Grzimek's ANIMAL LIFE ENCYCLOPEDIA

Volume 1

LOWER ANIMALS

Volume 2

INSECTS

Volume 3

MOLLUSKS AND ECHINODERMS

Volume 4

FISHES I

Volume 5

FISHES II AND AMPHIBIA

Volume 6

REPTILES

Volume 7

BIRDS I

Volume 8

BIRDS II

Volume 9

BIRDS III

Volume 10

MAMMALS I

Volume 11

MAMMALS II

Volume 12

MAMMALS III

Volume 13

MAMMALS IV

Grzimek's
ANIMAL LIFE ENCYCLOPEDIA

Editor-in-Chief

Dr. Dr. h.c. Bernhard Grzimek

Professor, Justus Liebig University of Giessen
Director, Frankfurt Zoological Garden, Germany
Trustee, Tanzanian National Parks, Tanzania

VAN NOSTRAND REINHOLD COMPANY

New York Cincinnati Toronto London Melbourne

First published in paperback in 1984

Library of Congress Catalog Card Number 79-183178

ISBN 0-442-23044-3

Printed in Federal Republic of Germany

Van Nostrand Reinhold Company Inc.
135 West 50th Street
New York, New York 10020

Van Nostrand Reinhold Company Limited
Molly Millars Lane
Wokingham, Berkshire RG11 2PY, England

Van Nostrand Reinhold
480 Latrobe Street
Melbourne, Victoria 3000, Australia

Macmillan of Canada
Division of Gage Publishing Limited
164 Commander Boulevard
Agincourt, Ontario M1S 3C7 Canada

16 15 14 13 12 11 10 9 8 7 6 5 4 3 2 1

EDITORS AND CONTRIBUTORS

Editor-in-Chief
DR. DR. H.C. BERNHARD GRZIMEK
Professor, Justus Liebig University of Giessen, Germany
Director, Frankfurt Zoological Garden, Germany
Trustee, Tanzanian National Parks, Tanzania

DR. MICHAEL ABS Curator, Ruhr University	BOCHUM, GERMANY
DR. SALIM ALI Bombay Natural History Society	BOMBAY, INDIA
DR. RUDOLF ALTEVOGT Professor, Zoological Institute, University of Münster	MÜNSTER, GERMANY
DR. RENATE ANGERMANN Curator, Institute of Zoology, Humboldt University	BERLIN, GERMANY
EDWARD A. ARMSTRONG Cambridge University	CAMBRIDGE, ENGLAND
DR. PETER AX Professor, Second Zoological Institute and Museum, University of Göttingen	GÖTTINGEN, GERMANY
DR. FRANZ BACHMAIER Zoological Collection of the State of Bavaria	MUNICH, GERMANY
DR. PEDRU BANARESCU Academy of the Roumanian Socialist Republic, Trajan Savulescu Institute of Biology	BUCHAREST, RUMANIA
DR. A. G. BANNIKOW Professor, Institute of Veterinary Medicine	MOSCOW, U.S.S.R.
DR. HILDE BAUMGÄRTNER Zoological Collection of the State of Bavaria	MUNICH, GERMANY
C. W. BENSON Department of Zoology, Cambridge University	CAMBRIDGE, ENGLAND
DR. ANDREW BERGER Chairman, Department of Zoology, University of Hawaii	HONOLULU, HAWAII, U.S.A.
DR. J. BERLIOZ National Museum of Natural History	PARIS, FRANCE
DR. RUDOLF BERNDT Director, Institute for Population Ecology, Heligoland Ornithological Station	BRAUNSCHWEIG, GERMANY
DIETER BLUME Instructor of Biology, Freiherr-vom-Stein School	GLADENBACH, GERMANY
DR. MAXIMILIAN BOECKER Zoological Research Institute and A. Koenig Museum	BONN, GERMANY
DR. CARL-HEINZ BRANDES Curator and Director, The Aquarium, Overseas Museum	BREMEN, GERMANY
DR. DONALD G. BROADLEY Curator, Umtali Museum	UMTALI, RHODESIA
DR. HEINZ BRÜLL Director; Game, Forest, and Fields Research Station	HARTENHOLM, GERMANY
DR. HERBERT BRUNS Director, Institute of Zoology and the Protection of Life	SCHLANGENBAD, GERMANY
HANS BUB Heligoland Ornithological Station	WILHELMSHAVEN, GERMANY
A. H. CHRISHOLM	SYDNEY, AUSTRALIA
HERBERT THOMAS CONDON Curator of Birds, South Australian Museum	ADELAIDE, AUSTRALIA
DR. EBERHARD CURIO Director, Laboratory of Ethology, Ruhr University	BOCHUM, GERMANY

DR. SERGE DAAN Laboratory of Animal Physiology, University of Amsterdam	AMSTERDAM, THE NETHERLANDS
DR. HEINRICH DATHE Professor and Director, Animal Park and Zoological Research Station, German Academy of Sciences	BERLIN, GERMANY
DR. WOLFGANG DIERL Zoological Collection of the State of Bavaria	MUNICH, GERMANY
DR. FRITZ DIETERLEN Zoological Research Institute, A. Koenig Museum	BONN, GERMANY
DR. ROLF DIRCKSEN Professor, Pedagogical Institute	BIELEFELD, GERMANY
JOSEF DONNER Instructor of Biology	KATZELSDORF, AUSTRIA
DR. JEAN DORST Professor, National Museum of Natural History	PARIS, FRANCE
DR. GERTI DÜCKER Professor and Chief Curator, Zoological Institute, University of Münster	MÜNSTER, GERMANY
DR. MICHAEL DZWILLO Zoological Institute and Museum, University of Hamburg	HAMBURG, GERMANY
DR. IRENÄUS EIBL-EIBESFELDT Professor and Director, Institute of Human Ethology, Max Planck Institute for Behavioral Physiology	PERCHA/STARNBERG, GERMANY
DR. MARTIN EISENTRAUT Professor and Director, Zoological Research Institute and A. Koenig Museum	BONN, GERMANY
DR. EBERHARD ERNST Swiss Tropical Institute	BASEL, SWITZERLAND
R. D. ETCHECOPAR Director, National Museum of Natural History	PARIS, FRANCE
DR. R. A. FALLA Director, Dominion Museum	WELLINGTON, NEW ZEALAND
DR. HUBERT FECHTER Curator, Lower Animals, Zoological Collection of the State of Bavaria	MUNICH, GERMANY
DR. WALTER FIEDLER Docent, University of Vienna, and Director, Schönbrunn Zoo	VIENNA, AUSTRIA
WOLFGANG FISCHER Inspector of Animals, Animal Park	BERLIN, GERMANY
DR. C. A. FLEMING Geological Survey Department of Scientific and Industrial Research	LOWER HUTT, NEW ZEALAND
DR. HANS FRÄDRICH Zoological Garden	BERLIN, GERMANY
DR. HANS-ALBRECHT FREYE Professor and Director, Biological Institute of the Medical School	HALLE A.D.S., GERMANY
GÜNTHER E. FREYTAG Former Director, Reptile and Amphibian Collection, Museum of Cultural History in Magdeburg	BERLIN, GERMANY
DR. HERBERT FRIEDMANN Director, Los Angeles County Museum of Natural History	LOS ANGELES, CALIFORNIA, U.S.A.
DR. H. FRIEDRICH Professor, Overseas Museum	BREMEN, GERMANY
DR. JAN FRIJLINK Zoological Laboratory, University of Amsterdam	AMSTERDAM, THE NETHERLANDS
DR. DR. H.C. KARL VON FRISCH Professor Emeritus and former Director, Zoological Institute, University of Munich	MUNICH, GERMANY
DR. H. J. FRITH C.S.I.R.O. Research Institute	CANBERRA, AUSTRALIA
DR. ION E. FUHN Academy of the Roumanian Socialist Republic, Trajan Savulescu Institute of Biology	BUCHAREST, RUMANIA
DR. CARL GANS Professor, Department of Biology, State University of New York at Buffalo	BUFFALO, NEW YORK, U.S.A.
DR. RUDOLF GEIGY Professor and Director, Swiss Tropical Institute	BASEL, SWITZERLAND

DR. JACQUES GERY	ST. GENIES, FRANCE
DR. WOLFGANG GEWALT Director, Animal Park	DUISBURG, GERMANY
DR. DR. H.C. DR. H.C. VIKTOR GOERTTLER Professor Emeritus, University of Jena	JENA, GERMANY
DR. FRIEDRICH GOETHE Director, Institute of Ornithology, Heligoland Ornithological Station	WILHELMSHAVEN, GERMANY
DR. ULRICH F. GRUBER Herpetological Section, Zoological Research Institute and A. Koenig Museum	BONN, GERMANY
DR. H. R. HAEFELFINGER Museum of Natural History	BASEL, SWITZERLAND
DR. THEODOR HALTENORTH Director, Mammalology, Zoological Collection of the State of Bavaria	MUNICH, GERMANY
BARBARA HARRISSON Sarawak Museum, Kuching, Borneo	ITHACA, NEW YORK, U.S.A.
DR. FRANCOIS HAVERSCHMIDT President, High Court (retired)	PARAMARIBO, SURINAM
DR. HEINZ HECK Director, Catskill Game Farm	CATSKILL, NEW YORK, U.S.A.
DR. LUTZ HECK Professor (retired), and Director, Zoological Garden, Berlin	WIESBADEN, GERMANY
DR. DR. H.C. HEINI HEDIGER Director, Zoological Garden (retired)	ZURICH, SWITZERLAND
DR. DIETRICH HEINEMANN Director, Zoological Garden, Münster	DÖRNIGHEIM, GERMANY
DR. HELMUT HEMMER Institute for Physiological Zoology, University of Mainz	MAINZ, GERMANY
DR. W. G. HEPTNER Professor, Zoological Museum, University of Moscow	MOSCOW, U.S.S.R.
DR. KONRAD HERTER Professor Emeritus and Director (retired), Zoological Institute, Free University of Berlin	BERLIN, GERMANY
DR. HANS RUDOLF HEUSSER Zoological Museum, University of Zurich	ZURICH, SWITZERLAND
DR. EMIL OTTO HÖHN Associate Professor of Physiology, University of Alberta	EDMONTON, CANADA
DR. W. HOHORST Professor and Director, Parasitological Institute, Farbwerke Hoechst A.G.	FRANKFURT-HÖCHST, GERMANY
DR. FOLKHART HÜCKINGHAUS Director, Senckenbergische Anatomy, University of Frankfurt a.M.	FRANKFURT A.M., GERMANY
FRANCOIS HÜE National Museum of Natural History	PARIS, FRANCE
DR. K. IMMELMANN Professor, Zoological Institute, Technical University of Braunschweig	BRAUNSCHWEIG, GERMANY
DR. JUNICHIRO ITANI Kyoto University	KYOTO, JAPAN
DR. RICHARD F. JOHNSTON Professor of Zoology, University of Kansas	LAWRENCE, KANSAS, U.S.A.
OTTO JOST Oberstudienrat, Freiherr-vom-Stein Gymnasium	FULDA, GERMANY
DR. PAUL KÄHSBAUER Curator, Fishes, Museum of Natural History	VIENNA, AUSTRIA
DR. LUDWIG KARBE Zoological State Institute and Museum	HAMBURG, GERMANY
DR. N. N. KARTASCHEW Docent, Department of Biology, Lomonossow State University	MOSCOW, U.S.S.R.
DR. WERNER KÄSTLE Oberstudienrat, Gisela Gymnasium	MUNICH, GERMANY
DR. REINHARD KAUFMANN Field Station of the Tropical Institute, Justus Liebig University, Giessen, Germany	SANTA MARTA, COLOMBIA

DR. MASAO KAWAI
Primate Research Institute, Kyoto University
KYOTO, JAPAN

DR. ERNST F. KILIAN
Professor, Giessen University and Catedratico Universidad Austral, Valdivia-Chile
GIESSEN, GERMANY

DR. RAGNAR KINZELBACH
Institute for General Zoology, University of Mainz
MAINZ, GERMANY

DR. HEINRICH KIRCHNER
Landwirtschaftsrat (retired)
BAD OLDESLOE, GERMANY

DR. ROSL KIRCHSHOFER
Zoological Garden, University of Frankfurt a.M.
FRANKFURT A.M., GERMANY

DR. WOLFGANG KLAUSEWITZ
Curator, Senckenberg Nature Museum and Research Institute
FRANKFURT A.M., GERMANY

DR. KONRAD KLEMMER
Curator, Senckenberg Nature Museum and Research Institute
FRANKFURT A.M., GERMANY

DR. ERICH KLINGHAMMER
Laboratory of Ethology, Purdue University
LAFAYETTE, INDIANA, U.S.A.

DR. HEINZ-GEORG KLÖS
Professor and Director, Zoological Garden
BERLIN, GERMANY

URSULA KLÖS
Zoological Garden
BERLIN, GERMANY

DR. OTTO KOEHLER
Professor Emeritus, Zoological Institute, University of Freiburg
FREIBURG I. BR., GERMANY

DR. KURT KOLAR
Institute of Ethology, Austrian Academy of Sciences
VIENNA, AUSTRIA

DR. CLAUS KÖNIG
State Ornithological Station of Baden-Württemberg
LUDWIGSBURG, GERMANY

DR. ADRIAAN KORTLANDT
Zoological Laboratory, University of Amsterdam
AMSTERDAM, THE NETHERLANDS

DR. HELMUT KRAFT
Professor and Scientific Councillor, Medical Animal Clinic, University of Munich
MUNICH, GERMANY

DR. HELMUT KRAMER
Zoological Research Institute and A. Koenig Museum
BONN, GERMANY

DR. FRANZ KRAPP
Zoological Institute, University of Freiburg
FREIBURG, SWITZERLAND

DR. OTTO KRAUS
Professor, University of Hamburg, and Director, Zoological Institute and Museum
HAMBURG, GERMANY

DR. DR. HANS KRIEG
Professor and First Director (retired), Scientific Collections of the State of Bavaria
MUNICH, GERMANY

DR. HEINRICH KÜHL
Federal Research Institute for Fisheries, Cuxhaven Laboratory
CUXHAVEN, GERMANY

DR. OSKAR KUHN
Professor, formerly University Halle/Saale
MUNICH, GERMANY

DR. HANS KUMERLOEVE
First Director (retired), State Scientific Museum, Vienna
MUNICH, GERMANY

DR. NAGAMICHI KURODA
Yamashina Ornithological Institute, Shibuya-Ku
TOKYO, JAPAN

DR. FRED KURT
Zoological Museum of Zurich University, Smithsonian Elephant Survey
COLOMBO, CEYLON

DR. WERNER LADIGES
Professor and Chief Curator, Zoological Institute and Museum, University of Hamburg
HAMBURG, GERMANY

LESLIE LAIDLAW
Department of Animal Sciences, Purdue University
LAFAYETTE, INDIANA, U.S.A.

DR. ERNST M. LANG
Director, Zoological Garden
BASEL, SWITZERLAND

DR. ALFREDO LANGGUTH
Department of Zoology, Faculty of Humanities and Sciences, University of the Republic
MONTEVIDEO, URUGUAY

LEO LEHTONEN
Science Writer
HELSINKI, FINLAND

BERND LEISLER
Second Zoological Institute, University of Vienna
VIENNA, AUSTRIA

DR. KURT LILLELUND Professor and Director, Institute for Hydrobiology and Fishery Sciences, University of Hamburg	HAMBURG, GERMANY
R. LIVERSIDGE Alexander MacGregor Memorial Museum	KIMBERLEY, SOUTH AFRICA
DR. DR. KONRAD LORENZ Professor and Director, Max Planck Institute for Behavioral Physiology	SEEWIESEN/OBB., GERMANY
DR. DR. MARTIN LÜHMANN Federal Research Institute for the Breeding of Small Animals	CELLE, GERMANY
DR. JOHANNES LÜTTSCHWAGER Oberstudienrat (retired)	HEIDELBERG, GERMANY
DR. WOLFGANG MAKATSCH	BAUTZEN, GERMANY
DR. HUBERT MARKL Professor and Director, Zoological Institute, Technical University of Darmstadt	DARMSTADT, GERMANY
BASIL J. MARLOW, B.SC. (HONS) Curator, Australian Museum	SYDNEY, AUSTRALIA
DR. THEODOR MEBS Instructor of Biology	WEISSENHAUS/OSTSEE, GERMANY
DR. GERLOF FOKKO MEES Curator of Birds, Rijks Museum of Natural History	LEIDEN, THE NETHERLANDS
HERMANN MEINKEN Director, Fish Identification Institute, V.D.A.	BREMEN, GERMANY
DR. WILHELM MEISE Chief Curator, Zoological Institute and Museum, University of Hamburg	HAMBURG, GERMANY
DR. JOACHIM MESSTORFF Field Station of the Federal Fisheries Research Institute	BREMERHAVEN, GERMANY
DR. MARIAN MLYNARSKI Professor, Polish Academy of Sciences, Institute for Systematic and Experimental Zoology	CRACOW, POLAND
DR. WALBURGA MOELLER Nature Museum	HAMBURG, GERMANY
DR. H.C. ERNA MOHR Curator (retired), Zoological State Institute and Museum	HAMBURG, GERMANY
DR. KARL-HEINZ MOLL	WAREN/MÜRITZ, GERMANY
DR. DETLEV MÜLLER-USING Professor, Institute for Game Management, University of Göttingen	HANNOVERSCH-MÜNDEN, GERMANY
WERNER MÜNSTER Instructor of Biology	EBERSBACH, GERMANY
DR. JOACHIM MÜNZING Altona Museum	HAMBURG, GERMANY
DR. WILBERT NEUGEBAUER Wilhelma Zoo	STUTTGART-BAD CANNSTATT, GERMANY
DR. IAN NEWTON Senior Scientific Officer, The Nature Conservancy	EDINBURGH, SCOTLAND
DR. JÜRGEN NICOLAI Max Planck Institute for Behavioral Physiology	SEEWIESEN/OBB., GERMANY
DR. GÜNTHER NIETHAMMER Professor, Zoological Research Institute and A. Koenig Museum	BONN, GERMANY
DR. BERNHARD NIEVERGELT Zoological Museum, University of Zurich	ZURICH, SWITZERLAND
DR. C. C. OLROG Institut Miguel Lillo San Miguel de Tucuman	TUCUMAN, ARGENTINA
ALWIN PEDERSEN Mammal Research and Arctic Explorer	HOLTE, DENMARK
DR. DIETER STEFAN PETERS Nature Museum and Senckenberg Research Institute	FRANKFURT A.M., GERMANY
DR. NICOLAUS PETERS Scientific Councillor and Docent, Institute of Hydrobiology and Fisheries, University of Hamburg	HAMBURG, GERMANY
DR. HANS-GÜNTER PETZOLD Assistant Director, Zoological Garden	BERLIN, GERMANY

DR. RUDOLF PIECHOCKI
Docent, Zoological Institute, University of Halle — HALLE A.D.S., GERMANY

DR. IVO POGLAYEN-NEUWALL
Director, Zoological Garden — LOUISVILLE, KENTUCKY, U.S.A.

DR. EGON POPP
Zoological Collection of the State of Bavaria — MUNICH, GERMANY

DR. DR. H.C. ADOLF PORTMANN
Professor Emeritus, Zoological Institute, University of Basel — BASEL, SWITZERLAND

HANS PSENNER
Professor and Director, Alpine Zoo — INNSBRUCK, AUSTRIA

DR. HEINZ-SIBURD RAETHEL
Oberveterinärrat — BERLIN, GERMANY

DR. URS H. RAHM
Professor, Museum of Natural History — BASEL, SWITZERLAND

DR. WERNER RATHMAYER
Biology Institute, University of Konstanz — KONSTANZ, GERMANY

WALTER REINHARD
Biologist — BADEN-BADEN, GERMANY

DR. H. H. REINSCH
Federal Fisheries Research Institute — BREMERHAVEN, GERMANY

DR. BERNHARD RENSCH
Professor Emeritus, Zoological Institute, University of Münster — MÜNSTER, GERMANY

DR. VERNON REYNOLDS
Docent, Department of Sociology, University of Bristol — BRISTOL, ENGLAND

DR. RUPERT RIEDL
Professor, Department of Zoology, University of North Carolina — CHAPEL HILL, NORTH CAROLINA, U.S.A.

DR. PETER RIETSCHEL
Professor (retired), Zoological Institute, University of Frankfurt a.M. — FRANKFURT A.M., GERMANY

DR. SIEGFRIED RIETSCHEL
Docent, University of Frankfurt; Curator, Nature Museum and Research Institute Senckenberg — FRANKFURT A.M., GERMANY

HERBERT RINGLEBEN
Institute of Ornithology, Heligoland Ornithological Station — WILHELMSHAVEN, GERMANY

DR. K. ROHDE
Institute for General Zoology, Ruhr University — BOCHUM, GERMANY

DR. PETER RÖBEN
Academic Councillor, Zoological Institute, Heidelberg University — HEIDELBERG, GERMANY

DR. ANTON E. M. DE ROO
Royal Museum of Central Africa — TERVUREN, SOUTH AFRICA

DR. HUBERT SAINT GIRONS
Research Director, Center for National Scientific Research — BRUNOY (ESSONNE), FRANCE

DR. LUITFRIED VON SALVINI-PLAWEN
First Zoological Institute, University of Vienna — VIENNA, AUSTRIA

DR. KURT SANFT
Oberstudienrat, Diesterweg-Gymnasium — BERLIN, GERMANY

DR. E. G. FRANZ SAUER
Professor, Zoological Research Institute and A. Koenig Museum, University of Bonn — BONN, GERMANY

DR. ELEONORE M. SAUER
Zoological Research Institute and A. Koenig Museum, University of Bonn — BONN, GERMANY

DR. ERNST SCHÄFER
Curator, State Museum of Lower Saxony — HANNOVER, GERMANY

DR. FRIEDRICH SCHALLER
Professor and Chairman, First Zoological Institute, University of Vienna — VIENNA, AUSTRIA

DR. GEORGE B. SCHALLER
Serengeti Research Institute, Michael Grzimek Laboratory — SERONERA, TANZANIA

DR. GEORG SCHEER
Chief Curator and Director, Zoological Institute, State Museum of Hesse — DARMSTADT, GERMANY

DR. CHRISTOPH SCHERPNER
Zoological Garden — FRANKFURT A.M., GERMANY

DR. HERBERT SCHIFTER Bird Collection, Museum of Natural History	VIENNA, AUSTRIA
DR. MARCO SCHNITTER Zoological Museum, Zurich University	ZURICH, SWITZERLAND
DR. KURT SCHUBERT Federal Fisheries Research Institute	HAMBURG, GERMANY
EUGEN SCHUHMACHER Director, Animals Films, I.U.C.N.	MUNICH, GERMANY
DR. THOMAS SCHULTZE-WESTRUM Zoological Institute, University of Munich	MUNICH, GERMANY
DR. ERNST SCHÜT Professor and Director (retired), State Museum of Natural History	STUTTGART, GERMANY
DR. LESTER L. SHORT, JR. Associate Curator, American Museum of Natural History	NEW YORK, NEW YORK, U.S.A.
DR. HELMUT SICK National Museum	RIO DE JANEIRO, BRAZIL
DR. ALEXANDER F. SKUTCH Professor of Ornithology, University of Costa Rica	SAN ISIDRO DEL GENERAL, COSTA RICA
DR. EVERHARD J. SLIJPER Professor, Zoological Laboratory, University of Amsterdam	AMSTERDAM, THE NETHERLANDS
BERTRAM E. SMYTHIES Curator (retired), Division of Forestry Management, Sarawak-Malaysia	ESTEPONA, SPAIN
DR. KENNETH E. STAGER Chief Curator, Los Angeles County Museum of Natural History	LOS ANGELES, CALIFORNIA, U.S.A.
DR. H.C. GEORG H. W. STEIN Professor, Curator of Mammals, Institute of Zoology and Zoological Museum, Humboldt University	BERLIN, GERMANY
DR. JOACHIM STEINBACHER Curator, Nature Museum and Senckenberg Research Institute	FRANKFURT A.M., GERMANY
DR. BERNARD STONEHOUSE Canterbury University	CHRISTCHURCH, NEW ZEALAND
DR. RICHARD ZUR STRASSEN Curator, Nature Museum and Senckenberg Research Institute	FRANKFURT A.M., GERMANY
DR. ADELHEID STUDER-THIERSCH Zoological Garden	BASEL, SWITZERLAND
DR. ERNST SUTTER Museum of Natural History	BASEL, SWITZERLAND
DR. FRITZ TEROFAL Director, Fish Collection, Zoological Collection of the State of Bavaria	MUNICH, GERMANY
DR. G. F. VAN TETS Wildlife Research	CANBERRA, AUSTRALIA
ELLEN THALER-KOTTEK Institute of Zoology, University of Innsbruck	INNSBRUCK, AUSTRIA
DR. ERICH THENIUS Professor and Director, Institute of Paleontology, University of Vienna	VIENNA, AUSTRIA
DR. NIKO TINBERGEN Professor of Animal Behavior, Department of Zoology, Oxford University	OXFORD, ENGLAND
ALEXANDER TSURIKOV Lecturer, University of Munich	MUNICH, GERMANY
DR. WOLFGANG VILLWOCK Zoological Institute and Museum, University of Hamburg	HAMBURG, GERMANY
ZDENEK VOGEL Director, Suchdol Herpetological Station	PRAGUE, CZECHOSLOVAKIA
DIETER VOGT	SCHORNDORF, GERMANY
DR. JIRI VOLF Zoological Garden	PRAGUE, CZECHOSLOVAKIA
OTTO WADEWITZ	LEIPZIG, GERMANY

DR. HELMUT O. WAGNER Director (retired), Overseas Museum, Bremen	MEXICO CITY, MEXICO
DR. FRITZ WALTHER Professor, Texas A & M University	COLLEGE STATION, TEXAS, U.S.A.
JOHN WARHAM Zoology Department, Canterbury University	CHRISTCHURCH, NEW ZEALAND
DR. SHERWOOD L. WASHBURN University of California at Berkeley	BERKELEY, CALIFORNIA, U.S.A.
EBERHARD WAWRA First Zoological Institute, University of Vienna	VIENNA, AUSTRIA
DR. INGRID WEIGEL Zoological Collection of the State of Bavaria	MUNICH, GERMANY
DR. B. WEISCHER Institute of Nematode Research, Federal Biological Institute	MÜNSTER/WESTFALEN, GERMANY
HERBERT WENDT Author, Natural History	BADEN-BADEN, GERMANY
DR. HEINZ WERMUTH Chief Curator, State Nature Museum, Stuttgart	LUDWIGSBURG, GERMANY
DR. WOLFGANG VON WESTERNHAGEN	PREETZ/HOLSTEIN, GERMANY
DR. ALEXANDER WETMORE United States National Museum, Smithsonian Institution	WASHINGTON, D.C., U.S.A.
DR. DIETRICH E. WILCKE	RÖTTGEN, GERMANY
DR. HELMUT WILKENS Professor and Director, Institute of Anatomy, School of Veterinary Medicine	HANNOVER, GERMANY
DR. MICHAEL L. WOLFE Utah State University	UTAH, U.S.A.
HANS EDMUND WOLTERS Zoological Research Institute and A. Koenig Museum	BONN, GERMANY
DR. ARNFRID WÜNSCHMANN Research Associate, Zoological Garden	BERLIN, GERMANY
DR. WALTER WÜST Instructor, Wilhelms Gymnasium	MUNICH, GERMANY
DR. HEINZ WUNDT Zoological Collection of the State of Bavaria	MUNICH, GERMANY
DR. CLAUS-DIETER ZANDER Zoological Institute and Museum, University of Hamburg	HAMBURG, GERMANY
DR. DR. FRITZ ZUMPT Director, Entomology and Parasitology, South African Institute for Medical Research	JOHANNESBURG, SOUTH AFRICA
DR. RICHARD L. ZUSI Curator of Birds, United States National Museum, Smithsonian Institution	WASHINGTON, D.C., U.S.A.

Volume 10

MAMMALS I

Edited by:

WALTER FIEDLER
WOLFGANG GEWALT
BERNHARD GRZIMEK
DIETRICH HEINEMANN
KONRAD HERTER
ERICH THENIUS

ENGLISH EDITION

GENERAL EDITOR:
George M. Narita

SCIENTIFIC EDITOR:
Erich Klinghammer

TRANSLATORS:
Renata Geist
Erich Klinghammer

ASSISTANT EDITORS:
Lori Burghardt
Frances A. Wilke

PRODUCTION DIRECTOR:
James V. Leone

EDITORIAL ASSISTANTS:
Detlef K. Onderka
Karen Boikess

ART DIRECTOR:
Lorraine K. Hohman

INDEX:
Suzanne C. Klinghammer

CONTENTS

For a more complete listing
of animal names, see systematic classification or the index.

1 The Mammals

Distinguishing characteristics

The Class *Mammalia* are characteristically homoiothermal, hair-covered vertebrates (there is a vestigial hair coat development in certain forms; some groups have secondarily adapted to an aquatic mode of life). The female mammals nourish their young with milk which is a secretion from specialized skin glands. The HRL may range from approximately 6.5 cm in the shrew to approximately 30 meters in the blue whale. Scales are occasionally found along with hair. There are many types of skin glands: sebaceous, sweat, milk, and odoriferous glands. Toes and fingers terminate in nails, claws, or hooves. A secondary jaw joint is found between squamosal and dentary. The articular and quadrate along with the primary jaw are reduced and form the auditory ossicles located in the mammalian middle ear. The skull is firm and unmovable with a secondary bony palate and two occipital condyles. Usually only seven vertebrae are found. The thoracic vertebrae support the ribs. The coracoid is reduced and is fused with the scapula (except in the *Prototheria*). Dentition consists of incisors, canines, premolars, and molars. The number and shape of the teeth vary greatly with the mode of feeding. The roots of the teeth are usually anchored in alveoli. Rootless teeth also occur. There is only one replacement of teeth. Dentition and an effective digestive system ensure efficient food utilization. The lungs are divided into many very small chambers. The inner surface area of all lung alveoli is over 100 square meters in the larger mammals. The diaphragm separates the thoracic and abdominal cavities. Inhaled air is pre-warmed in the nasal cavities. The pulmonary circulation and that of the rest of the body is totally separated. The heart consists of two separated ventricles and auricles. Sensory organs are highly developed, although sensory performance varies greatly in individual groups. The cerebral hemispheres predominate the other parts of the brain in the more highly developed mammals. Monotremata lay large-yolked eggs. All other mammals bear live young. Embryos are nourished within the uterus of the mother; in the case of the marsupials this is done with a uterine secretion. In the higher mammals there is a direct exchange of metabolites via the placenta.

Class: Mammals by D. Heinemann

Mammals enjoy a world-wide distribution on land, water, and air. There are three subclasses: 1. EGG-LAYING MAMMALS *(Prototheria);* one order; 2. POUCHED MAMMALS *(Metatheria);* one order; 3. HIGHER MAMMALS *(Eutheria);* 16 orders and a total of 4000–5000 species.

Out of more than one million species of animals, only a mere 5000 belong to the mammals. This is not even one half of one percent. Nevertheless, this vertebrate class is of the utmost importance to man because he belongs to this group himself. The quadrupedal development started over 400 million years ago with the transition from the finned fish to the ancestral amphibian which crawled to the shore and finally reached its peak in the mammals. Only very few have given up the quadrupedal stance. Some, like the whales and sea cows, have returned to the water. Only a few vestigial bones of the former hind legs have remained. In flying mammals the frontal appendages evolved into wings. Kangaroos, elephant shrews, jumping mice, the springhaas, and man have become bipedal. The majority of mammals have remained faithful to the quadrupedal type of locomotion.

Regulation of body temperature

The ability to maintain a constant body temperature independent of outside temperatures differentiates the mammal from the typical reptile from which it originated. Although the internal temperature of all animals (and plants) is usually somewhat higher than the external temperature, this is only a by-product of normal metabolic processes. The poikilotherms respond to a drop in the outside temperature by lowering their body temperature and by slowing down their metabolic rate. This response to the environment is different in the warm-blooded or homoiothermic mammals. A drop in the environmental temperature results in an increased metabolic rate to produce an optimal amount of body heat for the normal functioning of body processes which, of course, requires additional energy. Homoiothermic animals require a greater food supply than do poikilotherms of comparable size. The food requirement increases with a decrease in the temperature. This is also the reason why we have a greater appetite after a walk on a cold winter's day than on a hot summer day. This increased energy consumption which requires extra food is a definite disadvantage of the otherwise so practical temperature regulation. Animals that were successful in maximizing this excess energy demand had a better chance of survival in the future than those whose food demand continued to increase. "Rationalization" in the free interplay of forces in living nature is just as important as in the economy. Rationalization in the temperature regulatory mechanism means, however, the necessity of saving fuel by means of heat-conserving mechanisms. This important function is fulfilled by the mammalian pelage. A similar type of fur served this purpose in the extinct flying reptiles; the plumage in birds also serves a similar function. The "invention" of a steady body temperature (homoiothermia) developed independently three times in vertebrates which originated from the reptilian stem during the Mesozoic era. It occurred in the ancestral stem of the mammals, the flying reptiles, and the birds.

Skin structures: hairs and scales

The bird feather developed from a converted reptilian scale. The mammalian hair, on the other hand, is a totally new structure. Certain species of mammals are not only covered by hair but certain parts of their bodies are still covered by reptile-like scales. We find such scales on the dorsal side of the pangolins, for example, or on the tail of the rat. As a rule there are several hairs underneath, i.e., behind, each scale, usually one thicker central hair and two smaller side hairs. The hair grouping in mammalian skin that is not covered by scales is similar and the hairs appear as if they were arranged between two rows of scales. For this reason, it has been assumed that the reptile-like mammalian ancestors developed a hair coat before the scales disappeared. The side hairs, in turn, are often surrounded by several additional hairs which in many mammals make up the thick warm underwool. The main hairs protrude out of this undercoat as long smooth guard hairs.

An individual hair is an epidermal structure. The lower part, the hair root, is embedded in an epidermal pocket, sunk deeply into the skin, and grows from a small papilla. The hair consists of three layers of atrophied cornified epithelial cells: the inner medulla, an outer cortex, and the cuticle. The color of the hair depends on the pigment inside the medulla. Air bubbles trapped inside the hair will lighten the color, sometimes to a brilliant white. Hair is shed from time to time and is replaced. In many mammals hair is shed seasonally, while in others it is gradually shed throughout the year.

Nails, claws, hooves, and horns

Nails, claws, and hooves tipping the toes and fingers of mammals consist of the same keratinized epidermal material as hair. The same is true for the horns of the ruminants and the nasal horns of the rhinoceros which the animal uses as a weapon.

The hair follicles usually lie at a slant within the dermis. Hairs are arranged in tracts which slope in the same direction. Each hair follicle is surrounded by a small muscle (arrector pilli muscle) which contracts involuntarily. Thus the hair follicle becomes stiff and the hair is erected. This results in an increase in the insulating air layer trapped between the skin and the fur surface. When we get cold our tiny body hairs are also erected and we get "gooseflesh"; that is, the body attempts to erect the hair coat which was already lost by our ancestors one hundred thousand years ago. In situations where we experience fright, anger, surprise, or enthusiasm, our hair too "stands on end," just as in other mammals in a similar situation.

Skin circulation and skin glands

The homeostatic mechanism which controls the temperature of a mammalian body not only prevents it from losing heat in extreme cold but also allows for ventilation at high ambient temperatures. In such a case, blood flow to the skin increases and heat can be dissipated. Under extreme conditions the body relies on vaporization to rid itself of excess heat. The animal either pants (i.e., rapid breathing) or sweats (i.e., the skin is moistened by the watery secretion of the sweat glands). True sweat glands, however, are found only in primates, but many other mammals have similar glands which fulfill the same function. There is another skin gland which is closely associated with the hair coat. This gland, the sebaceous gland, empties an oily secretion into the hair follicle which protects and lubricates the hair and skin against moisture.

The mammary glands

The most important skin glands of the mammals, however, are the mammary glands after which this class of animals is named. The egg-laying mammals possess two glandular regions which consist of several discrete glands which discharge into depressions in the abdominal surface. In all other mammals the gland openings are concentrated in teats or nipples. The marsupials have mammae which are arranged in four rows or in a circle, and in the higher mammals they are always in two rows. Originally mammae were on the right and left side of the trunk starting at the armpit down to the lower abdomen; they developed from the "mammary ridges" of the embryo. Mammals with a smaller number of young have fewer nipples, sometimes only one pair. In this case the nipples are either found at the anterior or posterior location of the original ridges. In the monotremes and the marsupials only the females possess mammary glands. In the higher mammals vestigial mammary glands and teats are also found in the males, and in exceptional cases minute amounts of milk may be secreted.

Feeding the embryo

The mammalian mother does not merely supply milk for her young. Already before birth most offspring have been better cared for than the embryos of other animal groups. However, bearing of living young is not only the prerogative of the mammals. Examples of viviparity can be found in almost all the invertebrate phyla, and also in the fishes, amphibians, and reptiles. Yet in the more evolved mammalian groups, protection and nourishment of the developing embryo reaches its highest perfection. In contrast to the reptilian and avian embryo, the mammalian embryo, aside from the monotremes, does not derive its nourishment from the large yolk supply within the egg but directly from the maternal body. The mammalian egg contains very little yolk. Upon fertilization the egg is implanted into the wall of the uterus. The uterus, an enlargement of the oviduct, is ideally suited for the reception of the blastocyst because of its highly vascularized endometrium. Just like the embryos of the reptiles and birds, the mammalian embryo forms have, besides the yolk sac which is characteristic of all vertebrates, three additional bladder-like membraneous appendages. These are the amnion, chorion, and allantois. In the marsupials and higher mammals the amnion takes over the role of nourishing the embryo. However, this type of nourishment is still not fully developed in most marsupials. The uterine wall of the marsupial exudes a mixture of gland secretions, fat droplets, and disintegrated white blood cells which are absorbed by the amniotic membrane of the embryo. In the higher mammals the outer membrane develops into the highly convoluted chorionic villi which internally come to lie next to the allantois and at this point of contact are penetrated with blood vessels. The chorionic villi project into the uterine tissue. In this manner the placenta, which supplies the embryo with food, develops out of embryonic and maternal components. There is great variability in the manner in which maternal and embryonic tissues interlock. The shape of the placenta differs greatly in the individual mammalian orders.

The form of the reproductive organs

There is also great variation in the shape of the uterus. In the primitive state each of the two ovaries enlarged into a uterus (uterus duplex). This type is still present in the marsupials and many rodents. In the insectivores, carnivores, prosimians, whales, and ungulates, the distal parts of the uteri are fused, giving rise to a bicornuate uterus. In the bats, monkeys, and man, both uterine halves are joined to a uterus simplex.

The male genital organ, the penis, lies within the cloaca of the egg-laying mammals when in the inactive state. In several other mammals, for example, the hedgehog, the penis lies within a penis sheath. However, in most mammalian groups the penis extends outside the body. The urethra passes through the penis and in many mammals is supported by a bony structure, the os penis. Usually the testes lie permanently or temporarily outside the body cavity within a scrotal sac. In certain species, elephants, for example, they are permanently withdrawn inside the body cavity.

A more efficient food utilization: the digestive organs

The increased energy demand which is essential for the maintenance of homeothermy is not met by an ever-increasing food uptake but by a more efficient method of utilizing food. No other vertebrate class is endowed with such diverse and specialized teeth and, not counting the birds, such highly efficient digestive organs as the mammals. The basic structure of the stomach does not differ greatly from other vertebrates but the stomach of certain herbivores is highly complex (compare ruminants, Vol. 13). The small intestine is highly coiled but otherwise is not very different from that of other vertebrates. The large intestine is immense. At the point where the small intestine joins with the colon a spacious cecum is found in many mammals. This area, which is a reservoir for intestinal bacteria, provides room for additional colonic activity where cellulose and other carbohydrates which are hard to digest are broken down and are thus made available to the body. The bacterial flora in the colon and cecum often play a significant role in the production of vitamins. Domestic rabbits can occasionally be observed to eat some parts of their fresh feces. If one prevents rabbits from eating their feces by putting wire mesh on the bottom of their cages they develop typical vitamin deficiency symptoms and eventually die as a result. These fecal fragments contain essential vitamin components which are produced in the cecum by the bacterial activity. The cecum and colon of the rabbit seem incapable of absorbing vitamins and transporting them into the blood stream. The animals have to eat this "cecal food" so that the vitamins can be absorbed by the small intestine and thereby utilized by their metabolism. Caecotrophy is also practised by other herbivores.

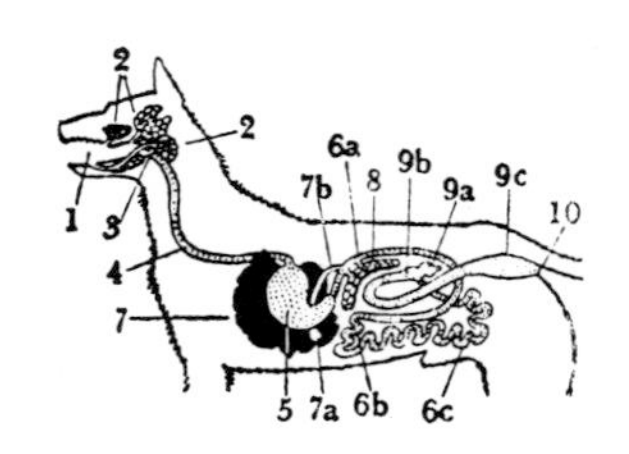

Mammalian teeth: the types of teeth

Normally the mammalian dentition consists of four types of teeth. In the front are the conical or chisel-like incisors. These are followed by a canine tooth which is basically long and pointed. Next is the molar series with the anterior premolars, and in the back are the molars with their relatively broad chewing surfaces. This basic mammalian dentition was greatly modified in the adaptations to the various modes of feeding. In the carnivores (meat eaters) the incisors remained small but the canines developed into large, pointed daggers and the molars into sharp-crested shears. Fish eaters, like many toothed whales, which swallow their food unchewed, possess very many sharp teeth in order to hold on to the slippery prey. The number of their teeth increased due to the division of the tooth buds, and all the various tooth types have a uniform, canine-like shape. Grazing and foraging animals developed wide-crowned molars, and even their premolars became "molarized." The herbivores also lost the canines, or these teeth came to resemble the incisors. However, in the musk deer and the Chinese water deer, the males have retained the canines which function as display weapons. In the rodents and lagomorphs the first pair of incisors evolved into enormous, rootless chisel-like gnawing teeth which grow throughout the animals's life. In the true plant-eating ruminants *(Pecora)* the upper row of incisors disappeared and was replaced by a hard horny plate. Some animals that are highly specialized to feed on one type of food lost all their teeth in the course of their evolutionary history. The right whales replaced theirs with baleen plates which are smooth on the outside but fringed on the inside. This structure forms an effective strainer for the enormous amounts of minute crustaceans which the whale requires. Some insectivorous mammals also lost their teeth, such as the South American anteater, the Old World pangolins *(Pholidota)*, and the Australian echidnas. Some possess teeth only as juveniles, such as the duck-billed platypus.

The dental formula as a distinguishing characteristic

The form and number of teeth serve as an important distinguishing feature even within individual mammalian orders and families. A succinct formula which expresses the number of teeth present in a

◁Fig. 1-1.
The Digestive Organs of the Dog

1-5: Foregut: 1. Oral Cavity *(Cavum oris)*. 2. Salivary Glands *(Glandulae salivales)*. 3. Pharynx. 4. Esophagus. 5. Stomach *(Ventriculus)*.

6-8: Midgut and Digestive Glands: 6. Small Intestine: a. Duodenum, b. Jejunum, c. Ileum. 7. Liver *(Hepar)* with: a. Gall bladder *(vesica fellea)*; b. Bile duct *(Ductus choledochus)*. 8. Pancreas with two pancreatic ducts.

9-10: Hindgut and Anus: 9. Large Intestine: a. Cecum, b. Colon, c. Rectum. 10. Anus.

mammal serves as an aid in distinguishing the great diversity of mammalian types. Above the fractional line the teeth of one upper jaw half are written, starting with the incisors, the canine, the premolars, and molars. Below the line the same sequence of the teeth in the lower jaw half is written. The following dental formula—$\frac{3 \cdot 1 \cdot 4 \cdot 3}{3 \cdot 1 \cdot 4 \cdot 3}$—indicates that each upper and lower jaw half consists of three incisors, one canine, four premolars, and three molars. This is the formulation of the primitive placental dentition of the higher mammals. All other types of dentition can be derived from it.

In most fishes, amphibians, and reptiles, lost or worn-out teeth are continually replaced; however, in the mammals there is only one tooth renewal. In most species the "milk teeth" of the young animal are replaced by the "permanent teeth" in a precise order. Molars are not present in the milk dentition since there is not room for them in the small jaws of the young animal. The "milk molars" do not precede the molars but rather the premolars of the permanent dentition. A highly developed dentition is concomitant with a similarly highly developed jaw apparatus, particularly a functionally "constructed" jaw articulation. In the fishes, amphibians, reptiles, and birds, the lower jaw still consists of several bone components. One of these, the articular, forms the connection with the upper jaw. The other half of this "hinge" is the quadrate which is a component of the palate. This structural arrangement is totally sufficient if the toothed jaw is mainly used for holding a prey animal or ripping off parts of plants, as is the case in the reptiles. However, the greatly diversified chewing, grinding, cutting, and gnawing motions which the true mammals "invented" for the more efficient utilization of their food required a more advanced form of jaw articulation than the one found in the reptiles. The "reconstructed" lower jaw had to evolve concurrently with the development of the typical mammalian dentition. In the ancestral group of the mammals, a new component originated next to the primary jaw articulation, and this is the tooth-bearing dentary bone of the lower jaw which connects to the skull. Henceforth, only this new "secondary" jaw remained in use in all mammals, and the dentary bone became one single lower jaw bone. The quadrate and articular bones became superfluous as jaw articulations, and the former "hinges" of the reptilian jaw took on a new function. These bone elements are found in the auditory bulla of the middle ear as incus and malleus. The single auditory ossicle found in the reptiles, the columellar apparatus, is retained as the stapes. We still possess the primary jaw of our distant reptilian ancestors in our ears, and every mammalian and human embryo repeats (recapitulates) this same transformation. During embryonic development the articulation between hammer and anvil is formed in the early stages, and later moves into the middle ear cavity.

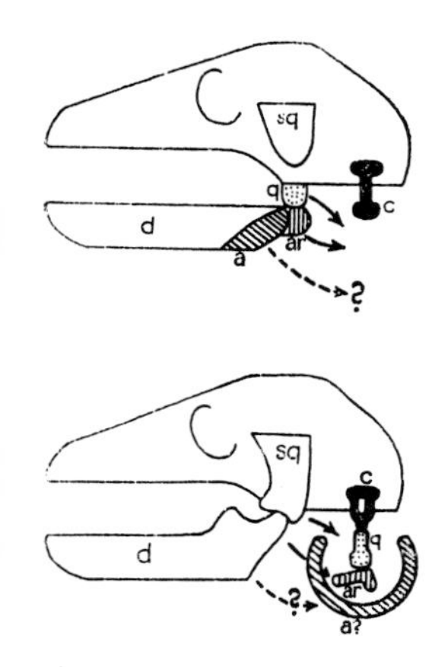

Fig. 1-2.
Transformation of the jaw during the evolution of mammals (top: reptile; bottom: mammal). a—timpanic ring; ar—malleus; c—stapes, q—incus, d—dentary, sq—squamosal bone.

The Anatomy of a Mammal (normal sized dog)

Upper Plate—The Superficial Musculature (Abbreviations: M. = Musculus, Muscle; Mm. = Musculi, Muscles)

1. Nasolabial elevator muscle (M. levator nasolabialis); 2. Jaw—nasal muscle (M. caninus); 3. Orbicularis muscle (M. orbicularis oris); 4. Elevator of the upper lip (M. levator labii maxillaris propri); 5. Cheek muscle (M. malaris); 6. Zygomatic muscle (M. zygomaticus); 7. Masseter muscle (M. masseter); 8. Orbicularis oculi muscle (M. orbicularis oculi); 9. External elevator of the upper eyelid (M. superciliaris); 10. Retractor of the temporal angle of the eye (M. refractor anguli temporalis oculi); 11. Temporal part (Pars temporalis); 12. Frontal part of the frontoscutularis muscle (Pars frontalis d. M. frontoscutularis); 13. Interscutularis muscle (M. interscutularis); 14. Cervicoscutularis muscle (M. cervicoscutularis); 15. Major deep cervicoauricularis muscle (long outward rotator of the ear) (M. cervicoauricularis prof. major); 16. Minor deep cervicoauricularis muscle (short rotator of the ear) (M. cervicoauricularis prof. minor); 17. Superficial cervicoauricularis muscle (long elevator of the ear) (M. cervicoauricularis superf.); 18. Ventral inward rotator of the ear (M. auricularis ventr.); 19. Dorsal superficial scutuloauriculus muscle (M. scutuloauricularis superf. dors.); 20. Transverse and oblique muscles of the auricle (Mm. transversi et obliqui auriculae); 21. Sternohyoideus muscle (M. sternohyoideus); 22. Mylohyoideus muscle (M. mylohyoideus); 23. Biventer mandibulae muscle (M. biventer mandibulae); 24. Straight abdominal muscle (M. rectus abdominis); 25. External oblique abdominal muscle (M. obliquus abdominis ext.); 26. External intercoastal muscles (Mm. intercostales ext.); 27. Internal oblique abdominal muscle (M. obliquus abdominis intern.); 28. Broad back muscle (M. latissimus dorsi); 28'. Lumbo-dorsal fascia; 29. Deep pectoral muscle (M. pectoralis profundus; 30. Sternocephalicus muscle (M. sternocephalicus); 31. Brachiocephalicus muscle (M. brachiocephalicus); 32. Cleidocervicalis muscle (M. pars cervicis); 32'. Trapezius thoracis muscle—Trapezius muscle (Pars thoracis d. M. trapezius); 33. Serratus dorsalis cranialis muscle (M. serratus anterior); 34. Supraspinatus muscle (M. supra spinatus); 35. Omotransversarius muscle (M. omotransversarius); 36. Scapular part (Deltoideus) (Pars scapularis); 36'. Acromial part—Deltoideus muscle (Deltoideus) (Pars acromialis d. M. deltoideus); 37. Brachialis muscle (M. brachialis); 38. Long head of the triceps muscle (Caput longus); 38'. Lateral head of the triceps brachii muscle (Caput laterale d. M. triceps brachii); 38''. Anconeus muscle (M. anconeus); 39. Extensor carpi radialis muscle (M. extensor carpi radialis); 40. Common digital extensor muscle (M. extensor digit. communis); 41. Lateral digital extensor muscle (M. extensor digit. lateralis); 42. Extensor carpi ulnaris muscle (M. extensor carpi ulnaris); 43. Flexor carpi ulnaris muscle (M. flexor carpi ulnaris); 44. Abductor pollicis longus muscle (M. abductor pollicis longus); 45. Deep digital flexor muscle (M. flexor digit. prof.); 46. Brachioradialis muscle (M. brachioradialis); 47. Extensor carpi radialis muscle (M. extensor carpi radialis); 48. Pronator teres muscle (M. pronator teres); 49. Flexor carpi radialis muscle (M. flexor carpi radialis); 50. Superficial digital flexor muscle (M. flexor digit. superfic.); 51. Ulnar head (Caput ulnare); 51'. Humeral head (Caput humerale), Flexor carpi ulnaris muscle (M. flexor carpi ulnaris); 52. Interosseus muscle (M. interosseus); 53. Humeral head (Caput humerale); 53'. Radial head (Caput radiale), Deep flexor muscle (M. flexor profundus); 54. Sartorius muscle (M. sartorius); 55. Tenson fascia lata muscle (M. tensor fascia latae); 55'. Fascia lata (Fascia lata); 56. Middle gluteal muscle (M. gluteus medius); 57. Superficial gluteal muscle (M. gluteus superf.); 58. Dorsal sacrococcygeus muscle (M. sacrococcygeus dors.); 59. Lateral coccygeus muscle (M. coccygeus lat.); 60. Ventral sacrococcygeus muscle (M. sacrococcygeus ventr.); 61. Biceps femoris muscle (M. biceps femoris); 62. Semitendinosus muscle (M. semitendinosus); 63. Lateral gastrocnemius muscle (M. gastrocnemius lat.); 64. Superficial digital flexor muscle (M. flexor digit. pedis superf.); 65. Anterior tibialis muscle (M. tibialis ant.); 66. Long digital extensor muscle (M. extensor digit. pedis longus); 67. Long fibular muscle (M. fibularis longus); 68. Flexor hallucis longus muscle (M. flexor hallucis longus); 69. Lateral digital extensor muscle; 70. Short digital extensor muscle (M. extensor digit. pedis lat. brevis); 71. Interosseus muscle (M. interosseus); 72. Long digital flexor muscle (M. flexor digit. pedis longus).

Middle Plate—A. Skeleton

1. Upper skull; 2. Mandible (Mandibula); 3-3'. Neck vertebrae; 4-4'. Thoracic vertebrae; 5-5'. Lumbar vertebrae; 6. Sacrum (Os sacrum); 7-7'. Caudal vertebrae; 8. Scapula (Scapula); 9. Upper arm (Humerus); 10. Ulna (Ulna); 11. Radius; 12. Carpus (Ossa carpi); 13. Metacarpus (Ossa metacarpalia); 14. Digital skeleton

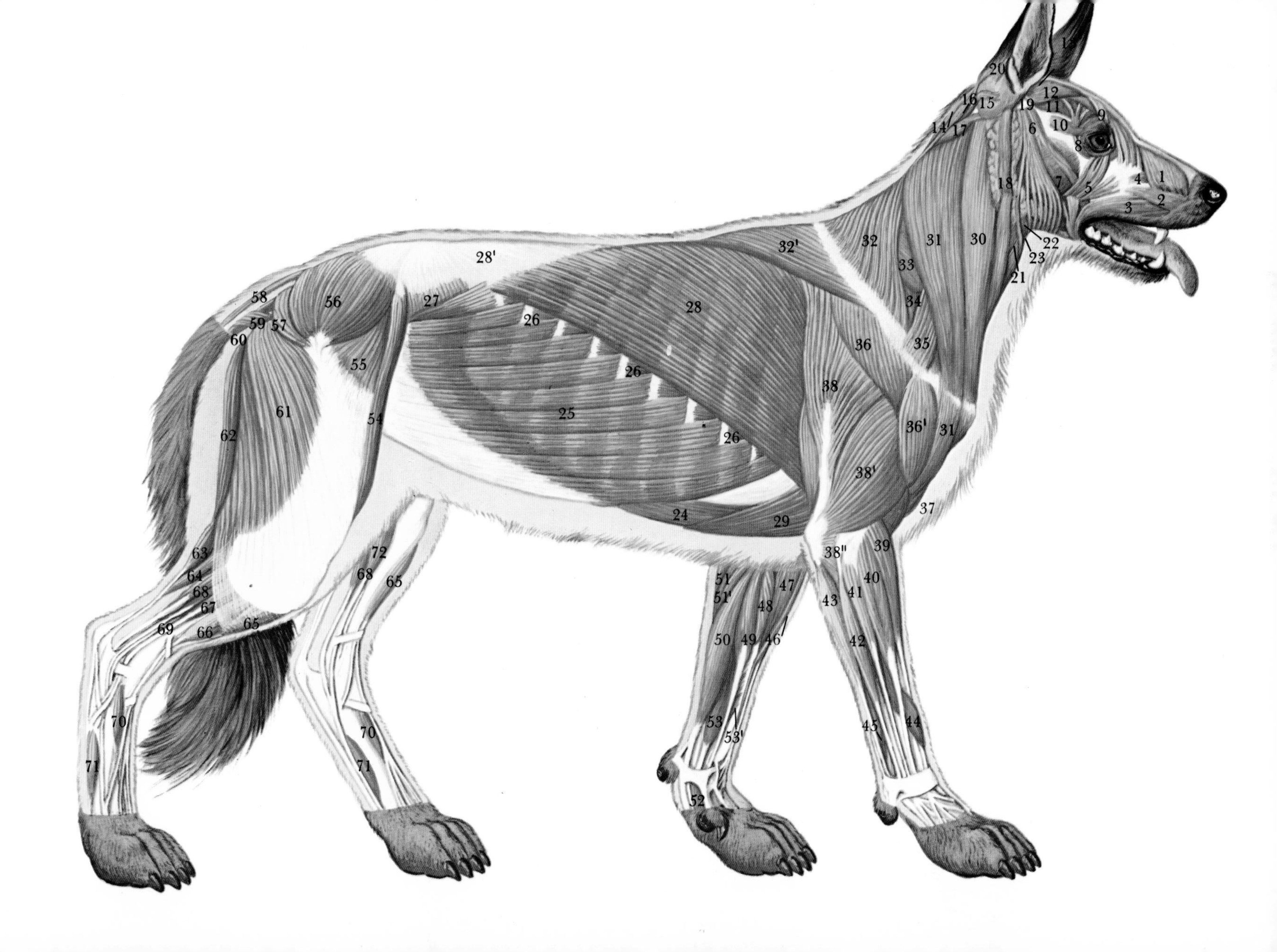

1
2
3
4
5
6
7
8
9
10
11
12
14
15
16
17
18
19
20
21
22
23
24
25
26
27
28
28'
29
30
31
32
32'
33
34
35
36
36'
37
38
38'
38"
39
40
41
42
43
44
45
46'
47
48
49
50
51
51'
52
53
53'
54
55
56
57
58
59
60
61
62
63
64
65
66
67
68
69
70
71
72

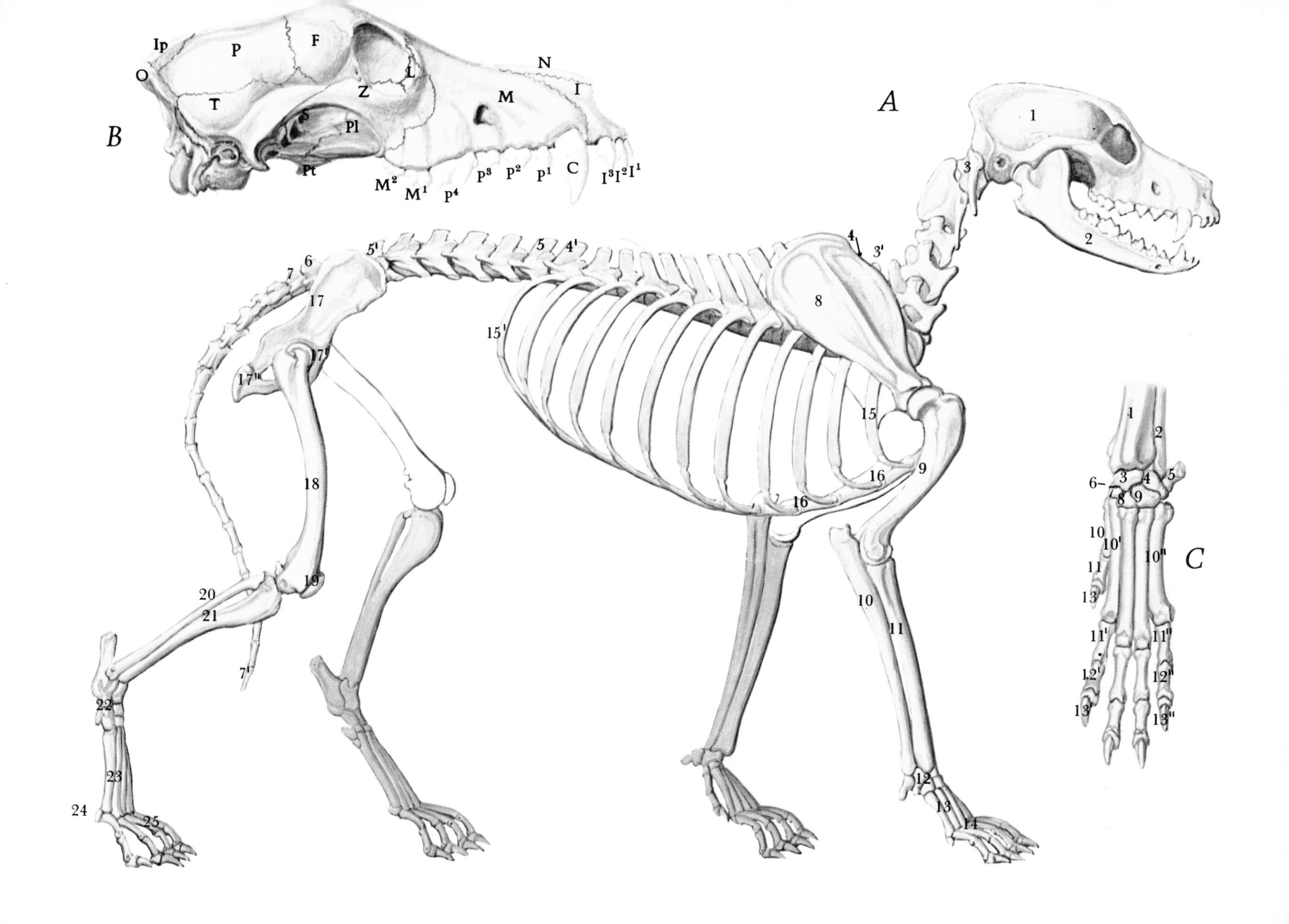

A
B
C
Ip
O
P
F
T
Z
L
S
Pl
Pt
N
I
M
M²
M¹
P⁴
P³
P²
P¹
C
I³I²I¹
1
2
3
3'
4
4'
5
5'
6
7
7'
8
9
10
11
12
13
14
15
15'
16
17
17'
17''
18
19
20
21
22
23
24
25
6–
10'
10''
11'
11''
12'
12''
13'
13''

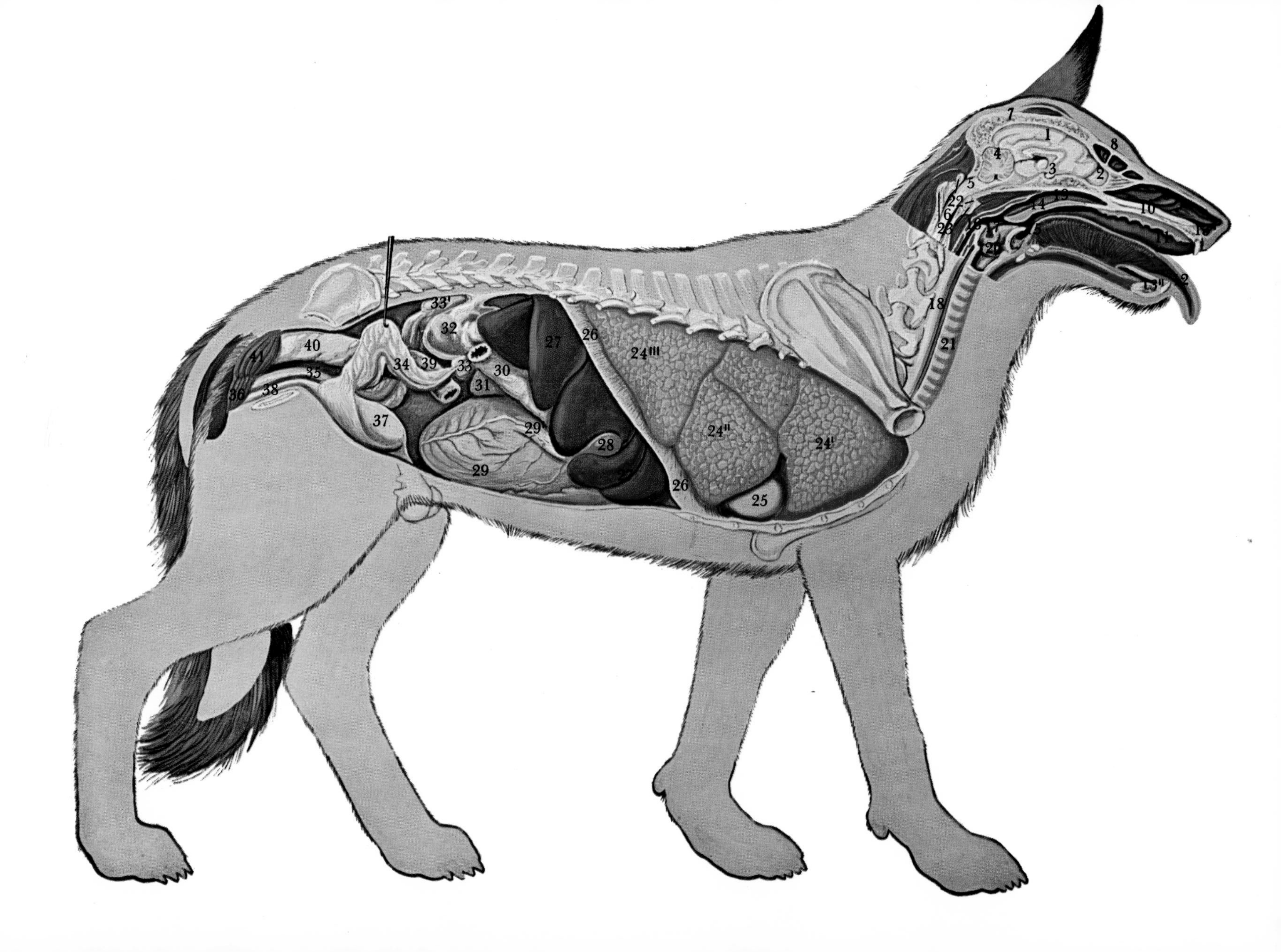
1
2
3
4
5
6
7
8
10
13''
14
18
21
22
23
24'
24''
24'''
25
26
27
28
29
29'
30
31
32
33
33'
34
35
36
37
38
39
40
41

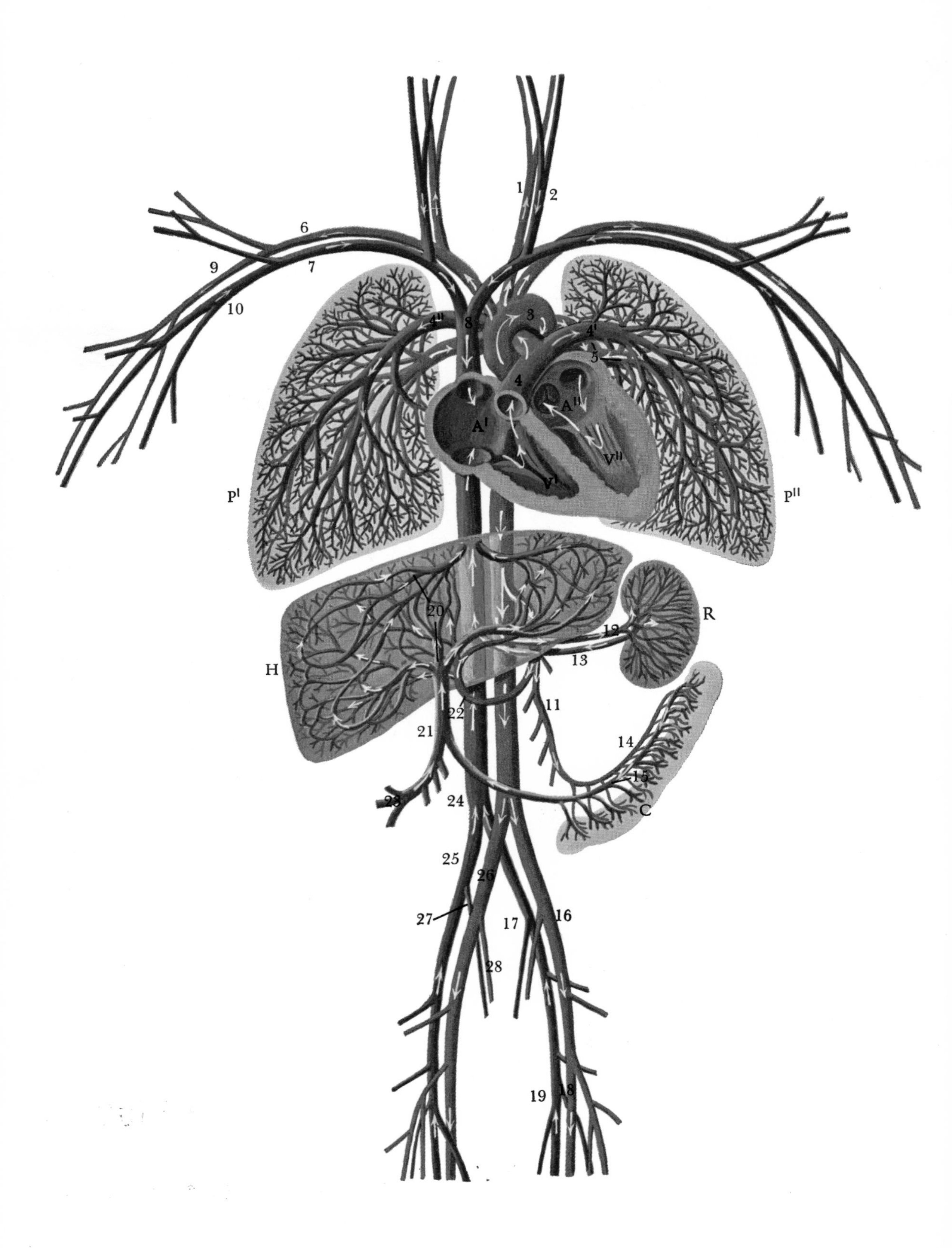

1
2
6
7
9
10
3
4''
8
4'
5
4
A'
A''
V'
V''
P'
P''
20
H
R
12
13
11
22
21
14
15
C
23
24
25
26
27
17
16
28
19
18

(Ossa digit manus); 15-15'. Ribs; 16. Sternum; Pelvis: 17. Ilium (Os ilium), 17'. Pubis (Os pubis), 17''. Os ischium; 18. Femur; 19. Patella; 20. Fibula; 21. Tibia; 22. Tarsus (Ossa tarsi); 23. Metatarsus (Ossa metatarsalia); 24. Calcaneum (Tubor calcanei); 25. Digital skeleton (Ossa digit pedis).

B. Skull and Dentition

Skull components: F. Frontal bone (Os frontale); I. Premaxilla (Os incisivum); Ip. Interparietal bone (Os interparietale); L. Lacrimal bone (Os lacrimale); M. Maxilla (Os maxillare); N. Nasal bone (Os nasale); O. Occipital bone (Os occipitale); P. Parietal bone (Os parietale); Pl. Palatine bone (Os palatinum); Pt. Pterygoid bone (Os pterygoides); S. Sphenoid bone (Os sphenoides); T. Temporal bone (Os temporale); Z. Zygomatic bone (Os zygomaticum).

Dentition: I. Incisors (Incisivi); C. Canine (Caninus); P. Premolars (Praemolaren); M. Molars.

C. The Hand (front foot)

1. Radius; 2. Ulna; Carpus: 3. Fused scaphoid and lunar bones (Lunatum); 4. Cuneiform bone (Cuneiforme); 5. Pisiform bone; 6. Trapezium; 7. Trapezoid; 8. Magnum (Capitaltum); 9. Unciform (Hamatum); 10. Metacarpal of the first finger; 10'. Metacarpal of the second finger; 10''. Metacarpal of the fifth finger; 11. First phalanx of the first finger; 11'. First phalanx of the second finger; 11''. First phalanx of the fifth finger; 12'. Second phalanx of the second finger; 12''. Second phalanx of the fifth finger; 13. Third phalanx of the first finger; 13'. Third phalanx of the second finger; 13''. Third phalanx of the fifth finger.

Lower Plate—The Digestive System

(Medical section of the head; the musculature, ribs, diaphragm, and pleural membrane have been removed from the right side. In the region of the duodenum and colon the intestine has been cut and removed.)

1. Cerebrum; 2. Olfactory bulbs (Bulbus olfactorius); 3. Hypophysis; 4. Cerebellum; 5. Medulla oblongata; 6. Spinal cord (Medulla spinalis); 7. Roof of skull; 7'. Base of skull; 8. Frontal sinus (Sinus frontalis); 9. Nasal cavity (Cavum nasi); 10. Bony palate (Palatum durum); 11. Oral cavity (Cavum oris); 12. Tongue (Lingua); 13'. Lower jaw; 13''. Lower jaw with first incisor; 14. Soft palate (Palatum molle); 15. Isthmus faucium; 16. Vestibule of the esophagus (Vestibulum oesophagi); 17. Trachynx; 18. Esophagus (Oesophagus); 19. Pharynx (Pharynx respiratorius); 20. Larynx; 21. Trachea; 22. First cervical vertebra (Atlas); 23. Second cervical vertebra (Axis); 24'. Apex lobe (Lobus apicalis); 24''. Cardiac lobe (Lobus medius); 24''. Diaphragmatic lobe (Lobus diaphragmaticus of the right lung); 25. Pericardial cavity with the heart; 26. Diaphragm (Diaphragma); 27. Liver (Hepar); 28. Gall bladder (Vesica fellea); 29. Stomach (Ventriculus); 29'. Pylorus; 30. Duodenum; 31. Pancreas; 32. Left kidney (Ren sin) underneath the peritoneum; 33. Right ovary (Ovar dex.); 33'. Left ovary (Ovar sin.); 34. Right uterine horn; 34'. Left uterine horn; 35. Vagina; 36. Constrictor vestibuli muscle (M. constrictor vestibuli); 37. Urinary bladder (Vesica urinaria); 38. Urethra; 39. Colon; 40. Rectum; 41. External anal sphincter muscle (M. sphincter ani ext.).

The Heart and Circulation

A.' Right atrium (Atrium cordis dex.); A''. Left atrium (Atrium cordis sin.); C. Colon; V'. Right ventricle (Ventriculus cordis dex.); V''. Left ventricle (Ventriculus cordis sin.); P'. Right lung (Pulmo dex.); P''. Left lung (Pulmo sin.); H. Liver (Hepar); R. Kidney (Ren); 1. Carotid artery (Art. carotis communis); 2. Jugular vein (Vena jugularis interna); 3. Aortic arch (Arcus aortae); 4. Pulmonary trunk (Truncus pulmonalis); 4'. Left, 4'' right pulmonary arteries (Art. pulmonalis); 5. Pulmonary vein (Vena pulmonalis); 6. Subclavian artery (Art. subclavia); 7. Subclavian vein (Vena subclavia); 8. Anterior vena cava (Vena cava superior); 8'. Posterior vena cava (vena cava inferior); 9. Axillary artery (Art. axillaris); 10. Axillary vein (Vena axillaris); 11. Anterior mesenteric artery (Art. mesenterica sup.); 12. Renal artery (Art. renalis); 13. Renal vein (Vena renalis); 14. Intestinal artery; 15. Intestinal vein; 16. External iliac artery (Art. iliaca ext.); 17. External iliac vein (Vena iliaca ext.); 18. Femoral artery (Art. femoralis); 19. Femoral vein (Vena femoralis); 20. Hepatic veins (Venae hepaticae); 21. Portal vein (Vena porta); 22. Hepatic artery (Art. hepatica); 23. Anterior mesenteric vein (Vena mesenterica sup.); 24. Posterior vena cava (Vena cava inf.); 25. Common iliac vein (Vena iliaca communis); 26. Common iliac artery (Art. iliaca communis); 27. Internal iliac vein (Vena iliaca interna); 28. Internal iliac artery (Art. iliaca interna.).

The ear: sense organ of hearing

The malleus is in close contact with the tympanic membrane which is an elastic membrane that separates the external and middle ear and acts as a receptor of sound waves from the outside atmosphere. Malleus, anvil, and stapes conduct the vibrations to the inner ear, and via the endolymph to the actual sensory organ, the organ of Corti which is endowed with sensory cells. The organ of Corti with its stretched-out basilar and tectarial membranes is found within a cavity which in the mammals is extremely elongated and coiled. This is the cochlea. Here the incoming vibrations are decoded according to wave length and are distributed to the countless sensory cells where the stimulus is transduced and conducted via the auditory nerve to the brain. Here, depending on the number and location of sensory cells stimulated by the various sound intensities, pitch and sound qualities are "understood."

Many mammals have an extremely acute sense of hearing. They can perceive the slightest whisper or rustling which our ear is incapable of detecting. In most mammals the ear pinnae, which often are movable, act as a sound funnel and a "directional aerial." Certain mammals—perhaps quite a number of species—perceive sounds which are too high for our ears because the human organ of Corti does not have receptor zones for these high frequencies. In Volume XI ultrasonic sounds are discussed more fully in connection with bats.

The sense of equilibrium and other senses

Aside from the auditory apparatus, the internal ear also contains the organ of equilibrium. In mammals this sense is essentially not different than in the other vertebrates. The other mammalian sensory organs, also, are structurally not very different from those of the other higher vertebrates; however, the performances vary greatly in the individual mammalian groups. The olfactory sense, more so than the senses of touch, temperature, balance, pain, and taste, is truly amazing in some animals, at least if compared to man who has a rather poorly developed sense of smell. In the carnivores, chemical distance reception reaches its peak of development. These animals follow the tracks of their prey with their noses. Herbivores are dependent on their sense of smell for the early detection of their enemies. In these species the olfactory sense plays an important role in the social and reproductive life. Many possess olfactory glands which secrete substances by which the animals recognize each individually as well as their mutual territory. Mammals with highly developed olfactory capacities, the macrosmatic animals, have large nasal cavities. The interior of the nasal chamber consists of highly folded conchae which are lined with sensory epithelium. In mammals, including man, which have a less highly developed sense of smell, the microsmatic animals, the nasal chambers are smaller and the conchae are less folded. Whales which lost their olfactory sense in their adaptation to an aquatic existence are known as anosmatic animals.

Olfaction and olfactory organs

The mammalian eye and its capacities

For man the eye is the most important sensory organ; however, this is not true for all mammals. Even animals with large, highly functional eyes, for example the dog, often rely more on other sense organs. Only a few mammals which live below the ground are totally blind: certain moles, the golden mole, some burrowing moles, and the marsupial mole. Other mammals, such as many bats, have very poor vision. In most herbivores, including the ungulates, kangaroos, rodents, and lagomorphs, the eyes are in a lateral position in the head. This provides for a wide field of vision but the animals are either unable or are only slightly able to fixate binocularly; thus they cannot view an object with both eyes. Here eye sight is mainly adapted to perceive movement. Stationary objects are only poorly perceived. Bifocal fixation, however, is the prerequisite for spatial vision. Only very few mammals are capable of stereoscopic vision; these include the cats, the primates, and a few primitive forms. Color vision in most mammals is less well developed than in most fishes and reptiles, and even in most birds. Only the primates are fully capable of utilizing colors. Almost all monkeys can differentiate colors just as well as man. Carnivores and herbivores are much less able to do so, but even decidedly nocturnal animals still have remnants of color vision, which could indicate that this capacity was lost secondarily in many mammals.

The circulatory system and associated organs

The circulatory system of mammals (Color plate p. 26) is also more complete than in the reptiles. Just as in birds, the two heart chambers are totally separated by a septum, so that the freshly aerated (arterial) blood does not, as in the amphibians and reptiles, mix with the deoxygenated (venous) blood. The aorta comes out of the left atrium and arches over to the left. This corresponds to the left aortic trunk of the reptiles. In the birds the aorta, which also comes out of the left atrium, bends to the right side. This corresponds to the right aortic trunk of the reptiles (see circulatory system of the reptiles, Vol. VI).

The nervous system, brain, and intelligence

In the higher primates and particularly in man, the central nervous system reaches the highest peak of development in the entire animal kingdom. Also the functions of the cerebrum of the more advanced forms of the other mammalian orders far exceed the performance of other animal classes. This is closely related to the development of homeothermy. A constant internal temperature is very advantageous for the efficient functioning of the extremely complex processes within the brain. The "investment" in the expansion and elaboration of the cerebrum only "paid off" for those animals which were already capable of maintaining a constant internal temperature.

In the evolution of the mammalian brain the cerebrum plays a significant role. The gradual enlargement of the neopallium from the primitive to the more evolved forms is very conspicuous. In the most highly evolved species the surface of the cerebrum is folded and

convoluted which greatly increased the amount of grey matter in the cerebral cortex. This grey cortical matter contains the most complicated neural mechanisms that are necessary for a successful utilization of personal experiences and purposeful action. The innate, predominantly instinctive behavior is influenced by relatively few key stimuli, and the neural connections are much simpler. The behavior of mammals, in particular in the more highly developed forms, is therefore less rigid than in other animals. They are less dependent upon innate behavior patterns. Learning by experience plays an even greater role in their life than in other animals. Yet the most highly developed ones already show the first signs of insight behavior. This capacity is most highly developed in man. However, even man does not make full use of this capacity for insightful behavior. For too often he is moved by drives, prejudices, or primitive training which is ill-adapted to his man-made environment. Each one of us has experienced these innate urges in our own personal or social behavior from time to time.

Mammals as predators and prey

A leopard *(Panthera pardus)* has pursued a chacma baboon *(Papio ursinus)* and is confronting it in the open plain. The desperate attempt by the baboon to defend itself is of no avail. A few minutes later, the leopard overpowered its prey. Strong baboon males, however, often successfully defend themselves against leopards and on occasion have even killed the predator.

A "trend" (i.e., developmental tendency) for an enlargement of the cerebrum and a concomitant improvement in its capacity is found in all developmental lines of mammals. In the constant struggle for survival, an efficient brain is advantageous for carnivores and herbivores alike, just as it is for a rodent or a monkey. Consequently, aside from the human brain, one not only finds highly developed brains in our closest relatives, the great apes, but also in the dolphins which belong to the toothed whales and presumably evolved from the earliest ancestors of the carnivores.

The skeleton, musculature, and the means of locomotion

The skeleton and musculature of the mammals are unique in a variety of ways but only the most important characteristics will be mentioned here. The transformation of the jaw articulation was previously discussed (see p. 23). The oral cavity is separated from the nasal cavity by a secondary palate enabling the animal to eat and breathe at the same time. Ribs are only present on the thoracic vertebrae in mammals. Most arch towards the sternum but the posterior ones are suspended as so-called "floating ribs." The limbs in the mammalian ancestors, the mammal-like reptiles, did not extend far out from the sides of the body as in other reptiles but were more or less in a vertical position beneath the body. This basic structural design was maintained in most mammals: the knee joint points forward, elbows and heels point to the back. In marsupials and higher mammals the coracoid fused with the scapula, and is recognizable as the acromion (Processus coracoides). Otherwise, the limbs became markedly modified in the various evolutionary lines of mammals. Only four modifications are mentioned here. In many tree-climbing mammals, the hands and feet were adapted as grasping organs. In most forms the thumbs and first toes are opposable to the remaining fingers and toes. In most higher monkeys, opposability of the first digits occurs unless the thumbs or

Movement and the structure of the limbs

the capacity of opposability has not been lost secondarily. Others, like the koala, certain prosimians, and new world monkeys, oppose thumb and index finger to the remaining fingers. In fast runners, particularly in animals that inhabit the hard steppe ground, the number of fingers and toes was reduced while at the same time the central digits became enlarged and elongated. In the horse, finally only the central finger and middle toe remained. As a running apparatus this type of limb only swings in a logitudinal direction which, in these forms, resulted in the disappearance of the redundant clavicle in the shoulder girdle. Concomitantly the hand and foot became more erect. The plantigrade animal which put down the entire palm and sole up to the wrist and heel evolved into a digitigrade animal which only placed down the ventral surface of the digits and finally the one-toed animals which only touch the ground with the tips of the digits. In the bats the forelimbs were modified into wings, and in the seals, whales, and sea cows, into flippers. The whales and sea cows completely lost their hind limbs while their tails broadened and took over as main locomotory organs.

The evolutionary origin of mammals by E. Thenius

Despite the great diversity in the mammals, a zoologist will never have any doubt as to which vertebrate class he would assign a living or freshly-killed mammal. Even whales, duck-billed platypuses, or other abnormally appearing forms clearly betray their mammalian nature: They are homeothermic, they possess a hair coat even if there are just a few vestiges left as in the whales, and most important of all, the females always have mammary glands. Additional characteristics are the four-chambered heart, the unique shape of the new lower jaw joint, the translocation of the primitive lower jaw into the middle ear, and many other features which are shared by all mammals living today.

The paleontologist who studies prehistoric creatures has more difficulties. He is usually confronted only with horns and teeth, and often mere bone fragments. On the basis of these fossil remains, he has to infer indirectly the shape and function of the soft parts. Today's living mammals can clearly be differentiated from other vertebrates and even present-day reptiles on the basis of their unique skeletal characteristics. These unique features include the transformation of the lower joint of the jaw and the bones of the middle ear, the newly formed bony palate, and other specializations. Even the shape of teeth and the way they are attached differ in the mammals and present-day reptiles. Although we know that the mammals arose from the ranks of reptiles, it is impossible to derive them from the reptilian forms living today.

Fossil remains of a very diverse group of rather unique, but often quite impressive, animals were discovered in ancient stone deposits of South Africa and the Soviet Union. These were mammal-like reptiles *(Therapsida)*. They were undoubtedly reptiles, and yet they possessed a whole series of characteristics which are only common in mammals today. These mammal-like reptiles lived during the Permian and Triassic periods, which was at the turning point of the Paleozoic and Mesozoic eras, about 270 to 180 million years ago. This was long before the mighty dinosaurs dominated the earth. The various stems of mammal-like reptiles *(Theriodontia)* with mammal-like teeth belonged to this group. The theriodonts clearly showed the trend towards the development of mammal-like characteristics. The shape and size of their nasal conchae, the structure of the rib cage, and other skeletal characteristics indicate that many mammal-like reptiles were already homeothermic. The nasal conchae are not only associated with the olfactory sense but are also important for prewarming the air the animal breathes. From the shape of the ribs it is evident that a diaphragm was utilized in breathing as is common for all homeothermic animals. In this respect it is impossible to draw any distinct lines between reptile and mammal. Many therapsids had crossed the dividing line to the mammals.

What is a mammal, and what is a reptile?

The discovery of the mammal-like reptiles forced scientists to find a new definition for the concept "mammal" as distinguished from a reptile. Wherever one comes across transitional forms which evolved from one animal form or group to another it is impossible to establish distinct identifying criteria. The many characteristics which distinguish today's mammals from their reptilian ancestors evolved in the course of millions of years. This development was a step-by-step, continuous process, and there were forms that represented the various transitional stages. Man, who is greatly fond of categorizing, was often forced to draw artificial dividing lines in the realm of the living where in reality there were gradual transitions. It is particularly difficult to draw boundaries between fossilized reptiles and mammals since typical mammalian characteristics occurred side by side in the various therapsid lines, and developed in a different sequence and at a different rate. This "mosaic" method of evolution made it necessary to pick a random set of characteristics and to use them as criteria for the artificial separation between the reptiles and the mammals. The American paleontologist G. G. Simpson suggested that only those fossil remains be considered as mammals where the tooth-bearing lower jaw bone is directly connected to the skull by the secondary jaw joint. Consequently, a number of related forms which still possess the primary (primitive) jaw joint between the articular and quadrate are classified as mammal-like reptiles even though they may have many other mammalian characteristics. This distinguishing characteristic is clear and is easily applied but subsequently certain fossil remains that were formerly regarded as mammals are now classified among the reptiles. For example, the theriodont genus *Tritylodon* belongs to this group. The molars of these animals have been found in South Africa and more recently in England. They are multi-cuspate and have several

Fig. 1-3.
Lycaenops ornatus is a mammal-like reptile which lived in South Africa over 200 million years ago (upper Permian).

roots just as in the mammals. The gap between the incisors and molars, and the secondary bony palate are mammalian characteristics also found in these animals. However, the lower jaw still consists of several bones as in the reptiles where the articular and quadrate are still located between the primary jaw. It is probable that such forms were homeothermic and had a hair coat but according to Simpson's definition they are not true mammals.

The mammalian stem originated from several roots

The secondary jaw joint did not evolve only once but several times in various related stems. This also occurred with other skeletal characteristics that separate the reptilian from the mammalian class. Seen from this aspect the vertebrates known as "mammals" emerged from various stem roots. Most of these forms vanished, but probably today's monotremes developed from a different line of theriodonts than did the marsupials and higher mammals. The evolution of one stem from several root forms is known as polyphyletic. The general trend in the therapsids to develop mammal-like characteristics make it extremely difficult to state with certainty which of these known forms is the actual stem group of the mammals. The experts have many different opinions. Some scientists consider the dog-toothed animals *(Cynodontia)* as possible ancestors of the mammals. Others believe that the *Bauriamorpha* are the ancestral group. Both of these groups were from the Triassic deposits in South Africa. Another group of paleontologists assumes that the *Ictidosauria* are the ancestral mammalian stock. The ictidosaurs were widely distributed throughout Africa, Asia, and Europe during the Triassic period.

For some time many experts doubted the theory about the transformation of the lower jaw. At first it is rather difficult to believe that the original lower jaw joint within one animal group should gradually lose its function in the course of countless generations, and to be finally found in greatly reduced size inside the middle ear. The formation of a new lower jaw out of two new bone components is just as difficult to comprehend. Such a modification is a slow, step-by-step process. All intermediate steps have not only to be viable but must also be superior to earlier adaptations. Otherwise these new innovations cannot maintain themselves in the struggle for existence. However, how did the intermediate viable form look that was somewhere between the reptile with its primitive lower jaw and the mammal with the newly-formed lower jaw? This whole theory about the origin of the mammals would almost have broken down because of the impossibility of imagining such transitional forms.

Fortunately, however, bone fragments of an animal were discovered which demonstrated the existence of these transitional forms. This fossil was called *Diarthrognathus,* meaning "double-jointed jaw." This animal had the primitive reptilian and also the newly-formed mammalian jaw articulation on each side of the head.

The first mammals lived 200 million years ago

The first indisputable mammals occurred in the late Triassic, approximately 200 million years ago. It was in this period that the reptiles began to evolve and to diversify into higher, more efficient forms. The age of the dinosaurs was dawning. At first the mammals were not equipped to compete with these reptiles. The mammals survived the age of the large reptiles as small mouse to rat-sized creatures. Finally, at the end of the Mesozoic period, the time had come for the mammals because of the extinction of the large reptiles. The great epoch of the reptiles was followed by the dominance of the mammals in the dawning of the Tertiary period. During the Tertiary the mammals underwent an explosive evolution and they adapted to a great variety of ecological niches which had previously been occupied by the reptiles.

Already the small mammals of the Mesozoic era had become diversified into many different evolutionary lines. It seemed as if nature had experimented in order to try out the best possible tooth construction. Of all these groups—the *Triconodonts, Symmetrodonts, Docodonts, Pantotheria,* and *Multituberculates*—only two survived. These are the ancestors of the egg-laying mammals *(Prototheria),* which have remained unknown to date, and the ancestors of the marsupials and higher mammals which have been combined in one subclass as the true mammals *(Theria).* There are several concepts concerning the stem group of the true mammals. The panthotheria may possibly have been the stem of the mammals. The panthotheria occurred in a variety of forms during the Jurassic period and also probably during the late Triassic. The appearance of these animals probably could be compared with today's opossums, phascogales, or tree shrews.

There are not many preserved fossils from the tiny Mesozoic mammals. Usually only teeth have been discovered, and occasionally a jaw fragment. Therefore many details about the origin of the tree mammals are still shrouded in darkness. However, today one can state with certainty that the higher mammals did not evolve from the marsupials which might be assumed on the basis of the latter's body structure. Both groups go back to a common ancestor, and their evolutionary development followed separate lines at an early stage. Already in the Cretaceous period there were marsupials as well as higher mammals. Various marsupials (i.e., *Eodelphis, Didelphodon,* and *Pediomys*) lived in North America during the upper Cretaceous, about 100 million years ago. In Manchuria an insectivore-like higher mammal (*Endotherium*) which was several million years old was found. It had lived during the lower Cretaceous. Recently remains of early ungulates (*Arctocyonidae*) were found in upper Cretaceous deposits. The diversification of the mammals had already begun by then.

Adaptation, destruction, and preservation by D. Heinemann

The ability to maintain a constant body temperature enabled the mammals to be more independent of the external environment. It endowed them with an increased vigor and adaptibility. Homeothermy also facilitated the gradual freeing from the rigidity of innate behavioral organizations. Mammalian behavior could become more diverse and more adaptable. Subsequently the mammals became the rulers of the earth during recent times and have occupied a great variety of ecological niches. They occupy the canopy of the forests as climbers, jumpers, and gliders. They are the secretive inhabitants of underbrush and thicket. They freely gallop over the wide open plains or they climb along treacherous paths in the cliffs of the mountains. Mammals can be found burrowing deeply under the surface of the earth or swimming in competition with the fish in the oceans of the world or flying through the air as shadows of the night. The habitat of a mammal may be the hot and humid rain forest of the tropics, the tropical sand desert, or the icy wasteland of the arctic. Mammals are also superbly adapted to the most unfavorable climatic conditions. Only one being, man, challenges the hitherto undisputed dominance that the other mammals have held for millions of years. He is constantly displacing them so that he can provide more space and food for millions of his own kind and his domesticated animals, which are likewise mammals. He kills millions of them, not only because of necessity but because of his greed or for the mere joy of killing. In a few decades he has annihilated species which took millions of years to evolve. These were creatures that man did not create, and will never be able to re-create. Our ancestors exterminated the Aurochs from the forests of Central Europe. The bison and the Alpine ibex would almost have shared the same fate. The Boers shot hundreds of thousands of quaggas in order to make sacks of their hides. Burchell's zebra, blaubok, Schomburg's deer, Falkland Island dogs, Steller's sea cow, the Cape and the Atlas lions were exterminated by man just as were several species of prosimians, whales and marsupials. The wild horses, bontibokke and blesbok, white-tailed gnu, bison and Père David's deer were saved practically only at the last minute. Today the orang-utan, the Javan and Sumatran rhinoceroses, the Andean deer, the Persian fallow deer, and many other species are in immediate danger of extinction, as, presumably, is the blue whale, the largest of all mammals. The mammals suffer most by man's misuse of his power which he obtained over his fellow creatures because of his highly developed brain. Protection of the mammals should have first priority in the preservation of animals because man is himself a mammal.

Dietrich Heinemann
Erich Thenius

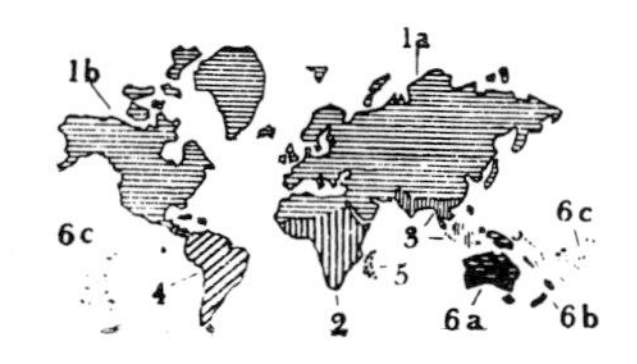

Fig. 1-4.
Zoo-geographical zones

1. Holarctic region:
a. Palearctic, b. Neartic.
2. Ethiopian region.
3. Oriental region.
4. Neotropical region.
5. Malagasy region.
6. Australian-oceanic region: a. Australian region, b. New Zealand, c. Oceania.
In each of these regions there is a distinct mammalian fauna.

2 Egg-laying Mammals

Distinguishing characteristics

Subclass *Prototheria*. There is only one order, the *Monotremata*. The animals are plump, stout, short-legged and short-tailed. The HRL is 40–80 cm. The body covering may be thick fur or hair and spines, which may be interspersed with irregular horny papillae reminiscent of reptilian scales. Hair follicles are associated with alveolar sebaceous glands. Sweat glands are located on the hand surfaces and the soles of the feet. There are two teatless mammary glands in the form of densely arranged tubular glands which ooze milk into mammary depressions. The ♂♂ possess a hollow horny spur on the ankle joint. The eyes are small, and the ear pinnae are small or missing. The arms are stocky and the broad hands have the ability to grasp. There are five fingers and five toes which terminate in strong, flattened claws. Occasionally a reduction of the last toe digits occurs. The horny snout is elongated, and the skull is flat and almost reptile-like. The inner ear is flat but mammal-like. The larynx is without vocal chords. Only the juveniles have teeth. The shoulder girdle consists of a large independent coracoid and epicoracoid, just like in the reptiles. The epipubic or "marsupium" is associated with the pelvis. The brain is large and mammalian with some reptilian elements. The mammalian heart and circulatory system has a few reptilian characteristics (incomplete right atrio-ventricular valve). The excretory and reproductive ducts plus the anus open into the cloaca. The penis is attached to the ventral wall of the cloaca. In the female monotremes, the oviducts open separately into the excretory-genital canal, whose end serves as the vagina. There one to two (rarely three) large yolked eggs from the left ovary only are fertilized and covered with a soft shell. There are two families: 1. SPINY ANTEATERS *(Tachyglossidae)* and 2. DUCK-BILLED PLATYPUS *(Ornithorhynchidae).* There are a total of three genera and six species.

Order: Monotremes by B. Grzimek

Professor Dr. Wilhelm Haacke, who was my predecessor as the director of the Frankfurt Zoological Garden during the last part of the previous century, discovered on his trip to Australia in 1884 that the spiny anteater, a mammal, laid eggs. At the same time, the Australian W. H. Caldwell discovered the same thing in the duck-billed platypus of Queensland, Australia. This coincidence settled an ongoing dispute which had been fought by zoologists in England, France, and Germany during 1789. The controversy centered around the taxonomic position of these egg-laying mammals. Where should the Monotremata of cloacal animals be classified in the animal kingdom? The order consisted of only two families, the spiny anteaters and the duck-billed platypuses. They occurred only in East Australia, New Guinea, and Tasmania. Remains of the extinct ancestral form have not been found anywhere else.

The first duck-billed platypus was thought to be a hoax

The first complete platypus skin which arrived at the London British Museum in 1798 was believed to be a hoax. The beaver-like fur and tail and the genuine dried duckbill was a phenomenon too strange to be real. However, reasons for doubting were not completely unjustified because the freighter which had brought the skins had sailed through the Indian ocean. Ports on this ocean had passed on many exotic and stuffed relics for many gullible sea captains who paid dearly for them. Fantastic "new species of birds of paradise" assembled from bits of skin and feathers, and even stuffed "mermaids," carefully sewn and glued from some dried monkey heads and the scaly posterior skin of large fish, were brought back by the sea captains.

Four years later, however, a complete duck-billed platypus carcass arrived in England and was dissected by the great Scottish anatomist Sir Everald Home. So it was shown that this mysterious animal indeed existed. Consequently, it was discovered that the bill was not as hard as that of a duck. The first dried skins had fooled the zoologists. However, controversy continued about whether the animals were mammals or a special class of vertebrates. In 1824 medical professor Johann Friedrich Meckel of Halle, Germany, and, incidentally, a corresponding acquaintance of the poet Goethe, discovered mammary glands on the body of a female platypus. The French school under the leadership of Geoffroy Saint-Hilaire maintained that these glands were a type of sebaceous gland because they thought it impossible that young platypuses could suck milk with a duckbill. Sir Everald Home and the famous paleontologist Richard Owen advocated the view that monotremes lay eggs, but that the young are born alive and without a shell. In fact, they would hatch inside the mother's body. After all, such phenomena were well documented for the reptiles. A medical doctor by the name of John Nicholsen from Victoria, Australia, wrote to Richard Owen that some gold diggers had caught a platypus and tied it up inside an empty whiskey case. The next morning they found two soft white eggs, which were without a shell and could be squeezed. "Miscarriage caused by fear," was R. Owen's sympathetic comment and he never wavered from this opinion.

Haacke and Caldwell discovered that mammals lay eggs

On September 2, 1884, two important telegrams arrived at approximately the same time. One was sent by Haacke to the Royal Society of Australia, and the other was sent by Caldwell to the members of the British Zoological Society who were holding a conference in the Canadian city of Montreal.

In Adelaide, Haacke had kept a few spiny anteaters which had been caught on nearby Kangaroo Island. Since he was well aware of the controversy about the reproduction and taxonomy of these animals, he asked an institute assistant to hold up one of the females by the hind leg so that he could examine her abdomen. Here are his own words describing the event: "Only an animal expert could comprehend my dismay when I pulled an egg from the abdominal pouch, the first egg laid by a mammal. This unexpected discovery left me dumbfounded to such an extent that I committed the stupid act of firmly squeezing the egg between my thumb and index finger to crack it. The egg contained a thin fluid which already had started to disintegrate because of the upsetting circumstances the mother had experienced. The elliptical egg was 15 mm long and 13 mm wide. The shell was tough and parchment-like, similar to reptile eggs."

Egg-laying mammals

1. Duck-billed platypus *(Ornithorhynchus anatinus)*.
2. Bruijn long-nosed spiny anteater *(Zaglossus bruijni)*.
3. Tasmanian short-nosed spiny anteater *(Tachyglossus setosus)*.
4. Australian short-nosed spiny anteater *(Tachyglossus aculeatus)*.

At approximately the same time, on August 14, on the Burnett River, Caldwell shot a female platypus which had just laid an egg. He opened up the animal and found a second egg in the oviduct, almost ready to be laid. The embryo was developed to the same stage as a chicken egg incubated for three days and, just like the chick embryo, it was lying flatly against the yolk-sac with its open abdominal cavity. This type of embryonic development, which is characteristic of large-yolked eggs, is called "meroblastic." Since telegrams sent from Australia to Canada are not exactly cheap, Caldwell summarized his discovery in these now famous four words: "Monotremes oviparous, ovum meroblastic." Caldwell was unable to mail this message for five days to a friend in Sydney who then quickly sent off the exciting telegram.

Family spiny anteaters

The family of the SPINY ANTEATERS *(Tachyglossidae)* is characterized by the ability to dig in rapidly when in danger and by a diet of ants, termites, and other insects which they pick up with their sticky tongues. The body is plump, compact, and short-tailed. The HRL is 40–80 cm. The long snout is beak-like, round when seen in cross sections, and covered with a horny material. The mouth opening is very narrow and the tongue is long and worm-like. The body is covered by fur with dorsally located spines. Teeth are completely missing, and instead there are horny ridges on the palate to aid in grinding the food. The ♀ has a marsupium or temporary pouch in the central region of the abdomen. The milk flows from the mammary glands along several hair tufts into the pouch.

Distinguishing characteristics

There are two genera and five species. 1. The SHORT-NOSED SPINY

1
2
3
4

Top: Australian spiny anteaters (*Tachyglossus aculeatus;* see p. 43). Spiny anteaters, which enter the water occasionally, swim quite well.
Bottom: The duck-billed platypus glides through the water like a submarine. The animal probes with its duck-like bill for snails and other aquatic animals.

ANTEATER (*Tachyglossus*) has an HRL of 40–50 cm. This genus has a flat body appearance. The snout is short and straight. The external ear is absent, and the tail is stump-like. The second foot claw is specialized for grooming. There are two species, the AUSTRALIAN SHORT-NOSED SPINY ANTEATER (*Tachyglossus aculeatus;* Color plate p. 41), which is found in Australia and New Guinea, and the TASMANIAN SHORT-NOSED SPINY ANTEATER (*Tachyglossus setosus;* Color plate p. 41), which is found in Tasmania and the Bass Strait Islands. 2. The LONG-NOSED SPINY ANTEATERS (*Zaglossus*) have an HRL of 55–78 cm. The legs are longer, and the snout is long and slightly curved. The external ear is small. The spines are longer and less dense. The fourth and fifth hand and foot claws are partly reduced. There are three species. The BARTON LONG-NOSED SPINY ANTEATER (*Zaglossus bartoni*) has short white spines covered by long black hair. The ventral side is free of spines. There are five claws. This anteater is found in Northeast and South New Guinea. The BRUIJN LONG-NOSED SPINY ANTEATER (*Zaglossus bruijni;* Color plate p. 41), with three or four claws, is found in West New Guinea and the Salawati Islands. The BUBU LONG-NOSED SPINY ANTEATER (*Zaglossus bubuensis*) has short, white spines which protrude from the brown dorsal hair. There are five claws. It is found in the Bubu river region of New Guinea.

Along with man, the spiny anteaters belong to the few mammals which have a life expectancy of over fifty years. One anteater in the London Zoo lived for thirty years and eight months. Another one in Berlin reached the age of thirty-one or possibly even thirty-six years. The exact records were lost during the air raids of World War II. One Australian anteater made its home in the Philadelphia Zoo from 1903 until 1953. He occupied a small enclosure with a hiding box, and here he lived for forty-nine years and five months. If his age prior to his zoo existence were added, it could be said that he reached the average age of civilized man.

Reproduction of the anteater

The anteater female puts the freshly laid eggs into her abdominal pouch and carries them around for seven to ten days, in a manner reminiscent of kangaroos or other marsupials. The young, when hatched, are only 12 mm long. They lick the thick yellowish milk which trickles from the hair close to the mammary gland. The small anteaters stay inside the mother's pouch for six to eight weeks until their spines develop. In the meantime, they reach lengths of nine or ten centimeters. At this point, the mother hides the babies in a nest-like structure. They are sexually mature at the age of one year and weigh about two and one half to six kilograms. The spines are up to six centimeters long. The pouch in the female is only prominent during the breeding season. In the Prague Zoo, it was observed that even some anteater males developed an abdominal pouch every twenty-eight days.

Anteaters have reproduced only twice in captivity, one time being in Berlin in 1908, but the one young survived for only three months. In 1955 a cold young weighing 83 g and measuring 12.5 cm in length was found in the Basel Zoo. After being warmed, the young started to move but it did not survive beyond two days.

Although the spiny anteaters are not known to climb trees in nature, they are quite able to clamber up the sides of a wire cage. When they reach the top, they do not know how to get down again, and they often drop to the floor, injuring themselves. The animals are silent except for some snorting sounds. Spiny anteaters have the amazing ability to dig rapidly vertically down into hard soil with great speed. Cansdale timed one animal until it disappeared; it took nine minutes. Despite this ability, these animals do not build their own subterrestrial burrows but use those of other animals. Usually spiny anteaters dig until half of their body, which is protected dorsally by spines, is submerged. It is impossible to lift up an animal from this position because it digs in with its large claws and pulls down its lateral spines so that they point to the ground. Consequently, anyone trying to touch its abdomen will be pricked by the spines.

The European hedgehog and the spiny anteaters share one advantage and one disadvantage. They are able to roll themselves into a ball, but they have difficulty in keeping the fur between the spines free from parasites. Therefore, they have to scratch themselves frequently with a specialized long and slightly curved grooming claw found on the second toe. Their eye sight is poor but, on the other hand, they are able to detect the slightest earth tremor. The tube-like, toothless mouth and the long tongue indicate that they exist primarily on ants, termites, and other insects. Anteaters which are kept in captivity will readily eat anything that will fit through their mouth opening, such as bread soaked in milk, fresh or soft-boiled eggs, and minced meat. Spiny anteaters, quite in contrast to their closest relatives the platypuses, are able to fast up to one month. In the southern part of their range, Victoria and Tasmania, these creatures hibernate.

These little fellows are amazingly strong. Captive anteaters have ripped off meshed wire which was nailed over their box and they have lifted off a box lid weighted down with heavy objects. When searching for food in the open, they will tip over stones and boulders twice their weight with comparative ease. A zoologist in Adelaide, who kept a spiny anteater in his kitchen overnight, found the next morning that the animal had moved the heavy cupboard, the table, chairs and boxes from the wall into the center of the room. In contrast to the platypus, spiny anteaters are often very active during the day, particularly during warm weather.

These animals are also able to walk on two legs. The zoologist Michael Sharland surprised a juvenile on a trail in Tasmania. The animal sniffed about and, when it felt tremors from the man's footsteps, became alarmed and reared on his hind legs. He remained like that for a few moments and then fled into the bush on two legs.

Family
Ornithorhynchidae

Distinguishing characteristics

As the spiny anteaters are adapted for searching and consuming insects on the land, so the DUCK-BILLED PLATYPUS (*Ornithorhynchus anatinus;* Color plate p. 41) is specialized for hunting small animals under water. There is only one species in the Family *Ornithorhynchidae.* The HRL is approximately 45 cm and the TL is 15 cm. The ♂♂ are larger than the ♀♀. The snout is elongated into a broad flat bill. The nostrils are situated above the tip of the bill. The eyes are small, and there are no ear pinnae. The evenly dense fur has large hair bristles and a very fine woolly underfur. Only the juveniles have teeth. There are 34 tooth buds: $\frac{0 \cdot 1 \cdot 2 \cdot 3}{5 \cdot 1 \cdot 2 \cdot 3}$, but only 12 breakthroughs: $\frac{0 \cdot 0 \cdot 1 \cdot 2}{0 \cdot 0 \cdot 0 \cdot 3}$. These teeth are later replaced by horny plates that have cross wrinkles and ridges near the front of the mouth to facilitate the crushing and straining of food. The five fingers and toes are cleaved, and the hands and feet are webbed. In the hands, the webbed skin extends beyond the claws. The brood pouch is absent. There is one species and four subspecies.

The duck-billed platypus is found in Australian and Tasmanian waters up to an elevation of 1650 meters. They are active at dawn and dusk. The animals will float on the water surface like a dry piece of wood, and a splash can be heard when the animal dives underwater with a flick of the tail. The animal is able to close a protective skin fold over its eyes and ears. Underwater the platypus depends heavily on the tactile sensory organs which are concentrated on the soft skin of the bill. The average time spent underwater is one minute, but in case of danger the animal can stay submerged for up to five minutes without breathing.

In the creeks and rivers, the platypus catches larvae, crustaceans, snails, and occasionally small fish. Smaller prey is temporarily stored in the external cheek pouches along with grit which aids in the grinding up of food. Crayfish and larger animals caught on land are transported in the bill. The animals swim almost silently while they sift the water for food. Only rarely does one hear slight humming sounds.

The burrows of the platypus have elaborate channels. The breeding burrow of the female opens one third of a meter above the water, extends upward into the bank up to seven meters above the water line, and horizontally into the bank as far as eighteen meters. The water may enter the burrow during flooding; thus a platypus, upon leaving the water, may enter its burrow very wet, but when it emerges moments later, its coat is dry and glossy. The tunnel walls seem to absorb the water like a sponge.

The female carries bundles of wet leaves under her tail, which is folded forward into the nest chamber. The entrance to the nest is plugged with soil in one or several places. The female usually lays two eggs, sometimes one or three. The animal curls around the eggs while they are incubating, or she may lie on her back and incubate the eggs on her abdomen, according to the observation of some zoologists. The temporary brood pouch, which is a characteristic feature in the anteaters, is not developed in the platypus. The eggs, which are the size of sparrow eggs, measure 1.6 to 1.8 cm, but are a little rounder. Incubation time is between seven and ten days. The incubating mother does not leave the burrow for days at a time, and then only briefly to defecate or to wet her fur. After she returns, she again plugs up the tunnel. A newly hatched platypus is about 2.5 cm long, blind, and naked. After about four months, when fully covered by hair and about 35 cm in length, the young emerge from the burrow. Soon they probe the mud for snails and larvae just like their parents.

Fig. 2-1.
A platypus can stand up on his hindlegs like a dog.

Platypuses belong to the few "poisonous" mammals. Both ankles of the male's hind limbs have inwardly-directed hollow spurs which are connected to venom glands. One male attacked his mate in captivity, driving his spurs into her flanks and the female nearly died as a result. A warden, spurred by a platypus, fell to the ground in intense pain. His hand and arm became greatly swollen and he suffered for months from the poison in his system and from loss of control in the injured hand. Females too have spurs but only as juveniles.

Up until now platypuses have been taken from their home range to another continent only three times, and each time it was to the Bronx Zoo in New York City. The keeping of platypuses in America has an interesting history. As early as 1910, the Australian zoologist Harry Burrell had designed and built a portable water tank for platypuses which could be connected to a labyrinth of tunnels. The inside of the tunnels was lined with rubber welts through which the animal had to squeeze itself to reach the nest. This helped to squeeze the water from the animal's fur. This provision was necessary since free-living platypuses dry their fur in earth tunnels. Burrell's first captive platypus escaped after sixty-eight days. One of the remaining four was displayed in the Sydney Zoo for three months. At this point, Burrell had enough of the platypuses. To provide the necessary food for the five animals, Burrell had to spend six hours daily to either dig or fish approximately two pounds of earthworms, crustaceans, beetle larvae, or water snails. When he kept only one platypus, he soon discovered that this one could easily devour the same amount all by itself.

Fig. 2-2.
A platypus grooming his fur.

Following World War I, Ellis S. Joseph, a well known animal collector, persuaded his friend Harry Burrell to again keep platypuses. Joseph was very eager to import a living platypus to America. Burrell obliged and provided him with more than just one animal. On May 12, 1922, Joseph started his return journey from Australia to San Francisco with many Australian animals, including five duck-billed platypuses, which were housed in the artificial Burrell burrow, and of course a large supply of earthworms. After forty-nine days at sea, the

Fig. 2-3.
This is how the duck-billed platypus swims.

ship arrived in San Francisco with one surviving platypus but not a single earthworm. It took Ellis Joseph several days to collect earthworms before he could continue on to New York via train.

In the Bronx Zoo, the platypus was displayed to visitors for one hour every afternoon. Long lines of people slowly filed past the open tank. Director William T. Hornaday complained that he had to spend $4.00 to $5.00 daily to provide half a pound of earthworms, forty shrimps, and forty grubs for one such small animal. Actually, this amount of food was quite insufficient, as we know today. Nevertheless, he wrote: "Really, it is hard to believe that such a small animal needs such large amounts of food. I know of nothing comparable among the mammals." The animal died after forty-seven days on August 22, 1922. This was taken as a quite acceptable life span in captivity.

Robert Eadie, director of the Sir Collin Mackenzie Sanctuary in Healesville, Melbourne, Australia, was more successful in keeping platypuses. The famous "Splash" lived for four years and one month (1933–1937) in an artificial burrow based on Burrell's design. When David Fleay became director of the same zoo, which was located in a forest, he introduced "Jack and Jill" in 1938. The enclosure was so situated that the female was able to dig in a wall of soil. When Jack was approximately six years old, he started to court Jill. The male nibbled the hairless beaver-like tail of his mate and the two swam slowly in a circle. This is the way platypuses seem to express their love. Mating took place in the middle of October, and Jill retired to her burrow on the 25th. After four months, the young appeared for the first time. They enjoyed playing with each other, and humans could also entice them to play. Jill, the mother, lived for almost ten years, and Jack, the father, reached the age of seventeen.

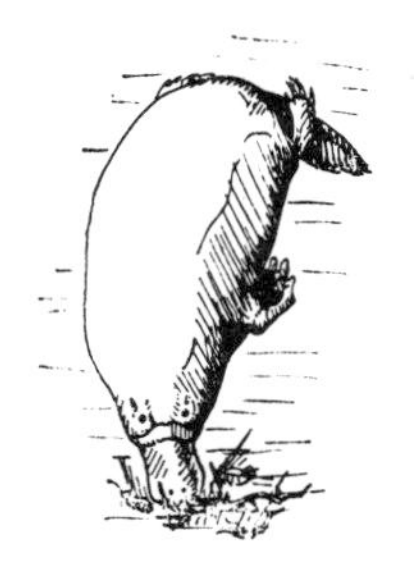

Fig. 2-4.
A duckbilled platypus searches for food on the river bottom.

The great success in Healesville made the people of the Bronx Zoo very anxious to get their own platypuses. Consequently, on March 29th, 1947, David Fleay and his wife, with one male and two female platypuses, boarded a ship for Boston. This time the journey did not take forty-nine days but only twenty-seven; nevertheless, the worm supply had to be supplemented twice during this trip. The animals arrived in Boston in good shape. They had been kept for one year previously in Healesville. From Boston the journey continued by car to New York where the animals were exhibited to the general public for three days. It was here that I saw my first platypus. I learned that an animal of 1½ kg weight could consume 540 earthworms daily, plus twenty to thirty shrimp, 200 mealworms, two small frogs, and two eggs. It was possible to discover numerous new facts about this Bronx Zoo platypus. The platypuses would preferably enter the water only if it were above 15°C; at less than 10°C, they stayed on land. Of course, the daily food bill for one single platypus was much more than four or five dollars. In the winter, large quantities of earthworms had to be

Fig. 2-5.
This is how the duck-billed platypus digs its den.

flown in from Florida. Two out of the three animals in New York lived more than ten years; thus their average life span was at least eleven years.

In the meantime, David Fleay had moved to the vicinity of Brisbane in the warm climate of Queensland. Here in his private zoo at West Burleigh, I heard the story about the "flying platypuses," which were to liven up the empty facilities at the Bronx Zoo. In 1946, it had been fairly easy to catch the first three platypuses. They had been caught within three weeks in the vicinity of Healesville. Actually nineteen animals had been caught but only the three best were selected. This time it was a different matter. Fleay was required by the government of Queensland and the Australian Commonwealth to have a permit to catch and to export platypuses, because these animals are now one of the most protected species. When permission was finally granted, the rainy season had not started and this was the cause of the low water level in the creeks and rivers. Many rivers had been reduced to small ponds or muddy fields. It was obviously a bad year for the platypuses. The females had not even dug their breeding burrows.

David Fleay started to search for platypuses in very rough terrain full of hills and deep gorges. The catchers who aided him nearly succumbed to the burning heat and mosquitoes because they had to be absolutely quiet when they spotted a platypus. After three months in the bush, 13,000 km logged by car, and repeated telegrams of warning, David finally caught three juveniles, one pair and one female. It was decided to transport the animals by plane. This decision was tested in a return flight to Brisbane. The animals were put into boxes cushioned with freshly cut grass. One female became very sick from over excitement, and had to be released again.

Winter was approaching in Queensland, and it was no great pleasure to wade and to swim in rivers to set up traps while looking for that one platypus. By chance, four weeks prior to the flight to the United States, an additional female was caught in a cow pasture. Now everything was ready for the flight. Five thousand earthworms and 5000 mealworms preceded the platypus' flight to Hawaii. The worms travelled in sterile plastic material because it was forbidden to bring earth to Hawaii to guard against the introduction of foreign plant diseases. In West Burleigh, the platypus would not accept washed earthworms. This necessitated a flight for the earthworms one week earlier so that they could be conditioned to Hawaiian soil.

All was ready and the platypus babies travelled from Brisbane to Sydney with some 10,000 earthworms, 2500 mealworms, and 550 shrimp. This simply did not last, however, because the plane to New York was delayed by two days. A telegram was dispatched to West Burleigh asking for an additional supply of several thousand more earthworms and fifty shrimp. Two hours after departure, the platypuses became very excited from the noise of the four engines. They swam around madly and tried to climb the corners of their container. At the first stop-over in Fiji, all three platypuses had disappeared in their hiding places. In Hawaii, the Fleays went through customs and health inspection. In the meantime, the health inspectors had transported the container from the plane and had tipped it in such a manner that all the nesting places of the platypuses were flooded. This required that the wet cushions were replaced with dry ones. The animals, however, were fine. Sunday morning at 7:30, the executives of the Bronx Zoo waited at Idlewild Airport, New York, to meet the Australian transport. Unfortunately, the three platypuses only lived eight months.

The spiny anteater and the duck-billed platypus have few enemies

Neither the spiny anteater nor the duck-billed platypus are particularly threatened species in Australia today. They have few natural enemies, except perhaps snakes, foxes, or the Tasmanian devil. However, platypuses still suffocate in fish traps today. They enter them underwater, but they cannot find their way out and hence are unable to surface for air. Unfortunately, there is no law in Australia which makes it mandatory for fishermen to provide an escape hatch on the top of their wire fish trap baskets.

Spiny anteaters have enjoyed full protection since 1905, and their numbers have increased since then. They are particularly numerous in Tasmania, and can be seen occasionally on the outskirts of Hobart, the Tasmanian capital. The zoologist Sharland thinks it is quite probable that an occasional burrow may be located right beneath the streets of some suburbs. This does not mean that the average pedestrian will see these shy and nocturnal animals. Occasionally, anteaters are run over by cars, just like hedgehogs in Europe. Spiny anteaters are more widely distributed than the platypuses. I believe that the former belongs to the most widely distributed wild animals of Australia, but I seriously doubt that this is due to Australian protective laws. I gained the impression that these laws are generally not heeded because any person is free to buy a gun and to shoot at any animal that moves only five miles outside the city limits. Platypuses and anteaters are not greatly endangered by all this shooting because of their secretive habits, worthless fur, and little meat which does not taste good at that. Significantly, not even the most ignorant sheep rancher can claim that these creatures kill his lambs, or eat his sheeps' food, as any wild animal that competes with sheep in any way is not safe although the law says it should be.

Bernhard Grzimek

3 The Marsupials

Distinguishing characteristics

Order *Marsupialia.* General characteristics: There are great size variations from the shrew-like animals to the giant kangaroos. The HRL is 10–200 cm, and dense fur is the rule. The toes and fingers terminate in claws. The first toe is clawless, except in the mouse opossums and marsupial moles. The sebaceous and sweat glands are well developed. There may be 2-27 teats, which are often arranged in two or four rows or sometimes in a circle. The opossum *(Didelphis)* has an uneven number of teats. The teats are long and serve as a means of attachment for the underdeveloped young (Color plate p. 146). A teat muscle (compressor mammae) controls the injection of milk into the mouth of the young. The skull is completely mammal-like. The angular process (processus angularis) of the lower jaw is curved to the inside (except in the honey phalangers, *Tarsipes).* There are 18–56 teeth. There is no replacement of teeth except for the fourth premolar. The collar bone is present, but it is reduced or absent in the bandicoots. The pubic bone of both sexes is equipped with a pair of "marsupial bones" (Ossa marsupialia) which are reduced only in the marsupial wolf *(Thylacinus).* The brain is primitive. The cerebral cortex is small, flat, almost without convolutions, and does not cover the brain stem. The body temperature, which is lower than in higher mammals (34–36°C), is independent of external temperatures. For young marsupials the nipple completely fills the oral cavity. The larynx is situated immediately behind the oral cavity just as in whales. This results in keeping the digestive and respiratory tubes completely separated. The penis, which lies behind the scrotum, is surrounded by a penis sheath. The anus and genital duct open into the cloaca which is surrounded by a sphincter muscle. Originally there were two completely separated uteri and vaginas, and the more recent marsupials show varying degrees of fusion of these structures. Marsupial distribution is limited to the faunal regions of Australia and South America, except for a few opossums which advanced into North America during the ice ages.

When the young koala (*Phascolarctos cinereus;* see p. 119) has outgrown its mother's pouch, it is carried about on her back for still another year.

Petaurus
Potorous
Macropus
Dendrolagus
Macrotis
MACROPODIDAE
Phascolarctos
Phalanger
Tarsipes
Diprotodon
Vombatus
PHALANGERIDAE
Myrmecobius
Notoryctes
Procoptodon
PERAMELIDAE
Thylacinus
Thylacoleo
VOMBATIDAE
PHASCOLARCTIDAE
DIPROTODONTIDAE
Dasyurus
NOTORYCTIDAE
MYRMECOBIIDAE
Phascogale
THYLACININAE
DASYURIDAE
Didelphis
DIDELPHIDAE
?
Chironectes
BORHYAENIDAE
Thylacosmilus
Prothylacinus
Caenolestes
CAENOLESTIDAE
POLYDOLOPIDAE

There are nine families: 1. OPOSSUMS *(Didelphidae)*; 2. CARNIVOROUS MARSUPIALS *(Dasyuridae)*; 3. MARSUPIAL ANTEATERS *(Myrmecobiidae)*; 4. MARSUPIAL MOLES *(Notoryctidae)*; 5. BANDICOOTS *(Peramelidae)*; 6. CAENOLESTIDS *(Caenolestidae)*; 7. PHALANGERS *(Phalangeridae)*; 8. WOMBATS *(Vombatidae)*; 9. KANGAROOS *(Macropodidae)*; with a total of 71 genera and 241 species.

The word marsupial brings the image of a kangaroo to most people's minds. Kangaroos and wallabies are only one specialized group within an extraordinarily diversified order. Besides kangaroos, there exist many weasel-like, marten-like, and wolf-like carnivorous marsupials. There are climbing marsupials of squirrel, flying squirrel, and teddy-bear shapes. There are marsupials that resemble badgers, moles, rats, mice, elephant shrews, and many other types.

Order: Marsupials
by D. Heinemann

All members of the marsupialia are characterized by a unique type of reproduction. The embryos are nourished inside the uterus for only a brief period. The chorio-allantoic placenta is found only in the long-nosed bandicoots *(Perameles)*. All other marsupials have a transient "yolk-sac placenta" which is formed from the chorion (p. 21) and endometrium. The embryos stay inside the uterus for only a brief period, usually 8–42 days depending on the species. The fetus is very small at birth, measuring from ½ to 3 cm. Their sensory organs are underdeveloped, the primitive kidney is still functional (see Vol. III), and the hind legs are bud-like. One can hardly compare these embryos with those of other mammalian young. Even the naked and blind altricial young of mice and hedgehogs seem highly developed in comparison. Young marsupials are actually premature embryos whose gestation period was merely interrupted by the birth process and not terminated by it as in other mammals. Marsupial babies require the immediate protection of the maternal body. This function is fulfilled by the pouch which the young leaves for the first time after many weeks. This is, so to speak, their second birth.

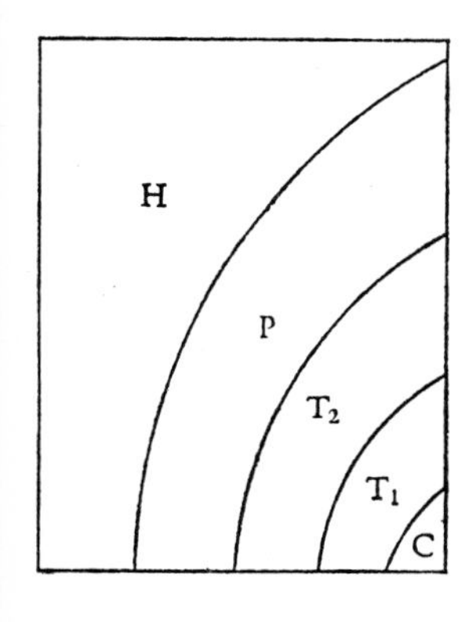

The phylogeny of the marsupials

C—Cretaceous, approximately 120–70 million years ago. T^1—Paleocene, approximately 70–25 million years ago. T^2—Pliocene, approximately 25–1 million years ago. P—Pleistocene, 1 million–20,000 years ago. H—Holocene, recent–20,000 years ago. The dotted lines indicate animal groups for which no fossil records have yet been found. Living animals are drawn in color, extinct ones in gray.

There are great variations in the pouch. The simplest version of the pouch consists of many teats which are individually surrounded by a specialized skin fold serving as protection for the embryo attached to the teat. A more advanced form is a pouch which opens posteriorly and is enclosed by a circular skin fold. The teats are inside the pouch. The most advanced form is the closed pouch. The pouch, which points either anteriorly or posteriorly, is closed by a sphincter muscle. Pygmy marsupial mice *(Marmosa)*, woolly opossums *(Caluromys)*, marsupial shrews *(Monodelphis)*, marsupial anteaters *(Myrmecobius)*, and rat opossums *(Caenolestidae)* have either no pouch or only a rudimentary one.

The unique mode of reproduction and the other previously mentioned peculiarities in the body differentiate the marsupials from all other mammals to such an extent that they have been grouped separately as a special subclass, the *Metatheria*, marsupial mammals.

Phylogeny of the marsupials by E. Thenius

The oldest marsupial fossils were recorded in the upper Cretaceous period in North America. During that period, various forms (i.e., *Eodelphis*, among others) lived which greatly resembled the present-day American opossum. This suggests that marsupials may have originated in the New World. However, this question has not yet been settled, and it can only be confirmed by additional fossil discoveries, although the absence of fossilized marsupials in Asia and most recent research results in Paleogeography point to the origin of marsupials in South America rather than in Asia. Recent geological tests on the position of the former magnetic fields indicate that South America and Australia were connected via Antarctica in earlier geological times.

Fossil records have made it possible to explain the distribution and phylogeny of the marsupials outside Australia. The origin of the Australian marsupial, on the other hand, is still shrouded in mystery because of few fossil discoveries from the Tertiary period. Opossums (*Peratherium*, etc.) were distributed throughout North America and Europe during the Tertiary. Ancestral forms from the Cretaceous period gave rise to the marsupial stock of South America. Marsupials flourished in South America during the Tertiary. One would not have surmised this by looking at today's South American marsupial fauna. Only opossums *(Didelphidae)* and Caenolestids *(Caenolestidae)* with twelve genera are found in South America today. Aside from the vastly different yapoks, most South American marsupials have a rather mouse- or rat-like appearance. In the Tertiary, herbivorous marsupials, with highly specialized teeth, and numerous carnivorous marsupials *(Borhyaenidae)* existed. The sabre-toothed marsupial *(Thylacosmilus)* of the pliocene was one of the more remarkable extinct marsupials. It is astonishing to what extent the teeth and the skull (except the decidedly smaller brain) resembled that of the extinct sabre-toothed tigers (*Machairodus*, etc.). However, the two forms are in no way related. The sabre-toothed tigers were true feline carnivores, while *Thylacosmilus* was undoubtedly a marsupial. This similarity between a marsupial and a higher mammal is one of the most interesting examples of convergence, that is, similar adaptations in response to similar environmental conditions. A comparable similarity can be seen between the marsupial and the golden mole, or the flying phalanger and the flying squirrel.

The BORHYAENIDS appeared as opossum-like forms *(Eobrasilia)* already in the late Paleocene of North America. In the Miocene marten-like Borhyaenids, like *Cladosictis* and *Thylacodictis*, were found along with dog-like forms such as *Prothylacinus* and *Borhyaena*. The latter were often regarded as close relatives of the Australian carnivorous marsupials *(Dasyuridae)* which some of them strongly resembled. The investigations of the American paleontologist G. G. Simpson have shown, however, that the similarity between South American and

Australian marsupials is based on parallel evolution and that both stocks rose independently from opossum-like animals.

The carnivorous marsupials branched out into many forms in South America because in the Tertiary the true carnivores were originally not present there. During most of the Tertiary, South America was separated from Central and North America by ocean channels. A land bridge was formed only at the end of the Tertiary, which in the late Tertiary and during the glacial periods enabled insectivores, ungulates, proboscidians, rodents, lagomorphs, and the ancestors of present-day South American carnivores to migrate across this land bridge from North to South America. During the Tertiary, the Borhyaenids occupied the ecological "niches" that the more recent carnivores occupy in South America today. In a similar manner, the true ungulates and lagomorphs took over the roles of the notoungulates and the litopterna. These became extinct at the end of the Tertiary. Only three remaining species of the Caenolestids, which in the lower Tertiary were represented by many species, have survived to the present, represented by three species. Opossums became extinct in North America and Europe during the late Tertiary. Opossums found in North America today are Pleistocene or ice age immigrants from South America.

Until a few years ago, only one Tertiary marsupial *(Wynyardia bassiana)* from the lower Tertiary was known in Australia. It was found in Tasmania. Recent excavations performed under the direction of Prof. R. A. Stirton (University of California) in South Australia have led to the discovery of additional Tertiary marsupials. For the time being, this gives us only an initial glimpse of the former marsupial fauna of the Fifth Continent. The fossil remains discovered up to now have not clarified the origin or the phylogenetic relationship of the individual marsupial stems. Nevertheless, Stirton's excavations have shown that the various marsupial stems, which died out during or after the ice ages, had existed during the Oligocene/Miocene period.

The Australian marsupials developed into regular giants during the ice ages, a phenomenon which occurred in several higher mammals in other continents. When the climate was still moist, these Australian animal giants occupied the regions that are grass lands and semi-deserts today. In this area particularly gigantic kangaroos, some with short snouts and short tails *(Sthenurus, Procoptodon)*, one giant koala *(Phascolarctos ingens)*, and one giant wombat *(Phascolomis gigas)* were found. In addition to these forms, there were herbivorous giant marsupials the size of rhinoceros, nototherions or diprotodonts *(Diprotodon, Euowenia, Euryzygoma,* and *Palorchestes)* with broad wombat-like skulls. The large leopard-sized marsupial lion *(Thylacoleo carnifex)* was probably a highly specialized herbivore. The few greatly enlarged, sharp molars are indicative of a meat-eating diet, but the frontal teeth are not at all similar to those of a carnivore.

Some of these extinct giant marsupials were still living 10,000 years ago, a period when man had already reached the Australian mainland. Extinction was not cause by humans but was probably due to climatic changes which ultimately resulted in changes in vegetation and the drying up of extensive land regions of Australia. Further extinction of species continued with the introduction of the dingo by the early Australians and, finally, the introduction of foxes and rabbits by the white man. Of course, man was further responsible for large deforestation and utilization of large land tracts solely for the purpose of raising sheep. The Tasmanian wolf, the Tasmanian devil, and several other marsupial species were completely exterminated from the Australian mainland due to the actions of humans. Even today the destruction has not ceased.

Dietrich Heinemann
Erich Thenius

4 Opossums

Family: *Didelphidae*
by W. Gewalt

Distinguishing characteristics

Family *Didelphidae:* The animals vary in size from that of a mouse to that of a cat. The HRL is 7–45 cm and the TL is 4–45 cm. The facial part of the skull, which is long and pointed, seems rat-like. The small to medium-sized hairless ears occasionally can be folded. The eyes ("button eyes") frequently protrude in the nocturnal members. The vibrissae are long. The tail is frequently longer than the body and the tail base is cylindrically enlarged for fat deposits in certain species. The tail, which is frequently prehensile, is either naked or only partially covered with hair. There are five fingers and five toes on the hands and feet. There are claws on all digits except the first toe which is opposite to the other toes while the animal is grasping. The fur is smooth in the mouse opossums and woolly in the woolly opossums. Some genera have long, projecting cover hairs (i.e., the American opossum). The hair coloration may be grey, yellow, or brown, and a dark mask around the eyes is found in some species.

Some water opossums have broad dark crossbands. Didelphids are commonly omnivorous, although meat predominates in the diet. There are 50 teeth with the following dental formula: $\frac{5 \cdot 1 \cdot 3 \cdot 4}{4 \cdot 1 \cdot 3 \cdot 4}$. The molars are multicuspate and there may be extra teeth. There are great variations in the pouch structure, which may be absent altogether, merely indicated by two lateral abdominal folds, or fully developed with a posterior opening (water opossums) or open towards the front as in all other forms. There are two to twenty-seven teats, which are arranged in several rows or in a circle. The gestation period is short and there are many young per litter. The animals are distributed over South America, certain regions of North America, and a few other places. There are 12 genera, 76 species, and 163 subspecies.

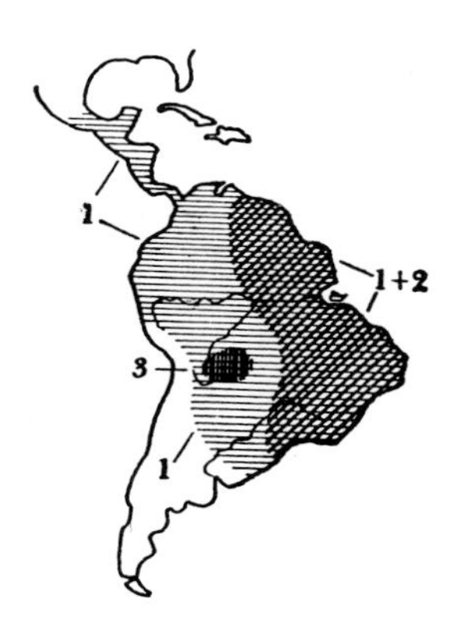

Fig. 4-1.
Distribution of: 1. Red woolly opossum *(Caluromys laniger)*. 2. Yellow woolly opossum *(Caluromys philander)*.
3. Black-shouldered woolly opossum *(Caluromys irrupta)*.

A. WOOLLY OPOSSUMS (Genus *Caluromys*). The HRL is 22–30 cm and the TL is 22–40 cm. The ears and ⅓ to ⅔ of the prehensile tail are naked. The fur is soft and woolly, much like a "teddy bear's." Their diet consists of fruit, insects, and other small animals. There are three species and twenty subspecies: 1. The BLACK-SHOULDERED WOOLLY OPOSSUM *(Caluromys irrupta)*. Externally the animals have a dark dorsal stripe and a dark collar stripe which extends to the front legs. Only two specimens have been found up to now. The other species are uniformly red, grey, or brown on the dorsal side and yellow on the ventral side. 2. YELLOW WOOLLY OPOSSUM *(Caluromys philander)*. 3. RED WOOLLY OPOSSUM *(Caluromys laniger;* Color plates p. 61 and p. 62).

Fig. 4-2.
Bare-tailed opossums
1. Short bare-tailed opossum *(Monodelphis domestica)*. 2. Striped bare-tailed opossum *(Monodelphis americana)*.
Bushy-tailed opossums
3. *Glironia venusta*.
4. *Glironia aequatorialis*.

B. SHORT BARE-TAILED OPOSSUMS (genus *Monodelphis*). The HRL is 7–16 cm and the TL is 4–8 cm. The eyes are small and regressed like those of the shrews. The tail is naked, and the pouch is not developed. The teats are arranged in a circle on the abdomen, and there are eight to fourteen young. There are ten species, two of which are: 1. SHORT BARE-TAILED OPOSSUM *(Monodelphis domestica)*. The species, which is frequently found in human settlements, is welcomed as an exterminator of insects and rodents. 2. STRIPED SHORT BARE-TAILED OPOSSUM *(Monodelphis americana;* Color plate p. 62).

C. BUSHY-TAILED OPOSSUMS (Genus *Glironia*). The HRL is 16–21 cm and the TL is 20–23 cm. The genus resembles the murine and "four-eyed" opossums, but in contrast, the tail is well covered with fur and is bushy to the tip. Only four specimens have been found. There are two species: 1. *Glironia venusta;* 2. *Glironia aequatorialis.*

D. MURINE OR MOUSE OPOSSUMS (genus *Marmosa*). The naked ears are delicate and can be folded during sleep. The smooth and shiny fur is from yellowish grey to chestnut brown dorsally and whitish on the ventral side. The eyes are often surrounded by a dark mark. The tail base is covered by hair, while the rest of it is naked and scaly. In some species the tail base is swollen by fat deposits. The pouch is not developed. There are nine to nineteen teats, which are arranged in two rows or in a circle on the abdomen. There are over forty species and approximately 150 subspecies. Three species are listed here. 1. MEXICAN MOUSE OPOSSUM *(Marmosa mexicana);* with an HRL of 13–18 cm and a TL of 15–20.5 cm. 2. MURINE OPOSSUM *(Marmosa murina)* has an HRL of 10–16 cm and a TL of 16–19 cm. 3. ASHY OPOSSUM *(Marmosa cinerea)* with an HRL of 15–18 cm and a TL of 20–25 cm.

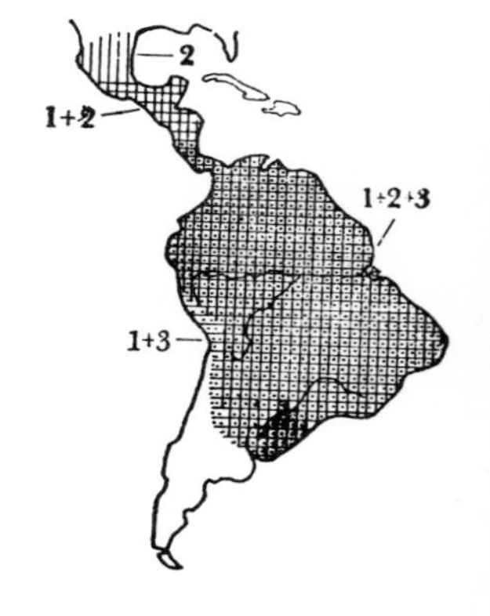

Fig. 4-3.
1. Mouse opossums (genus *Marmosa*).
2. "Four-eyed" opossum *(Metachirops opossum)*.
3. Rat-tailed opossum *(Metachirus nudicaudatus)*

E. "FOUR-EYED" OPOSSUMS (genera *Metachirops* and *Metachirus*). The HRL is approximately 30 cm and the TL is approximately 35 cm. There is a distinct light spot above each eye. Each genus has one species, the "FOUR-EYED" OPOSSUM *(Metachirops opossum;* Color plate p. 62), with a well developed pouch which opens to the front. There are twelve subspecies. There is also the RAT-TAILED OPOSSUM *(Metachirus nudicaudatus)*, whose pouch is indicated only by lateral skin folds. There are twelve subspecies.

F. LITTLE WATER OPOSSUM OR THICK-TAILED OPOSSUM (genus *Lutreolina*). The HRL is approximately 30 cm and the TL is approximately 35 cm. The ears and legs are short, and the tail is thickly covered with fur

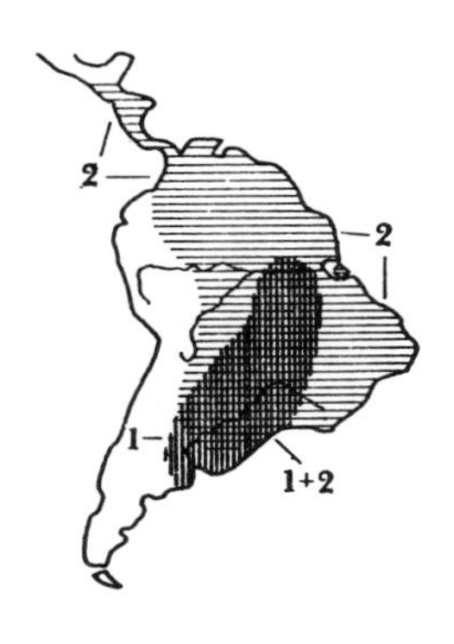

Fig. 4-4.
1. Thick-tailed opossum *(Lutreolina crassicaudata).*
2. Water opossum or yapok *(Chironectes minimus).*

at the base and conspicuously enlarged by fat deposits. The hair on the terminal part of the tail is sparse and the tip is not as prehensile as in other Didelphids. There is no pouch. The animal is an excellent swimmer and seems to prefer water. It is predominantly found in swamps and river banks, although it also occurs in forests and on the treeless pampas. There is one species, the THICK-TAILED OPOSSUM (*Lutreolina crassicaudata*, which translated literally is the thick-tailed fish otter; Color plate p. 62).

G. WATER OPOSSUM OR YAPOK (genus *Chironectes).* The HRL is approximately 35-40 cm and the TL is approximately 40-50 cm. The animal is adapted to an aquatic existence (see p. 65). There is only one species, the WATER OPOSSUM OR YAPOK *(Chironectes minimus;* Color plate p. 62). The tail is naked, scaly, and flat.

H. LARGE AMERICAN OPOSSUMS (Genus *Didelphis).* The HRL is 40-45 cm and the TL is 30-35 cm. There are two species, the LARGE AMERICAN OR NORTHERN OPOSSUM *(Didelphis marsupialis;* Color plate p. 62) and the AZARA'S OR SOUTHERN OPOSSUM *(Didelphis paraguayensis).* This species is somewhat smaller in size, with longer legs and distinctly dark head markings (usually several dark streaks). Both species have many subspecies. There are three additional genera.

In addition to the popular kangaroos and koala bears, the opossums are, at least because of their names, the best known marsupials, although this animal group is not even a typical representative of this unique mammalian order. Their home range is not located in the classical marsupial land of Australia, but in South, Central, and North America. Some opossums do not even possess the characteristic pouch. The young of the mouse opossum of the genus *Marmosa* are only the size of grains of rice. A single opossum litter may contain up to eighteen young, or, according to some records, twenty-one young. Normally, however, only half a dozen of these develop into adults. The pouchless species carry their young in a "bundle" either between the hind legs or on the back of the mother.

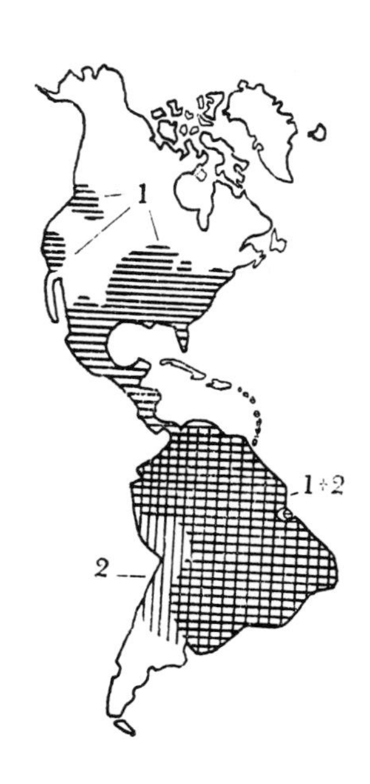

Fig. 4-5.
1. Common opossum *(Didelphis marsupialis).*
2. Azara's opossum *(Didelphis paraguayensis).*

The numerous sharp teeth, the five fingers and toes on all four paws, and various other morphological characteristics indicate that the Didelphids are at the bottom of the marsupial phylogenetic tree. The opossums lack the specialized adaptations that characterize the more highly evolved "modern" marsupials. Examples of these specializations are the broad grinding molars of the herbivores and the enormously elongated hind legs of the runners and jumpers where the number of toes has been reduced to two fully developed and very powerful toes. However, the opossums constitute a relatively uniform basic type which still contains all the original phylogenetic components. Most species have a prehensile tail, although none of the Didelphids can use this "fifth hand" with anywhere near the same mastery as the South American spider monkeys. Nevertheless, certain ground-dwelling opossums, such as the mouse opossums, still use their tails for grasping and even transporting material. Walker observed a mouse opossum in a Guiana Indian hut which curled its tail around a piece of paper, then placed the tail on its back and carried the paper off, presumably to its nest.

In South America the smaller species of the Didelphids occupy the niche of the order of insectivores which are almost completely absent there. The larger members consume eggs, small vertebrates, fruit, and carrion. Some are efficient hunters, like the striped mouse opossum called "Gatita" by the South Americans after their relatives, which attacks birds and mammals hardly smaller than itself. It is no wonder that Goeldi, in the anthropomorphic manner of speaking of his time, accused them of "blood thirst and a blind pleasure in cruelty, imprudent restlessness, and insatiable gluttony."

Most species are arboreal and live in forest and parkland regions. Snethlage found the yellow woolly opossum in trees along the traffic-jammed streets of South American cities. In recent times man has extended the range of opossums by introducing certain species to some islands in the West Indies. Opossums have further been introduced to Madagascar, New Zealand, and Pemba, along the East African coast.

Apart from the peculiarities in body structure and development, the opossums fill a special position within the marsupials in other respects as well. While many Australian marsupials are threatened with extinction today because of changes in the environment and the introduction of foreign species, the American opossums have demonstrated that they can compete with the "modern" mammals on their continent. Many Australian marsupial species have been pushed back to small centers of their range or into special sanctuaries because of their very specialized requirements. In contrast, the opossum is still extending its range today. The large American opossum has successfully competed for food and living space with raccoons and lynx, the threat of dogs and rats, technology, and man. The most northerly distributed marsupial, it has already crossed the border into Canada. The surprising adaptability which allows the opossum to utilize any dark hole as its den, be it behind the bark of a jungle tree or below the floor boards of a modern weekend house, was also discovered by H. Krieg in a similar form in the South American pygmy marsupial mouse. These delicate creatures, which are so reminiscent of deer mice, were undoubtedly "made" for an arboreal existence. However, Krieg also found them living in stone rubble on completely flat terrain. Another factor contributing to their great adaptibility is the omnivorous feeding habits of the murine opossums, which can live equally well on insects, fruit, lizards, earthworms, or carrion.

A condition of "lethargy," similar to hibernation in higher mammals, is an additional factor for the Didelphid's success. The

The woolly opossums (*Caluromys laniger*; see p. 58) are nocturnal and arboreal. The agile, quick movements are reminiscent of our squirrels. The pouch is vestigial and is only indicated by two lateral skin folds.

1
2
3
4
5
6
7
8
9
F·66

animals are capable of "sleeping" through cold, dry, or hungry periods. The body temperature and the rate of respiration and heart beat decreases and the body's metabolic rate is slowed down until conditions are favorable again. During such a fasting period the fat deposits under the skin are utilized.

It is well known that occasionally tarantulas, tropical insects, or small snakes find their way into European harbors and market halls with goods brought in by some banana freighter. On several occasions I (Gewalt, W.) have received mouse opossums and "four-eyed" opossums which reached Germany as stowaways. These animals had survived for weeks in cold storage among green bananas that were not palatable to them. In the Berlin Zoo I once received a Mexican mouse opossum *(Marmosa mexicana)*. Arriving with a load of fruit from Central America, it had been caught, supposedly as a "night monkey," in the large municipal market hall. The animal was barely as large as a golden hamster and noticeably slimmer. Only when we placed her into the hurriedly assembled terrarium did I notice a cluster of eleven young between her hind legs. The youngsters, which were about the size of one finger segment, were attached by their mouths to the whitish millimeter-long teats. In a manner of speaking, they were literally glued to the mother's abdomen. The newly-born kangaroo also grows attached to the mother's teat in a similar manner, but in this case the youngster is snug and well-protected inside of the pouch. The young of the pouchless mouse opossums, on the other hand, hang freely and seem to be in constant danger of being dropped. Although still in the infant stage, the *Marmosa* youngsters demonstrate the amazing aerial feats that some aerial acrobats perform when they hang suspended in mid-air by holding on to a leather strip with their teeth. My heart beat faster whenever I observed the mouse opossum mother swinging recklessly through the twigs with her eleven appendages. As a result of her hasty movements, the youngsters dangled in all directions. Some would bounce against twigs, become caught in a forked twig, or even fall to the ground. A lost baby would immediately scream piercingly, a sound well within the ultrasonic region and not audible to human ears. The mother responded quickly to the cries of her lost young.

As soon as the young have opened their eyes, they move from the abdomen first to the flank and finally to the back of the mother. The manner in which the murine opossum mother transports her children *(Marmosa murina)* served as the example of the "Aeneas rat" picture done by Maria Sybilla Merian (see p. 67).

The Didelphids, and almost all the other marsupials, did not fare too well in earlier times in the assessment of their mental abilities. During the times of Brehm and Bölsch, it was customary to anthropomorphize the characteristics and abilities of an animal. The opossum

Opossums

1. Four-eyed opossum *(Metachirops opossum)*.
2. Red woolly opossum *(Caluromys laniger)*.
3. Thick-tailed opossum *(Lutreolina crassicaudata)*.
4. North American opossum *(Didelphis marsupialis virginiana)*.
5. Yapok *(Chironectes minimus)*.
6. Short bare-tailed opossum *(Monodelphis americana)*.

Marsupial mice

7. Yellow-footed marsupial mouse *(Antechinus flavipes)*.
8. Narrow-footed marsupial mouse *(Sminthopsis crassicaudata)*.
9. Central jerboa marsupial *(Antechinomys spenceri)*.

was thought to be "stupid," "sluggish," "lazy," "somnolent," or even "ugly" and "repulsive." In German the term "marsupial stupidity" *(Beuteltierstumpfsinn)* was coined to show that a marsupial was more stupid than any other comparable higher mammal. This misconception was probably the result of observing a marsupial under completely improper conditions. It is possible that people had kept some secretive nocturnal animal in a bare, brightly-lit cage or some other biologically inappropriate arrangement. Hediger has repeatedly emphasized the importance of keeping and observing wild animals in accordance with their natural living conditions. The opossum had always been regarded as an untameable animal. However, Hediger demonstrated without great difficulty that the opossum can be friendly and playful and can even learn and be trained. Nevertheless, it seems that the Didelphids do have rather primitive behavior patterns, such as their relatively little evolved facial expressions, their amazing insensitivity to pain, their simple brain structure, and their relatively few specialized adaptations. Hediger rightly cautions against interpreting these peculiarities as "stupidity" or "dullness." We do not know enough about the habits of most marsupials to judge their real performances.

For centuries man has thought of the Didelphids as vermin that had to be exterminated, as prey that had to be hunted, or as curiosities. Yet he was never interested in understanding their true nature. The animals are extremely easy to catch. It makes little difference if the traps are baited with meat or fruit, camouflaged, or left in the open. During his explorations in South America, H. Krieg was repeatedly annoyed by finding hissing and spitting Didelphids in his traps in the morning. He had been hoping for something "better."

A cornered opossum, with its open mouth displaying rows of teeth, is a fearsome sight. The mouth can open beyond 90° and, if necessary, can remain open for fifteen minutes. They share this capacity with only a few other carnivorous marsupials and the tenrecs of the order of insectivora. This behavior is displayed by young mouse opossums when they are not even the size of a May bug. If they are disturbed in their hiding place, twelve gaping and hissing mouths may confront the intruder.

Not all mouse opossums are as small as their name indicates. There are some species, like the GREY MOUSE OPOSSUM *(Marmosa cinerea)*, which have an HRL of 50 cm. It is very difficult, aside from size, to differentiate between the various species of the genus *Marmosa*. Even in technical books, photographs of mouse opossums were often only labelled with the genus name *(Marmosa)* because it was difficult to ascertain the species. The mouse opossum, like all its relatives, lives solitarily. Although the usual habitat of the mouse opossums is the forest, parkland, and fruit plantations, they have also moved into barren boulder fields. In their natural habitat, the mouse opossums move about comparatively slowly through the branches while searching for insects, fruit, bird nests, and similar food.

One can easily keep murine opossums on a diet of meal worms, grasshoppers, pears, bananas, raw meat, eggs, and milk. The animals are charming to watch and their unique parental care actually makes them ideal pets. It seems there is no end to the fascinating observations one can make; for example, they will quickly catch a flying insect in the air. It is particularly charming to observe one whole litter of *Marmosa* young reaching for one elderberry each and placing them in their mouths. Initially wild, the animals eventually stop hissing with opened mouths at their keeper and become quite tame in captivity.

The water opossum (Yapoks)

The WATER OPOSSUM or YAPOK *(Chironectes minimus)* is even better adapted to an aquatic existence than the thick-tailed opossum. Most other opossums, which have remained very flexible in their choice of habitat, can live equally well in trees, bush or forest, on the ground, in grass, between boulders, or in house attics. The yapoks, however, prefer to live in the vicinity of creeks, ponds, or rivers. The body is covered by a short, firm, dense fur. A glance at the yapok's hind feet betrays his aquatic mode of life; they have well developed webs between the toes. The fingertips, however, are thickened into pads which are reminiscent of the fingerpads of tree frogs.

The long, coarse facial bristles, which are arranged around the snout of the water opossum, are reminiscent of those of otters and seals. The facial bristles aid in detecting a prey under water. The small ears can be tightly folded and the eyelid glands protect the eyes against the water.

When we think of such perfect swimmers as otters, sea lions, or dolphins, and compare them to the water opossum, then its aquatic adaptations seem to be inadequate. Their swimming abilities are only the mere beginnings, however. The yapok actually "runs" in the water, aided by the undulations of the tail. While slowly probing through the mud, the yapok collects shellfish, crayfish, and above all, spawn. Only rarely does he catch a quickly moving fish. The animal digs a burrow into the bank and only leaves it at night. This is one of the reasons why so little is known about this animal. Although the yapok has a large range, it is only rarely seen. On his South American expedition, Krieg considered it an extraordinary catch when a water opossum became tangled in one of the weir-baskets, because so very few specimens of these interesting animals are available even in museums. It was formerly believed that yapoks possessed cheek-pouches for transporting food, but, according to Haltenorth, this is not so. Both sexes, however, have a pouch, although in the male the scrotum is situated in the marsupium. The female may carry up to seven young in her pouch.

Brood-care behavior by the water opossum has long remained a puzzle. What happens to the young when the mother dives underwater? Older authors like Snethlage assumed that female yapoks stay out of the water as long as they are raising young. Walker has observed, however, that females are able to swim and dive with the youngsters in their pouch. The pouch opening points to the rear and can be completely closed by a well developed sphincter muscle to make it watertight. The result is a small, air-filled nursery with a limited supply of oxygen. When the young are larger, they are probably left in the nest while the mother goes hunting, or possibly they are clustered in the fur on brief diving maneuvers. There is always some oxygen present in the spaces between the dense fur and within the clump of young. Water opossums have rarely been displayed in zoos. The animal, which requires an aquarium type of enclosure, probably would make a highly interesting member of any nocturnal animal house in a modern zoo.

The opossums by B. Grzimek

The cat-sized OPOSSUMS (Genus *Didelphis*) are the best-known and by far the most impressive representatives of the entire family. The American opossum is still generally regarded as *the* Didelphid. Before Europeans had set foot upon Australian soil and become acquainted with kangaroos, koalas, wombats, and carnivorous marsupials, it was also known as *the* marsupial. When zoologists examined the first opossum female, they not only discovered the pouch but also the paired uteri (Uterus duplex) with the two separate oviduct exits. The name given to the genus, *Didelphis,* means two vaginas. At this time, it was not known that this condition exists in many other marsupials as well. The majority of marsupials were not known before the discovery of Australia.

Opossums almost always appear dishevelled as well as colorful. The long guard hairs are multicolored and, depending on the hair texture, may produce different color effects from the light striking them. There are black, white, creamy yellow, and brown forms as well as spotted or mottled forms in all possible combinations. The opossum folds its ears while sleeping. Krumbiegel describes the ears as "peculiarly crinkled, but after the animals awake, they quickly regain their former state of elasticity and surgidity." At the North American limit of their range, opossums may lose their hairless ears and parts of the tail from frostbite, since the animals will travel through snow and ice because of hunger or for some other reason. The opossum possesses an extraordinarily acute sense of hearing. Even a moderate sound causes the ear pinnae to flinch or the entire animal may jerk as if in pain. At the time that Hediger was trying to refute the theory that the "stupid" and "malicious" opossums were untameable, he very carefully approached his charges with crepe-soled shoes, because the animals would easily become highly excited by the mere sound of crawling mealworms or a piece of paper crackling or the clicking of a light switch. Such an acute sense of hearing is highly advantageous in detecting the slightest motion of prey animals. The visual and olfactory senses of the opossum are only moderately developed.

The picture by Maria Sibylla Merian

"My school biology text had a beautiful picture of an opossum mother that I will never forget," says Grzimek. "She had her tail bent up horizontally above her body. The youngsters were sitting on her back and they all had their tails wrapped around their mother's larger one which was hanging above them. It seemed obvious that the youngsters could never fall off mother if they held on in this way. I only discovered recently that the picture was incorrect. The original had been painted by the famous Frankfurt artist Maria Sibylla Merian some 250 years ago. Through the years her picture served as the basis for a slowly evolving invention. From 1699 to 1701 she spent some time in Surinam in the north of South America where she sketched flower and insect pictures which still delight people today. The last page of her book shows some insects on the ground and an opossum mother carrying her youngsters on her back. The tails of the youngsters point posteriorly and are wrapped around the mother's tail which is also pointed straight back. One can observe how this opossum picture slowly changed through the decades and centuries and how it was constantly improved until finally the mother's tail became a sort of holding bar positioned over her back and the youngsters were holding onto it. Even photographs exist, but these originated from stuffed museum specimens."

Ever since the Europeans discovered the opossum in 1520, this animal has stimulated our imagination and has been the source for the most incredible stories. In America the opossum was intensely studied during the last decade like no other marsupial on earth. The studies revealed facts which everyone had previously considered to be fables. An opossum mother with young in her pouch, which the discoverer Pinzón brought back from the newly-discovered Brazil, astonished the King and Queen of Spain. Ferdinand and Isabelle put their royal fingers on the pouch and were astounded that such a peculiarity should occur in nature. Marsupials are found only on the American continent, Australia, and the surrounding islands. In North America it seemed for a while that the opossum had lost ground to the other animals. It occurred infrequently and it seemed to have completely disappeared from most areas. Since 1920 the number of opossums has increased in North America, and they have extended their range well into the north. Two to three million opossum pelts are handled commercially every year in the U.S.A.

The small brain of the opossum

It was generally believed that the opossum was simply not clever enough in comparison to other mammals. A comparison of an opossum and a cat of equal weights will show that the opossum only has 1/5 the brain mass of the cat. Books about America always emphasized the dreadful smell of the opossum. It was said that the smell was so powerful that it would penetrate through wood, stones, and would even make Indians fall over dead. If an opossum came close to the site of a village, everyone would have to leave. The one item that contained a grain of truth was the fact that opossums do not taste good to other animals. One rarely finds opossum remains in the dens of foxes. A dog that catches an opossum will shake it several times and then leave the carcass. On the other hand, he will partially eat and bury the remainder of a wood chuck. Nevertheless, in the southern United States opossum meat is a sort of culinary specialty served with sweet potatoes. Some people, however, recommend that one should concentrate on the potatoes when eating such a meal. There are even photographs of President Franklin D. Roosevelt sitting down to a roast of opossum.

The opossum's success in life, in spite of its small brain, may not be solely due to its bad taste, but also to the fact that it is hard to kill. An examination of a collection of opossum skeletons in a museum reveal an unusual number of well-healed broken bones which other mammals of similar size could not have survived, especially in such large numbers. In addition, the opossum has the ability to "feign death." "To play 'possum" is an American expression which means "to play dead." This type of behavior is widespread throughout the animal kingdom. Birds, reptiles, amphibians, and some arthropods are able to feign death, but none of these are as well-known and as popular as the opossum in America. However, opossums, like all other mammals, cannot really "feign death" and suddenly become motionless due to some innate protective behavior. In mammals the temporary paralysis is usually caused by the pressure of the carnivore's teeth on the breathing center in the brain, from beatings, or from unconsciousness. A paralyzed opossum lies on its side, its eyes closed, and its mouth partially opened with the tongue hanging out. A boy once carried one of these "dead" opossums around with him for two hours. He held the animal by the tail and when he climbed over a fence, he suddenly noticed that his "dead" animal held on.

Although the tail of the opossum is not used as a holding bar for her young's tails, it is highly functional for transporting leaves and grass into the burrow. The animal bends the tail forward, below the body, and then it stuffs the nest material with its mouth between its tail and abdomen. In this manner six to eight mouthfuls can be transported at one time. The opossum never winds its tail tightly around branches the way that spider monkeys do. The opossums use the tail as a loose anchor while they are climbing. They hang by their tails only for short periods, and then they drop down. The gestation period of the opossum is only thirteen days. The embryonic newborn weigh only 1/6 gram. Twenty-four newborn young fit into one teaspoon. The births of as many as twenty-five young may take only five minutes.

How do the new-born opossums reach the mother's pouch?

For centuries it remained a puzzle as to how these underdeveloped, tiny creatures made their way from the vagina to the mother's pouch. As with the kangaroos, many theories and speculations abounded. Many believed they had seen the mother pick the young up with her teeth or lips and put them into the pouch. Others assumed that the female bent her back so that the vagina was next to the opened pouch and the young only had to grasp one of the teats. Dr. Carl G. Hartman of the University of Texas finally settled this question. He observed how the tiny opossum embryos made the eight centimeter journey to the pouch unaided by the mother. The young used the well-developed front legs to climb along the mother's abdomen. At this stage the embryos looked more like worms than mammals. It is essential for the embryo to reach the pouch. Apparently only about half of the young succeed, while the others perish on the way.

When the opossum baby has finally reached the warm pouch of the mother, it has mastered the first and greatest difficulty in its life. There are thirteen teats inside the pouch which the female can voluntarily open and close. A closed pouch may contain up to 6% carbon dioxide. It is incredible that the young are able to tolerate this high concentration. After ten weeks the young climb out of the pouch quite dexterously and without the aid of the mother. At three to four months they are independent. Two-thirds of their diet consists of animal matter. Insects, mice, grubs, earthworms, and even toads, which most other small carnivores avoid, are eaten. The life span of the opossum is approximately two years, and to maintain their numbers they have to reproduce enormously. A female has two litters per year. Glen Sanderson in Illinois was successful a few years ago in individually marking a litter of youngsters while they were still in the mother's pouch. Beneath a magnifying glass, he clipped a small piece from one of the ten toes of the hind legs. Each individual was clipped differently. This enabled him to obtain some information about the animal's life history. For example, he discovered that there were approximately an equal number of male and female young, and that approximately one hundred individuals lived on one square kilometer.

Introduction of opossums into New Zealand

The opossums have repeatedly demonstrated that they are extremely well adapted to life in this world. Not only have they been successful in the United States in the last four centuries, but recently they have arrived in New Zealand. Europeans living in New Zealand have always had a great passion for importing various animals from all over the world. The American opossum that was introduced there has increased to twenty million and is yearly increasing by six to seven million.

Wolfgang Gewalt
Bernhard Grzimek

5 Marsupial Carnivores, Marsupial Anteaters, and Marsupial Moles

Family: Marsupial Carnivores by D. Heinemann

Distinguishing characteristics

Family: *Dasyuridae.* The size may vary from the size of a mouse to the size of a dog. The total body length is between 4.5–110 cm. The ears are small or medium-sized. The legs are short to medium long. There are five digits on the front feet. The hind feet have four or five toes. If the first toe is present, it is small and nailless. Some animals walk on the soles of their feet, and others walk on their toes. The long-legged jumping marsupials *(Antechinomys)* have elongated hind legs. The tail, which is medium to very long, can be naked or furred and is often bushy. In certain species the tail is thick and serves as a fat storage. There are forty to sixty teeth. The canines and the premolars have sharply pointed ridges and the molars are strong. The females have four to twelve teats located on the abdomen. Usually the pouch is only conspicuous when it contains young, while at other times it is flat. It is indicated by a posteriorly or ventrally open skin-covered sphincter muscle. In some forms it consists of a pair of lateral skin folds. The pupils are round, although in some marsupial martens they are slit-like.

There are three subfamilies: (1) MARSUPIAL MICE *(Phascogalinae);* (2) MARSUPIAL MARTENS *(Dasyurinae);* and (3) MARSUPIAL WOLVES *(Thylacininae).* There are a total of sixteen genera and forty-six species in the Australian faunal region.

Of all Australian marsupials, these insect eaters and carnivores seem to be the group most closely related to the American opossums *(Didelphidae).* The Didelphids, which are the most primitive marsupials living today, are closest to the root of the entire marsupial family tree. The family *Dasyuridae* includes the smallest marsupials, small rodent-like, insect-eating marsupial mice and also the largest marsupial carnivore, the Tasmanian wolf.

The marsupial carnivores *(Dasyuridae)* are characterized by a carnivorous or omnivorous dentition and the primitive hind foot with three separate, independent digits. These characteristics are also

Dasyures

1. Tiger cat *(Dasyurus maculatus;* see p. 79)
2. Eastern dasyure *(Dasyurus quoll;* see p. 79)
3. *Myoictis melas* (see p. 81)

Marsupial wolf and m

4. Marsupial wolf *(Thylacinus cynocephalus;* see p. 83)
5. Marsupial mole *(Notoryctes typhlops;* see p. 93)

1
2
3
5
4

1
2
3
4
5
6
7
8
9
10
11
12
13

Bandicoots

1. New Guinea bandicoot (*Peroryctes raffrayanus*). 2. Long-tailed New Guinea bandicoot (*Peroryctes longicauda*). 3. Long-nosed bandicoot (*Perameles nasuta*). 4. Tasmanian barred bandicoot (*Perameles gunni*). 5. Eastern barred bandicoot (*Perameles fasciata*). 6. Mouse bandicoot (*Microperoryctes murina*). 7. Southern short-nosed bandicoot (*Thylacis obesolus*). 8. Brindled bandicoot (*Thylacis macrourus*). 9. *Echymipera clara*. 10. *Echymipera kalubu*. 11. Pig-footed bandicoot (*Chaeropus ecaudatus*). 12. Rabbit bandicoot (*Macrotis lagotis*).

Rat opossums

13. Ecuador rat opossum (*Caenolestes fuliginosus*; see p. 108)

present in the *Didelphidae*. While the bandicoots (*Peramelidae*) possess all of the incisors, the second and third toes are joined by tissue so that it appears to be one single toe with two claws, just like those of the phalangers and kangaroos.

The teeth formation of this family is particularly interesting. It is possible to trace the gradual transition from the primitive needle-sharp teeth of the insectivorous forms, such as marsupial mice, to the teeth of the marsupial marten which is adapted to an omnivorous diet, and finally to the specialized teeth of the meat-eating Tasmanian wolf.

The differences in food habits are also paralleled by adaptations to a particular environmental niche and certain behavioral patterns. Thus the external appearance and mode of life of the marsupial mice resemble that of small rodents and insectivorous shrews of other continents. The spotted and striped marsupial marten (*Dasyures*), the Tasmanian devil, and the Tasmanian wolf are reminiscent of weasels, martens, cats, and dogs.

The marsupial carnivores are lithe and lively animals. While almost all of them are capable of climbing trees, they are not truly arboreal. Most are quick and intelligent. They attack anything alive which they can overpower. They are primarily nocturnal and they sleep in caves, trees, and holes in the ground when they are not active. The Tasmanian devil and the Tasmanian wolf emit hoarse, coughing, growling sounds. This sound repertoire does not have any greater expression and tonality than that of most marsupials. The mouse-like species very rarely emit sounds.

Subfamily *Phascogalinae*

Distinguishing characteristics

The MARSUPIAL MICE (Subfamily *Phascogalinae*) include a number of the smaller marsupial carnivores. In contrast to the true rats and mice, these animals have very pointed, conical snouts. The size variation ranges from mouse-like to rat-like. The HRL is between 4.5 cm and 30 cm, and the TL is 5–23 cm. The animals are mouse-, rat-, weasel-, and shrew-like. The snout is pointed, and the small to medium-sized ears are almost naked or covered by short hair. There are 44–46 teeth: $\frac{4 \cdot 1 \cdot 3 \cdot 4}{3 \cdot 1 \cdot 3(2) \cdot 4}$. The females have four to twelve teats. The marsupium is either fully developed, poorly developed, or absent. There are nine genera with thirty-nine species.

BROAD-FOOTED MARSUPIAL MICE (Genus *Antechinus*) have an HRL of 6.7–17 cm, and a TL of 6–14.5 cm. The feet are short and broad. The short-haired tail appears to be almost naked and the end never has a hair crest or tassel. The marsupium may be absent, poorly, or well-developed. They are insectivorous or carnivorous. It is questionable whether or not they consume plant matter. There are eleven species. The YELLOW-FOOTED MARSUPIAL MOUSE (*Antechinus flavipes*; Color plate p. 62) has an HRL of 9-17 cm and a TL of 8-12 cm. Despite great color variation of the fur, the animal is always recognizable by the brownish-yellow or reddish-brown feet. The PYGMY MARSUPIAL

Fig. 5-1. Broad-footed marsupial mice (Genus *Antechinus*).

MOUSE *(Antechinus maculatus)* has an HRL of 6.7–7.2 cm and a TL of 6–6.5 cm. The FAT-TAILED MARSUPIAL MOUSE *(Antechinus macdonnellensis)* has an HRL of 9.2–12 cm and a TL of 17–17.7 cm. The thick tail serves as fat storage. The SPECKLED MARSUPIAL MOUSE *(Antechinus apicalis)* has an HRL of 11–12 cm and a TL of 8–9 cm. The dorsal side is covered with reddish-brown spots. There are seven additional species. The last premolars in the fat-tailed and speckled marsupial mice are atrophied or absent. Therefore, these animals are placed into separate subgenera: *Pseudantechinus* and *Parantechinus.*

Fig. 5-2.
Planigale (Genus *Planigale*).

The yellow-footed marsupial mouse *(Antechinus flavipes)* can be found in dense forest as well as in sparsely forested land. It is most frequent along creeks in rocky terrain. The animal hides in crevices, holes in the ground, and in trees. One can often find them in wind-eroded caves in sandstone, which have been packed with interwoven Eucalyptus leaves that are shaped into nests with an entrance facing to the back. The furrowed foot pads and long claws enable the animal to hold on firmly while climbing trees and vines. This marsupial mouse is even capable of running along the ceiling of a cliff for a short distance if it has the proper momentum. This is advantageous because the animal can then build its nest beyond the reach of larger enemies. Despite its agility, it is very difficult for the tiny animal to carry the many leaves to its nest. In the National Park close to Sidney, *A. flavipes* often use the abandoned nests of lyre birds *(Menura novaehollandiae)* which are high up in cliffs.

Fig. 5-3.
Crest-tailed marsupial mouse (Genus *Dasycercus*; see p. 76).

Their diet consists mainly of insects. Traps baited with bug larvae will catch the yellow-footed marsupial mouse. Meat bait also succeeds; hence, the animals must like the taste. Even raisins may attract the "mice"; therefore, their menu must be quite varied. This animal is very useful as an insect exterminator as are most other marsupial mice. These marsupials even attack the house mice that had been introduced by the white settlers. The well-known naturalist Gould reported the frequent occurrence of the yellow-footed marsupial mouse in the first settlement of New South Wales. He describes how these animals would cross fallen trees with quick squirrel-like jumps. The female does not possess a pouch. Instead, she has a shallow skinfold around the eight teats.

The FLAT-SKULLED MARSUPIAL MICE (genus *Planigale*) are usually even smaller than the smallest broad-footed marsupial mice. The HRL is 4.5–10 cm, and the TL is 4–8 cm. The tail is short-haired and not enlarged. There are four species, one of which is the NORTHERN PLANIGALE *(Planigale ingrami).* The HRL is approximately 8 cm and the TL is approximately 6 cm. The KIMBERLEY PLANIGALE *(⊕Planigale subtilissima)* is the smallest known marsupial. The HRL is approximately 4.5 cm and the TL is approximately 5 cm.

These animals are truly terrestrial, and this is the reason for the soft,

Fig. 5-4.
Northern Planigale *(Planigale ingrami).*

smooth foot pads. The skull is extremely flattened, which is very unusual for a mammal but is characteristic of lizards. The tiny marsupials probably slip through narrow cracks, and scurry between the tough grass clumps in a manner similar to the lizards. The flat skull structure is probably quite advantageous for this mode of life. Grasshoppers, found by the thousands in the grass, make up the major part of their diet.

The largest member of this subfamily is the BLACK-TAILED PHASCOGALE *(Phascogale tapoatafa).* The HRL is 20–24 cm and the TL is 18–22.5 cm. The terminal half of the tail is covered with a hairy brush. There are two additional species of this genus. One is the RED-TAILED PHASCOGALE *(⊖Phascogale calura)*, with an HRL of approximately 12.5 cm and a TL of approximately 14.5 cm. The other is the *Neophascogale,* a special subgenus which includes the NEW GUINEA PHASCOGALE *(Phascogale lorentzi).* The HRL is 17.5–23 cm and the TL is 17–21 cm. The fur is coarser. The white tail end is not bushy.

Fig. 5-5.
Black-tailed Phascogale *(Phascogale tapoatafa)*
The ancestral mammals must have looked similar to this animal. The first mammals originated in the Mesozoic about 200 million years ago.

The BLACK-TAILED PHASCOGALE or BUSHY TAILED MARSUPIAL MOUSE *(Phascogale tapoatafa)* is also called TUAN. This animal is more carnivorous than the smaller species. These active animals are rarely seen by humans, except perhaps when one surprises one during its stalk of the chicken coop.

The first settlers in West Australia mistook these squirrel-like, agile tree-climbers with the bushy tails for squirrels. However, the mode of hunting and diet are more reminiscent of the martens. In one of his reports about pioneer times, Gould says that these little fellows, which would raid caches and chicken houses, were a real nuisance. Yet Gould's examination of the stomach contents of the tuans only revealed the remains of bugs and something resembling parts of mushrooms. Gould relates how wild captive tuans behave, and how desperately they fight for their freedom. Even natives avoided coming close to the bushy-tailed marsupial "mouse," which were said to inflict horrible bites. The Australian Museum in Sydney once received an animal which had bitten through the larynx of two ducks and mutilated their wings. The animal, which was surprised while it was biting the duck, was so intent on this task that one could easily have killed it with a stick. The tuan, which also preys on rats and mice, is therefore a very useful animal.

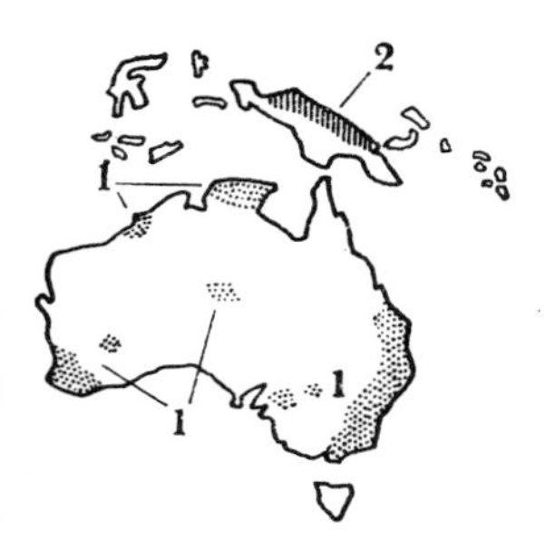

Fig. 5-6. 1. Phascogales (Genus *Phascogale*); 2. Neophascogale (Subgenus *Neophascogale*).

The bushy tailed "mice" build their nests of leaves or grass in tree holes or on the ground. They can also utilize bits of cloth or paper for nest-building. A logger in a camp once lost a pound note. This resulted in a great deal of mutual suspicion until the lost note was discovered in the nest of a tuan in a freshly-cut tree close to the camp.

There are two marsupial mouse genera on New Guinea which greatly resemble the bushy-tailed mice in appearance and mode of life. However, their tails are short-haired and without a brush. These are the NEW GUINEA MARSUPIAL MICE *(Murexia)* and the *Phascolosorex*, each with two species.

Two other genera on the Australian mainland are characterized by hair crests or hair brushes on their tails. They are terrestrial and inhabit the desert and semi-desert regions: 1. CRESTED-TAILED MARSUPIAL "RAT" *(Dasyuroides byrnei)*. There is only one species. The HRL is 14.5–18 cm and the TL is 13–14 cm. The body form is strong and they have short legs. The first toe is lacking. The tail is only slightly thickened with a dorsal and ventral hair crest. 2. CRESTED-TAILED MARSUPIAL "MOUSE" (genus *Dasycercus*). The HRL is 13–15 cm and the TL is 8.5–10 cm. The first toes are present. The base of the tail is enlarged for fat deposits, and the tip has a dorsal hair crest. There are two species.

Fig. 5-7. 1. New Guinea marsupial mice (Genus *Murexia*; see p. 76). 2. Crest-tailed marsupial mouse (*Dasyuroides byrnei*; see p. 76).

The great difficulty encountered when studying small mammals in their natural habitat is demonstrated in the history of the discovery and study of the crested-tailed marsupial mouse or MULGARA *(Dasycercus cristicauda)*. In 1867 Krefft described this species on the basis of one single specimen found in an Australian museum. It was more than fifty years later when Wood Jones published his highly interesting observations on captive crested-tailed marsupial mice. According to Jones' report, the mulgara is one of the most fearless and most clever marsupials. Although it is a highly competent carnivore for its size, it nevertheless can be friendly and trusting, provided it is not teased. A mulgara will kill a mouse if it is hungry, but if it is not, it will even permit the mouse to share its nest. When a mulgara is hunting, its body stiffens suddenly, the tail twitches like that of a lizard, and the prey is quickly seized by the neck and killed instantly. Prior to consuming the killed prey, the mulgara grooms itself most carefully. It really is a remarkably clean little animal.

The crested-tailed marsupial mouse starts to eat at the nose tip of the mouse. It detaches the skin, turns it over, bites the roof of the skull, and eats the brain. Then the rest of the head, body, and tail is consumed, often without any damage to the skin of the prey. Three hungry mulgaras once consumed an entire rat, but left behind an almost undamaged, turned-out skin which could have competed with the work of an experienced taxidermist. Not one bone was attached to the rat skin.

Mulgaras are not nocturnal. They like to stretch out like lizards in the hot sun. For mammals, they can tolerate an amazing amount of heat. In the region of Ooldea, the animals bear their young between June and September with an average litter size of seven. The pouch area consists of only slightly developed lateral skin folds which protect the youngsters. The young remain attached to the nipples for about one month. It is quite a sight to see a mother with seven sizeable youngsters hanging on to her teats while she is hunting for insects or killing a mouse.

Fig. 5-8.
Narrow-footed marsupial mice (Genus *Sminthopsis*).

The NARROW-FOOTED MARSUPIAL "MOUSE" (genus *Sminthopsis*) has an HRL of 7-12 cm and a TL of 5.5-20 cm. These animals are slender, narrow-footed, and have a pointed snout. The ears are medium-sized. The marsupium is relatively well-developed. They are terrestrial and inhabit deserts, semi-deserts and the savannah. There are twelve species, which include the MOUSE SMINTHOPSIS (*Sminthopsis murina*) and the FAT-TAILED SMINTHOPSIS (*Sminthopsis crassicaudata*; Color plate p. 62) with a tail thickened at the base for fat storage, particularly in the Southern subspecies *Sminthopsis crassicaudata crassicaudata*. These animals also have large ears.

The narrow-footed marsupial mice belong to the most graceful of all smaller marsupials. The feet are slender, and the hind part of the soles, which lack pads, are partially covered by hair. This characteristic distinguishes them from all of the other discussed genera. The formation of the foot is an adaptation to their mode of life. They hop bipedally on the ground. The tail may swell extremely, depending on the season. Most species take shelter in crevices, beneath tree stumps, in ground burrows, hollow logs, dead branches, or garbage piles. Their diet is mainly insectivorous, but they also hunt small mammals, such as the house mouse (*Mus musculus*) which was introduced into Australia. Once two narrow-footed marsupial mice were shipped in one box. Upon arrival only one animal was alive and well, and only a few remains of the other were found. Of course, this case of cannibalism was brought about by the unnatural circumstances. However, it does show that the animal would be capable of dealing with a common mouse in a similar manner.

Under favorable conditions the smaller species are able to produce ten young, which is actually quite a number for such a small animal. After the young leave the pouch, they often still hang onto the mother's sides. Once a farmer from New South Wales, while plowing, opened up the subterranean nest of a mother and her ten young. When he first noticed the animals, the mother was trying to flee with all ten young holding on to her. When she lost some, she would halt. When the babies squeaked, she twisted all over her brood until all were attached again. She finally disappeared with her children beneath a clump of earth.

Narrow-footed marsupial mice kept by man are extremely lively and they are able to consume astonishingly large amounts of fresh meat and insects. A mouse Sminthopsis (*Sminthopsis murina*) with a body weight of barely 21 grams consumed five large bug larvae and three small lizards, including the bones, skin and tails, with a total weight of 28 grams, in one night.

The beautiful JERBOA MARSUPIALS (genus *Antechinomys*) resemble narrow-footed marsupial mice somewhat, but the body form and limb construction indicate a kangaroo-type of locomotion. The HRL is 8-11

Fig. 5-9. 1. Jerboa marsupials (Genus *Antechinomys*; see p. 77). 2. Striped marsupial mice (Genus *Phascolosorex*).

cm and the TL is 11.5–14.5 cm. The forelimbs, lower legs, and tarsals are strongly elongated. The first toes are missing and the surfaces of the soles are without thick pads. There are two species: EASTERN JERBOA MARSUPIAL (◇*Antechinomys laniger*) and CENTRAL JERBOA MARSUPIAL (*Antechinomys spenceri*; Color plate p. 62).

The jumping locomotion of the jerboa marsupials makes them appear superficially similar to the kangaroo mice (genus *Notomys*) which belong to the rodents and are found in the same regions in greater numbers. The long-legged jumping marsupials eat insects, small lizards, and mice. Some have claimed that the Jerboa marsupials do not have a pouch. However, a specimen in the British Museum of London has a distinctly developed pouch with skin folds on the front and sides, although there were no young inside. The pouch opens to the rear while in the kangaroos, which have a similar mode of locomotion, the pouch opens to the front.

In 1865 Krefft received two *Antechinomys laniger* from some natives. One of the animals, which remained alive for several weeks, greedily consumed meat and immediately attacked the frightened mice which were put in its cage. Both species of *Antechinomys* are very rare. *A. spenceri*, which is a little larger in body size, prefers sandy hills covered with clumps of grass and other sparse vegetation in an otherwise desolate landscape. There one can find them in ground burrows, which they often share with the more numerous kangaroo mice.

In his book *Across Australia*, Baldwin Spencer describes how he observed one of these graceful little animals during its hunting activities on a bright moon-lit night on the stony flats close to Charlotte Waters. Occasionally the jerboa marsupial would cease jumping and curiously peer at the intruder. It would stand very erect on its stilt-like legs and the tail would swing up in an arch. It looked as if the animal was barely touching the ground. The individual jumping distances were considerable compared to its body size. The movements were very fast and difficult to follow. At one moment the animal would sit up on a rock, and then it would dash off to a distance of two meters.

Subfamily *Dasyurinae*

DASYURES (Subfamily *Dasyurinae*) range in size from weasel to racoon. The HRL is 17.5–75 cm, and the TL is 14.5–35 cm. Most of the animals are weasel or marten-like. The short to medium-long snout is pointed or blunt. The small ears are almost naked or covered by short hair. The tail is usually long and either bushy or short-haired. There are 42 teeth: $\frac{4 \cdot 1 \cdot 2 \cdot 4}{3 \cdot 1 \cdot 2 \cdot 4}$. This is a carnivore dentition with the canines and molars particularly well-developed. There are four to eight teats. The pouch, which opens to the rear or down, is usually fully developed only during the breeding season. There are three genera and a total of six species.

Distinguishing characteristics

In the last century many Europeans who settled in far-off continents

Fig. 5-10. 1. Eastern dasyure *(Dasyurus quoll).* 2. Little northern dasyure *(Dasyurus hallucatus).*

Fig. 5-11. 1. Western dasyure *(Dasyurus geoffroyi).* 2. Spotted-tailed dasyure *(Dasyurus maculatus).* 3. Striped dasyure (*Myoictis melas;* see p. 81).

frequently met a totally foreign fauna in their new homelands. Very often the animals had unfamiliar shapes, such as the giraffes of Africa or the kangaroos of Australia. Quite often the immigrants would give names to these exotic creatures that they had learned in school, at home, or the names that the natives had given them. Many animals reminded the immigrants of similar-looking creatures of their former homeland and so without further thought they gave these familiar names to the new animals. The South African Beisa antelope became the "Gemsbock," the giant rodents of the Amazon region were called "water pigs," and the plump wombats were a kind of "wild boar" that could be beaten to death and eaten. The white settlers of Australia soon became acquainted with cat or marten-like marsupial carnivores that raided their chicken coops at night. They believed that these nocturnal robbers were a type of cat and so, to differentiate these from the domestic cats brought back from Europe, they called them "native cats."

These "native cats" are, of course, like almost all Australian mammals, marsupials. They are not related in any way to the true cats or martens which belong to the Carnivora, a group of more highly evolved mammals. These carnivorous marsupials are classified under the subfamily *Dasyurinae* (Dasyures). Aside from the white-spotted weasel and marten-like "native cats," this group also includes the small striped dasyures from New Guinea and the plump, big, black Tasmanian devils from Tasmania. Although all these animals look very different, they are, nevertheless, closely related. The shape of the ear, the design of the snout, and the hand and foot surfaces are similar in all the species. The teeth are specialized to varying degrees for a carnivorous diet. The Tasmanian devil has the most specialized teeth. Most dasyures are more or less arboreal, but the Tasmanian devil is strictly terrestrial.

The EASTERN AUSTRALIAN NATIVE "CAT" (genus *Dasyurus*) is marten-like and slender. The fur is dark and covered with white spots. There are four subgenera with one species each (some zoologists consider each a separate genus): 1. Subgenus *Dasyurus* (i.n.s.): EASTERN AUSTRALIAN NATIVE "CAT" (*Dasyurus quoll;* Color plate p. 71). The HRL is 40–45 cm and the TL is 20–30 cm. The first toes are missing and the foot pads are not striated. The coloration is as seen in the Color plate p. 71 or dark-brown with white spots. 2. Subgenus *Dasyurinus:* WESTERN AUSTRALIAN NATIVE "CAT" (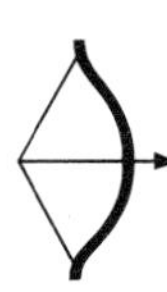 *Dasyurus geoffroyi).* The HRL is 35–45 cm, and the TL is 27–35 cm. The tail is not bushy and the terminal part is black. 3. Subgenus *Dasyurops:* LARGE SPOTTED-TAILED NATIVE "CAT" (*Dasyurus maculatus;* Color plate p. 71). The HRL is 35–75 cm and the TL is 25–35 cm. The tail is also spotted. The first toes are present, and the foot pads are well striated. The animals are excellent climbers. The teeth are more specialized for a meat diet than in the other species of the genus.

4. Subgenus *Satanellus;* LITTLE NORTHERN NATIVE "CAT" *(Dasyurus hallucatus).* The HRL is 25–30 cm, and the TL is 22–30 cm. The black tail is not bushy. The first toes are present.

The name "native cat" best fits the Eastern Australian native "cat" since its size and appearance are reminiscent of a domestic house cat. The animal, which rarely climbs trees, is found in many different types of habitat, frequently in the vicinity of human dwellings. Formerly, the "native cat" was found close to the larger cities of Southeast Australia, particularly near the coast. It is still found in the suburbs of Sydney. Twenty years ago it was not uncommon in Adelaide, but after that it was only found in the vicinity of Melbourne. The first settlers hated the native "cat" because of its alleged "blood-thirsty" attacks on the chicken yards. Today many a rancher would be happy to have some of these "cats" around to help him cope with the mouse plague. Unfortunately, native cats have either been exterminated from most regions or have been reduced to very small numbers. The female may bear up to twenty-four young at one time, but only six teats are present. The surplus young are doomed.

It is quite a fearless little marsupial displaying the undaunted intelligence of the true carnivores. The "native cat" makes an engaging pet. It gets used to its human keeper very quickly, and it is not very difficult to keep. The nocturnal animal sleeps in the daytime with its ears folded slightly to shut out noises. It holes up between stones or hollow tree stumps. At night it hunts insects, lizards, fish, small birds and mammals. Without a doubt, these "native cats" could play a very useful role in the extermination of rats, mice, and young rabbits. However, during 1901–1903 most "native cats," along with many other marsupials, died in large numbers due to an epidemic. This species disappeared from most of its known range of distribution, except for a remnant pocket around Sydney. The large spotted-tailed native "cat" was formerly very frequent in the forested regions of the South coast and the close mountain ranges. This tiger "cat" can occasionally overpower small kangaroos and large birds. The usual prey animals consist of reptiles, birds, and small mammals including rabbits; tiger "cats" will also eat eggs.

Close to the Hawkesbury River, north of Sydney, a tiger "cat" was observed carefully creeping up to a heron in cat-like fashion. The heron was searching for food in one of the swampy puddles. Every time the bird lowered its head, the "cat" would run forward; whenever the bird lifted its head, it would halt. On a different occasion another "tiger cat" killed hens from a chicken house on several rainy nights in succession. In June, 1933, one of these "night raiders" was caught in Woodford in the Blue Mountains. Of twenty-four guinea hens roosting in high trees, the tiger "cat" had killed twenty-two in a very short time. Of yet another tiger "cat" it was reported that it kept two

Irish terriers at bay and yet another animal killed a large male house cat in a desperate battle. A farmer in Tasmania surprised one large spotted-tailed native "cat" in the process of eating the remains of a wallaby in bright daylight. As the man approached with an axe, the "native cat" jumped aside. But when the farmer just stood there with the lifted axe, it returned to its prey as if nothing had happened.

The smallest dasyures (*Myoictis melas;* Color plate p. 71) have an HRL of 17.5–21.5 cm and a TL of 14.5–20 cm. It is found in New Guinea and the Aru Islands. Some zoologists do not classify this species with the dasyures but with the marsupial mice. This animal was named "melas" (black) because the first specimen known to science (1840) was a melanistic individual. Very little is known about the mode of life of this small carnivorous marsupial. A. R. Wallace observed in 1858 on the Aru Islands that these animals "destroy anything edible in the house just like rats."

The Tasmanian devil by B. Grzimek

Aside from the large spotted-tailed native "cat," the TASMANIAN "DEVIL" (*Sarcophilus harrisi;* Color plate p. 102) is the largest dasyure (the HRL is approximately 50 cm and the TL is approximately 25 cm). It is heavier and stronger than its spotted relative. The first toes are absent. The pouch, which is hoof shaped and opens to the rear, can be completely closed. There are four teats.

Their alleged wildness and viciousness

The Tasmanian devil received its unflattering name from the white settlers on Tasmania, south of Australia, because the animals had the reputation of being vicious and violent. However, if one picks the animal up by the tail, one shouldn't be surprised if it spits, growls, bites, and foams with rage. The tale of the fiendish disposition and constant bad mood goes back to zoologist Harris who discovered the animals in 1808: "They seem to be uncontrollably wild. They bite frequently and snarl and bark loudly." The pair of Tasmanian devils that he kept started to fight as soon as darkness fell (they slept all day) and they continued throughout the night without interruption. They emitted a "hollow" bark, which is not so astonishing since for several months Harris kept the poor animals chained to each other inside a barrel. It is no wonder that the animals were desperate and vicious. This attitude towards wild animals is typical of that era. The animals were tormented by their keepers, who even described these conditions in a rather unperturbed manner. During Harris's time, these big black marsupial carnivores were still quite numerous around the Tasmanian capital of Hobart. At night the animals would catch poultry and small animals from the houses. On the other hand, they were a welcome source of fresh meat because they could be caught very easily with meat bait. The bush, which formerly surrounded the settlement, disappeared gradually and so also did the "little devils."

Fig. 5-12. Tasmanian devil (*Sarcophilus harrisi;* see p. 81). +— extinct

Zoos construct large enclosures without bars for lions and tigers, and spacious cages for leopards, but hyenas and smaller wild cats do not fare as well. Such inconspicuous animals as Tasmanian devils, which are also nocturnal, are generally not housed very elaborately. In dull surroundings, the animal's behavior is often sullen and bored. A little friendliness and interest, however, will reveal a very different animal. Mrs. Mary Roberts, who kept and bred Tasmanian devils in Beaumaris, Tasmania, has done much to clarify the animal's true disposition. The little devils that she raised were engagingly affectionate, playful, and mischievous. Even wild adult animals could be tamed so that they could be handled easily. The animals are very clean, and they love to bathe and bask in the sun. They used their front paws to clean their faces. They would cup their hands and lick them thoroughly and then rub their heads. A farmer in Tasmania tamed two devils and taught them to walk on a leash. He could even lead them around on the streets of Melbourne. These dachshund-sized animals love the water. If they are pursued, they will enter the water, dive under and swim beneath the water to a place, preferably below plant cover, where they re-emerge silently.

The Tasmanian devil, which lived for a number of years in the Frankfurt Zoo, could "sing" loudly and persistently if it were encouraged to do so. One only had to stand in front of it while cleaning the cage and produce the right note, upon which it would join in and hold the note for quite a while with its mouth open. In a similar manner, I could induce my wolves to sing. The normal life span of these carnivorous marsupials seems to be between five and seven years. One Tasmanian devil lived in the Basel Zoo for six years and fifteen days. Tasmanian devils kept in zoos are notorious for escaping, which they probably accomplish with the aid of their strong teeth and powerful masseter muscle. A newly-arrived devil in the Vienna Zoo escaped during the first night. It had bent the sturdy bars of the cage and gotten through an opening only 7½ cm wide. It was found jammed into an eight centimeter wide crack behind a box. During a state of excitement, the pale ears of the Tasmanian devil will gradually turn red.

In Tasmania, Tasmanian devils occur frequently

Today Tasmanian devils are not uncommon on the Island of Tasmania, although there are none on the Australian mainland. In 1912 one devil was shot ninety kilometers outside of the city of Melbourne, but this was probably an animal which had escaped from a zoo or from a private owner. The bones of extinct devils have been frequently found on the mainland. Occasionally, someone claims that the devils are present in remote areas of Australia. The animals probably became extinct on the mainland because of the introduction of the dingo *(Canis familiaris dingo)*, which never reached Tasmania.

Reproduction and care of the young

Tasmanian devils breed at the end of the southern summer between April and May. Towards the end of May or at the beginning of June the 12 mm long young are born. After seven weeks inside the pouch which can be closed completely, the young have grown to a length of seven centimeters. By fifteen weeks they can be detached from the teats to which up to now they had been holding on constantly. At this stage they have hair and their eyes are open. Towards the end of September, the Tasmanian spring, one can occasionally observe a leg or tail poking out of the pouch. In nature both mother and father build the nest for the youngsters. They choose a hollow tree stump, a place underneath rocks, or the burrow of a wombat. The nest is lined with vegetation to provide a soft nest. There are never more than four young because the mother has only four teats. Small devils, which are suckled for at least five months, apparently do not breed until their second year. In Tasmania the Tasmanian devils are protected by law. Only once did a *Sarcophilus* female bear young outside of Tasmania. This occurred in the Basel Zoo in Switzerland. Three weeks after the devil had been caught in a rat trap close to Hobart, a little flesh-colored tail protruded from the pouch. Five days later there were two tails. Since the female was very aggressive and would stand up on the bars of the inside cage to bite, it was possible to look inside the pouch opening. After two weeks, I personally saw two little rear ends hanging out of the pouch and slowly crawling to the inside. By now the youngsters were already colored and were a little larger than house mice. Two days later the pouch was empty. Movement and soft noises under a pile of hay, plus the fact that the female would stand guard over this pile, showed where the young had gone. Unfortunately, everything was silent the next day and there were no more "signs of the young," wrote Basel Zoo director Dr. E. Lang.

Subfamily
Thylacininae
by B. Grzimek

The TASMANIAN POUCHED "WOLF" (⊕ *Thylacinus cynocephalus;* Color plate p. 71) is the largest carnivorous marsupial living today. The animal deviates so greatly from all of the other forms of this diversified family that it is classified as one species and genus under its own family *(Thylacinidae),* or at least as a separate subfamily *(Thylacininae)* within the *Dasyuridae.* The general characteristics are as follows. The animals have the appearance and size of dogs. The HRL is 100–110 cm, and the TL is 50 cm. The head is pointed like a dog's, but the mouth can be opened much farther than in a true dog. The ears are short, rounded, and erect. They walk on their toes, the first of which is absent. The tail is covered with short hair. The pouch is a flap of skin which opens to the rear. There are four mammae. There is only one, possibly extinct, species on Tasmania. *Thylacinus* became extinct on the Australian mainland quite recently, possibly due to the dingo. Fossils of thylacines have been found on New Guinea. "He has reached the point of no return, and the most lavish attention will not save him," writes Michael Sharland, the leading zoologist of Tasmania. The fate of the Tasmanian wolf is a cause for despair. If this largest of all marsupial carnivores had been created and constructed by the hand of man and not by God or nature in a process that took millions of years, it would have been saved as a matter of course. Is it not true that the U.N. and wealthy people have donated millions of dollars to preserve the stone monuments of Abu Simbel from submersion in Lake Aswan? These statues have been on the left bank of the Nile for over 3000 years as a glorification of King Ramses II in four twenty-meter tall versions. The preservation of the Tasmanian wolf, one of the most exciting and rare animals on the face of this earth, would require only a fraction of all this money.

Fig. 5-13.
Tasmanian wolf *(Thylacinus cynocephalus)*, + locations where thycaline fossils were found.

However, the pouched "wolf" lives at the end of the world on a heavily-forested, rugged, scarcely-populated island where European immigrants also permitted the extinction of the original woolly-haired natives without a thought. If Tasmanian "wolves" existed in the United States, somewhere in Europe, or the Soviet Union, there would be more publicity and perhaps the cost for preservation would not be shunned. No efforts are carried out in Australia, on the other side of the globe. What will our grandchildren write about this in the year 2020? I think I already know.

The famous naturalist John Gould predicted the troubled future of the Tasmanian "tiger," as the "wolf" is also called; it was so absurdly named this because of the transverse bands across its body. When he first visited the forested mountainous island of Tasmania one hundred years ago, he wrote: "Once this relatively small island is more densely populated and roads have been built through the virgin forests from the east to the west coast, the numbers of this unique animal will quickly decrease. The animal will be largely exterminated and, like the wolf in England and Scotland, will become an animal of the past." Fortunately, wolves are found in other places on earth besides England and Scotland. Pouched "wolves," however, are found only on Tasmania. The European settlers did not wait for the extermination of the marsupial wolf until roads had crossed Tasmania or the island was more densely populated. Tasmania has a surface area of 63,000 square kilometers and is three quarters as large as Ireland. Even today there are only 300,000 inhabitants, of which every third person lives in the capital of Hobart. One hundred years ago there was a $25.00 bounty on each pouched "wolf" because the "wolves" had not restricted their diet to kangaroos but had also attacked sheep. For over twenty years now, there has been a fine of $500.00 imposed for killing a Tasmanian wolf. This measure, however, cannot save this animal species anymore.

It took a relatively long time for the Tasmanians to realize how near to extinction the largest marsupial carnivores really were. At first, hundreds of animals were killed yearly. Pouched "wolves," which were accidentally caught in kangaroo traps and did not choke in them and were not eaten by Tasmanian devils were usually delivered to the zoo in Hobart which closed its doors in 1940. It contained a total of nine to ten marsupial "wolves," most of which were caught in the Florentine Valley on the west coast of Tasmania. The last one was caught in 1933. The zoo traded the "wolves" for other foreign species. Thus they obtained a pair of lions, a polar bear, an elephant, and finally a whole series of exotic birds. Other "wolves" were sold directly overseas. The Cologne Zoo received one animal, and the zoos of Antwerp and London all together "at least a dozen," over the years. The last Tasmanian wolf died in London in August, 1931. In New York alone, four animals were kept during 1908-1919. The people in Hobart were very active in trading off their Tasmanian wolves because they believed that there was an unlimited supply of them. In the end, they only had one lame animal which finally died after a lonely existence. Since 1933 no Tasmanian wolf has been caught.

Tasmanian wolves in zoos

The zoo animals were relatively insensitive to the cold and showed no tendencies towards "nocturnal activity," as was widely believed. No special care was given to the animals. They were kept on horsemeat and beef and an occasional small mammal. Despite this dreary existence, one Tasmanian wolf lived in the London Zoo for eight years and four months, and in Washington one lived for over seven years. The pouched "wolves" could jump up two to three meters. Privy Councillor Ludwig Heck, then the director of the Berlin Zoo, wrote in 1912 that it was still possible to obtain a pair of Tasmanian wolves on the market for about $500.00.

The scientific name of this marsupial carnivore means "marsupial dog with a wolf's head." This seems to be an appropriate term, although the animal is a marsupial which is in no way related to the dogs or wolves. The external appearance is very dog-like. The jaws are remarkable because of the width to which they may be opened. Many marsupials are able to open their mouths at a wide angle. It has been claimed that the Tasmanian wolf can open his to almost 180°. The hind end and tail look the least dog-like. The base of the tail is thick and is more reminiscent of a kangaroo tail than that of a dog. Unlike a wolf or dog, the Tasmanian "tiger" ("wolf") does not express his moods by different positions of the tail. There is never a friendly wagging or a dejected pulling in of the tail. Some people have claimed that a pouched "wolf" could not bite one's hands if the tail were held.

Meager observations on free living animals

All the information regarding observation on free-living pouched "wolves" which has been compiled during the last decades has to be interpreted cautiously. After all, nobody took the trouble to observe the animals while they still lived in the wild, and even captive "wolves" in zoos received a minimum of close observation. It has always been claimed that Tasmanian wolves are extremely blood-thirsty, sucking blood from the jugular vein of sheep and kangaroos, and at most ate only the blood-filled nasal membranes, liver, and kidney fat. It was also said that they never returned to a kill and that they never touched carrion. It is highly probable that all this "slander" originated with the sheep ranchers. The first two "wolves," caught by G. Harris in 1824, had been trapped with kangaroo meat as bait. The one specimen that was used for a description of the entire species contained a spiny anteater in its stomach. It is said that Tasmanian wolves do not run as swiftly and nimbly as dogs but instead trot along. They are reported to have shown no particular fear of dogs, and even several dogs would not dare to attack a Tasmanian wolf. If relentlessly pursued, the Tasmanian wolf will eventually face its foe on its hind legs, similarly to the kangaroo. The body structure of the pouched "wolf" could substantiate this claim. They are said to trot relentlessly after their prey until the victim is exhausted, and then they close in with a rush. They growl hoarsely when excited.

Everyone agrees, however, that a pouched "wolf" has never attacked a human. Only once, in 1900, did a Tasmanian wolf bite a Miss Priscilla Murray in her right arm. The lady lived in an isolated house, and when the attack occurred she was doing her laundry in the river. Fortunately, she was dressed in heavy winter clothing and the bite did not harm her. However, when she tried to chase the animal away, it again bit her left hand. In an effort to reach for a garden rake, she stepped on the animal's tail, whereupon it let go of her hand and ran away. Miss Murray reported that the animal had one eye missing, and she assumed that the animal was starved because it was winter. It is possible that the half-blind "wolf" mistook the human arm for a bird or another small vertebrate. Another point that discredits the claim that pouched "wolves" are bloodthirsty is the fact that farmers formerly set out poisonous bait for Tasmanian wolves and devils along side of their kangaroo snares, which they obviously must have eaten. The brood care of the Tasmanian "wolf" is similar to that of the Tasmanian devils. The young were carried in the flat pouch, which opened to the rear, for three months. When they were more independent, the young were placed in a padded nest. The female had up to four young, which would later join her for some time in hunting.

About the characteristic dullness of marsupials

In 1902 the Berlin Zoo received the last pair of Tasmanian wolves. The male survived for six years. Ludwig Heck describes his encounter with the animals in the following manner: "Making allowance for the general marsupial dullness, the animals were trusting and would restlessly sniff along the bars if one stood in front of their cage. They were constantly greedy for food when they were not sleeping, and with persistent stupidity they would try to bite through the iron bars. It is difficult to awaken the animals when they are sleeping on the soft straw in the semi-dark cage. However, they are not angry if one attempts this. The clear, dark brown eyes stare vacantly at the observer. The expressiveness of the true carnivore is totally lacking." Such reports of animals kept in very small cages are sceptically scrutinized today. Many so-called "stupid" animals have turned out to be quite responsive, lively, and interesting provided one takes the time to work with them and care for them.

We probably will never know the true behavior of the Tasmanian wolf. One of the last "wolves" was killed in 1930 on the Northwest coast of Tasmania close to Mawbanna. Three years later another one was caught in a snare. Since then all searches for them have been unsuccessful. In 1937 Roy Marthick, who searched for three weeks, maintained that he had found the tracks of twenty animals and had even seen them during dusk. Since then more expeditions have sought the elusive "wolves." The last took place in 1945; all trips were without success. Occasionally reports about Tasmanian "tiger"sightings reach the newspapers. Usually people (i.e., telegraph construction crews) that have to work in the more remote regions of Tasmania are responsible for these reports which could never be verified. A report of a helicopter crew which had seen and followed a marsupial "wolf" on the west coast in 1957 caused quite a stir. The photograph which they showed as proof was identified by the authorities as a dog.

The "Mercury," the Hobart newspaper, reported the following story in August, 1961. Bill Morrison and Laurie Thompson were out on a fishing and tenting trip on the West Coast. One night the two men heard the noise of some animal that was trying to reach a basket of fish bait. Thompson got up and grabbed a piece of wood to chase the animal away. He saw a shadowy figure at the basket and repeatedly hit it with the stick. The animal disappeared into the night. The next morning they found a dead male juvenile Tasmanian "wolf" close to the tent. At least, this is what they said. They put the carcass inside the tent with the intention of taking it to a museum after the trip was finished. When they returned to camp the carcass was gone. Somebody obviously had stolen it. The two men were greatly upset over the disappearance of this important piece of evidence, and reported the whole incident as soon as they returned home. Nevertheless, they had collected hair remains and dried blood which had seeped into the sand. Authorities in Hobart who examined these clues confirmed that these were indeed from a Tasmanian "wolf." The carcass of the animal was never tracked down.

Reports by laymen who have seen a lone Tasmanian wolf were often followed up. The authorities were never able to see an actual animal although they did find fresh tracks and hair which had been shed. In 1966 Dr. Eric Guiler identified some hair from a Tasmanian "wolf." It seemed that this animal had regularly used an old cooking stove as its resting place, and it lost some hair while slipping in and out of it. This sighting was close to Mawbanna in Northwest Tasmania. A trap set out for the animal was left untouched. The tracks indicated that this animal was a female with young.

Mammals: 1. Thin-tailed dormouse possum *(Cercaërtus concinnus)*; 2. Grey kangaroo *(Macropus giganteus ocydromus)*; 3. Marsupial anteater *(Myrmecobius fasciatus)*; 4. Quokka *(Setonix brachyurus)*; 5. Kangaroo mouse *(Notomys richardsonis)*; 6. Echidna *(Tachyglossus aculeatus)*. Birds: 7. White-faced heron *(Notophoyx novaehollandiae)*; 8. Wedge-tailed eagle *(Uraëtus audax)*; 9. White-tailed cockatoo *(Calyptorhynchus baudinii)*; 10. Spotted pardalote *(Pardalotus punctatus)*; 11. Bauers' rosella *(Platycercus zonarius zonarius)*; 12. Pale silvereye *(Zosterops albiventris)*; 13. Zebra finch *(Taeniopyga guttata castanotis)*; 14. Western magpie *(Gymnorhina dorsalis)*; 15. Banded wren *(Malurus elegans)*; 16. Western rosella *(Platycercus icterotis)*; 17. Laughing kookaburra *(Dacelo gigas)*; 18. Tawny-crowned honeyeater *(Gliciphila melanops)*; 19. Brush bronzewing pigeon *(Phaps elegans)*; 20. Straw-necked ibis *(Thresciornis spinicollis)*; 21. Musk duck *(Biziura lobata)*; 22. Maned goose *(Chenonetta jubata)*; 23. Black swan *(Cygnus altratus)*. Reptiles: 24. Bandy-bandy *(Vermicella annulata)*; 25. Cone-lizard *(Tiliqua rugosa)*. Amphibians: 26. Golden frog *(Hyla aurea)*.

Presently there are still some Tasmanian "wolves" alive in the remote forested mountain regions of Tasmania. Although the animals have been under complete protection since 1938, their chances for survival are extremely slim. The forest region is unsuitable for the animals because there is little food for them. The Tasmanian "tiger's" natural habitat is the open plain, the only place where it can hunt sufficient numbers of kangaroos and wallabies. However, the sheep ranchers and farmers have taken over this habitat for their own purposes, and the "wolves" were forced to retreat deeper and deeper into the forest. Although no one harms the "wolves" there, they cannot maintain themselves. Sincere efforts to save the species would mean providing them with open plains, which contained other marsupials or perhaps simply sheep. But who wants to go to the trouble of providing all this for some "burdensome vermin"!

Certain regions in Southwest and South Australia are inhabited by a unique ant and termite-eating marsupial. The creature has fifty small, delicate, regressed teeth and a long, thin, extendable tongue. Dental regression, usually due to an increase in the number of teeth, and the development of a long extendable tongue are characteristics found in other ant- and termite-eating mammals as well. These morphological features are found in the anteaters of the order *Edentata,* aardvark, pangolins , and the Australian echidnas.

Family *Myrmecobiidae* by D. Heinemann

The MARSUPIAL ANTEATER (*Myrmecobius fasciatus;* Color plate pp. 89–90) serves to illustrate the way in which the diversified order of *Marsupialia* has utilized this particular means of making a living. This group is distantly related to the *Dasyuridae,* but its many unique features justify assignment to a distinct family *(Myrmecobiidae).*

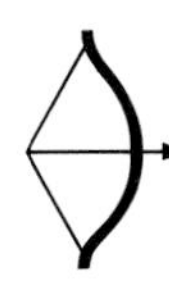

Distinguishing characteristics

The HRL is approximately 23 cm and the TL is 17 cm. The head is flat, slender, and pointed. The eyes are large and the pointed ears are medium-sized. The tail is bushy, and the fur is long-haired. Above the sternum is a collection of sebaceous and sweat glands. There are five fingers and four toes on each hand and foot respectively. There are 50 teeth: $\frac{4 \cdot 1 \cdot 4 \cdot 4}{3 \cdot 1 \cdot 4 \cdot 4}$. The teeth are small, weak, and regressed. Their size is not constant; the molars on the right and left sides of the skull vary in length and width. Extra teeth occur. The tongue is long and thin. There are four teats, and there is no pouch. There is only one species.

In Southwest Australia the marsupial anteater is often referred to by the name "numbat" which is given by the natives. From an early age, the numbat is always recognizable by the distinct stripes on its fur. The rat-sized animal has the appearance of a squirrel because of its bushy tail which is bent over its back during a state of excitement. However, the snout is much more pointed than in a rodent. The numbat is able to project its sticky tongue ten centimeters out of its mouth. It extracts termites, its main source of food, from crevices in the wood. The body of the numbat is much broader than that of most other small mammals.

The female is considerably smaller than the male. Although it is a marsupial, the female lacks any suggestion of a pouch. The region around the four nipples is surrounded by long hair, which may serve as some protection for the "marsupial" embryos. As long as the young are attached to the nipples, they are dragged about beneath the mother as she moves about. Very little is known about the life of these shy, small marsupials.

Fig. 5-14. Marsupial anteater (*Myrmecobius fasciatus*; see p. 91).

Years ago, David Fleay, who was at that time the director of the natural conservatory in Healesville, received a young female which he kept alive for two months. He named her "Miss Numbat." It had been known that small marsupials will occasionally fall into a state of temporary stupor. However, Fleay was concerned when he found Miss Numbat lying on the floor rigid and lifeless. He remembered how marsupial anteaters in general did not do well in captivity. Yet the following morning, Miss Numbat was active once more, and she started to eat again in the afternoon. Fleay kept her on a diet of termites, ants, ant larvae, meal worms, bugs, insect larvae, earthworms, raw eggs, honey, and bread soaked in milk. Miss Numbat, however, preferred the termites which she picked up individually if they fell within the range of her vision or her acute sense of smell. Her appetite for termites seemed insatiable. The smaller varieties were swallowed whole, and the larger "soldiers," which can bite with their powerful mandibles, were chewed quickly and noisily. Miss Numbat eagerly anticipated her breakfast pail which contained termite-infested pieces of wood. She would jump into the pail and project her tongue into the crevices to extract the soft insects. She preferred picking termites out of wood to collecting those which were running around freely on the floor. The animal became fearful when sawdust stuck to her tongue and prevented her from extracting the termites. The tongue is manipulated with great speed and dexterity. The extended tongue flicks into all corners and tunnels of the termite-infested wood. The numbat could turn a piece of wood with its mouth into a more favorable position for picking out the hidden insects. Rotten wood was worked over with the strong front claws. When the animal was startled while searching for termites, she would emit deep humming sounds. If she was lifted up, she would not attempt to bite. Sometimes when she was frightened, she would bolt upright just like a little kangaroo.

How they collect and eat termites

Marsupial anteaters do not construct ground burrows. Contrary to older reports, Fleay observed that the animal is a good climber and often probes for termites in high, dead trees. After a plentiful meal of termites, Miss Numbat would stretch out on her favorite branch. The ultimate sign of utmost comfort was the extended pink tongue which would dangle out of her mouth in a graceful arch. Unlike most

Fig. 5-15. 1. Marsupial mole (*Notoryctes typhlops*), 2. Northwestern marsupial mole (*N. caurinus*; see p. 93).

Marsupial anteaters are diurnal

marsupials, the numbat is active during the day. It eats only during the day and sleeps through the night. Miss Numbat had selected a hollow log and had made a comfortable bed of leaves and grass in it. The teeth were used to pull individual blades from old grass clumps. "After dusk Miss Numbat would not venture out of her snug boudoir," related Fleay. He emphasized the fact that marsupial anteaters are dependent on hollow logs for their shelter. The systematic application of brush fires has probably contributed just as much to the near extinction of these unique, beautiful creatures as has the introduction of the fox into Southwestern Australia. In certain regions of Australia, excellent sanctuaries for the threatened fauna have been set aside. Many of the original, yet often highly specialized, marsupials do not have a chance against the fox or the dingo. There is only one way to protect this unique fauna which flourished as the result of millions of years of separation from the rest of the world. The solution would be to either exterminate the introduced foxes and dingos or to keep them away from the marsupials by suitable fences.

Family *Notoryctidae* by D. Heinemann

Australia also has its very own mole-like, insect- and worm-eating burrowing marsupials, just as other continents have their golden moles or true moles. These Australian "moles" are not members of the order *Insectivora*, like the golden moles or moles, but like most Australian mammals they belong the the order *Marsupialia*. There is a possibility that the MARSUPIAL OR POUCHED "MOLES" (genus *Notoryctes*) are related to the *Dasyuridae* or *Peramelidae*, but since not enough data is available to clarify this point, the "moles" are classified as a separate family (*Notoryctidae*).

Distinguishing characteristics

The HRL is 9–18 cm and the TL is 1–2.5 cm. The head is short and thick and there is a horny shield on the nose. The eyes are vestigial and hidden under skin and muscle. The ear pinnae are missing, and the small ear opening can be closed off. The body is plump, stout, and cylindrical. The stumpy tail is naked and marked by a series of distinct rings. The limbs are short and thin. There are claws on all five fingers and toes. The arms are very short, and the hand protrudes directly from the body fur. The forefeet have two large claws on digits three and four which are modified and enlarged for display; the remaining digits are small. The first segment of the first toe is supported by a lateral bony plate which increases the sole surface. The fur is short, dense, silky, and, in parts, shiny. There are 40–42 teeth: $\frac{3\text{-}4 \cdot 1 \cdot 2 \cdot 4}{3 \cdot 1 \cdot 2 \cdot 4}$; the incisors are small and the premolars and molars are cuspate. The teeth of the upper jaw are smaller than those in the lower jaw. The epi-pubic bones are reduced. The flat marsupium opens to the rear, and there are two teats. There is no scrotum. There are two species: (1) MARSUPIAL MOLE (*Notoryctes typhlops;* Color plate p. 71). The HRL is 15-18 cm and the TL is 2-2.5 cm. (2) NORTHWESTERN MARSUPIAL MOLE (*Notoryctes caurinus*). The HRL is approximately 9 cm and the TL is approximately 1 cm.

The discovery of the marsupial mole

One day in 1888 the manager of a cattle ranch, William Coulthard, discovered a strange, mole-like creature underneath a clump of Spinifex grass *(Triodia irritans).* Coulthard had lived in Central Australia between Charlotte Waters and Alice Springs for many years, but he had never encountered such an animal. He picked up the animal and forwarded it to the zoologist Stirling. Among the mammalogists the discovery of this mole-like creature in Australia caused quite a sensation, comparable only to the discovery of the duck-billed platypus.

The marsupial "moles" more closely resemble the golden mole (*Chrysochloridae,* see p. 199) in general body form than the typical mole *(Talpidae).* There is such a striking similarity between the two families that the famous anatomist Cope first believed that they originated from the same ancestoral stock. In fact the similarity is the result of convergent evolution due to a similar way of life. Adaptation to a burrowing mode of life resulted in similar morphological changes in both the golden mole and the marsupial "mole." Both families (*Notoryctidae* and *Chrysochloridae*) are so highly specialized for the subterranean, burrowing life that they cannot exist in any other way.

How the marsupial mole burrows into the ground

When burrowing either for food or to escape from an enemy, the marsupial "mole" moves along at no more than about eight centimeters below the surface. The marsupial "mole" does not leave behind a permanent burrow as do the true moles. The tunnel always caves in behind it. The animal emerges after it has burrowed for several meters. On the surface it is able to breathe more freely. It shuffles along the surface leaving a triple track made by the body and limbs. The shielded snout penetrates the earth, the foreclaws function as scoops, and the hind feet throw up the sand. It does not take long and the "mole" disappears below the surface.

It is very difficult to keep such highly specialized animals in captivity. Professor Wood Jones, who observed marsupial moles in captivity, was impressed by their feverish drive for activity. The animals gave the impression of being intensely "nervous." Because of this high degree of activity, the animal requires a great amount of food in relationship to its body size. Food shortage can quickly end in death. It is possible that they can get through a fasting period by falling into a deep sleep. A. G. Bolam, a stationmaster for the Trans-Australian Railway in Ooldea who is also an avid naturalist, is of the opinion that marsupial "moles" are active in loose soil only after a rainfall and that they consume insects and worms. One *Notoryctes* that Professor Wood Jones received from Mr. Bolam fed ravenously on earthworms a few minutes after being unpacked from its shipping box. The feverish restlessness of the marsupial mole was reminiscent of the European mole's. After quickly feeding, the "mole" continued its rapid shuffling through the box where he regularly somersaulted at the corners. This was because its nose was always pointed down. The "mole" would suddenly fall asleep, only to awake just as suddenly to resume its feverish activity. It did not seem to resent handling, and it would even consume milk rapidly while being held; then it would suddenly fall asleep again. Wood Jones' captive "moles" made chirping sounds which were repeated two to three times.

Nothing is known about the breeding habits of the marsupial "mole," but it has been suggested that the female makes a permanent burrow deep in the ground for raising her young. Occasionally, a marsupial "mole" has been caught in one of their permanent burrows.

Bernhard Grzimek
Dietrich Heinemann

6 Bandicoots and Rat Opossums

Family *Peramelidae*
by W. Gewalt

Distinguishing characteristics

The family *Peramelidae* are also called bandicoots. These animals range in size from that of a rat to that of a badger. The HRL is 17.5–50 cm and the TL is 7–26 cm. The general body form resembles that of the kangaroo. The front limbs are shortened and the hind limbs are longer. The snout is long and pointed, and the ears, which may be covered with hair or may be naked, range in size from small to large and narrow. The tail is of variable length. The hand has two to four fingers with well-developed claws; in the pig-footed bandicoot, the fourth finger is vestigial and clawless. The first and fifth fingers are short and clawless; they are lacking in the pig-footed bandicoot. The hind feet vary in width in bandicoots. The main digit is the fourth toe. The first toe, which is vestigial and lacks a nail, is absent in rabbit bandicoots and pig-footed bandicoots. The second and third toes are fused and are used for grooming. The nails are separate. There are 46–48 teeth: $\frac{5\text{-}4 \cdot 1 \cdot 3 \cdot 4}{3 \cdot 1 \cdot 3 \cdot 4}$. The canines are long in the rabbit bandicoots and tusk-like in the *Echymipera clara*. The molars are four-sided or triangular in outline with four external cusps. The well-developed pouch opens downward and backward. There are six to ten teats. Unlike all other marsupials, the long-nosed bandicoots have a true placenta consisting of chorion and allantoic membranes. Generally there are two to six young per litter. There are a total of eight genera and nineteen species.

Fig. 6–1.
1. Long-nosed bandicoot (Genus *Perameles*). 2. New Guinean bandicoot (Genus *Peroryctes*).
3. Ceram Island long-nosed bandicoot (Genus *Rhynchomeles*).

A. LONG-NOSED BANDICOOTS (genus *Perameles*; Color plate p. 72). The HRL is 21–43 cm and the TL is 9–17 cm. The animals are slender with a particularly long, pointed, almost trunk-like muzzle and medium long ears. The pelage may be soft or coarse. A part of their diet consists of vegetable matter such as onions, tubers, and roots. There are five species which include: 1. LONG-NOSED BANDICOOT *(Perameles nasuta)*. This insectivorous animal also consumes small vertebrates. 2. EASTERN BARRED BANDICOOT *(♁Perameles fasciata)* and 3. TASMANIAN BARRED BANDICOOT *(Perameles gunni)*.

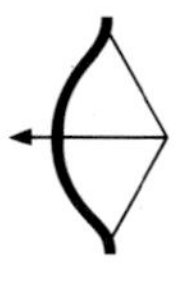

Fig. 6-2.
1. New Guinean spiny bandicoot (Genus *Echymipera*). 2. Rabbit bandicoot (Genus *Macrotis*).

B. NEW GUINEAN BANDICOOTS (genus *Peroryctes,* Color plate p. 72). The HRL is 17–50 cm and the TL is 14–26 cm. The animals live in the dense mountain forests. There are three species which include the (1) NEW GUINEA BANDICOOT *(Peroryctes raffrayanus);* and (2) the LONG-TAILED NEW GUINEA BANDICOOT *(Peroryctes longicauda).*

C. MOUSE BANDICOOTS (genus *Microperoryctes;* Color plate p. 72). There is only one species *(Microperoryctes murina).* The smallest member of the bandicoots, they have an HRL of approximately 17.5 cm and a TL of approximately 11 cm. Only three specimens have been found.

D. NEW GUINEAN SPINY BANDICOOTS (genus *Echymipera;* Color plate p. 72). The HRL is 27–45 cm and the TL is 7–12 cm. The pelage is stiff and spiny. There are three species, including the WESTERN NEW GUINEAN SPINY BANDICOOT *(Echymipera clara)* and the *Echymipera kalubu.*

E. RABBIT BANDICOOTS OR RABBIT-EARED BANDICOOTS (genus *Macrotis;* Color plate p. 72). The HRL is 20–44 cm and the TL is 12–22 cm. The pelage is long, silky, and soft. The terminal half of the tail has a conspicuous dorsal crest of long light hair. The long ears are rabbit-like. There are two species: 1. The LARGE RABBIT BANDICOOT *(⊕Macrotis lagotis)* and 2. The SMALL RABBIT BANDICOOT *(Macrotis leucura).*

F. SHORT-NOSED BANDICOOT (genus *Thylacis;* Color plate p. 72). The HRL is 24–41 cm and the TL is 9–18 cm. The nose is only slightly elongated and the ears are short. The pelage is short-haired and coarse. There are three species, including the SOUTHERN SHORT-NOSED BANDICOOT *(Thylacis obesolus)* and the BRINDLED BANDICOOT *(Thylacis macrourus).*

G. CERAM ISLAND LONG-NOSED BANDICOOT (genus *Rhynchomeles).* There is only one species *(Rhynochomeles prattorum)* on the island of Ceram.

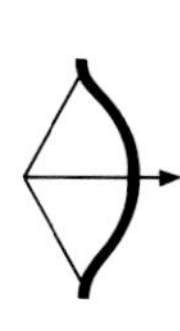

H. PIG-FOOTED BANDICOOTS (genus *Chaeropus;* Color plate p. 72). The HRL is approximately 25 cm and the TL is approximately 10 cm. The limbs are longer and more slender than in other bandicoots. Only two toes of the front feet and one toe of the hind feet touch the ground. There is only one species, the *⊕Chaeropus ecaudatus.*

"Miserable as a bandicoot"

Bandicoots are not well known even to zoologists, unless they happen to be specialists of the Australian fauna. This animal group is very diversified and is widely distributed in the Australian faunal regions. The animals are known by the term "bandicoot" in their home range. In the last century the saying "miserable as a bandicoot" was known throughout the English-speaking world. It is not clear why such an appealing animal, which looks like a cross between a jumping mouse, shrew, and a kangaroo, should be branded as "miserable." The origin of the word "bandicoot" is just as puzzling. One assumes that "bandicoot" is not of Australian origin and is a corruption of a word from the Telugu language, spoken by the people of the eastern Deccan plateau of India. This word, *Pandi-kokku,* means "pig-rat" and was originally applied to a group of large rats, *Bandicota indica.*

The tricuspate structure of their teeth is shared by the bandicoots

Legends to the following color plates

Phalangers (Plate I, p. 99)

1. Pygmy possum *(Eudromicia caudata).*
2. Grey cuscus *(Phalanger orientalis).*
3. Spotted cuscus *(Phalanger maculatus)* [Two color variations, a and b].
4. Bear phalanger *(Phalanger ursinus).*
5. Scaly-tailed phalanger *(Wyulda squamicaudata).*
6. Striped phalanger *(Dactylopsila trivirgata).*
7. Brush-tailed phalanger *(Trichosurus vulpecula).*
8. Short-eared brush-tailed phalanger *(Trichosurus caninus).*

Phalangers (Plate II, p. 100)

1. Honey phalanger *(Tarsipes spenserae).*
2. Honey glider *(Petaurus breviceps).*
3. Yellow-bellied glider *(Petaurus australis).*
4. Leadbeater's phalanger *(Gymnobelideus leadbeateri).*
5. Pen-tailed phalanger *(Distoechurus pennatus).*
6. Pygmy phalanger *(Cercaërtus nanus).*
7. Pigmy flying phalanger *(Acrobates pygmaeus).*
8. Queensland ring tail *(Pseudocheirus peregrinus).*
9. Brush-tipped ring tail *(Pseudocheirus lemuroides).*

Phalangers (Plate III, p. 101)

1. Greater flying phalanger *(Schoinobates volans).*
2. Koala *(Phascolarctos cinereus).*
3. Wombats: a. Common wombat *(Vombatus ursinus platyrrhinus);* b. Tasmanian wombat *(Vombatus ursinus tasmaniensis).*
4. Soft-furred wombat *(Lasiorhinus latifrons).*

Color plate p. 102

Top: Among the dasyures, the Tasmanian devil (*Sarcophilus harrisi;* see p. 113) is the species most dependent on a meat diet. These two are eating a large lizard. *Sarcophilus* means "meat loving."

Bottom: The spotted cuscus *(Phalanger maculatus;* see p. 113) may appear in different color variations. This is a particularly light colored animal. Some are completely white.

Color plate p. 103

Top: The Australian pygmy gliding possum (*Acrobates pygmaeus;* see p. 117) is actually only barely half as large as in this photograph. The feather-like tail serves as a rudder during jumping.

Bottom: The common wombat (*Vombatus ursinus;* see p. 139 is a plump terrestrial animal. In Australia the wombats fill the niche that the marmots fill in the northern zones.

Color plates p. 104 and p. 105

It is hard to believe that these two photographs represent the same animal. The animal shown is the cat-sized greater gliding possum (*Schoinobates volans;* see p. 119). While gliding the animal bends its underarm to the inside so that its hands lie close to its head. In this manner the great gliding possum can sail through the air for a distance of up to 100 meters.

Color plate p. 106

The Grey Cuscus *(Phalanger orientalis mimicus)* in the tropical rainforest of New Guinea. This genus contains more species than any other among the marsupials in the Papuan region. For the native Papuans the squirrel-sized animals are an important source of protein, since New Guinea has only few large animals. Hunters stir up sleeping marsupials which are hiding in hollow trees by poking sticks into the openings. They pull out the stick to see if any hairs cling to it which might betray the presence of an animal. Phalangers are formidable fighters. They defend themselves with loud threat calls and with bites with their long incisors as well as with the quick slashing thrusts of their front feet which have long, sharp claws.

1
2
3b
4
5
6
7
8

II
1
2
3
4
5
6
7
8
9

III
1
2
3a
3b
4
B BERTRAM

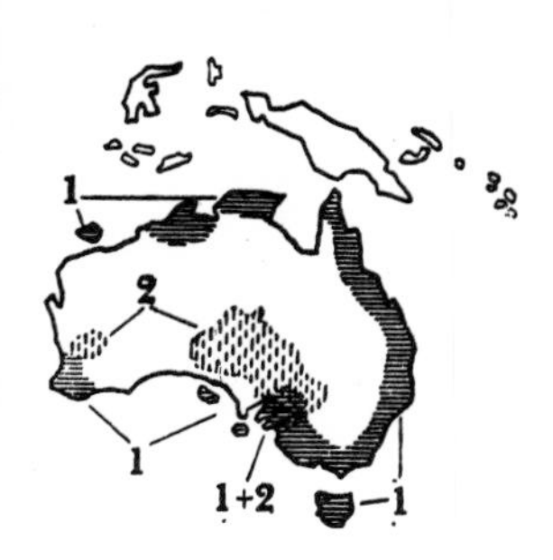

Fig. 6-3.
1. Short-nosed bandicoot (Genus *Thylacis*). 2. Pig-footed bandicoot (Genus *Chaeropus*).

and the carnivorous marsupials. The actual shape of the teeth changes throughout the animal's life. The young bandicoot has the needle-like teeth of the insectivores. The teeth become duller as they are worn down by the tougher meat diet. In the more mature animal, the teeth are reminiscent of the teeth of herbivores with the crowns frequently worn right down to the roots.

The term "nose" marsupials (*Nasenbeutler* in German) is rather appropriate because the muzzle is long and pointed, and occasionally the snout appears trunk-like. In certain species the ears are also long, narrow, and pointed. Generally bandicoots walk on all four legs, although the forelimbs are short and the hind limbs are longer. Nevertheless, the structure of the appendages indicates an adaptation to an existence in the savannah and semi-desert. Many animals that have adapted to either a running or jumping type of locomotion in response to living on plains, for example the zebras, kangaroos, and ostriches, show a reduction and modification in the foot digits. The number of toes is reduced and the body weight is consequently redistributed on the remaining toes or only on one main digit. It is remarkable to find these adaptive characteristics present in the hind feet of the bandicoot. On the basis of these modifications, it was believed that the bandicoots were the "transitional" group between the *Dasyuridae* and *Macropodidae.* However, the similarity between the hind foot of the bandicoots and the kangaroos is not based on a close phylogenetic relationship but on an adaptive response to similar environmental conditions.

All bandicoots are terrestrial and nocturnal. The animals are very secretive. Only occasionally do they betray their presence by loud squeaking sounds. They dig for insects, worms, and roots. Bandicoots construct grass nests or deep burrows in the ground with hasty scratching movements. It is said of some species that they can penetrate the ground with their clawed front feet much quicker than a man can with shovel and hoe. Often bandicoots are found near towns and cities where they quite frequently dig in gardens, looking for numerous grubs or mouse nests. Although this may seem like a useful service to the garden, people do not appreciate the occasional scratched up flower bed. Consequently, the animals are pursued quite persistently. Several species were in danger of extinction and are now protected. In contrast to opossums and the dasyures, which spit and bite when they are threatened, bandicoots remain completely defenseless and passive, even when in extreme danger to themselves. This could be the reason why they were called "miserable."

The rabbit bandicoots are particularly delightful creatures. They look like Morgenstern's Nasobem, an appealing mixture of springhaas, desert fox, aardvark, and kangaroo, come alive. Although the rabbit bandicoots look very graceful, they are the most avid burrowers among the bandicoots. The burrow usually descends from a single opening in a fairly steep, ever-widening spiral to a depth of 1.5 meters. The Australian aborigines highly value the meat of the "bilbies" and they use bundles of rabbit bandicoot tails as decorations. The aborigines put their ears on the ground to detect the scratching of the rabbit bandicoots. The animals are highly "useful" by human standards because they consume large quantities of insect larvae and mice. Nevertheless, the bandicoots' numbers are decreasing due to senseless persecution. Fur-trapping and sport-hunting has contributed greatly towards diminishing the remaining populations.

Fig. 6-4.
As in most burrowing marsupials, the pouch of the bandicoot opens to the rear. A young long-nosed bandicoot *(Perameles)* is about to enter the mother's pouch.

The most unique member of the bandicoots, the pig-footed bandicoot, is near extinction. This slender, rabbit-sized animal with the trunk-like nose and pointed ears looks like a normal bandicoot that has been placed on the thin, long legs of an ungulate. In a way it looks like a "bandicoot on stilts." The legs are long and thin. The second and third digits of the forefeet are fused and all the other digits are absent. The footprints are similar to those of a small deer. The third digit of the hind foot is functional. The structure is more kangaroo-like than deer-like. It seems apparent that the pig-footed bandicoot is not a burrowing animal but a plains runner, and it constructs its own grass nests. This bandicoot is not as exclusively nocturnal as are the other members of the *Peramelidae.*

General Major Mitchell was the first European to find a pig-footed bandicoot on his famous expedition to the Darling and Murray Rivers in 1836. The animal had fled into a hollow tree from which Mitchell's native guides pulled it out. During this process the bandicoot probably lost its tail. This tailless specimen, which was then presented to zoologists, was termed *Chaeropus ecaudatus,* which means "tailless pig-footed bandicoot." Although additional trapped pig-footed bandicoots proved this to be a misnomer, the scientific name remained because of the strict rules of priority in classifying animals.

Rat opossums:
Family *Caenolestidae*
by W. Gewalt

During the second half of the nineteenth century, fossil remains of small marsupials from the lower Tertiary period were discovered in Argentina. At the turn of the century, the living counterpart to the extinct ancestral marsupials was found in the cool rain forests of the Andes. These are the rat opossums (family *Caenolestidae*). About the size of a shrew, they have an HRL of 10.5–13.5 cm and a TL of 6.5–12.5 cm. The head is pointed. The tail, which is sparsely covered with hair, occasionally has a thickened base. The pelage is dense and soft. There are five fingers and five toes on the front and hind limbs. There are 46 teeth: $\frac{4 \cdot 1 \cdot 3 \cdot 4}{3 \cdot 1 \cdot 3 \cdot 4}$. The incisors are weak and small except for the large first lower incisors. The pouch is absent. There are four to five teats. There are three genera, each represented by one species, in each of the Andes' countries starting in Venezuela and extending to South

Chile. 1. ECUADOR RAT OPOSSUM (*Caenolestes fuliginosus;* Color plate p. 72); 2. PERU RAT OPOSSUM *(Lestoros inca);* 3. CHILE RAT OPOSSUM *(Rhyncholestes raphanurus).*

The last survivors of an ancient tribe

At first the relationship between the fossilized and living caenolestids was not recognized. The cuspate molars of the rat opossums seemed to indicate a connection with the didelphids. Additional information, however, indicated that the long, vertically rooted incisors in the lower jaws are very similar to those of the Australian phalangers, wombats, and kangaroos. For a time, the caenolestids were considered to be a transitional form between the predominantly carnivorous and insectivorous "multi-incisoral" didelphids, dasyures, and bandicoots on the one hand and the "bi-incisoral" herbivorous forms on the other. However, the shape and size variation in teeth is usually dictated by the type of food consumed, and only very rarely does it serve as a clue to phylogenetic relationships. Thus, these probably very primitive marsupials are classified separately. The animals flourished during the Tertiary and were represented by many different species. Some of the species were more highly developed than the rat opossums living today. Very little is known about the life history and brood care of the few species that have survived to present times.

Wolfgang Gewalt

7 Phalangers

Distinguishing characteristics

The PHALANGERS *(Phalangeridae)* are the most diversified group of all the marsupial families. The size variations are from mouse- to fox-size. The HRL is 7–82 cm and TL is 0–47 cm. The snout may be blunt or trunk-like, and the ears are small to medium-sized. The usually long tail may either be covered by hair or it may be naked. The tail is often prehensile, or it may be absent as in the koala. All four limbs have five digits and all the digits, except for the big toe, have claws. The second and third toes are joined by skin at the top joint, and they are smaller than the fourth and fifth toes. The pelage is dense, soft, and occasionally long-haired. There are twenty-four to forty-two teeth. The central lower incisors are well-developed and strong. The lower canines are missing. The pouch is well-developed and opens forward, except in the koala where it opens to the rear. There are two to four mammae in the female phalangerids. Young are born once a year or every other year (koala). There is usually only one young, rarely two, three, or even six.

There are three subfamilies: 1. PHALANGERS *(Phalangerinae)*; 2. HONEY PHALANGERS *(Tarsipedinae)*; 3. KOALAS *(Phascolarctinae)*. There are fourteen genera and forty-one species distributed in Tasmania, Australia, New Guinea, west to the Celebes, Timor, Ceram, and adjacent islands, and east to the Bismarck Archipelago and the Solomon Islands.

Family: Phalangers by W. Gewalt

The phalangers are arboreal and, in addition to the tree kangaroos, are the only predominantly herbivorous tree animals in the Australian faunal region. They occupy the niche that is filled by monkeys, prosimians, squirrels, flying squirrels, giant gliders, and sloths on other continents. This is the reason that they occur in such a variety of body forms. There are extremes in sizes, from mouse-like to fox-like. The mode of locomotion also varies greatly from the plump, slow-moving koalas to the flying possums which can glide through the air for nearly 100 meters. Then again there is a variety of tail forms. Phalangers may have rat-like scaly tails, prehensile tails, bushy squirrel-like tails, or only vestigial tails. Gliding marsupials have evolved independently three times in Australia. The pygmy flying phalangers, the flying phalangers, and the greater flying phalangers are not closely related to each other. Each group evolved their gliding syndrome independently of the others. Although all are marsupials, each group of gliders is closely related to species without the gliding membrane. For example, the smallest glider, the pygmy flying phalanger, is closely related to the pin-tailed phalanger, which does not have a flying membrane. The three Leadbeater's possums closely resemble the sugar glider but lack the gliding membrane. The greater flying phalanger belongs to a completely different subfamily which includes the distantly related koalas, although it resembles the ring-tailed possums in its anatomy.

Subfamily True Phalangers

The TRUE PHALANGERS (subfamily *Phalangerinae*) are heavy and rather powerfully built animals. The head is usually short with an elongated snout. The tail, which is covered with hair to varying degrees, often has a naked and horny underside which permits grasping. There are ten genera and twenty-three species.

The BRUSH-TAILED POSSUMS (genus *Trichosurus;* Color plate p. 99) vary in size from that of the rat to that of the fox. The HRL is 32–58 cm and the TL is 24–38 cm. The medium-sized, triangular, and naked ears can be folded. The fur is dense and soft with a great variation in coloration. The prehensile tail is bushy, but the terminal part is covered with scales and lacks hair. There are two species: 1. BRUSH-TAILED PHALANGER *(Trichosurus vulpecula).* The ears are large and pointed and the face looks fox-like. 2. SHORT-EARED BRUSH-TAILED PHALANGER *(Trichosurus caninus).* The ears are smaller.

The SCALY-TAILED PHALANGER (⊕ *Wyulda squamicaudata;* Color plate p. 99) is similar to the cuscuses but the tail is naked and scaly like a rat's tail. The animals are about the size of squirrels. Only four specimens have been found up to now.

The brush-tailed possums are nocturnal arboreal animals although some of them descend to the ground. Their diet consists of eucalyptus leaves, bark, tree sprouts, and other vegetation. It may also include eggs and small animals. The fur coloration has great variability. There is color to suit every lady, and, of course, the furriers were greatly attracted. Up to very recently, the cuscus was senselessly exploited for its fur. In one season alone (1931–1932), over one million brush-tailed possum pelts were exported from Australia under phony fashionable names like "Adelaide-Chinchilla," "Skunk," or "Australian Beaver." Once in South Australia over 100,000 brush-tailed possums were killed in the short period of three months. These statistics not only show what man has done to the marsupials in Australia, but it also indicates that the brush-tailed possum is one of the most frequently found marsupials on the Fifth Continent. The animal is distributed over the entire continent except in the desert regions. White settlers introduced the phalanger into New Zealand. The Australian mammalogist E. Throughton praises the brush-tailed phalanger for the same qualities that the Americans attribute to the opossum: "One can regard it (the brush-tailed phalanger) as the most adaptable of all mammals." Although the brush-tailed phalanger is arboreal by nature, it can thrive equally well in sparsely forested or treeless regions. They can hole up equally well in the crown of a huge eucalyptus tree, a rabbit burrow, or a river bank. Brush-tailed phalangers often live within the proximity of people, in spite of being hunted extensively by them. The animals seem to prefer roofs and sheds as hiding places. They have even been found in big cities. Landlords and gardeners are not too fond of the noises they make at night, nor of the dark spots they leave behind on white ceilings. Equally unpopular is the damage they cause to rose bushes and fruit trees. Brush-tailed possums even have a reputation as "forest vermin" because they debark young evergreens. The only way to keep the animals away from these trees is to spray them with chemicals.

Fig. 7-1. 1. Short-eared brush-tailed Phalanger *(Trichosurus caninus)*. 2. Scaly-tailed phalanger *(Wyulda squamicaudata)*. 3. Spotted cuscus *(Phalanger maculatus)*.

Since the brush-tailed phalangers are well hidden in their shelters during the day, they have very few natural enemies. The occasional poorly camouflaged animal may fall prey to the wedge-tailed eagle. A more frequent enemy is the monitor lizard which will climb up a tree to root out the brush-tailed possum from its hiding place. When the phalanger hears the scratching claws of the monitor, it shrieks loudly but it does not try to escape. The Australian aborigines made use of this reaction. They imitated the climbing sounds of the reptile by scratching a stick on the base of a tree which might hold a "possum." The frightened animal would betray its presence with a scream.

A single young is usually produced and the reproductive rate is low. The gestation period is seventeen days. The young possum does not leave the pouch until it is four to five months old. After leaving the pouch, it remains with its mother and rides about on her back. The young is weaned and becomes independent in about six months. Winkelsträter, who observed the behavior of a group of vulpine phalangers in the Zurich Zoo, once recorded the birth of a pair of twins. A possum female in the Frankfurt Zoo gave birth to one young approximately every five months between 1965 and 1967. The rate of population increase is not low, however, particularly since the young become sexually mature very early.

The CUSCUSES (genus *Phalanger;* Color plate p. 99) range in size from that of a rat to that of a cat. The HRL is 27-65 cm and the TL is 24-60 cm. The head is round with a pointed snout and large, nocturnally-adapted eyes. The ears are small and, in some species, are buried inside the fur. The first and second fingers are opposable to the remaining ones (forcipated claw). In the foot which is modified for grasping, the big toe is opposed to the others. The small second and third toes are fused. The tail is strongly prehensile and the terminal end is naked. The underside of the tail tip is striated with callouses. The pelage is dense and woolly. There are six species and twenty-three subspecies, including: 1. The GREY CUSCUS (*⊕Phalanger orientalis*); 2. The SPOTTED CUSCUS (*Phalanger maculatus*); 3. The BEAR PHALANGER (*Phalanger ursinus*).

Cuscuses are arboreal and nocturnal. The large night-adapted, orange rimmed eyes are red, yellow, or even bluish-green. The cuscus belongs to the most colorful mammals. There are countless color variations and patterns ranging from white, rust-brown, yellow, greyish-green, to almost totally black. In the spotted cuscus, the female is usually grey while the male is covered by white spots of varying sizes. However, some females are honey-blonde with rust-brown dots. The fur coloration varies with age, locality, and state of health. Even members of the same litter show remarkable color variations. However, the bear phalanger, which is limited to the island of Celebes, is always a dark chocolate brown.

Cuscuses, which are slow-moving and somewhat sluggish, resemble the South American sloths. Like the sloths, their diet consists mainly of the leaves of the virgin forest in addition to small animals. The round face with the short snout and protruding eyes and the animal's sluggish movements resemble similar characteristics in certain prosimians, like the slow loris, African pottos, and angwantibos. In captivity the cuscus, like the prosimians, can exist on a diet of eggs and meat. Formerly, the cuscuses on Cape York were occasionally mistaken for prosimians, and often they were the source for sensational reports announcing the presence of monkeys in the virgin forests of Australia! The island habitat of the cuscus is free of tree-climbing predators that would prove threatening to them. This is the reason why the phalangers can afford such a clumsy, slow mode of locomotion. The native Papuans relish cuscus flesh and the beautiful, dense fur for caps and capes. Other enemies, besides man, are the python and the monitor lizard.

I (W. Gewalt) kept a spotted cuscus from the Berlin Zoo in my apartment for some time. It made a very interesting pet. In the beginning, it snarled, barked, and struck out with its clawed paws, just like a hamster. Gradually, however, the animal became tame. It did not mind being petted, and it was even active in the day during certain times. If released from its cage, the cuscus would investigate the rooms with great interest. The animal provided amusement because visiting zoologists had difficulty identifying our strange house guest. Indeed, the cuscus is one of the most unusual animals. It is looked at with more astonishment than any other animal in the nocturnal display house in the zoo.

The STRIPED POSSUM or STRIPED PHALANGER (genus *Dactylopsila*) has the appearance and size of a squirrel. The bushy tail is well covered with hair. The animal secretes a penetrating odor of glandular origin which is non-volatile, as a defensive weapon. The black stripes serve as a "warning signal" (similar to the skunks; see Vol. XI). The slender, elongated fourth finger is an adaptation for dislodging bug larvae and the like from narrow holes and crevices (similar to the third finger of the aye-aye; Color plate p. 247). The animal is arboreal and nocturnal. There are two species: 1. STRIPED PHALANGER (*Dactylopsila trivirgata;* see Color plate p. 99). The HRL is 17–32 cm and the TL is 24–40 cm. 2. LONG-FINGERED STRIPED PHALANGER (*Dactylopsila palpator*). The HRL is 20–27 cm and the TL is 20–24 cm.

Fig. 7-2. 1. Brush-tailed phalanger (*Trichosurus vulpecula*).
2. Grey cuscus (*Phalanger orientalis*).
3. Bear phalanger (*Phalanger ursinus*).

The relatives of the following genera correspond in size, appearance, and, to a certain degree, mode of life to the European dormice (*Glis glis*). They are called pygmy phalangers. Their diet consists of insects, vegetation, and honey. The animals are arboreal and nocturnal.

A. PYGMY POSSUMS (genus *Eudromicia*). The HRL is 7–12 cm and the TL is 7.5–14.5 cm. The fur is soft and fine. The prehensile tail is covered with hair only at the base and is bare otherwise. Fat is deposited in the tail and body for periods of hibernation in the temperate regions. There are three species, including the NEW GUINEA PYGMY POSSUM (*Eudromicia caudata;* Color plate p. 99).

B. "DORMOUSE" POSSUM (genus *Cercaërtus*). The HRL is 8–10 cm and the TL is 8–11 cm. The appearance and mode of life resembles that of the pygmy possums. There are two species: 1. The THICK-TAILED "'DORMOUSE" POSSUM (*Cercaërtus nanus;* Color plate p. 100), which has a pronounced fat deposit in the tail. 2. The THIN-TAILED "DORMOUSE" POSSUM or SOUTHWESTERN PYGMY PHALANGER (*Cercaërtus concinnus;* Color plate pp. 89-90) has no fat deposit in the tail.

Fig. 7-3. 1. Striped phalanger (*Dactylopsila*). 2. Pygmy phalanger (*Cercaërtus*).

In August 1966 two skiers returned from a cabin in the mountains of Victoria with a living dormouse possum. The zoologists of the Melbourne Fish and Wildlife Department discovered to their surprise that this rat-sized creature belonged to a genus that up to that time had been known only on the basis of fossilized bone remains. It had been thought that this species (*Burramys parvus*) had been extinct for at least 20,000 years. Nothing is known of the distributional range or mode of life of this newly-discovered animal. A second specimen has not yet been found.

The LEADBEATER'S POSSUM (♰*Gymnobelideus leadbeateri;* Color plate p. 100) is a great zoological rarity. Up to 1960 only five specimens were known. The species was believed to be extinct, but in 1961 a small population was rediscovered. The region was consequently declared a sanctuary. The leadbeater's possum has a dark mid-dorsal stripe which extends from the forehead to the base of the tail. The appearance is similar to that of the sugar glider but, unlike the latter, it cannot glide. In a way it is "a gliding possum without gliding membranes."

Gliding possums by W. Gewalt and B. Grzimek

The closest relatives, the gliding possums (genus *Petaurus;* Color plate p. 100), are characterized by the gliding membranes. There are three species which progressively increase in size.

1. HONEY GLIDER *(Petaurus breviceps).* The HRL is 12–17 cm and the TL is 15–20 cm. The weight is 90–130 g. The head is rounded. 2. LESSER GLIDING POSSUM *(Petaurus norfolcensis).* The HRL is 21–25 cm and the TL is 25–28 cm. 3. YELLOW-BELLIED GLIDER (*Petaurus australis,* not to be confused with the greater flying phalanger [*Schoinobates volans,* see p. 120 cont'd.]). The animal is almost the size of a cat. The HRL is 30–32 cm and the TL is 42–48 cm. The abdomen is yellow.

The gliding membrane of this gliding possum extends from the forelimbs to the hindlimbs. It spreads out when the legs are stretched. The dorsal surface of the membrane is covered with a normal amount of hair, but the lower side is more sparsely covered. When the animal climbs, runs, or sleeps, the membrane is carefully draped along its flanks. The gliding membrane is hardly noticeable except for the rather roundish appearance of the animal. Just prior to jumping, the animal extends all four legs, which thereby stretches out the membrane. At this point, as Ludwig Heck describes, "they are a strange sight; it seems as if the animal has lost all its body weight and turned into towel." Then the gliding phalanger sails through the air in a shallow arch. The gliding membrane has frequently been compared to a parachute, but this comparison is not quite accurate. The gliding phalangers do not drop down vertically and the membrane does not function as a brake against falling. Rather, it provides an increased gliding surface so that the animal can cover as much distance as possible with the least loss of altitude on a type of "air cushion." A similar principle is involved when a skier tries the ski jump. The gliding membrane could be more appropriately compared to the paper planes that school children fold from their notebook pages and let glide out of the classroom window. Of course, the "sailing" of the gliding phalanger is not such a leisurely process. The stretching of the gliding membrane is at the start. Navigation through the air and the folding of the membrane at the point of landing occur in such rapid succession that it is impossible to examine all the details, particularly during the shorter jumps. Anyone who has ever observed the restless motions of the honey glider and lesser gliding possum in the nocturnal display house in zoological gardens can easily convince himself of the rapidity of their movements. Honey gliders have been observed catching moths in flight. According to Simon, "yellow-bellied gliders are even able to change course in mid-flight and to land on a different tree than the one at which they had originally aimed."

Fig. 7-4. 1. Pygmy possum *(Eudromicia)* 2. Leadbeater's possum *(Gymnobelideus).*

Fig. 7-5. 1. Gliding possum *(Petaurus).* 2. Pen-tailed phalanger *(Distoechurus penatus).*

The bushy tail of the gliding possums is not only used as a steering device during flight but also for collecting nest material. The animals suspend themselves by their hind legs, pass the leaves from their forefeet to the hind feet, and, in turn, to the tail which then coils around the nesting material. It is obvious that the animal cannot glide with this load, but must run along the branches to its nest.

The gliding phalangers probably are the most common Australian mammals. However, this does not mean that the animals are seen frequently; just the opposite is true. A medical doctor in Hobart, Tasmania found a dead animal in front of his house. It had probably sailed against the white wall of the house, mistaking it for the sky. Only after this incident did the doctor realize that the animals were living in his vicinity. It can happen that two wrestling possum males may drop from a tree at the feet of a person walking through the woods at night. Unperturbed by their fall, they will continue fighting. These "sugar gliders" or "honey gliders," as the Australians call the smaller species, are very fond of sweets, such as honey, cake, or fruit, when they are kept as pets. In the wilds, they will also kill insects and small birds.

It has been claimed that gliding possums can glide for as much as fifty meters. David Fleay thinks that this is an exaggeration. He let possums glide between two poles which were seven meters apart. The animals had great difficulty navigating this distance and never went beyond it. Nevertheless, Grzimek believes that a fifty meter glide is possible, depending on the height from which the downward flight is begun. A pole within a fenced meadow is probably much too low. When a gliding possum lands on a tree, it jumps up into the air again in order to land with its head right side up. It then proceeds to coil its tail around the tree trunk and to climb in spirals to the top. The young of a female sugar glider will sit on her back while she glides through the air. At this time she also may carry two new small infants in her pouch.

The gliding possum youngsters are incredibly tough. A Miss Ivey, who rescued a small youngster from the teat of its dead mother, fed it with milk and kept it alive for a considerable time. At first only two drops of milk administered with an eye dropper were needed to fill the little worm-like creature so that it looked round and turgid. This dose of warm milk and sugar was given five to six times daily. Only after three weeks was the possum youngster able to lap up the milk by itself. Another possum youngster, which was rescued in a similar manner from the cold body of its mother which had been killed by a cat, lived to the age of ten years.

One of these tame "sugar gliders" wakes up only briefly during the daytime to eat a morsel of cake and then goes back to sleep with its tail curled around its face and body. Some of these pets are carried about in their owner's jacket pocket during the entire day. At dusk the animals become active. They run up and down the window drapes and glide from objects to persons much as they would from tree to tree in the wild. People cannot smell any odor on the sugar glider itself. However, their nests in the hollows do smell quite strongly because they urinate on the nesting material. Th. Schultze Westrum found a series of glands on the forehead, chest, and anal regions of the PAPUAN HONEY GLIDER *(Petaurus breviceps papuanus)*, each of which produced a different odor. Members of one tribe recognize one another by their common odor although they do not know each other individually. However, the odor of a foreign tribe alone does not elicit attacks. Sometimes one can find half a dozen or so individual honey gliders of the same tribe in one nest. Prior to nocturnal gliding, the animals emit deep, loud calls which sound like groaning. Sometimes they shriek loudly. Many Australians raise honey gliders in captivity. Often one can observe how glider possum mothers open their pouches with their hands to check their young. There may be one to three of them, but usually there are two. Honey gliders kept in the London Zoo repeatedly had young.

The close relationship that exists between the non-gliding Leadbeater's possums and the gliding possums is similar to that found between the non-gliding pen-tailed phalangers and the pygmy gliding possums.

PEN-TAILED OF FEATHER-TAILED PHALANGER (*Distoechurus pennatus;* Color plate p. 100). The HRL is 10.5 cm and the TL is 10–15 cm. It resembles the true dormouse in appearance and life habits. The tail is fringed laterally with long, relatively stiff hairs; perhaps it serves as a rudder during jumping.

Fig. 7-6. 1. Pygmy gliding possum *(Acrobates)*.
a. Pygmy flying phalanger *(Acrobates pygmaeus)*;
b. New Guinea pygmy flying phalanger *(Acrobates pulchellus)*. 2. Honey phalanger *(Tarsipes spenserae)*.

PYGMY FLYING POSSUMS OF FEATHER-TAIL GLIDERS (genus *Acrobates*). The HRL is 6–7 cm and the TL is 6–9 cm. The animal greatly resembles the feather-tailed phalanger, especially in the feather-like tail. However, the differentiating feature is the gliding membrane of *Acrobates.* There are two species: 1. PYGMY FLYING POSSUM (*Acrobates pygmaeus;* Color plate p. 100) and 2. NEW GUINEA FLYING PHALANGER (*Acrobates pulchellus).*

The scientific name given to the more common Australian species (pygmy gliding possum) is really appropriate this time. *Acrobates pygmaeus* literally means "pygmy acrobat." Although these little fellows, the size of a delicate mouse, are not uncommon, the average Australian will rarely see one and probably will not even suspect that they inhabit his neighborhood or even his backyard. We too are aware of the presence of rats and mice more because they nibble on our supplies than because we actually see the animals. Since these little acrobats live on insects and the nectar of flowers and glide through the air at night, they may elude us for our whole life. Only by accident will an Australian discover these dwarfs in his garden. In a suburban area north of Sydney, a cat had specialized in catching feather-tail gliders. The little creatures usually remained unharmed and the cat would give them up for a bowl of milk and some meat. After a while, the cat just placed the little pygmy gliding possums on the pillow of her sleeping masters. Domestic cats, which were introduced into Australia by the Europeans, have done extensive damage to these tiny gliders and other marsupial wildlife as well.

Trying to catch one of these little pygmy acrobats is almost as difficult as looking for a needle in a hay stack. Harry Frauca, a man with a great deal of experience with Australian animals, lived for years in a region inhabited by pygmy gliding possums. Yet he only occasionally rescued one from a logged tree. In many years, he only saw gliding acrobats four times. Finally he posted a reward with different logging firms for anyone that could catch one. After three months he finally received one animal in a cardboard box. It was a little female with one blind young which alternately sat on her back or inside the pouch. Before the young could see, it climbed on branches or underneath leaves on the ground on its own. Both animals licked the drops of water and honey that Frauca spread on the branches.

The gliding membrane of the pygmy gliding possum extends from the wrist to the ankle. The membrane is not nearly as large as that of the American flying squirrels, a small gliding rodent. The female of the feather-tail gliders builds a relatively large nest in knot holes or hollow trees, often at a height of fifteen meters or more. The nest is lined with eucalyptus leaves and bits of bark. The litter size is between three and four. More would not survive because the female has only four teats and a pouch that would not hold more than four youngsters. The juveniles frequently stay with the parents. These family groups may contain up to sixteen members. The life expectancy of these creatures in the wilds is not known, but one animal kept in the London Zoo reached almost four years.

Subfamily *Tarsipedinae* by W. Gewalt

Among the marsupials exists a species that could be considered the "hummingbird of the marsupials." It is a very small, slender, nocturnal and long-snouted creature which feeds on honey and pollen. The long, narrow head and the thin, bristled tongue are adaptations for probing flowers and licking up honey and pollen.This animal is the HONEY POSSUM (*Tarsipes spenserae;* Color plate p. 100). The HRL is 7–8 cm and the TL is 9–10 cm. Quite apart from the rather unusual mode of life for a mammal, the honey possum has several other distinct characteristics. On the basis of its uniqueness, the animal is classified into a special subfamily, the LONG-SNOUTED PHALANGERS *(Tarsipedinae).* The honey possum has small eyes and medium-long, roundish ears. The tongue can be extended approximately three centimeters beyond the nose. The long, whip-like prehensile tail is almost hairless and has a naked tip. Usually honey possums are solitary or live in pairs but during the blossom time of certain trees, they congregate in large numbers to lick honey, pollen, and insects. Captive honey possums have been observed very skillfully catching flies out of the air. Honey possums build globular nests similar to those constructed by hazel mice. However, they will also seek shelter in deserted birds' nests. Females have been found with up to four youngsters in the pouch.

Subfamily *Phascolarctinae* by W. Gewalt

The last subfamily of the phalangers is composed of many different animal forms which at first glance do not seem to be related at all. This group consists of the koala bears and related forms, the subfamily *Phascolarctinae*. The ears are round, short or large, and covered with dense fur. The first and second fingers are opposable to the other three fingers (forcipated claw). There are 28–40 teeth: $\frac{2\text{-}3 \cdot 1 \cdot 1\text{-}3 \cdot 3\text{-}4}{1\text{-}2 \cdot 0 \cdot 1\text{-}3 \cdot 4}$. The diet consists of leaves, fruit, and flowers. The animals are nocturnal or active at dusk and dawn. They are predominantly arboreal. There are three genera (or groups of genera) and seventeen species:

A. RING-TAILED POSSUMS (genus *Pseudocheirus*). Some zoologists have split this group into four genera on the basis of dentition and skull differentiations. In this volume, they are treated as four subgenera. The size ranges from that of the squirrel to that of the marten. The HRL is 19–45 cm and the TL is 17–37 cm. The animals resemble martens with their short, round ears. The long, prehensile tail is usually completely naked on its undersurface. 1. Subgenus *Pseudocheirus* in a narrower sense: QUEENSLAND RING TAIL (*Pseudocheirus peregrinus;* Color plate p. 100) and eight additional species. 2. Subgenus *Pseudocheirops:* STRIPED RING TAIL *(Pseudocheirus archeri)* and three additional species. 3. Subgenus *Petropseudes:* ROCK-HAUNTING RING TAIL *(Pseudocheirus dahli)*. The HRL is approximately 45 cm and the TL is approximately 27 cm. 4. Subgenus *Hemibelideus:* BRUSH-TIPPED RING TAIL (*Pseudocheirus lemuroides;* Color plate p. 100). This subgenus deviates most from the other subgenera. The animals are like lemurs. The tail is covered by thick fur. On the lateral body sides there are indications of gliding membrane supports. The brush-tipped ring tails inhabit the rain forests.

B. GREATER GLIDING POSSUM (genus *Schoinobates*). There is only one species, *Schoinobates volans* (Color plates p. 100 and pp. 104/105). The body weight is 1–1.5 kg. The large round ears are covered by thick fur. The gliding membrane, which is covered by fur, extends from the elbow to the knee. The fur is very long and dense. The long tail is prehensile and evenly furred, except for the naked underpart of the tip. The coloration is variable. The diet consists only of the leaves and the buds of the eucalyptus tree.

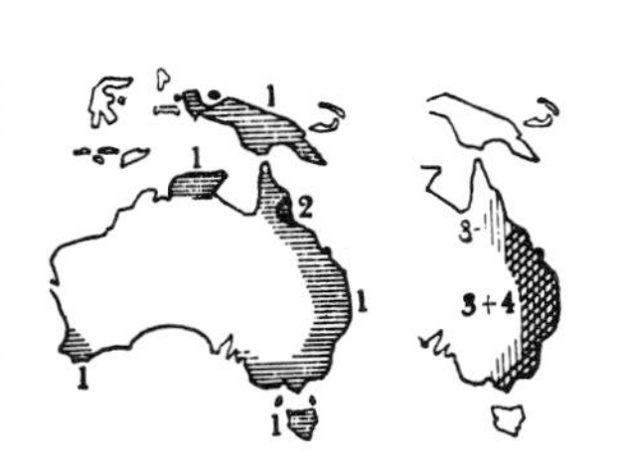

Fig. 7-7.
1. Ring-tailed phalangers (Genus *Pseudocheirus*).
2. Brush-tipped ring tail *(Pseudocheirus lemuroides)*.
3. Greater gliding possum *(Schoinobates volans)*.
4. Koala *(Phascolarctos cinereus)*.

C. KOALA OR "NATIVE BEAR" (genus *Phascolarctos*). There is only one species, *Phascolarctos cinereus* (Color plates p. 51 and p. 101). It is the largest of the phalangers. The HRL is 60–82 cm. There is no tail. The body weight can reach 16 kg. The big round ears are covered by thick fur and the surface of the nose is free of hair. The fur is soft and dense. The pouch opens in the rear which is unique in the phalangers. The animal is slow, nocturnal, and arboreal. The diet consists predominantly of eucalyptus leaves. The animals sleep in the forked branches.

In the ring-tailed possums, the prehensile end of the long and tapered tail is usually curled into a ring, which accounts for the common name of these possums. Ring-tailed possums are nocturnal. By day they shelter in nests of interwoven leaves and branches located on tree branches, or in hollow trees. The diet consists of fruit, leaves, flowers, insects, and small vertebrates. The pouch opens forward. Although four teats are present, only two young are born per litter. The ring-tailed possums inhabit regions where forest and parkland vegetation predominate. Only the rock-haunting ring tail inhabits the treeless rocky plateau of North Australia. All ring tails are solitary and relatively unsociable. They are not uncommon in suburbs of cities, but they are not conspicuous because of their nocturnal activities. Only accidentally does one become aware of the presence of a ring-tailed possum by finding an odd hollowed-out grapefruit beneath a tree. Usually such an incident is unjustly blamed on some rat. Countless ring tails fall victim to indiscriminate hunting because of their valuable furs.

The largest marsupial glider by B. Grzimek

The greater gliding possum is the third marsupial glider, along with the pygmy gliding possum *(Acrobates)* and the sugar glider *(Petaurus)*. Even today this peculiar animal is relatively well distributed throughout the mountainous timbered country of eastern Australia. The animals occur singly or in pairs in the rolling and mountainous eucalyptus forests. They are never found in large numbers. The gliding membrane of the greater gliding possum extends from the elbow to the ankle, while in the sugar possum it extends from the outside finger. This difference results in two varied gliding shapes. The greater gliding possum appears more or less triangular while floating. The front end looks tapered. The sugar glider, on the other hand, looks more rectangular. Even at night one can differentiate between the lesser and greater gliding possum. The eyes of the *Schoinobates volans* glow in the glare of a search light, while the eyes of the *Petaurus* reflect the light only weakly. A greater gliding possum has never been kept in a European zoo because of its highly specialized food requirements, similar to those of the koala. The animals are almost exclusively restricted to the narrow peppermint-scented leaves and young shoots of the eucalyptus trees (*Eucalyptus australiana* and *Eucalyptus elaephora*). Some specimens, which were shot in an orchard, had only leaves and blossoms of the eucalyptus in their stomachs. They had not touched the peaches or apricots.

They eat eucalyptus leaves almost exclusively

The usual color of the greater gliding possum is a brownish-black. However, there are many variations ranging from pure black, cinnamon, red, grey, and yellow to pure white. Although this large glider weighs between 1–1.5 kg, it can glide for great distances. One animal flew over ½ km in six successive leaps. The first jump went from the top of a tree higher than 30 m to the tree trunk of another which was almost 70 m away. As soon as the glider had landed, it galloped vertically up the tree trunk in great jumps, unlike the spiral climb of squirrels or sugar gliders. In this manner, it climbed up to the top of the next tree and sailed off to another tree almost 80 m away, and then on to other trees which were 100, 110, and 82 m from each other. Finally, it glided an additional 110 m. Every time this greater gliding possum climbed up a tree, it emitted a shrieking sound. Closely spaced trees that have been surrounded by barbed wire as a protective measure against debarking by domestic animals are often the cause of death for greater gliding possums. The gliding membrane is pierced by the barbs and the unfortunate creatures slowly die. Other enemies of these marsupials are a species of large owls, introduced foxes, and bush fires.

They glide more than 100 meters

There are two mammae inside the pouch, but usually only one young. For the first six weeks, it is firmly attached to the teat. Later it opens its eyes and at four months of age emerges from the pouch. Even then it still rides on its mother's back for a certain time period.

The easiest way to obtain living greater gliding possums is to be present during logging operations. David Fleay received one pair of animals this way. For two and a half years he kept the pair on a diet of eucalyptus leaves and honeybread soaked in water. The possums slept all day and were active only at night. They tolerated petting, but they did not climb around on people like the sugar glider. This would be highly uncomfortable in any case because their claws are as sharp as steel hooks. The male of the captive pair escaped one day through a crack in the door of the cage. However, he returned after several days through the same hole.

In 1789 the greater gliding possum was described as the "black flying opossum". The animal was mentioned in a report on the journey of Governor Phillip to Botany Bay, the first British penal colony and today the location of Sydney, a city with over one million inhabitants. "The fur is wonderful, and probably will be a valuable export item, provided one can hunt more of these animals," wrote the author at that time. Fortunately, the furriers did not share this opinion. The fur of the lesser and greater gliding possums is long, but it is rather loose and soft, making it difficult to work with. If the fur were different, not many of these gliders would be alive today.

The koala by B. Grzimek

Apart from the kangaroos and the opossums, the koala is the best known marsupial in Europe and America, although in Europe not one koala is present in a zoo. There are two good reasons why as late as 1952 not one of these appealing, popular animals existed in a zoo outside of Australia. The first reason is the strict Australian law which prohibits the exportation of the koala. Similar restrictions have also been established in other countries. It is believed that preservation of an animal is assured by prohibiting the exportation of living animals. This is only true for a very few animals, such as the orang-utan, which is threatened with extinction because of the uncertain political situation in its home range, and the primitive fashion which is employed to meet the demand of zoos. In 1963 the game department of Kenya deplored the fact that it had to grant permission to export 235 animals while at the same time it had issued close to 8000 hunting licenses to foreigners.

Only 100 years ago, millions of koalas lived in Australia

One hundred years ago, millions of these "cute little teddy bears" still existed in Australia. It used to be a popular sport with the younger folks to shoot koalas from the trees. The animals were an easy target sitting in the bare branches of the eucalyptus trees. However, it usually required several shots to kill the tenacious koalas. Of course, this "sport" was not very suitable for "tender hearted souls" because the wounded animals uttered loud cries which were painfully reminiscent of the crying of a helpless child. Another distressing fact was the way the seriously wounded animals desperately held onto the branches, sometimes only with one limb or one foot. The hands are especially well adapted to grasping branches. The fingers and toes are strongly clawed, and the first and second fingers are opposable to the other three fingers, thus facilitating a good grasp on twigs and branches.

Almost all were exterminated

Even the annual fires set in the eucalyptus forests by the Australians would not have threatened the koala bear species as such, although millions burned to death. It was the beautiful, silver-grey, soft, and durable fur of the koala which sealed its fate. In 1908, 57,533 koala hides were traded on the market in Sydney alone. In 1924, over two million furs were exported from the eastern states in Australia. At this point, the United States was the first country to prohibit the import of koala furs. Although these appealing animals had virtually disappeared from Victoria and New South Wales during 1927, Queensland, the last state where they were still present in reasonable numbers, declared "open season." In that year alone, 10,000 licenses were issued to hunters, and 600,000 furs of these animals were exported. The Australian zoologist Ellis Troughton wrote: "It is almost unbelievable that a civilized community permits the pitiless slaughter of such a harmless, indigenous animal solely for the purpose of selfish trade and gain." In the years 1887–1889 and 1900–1903, widespread epidemics eradicated millions of koalas. The animals died of eye diseases, meningitis, kidney infections, and intestinal parasites. A large population can absorb such losses. However, a new epidemic could completely wipe out the greatly reduced population.

In the last thirty years, the Australians have awakened to the plight of the last of these helpless and pretty animals. Quite in contrast to many other wild animals, it is very easy to observe the koalas in their natural habitat, the dry open forest and the parkland region. At night the koalas can easily be located by the sounds they make. Particularly the male is very vocal during the breeding season. His voice sounds like a handsaw going through a thin board. The greater gliding possum is the only other animal that is equally noisy. One can shine headlights on the koalas and they do not become disturbed. Even during the day, the animals do not pay any attention to people. They look down on people just as curiously as the latter look up at them. This is one reason why the aborigines could so easily kill them with their boomerangs.

Protective measures and resettlement

After the Australians had finally discovered that living koalas are just as beautiful and precious as the lifeless furs, they introduced protective laws. There are no koalas left in South and West Australia. In Queensland the former population of millions of animals has been decimated to a few thousand. In the last decade, attempts have been made to re-introduce koalas into the forests of Victoria. Most of these animals originated from the island of Philipp where very sophisticated capture methods were employed. With the help of a long stick, a noose is slipped over the koala as it sits up in a tree. A knot in the cord prevents choking the animal. The koala is forcefully pulled from the branch and caught in a stretched-out cloth to prevent any injuries from the high fall. Animals caught in this fashion have been reintroduced to more than fifty locations in Victoria. The wardens who transport the animals to their new habitat stop off at every school along the way to impress upon the children what pretty and harmless creatures the emblem animals of Australia really are. This is the most effective way of preventing the senseless shooting of koalas in the wild. The danger of shooting does exist because every Australian seems to own a gun. In the future there may be a "come back" of the koala in Australia.

However, the population does not increase too rapidly because of the low reproductive rate. It is thought that the animals become sexually mature only at three or four years. A mature male usually has a small harem, which he jealously guards. The gestation period is 25–30 days. The young, which weighs five and one half grams at birth, remains in the mother's pouch for six months. There is usually only one young, and twins are very rare. There are only two mammae, and thus more than two youngsters could not survive.

Koalas eat eucalyptus leaves exclusively

The various factors discussed on the preceding pages would theoretically not prevent the keeping of koalas in European zoos. The second and decisive reason for this impossibility is the fact that we are unable to feed the animals. The koalas are specialized to a higher degree than any other mammal. They are herbivorous and they only consume leaves. The caecum of the koala, which is 1.8–2.5 m long, is three to four times as long as the entire animal. This is an adaptive response to the difficulty of digesting the tough vegetation. The caecum of many animals, like chickens and horses, is specialized to digest crude fibers. Howler monkeys and leaf monkeys are other groups of animals which are specialized in eating only leaves. However, koalas are not indiscriminate eaters of leaves. They have limited themselves to the eucalyptus leaves. Of the 350 different species of eucalyptus in Australia, the koala eats only twenty, and from these he prefers five species. The favorite species are the mana or sugar eucalyptus *(Eucalyptus viminalis)*, the spotted eucalyptus *(Eucalyptus maculata)*, and the red eucalyptus *(Eucalyptus rostrata)*. The koala consumes a daily quota of two and one half pounds of leaves which it leisurely chews. Koalas from different localities have different food preferences. Two koala bears from the southern state of Victoria, which were kept in a small zoo near Brisbane, Queensland, would not touch the leaves of the "blue" and "grey" gum trees which were the favorite food of the Queensland koalas. For six months mana leaves had to be shipped by train from Victoria to Brisbane before the two "native bears" became accustomed to the leaves of the grey and blue eucalyptus tree. The koala even discriminates between the leaves of the mana tree. The animal often ignores certain bunches of leaves and reaches for an alternate twig. As we shall find out later, there are good reasons for this. Nevertheless, it becomes clear from the preceding discussion why it is very difficult to keep koalas in the zoo.

The youngest leaves are lethal

Mr. Ambrose Pratt, President of the Zoological Society in Victoria, observed that seemingly healthy and happy koalas kept in the Melbourne Zoo would be dead the next day. Different types of medical treatment did not help, and the autopsies revealed nothing. At approximately this same time, chemists and pharmacists showed an interest in eucalyptus leaves. They found that the mana tree, the koala's favorite tree, seasonally produced hydrocyanic acid in its leaves and shoots. A greater percentage of hydrocyanic acid was produced in winter than in summer and more in the young leaves and shoots than in the more mature leaves. In the wild, the koalas change over to other trees and avoid the fresh sprouts. When captive animals are given, with the best of intentions, only these tender and juicy shoots, the koalas will sooner or later eat them. Some of the analyzed samples contained as much as .09% hydrocyanic acid. This is an extraordinarily high amount. Just twenty-five grams of these leaves will kill one sheep. The zoo keepers had unknowingly forced their koalas to commit suicide.

Most eucalyptus trees contain substances which are highly important to the health of the koala. Two significant substances are cineol and phellandren. The former decreases blood pressure and body temperature and relaxes the muscles. An overdose of cineol stops respiration. Phellandren seems to increase the body temperature. The smaller koalas in the warm north of Queensland avoid the eucalyptus species which contain phellandren and prefer those with cineol. The larger koala of cooler South Australia chews eucalyptus leaves which contain phellandren and so avoids cineol.

It takes extensive scientific knowledge to keep koalas in captivity.

One way to solve the intricacies of finding the proper diet is to present the animals with a great variety of eucalyptus leaves and let them choose. This is the real reason for the absence of koalas in European zoos. How could one provide a continuous and abundant selection of fresh leaves from the tropical eucalyptus trees?

The koalas are "super saturated" with the essential oils found in the gum tree leaves. They smell just like cough drops. This may be one reason why no ectoparasites are found in their beautiful soft fur. In 1933 Keither Minchin visited a "koala farm" near Adelaide in South Australia and quite coincidentally observed the manner, which is disgusting to us, in which koala youngsters are weaned from their mothers. A koala female, which may become twenty years old, bears a young every two years. She carries this youngster on her back for a whole year. The following observation was made on a female with a young which had appeared out of the pouch for the first time five weeks previously. The female was crouching on a forked branch and the head and arms of the young were protruding from the posteriorly pointing pouch. The entire face of the youngster was covered with yellowish-green mucus, and the young was in the process of pushing its nose into the anus of the mother, the opening of which it was attempting to widen. It greedily ate the intestinal content for about one hour. The female was not suffering from an attack of diarrhea because there was evidence of normal fresh feces on the ground. The koala young was consuming predigested eucalyptus leaves. It seems that the maternal intestine is cleared of fecal material and the partially digested food of the anterior intestine is quickly passed through. It could also be possible that the content of the caecum is separately excreted. Obviously, this is a transitional mode of nourishment to adapt the koala young gradually away from the mother's milk to the diet of the tough gum tree leaves.

Predigested food from the intestine

It was observed that the koala young received the predigested food every second or third day for approximately one month. Usually feeding occurred between three and four o'clock in the afternoon, although it is possible that the youngster also feeds at night. This mode of feeding also explains why the opening of the pouch faces the maternal anus. In all other phalangers and kangaroos, the marsupium opens anteriorly. In the wombats the pouch also opens to the rear, but this is an adaptation for preventing soil from entering the opening. However, these animals are burrowers. Before Minchin had observed the weaning process in the koalas, no one had been able to explain a rear pouch opening in this arboreal species. Minchin has records of fifteen additional koala youngsters being weaned in this unique manner.

Raising koalas in captivity

Small koalas raised by humans quickly adapt to their foster parents. "Teddy," a much-travelled koala bear, was widely known to many people. The Faulkners of North Queensland received the three month old animal wrapped in a piece of fur. The small female cried every night and had to be comforted constantly until the Faulkners gave her a koala fur tied around a pillow. This artificial substitute mother seemed to comfort the youngster when she had to be left alone. In the beginning Teddy received cow's milk which she slowly lapped like a kitten. Later she ate fresh blue eucalyptus leaves. Four weeks later she went on her first journey to West Australia. She slept in a basket clinging to a stuffed toy bear. In West Australia she consumed the leaves of specific eucalyptus trees which had probably been the food of the exterminated koalas in that region. In addition, Teddy was given some milk and a few small peppermint lozenges. Frequently, she would swallow soil and small pebbles. Teddy reached the age of twelve years, which is probably the longest time that a koala has lived in captivity.

As soon as a koala is used to humans, it does not want to be left alone and loves to be carried around. It has no desire to run away. Compared to other marsupials, the koala is remarkably intelligent. One captive koala showed great interest in its image in the mirror and even walked behind it to look for the "other" one. Today it is forbidden to keep koalas privately, however.

In many books, one can read that "koala" in the native language means "does not drink." However, almost all captive koalas were known to take milk and water. They sip it noisily like dogs.

Discovery and first zoological reports

The explorer Captain Cook did not report seeing any koalas in 1770, and the first penal colonies around Sydney were not aware of these animals. Finally in 1798 a young man making an expedition to the Blue Mountains west of Sydney reported on "another animal which the aborigines called 'cullwine' and which was reminiscent of the sloths in America." In 1802 an interested young French investigator, Ensign F. Barallier, traded spears and one tomahawk to the aborigines for "parts of a monkey which they called Colo." Unfortunately, he only received the feet which were shipped inside a bottle of cognac to the governor. Yet only one year later, the first living koala female with a pair of twins reached Governor King in Sydney. It did not take long to realize the difficulty of feeding koalas outside Australia. The first koala reached Europe much later. On April 28, 1880, the London Zoo bought a koala from "a dealer." The animal must have been a particularly tough little fellow because at first he was fed only the dry eucalyptus leaves sent with him from Australia. Later he received fresh leaves. It is hard to believe that he stayed alive for fourteen months! He perished not because of the diet but because he accidentally caught his head under the heavy cover of a washing stand as he was freely roaming in the director's room.

Subsequent attempts to keep koalas in the London Zoo were less successful. In 1908 the curator of the aquarium brought some koalas by boat from Australia. The animals, however, refused to eat the eucalyptus leaves which had wilted during the journey. Yet they eagerly ate bread, milk, honey, and even eucalyptus cough candies. When the boat reached the more temperate regions, the koalas caught cold and died. Another koala, which reached New York on October 20, 1920, only survived for five days. This was the fate of additional koalas that periodically were exported from Australia.

In December, 1927, a sailor with a sack in his hand appeared in the office of the London Zoo. He opened the sack on the desk of the bookkeeper and, to everyone's surprise, out came two pretty and lively small koalas. Both animals were very tame. One climbed up on the shoulder of the sailor and started playing with his hair. The sailor said that he had "brought these home" with him. But what about food? Oh yes, he had kept a fresh supply of eucalyptus leaves in the cooling room of the boat. He just wanted to inquire if the zoo was interested in buying them. The answer, of course, was yes, because this opportunity was too good to pass by. This was the way that the London Zoo acquired koalas "across the counter," so to speak. Unfortunately, the supply of eucalyptus leaves did not last too long. Emergency calls for eucalyptus leaves were printed in newspapers but only a meager amount from a botanical garden in the province arrived. The supply, which did not meet the demands of the two koalas, stopped altogether when it grew colder. After four weeks the two charming animals were dead. Many Londoners had been able to view them, however.

The sad experiences accumulated by keeping "deported" koalas finally took a turn for the better in the winter of 1952. The Paramount film industry had imported some koalas to Hollywood to use in a film about the Australian penal colonies. Belle Benchley, the director of the San Diego Zoo at that time, was, of course, much more excited about the koalas than the film drama about Governor Arthur Phillip and his band of prisoners. It had been her secret wish for many years to keep some koalas. She found a suitable place for the four male and four female koalas in the San Diego Zoo. The zoological gardens in California are more fortunate than European zoos, particularly in San Diego where they had planted and raised eucalyptus trees for over twenty-eight years in a part of the zoo where the Australian fauna was kept. The semi-arid landscape and climate of California are very similar to those of Australia, and the eucalyptus trees do very well there. Of two koalas brought to San Diego in 1925 by director Faulconer from an Australian trip, one lived for two years. An autopsy of this animal revealed dried eucalyptus leaves in the caecum. However, since that time there are a greater variety of eucalyptus trees in San Diego.

The koalas which had arrived in 1952 did have a sufficient supply of eucalyptus leaves. Two males lived for five years and one female almost reached age seven. In April 1959 the zoos of San Diego and San

Francisco each received three koala bears. In each group was a female which carried a young in her pouch. Both zoos were even successful in raising their own koalas.

There is hardly an American newspaper or magazine which does not have several pictures of these "photogenic" animals. They constantly star on television shows. The cautious and friendly manner of these animals and the great human urge to cuddle and to hold them has subconsciously done more for Australian public relations in America than any consulate, tourist office, or advertisement could ever have done. Fortunately, the Australians have come to realize that these endearing bears are real treasures, although at one point they had nearly been exterminated.

Wolfgang Gewalt
Bernhard Grzimek

8 Wombats

Family: Wombats
by B. Grzimek

Many of the marsupials that have been discussed up to now have been similar in appearance or had certain characteristics found in various higher mammals, although there was no phylogenetic relationship between these groups. The WOMBATS (Family *Vombatidae*) are rodent-like and "are adapted to a digging and root-eating existence," as the English mammalogist Thomas describes them. The incisors of these stout marsupials are rootless and grow continually just as in the rodents and the lagomorphs. The HRL is 67–105 cm and the tail is vestigial. The body form is low and plump. The head is thick and broad and the ears are medium long. The limbs are short and powerful. There are five fingers on each hand and five toes on each foot. The short fingers have powerful claws. The second and third fingers are fused at the base. There are 24 teeth: $\frac{1 \cdot 0 \cdot 1 \cdot 4}{1 \cdot 0 \cdot 1 \cdot 4}$. All the teeth are rootless and grow continuously to compensate for wear. The two large and powerful rootless incisors have enamel on their front surface only, which provides a continual sharp cutting edge as in the rodent's teeth. The pouch opens to the rear. There are two mammae, but only one young is produced.

Distinguishing characteristics

In Southeastern Australia two genera are found, each with one species, on the islands of the Bass Strait and Tasmania: 1. COMMON WOMBAT (genus *Vombatus*). The HRL is 67–105 cm. The surface of the nose is hairless and black. The short and coarse fur has little or no underfur. The head is thick and round. The rounded ears are slightly pointed at the tip. The back is slightly raised. 2. HAIRY-NOSED WOMBAT (genus *Lasiorhinus*). The HRL is 87–102 cm. The surface of the nose is covered with either white or brown fur. The long and silky fur has a well-developed underfur. The head is more slender and angular and the ears are triangular. The back flattens out more and tapers off abruptly at the croup (Color plates p. 101 and p. 102).

How the wombats were discovered

The first Europeans to see wombats were shipwrecked on one of the islands in the Bass Strait between Australia and Tasmania. They be-

Legends to the following color plates

The family of kangaroos *(Macropodidae)* is comprised of 51 species of which 43 species are represented by 49 subspecies in the following color plates.

Musky rat kangaroos (Plate I, p. 131)

1. Musky rat kangaroo *(Hypsiprymnodon moschatus).*

Rat kangaroos

2. Broad-faced rat kangaroos (*Potoroops platyops*—extinct).
3. Long-nosed rat kangaroo *(Potorous tridactylus).*
4. Gilbert's rat kangaroo *(Potorous gilberti).*
5. Desert rat kangaroo *(Caloprymnus campestris).*
6. Rufous rat kangaroo *(Aepyprymnus rufescens).*
7. Gaimard's rat kangaroo *(Bettongia gaimardi).*
8. Tasmanian rat kangaroo *(Bettongia cuniculus).*
9. Lesueur's rat kangaroo *(Bettongia lesueur).*

Banded hare wallaby (Plate II, p. 132)

1. Banded hare wallaby *(Lagostrophus fasciatus).*

Hare wallabies

2. Brown hare wallaby *(Lagorchestes leporoides).*
3. Western hare wallaby *(Lagorchestes hirsutus).*

4a. Spectacled hare wallaby *(Lagorchestes conspicillatus conspicillatus).*

4b. Leichardt's spectacled hare wallaby *(Lagorchestes conspicillatus leichardti).*

Rock wallabies (Plate III, p. 133)

1. Plain rock wallaby *(Petrogale inornata).*
2. Brush-tailed rock wallaby *(Petrogale penicillata).*
3. Short-eared rock wallaby *(Petrogale brachyotis).*
4. Ring-tailed rock wallaby *(Petrogale xanthopus).*
5. Little rock wallaby *(Peradorcas concinna).*

Nail-tail wallabies (Plate IV, p. 134)

1. Northern nail-tail wallaby *(Onychogalea unguifer).*
2. Crescent nail-tail wallaby *(Onychogalea lunata).*
3. Bridled nail-tail wallaby *(Onychogalea fraenata).*
4. Northern New Guinea wallaby *(Dorcopsis hageni).*
5. New Guinea mountain wallaby *(Dorcopsis macleay).*

Tree kangaroos (Plate V, p. 135)

1. Lumholtz's tree kangaroo *(Dendrolagus lumholtzi).*
2. Black tree kangaroo *(Dendrolagus ursinus).*
3. Doria's tree kangaroo *(Dendrolagus dorianus bennettianus).*
4. Matschie's tree kangaroo *(Dendrolagus matschiei).*

Pademelons (Plate VI, p. 136)

1. Red-legged pademelon *(Thylogale stigmatica).*
2. Red-necked pademelon *(Thylogale thetis).*
3. Rufous-bellied pademelon *(Thylogale billardierii).*
4. Bruijn's pademelon *(Thylogale bruijni.)*

Wallabies

5. Tammar *(Wallabia eugenii)*
6. Parma wallaby *(Wallabia parma)*

Quokkas

7. Quokka *(Setonix brachyurus)*

Brush wallabies (Plate VII, p. 137)

1. Pretty-face wallaby *(Wallabia canguru).*
2. Black-tailed wallaby *(Wallabia bicolor).*
3. Black-gloved wallaby *(Wallabia irma).*

4a. Red-necked wallaby *(Wallabia rufogrisea rufogrisea).*

4b. Tasmanian red-necked wallaby *(Wallabia rufogrisea frutica).*

5. Black-striped wallaby *(Wallabia dorsalis).*
6. Sandy wallaby *(Wallabia agilis).*

Kangaroos (Plate VIII, p. 138)

1. Wallaroo *(Macropus robustus)* a. Southeastern wallaroo *(Macropus robustus robustus)* b. Deer wallaroo *(Macropus robustus cervinus)* c. Antelope wallaroo *(Macropus robustus antilopinus).*
2. Great gray kangaroo *(Macropus giganteus)* a. Northeastern great gray kangaroo *(Macropus giganteus giganteus)* b. Tasmanian great gray kangaroo *(Macropus giganteus tasmaniensis).*
3. Red kangaroo *(Macropus rufus)* a. Northwestern Australian red kangaroo *(Macropus rufus pallidus)* b. Western Australian red kangaroo *(Macropus rufus dissimulatus).*

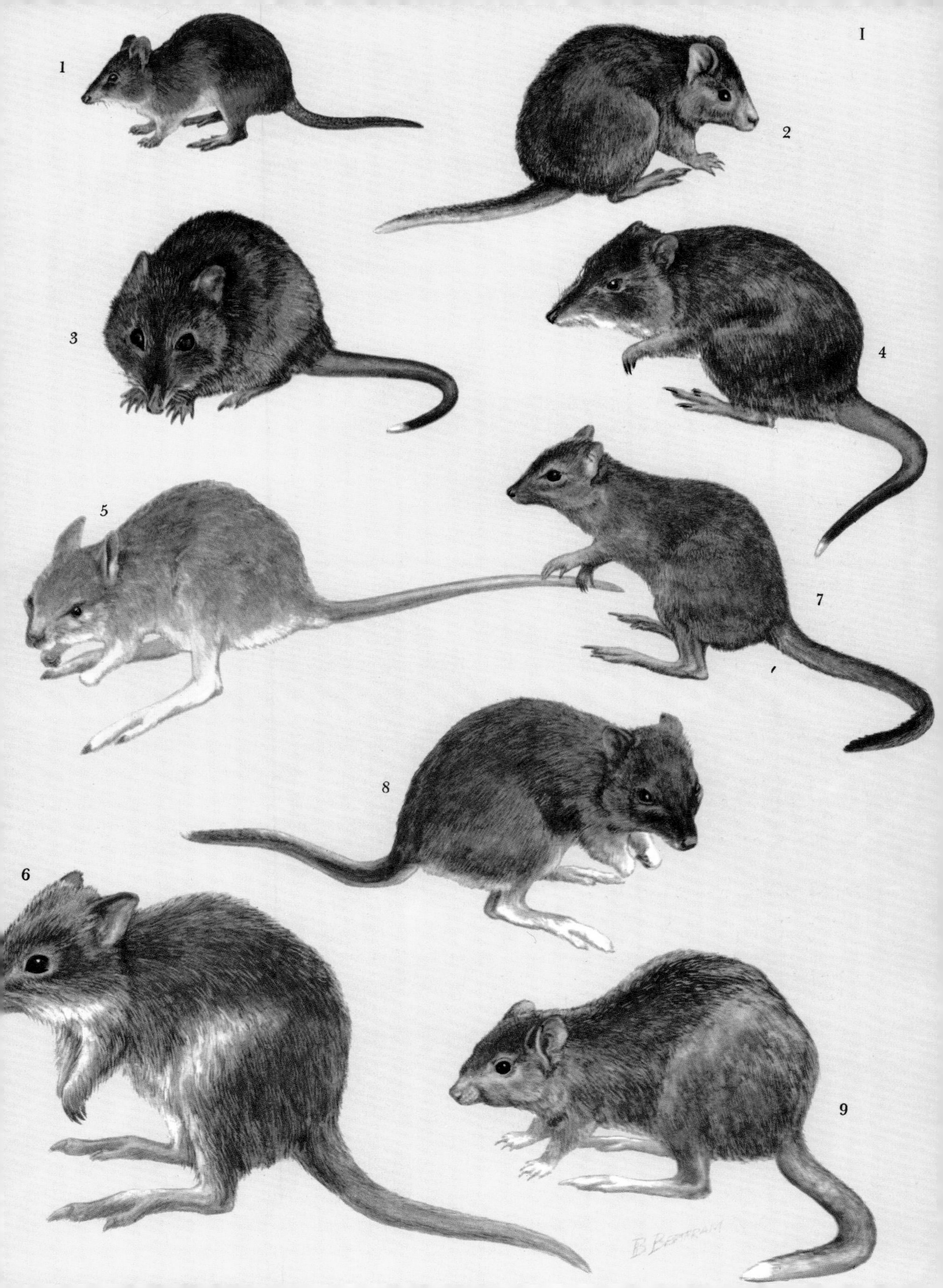
I
1
2
3
4
5
7
8
6
9

II
1
2
3
4a
4b

2
5
3
4
B. BERTRAM

IV
1
2
3
4
5

V
1
2
3
4

VI
1
2
3
4
5
6
7

1
2
4 a
4 b
3
6
5
B BERTRAM

VIII
♂
1a
♀
1b
1c
2b
2a
3a
3b

Fig. 8-1.
1. Common wombat *(Vombatus ursinus).*
2. Hairy-nosed wombat *(Lasiorhinus latifrons).*

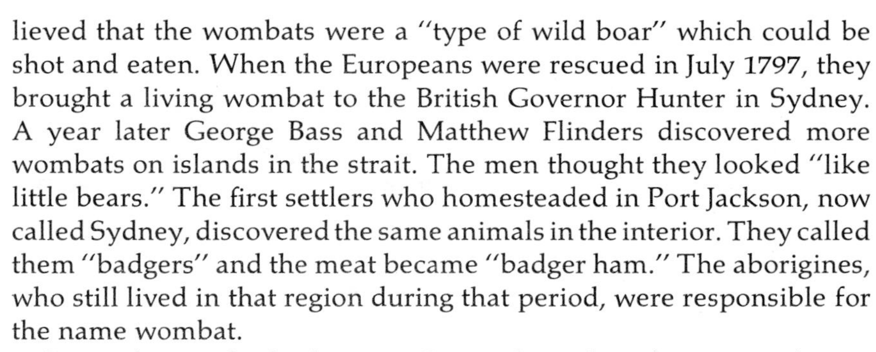

lieved that the wombats were a "type of wild boar" which could be shot and eaten. When the Europeans were rescued in July 1797, they brought a living wombat to the British Governor Hunter in Sydney. A year later George Bass and Matthew Flinders discovered more wombats on islands in the strait. The men thought they looked "like little bears." The first settlers who homesteaded in Port Jackson, now called Sydney, discovered the same animals in the interior. They called them "badgers" and the meat became "badger ham." The aborigines, who still lived in that region during that period, were responsible for the name wombat.

Soon afterwards the botanist Brown brought a living wombat to London. For two years, Clift, a professor of surgery, kept the wombat in his house. The animal was very friendly towards people. It would sit up and put its front legs on a person's knee or go to sleep in his lap. Even children could carry it around. Occasionally the wombat would bite, but never very seriously. In 1808 Everard Home described the living habits of these animals. He characterized them as passionate burrowers which were very restless at night and which fed exclusively on vegetative matter.

In the following century not much else was added to our knowledge about the living habits of the wombat, according to Wood Jones in 1924. However, in accordance with the practices of the last century, many skeletons and skins were sent to museums, and a large number of different species of wombats were described. Today we know that these animals vary in size and fur coloration depending on the region in which they live. Only two genera, each with one species, are recognized today: the COMMON WOMBAT (◊ *Vombatus ursinus)* and the HAIRY-NOSED WOMBAT (◊ *Lasiorhinus latifrons).* The various "species" described formerly are regarded today as a subspecies of either one of the two genera.

The settlers not only welcomed the wombats as "badger hams," but they also hated them intensely as "vermin." These powerful animals were capable of tearing large holes in the wire fences of the settlers. Furthermore, the burrows of the wombats were used by the newly introduced rabbits, which in turn gradually plagued the countryside. Extermination procedures directed against the rabbits also took their toll of wombats. Occasionally, the burrows of the wombats were blamed for a horse or cow breaking a leg. The wombats were shot or caught alive in box traps. In certain regions, a "bounty" was paid for an animal. In 1958 premiums were still paid for 4180 dead wombats.

They build spacious burrows

Some wombat burrow systems are of such a size that children can crawl through the tunnels right into the nest chamber. However, if the burrow is occupied, going inside is not recommended even for dogs. Wombats strike out with their powerful hind legs, particularly if they are grasped on the back. A dog has difficulty holding on to a wombat inside its burrow, because there is no tail and the skin on the rear is tough and very difficult to penetrate even by dog's teeth. Furthermore, the animals have the habit of planting the short legs against the opposite wall and pushing any intruder against the ceiling or wall of the burrow with great suddenness. This force may result in a broken nose for a dog or a badly crushed human arm.

Wombats are solitary except during the breeding season, although the burrows of neighboring animals are often interconnected. Once a colony which extended over an area of eight hundred by sixty meters was found. It is not known whether there might be some sort of social relationship after all. Often a shallow resting place is excavated against a tree near the mouth of the burrow, as a site for sunbathing. The hard-trodden paths of the wombats often extend for many kilometers. In Tasmania where the wombats are now protected, the animals often settle in the vicinity of towns and streets. The wombats are not seen too frequently because they are nocturnal and most people do not even suspect the presence of their subterranean neighbors.

At the time when wombats were still found on the islands of the Bass Strait, fishermen often tamed them and kept them like dogs. These tame wombats would spend the day in the forest and return to the house in the evening. They had changed their normal daily rhythm to suit their human keepers. In the year 1798, George Bass found numerous wild wombats on all the islands in the strait, which was named after him. Ninety years later, they were totally eradicated. In 1908 the zoologist Charles Barrett discovered a few on Flinders Island.

Wombats do well in captivity. For this reason, they have frequently been exported to zoos overseas, although they are not too interesting to viewers because of their nocturnal habits. The hairy-nosed wombat of the London Zoo reached the age of seventeen, and the common wombat grew to be twenty years old. "Wanda," of the MacKenzie Sanctuary near Melbourne, was a bit of a celebrity. In the zoo she freely ran around and even permitted visitors to carry her about. When the new zoo opened in Rome in 1910, director Knottnerus-Meyer received a pair of wombats. The male had badly bitten the female during the long sea voyage. This is not surprising since the animals are solitary except during the breeding season. One day in November, the warden forgot to lock the animals into the cage with the cement floor. The wombats lost no time in digging beneath the fence of the outside enclosure and disappearing. They lived unnoticed in the park bordering the Villa Borghese, which was closed at night. Finally in March of the following year, workers were able to catch the pair.

It was customary in zoos in those days to associate ground-dwelling animals and monkeys. This is not a very enviable existence for the poor ground dwellers like turtles or armadillos. In the Rome Zoo, the wombat was very indifferent to the baboons with whom it shared the enclosure. Knottnerus-Meyer describes the coexistence of the two species in the following manner: "At feeding time, the wombat would very quietly and determinedly take his place among the baboons and peacefully eat his food. Yet it must have seemed to the baboons that he ate too much. At first their eyebrows lifted and they started to murmur, which is an initial sign of excitement. Then they would hit their hands on the ground, back up, and run towards the wombat in a threatening posture. All of this was to no avail. The wombat was not impressed with any of these antics. At this point one of the baboons would softly and boldly push against the wombat's broad forehead, but this only resulted in a more vigorous retaliatory push and the hissing wombat continued his meal. Next the baboons tried to snatch food from the mouth of the wombat while being very careful not to approach the sharp teeth too closely. The wombat would rush and hiss at the opponent, although he was never able to catch the agile monkeys. Finally the drama ended when all the baboons joined together in some terrible howling. They bombarded the wombat with sand while he tried to continue eating, emitting hissing sounds. He could kick out with his hind legs like a donkey and throw sand at the baboons. There was never an exchange of bites. There was no place to penetrate the thick, rounded hide of the wombat. The legs and ears were relatively short, and the wombat was able to bite back severely."

There are only two records of wombat births in zoos

In captivity wombat pairs occasionally formed, but only two normal births of young have been recorded in zoological gardens up to now. A pair of hairy-nosed wombats in the Halle Zoo had one youngster in April 1914. The young was born in the same time period, between April and June, as in its natural habitat, and it also stayed in the mother's pouch until December. The second breeding success occurred in 1931 in the Whipsnade Zoo. It is rather a surprise to see the little wombat looking out of the backward-pointing pouch from underneath its mother's rear between her hind legs. The majority of people think that most marsupials have a pouch similar to the kangaroos'. The pouch is attached to the abdomen rather in the manner of a clothes-pin apron that some housewives wear, and the young peeks out of it like we might look out of a window. Yet in an animal that walks bipedally, it is more of an advantage to have the opening of the pouch be forward and up, in the direction of the head.

The situation, in reference to the pouch opening, is different in burrowing marsupials, where the pouch opens to the rear. The advantage of this arrangement seems very obvious. A pouch that opens forward would soon fill up with sand and dirt while the animal is digging or running. The arboreal koala, which does not have a sand and dirt problem, also has a pouch opening to the rear, but the reasons for this were discussed on p. 125.

The newly-born wombat and koala young have a much shorter and less dangerous trip into the pouch than does the kangaroo young. It is relatively easy for the newly born to crawl from the vagina to the posteriorly pointing pouch. However, there is one disadvantage. Unlike the kangaroo, the wombat mother is not able to open the pouch with her hands to clean it out.

W. Staudinger wrote about the wombat young born in Halle as follows: "Often one foot, or sometimes a hind and front foot, would protrude at the same time from the pouch. The light pink color of the sole of the feet indicated that the little one had not been outside yet. Approximately three weeks later, the young was first observed leaving the pouch for the first time. It would run beside its mother for short periods and, if it was disturbed at all, it would quickly disappear into the pouch again. At a later stage the rabbit-sized youngster could be observed when the mother left the burrow. It returned to the pouch with decreasing frequency. If it needed protection, the wombat young simply crawled underneath its mother, which covered it lovingly, like a hen her chickens."

Bernhard Grzimek

▷

A gray kangaroo young *(Macropus giganteus)* looking out of its mother's pouch. Although the young is able to jump around on its own and to eat grass, it nevertheless returns to the warmth and protection of its mother's pouch for resting.

▷▷

In an enclosure of the Wildlife Research Station of the CSIRO a red kangaroo male *(Macropus rufus)* stands up to more than a man's size in a threatening posture.

▷▷▷

These two appear as if they were in a tender embrace. However, in the next moment these two red kangaroo males will kick each other in the abdomen with great force.

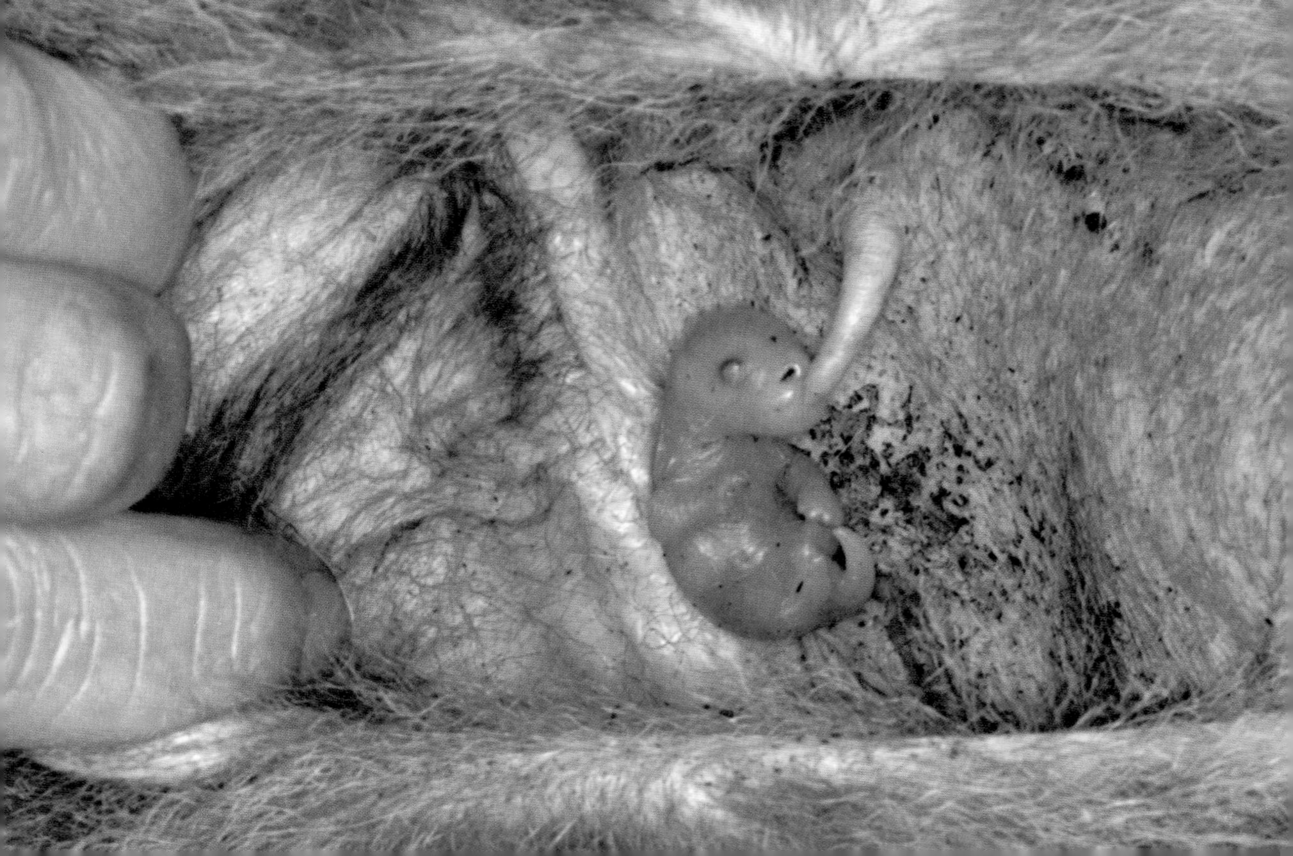

9 Kangaroos

Family: Kangaroos
by B. Grzimek

Distinguishing characteristics

Albino red-necked wallaby (*Wallabia rufogrisea*; see p. 169). Twins are rare in kangaroos. Offspring of different mothers may visit each other in the maternal pouch, however. Occasionally albinos occur in kangaroos, as they do in almost all other mammals.

The tiny kangaroo infant is firmly attached to the mother's teat inside the pouch. The offspring is not yet able to suckle by itself so the milk is injected. The way in which this embryo-like, immature being reaches the pouch is explained on p. 155.

Family *Macropodidae*. The animals vary in size from that of a rat to that of man. The HRL is 25.5–166 cm and the TL is 16.5–107 cm. The body posture of this group is often upright. The slender head is either long-snouted or short-snouted. The ears may be round, oval, or long. There is great variation in the fur, but usually it is dense. The tail is covered with the same type of fur as the rest of the body except in the musky rat kangaroo *(Hypsiprymnodon moschatus)* where the tail is naked. The muzzle is more or less naked, and serves as a good differentiation characteristic for the various species. The front limbs are usually small and short with heavily-clawed, almost equal digits. The hind feet are lengthened and are often very powerful. The first digit of the hind foot is lacking in all genera except *Hypsiprymnodon;* the small digits two and three are united by skin and the two separate claws are used for grooming; digit four is long and strong with a large nail; and digit five is moderately long. The tail is usually long and strong and often very powerful.

The kangaroos are predominantly herbivorous, except for the musky rat kangaroo which is also insectivorous. The upper incisors, particularly the middle ones, are long and strong. Only the middle teeth of the lower incisors are present. The lower incisors are very long and chisel-like and project forward in a horizontal manner. The upper canines are small or absent; the lower ones are always absent. The one to two premolars drop out prematurely after the second set of teeth has grown. The four to seven broad molars move forward gradually. The stomach consists of a long, jointed glandular portion. The caecum is present.

The family is found in Tasmania, Australia, New Guinea, the islands of Aru and Kei, and parts of the Bismarck Archipelago. The members of this family occupy the steppe and bush regions. The genus *Dendrolagus* is arboreal.

The pouch is always well developed and opens to the front. There are four mammae but usually only two secrete milk. Usually one young is born per year, during the rainy season. Two young are occasionally born, but three are very rare. The gestation period is thirty to forty days, but the young stay in the pouch for a long time.

There are three subfamilies: 1. MUSKY RAT KANGAROOS *(Hypsiprymnodontinae)*; 2. RAT KANGAROOS *(Potoroinae)*; 3. KANGAROOS *(Macropodinae)*. There is a total of seventeen genera, fifty-one species, and ninety-three subspecies.

How kangaroos were discovered

It is generally believed that the great British explorer James Cook discovered the kangaroos in the eighteenth century. However, 140 years prior to this time the Dutch captain Francisco Pelsaert, who was stranded near the Wallaby Islands on the west coast of Australia in 1629, discovered the small tammars *(Wallabia eugenii)*. He also observed the minute young hanging with its mouth on one of the mother's teats, but he believed that the embryo grew out of the mammary gland. This report did not cause any excitement and was soon forgotten.

On June 11, 1770, James Cook's boat "Endeavour" was stranded on a coral reef off the northern coast of Australia. Fortunately, he was able to get free but he had to stay in a bay until the beginning of August in order to repair his boat. He had docked close to the Cape York Peninsula which on the map looks like the long pointed finger of the Australian continent directed towards the island of New Guinea. In Cook's memory a town at the approximate location has been named Cooktown. It is a city today. Cook and his crew used this opportunity to explore the surrounding area and the roaming aborigines. On June 22 Cook sent some of his men on land to shoot some pigeons for the sick sailors. When the men came back they reported having seen an animal which was smaller than a greyhound, slender in build, had a mouse gray fur and was extremely fast. Cook saw the animal himself two days later. He said: "I would have said it was a greyhound, but when moving it hopped like a rabbit or deer." Again another two weeks passed before the naturalist Joseph Banks and four companions conducted a more extensive exploratory land trip tween July 6 and July 8. Cook wrote about this trip: "After a march of many miles they discovered four of the same animals. Two of these were hotly pursued by Bank's greyhounds but they escaped when they jumped over the dense, high grass which proved an obstacle to the dog." On July 14th Lieutenant Gore shot one of these animals which was of a smaller variety. "It hopped and jumped with its hindlegs," reported Cook. "The animal does not resemble any European animal that I have seen. At best it resembles a jumping mouse but it is much larger." Following the rather vague information of the aborigines, Cook named the animal "kangaroo."

The rediscovery of the strange kangaroos caused a great sensation. Their body form was vastly different from that of all previously known animals. Three years after the first English fleet unloaded their first load of prisoners in Port Jackson (it is Sydney today), the first living kangaroo was sent to England and presented to King George III. Just to make sure Governor Phillips sent a second one over to England on another ship, the Londoners were greatly excited about this uniquely shaped animal from the new continent, and consequently other kangaroos followed. An old handbill from those days had the following message printed on it: "The wonderful kangaroo of Botany Bay, a truly remarkable, beautiful and tame animal! It is approximately 1.50 meters high. It comes from the Southern Hemisphere and seems almost unbelievable." For one shilling, which in those days was a lot of money, one could admire a kangaroo at the Haymarket.

When we speak of kangaroos today, we are usually thinking of the almost man-high reddish or gray erect figures from the zoo which have a well-developed rump and powerful hind legs, yet small arms, a narrow chest, and a small, rabbit-like head. It seems a rather odd assemblage. Aside from these large kangaroos, a few similar, rabbit-sized animals may come to mind. Actually, however, the kangaroos and wallabies *(Macropodidae)* comprise a family of very many species. However, the animals are limited to a small section of the globe: Australia, Tasmania, New Guinea, and the islands of the Bismarck Archipelago. Recently they have been introduced by man into New Zealand. The body length (not including the tail) of the smallest kangaroo is 23 cm, and the largest measure 1.60 m. In most species the male is larger than the female.

In the English-speaking world the term "kangaroo" is usually limited to the three species of giant kangaroos: the great gray kangaroo, the red kangaroo, and the wallaroo or euro. The numerous smaller species are called "wallabies."

Before the introduction of European foxes into Australia, RAT KANGAROOS *(Potoroinae)* were fairly numerous. In the year 1904 dealers in Adelaide still offered them a dozen at a time for a few pennies. People amused themselves on Sundays by letting the animals race. Recently, at least two species of this family have been exterminated. In western Australia they are a little more common. In the New York Zoo a long-nosed rat kangaroo lived for almost two years. A Tasmanian rat kangaroo kept in the London Zoo used to wrap its tail around a bundle of straw and hold this in front of its abdomen. With this bundle the animal used to hop around for hours at night. Other rat kangaroos also carry their nest material via the tail. The larger kangaroo species are not able to do this because their tail is stiffer and serves more as a balancing organ.

Some kangaroos beat on the ground with their hind feet when in danger. This alarm signal is loud and carries over a long distance. It is similar to the alarm signal given by rabbits. The narrow, elongated feet enable the animal to make bouncing and remarkably long jumps. The TRUE KANGAROOS *(Macropodinae)* particularly demonstrate this jumping capacity. The scientific term of this family, *Macropodidae*, means "large feet." The fused second and third toes are shaped into a grooming claw with which the kangaroo combs its fur and scratches behind the ears. The front feet have five claws and are also used as a comb.

Since the Europeans came to the Australian continent, four species of kangaroos have been exterminated. Ten additional species of small, often particularly pretty and interesting kangaroos, are in danger of extinction. These shy, inconspicuously living animals are not hunted directly; however, the individual species has only a small distribution range. Man has limited the living conditions of these jumping marsupials by either changing the vegetation or by introducing domestic animals. The three species of man-sized giant kangaroos have been almost totally eradicated from certain regions of Australia by the European immigrants, but in other areas the living conditions for the kangaroos were improved by agriculture and their populations have increased.

Heat, drought, and hunger are the biggest hazards for the kangaroos and not so much the predators, although many species get along with small amounts of water and meagre food. Undoubtedly the persistent eradication of the dingo *(Canis familiaris)* by the Europeans has been of some aid to the kangaroo populations. Some snakes, including pythons *(Morelia argus)*, occasionally catch the smaller species of kangaroo or the young of the larger species. The wedge-tailed eagle *(Uraëtus audax)*, which has defied all human persecution, may capture an occasional small kangaroo. The great kangaroos try to escape pursuing dogs by entering water up to their chests. They will stay in one spot and try to grab the dogs with their hands and hold them underwater until they have drowned. If a body of water is not nearby, which is frequently the case in Australia, a pursued kangaroo will try to get its back against a tree and will kick the opponent with the hind legs. The wounds caused by such a kick can be devastating. A man can count himself lucky if only his belt and pants get ripped off. There are records of persons that have been gored by kangaroos or have died as a result of broken jaws and limbs. Fighting kangaroo males kick each other. They try to grab each other's hands and arms and then kick each other in the abdomen. According to E. H. M. Ealey, the wallaroos of Northwest Australia do not fight with their clawed hind feet but bite instead. This makes it a little easier to handle them in the zoo.

Fig. 9-1.
The sharp claws of an attacking kangaroo can not only rip the person's clothing but can also tear open his abdomen.

Sometimes one gets to view "boxing kangaroos," with boxing gloves tied over their hands, in circuses. Of course, this fight with the human opponent is only play, for if the animal would really fight with its hind legs, the man would be in serious trouble. Older, more mature and self-confident kangaroo males are not very suitable for such performances. In the Hagenbeck Zoo a red kangaroo, which was separated from a hippopotamus by a wall, one day jumped over and scratched and kicked this huge animal's nose with its front legs. The hippo was both surprised and curious and did not harm the intruder.

Kangaroos kept in enclosures with water do not avoid it. In fact, in the Rome Zoo the gray kangaroos bathed daily, but the red kangaroos did not enter the water at all. In Australia the water holes are often beleaguered by sand flies which attack the kangaroos and can be responsible for eye infections which may cause blindness.

Only in the last two decades have Australian scientists discovered the secret of the kangaroo's ability to survive in the most desolate regions. Kangaroos have ruminant-like bacteria in their stomach and the upper part of the small intestine to aid them in the breakdown of cellulose found in the vegetation. A female visitor at the Basel Zoo observed a kangaroo, which was settled in its stall for the evening, regurgitating and re-chewing cud. Since then the regurgitation of cud has been discovered in five additional genera of kangaroos. Domestic sheep take up the hard, dry, hard-to-digest Spinifex grasses only as a last resource. The millions of Australian sheep graze at first on those grasses and herbs which are more easily digested. As the more favored grasses and herbs of the sheep decrease, the Spinifex species increase. Thus unknowingly the sheep rancher has provided an excellent habitat for the kangaroos. A rancher will grimly endure this in certain regions. As soon as he doubles his number of sheep, the number of greater kangaroos quadruples as a matter of course. Although kangaroos require little in their nutritional demands, they nevertheless are dependent on certain basic food elements. All but three of the fifty-two great gray kangaroos imported to the Brookfield Zoo near Chicago in 1934 died a slow death. After the diet was changed to alfalfa, rolled oats, field produce, and an addition of calcium and other minerals, the number of kangaroos increased again. In 1949 the Brookfield kangaroo herd had increased from the original three to seventy-four. One can often read in older zoological texts that kangaroos died as a result of lump jaw or mysterious infections of the oral mucuous membranes or the larynx. These symptoms occur with greater frequency if the kangaroos are fed with straw or grains that possess acorns which can easily injure the mucous membranes inside the mouth and thus expose the area to pathogenic agents. However, since modern zoos have become aware of these hazards, kangaroos are fed on a more diverse diet containing soft hay and alfalfa. Kangaroo deaths due to nutritional deficiencies have now become rare.

Kangaroos are not as stupid as some people make them out to be, at least not if one compares them to other marsupials or even ruminants. D. H. Neumann of Münster, Germany, had gradually trained a red kangaroo and an American opossum *(Didelphis marsupialis)* to differentiate between two drawings on a piece of paper. Only one drawing was rewarded by food. The kangaroos learned to differentiate between seven different sets of drawings; the opossum only two. After 160 days the kangaroo still remembered six of the seven figure sets, while the opossum still remembered its two sets after two weeks but forgot them completely after four weeks.

Fig. 9-2.
When it is hot, certain kangaroos lick the underside of their arms in order to cool down.

The kangaroos "pant" like dogs and sheep in order to get cool. This means that they rapidly breathe in and out through the open mouth. In addition, kangaroos lick their arms and chest, occasionally even the hind legs, because the evaporated mucus helps to cool down the body. Wallaroos start to lick their bodies when temperatures rise to 31.5°C. But how do kangaroos manage without water for weeks and months? The west Australian scientist E. H. M. Ealey found that the kangaroos kept in an enclosure required a water supply of 5% of their body weight when fed on dry food. If the animals consumed vegetation which contained 30–50% water, they nevertheless lost one third of their body weight after seventy days. They had not taken any pure water at all. Free living wallaroos kept under the same conditions did not lose any weight at all; however, these animals actively dig for water, sometimes up to one meter deep. In the dry season in Africa the rhinoceros, antelope, snakes, and zebras are dependent for water on the water holes that the elephants dig in the soft sand of the dry river beds. In Australia wild pigeons, the pink cockatoo, dasyures, and even emus drink from the water holes dug by kangaroos. Many of the water places made available to sheep and cattle by the ranchers with the aid of deep drilling, artesian wells and electrical pumps, have opened up new habitats to kangaroos and certain species of wild birds. The wild animals have the same access to water as the sheep and cattle.

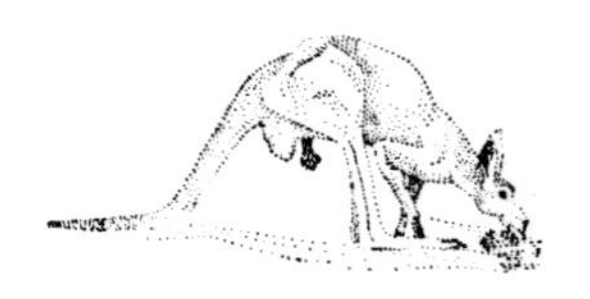

Fig. 9-3.
A drinking kangaroo.

Yet many of the wallaroos that Ealey observed in Northwestern Australia for five years never came to these drinking places on farms. Ealey fenced the various water troughs and caught the animals as they slipped through openings that he had left. He marked them with colored neck bands and numbers. The neck bands were made of a material that reflected at night when hit with a spotlight beam. Ealey also invented little gadgets which would squirt luminescent paint on each kangaroo as it tried to slip through a fence hole.

These marking methods helped him to discover that wallaroos or euros, as they are called in Australia, did not drink even when the temperature was 46°C in the shade. The animals conserved body water by hiding in hollows under granite boulders during the hottest part of the day. The temperature in these hollows never rose above 32°C.

Why do the euros avoid the beautifully clear water which is available in such close proximity in the paddocks? The answer lies more with the type of food that they eat than with their drinking habits.

Recent investigations which were carried out in Kenya, East Africa, have shown that cattle that drink a lot of water excrete more nitrogen. In 1963 similar tests were carried out on euros in laboratories of the University of Perth, the capitol of West Australia. The results were similar to those of Kenya. The smaller the amount of water an animal drinks, the less the percentage is of nitrogen that is lost. This provision enables the animal to make maximum use of the scanty, protein-poor vegetation of the semi-deserts.

Dr. Main of the Zoology Department of the University of Perth solved another kangaroo mystery. When I visited him he kept several tammars, which had been on sea water for 30 days, in cages. Sea water contains 3% salt, and is not tolerable for humans and most other animals. The tammars gained weight despite the fact that the other food only contained 10% moisture. These small kangaroos are almost completely exterminated on the mainland. They are still found on small coastal islands where no rain falls for months during the dry season, and where there is no source of fresh water. Dr. Main found that almost all the females on one island carried a young in their pouch; however, he did not find any yearlings at all. These rabbit-sized animals occupy tunnels in the high grass which they defend against other conspecifics. As soon as a juvenile reaches sexual maturity it is chased out of the territory of the parents and is forced to escape to the open plain. Here they are easy prey for the bald-headed eagles. The number of the tammars on this small island remains more or less the same—just enough animals that can support themselves. If the situation were different the animals would totally graze off all the grass and almost all of them would die or would be prey to the various predatory birds because of the lack of cover.

How old do kangaroos get? One day I visited G. B. Sharman at the experimental station of the Commonwealth Scientific and Industrial Research Organization close to Canberra, the capital of Australia. He told me that two thirds of the wild kangaroos that are shot are not older than four years. Over one quarter grow older than eight years. The life expectancy of the great kangaroos kept in zoos of America and Europe is at the most seventeen to eighteen years; for the wallabies it is twelve years.

Towards the end of the nineteenth century it was possible to see kangaroos instead of deer on the meadows of the Rhineland and Silesia (Germany). Generally the animals did well in these areas. The introduced species did not belong to the great kangaroos of the Australian mainland. The red-necked wallaby *(Wallabia rufogrisea frutica)* from Tasmania was particularly well suited to the central European climate because of its dense fur. Tasmania, the island situated at the southern tip of Australia, lies closest to the South Pole, and the climate is already quite cold although there is hardly any snow and ice.

In the year 1887 Baron Philipp von Böselager released two males and three females of this species in a forest area of five square kilometers near Heimerzheim in the district of Bonn (West Germany). Six years later there were thirty-five to forty animals, and they even survived the extremely cold winter of 1887/1888 when temperatures sank to -22°C. Unfortunately the baron's two game wardens died shortly thereafter, and by the time they were replaced, a horde of poachers had shot most of the animals at their feeding station, where they were such an easy target. A few kangaroos were still found in the area. One was even found in the Taunus, 100 kilometers away from Heimerzheim. Another one was shot in the Eifel in 1889 but by 1895 the entire population had been exterminated. "Many years later we did find out in which tavern the scoundrels had eaten the kangaroos," wrote the son of Baron Böselager.

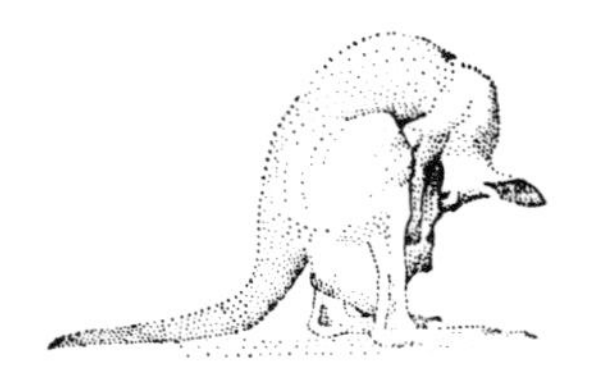

Fig. 9-4. A kangaroo mother uses her mouth to clean out the pouch.

In England several attempts to introduce kangaroos have met with success, for example the introduction of kangaroos at the castle of Tring at the turn of the century. The Baron of Rothschild kept red-necked wallabies and also the great grey kangaroos in complete freedom. The count of Witzleben, who had released red-necked wallabies on his property, Altdöbern, in the district Frankfurt/Oder, also had good breeding success with these animals, but he soon tired of the "hopping giant fleas" and shot them all because, in his view, they scared off his deer.

Prince Gerhard Blücher von Wahlstatt released red-necked wallabies on his island of Herm in the English Channel east of the island of Guernsey at the end of the nineteenth century. The population increased quite well on the island, which was 1.3 square kilometers. However, at the beginning of World War I, English soldiers occupied the islands, and all the wallabies gradually ended up in the kitchen. Some of the descendants of these channel-wallabies were released on the estate Kriblowitz in Silesia, another possession of the Blüchers. This herd increased to sixty to seventy head but these animals were shot by poachers in the first two years after World War I.

Fig. 9-5. This is how the newborn kangaroo independently crawls into the pouch (the offspring is drawn much larger in proportion to the mother so that it can be seen).

Professor Otto Koenig of Vienna also attempted the introduction of kangaroos. He did not keep the Tasmanian red-necked wallabies but the same species *(Wallabia rufogrisea banksiana)* from the mainland. These animals survived the winter on a diet of bark, buds, and dry grasses. The youngsters did not seem affected by the cold as long as they stayed inside the mother's pouch. Nevertheless, the wallabies were in need of shelter and hiding places away from the wind. Some wallabies broke through the ice and drowned but such mishaps also occur with endogenous species. In the European fauna the kangaroos would fill the niche that is occupied by the deer.

The pouch in the female of all kangaroos is well developed. It opens to the front, and there are four mammae but usually only two secrete

Fig. 9-6. A kangaroo young that is already several weeks old is shown attached to the teat inside the mother's pouch (the pouch is shown transparent in order to show the otherwise invisible young).

milk. Every year during the rainy season one young, occasionally two but rarely three, is born following a gestation period of thirty to forty days. Close to one hundred years ago people were puzzled about the way in which the newly born kangaroos reached the pouch of the mother. On very rare occasions somebody was present in a zoo when a kangaroo gave birth but was never close enough to observe the process. Approximately two hours before giving birth the female very carefully licks the inside of the marsupium. A little later she takes up an unusual position: She pulls the tail forward between her two hind legs and leans back, often supporting her back against a wall. Up to then it had been supposed that the mother would pick the newly-born from the vagina with either her teeth or her lips and put it inside the pouch. This assumption seemed quite logical since the fetus of the great red kangaroo only weighs about ¾ to 1 gram, approximately 1/30,000 the weight of the mother (20–30 kg). The eyes and ears of the newly born are completely undeveloped and it does not possess any hair. It has more the appearance of an undeveloped fetus from the uterus. The hind legs of the young are still short and not at all kangaroo-like. It is almost unbelievable that such a "premature infant" is able to crawl through the fur of the mother by itself into the warm pouch. Nevertheless, the young accomplishes this feat. The biologists G. B. Sharman and H. J. Frith of Canberra have filmed and photographed the birth process in the red kangaroo from close proximity. In this species the young is born thirty-three days after copulation. The young opens the amnion itself and climbs into the pouch even if the mother is anaesthetized and would not be able to help at all. During the first part of the journey to the pouch the young is still attached to the placenta by the umbilical cord, which is still blocking the vagina. The young moves in a snake-like motion and is inside the pouch in three to five minutes. The mother licks away the blood and mucus. Prior to these exact studies it had been assumed repeatedly that the mother licked the fur in the abdominal regions in order to prepare a smoother passage for the young through the coarse hair. Inside the pouch the young latches on to one of the teats and sucks it into its mouth. The end of the teat develops into a bulb-like structure. It is no wonder that the first investigators assumed that the young grew from the teat. Removing such a small young from a teat will cause slight bleeding.

Fig. 9-7. A large kangaroo young puts its head inside the mother's pouch in order to suckle (the pouch is drawn transparent).

Although the eyes and ears are still in the early stages of development, the embryo has wide open nostrils and the olfactory center in the brain is also well-developed. On the basis of this evidence it is assumed that the blind and deaf embryo finds its way to the teat within the pouch via the olfactory sense.

Sharman and Pilton reported on the behavior of a newly-born red kangaroo as follows: "The front feet and the five separate toes are well developed; each is equipped with a well-developed claw. The young continued with its crawling motions even after it had been removed from the fur of the mother. It took considerable force to remove the young from the mother's fur. It was observed that only the toes and claws of the front feet were used for grasping the mother's fur. Contrastingly, the hind feet are less well developed but each foot had four distinct toes but no claws. Part of the umbilical cord was still attached to the abdomen but otherwise the embryo was free of embryonic linings. The tail was positioned frontwards between the hind legs."

The red kangaroo youngsters stay inside the pouch for approximately 235 days, and at the end of this period weigh two to four kilograms. One wallaroo in the New York Zoo put its head out of the pouch for the first time five months and eleven days after it had crawled into it. In the weeks following it spent part of its time outside of the pouch and part of it inside until it had grown too large for it. Nevertheless, it still poked its head into the marsupium in order to suckle. During the "pouch period" of the young the female regularly cleans the inside of the pouch with her lips; often she holds it open with her hands.

Twins and triplets are rare in kangaroos. Of 219 kangaroo and wallaby birth records in the London Zoo, eleven were twin births and once triplets were born. A great gray kangaroo female in the Philadelphia Zoo, having borne twins, threw one of them out of the pouch; it was then raised with a bottle.

The research team of Canberra investigated many of the wild red kangaroos which were shot for meat. More than 75% of the females carried young inside the pouch, and of these another 20% suckled an additional young which was running free. Furthermore, 60–70% of all kangaroo females that carry a young inside the pouch are pregnant. This fact is very decisive in the desperate struggle for survival in the face of millions of competitive sheep introduced by man, although this way of reproducing of course preceeds the arrival of sheep and man. During the dry season more than 75% of the young that left the mother's pouch die. Nevertheless, the reproductive process continues. Soon after giving birth the female copulates again but the implanted ovum only develops to a stage of about 100 cells. Further embryonic development ceases until the young inside the pouch dies or has become independent. In this manner the female is able to bear another young in approximately four weeks without having to copulate. Unfavorable conditions thus do not bring about unnecessary delays in reproduction.

The number of kangaroos is increasing on grazing land

The sheep is also well adapted to living in hot temperatures. It can tolerate an increase in its body temperature up to 43°C, and can lose up to 25% of its body weight due to dehydration (man dies if he loses 12% of his body weight because of dehydration). The reproductive rate of sheep is higher than that of the great kangaroo because of the higher percentage of twinning in sheep. However, during a dry spell or a famine all young may die. Here the kangaroo has the advantage of having another young in reserve within the uterus and the pouch. For sheep it may take another year before they are again in estrus, copulate, and bear young.

In the Pilbara region of Northwestern Australia, the number of sheep has decreased by as much as 50% during the last twenty-five years, and over a dozen large ranches with eight million sheep were abandoned. Concurrently the number of kangaroos and wallaroos has greatly increased. Just how much they had increased was demonstrated by the poisoning programs. Between 1930 and 1935, 90,000 wallaroos were poisoned on a farm of 14 square kilometers. Another farm could only maintain 4000 sheep on a 10 square kilometer area. Here the meadows were of such poor quality that the undernourished sheep had ceased to bear young. However, approximately 30,000 wallaroos lived on this same area, and even reproduced. Most sheep ranchers have just as many great kangaroos grazing on their meadows as sheep, and very often the number of kangaroos is far greater. Yet the smaller species of kangaroos have long since disappeared from all sheep meadows.

In former times the number of great kangaroos was much smaller in Northwestern Australia than it is today. The aborigines hunted them regularly. The kangaroos were their mainstay of life. But then the Europeans killed most aborigines or sold them to the pearl fishers for slave labor. The number of aborigines in the Pilbara region has declined from 5000 to 1300. Today they usually live in closed settlements. Also many of the 3000 European gold diggers that roamed in this region and mostly lived off of kangaroo meat have almost totally disappeared since 1930; only a bare dozen are still around.

They do better than sheep

The government of West Australia bought two large farms in this region in order to study the apparent mystery of kangaroo population increase. Since 1955 the biologist E. H. M. Ealey has conducted a five year study. In this area the temperature rises to 50°C in the shade and only about 25–30 cm of rain falls during the entire year. The spiny grass species *Triodia spinifex* grows predominantly on this soil. It has very little nutritive value. The few other grass species, which have a slightly higher percentage of protein, are the cause for the desperate survival struggle between kangaroos and sheep. A sheep requires at least 6.5% protein in its dry fodder in order to stay alive, and it needs more to produce wool and to reproduce. The kangaroo uses this protein more efficiently because it rarely or never drinks. The farmers further worsen ranch conditions by burning the old dry grass during shearing time. Thus, they also burn all the grass seed and foster the spread of the more undesirable species. The meadows are continually grazed by sheep, and consequently there is no recovery period for the grasses.

The ranch becomes run down and thereby provides a better competitive food basis for the kangaroos. E. H. M. Ealey is of the opinion that the two million kangaroos in the Pilbara region will survive all methods of extermination.

Mass shootings and utilization

Actually one cannot claim that the mass shootings and the utilization of kangaroo meat will eradicate faunal species from Australia. Persons hearing about these shootings for the first time are repulsed and emotionally against it. The smaller species of kangaroos which are close to extinction were hardly ever shot. Nevertheless, it is hair-raising to read the occasional newspaper report which relates the vast extent to which the emblem animal of Australia is turned into dog food and leather for shoes. In Queensland alone between the years 1950 and 1960 approximately 450,000 kangaroo skins were traded annually. Recently an American manufacturer ordered $140,000 worth of kangaroo skins for making ski apparel. The state of Queensland alone issued 1800 licences for professional hunters. These men can make between $125.00 and $200.00 a week; in good areas this may even come to approximately $60.00 a day. I read in an Australian hunting magazine that twenty-five hides a day are a good average for a hunter; 140 hides obtained in six days was considered phenomenal. The animals are shot from a distance of 50 to 250 meters. The kangaroos have the habit of stopping and looking around after a short escape; of course, this spells disaster if the enemy is a man with a gun in his hand.

Between July, 1958, and June, 1962, 7500 ton of kangaroo meat was exported; still more was sold in Australia as pet food. Only the hindquarters of the animal are utilized. This means that one to two million kangaroos were shot to provide that amount of meat. At this period (according to Dr. Sharman and Dr. Frith), it is estimated that one million kangaroos will be shot annually in western New South Wales. One can of dog food sells for 23¢ in Australia. The hunters who are paid so much per hide shoot at everything in sight. In 1964, fifty tons of meat were exported weekly from the whole of Australia. In addition, ten tons of meat were used as dog and cat food within Australia itself. It takes 133 kangaroos to make up one ton of meat. This means that every week 7980 animals are shot, and when one counts all the young within the pouches the number climbs to approximately 10,000.

These numbers, which may have a rather depressing effect, have to be viewed in the light of the existing kangaroo populations. According to Sharman and Frith, over two million hides of red-necked wallabies were exported from Tasmania alone between the years 1923 and 1955. This implies that twice as many animals were shot, but today there are still large numbers of them. Two thousand great kangaroos shot on a mainland farm of 10,000 hectares did not seem to have made any dent on the size of the population. On the plantation Talga Talga in West Australia which extended over an area of 84,000 hectares, only

2300 sheep could be maintained. That means that each sheep required eight hectares. When water wells were poisoned for durations of one, three, and ten weeks, 12,834 dead wallaroos were counted. Just to bring things into perspective it should be mentioned that in small West Germany over one million hares and 550,000 deer are shot annually by hunters, and that more than 200,000 rabbits are run over by cars. Yet the populations of these species are in no way decreasing.

Kangaroo meat has an excellent taste and is suitable for man as well as pets, provided that it is handled just as hygienically as beef or mutton in Australia. Naturally wild game on the average contains more parasites than domestic animals but usually these are harmless for humans. All the rabbits, wild boars, and deer consumed in Europe contain just as many parasites as the Australian kangaroos. In these dry countries the kangaroo is by far a better meat producer than sheep and particularly cattle. Would it not be more advantageous to keep kangaroos and regularly crop them? There is one disadvantage—kangaroos do not grow wool.

Subfamily *Hypsiprymnodontinae* by D. Heinemann

However unusual and odd kangaroos with their powerful hind legs appear to us, the differences between the smallest species of the jumping marsupials and the average type of mammal as exemplified by the rat or among the marsupials the phalangers *(Phalangeridae)* is very small. Many of the jumping marsupials even resemble the phalangers in their internal structure, so much so that the renowned mammal taxonomist Thomas was in doubt if he should classify the

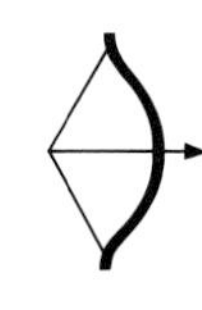

MUSKY RAT KANGAROO (◊*Hypsiprymnodon moschatus;* Color plate p. 131) with the kangaroos or the phalangers.

The hind foot of this smallest of kangaroo species still possesses the original five mammalian toes. All the other species of kangaroos only have four toes. The reduction in the number of toes in the many fleet-footed mammals has been a gradual development in the phylogenetic history. The presence of five toes is therefore considered a primitive condition. This is the reason why the musky rat kangaroos have been grouped into their own subfamily *(Hypsiprymnodontinae)*, separate from all the other rat kangaroos.

Fig. 9-8. 1. Musky rat kangaroo *(Hypsiprymnodon moschatus)*
2a. Broad-faced rat kangaroo *(Potoroops platyops)*
2b. Gilbert's rat kangaroo *(Potorous gilberti)*
2c. Long-nosed rat kangaroo *(Potorous tridactylus)*

The animals are the size of a rat. The HRL is 23.5–33.5 cm. The TL is 13–17 cm. The body weight is approximately 500 g. The snout is pointed. The ears are medium-sized and naked. The fur is dense and soft, and the tail is free of hair but covered with scales just like a rat's tail. The hind legs are only slightly elongated. The first toe does not have a nail. There are 32–34 teeth: $\frac{3 \cdot 1 \cdot 1\text{-}2 \cdot 4}{1 \cdot 0 \cdot 2 \cdot 4}$. For the distribution, see the map. There is only one species and no subspecies.

The first toe is short but well-developed, and movable; however, it is not opposable to the other toes. The second and third digits are fused to form a grooming claw which is found in all kangaroos. The fourth digit is longest. Musky rat kangaroos usually run on all fours; only when in flight do they hop on their hind legs like other kangaroos. The difference in length between the front and hind legs is relatively small. The scientific species name, *Moschatus,* refers to the musky odor emitted by both sexes.

The animals inhabit the undergrowth of the rain forests in northeast Queensland. Ramsey found musky rat kangaroos in the dense and moist vegetation along the Herbert River. The animal is more or less diurnal. It seeks its food by turning over plant matter and looking for insects, worms, and tuberous roots. The berries of a palm *(Ptychosperma alexandrae)* are frequently eaten by the musky rat kangaroo whereby it sits on its hips and holds the berries in its front feet in the manner of a brush-tailed phalanger. Sometimes they dig like bandicoots. They are usually found in pairs unless accompanied by young. They apparently breed during the rainy season, from February to May.

Subfamily rat kangaroos by D. Heinemann

In addition to the musky rat kangaroos, the TRUE RAT KANGAROOS of the subfamily *Potoroinae* are also very primitive kangaroos, although in many aspects they are reminiscent of the phalangers (family *Phalangeridae*).

Distinguishing characteristics

The animals are rat-like and about the size of a rabbit. The head is rather short and the snout is pointed. The ears are medium long and usually spoon-like with pointed tips. The hind feet are only moderately elongated, usually without the first toe. The animal moves predominantly by hopping. The tail is covered with hair and is partially prehensile. There are 34 teeth: $\frac{3\cdot1\cdot2\cdot4}{1\cdot0\cdot2\cdot4}$. The upper central incisors are very long and powerful. The animals occupy all types of habitats except rain forests. Rat kangaroos are terrestrial. During the day they rest in grassy nests, covered tunnels, or ground burrows. The animals are threatened with extinction because of the cats and dogs introduced by man. This was discussed on p. 150. There are five genera, eight species, and eleven subspecies.

The BROAD-FACED RAT KANGAROO (Genus *Potoroops*) has only one species, *Potoroops platyops* (Color plate p. 131). The face is short and broad. The tooth and skull characteristics indicate a relationship to the phalangers (*Gymnobelideus* and *Petaurus*). This species is extinct.

The LONG-NOSED RAT KANGAROOS (Genus *Potorous*) has two species: 1. The GILBERT'S RAT KANGAROO (*Potorous gilberti;* Color plate p. 131) became extinct in the last decade. 2. The LONG-NOSED RAT KANGAROO (⊕ *Potorous tridactylus;* Color plate p. 131) has two subspecies.

The DESERT RAT KANGAROO (Genus *Caloprymnus*) has only one species *(⊕ Caloprymnus campestris).* The HRL is 27–44 cm and the TL is 35–38 cm. The ears are large. The front legs are weaker, but the hind legs are longer and stronger than in other rat kangaroos. The pelage is soft. They are desert dwellers.

The SHORT-NOSED RAT KANGAROO (Genus *Bettongia*) has a prehensile tail supplied with a stiff, brush-like crest on the upper side. There are

at least three species with seven subspecies: 1. GAIMARD'S RAT KANGAROO (*Bettongia gaimardi;* Color plate p. 131) is rabbit sized. The HRL is 26–39 cm and the TL is 28–31 cm. It was widespread in former times but is almost extinct today. 2. LESUEUR'S RAT KANGAROO (*Bettongia lesueur;* Color plate p. 131). 3. TASMANIAN RAT KANGAROO (⊖*Bettongia cuniculus;* Color plate p. 131) has a tail crest which is less well pronounced. The tip of the tail is often white.

The RUFOUS RAT KANGAROO (Genus *Aepyprymnus*) has only one species (*Aepyprymnus rufescens;* Color plate p. 131). This is the largest species of the subfamily. The HRL is 38–52 cm and the TL is 35–40 cm. The fur is rough. The tail is completely covered with hair and is not crested.

How rat kangaroos live

Sharland writes that the Tasmanian subspecies of the long-nosed rat kangaroo with its long snout and nose and short hind legs is more reminiscent of a rat than a kangaroo. The manner in which the species apparently scurries along on all fours through the bush is very much like that of a large rat at first glance. The animal is very aggressive, alert and lively, but during the day time it is shy and hides in nests of grass or leaves dug into the ground. Its diet consists of roots, tender leaves and fruits. Often it will plunder gardens, and is responsible for destroying tomatoes and other domesticated plants. Tasmania is the last refuge of this genus.

Fig. 9-9. 1. Short-nosed rat kangaroo (Genus *Bettongia*)
+ means extinct

Despite its small size, the desert rat kangaroo reminds one in certain aspects of the true large kangaroo. In former times it enjoyed a much wider distribution in the central Australian desert than today. Prehistoric fossil remains of this animal have also been found in Western Australia. When the animal was first described in 1843 it was already very rare and only three specimens were found. Then nothing was heard about it for ninety years, and the genus was assumed to be extinct until recently rediscovered east of Lake Eyre.

Just as a true large kangaroo, the desert rat kangaroo utilizes the tail as a supporting organ while it slowly hops along. A peculiar feature of the hopping gait of *Caloprymnus* is that the feet are not brought down in line with one another, but rather the right toe mark is found well in front of the left toe mark. The animal seldom dodges or doubles back when moving rapidly and is noted for endurance rather than speed. *Caloprymnus* is a nest builder, not a burrower, and shelters in simple leaf and grass nests in shallow scratched-out depressions, despite the heat during the day.

Fig. 9-10. 1. Desert rat kangaroo (*Caloprymnus campestris*)
2. Rufous rat kangaroo (*Aepyprymnus rufescens*)

The short-nosed rat kangaroo is characterized by a stiff, brush-like crest which in most of these animals is found on the upper side of the tail. At one time Gaimardi's rat kangaroo was very frequent in the entire southern half of Australia and in the east it was distributed up into North Queensland. Today it is almost extinct except for a few places in Southwest Australia where it has survived in small populations.

All short-nosed rat kangaroos scratch depressions into the ground and utilize them as their nests. The burrow of Gaimardi's rat kangaroo is quite shallow, and the nest of grass is visible. However, often such nests are camouflaged with bunches of grass or bushes and are therefore very hard to discover, particularly since the rat kangaroo very carefully closes the opening. These animals use their prehensile tail to transport the nest material. The prehensile tail is responsible for the name "opossum rat" by which term they are known in Australia. These small kangaroos prefer a habitat of dry plateaus and hills with only scattered trees or bushes. They occasionally run some field produce, and for this unfortunate reason they are still trapped or poisoned in large numbers in the last larger refuge in Southwest Australia.

Fig. 9-11. 1. Hare wallabies *(Lagorchestes)*
2. Banded hare wallaby *(Lagostrophus fasciatus)*

The rufous rat kangaroos build semi-circular nests in high grass in which they hole up during the day, but nest-building may possibly be only a winter activity. At night the rufous rat kangaroo digs for roots and bulbs with its sharp front claws; however, it will also feed on grasses. Today the species is restricted to coastal Queensland and Northern New South Wales. The last record of finding one in Victoria was in 1905. It inhabits the open plains and open grass-tree forests.

Rufous rat kangaroos are generally solitary, although a female is often accompanied by a nearly full-grown offspring.

Subfamily true kangaroos *(Macropodinae)* by B. Grzimek and D. Heinemann

The TRUE KANGAROOS (subfamily *Macropodinae*) vary in size from that of the rabbit to that of man. The HRL is 35–160 cm and the TL is 30–105 cm. The head is elongated and the snout is usually pointed, although in some species it is rather broad. The ears are from medium long to long, spoon-shaped and tapering to a point. The diet consists of grasses, herbs, bark, and leaves. They inhabit either grass plains or rocky terrain in their range of distribution. The exceptions to this are the genera *Dorcopsis* and *Dendrolagus* which inhabit forest regions. There are eleven genera and forty-two species.

HARE WALLABIES (Genus *Lagorchestes;* Color plate p. 132) are the size of a rabbit. The HRL is approximately 35–50 cm and the TL is 30–46 cm. The nose is either totally or partially covered with hair. The central hind claw is long and strong and is not hidden by the fur of the foot. The members of this genus have thirty-four teeth, two more than other genera. There are three species: 1. The BROWN HARE WALLABY *(Lagorchestes leporoides)* is very rare and is only found in New South Wales. Formerly it was distributed in Victoria and South Australia. 2. The WESTERN HARE WALLABY *(⊕Lagorchestes hirsutus)* has long reddish hair on the lower back. The animal inhabits desert country in Western and South Australia. *Lagorchestes hirsutus bernieri* is found on the island of Bernier. *Lagorchestes hirsutus dorreae* is found on the island of Dorre. 3. The SPECTACLED HARE WALLABY *(Lagorchestes conspicillatus)* is smaller; the ears are shorter and the fur is more dense. It lives or lived in Northwest Australia *(Lagorchestes conspicillatus leichardti)* and on several islands off the northwest coast.

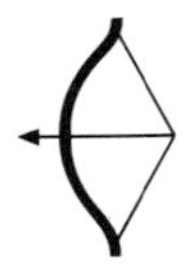

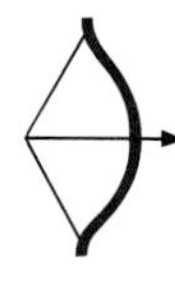

The BANDED HARE WALLABY (⦶*Lagostrophus fasciatus;* Color plate p. 132) has been assigned by some zoologists to the genus *Lagorchestes.* The HRL is approximately 40 cm and the TL is approximately 40 cm. The upper incisors are broad and brachyodont and the lower ones lie flatly under the tips of the upper ones. The nose is not covered by hair. The hind feet are covered by long, coarse hair which also covers the central claw.

The hare wallabies are the smallest species of the true kangaroos; however, they are among the most agile jumpers. The Australian explorer Gould told of one hare wallaby that jumped right over his head.

The hare wallabies resemble hares not only in their size but also in certain aspects of their life habits. Just like hares they rest by day in a depression scratched into the ground and they jump away at the last moment if approached by predators, hunters, or dogs. Gilbert reports that the western hare wallaby escapes into a tunnel with two entrances near the actual burrow when in danger.

The ROCK WALLABIES (Genus *Petrogale;* Color plate p. 133) have an HRL of 50–80 cm, a TL of 40–70 cm, and a body weight of 3–9 kg. They are slightly larger than the hare wallabies. The nose is naked. The tail base is hardly thickened. The long tail is primarily used as a balancing organ while jumping or climbing rocks rather than as a prop for sitting. They inhabit rocky terrain; they are rare or extinct over much of their range. The two species are: 1. The BRUSH-TAILED ROCK WALLABY *(Petrogale penicillata);* and 2. The RING-TAILED ROCK WALLABY *(⦶Petrogale xanthopus).* For additional species, see Color plate p. 133.

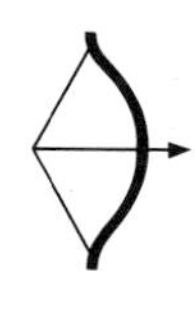

The climbing ability of these "Australian chamois" is astonishing; some of their leaps measure up to four meters horizontally. They can also scramble up cliff faces and leaning tree trunks with relative ease. However, they are not able to encircle branches with their arms as are the true tree kangaroos (Genus *Dendrolagus*). This is one reason why the rock wallabies are unable to climb up vertical tree trunks.

The animals hole up in inaccessible rock crevices. The ground and the rocks of the trails leading to the hiding places are often polished by the use of generations of rock wallabies. Occasionally these predominantly nocturnal animals emerge in the day to sun bathe on a rock or a tree trunk.

One of the best known brush-tailed rock wallabies *(⦶Petrogale penicillata penicillata)* live (or lived) in the mountainous region of eastern New South Wales. In one of his earlier reports Gould describes the agile, monkey-like movements of these animals. He assumed that this agility was ample protection against the pursuits of the dingo or hunting aborigines.

The ring-tailed rock wallaby is one of the most colorful kangaroos. Formerly it was widely distributed in the mountainous regions of

Queensland and southeastern Australia. However, the fur of these rock wallabies is very valuable because of its softness and beautiful coloration. For this reason the animals are not safe in the vicinity of humans. The ring-tailed rock wallabies are protected in South Australia, but are allowed to be hunted in the bordering states. There the trade of poached hides is flourishing.

Fig. 9-12.
3. Nail-tail wallabies *(Onychogalea)*

The LITTLE ROCK WALLABY (*Peradorcas concinna;* Color plate p. 133) from northwestern Australia is considered as a separate Genus because of its unique molar dentition. There are supplementary replacement molars behind the last regular molar. These teeth move forwards as the frontal molars become worn out and drop out. Seldom are there more than five molars in place at any one time, but studies suggest that as many as nine molars may erupt successively.

The NAIL-TAILED WALLABIES (Genus *Onychogalea;* Color plate p. 134) are characterized by a horny nail at the tip of the tail, similar to lions. The HRL is 46–65 cm and the TL is 33–65 cm. The tail is tapering, short-haired and as in the larger kangaroos and wallabies, operates as a lever during leaps. There are three species: 1. The NORTHERN NAIL-TAIL WALLABY *(Onychogalea unguifer)* is found in the northern regions of Australia from northern Western Australia eastward through the Northern Territory to Queensland. 2. The BRIDLED NAIL-TAILED WALLABY *(⦵Onychogalea fraenata)* is found in New South Wales. 3. The CRESCENT NAIL-TAIL WALLABY *(⦵Onychogalea lunata)* is present in parts of Western Australia and South Australia. The nail-tail wallabies move their arms in a rotary motion when hopping, which has led to the common name of "organ grinders." The significance of the nail on the tail is not known. Gould considers the northern nail-tail wallabies as the most elegant of all kangaroos, although it does not have the beautiful fur coloration of the other two species. At one time the bridled nail-tail wallaby was very common in New South Wales but just like the crescent nail-tail wallaby, it is threatened with extinction.

The TREE KANGAROOS (Genus *Dendrolagus;* Color plate p. 135) inhabit timbered areas in New Guinea and northwestern Queensland. The HRL is 52–81 cm and the TL is 43–94 cm. The hind legs are slightly shorter and less well-developed than the front legs. The arms and hands are more strongly developed than in other kangaroos. The animal possesses strong claws. There are seven species, including: 1. The BLACK TREE KANGAROO *(Dendrolagus ursinus)* which is found in northwestern New Guinea. 2. GOODFELLOW'S TREE KANGAROO *(Dendrolagus goodfellowi),* and 3. MATSCHIE'S TREE KANGAROO *(Dendrolagus matschiei)* are found in New Guinea. 4. DORIA'S TREE KANGAROO *(Dendrolagus dorianus)* is found in New Guinea and its subspecies BENNETT'S TREE KANGAROO *(Dendrolagus dorianus bennettianus)* is found in northern Queensland.

The order *Marsupialia* has its share of unique and unusual species. The tree kangaroos (Genus *Dendrolagus*) seem to belong to the most peculiar, almost paradoxical, types. The animals have the appearance of a "misconstruction." Already Alfred Russel Wallace, who together with Darwin discovered the principle of natural selection, wrote: "These animals are not essentially different in body form from the terrestrial kangaroos, but their movement is quite slow and they do not seem too steady on the branches." On another occasion he remarked: "The main difference from the ground-dwelling kangaroos seems to be the well-furred tail of nearly uniform thickness. The tail does not serve as a prop. The powerful claws on the front feet are used to grasp the tree bark and branches or to reach out for leaves which they eat. They progress by means of small leaps which does not seem very adaptive for climbing trees. The animal probably adapted to a mode of leaf eating in the extensive forests of New Guinea which are characterized by lush vegetation and which is a differentiating feature from Australia." Recently it has been discovered that the tree kangaroos travel rapidly from tree to tree by simply leaping downward from remarkable heights, up to 18 meters allegedly. This feat cannot cover up the fact that the tree kangaroos seem to be poorly adapted to their habitat. The explanations for this seemingly contrary fact appear to be quite simple: The selection pressure, which almost always has resulted in astonishingly perfect adaptations, has been almost minimal in the tree kangaroos. The animals do not have predators which can follow them into the tree branches. Only smaller and weaker animal species compete with them for the same source of food. The fruits and leaves of the trees literally grow right into the mouths of the animals. A tree kangaroo having more agile and faster climbing abilities would not enjoy any advantage. The chance to pass on these capacities to a numerous progeny are not any larger than in the slower conspecifics. Thus the tree kangaroos are barely able to climb and this seems quite sufficient. The animals spend a large part of the time on the ground, and probably only climb up trees in order to eat the leaves or to escape predators which is an observation confirmed by their behavior in zoos. In contrast to rock wallabies which occasionally climb trees, the tree kangaroos use their arms and hands to hold on to the tree trunk, especially those which are vertical.

Fig. 9-13
1. Rock wallabies *(Petrogale)*
2. Little rock wallaby *(Peradorcas concinna)*
3. Tree kangaroos *(Dendrolagus)*
4. New Guinean forest wallabies *(Dorcopsis)*

The thick fur on the nape and sometimes on the back grows in a reverse direction and apparently acts as a natural water-shedding device, as the animals sit with the head lower than the shoulders. This is an adaptation to the heavy tropical rainfalls which occur in their habitat.

Ground-dwelling kangaroos also occur in the rain forests of New Guinea. These are the NEW GUINEAN FOREST WALLABIES (Genus *Dorcopsis;* Color plate p. 134); the HRL is 49–80 cm and the TL is 30–55 cm. The nose is large, broad, and naked. The hind feet are hardly longer than those of the tree kangaroo. The arms are relatively well-developed. The fur of the nape faces forwards. There are five species, including: 1. The NORTHERN NEW GUINEA WALLABY *(Dorcopsis hageni)* is found in northern coastal New Guinea; 2. GOODENOUGH'S NEW GUINEA WALLABY *(Dorcopsis atrata)* inhabits oak forests on mountainous slopes on Goodenough Island. Subgenus *Dorcopsulus:* NEW GUINEA MOUNTAIN WALLABY *(Dorcopsis macleayi)* is smaller and has a denser pelage. The skull and dental features vary from those of the other species. It inhabits rain forests of mountains near Port Moresby.

In many aspects the New Guinean forest wallabies seem to represent a transitional form between the tree kangaroos and the remaining jumping marsupials. It is very easy to imagine the emergence of the tree kangaroo from the species similar to *Darcopsis.*

Formerly certain zoologists classified all other kangaroos and wallabies in one single genus *(Macropus).* Today, however, the kangaroos are classified in several genera according to differences in dental feature and also size. The smallest representative is the PADEMELON (Genus *Thylogale;* Color plate p. 136); the HRL is 53–77 cm and the TL is 32–47 cm. The pelage is soft and dense. There are four species: 1. The RED-NECKED PADEMELON *(Thylogale thetis)* has very colorful markings. Individuals vary greatly in coloration and markings. They inhabit the coastal regions of New South Wales and southern Queensland. 2. The RED-LEGGED PADEMELON *(Thylogale stigmatica)* is found fron New South Wales up to Cape York Peninsula, southeastern New Guinea. 3. BRUIJN'S PADEMELON *(Thylogale bruijni)* is found in New Guinea and the Aru and Bismarck Archipelago. 4. RUFOUS-BELLIED PADEMELON *(Thylogale billardierii)* is found in South Australia and Tasmania.

According to Murray's "New English Dictionary" the pademelon is a bowdlerized version of a name from the language of the aborigines. The pademelons inhabit the dense undergrowth of forests, or live in thickets of long grasses, ferns and bushes which are found in swampy areas. In such places one can recognize their tunnel-like runways but one cannot walk on them. If these beautiful, small animals are left undisturbed by dogs or other hazards one can observe them grazing on green grassy slopes or openings close to the thicket. They feed at dusk or dawn. At the smallest indication of danger they quickly hop back into the underbrush. The main portion of the diet consists of grass, but leaves and shoots are also eaten. Occasionally they ruin field crops and vegetables but otherwise they are harmless, if not present in large numbers. They make friendly and pleasant pets. They are without the aggressive tendencies of the mature males of wallabies and kangaroos.

Like all kangaroos the pademelons have, unfortunately, a low productive rate. Usually they only carry one young in the pouch, very rarely two. Thus many species which were formerly abundant are on a constant decline or have already completely disappeared from the mainland.

Thylogale bruijni is named after the Dutch painter DeBruijn who provided the first exact description of a kangaroo in 1714. Bruijn's pade-

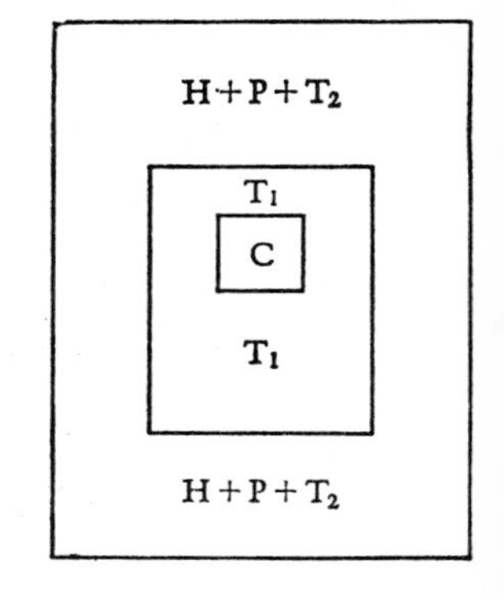

The evolution of the higher mammals: (C) Cretaceous, approximately 120–70 million years ago (T_1) Lower Tertiary, approximately 70–25 million years ago. (T_2) Upper Tertiary, approximately 25–1 million years ago (P) Pleistocene—the "ice ages"—approximately 1 million to 20,000 years ago. (H) Holocene, recent to 20,000 years ago. The broken lines represent groups of animals for which no fossils have been found during those time periods. Animals alive today are colored; extinct ones are in gray.

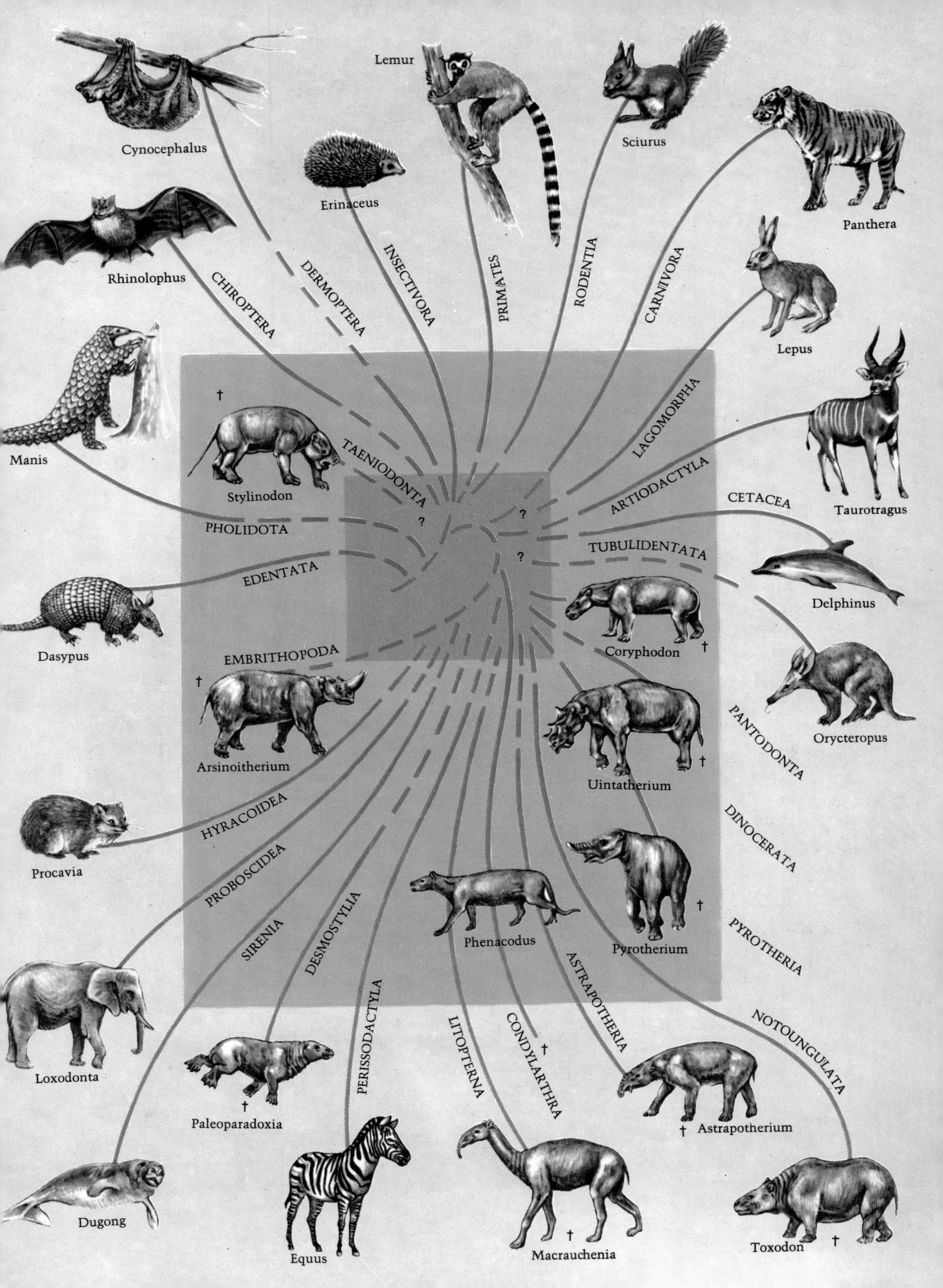

Cynocephalus
Lemur
Sciurus
Panthera
Erinaceus
Rhinolophus
CHIROPTERA
DERMOPTERA
INSECTIVORA
PRIMATES
RODENTIA
CARNIVORA
Lepus
Manis
†
Stylinodon
TAENIODONTA
LAGOMORPHA
ARTIODACTYLA
Taurotragus
PHOLIDOTA
?
?
?
CETACEA
TUBULIDENTATA
EDENTATA
Delphinus
Dasypus
Coryphodon †
EMBRITHOPODA
†
Arsinoitherium
Uintatherium †
Orycteropus
PANTODONTA
HYRACOIDEA
Procavia
DINOCERATA
PROBOSCIDEA
SIRENIA
DESMOSTYLIA
Phenacodus
Pyrotherium †
PYROTHERIA
ASTRAPOTHERIA
NOTOUNGULATA
PERISSODACTYLA
LITOPTERNA
CONDYLARTHRA †
Loxodonta
Paleoparadoxia †
Astrapotherium †
Dugong
Equus
Macrauchenia †
Toxodon †

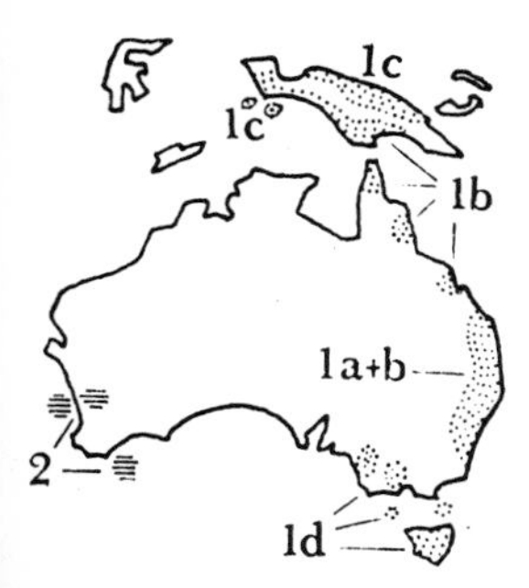

Fig. 9-14.
1. Pademelons *(Thylogale)* a. Red-necked pademelon *(Thylogale thetis)* b. Red-legged pademelon *(Thylogale stigmatica)* c. Bruijn's pademelon *(Thylogale brunii)* d. Rufous-bellied pademelon *(Thylogale billardierii)*
2. Quokka *(Setonix brachyurus)*

melon at one time was one of the most common wallabies.

Today Bruijn's pademelons are still found in large groups in Tasmania. They inhabit regions near water ditches, thickets, and grass lands where one can see their well-used trails.

The SHORT-TAILED SCRUB WALLABY or QUOKKA (*Setonix brachyurus;* Color plate p. 136) has an HRL of 47.5-60 cm and a TL of 25-35 cm. It was at one time classified in the genus *Thylogale*, but today it is considered to be its own genus. Quokkas used to be widely distributed throughout Western Australia but today only remnant colonies are found. Some colonies, for example, occur in swampy thickets on Rottnest Island, near Perth, and on Bald Island, near Albany. When moving slowly they do not use their thin and sparsely furred tails as a third prop for the rear end of the body, as do kangaroos and the larger wallabies. Some quokkas often bring the tail forward between the hind feet and sit on it. They make runways and tunnels through dense grass and undergrowth.

Ruminant-like digestion in the macropod marsupials was first demonstrated in *Setonix.* The pregastric bacterial digestion in the quokka is similar to that in sheep, for most of the fifteen or so species of bacteria present in the large stomach region of the quokka are comparable with those in the rumen of sheep. This wallaby seems to occupy a position between the ruminants and the non-ruminant herbivores.

The previously discussed delayed birth in marsupials was first demonstrated in the quokka. As long as the young remains in the pouch, the newly-fertilized egg proceeds only to the early blastocyst stage, and then becomes quiescent in the uterus. Should the pouch young die or leave the pouch the blastocyst resumes development. The Frankfurt Zoo has raised quokkas for years.

The medium-sized BRUSH WALLABIES *(Wallabia)* are a genus of kangaroos that include a great number of species. The HRL is 45-105 (120?) cm and the TL is 33-75 cm. The body weight may go up to 24 kg. The tail tapers off evenly and serves as a prop while the animal is sitting or hopping. There are about 11 species.

The TAMMAR (*Wallabia eugenii;* Color plate p. 136) is the smallest species of this genus and is counted as one of the pademelons by some scientists. There are several subspecies. It is found in southwestern and South Australia and the coastal islands. Some subspecies have become extinct.

The RED-NECKED WALLABY (*Wallabia rufogrisea;* Color plate p. 137) is the largest member of the wallabies. The HRL is 70-90 cm and the TL is 65-75 cm. The subspecies *Wallabia rufogrisea frutica* from Tasmania is sometimes even larger in body size than the other species. The subspecies *Wallabia rufogrisea banksiana* from the mainland and *Wallabia rufogrisea rufogrisea* from the King and Flinder islands are smaller.

The lesser hedgehog tenrec *(Echinops telfairi)* is a very skillful climber. The spiny hedgehog cannot do this.

The BLACK-STRIPED WALLABY (*Wallabia dorsalis;* Color plate p. 137) occupies brush thickets in the central regions of New South Wales and Queensland. The PRETTY-FACE WALLABY (*Wallabia canguru;* Color plate p. 137) is found on the hills and mountainous regions of eastern New South Wales. It lives in herds. The BLACK-TAILED WALLABY (*Wallabia bicolor;* Color plate p. 137) inhabits swampy thickets. The SANDY WALLABY (*Wallabia agilis;* Color plate p. 137) and the BLACK-GLOVED WALLABY (*Wallabia irma;* Color plate p. 137) occupy the open plains.

Brush wallabies generally spend the daylight hours resting in cover, although occasionally they feed in the latter part of the afternoon or until late in the morning. In common with the other kangaroos and wallabies, brush wallabies may lick their hands and forearms excessively during periods of nervous excitement or hot weather.

They are persecuted by cattle and sheep ranchers, who claim that they eat too much grass. These wallabies are also trapped for their fur, particularly the appealing pretty-face brush wallabies, who are shot in masses. Their soft fur is used for making artificial little koala bears which are sold as souvenirs or toys. The tammars were discussed on p. 153.

The red-necked wallaby was discussed on p. 153. It has been completely eradicated from certain regions. The pretty little black-striped wallaby which inhabits the brush thickets of central New South Wales and Queensland does not seem seriously threatened. It is the most frequent kangaroo species of the southern coastal regions of Queensland. It prefers to stay in the thick underbrush which is penetrated with vines and thorny bushes.

Fig. 9-15.
1. Brush wallabies (Genus *Wallabia*)
2. Red-necked wallaby (*Wallabia rufogrisea*)

The eastern variety of the black-gloved wallaby *(Wallabia irma greyi)* which is known in Australia by the aborigine's name of "toolache" formerly had only the wedge-tailed eagle as its enemy. Andrew Robson describes the courage of a black-gloved wallaby female in the presence of such an eagle: "A low flying eagle persistently followed a mother and her large offspring and attempted to separate the two. The female tried very hard to block this attempt. After a few minutes the eagle dived down from a height of one meter to grasp the offspring. At this moment the mother reared up and kicked the attacker with a peculiar tearing jump with both her hind legs, a mode of behavior common to all other members of the kangaroo family. The observer was not able to ascertain if the blow hit its mark. Nevertheless, the eagle seemed perturbed and retreated and the previously threatened offspring sought comfort in the maternal pouch."

The extermination of the "toolaches" by humans occurred at a rapid pace. In 1923 a small herd of approximately fourteen animals was left. An attempt was made to collect these alive and to transfer them to a protected area. Only four dead or dying animals plus one living female were the success of the operation, and this then was the remainder of this valuable last reserve of a formerly plentiful kangaroo species.

The last genus of the true kangaroo to be mentioned is the large

Fig. 9-16.
1. Red kangaroo *(Macropus rufus rufus)*
2. Western red kangaroo *(Macropus rufus dissimulatus)*
3. Northern red kangaroo *(Macropus rufus pallidus)*

kangaroo *(Macropus)* after which the entire family has been named—*Macropodidae.* This genus represents the largest marsupials living today and also the most highly specialized ones. Species of this genus are most often found in zoos and natural museums, and for this reason many people usually associate the concept of the marsupials with these larger kangaroos. The females of this group are usually considerably smaller than the males and in some species the sexes are further differentiated by a difference in fur coloration. The three species of kangaroos are classified under three different subgenera. Some scientists even consider the groups under three different genera.

1. Subgenus *Megaleia:* RED KANGAROO (*Macropus rufus;* Color plates p. 138, p. 144, and p. 145); the HRL of the ♂♂ is 130–160 cm, and of the ♀♀, 100–120 cm; the TL of ♂♂ is 85–105 cm and of the ♀♀ it is 65–85 cm. The body weight is 23–70 kg; the ♂ is approximately twice as heavy as the ♀. The upper part of the nose is covered by hair. The snout broadens frontally. There are three subspecies.

2. Subgenus *Macropus* (i.n.s.): GREAT GRAY KANGAROO (*Macropus giganteus;* Color plate p. 138); the HRL of ♂♂ is 105–140 cm, and of the ♀♀ it is 85–120 cm; the TL of ♂♂ is 95–100 cm, and of ♀♀ it is approximately 75 cm. The dorsal surface of the head and body are grayish-brown to reddish-gray. The ventral side is whitish-gray to white. There is an indication of a preorbital dark stripe and a lighter area above the eye. There are five subspecies; see the map.

3. Subgenus *Osphranter:* WALLAROO (*Macropus robustus;* Color plate p. 138); the HRL of ♂♂ is 100–140 and of ♀♀ is 75–100 cm; the TL of ♂♂ is 80–90 cm and of ♀♀ is 60–70 cm. There are seven subspecies.

In Australia the kangaroos occupy the niche that is represented in other continents by grass-eating ungulates such as the antelope, deer, zebra, and bison. Kangaroos are able to reach up to 88 kilometers per hour over short distances; however, like all wild animals they soon tire after such an exertion. Even a rider on a horse can finally catch up with a kangaroo. Of course a kangaroo is no match for a car even on the open plain. The tail serves as a support when the animal sits down, and predominantly acts as a balance and rudder when the animal is leaping. The slow jump of a kangaroo measures a distance of from 1.2 to 1.9 meters, but during flight the animal can jump 9 meters and more. One great gray kangaroo once jumped nearly 13.5 meters. The large kangaroos, if need be, can jump up as high as 3.3 meters; however, these are exceptions. It seems clear that the animals do have difficulties with fences of 1.5 meters as is demonstrated by their desperate running alongside the fence when they are chased or by the many dead animals that are found hanging in the barbed wire.

Fig. 9-17.
The great gray kangaroo can jump a distance of up to 13.5 meters and a height of up to 3.3 meters.

The fur coloration of the red kangaroo and the wallaroo differs greatly in the two sexes and the various subspecies. The smaller females usually have a gray fur (although rufous females also occur). The males of the red kangaroo and certain subspecies of the wallaroos are colored reddish like deer. In the west Australian subspecies *Macropus rufus dissimulatus* the females are also rufous while the northwestern Australian variety *(Macropus rufus pallidus)* is of a pale reddish color.

The breeding color of the large males of the red kangaroo is a brilliant red on the chest and back. The skin in the region of the larynx and chest secretes a powder-like rose-red substance. The animal also rubs this powder on his back with his hands. A white handkerchief turns pink if rubbed against such fur. This reddish color gradually fades in dried kangaroo skins.

A similar, but colorless, skin secretion occurs in the great gray kangaroo. If one pets a male of this species, as was possible in the Berlin Zoo, one's hands would smell strongly of Maggi *(Levisticum officinale)*.

The red kangaroos prefer the wide open plains with neither trees nor brush. In their attempt to provide more meadows for huge herds of sheep and cattle the immigrants burnt down forests and bush regions. Of course, concurrently, they created beautiful new habitats for the red kangaroos at the expense of other kangaroo species.

The great gray kangaroo *(Macropus giganteus)*, which is almost as large as the red one, is distributed with its five subspecies over most of Australia and Tasmania. It inhabits the open forest and dense bush, to which it retreats during the hot summer. This species is not as gregarious as the red kangaroo. Usually they exist as small family groups. These species were extensively discussed from p. 156 on.

The great gray kangaroo of Tasmania *(Macropus giganteus tasmaniensis)* which goes by the name of "forester" in its home range, is described by Michael Sharland in his book *Tasmanian Wild Life*: "The Tasmanian gray kangaroo may be just as large as the variety on the mainland to which it is closely related, but no other kangaroo has such a heavy body build, such a powerful appearance, or such long, coarse fur as the forester. The number of these great gray kangaroos has increased in the northeastern corner of their range of distribution, and the farmers have permission to shoot them. This kangaroo is predominantly a forest dweller which grazes in the natural forest clearings rather than fields or unfenced meadows. For this reason it does little harm and is in little danger of being exterminated."

These Tasmanian great gray kangaroos seem to have the greatest endurance of all kangaroos. There is a report of one kangaroo that was chased by a man on a horse for thirty kilometers out of its home range, and then still swam out into the sea for another three kilometers.

A deworming chemical, Phenothiazin, which was given in the Zurich Zoo had a disastrous effect on the great gray kangaroo mothers. They were not able to clean out the pouch. In less than twenty-four hours the offspring were floating in a mixture of feces and urine, and some of them died despite human efforts to save them.

The Zurich Zoo was very fortunate in having an extraordinarily tame

Fig. 9-18.
Great gray kangaroos

1. Northern great gray kangaroo *(Macropus giganteus giganteus)*
2. Southern great gray kangaroo *(Macropus giganteus major)*
3. Tasmanian great gray kangaroo *(Macropus giganteus tasmaniensis)*
4. *Macropus giganteus fuliginosus*
5. Great gray kangaroo *(Macropus giganteus ocydromus)*

Fig. 9-19.
Wallaroos

1. New South Wales wallaroo *(Macropus robustus robustus)*
2. Antelope wallaroo *(Macropus robustus antilopinus)*
3. Deer wallaroo *(Macropus robustus cervinus)*
Those without numbers represent the remaining subspecies.

Fig. 9-20.
This is how an older kangaroo young jumps into the mother's pouch, so that the head and feet still stick out (the pouch is drawn to appear transparent).

great gray kangaroo mother, which was the ancestress to a large herd, and which also permitted the examination of her pouch at any time. Professor Hediger writes about her: "We even dared to take her to the X-Ray room of the Kanton Hospital in order to find out the natural position of the young within her pouch. This enabled us to explain the manner in which the offspring jumps into the marsupium from the outside, and to document the entire movement pattern of this process. Larger offspring which repeatedly have left the pouch already will head for their mother's pouch at the sign of danger in a sort of head-first dive, and then roll over. Once inside the pouch the young has to turn its head around 180° in order to look out; otherwise it would be facing its mother's abdomen. A little later it shifts its body position so as to be able to see out better.

The wallaroo *(Macropus robustus)*, also known as "euro" by the Australians, occupies hilly country interspersed by rocky areas. It seems best adapted to heat and dessication. The species was discussed more fully on p. 152.

The wallaroos in the various regions of Australia deviate decidedly from this type. Particularly the males of the species which occupy the north and northwestern parts of the continent are often reddish brown or brilliant red, and also the fur of the females may have a reddish tinge. The northeastern subspecies of the Kimberly region belongs to these red types. This attractively colored, short-haired fur is reminiscent of certain species of antelopes. The animal is therefore named the antelope wallaroo *(Macropus robustus antilopinus)*. A subspecies found in southwestern Australia is known as the deer wallaroo *(Macropus robustus cervinus)*. There are four additional subspecies.

The kangaroos and wallaroos represent the highest degree of evolutionary development in the marsupials, and thereby the Metatheria. It is futile to speculate if the order Marsupialia would have given rise to more highly developed and more intelligent species if the higher mammals had not entered the existential competition in all continents but Australia. However, it is an almost certain prediction that nowhere on earth could such a highly specialized marsupial as the kangaroo have evolved if Australia had not been separated from the other continents and their higher mammals through millions of years. Man, not counting a few mice and bats, was the first to overcome this isolation of the fifth continent. He penetrated with his dogs, sheep, foxes, cats, and rabbits into the land of marsupials, and by this act sealed the fate of many, if not ultimately all, marsupials. The hope is to preserve at least some species and particularly the vigorous kangaroos in natural sanctuaries and national parks to serve as a reminder of a past era in the development of mammals.

Bernhard Grzimek
Dietrich Heinemann

10 The Higher Mammals (Placentals)

Subclass: Higher mammals by K. Herter

Distinguishing characteristics

Subclass *Eutheria.* The marsupium and the marsupial bones are absent. The uro-genital duct and the rectum have separate openings. The scapula has a scapular spine running down its length. The coracoid has receded to form the processa coracoideus. The angular process (Processus angularis) which projects from the lower jaw is never bent to the inside but is usually completely absent. The optic foramen (foramen opticum) is present and only absent in a few cases. The hard palate is occasionally perforated. The tympanum can be expanded inward to form part or all of the tympanic bulla (Bulla tympanica). Originally the number of teeth was 44: $\frac{3 \cdot 1 \cdot 4 \cdot 3}{3 \cdot 1 \cdot 4 \cdot 3}$. The number of teeth may vary from 0 to 250. The cerebral hemispheres are joined by a massive commissure, the corpus callosum.

The vagina is always simple. The penis lies in front of the scrotum. The young inside the uterus are nourished via the placenta and are born at a more mature stage than the marsupial offspring.

The higher mammals are distributed over the entire earth and oceans; however, only a few species reached Australia and New Zealand.

There are sixteen orders: 1. PRIMITIVE INSECT-EATING MAMMALS: MOLES and SHREWS *(Insectivora)*; 2. MONKEYS, APES, and MAN *(Primates)*; 3. FLYING LEMURS *(Dermoptera)*; 4. BATS *(Chiroptera)*; 5. ARMADILLOS, SLOTHS, and ANTEATERS *(Edentata)*; 6. SCALY ANTEATERS *(Pholidota)*; 7. RODENTS *(Rodentia)*; 8. WHALES *(Cetacea)*; 9. FLESH-EATERS *(Carnivora)*; 10. RABBITS and HARES *(Lagomorpha)*; 11. AARDVARKS *(Tubulidentata)*; 12. ELEPHANTS *(Proboscidea)*; 13. HYRAXES *(Hyracoidea)*; 14. SEACOWS *(Sirenia)*; 15. ODD-TOED HOOFED ANIMALS *(Perissodactyla)*; and 16. EVEN-TOED HOOFED ANIMALS *(Artiodactyla)*. Eleven additional orders are extinct.

What are higher mammals?

The first two orders of the class *Mammalia,* the egg-laying mammals (*Monotremata* or *Ornithodelphia*) and the pouched animals (*Marsupialia* or *Didelphia*) are two independent branches of the main mammalian stem. They evolved quite separately at an early stage. For this reason they are classified as separate subclasses (Primitive Mammals = *Prototheria* and *Marsupialia* = *Metatheria*). All other living and extinct mammals are classified under the third subclass of the higher mammals (*Eutheria* or *Monodelphia*).

The three subclasses are mainly differentiated on the basis of the reproductive organs of the female and the manner in which the embryo is nourished.

In the "true mammals" the embryo inside the uterus is connected and nourished by the mother's metabolic processes. Metabolic exchanges between embryonic and maternal tissue takes place via the placenta. Formerly the higher mammals were also known as *Placentalia* because of this new development. Occasionally placentas do occur in the marsupials, and therefore this subclass is today more appropriately known as *Eutheria*, which means "true mammals." As a result of the better conditions for the embryo within the mother's body the eutherian young are born at a more advanced stage than young marsupials. Two types of young may be differentiated: the altricial young and the precocial young. The altricial newborn is small, naked or barely covered with hair, and is not able to regulate its own body temperature entirely. Its eyes and ears are not yet functional. It cannot take up any nourishment except its mother's milk. This means that the altricial young are completely dependent on the mother for warmth and food. The precocial newborn, on the other hand, are completely covered by hair. All senses are functional, as well as the limbs for locomotion. They are able to regulate their own body temperature and are able to look for their own food shortly after birth.

Despite fluctuating outside temperatures, the higher mammals almost always maintain a steady body temperature. This ability allows some mammalian species to occupy the arctic, mountains above the snow line, or other cold regions. The majority of the higher mammals are terrestrial; however, members of several orders have more or less adapted to an aquatic way of life. Whales, sea-cows, and seals are dependent on water. Only bats are completely airborne, but gliding forms are found among several other orders, similar to the gliders found in the marsupials (see p. 115). The higher mammals have been highly successful in filling a great variety of ecological niches on land, fresh water, the oceans, and the air.

Konrad Herter

▷

Solenodons

1. Haitian solenodon *(Solenodon paradoxus)*
2. Cuban solenodon *(Atopogale cubana)*

Tenrecs

3. Tenrec *(Tenrec ecaudatus)*
4. Streaked tenrec *(Hemicentetes semispinosus)*
5. Large Madagascar hedgehog *(Setifer setosus)*
6. Small Madagascar hedgehog *(Echinops telfairi)*

11 The Insectivores

Order: Insectivores
by K. Herter

Distinguishing characteristics

Order *Insectivora:* these animals are small or very small. The HRL is 3.5 (Savi's pygmy shrew) to 44.5 cm (Moon rat). There are six types of insectivores: a. Shrew type: The body is elongated. The snout is long, pointed, and often trunk-like. It protrudes way beyond the mouth. The legs are short and the tail is usually long (shrews, solenodons, rice tenrecs); b. rat type: The body is more compact than in the shrews. The head is less pointed. The tail is long and naked (hairy hedgehog); c. otter type: The body is long. The head is relatively blunt. The long, laterally compressed tail is used for propulsion in the water (otter shrews); d. hedgehog type: The body is compact, and can roll itself into a ball. The tail is very short and is not visible on the outside. The back is covered with bristles and spines (gymnures, hedgehogs). e. mole type: The body is cylindrical. The nose is pointed and often trunk-like. The tail is usually very short or long and used as a rudder. The eyes and ears are small (golden mole, common mole); f. jerboa type: The nose is trunk-like. The eyes and ears are large. The tail is long. The two lower bones of the hind leg are united and are elongated (elephant shrew).

Sexual activity is often accentuated by odors given off by skin glands. There are usually five (sometimes four) clawed fingers and toes. The collar bone is absent only in the otter shrews. The skull is usually long, flat, and small. The zygomatic arch is absent in the solenodons, tenrecs and shrews. The dentition is very variable. The brain is primitive and small. The cerebral hemispheres of the brain are smooth, and do not extend backward over the cerebellum. They are pear-shaped, small and short. The olfactory bulbs are well developed and sharply demarcated. The visual sense varies in quality. The sense of hearing is well developed. The tactile and vibratory sense is very well developed and is vital for the animal's existence. Long, movable vibrissae are located around the nose, mouth and eyes, and also on the body. The olfactory sense is very acute. The nasal chamber, which

▷▷

Rice tenrecs

1. Ground tenrec *(Geogale aurita)*
2. Long-tailed tenrec *(Microgale longicauda)*
3. Rice tenrec *(Oryzorictes talpoides)*
4. Web-footed tenrec *(Limnogale mergulus)*

Otter shrews

5. Otter shrew *(Potamogale velox)*
6. Lesser otter shrew *(Micropotamogale lamottei)*

Golden moles

7. Cape golden mole *(Chrysochloris asiatica)*
8. African golden mole *(Amblysomus sp.)*
9. Giant golden mole *(Chrysospalax trevelyani)*

1
2
3
5
4
6

1
2
3
4
5
6
7
8
9
P.B.

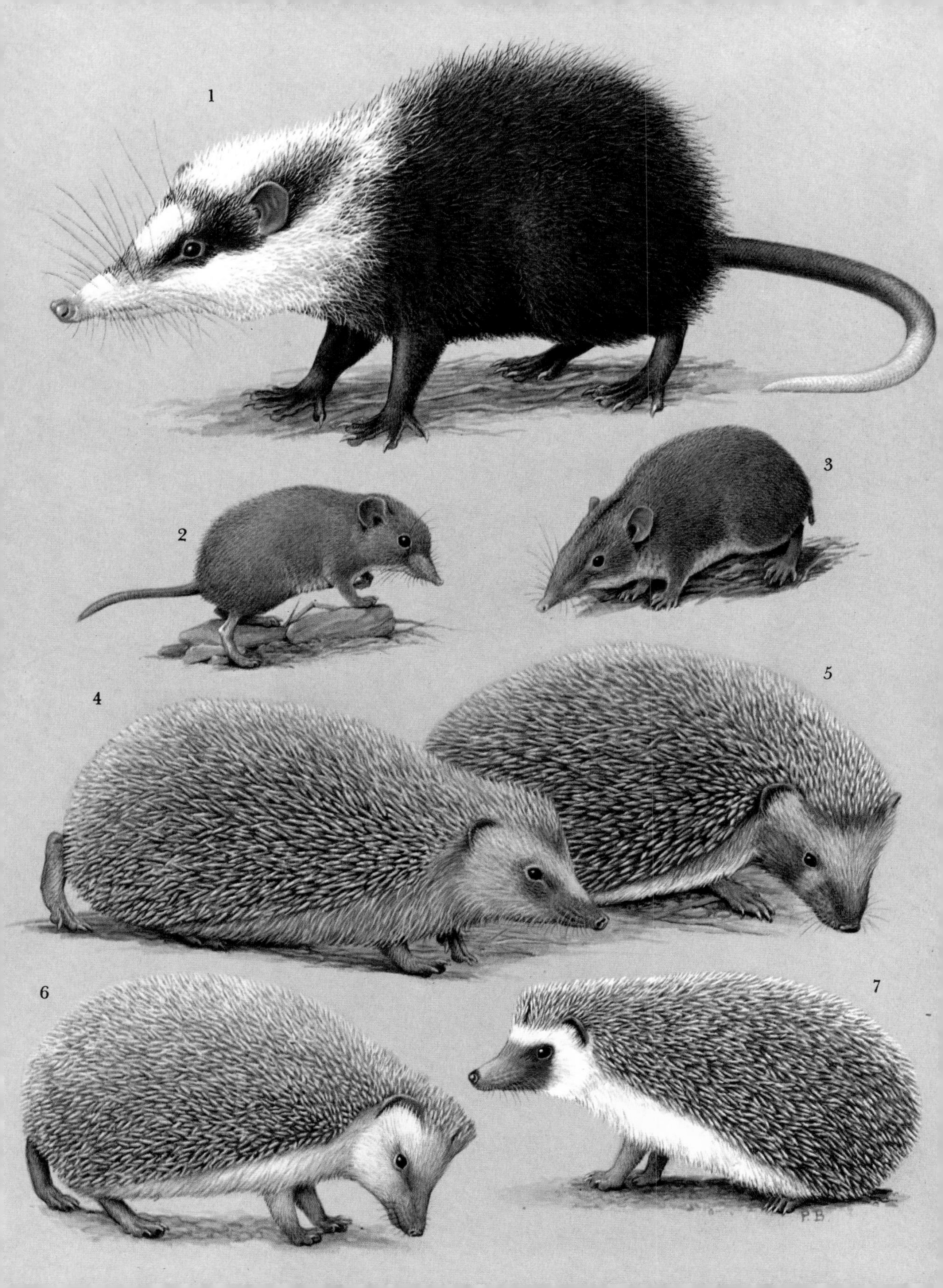
1
2
3
4
5
6
7
P.B.

1
2
3
4
5
6
P.B.

consists of coiled scrolls of bone, is covered with olfactory epithelium. The ethmoturbinals are large. The animals are also equipped with a heat sensory organ. The range of distribution of some families is very restricted—"insular." Solenodons are only found in Cuba and Haiti, tenrecs only on Madagascar. Otter shrews, golden moles and elephant shrews are limited to Africa. Hedgehogs, shrews, and moles are widely distributed.

There are five superfamilies and eight families: 1. Tenrec-like species *(Tenrecoidea)*: Solenodons, tenrecs, otter shrews; 2. Golden mole-like species *(Chrysochloroidea)*: Golden mole; 3. Hedgehog-like species *(Erinaceoidea)*: Hedgehog; 4. Shrew-like species *(Soricoidea)*: shrews, moles; 5. Elephant shrew-like species *(Macroscelidoidea)*: elephant shrew. There are approximately 370 species.

The insectivores are the most primitive of all living higher mammals today. In earlier evolutionary times an insectivore-like creature gave rise to all other orders of the mammals. Insectivore fossil remains have been dated back to the Cretaceous. Ancestral insectivores and also marsupials originated from still more ancient forms, namely the mouse to rat-sized insectivorous pantotherians which existed during the Jurassic and lower Cretaceous.

Fossils from the Tertiary and the Pleistocene not only reveal the geological age of the various insectivore stems but also disclose the way in which the various extinct and living forms are related to each other. It has been shown that the insectivores are not one phylogenetic unit but rather are a "cross-section type order." Present day insectivore specializations do not cover up ancestral characteristics. The animals represent a cross section of the early evolutionary stages of the higher mammals. It is probable that the hedgehog-like types *(Erinaceoidea)* and the shrew-like types *(Soricoidea)* are closely related and make up the "true" insectivores. Similarly, the tenrec-like animals *(Tenrecoidea)* and the golden mole-like species are related. Some paleontologists group these and the extinct forms under the *Zalambdodontia.* There is a possibility that the *Zalambdodontia* may be related to the "true" insectivores. The solenodons are generally assigned to the tenrec-like forms; however, the two groups evolved quite independently at an early stage.

The insectivores with the most highly evolved specializations, the elephant shrews, cannot be derived from either the fossilized or living forms. The elephant shrews are more closely related to the tree shrews *(Tupaiidae).* Formerly both groups originated from a common stock. Today most zoologists classify the tree shrews with the prosimians because of their primate-like characteristics, although new evidence now disputes this. This leaves the elephant shrews in a group by themselves. It is rather questionable if one should classify them under the insectivores at all.

◁◁
Hairy hedgehogs

1. Moon rat *(Echinosorex gymnurus)*
2. Shrew hedgehog *(Neotetracus sinensis)*
3. Lesser gymnure (*Hylomys suillus)*

Spiny hedgehogs

4. Western European hedgehog *(Erinaceus europaeus)*
5. Eastern European hedgehog *(Erinaceus roumanicus)*
6. Algerian hedgehog *(Aethechinus algirus)*
7. Pruner's hedgehog *(Atelerix pruneri)*

◁
Spiny hedgehogs

1. Long-eared hedgehog *(Hemiechinus auritus)*
2. Ethiopian hedgehog *(Paraechinus aethiopicus)*

Elephant shrews

3. North African elephant shrew *(Elephantulus rozeti)*
4. Forest elephant shrew *(Petrodromus sultan)*
5. Short-eared elephant shrew *(Macroscelides proboscideus)*
6. Checkered elephant shrew *(Rhynchocyon cirnei)*

Fossil remains have indicated that *Zalambdodontia* and hedgehog-like creatures were already present towards the end of the Cretaceous. Today's representatives of the other insectivore families originated in the Tertiary. Numerous remains of extinct insectivorous groups have been discovered. Some seem to be phylogenetically related to other mammalian orders. This shows the significance of the insectivores as the ancestral group in the evolution of the mammals.

Fig. 11-1.
Shrew-type
(shrew)

Echolocation, similar to the phenomenon found in bats, has been found in certain insectivores. Up to now these high frequency vocal sounds have been demonstrated in solenodons, various species of tenrecs, and also certain shrews. It is probable that this ability is widely distributed through the entire order.

Family *Solenodontidae*

Certain morphological characteristics are common to the solenodons, the tenrecs, and the otter shrews, and presumably they are distantly related. For this reason these three insectivore groups are combined under the superfamily *Tenrecoidea.* The solenodons *(Solenodontidae)* are a rather unique group since they do not have much in common with the other members of the order. Solenodons are large, somewhat antiquated looking insectivores which inhabit the islands of Cuba and Haiti. There are two species. These animals may have died out.

Distinguishing characteristics

The HRL of the shrew-type species is 28–32.5 cm and the TL is 17.5–25.5 cm. The body weight is approximately 1 kg. The elongated snout is supported at its base by a small round bone. The eyes are small, and the ears partially extend beyond the pelage. The front legs are shorter than the hind legs. There are five fingers and five toes. The underfur is short and dense; the guard hairs are long. The forehead is sparsely covered with hair. There are long vibrissae. The ears, legs, and scaly tail are almost hairless. The thighs and the lower back are naked but are covered by the long hair of the front of the body. There are scent glands in the armpits and groin. The salivary glands are very large. The two mammae are located in the inguinal region. There are 40 teeth: $\frac{3 \cdot 1 \cdot 3 \cdot 3}{3 \cdot 1 \cdot 3 \cdot 3}$. The second lower incisor is deeply grooved on the inside (*Solenodontidae* means grooved teeth). The duct of a lower jaw gland, which produces a possibly toxic saliva, opens into the groove. There are two genera, with one species each. The animals inhabit mountainous forests on Haiti and Cuba.

Fig. 11-2.
Rat-type
(Hairy hedgehog)

1. The HAITIAN SOLENODON (*Solenodon paradoxus;* Color plate p. 177) has a fur color that can occasionally turn red, independent of age, sex, or season. The light spot on the nape changes in size and shape. The animal is found in the high forests of Haiti.

2. The CUBAN SOLENODON (†Atopogale cubana; Color plate, p. 177) has hairs which are longer and softer. The secretion of the lateral glands are of a reddish color. They are assumed to be extinct since 1909. In North America fossils of three further genera from the middle Oligocene are known.

How solenodons live

In the daytime the Haitian solenodon shelters in stone crevices, or occasionally in hollow trees. Usually several solenodons are found in one hollow. Presumably the female bears young twice a year, independent of the season. The litter size is one to three. The offspring of two successive litters may stay with the parents. Up to eight animals have been found in one burrow. In the other insectivores the family groups break up soon after the young become independent. For a long time it had been assumed that the solenodons had been totally exterminated by the mongoose which was introduced in 1872 for controlling snakes, or by domestic dogs and cats. However, since 1935 the animals have been rediscovered in various regions on the island. Some living specimens have even reached Europe.

Fig. 11-3.
Otter-type
(Otter shrew)

Keeping and care of solenodons

The Frankfurt Zoo is presently very successfully raising solenodons. Also the zoologist Erna Mohr of Hamburg has kept fifteen of these antiquated animals. The animals quickly became familiar with their keeper and would climb up on her arm or shoulder. Usually one could hold the animals around the abdomen, but sometimes they could also bite quite severely. Some of the solenodons had names and were trained to come when called. In the daytime the animals only occasionally left their sleeping boxes and then only to eat a little, urinate, or defecate. In the latter part of the afternoon they became more active and then stayed up for twenty to forty minutes every two hours. Close to 3 A.M. all activity ceased abruptly, writes Erna Mohr, and the animals would only emerge if someone was working on their cage. If not interrupted, the animals would sleep through until 11 o'clock. The solenodons lined their sleeping boxes with hay, dried leaves, peat moss, or other material. With their hands they scratched the nesting material into a pile, and then transported this with either their hands or teeth into the nest while walking backwards. The animals also collected hard, solid objects such as coarse chunks of peat moss, the prickly hulls of beechnuts which they picked out of the leaves, empty water and food bowls, and even shovels and brooms left behind after the cleaning of the cage. They tried to drag all of these objects into their nests. In the burrows of free living solenodons one can frequently find an accumulation of snail shells.

Fig. 11-4.
Hedgehog-type
(Hedgehog)

Fig. 11-5.
Mole-type
(Common Eurasian mole)

Fig. 11-6.
Elephant shrew

The solenodons kept by Erna Mohr used to all sleep in a pile whereby the animals on top used to push their noses between the legs of those lying on the bottom. Outside the nest, too, the animals of different ages and sexes were usually agreeable with each other. Some animals were inseparable. Two mature males were hostile to each other outside the sleeping box, but inside they used to curl up back to back. Yet one of these males had killed three females during the sea journey.

Mode of locomotion and sounds

While walking or standing the solenodons put down the entire sole of the front feet, but only lift the hind feet to the base of the toes, so that the heels do not touch the ground at all. They stroll along leisurely, somewhat in a zigzag course. The tail is held stiffly and almost motionless. It is never bent to the side. The animals are incapable of jumping but are able to run straight ahead quite fast. Frequently they attempt climbing. They will hook their powerful front claws into rough wood or a crack and then pull themselves up by pushing with their hind feet and the propped-up tail. On the flat ground they prop up the tail in kangaroo fashion and sit up on their hind legs. They do a lot of scratching in the ground, but usually only with one hand. If they do use both hands they literally fall flat on their noses. When a solenodon scratches itself with one hind foot, it stands on three legs and uses the tail as an additional prop, thus enabling the animal to reach almost all of its body parts. It seems that scratching is the only body care activity. I have never seen them bathe, lick, or groom themselves.

Solenodons wheeze like hedgehogs, squeak like guinea pigs, or twitter like mice. Erna Mohr found that the most frequent sounds heard "were like plaintive whimperings coming from a nest of restless, still blind kittens." An excited solenodon will emit penetrating shrieking sounds. Contact calls between mother and young or between a pair of mates which had been separated for some time sound rather melodic, and reminded Erna Mohr of the song of a red-breasted robin.

Nutrition and feeding

The rural people of Haiti frequently kill solenodons because they are reported to eat sweet potatoes and other field crops. Probably the animals only damage the plants in their search for insects or other small animals. The solenodons kept by Erna Mohr refused to eat any vegetable food. Some solenodons kept in zoos have taken very small amounts of rice, bananas, or soaked white bread. The solenodons kept in the Frankfurt Zoo have been doing very well for years on a diet of minced meat (supplemented with vitamins, minerals, and calcium), horse meat, heart, mice, chicks, fish, meal worms, milk, as well as bananas and salad which they seem to eat incidentally. Mohr's animals refused grubs and a living frog. However, they preferred live animals to dead ones. The solenodons would remove the soil which was attached to an earth worm with their hands; otherwise, they only touched food with their mouths. When drinking they would bend back their noses as far as possible to keep them from touching the water. They lapped up the water with their tongues, and then held their heads back and swallowed with chewing motions.

The auditory and olfactory senses

The auditory and olfactory senses are most important to the solenodon. The animals will be startled and dash off aimlessly when they hear a sudden high or shrill sound, loud laughter, or smacking noises. The odors given off by the green oil-like secretion from the glands in the armpits and the groin play an important role in the sexual activities of these animals. Erna Mohr wrote: "All mature males that I handled were always green on the ventral side. The color would stain my hands. The mature females were never as damp as the males and at times they were completely dry." It is probable that the males are always ready to copulate while the females have "individual breeding times." When the female is not in estrus, the glands probably do not secrete at all or only very small amounts. A male put together with a female approached her immediately, and vigorously pushed his nose into her armpits. The female pushed back less vigorously. Then both animals rubbed certain parts of their bodies against each other, and eagerly sniffed each other. The male crawled underneath the female, encircled her abdomen while dragging his hind legs. He made several attempts to mount, but copulation did not take place.

Reproduction and life expectancy

Nothing is known about the gestation period or the birth process. Newly born solenodons are naked except for the vibrissae and a few guard hairs. The eyes are still closed. Within two weeks the youngsters have grown a fine, normal fur. A young female started to drink milk from a bowl at thirteen weeks of age. At eighteen weeks she ate minced meat and shortly thereafter, cockroaches. This young was suckled by her mother until twenty weeks of age. The life expectancy of the solenodon is not known. One in the Breslau Zoo reached an age of six and one half years. A Cuban solenodon which arrived in the Philadelphia Zoo in 1886 survived for more than five years.

Family: *Tenrecidae*

TENRECS (family *Tenrecidae*) are of the hedgehog and shrew types. The HRL is 4–39 cm. The nose is frequently trunk-like. The front legs are shorter than the hind legs. There are four to five fingers and five toes. Up to twelve pairs of mammae may be present. The rectum and urogenital ducts open into a common cloaca-like skin fold. The zygomatic arch is absent. There are thirty-two to forty teeth. The animals are found on Madagascar. On Madagascar fossil remains from the Pleistocene have been found, and also from the Oligocene in Africa. There are two subfamilies: 1. TENRECS (*Tenrecinae*); 2. RICE TENRECS (*Oryzorictinae*); there are a total of ten genera with thirty species.

The tenrecs *(Tenrecidae)* of Madagascar are probably the only forms which have not deviated much from the common ancestral group of all the higher mammals. In many respects they are comparable to the primitive opossums of South America in their degree of development. The tenrecs, however, possess all the morphological characteristics and mode of reproduction typical of the higher mammals (see p. 175). We should regard the tenrecs as the last branch of the prototype higher mammal which survived until today because of the isolated position of the ancient island of Madagascar.

The mammal with the largest number of young

Among the mammals, the tenrec produces the largest number of offspring per litter. In the uterus of a female tailless tenrec *(Tenrec ecaudatus)*, thirty-one embryos were found. The average litter size for this species seems to be twelve to fifteen. It appears that the twelve pairs of mammae are matched by the large number of young. The offspring are able to run about soon after birth. After approximately nine days they open their eyes and within four weeks they are completely independent. The female of the streaked tenrec *(Hemicentetes semispinosus)* has only four pairs of mammae, and probably has a maximum litter size of eight. The Madagascar hedgehogs (genera *Setifer, Echinops,* and *Dasogale*) have five pairs of teats.

Temperature regulation and states of torpidity

There are certain unique features in the temperature regulatory mechanism of the tenrecs that, as a rule, are not found in other "warm-blooded" animals. For example, the activity patterns of the streaked tenrec are strongly influenced by the outside temperature. At approximately 20°C the animals rest inside their shelter for most of the day, but in the afternoon and early evening hours become very active. After midnight they become more quiet and the periods of rest increase in length. In the morning they finally crawl into their nest for the long rest during the day. Around 16°-18°C the animals are much more restless and run around at night, and are also active for many hours during the day. This physical activity generates additional metabolic heat and the animal is thus able to maintain its body temperature of 26°-30°C. However, if the temperature falls below 16°C, increased activity will not suffice to maintain an even body temperature. The animals become numb, and may freeze to death.

The possibility of this occurring in Madagascar is very rare. The streaked tenrec usually inhabits the moist regions where the temperatures are warm and do not fluctuate too much. The tailless tenrec, on the other hand, prefers dry, sandy locations where outside temperatures vary, particularly during the dry season. Aside from a few permanently moist regions, hardly any rain falls in Madagascar from April or May to October or November. This "winter" in Madagascar is fairly cool. The vegetation dries up and the small animals hide. The tailless tenrecs survive this period of coolness and food shortage by rolling themselves up in their ground burrows, placing their noses between their front paws and closing their eyes. They feel cold to the touch, and have a breathing rate of approximately one respiration per three minutes. The breathing is very shallow and barely discernable. If touched, the animals snort and grunt, and sometimes snap at one's hand. If left undisturbed, the animals may stay in this torpid state for the length of the dry season. They do not take any food during this time. Animals dug out of their burrows had empty intestines, but their bodies were fat. The burrows were void of feces or nesting material. Tailless tenrecs kept in captivity in Madagascar were also in a torpid state from the beginning of May to the middle of October, even when it was warm and sunny outside. The normal body temperature of tailless tenrecs is 34°-35°C. A decrease in the outside temperature results in a lowering of the body temperature of these creatures, and they enter into the torpid state. Eisentraut in Stuttgart measured the lowest body temperature (13°C) in a tailless tenrec when the outside temperature measured 10°C. The Madagascan hedgehogs enter a state of torpidity not only during the unfavorable cool months but also during their daily resting periods. Their body temperature drops, the rate of respiration is decreased and all vital activities are slowed down. Only if it is warmer than 25°C does the body temperature not drop during the resting period.

Subfamily: *Tenrecinae*

Of the two subfamilies in the tenrecs, the *Tenrecinae*, which belong to the hedgehog-type, are best known. The HRL is 9.5–39 cm. The tail is not visible on the outside, or it may be a small bulb-like structure covered with spines. The ventral side is covered by hair interspersed with bristles and spines, or only spines. The main part of the diet consists of invertebrates or small vertebrates. The animals inhabit forests, brushlands, semi-deserts, and plains. There are five genera and six species.

The best-known species: tenrec

The TAILLESS TENREC (*Tenrec ecaudatus;* Color plates p. 177 and pp. 283–284) has an HRL of 26.5–39 cm and a TL from 1–1.6 cm. The tail is not visible. The nose is long and pointed. The fur is coarse and interspersed with bristles and long, thin spines. The mane, consisting of bristles and spines, can be erected. There are long tactile hairs on the back. Young animals have white spines arranged in longitudinal rows along the back; these are lost in the adults and replaced by hair. There are 38–40 teeth: $\frac{2 \cdot 1 \cdot 3 \cdot 3\text{-}4}{3 \cdot 1 \cdot 3 \cdot 3}$. The canines are long. The fourth molar does not erupt until after the first molar has been dropped.

Distribution and habits

The tailless tenrec is one of the best known tenrec species. Originally it only inhabited the low lands and mountains up to 1000 meters elevation in Madagascar. However, it is now also found on the Comoro Island of Mayotte, and man has introduced the tailless tenrec to the islands of Réunion, Mauritius, and the Island of Mahé in the Seychelles. The animal prefers warm, sandy areas in the brushlands, clearings of the drier forests, the highland plateaus and the cliffs along river beds. The tenrec is also found in the gardens of Tananarive, the capital of Madagascar. The animal is rarely seen in the arid southwest districts or the eastern rain forests. During the day the animals rest in rocky crevices, tree hollows, ground holes, or dug-out burrows. A burrow excavated in August revealed a tunnel one and one half meters in length lying just under the surface which then dropped down almost vertically for seventy centimeters. The duct ended in a little chamber of smooth walls which was barely large enough to house one rolled-up tenrec. Part of the tunnel was filled with soil.

The animals forage at dusk and at night. They dig with their snouts and claws to obtain food. They bore funnel-like holes 2½–5 cm deep. Apparently they will also enter shallow water and search for prey in the mud. The major part of the diet consists of worms, snails, arthropods, and particularly grasshoppers. They will also eat lizards, eggs from ground-nesting birds, roots, and fruits.

The mating season probably begins soon after hibernation ends. Pregnant females were found in October and November. From March to May female tenrecs with not yet half grown young were seen. The average litter size of the tailless tenrec is twelve to fifteen young, but litter sizes of over twenty have also been recorded.

Tailless tenrecs have frequently been kept in scientific institutions and zoos. In New York one animal lived for over twenty-five months, and in the Frankfurt Zoo one survived for fifty-nine months. The diet of these captive tenrecs consisted of raw minced beef or horse meat, and earth worms, plus a supplement of white bread soaked in milk, bananas, and boiled potatoes. The animals preferred the meat to the vegetative material. They would jump on the meat with snarling sounds, shake it and chew it for long periods of time. A juvenile tenrec sniffed a dead mouse only briefly and then suddenly grabbed it. While eating the tenrec held the mouse to the ground with its front paws, and ripped pieces of meat from it. The animals usually deposit their feces in a certain spot, and then first scratch with their front paws and then their hind paws.

If one approaches a tailless tenrec with a hand, the mane of spines on the upper back will erect. The animal jerks up frontally and tries to push the erected mane against the hand. The spines do not penetrate human skin. The animals quickly adapt to being touched by humans. They are compatible in a group of conspecifics. If one antagonizes the animal it will snort, grunt, spit, and hiss. A tied-up tenrec will squeak and whimper loudly. Juveniles shake the dorsal spines and the noise created in this manner helps the animals to maintain contact. A similar behavior is observed in the streaked tenrec. The inhabitants of Madagascar dig the animals out of their burrows or chase them with dogs. They eat the tenrec despite its rather unpleasant odor, unpleasant at least for the noses of Europeans. The spines are singed off and the bristles and top skin are scratched off. Then the animal is gutted and the very fat meat is cut into pieces and boiled. In former times soldiers were not allowed to eat tenrecs because it was believed that the meat of this fearsome animal would affect their courage. Although all Madagascan hedgehogs are protected by law, the tailless tenrec is, nevertheless, regularly sold alive or cooked at the weekly markets. The meat is very much in demand. Their high reproductive rate has forestalled their total extinction.

The streaked tenrec and blackheaded streaked tenrec

The STREAKED TENREC (*Hemicentetes semispinosus;* Color plates p. 177 and pp. 283-284) resembles the tailless tenrec; however, it is smaller, has longer legs and a larger head with an extraordinary, long, beak-like snout. The HRL is 16-19 cm and the body weight is 150 g. The body form is stout and the tail is not visible. The body is covered by bristles and spines. The ventral side is covered by soft hair. Long tactile hairs are located around the muzzle, behind and below the eyes and the back.

Red-toothed shrews

1. Common shrew *(Sorex araneus)* 2. Lesser shrew *(Sorex minutus)* 3. Laxman's shrews *(Sorex caecutiens)* 4. Alpine shrew *(Sorex alpinus)* 5. Northern watershrew *(Sorex palustris)* 6. American pygmy shrew *(Microsorex hoyi)* 7. European watershrew *(Neomys fodiens)* 8. Mediterranean watershrew *(Neomys anomalus)* 9. Short-tailed shrew *(Blarina brevicauda)* 10. Least shrew *(Cryptotis parva)* 11. Crawford's desert shrew *(Notiosorex crawfordi)*

White-toothed shrews

12. Bicolor white-toothed shrew *(Crocidura leucodon)* 13. Lesser white-toothed shrew *(Crocidura suaveolens)* 14. African giant shrew *(Praesorex goliath)* 15. Savi's pigmy shrew *(Suncus etruscus)* 16. Long-clawed shrew *(Feroculus feroculus)* 17. Pearson's long-clawed shrew *(Solisorex pearsoni)* 18. Mole shrew *(Surdisorex norae)* 19. Piebald shrew *(Diplomesodon pulchellum)* 20. Szechuan burrowing shrew *(Anourosorex squamipes)* 21. Himalayan watershrew *(Chimmarogale platycephala)* 22. Szechuan water shrew *(Nectogale elegans)*

Armored shrews

23. Armored shrew (*Scutisorex species:* see p. 233)

1
2
3
4
5
6
7
8
9
10
11
12
13
14
15
16
17
18
19
20
21
22
23

1
2
3
4
5
6
7
8
9
10
11
12

Shrew moles

1. Shrew mole *(Uropsilus soricipes)*

Desmans

2. Russian desman *(Desmana moschata)* 3. Pyrenean desman *(Galemys pyrenaica)*

Old World moles

4. Eastern mole *(Talpa micrura)* 5. Mediterranean mole *(Talpa caeca)*

American-Asiatic moles

6. Long-tailed mole *(Scaptonyx fuscicaudus)* 7. Japanese shrew mole *(Urotrichus talpoides)* 8. American shrew mole *(Neurotrichus gibbsi)* 9. Hairy-tailed mole *(Parascalops breweri)* 10. Townsend's mole *(Scapanus townsendi)* 11. Eastern American mole *(Scalopus aquaticus)*

Star-nosed moles

12. Star-nosed mole (*Condylura cristata*; see p. 244)

The sexes cannot be differentiated externally. There are 40 teeth: $\frac{3 \cdot 1 \cdot 3 \cdot 3}{3 \cdot 1 \cdot 3 \cdot 3}$. The incisors, canines, and first upper premolars are widely spaced. The teeth are hook-like and point backwards and serve to hold the prey. The remaining premolars and molars form a closed row. The teeth are laterally flattened, and multicuspate. The sharp edges aid in the mastication of soft prey animals.

The BLACKHEADED STREAKED TENREC *(Hemicentetes nigriceps)* is a close relative of the streaked tenrec. The head is black and the nape is yellowish-white. There is a short yellowish-white band between the eyes. The ventral side is also whitish. The spines are shorter and spaced more widely. The fur is softer and denser.

The streaked tenrec is widely distributed in Madagascar. It prefers moist regions with a ground cover in brushland and forest edges. It also occurs on the east coast, southwestern coast, and the rain forests of central Madagascar up to 950 meters elevation. The animal is nocturnal but occasionally single animals, which are highly active, have been observed in the early morning hours. The animals dig little burrows about 15 centimeters underneath the surface into the rich, dark soil of the manioc plantations. The burrows have solid, smooth walls. In grassland or thicket they dig tunnels which are only a few centimeters underneath the surface and are about 24–50 centimeters long. They shelter here during the daytime. The "ambikos," which is the native name for streaked tenrecs, search for food during the night. The animal sniffs around in the vegetation and investigates the ground with its long nose. If it finds something edible the ambiko will dig it out with its front paws, grasp the prey with its mouth and pull. The main part of the diet consists of earthworms. It probably also eats slugs, insects, and other ground-dwelling soft invertebrates which it dismembers with the sharp edges of its molars. The animals seem to be active throughout the year; however, they are somewhat less active during the cool dry season than in the Madagascar "summer." The animals, which were dug out on a relatively cold September night, were cold to the touch and were lethargic, although they were not in a state of torpidity. Traces of food were found inside the intestines, and this indicates that they had not been inactive for too long.

The raising of young in captivity

In May, 1961, H. Francke of the Berlin Zoo received a mature female streaked tenrec from the rain forest of Perinet. Presumably, this was the first streaked tenrec that had reached Europe alive. This female bore two young on July 18. The gestation period must have been at least fifty days. The newly born tenrecs were 2½ cm long, short-haired, and had the same markings as the mature tenrecs. The young were able to run quite well. On the sixteenth day one could see the tips of the spines. Both young were very agile and accompanied their mother on her foraging trips, and just like her, dug in the ground with their noses and front paws. When three weeks old, they measured 9 cm, and were able to feed independently. When the young were five weeks old I took over their care and the observation of their behavior. At first I fed them with live earthworms because up to that point the animals had refused all other food. Gradually I got them used to raw beef and poultry hearts, kidneys and chicken stomachs; however, I always had to mix in some pieces of earthworm to get them to accept this substitute. Only rarely did they eat the larvae of the mealworms. The hard chitinous cover was too difficult for them to chew. When I put the food into their cage in the evening the animals usually ate the greater portion of it right away. The remainder was eaten in smaller amounts during the night. Each animal consumed about 40–60 grams daily. This is about one third of their total body weight.

How a streaked tenrec catches earthworms

When the material covering the floor of the cage contained living earthworms, the streaked tenrec would suddenly interrupt its foraging, sniff briefly at a specific spot and push its long nose into the ground. By jerking its head up and stretching its front legs the animal would pull the earthworm to the surface. Usually the tenrec would grab the worm somewhere at the front end, briefly jerking up the head and shaking the worm. Then the animal would back up a few steps and hit at the worm with stretched out arms. Frequently it would step on the earthworm, and pull off a piece, gulping it down quickly. Then it would proceed to the next section of worm, tear it off and gulp it down also. In this way even large earthworms disappeared in a few minutes. This sequence of innate behavior patterns in the eating of large, live worms is very appropriate. These same behavior patterns are also carried out when the animal consumes pieces of meat. Frequently the tenrecs would regurgitate the contents of the meal right after they had eaten. However, they would then pick out the pieces of meat or earthworm segments from the contents and again swallow them.

Drinking habits, urination, and defecation

Several times during the day the streaked tenrecs would drink water or diluted canned milk. While drinking they would lift the tip of the nose upwards a little bit, and lap the fluid with their long tongues. If the animal had to defecate or urinate, it would almost always leave the sleeping box. Usually it would hurry into a corner of the cage. Often the same corner was used several days in succession. The tenrec would relieve itself into a little three centimeter deep hole which it had scratched with its front paws; however, the animal did not cover up the hole after it had finished.

Transport of nesting material

My streaked tenrecs would carry wood shavings, paper strips, or dry leaves in their mouths, and stuff this material into the sleeping box and all the entrances. A family of tenrecs would usually occupy one sleeping box, even though other boxes were available. The animals were very peaceful among themselves. Sometimes one would try to snatch an earthworm from another one, but aside from this there were no aggressive interactions. Generally they would share one corner of the resting box, all sleeping with their heads stretched between their front legs. Sometimes their sleep was deep, and they did not awaken when one opened their box. If one blew at them or touched them they would stand up immediately and erect the crest of spines on the nape. If one tries to grab a streaked tenrec it will jerk up its head and the front part of its body and thrust the erected spines into the hand. The very thin and fine spines of the back detach easily and stick to the human skin. Carelessly touching the animal regularly results in a finger covered with spines. This occurrence makes it a little painful to handle these otherwise very delightful animals. However, I must say that my charges never bit me.

How they use the spines on the nape

When my streaked tenrecs were awake they would usually run around stilt-leggedly and with stretched-out snouts, often for long periods at a time. When they were searching for food they walked slowly. Sometimes they tried, rather clumsily, to climb or jump. Occasionally they ran through the water bowl and got their bellies wet, but apparently this did not disturb them. In deep water they swam rapidly with strong rowing strokes; however, they always headed eagerly back to land.

The female died of pneumonia after five months. The other young died at two months of age. The remaining male was sexually mature at barely three months of age. He survived for one year and four months, which probably is the normal life expectancy for such a small insectivore with an extremely high rate of metabolism.

Three species of Madagascan hedgehogs

Formerly the genera *Setifer* and *Echinops* with one species each were classified in one genus, *Ericulus.* In the German language the animals are very appropriately termed "hedgehog tenrecs"; in English they are known as Madagascar hedgehogs because they resemble and behave like the spiny hedgehogs. The entire back is covered with spines. The tail is a flattened, short, broad spine-covered conical structure. The head, ventral surface, and legs are covered by fur. There are five powerfully clawed toes and fingers. There are five pairs of mammae. The sexes cannot be differentiated externally. The last premolars and the molars have specialized chewing surfaces for crushing hard insects. There are three genera, each with one species: 1. The LARGE MADAGASCAR HEDGEHOG (*Setifer setosus:* Color plate p. 177) has an HRL of 15–18.5 cm and a TL of 1.5–1.6 cm. The body weight is 180–200 g. There are 36 teeth: $\frac{2 \cdot 1 \cdot 3 \cdot 3}{2 \cdot 1 \cdot 3 \cdot 3}$. The spines are short and thin and more densely spaced than in the true hedgehogs. There are numerous widely-spaced long vibrissae growing out of wart-like elevations between the region of the nose and the eyes. The animal is found in the dry forests, brush country, and cultivated land in Northwest and South Madagascar and also on the east coast. 2. The LESSER HEDGEHOG TENREC or SMALL MADAGASCAR HEDGEHOG (*Echinops telfairi;* Color plate p. 177) has an HRL of 12–18 cm.

The body weight is 170–200 g. There are 32 teeth: $\frac{2 \cdot 1 \cdot 3 \cdot 2}{2 \cdot 1 \cdot 3 \cdot 2}$. The spines are longer and less densely spaced. The vibrissae are shorter and the tail is a little longer. The ear pinnae and eyes are larger and the claws are shorter than in the large Madagascan hedgehog. The animal is found in semi-deserts, thorny forests, and dry areas in the forests of Southwest Madagascar and on the west coast. 3. FONTOYNONT'S HEDGEHOG TENREC *(Dasogale fontoynonti)* has teeth and a skull that are a little differently shaped. Only one specimen is known.

Very little is known about the behavior of the Madagascar hedgehogs in their natural environment. They are nocturnal, and sleep during the daytime all curled up into a ball. The lesser hedgehog tenrecs usually shelter in small groups underneath boulders, tree roots, or tree holes close to the ground. These animals, like the tailless tenrecs, are believed to "hibernate" during the cool season. Like the hedgehogs, when disturbed they roll themselves into a protective ball of spines. Their diet consists of insects and other invertebrates. The natives of Madagascar do not like to eat the lesser hedgehog tenrec as much as the tailless tenrec.

Madagascar hedgehogs in captivity

In the summer of 1960 the first living lesser hedgehog tenrecs reached some of the zoos in Europe, and some of them came under my care. In 1961 I also received a large Madagascar hedgehog. In the evenings I would give my animals a mixture of finely chopped beef heart, chicken stomachs and hearts, and some living and freshly killed insects. The young ones were less fastidious than the older animals. The former would also eat minced meat, very young mice and commercial cat food. The youngsters consumed large amounts of bananas and boiled rice while the adults almost always refused vegetative food. During one night every adult lesser hedgehog tenrec would consume approximately twenty grams of food. Once an animal consumed ninety-four mealworm larvae in succession, while another ate ten large locusts in ten minutes. In contrast to the streaked tenrecs the small Madagascar hedgehogs chew their food well. A lesser hedgehog tenrec will stalk a living locust with a stretched-out, sniffing nose to within a few centimeters and then grab it with a quick thrust of the mouth. While making grinding motions the animal adjusts the prey with its hands until the front end is located between its teeth, and then chews it down. My lesser hedgehog tenrecs drank water or milk very rarely and then only in small amounts while the large Madagascar hedgehog drank a lot of water at quite frequent intervals. The lesser Madagascar hedgehogs inhabit very dry regions where no rainfall is recorded for months; therefore they are adapted to require less fluid than the large Madagascar hedgehogs which come from the more humid areas.

During their active periods the Madagascar hedgehogs would often walk about stiff-leggedly, and sometimes stand briefly with one raised front leg. In contrast to the spiny hedgehogs and the streaked tenrecs the small Madagascar hedgehogs are agile climbers (Color plate p. 168). They will hook into the bark of vertical tree trunks and prop themselves up with their powerful, wedge-shaped tails. They even manage to scale successfully the wire roof of their cages. Often the animals rest in a forked branch. Frequently they sit up and leisurely groom their faces with the spread-out fingers of both hands.

Olfactory and auditory senses

The olfactory and auditory senses are of great importance to the Madagascar hedgehogs. When one places a bowl of food inside a cage the animals respond by immediately sniffing the air with raised heads, and then proceed to the bowl in a straight line, then sniffing at each chunk of food individually. When they hear a sudden noise they partially curl up, even if the noises are of a low intensity. The animals locate and pursue jumping grasshoppers by the sound they make as they bounce off the ground; however, they also follow jumping grasshoppers or large crawling insects with their eyes. The small eyes are therefore not completely without function despite their generally poor vision. To a light touch the Madagascar hedgehogs respond by switching or erecting the spines at the stimulated spot. If one touches them harder, the animal erects the spines on the forehead and thrusts these like a spiny hedgehog, at the same time emitting threatening and defensive sounds which are reminiscent of the blowing snorts and "puffing" of the spiny hedgehogs. The blowing and snorting sounds can also be heard when two conspecifics meet. A subordinate animal which is attacked by a more dominant one sometimes shrieks and squeaks loudly. This probably is indicative of "fear" and defense. At first my lesser hedgehog tenrecs were shy and defensive with humans. When I picked them up they would snort, and grind their teeth in a threatening manner and then attempt to bite. The bite of their sharp teeth is very painful, and if one moves the hand they bite even harder. However, they soon became friendly and could be fed by hand. They would come to the cage door as soon as it was opened, and would take tidbits out of my hand.

The two males avoided each other and slept in separate boxes. The dominant male chased the other from the food or from the sleeping burrow. Madagascar hedgehogs of both sexes get along well and sleep in one burrow. Prior to giving birth or while raising young, the female is unsociable and does not tolerate the male in her nest. Long after the young have become independent they may still sleep in the same nesting box. Until recently nothing was known about the reproduction of these animals. However, recently the reproductive process was observed in the Zurich Zoo and in America. The male of the lesser hedgehog tenrec is stimulated by the odor of the female. His eyes become covered by a milky-white, viscous substance secreted by the Harderian gland. Occasionally this secretion covers both eyes

Copulation of Madagascan hedgehogs

completely and trickles down the cheeks and runs out of the nostrils; the animal then tries to lick it off. The precopulatory displays may go on for several hours. The male pokes the female with his nose or attempts to mount. She turns away from him and retreats into a corner, staying there with pulled-in head and tail. The approaches of the male are rebuffed by erection of spines, snorting, and biting. Finally the male grabs her by the nape and pulls her out of the corner, quickly scratching her back with his front and hind legs. She squeaks with an opened mouth and finally stretches her hind legs backwards and lifts her tail so as to permit the male to penetrate.

Birth and raising of young

The length of the gestation period is not known. Shortly before giving birth the female sometimes gathers nesting material or cleans out the nesting burrow. Two to ten young are born, usually during the evening hours or at night. The mother licks the newborn young dry, and eats the placenta. At first she rarely leaves the nest, and when she does she always covers the offspring with nesting material. She lies on her side while the young suckle. She eats the feces of the offspring. At first the young are naked but by the fifth day they are already covered with a gleaming blackish spiny pelage. On approximately the tenth day they open their eyes and periodically leave the nest. On the thirty-third day they are independent and weaned. A male of eight and one half months was observed with his first milky eye secretion, and a female of barely fourteen months bore her first litter. At five to six years of age, Madagascar hedgehogs are still able to reproduce. In the Frankfurt Zoo Madagascar hedgehogs have been raised for the past six years. Of four animals that had been acquired and their young, twenty-four young were raised during this time period (usually two to five young per litter).

Up to two months of age the male and female juveniles defecated and urinated into the same corner of the cage. They mark all unusual smelling spots by either urinating or defecating on them and then rubbing their abdomen on the ground. Next they stand up a little and rub their front paws in the odorous substance and in their own saliva which they then spread over the surface of their body. This impregnation with saliva, but without urination and defecation, was also observed by Eibl-Eibesfeldt: "It looked as if it wanted to impregnate my hand with its saliva and body odors, and true enough, my hand soon smelled like a Madagascar hedgehog. It is quite possible that this is odor marking, where the saliva dissolves certain odorous substances on the body of the animal. It is probable that the animals mark specific localities in their territory with these odorous substances, and also that "odor duels" are carried out in these disputed areas where one animal tries to mark over another's odor by adding its own secretion over it." This is similar to the situation in the Eurasian hedgehogs which perform this "self-anointing" behavior with saliva (see p. 213). This might be indicative of certain ancient phylogenetic relationships.

Subfamily rice tenrecs: *Oryzorictinae*

The RICE TENRECS are closely related to the spiny hedgehogs. The rice tenrecs are classified with the shrews and because of certain specialized features in the skeleton justify a separate subfamily *(Oryzorictinae)*. The HRL is 4–13 cm and the TL is 3–16 cm. The body is covered by fur. There are thirty-four to forty teeth. The diet consists mainly of invertebrates and small vertebrates, and a certain amount of vegetative matter. They inhabit the warm, moist forests, rice paddies, meadows, swamps, and banks. There are five genera and a total of twenty-four species.

Rice tenrecs and shrew-like tenrecs

In the rice tenrecs (genus *Oryzorictes*) the head is mole-like. The eyes and ears are small. The front legs are short and have powerful digging claws. The diet consists of insects and other invertebrates. Most of their activities take place underground in marshy areas. There are two species: 1. The MOLE-LIKE RICE TENREC or VOALAVONAROBO (*Oryzorictes talpoides:* Color plate p. 178) has an HRL of 8.5–13 cm and a TL of 3–5 cm. The fur is velvety. 2. The FOUR-FINGER RICE TENREC or VOALAVORANO *(Oryzorictes tetradactylus)* has a coarser hair texture. The first finger is absent. It inhabits the high plateaus.

The SHREW-LIKE TENRECS (genus *Microgale*) have an HRL of 4–12.5 cm and a TL of 3.5–19 cm. The digging claws are absent. The ears project well above the short, soft fur. There are approximately nineteen species, which include the LONG-TAILED TENREC *(Microgale longicauda;* Color plate p. 178) with an HRL of 6–7 cm and a TL of 15–16 cm. The tail is particularly long and has twenty-seven tail vertebrae.

At times both species of the genus *Oryzorictes* have been the cause of great damage done to rice paddies because they uproot the young plants by their burrowing activities, or destroy the water-retaining walls. Although the animals are widely distributed in the marshy areas of Madagascar very little is known about their life history because, like golden moles and common moles, they lead a subterranean life. Even less is known about the appearance and the habits of the quite shrew-like tenrecs *(Microgale)*.

Web-footed tenrecs and ground tenrecs

The WEB-FOOTED TENREC (*Limnogalemergulus;* Color plate p. 178) is the only aquatic species of tenrec. It is the largest rice tenrec (the HRL is 12.8 cm and the TL is 13.4 cm). It inhabits the banks of streams, lakes, and cold mountain creeks as well as the edges of swamps. The five fingers on the front feet are fringed with silver-white hairs, and the toes are connected by webbed skins. The laterally compressed tail is probably the main organ of propulsion in the water. The GROUND TENREC (*Geogale aurita;* Color plate p. 178) is another tenrec species that is even more closely related to the otter shrews than the other rice tenrecs. The *Geogale* is one of the rarest and least known of all Madagascan mammals.

Family otter shrews: *Potamogalidae*
Distinguishing characteristics

OTTER SHREWS (family *Potamogalidae*) are of the otter type. The HRL is 14.7–35 cm and the TL is 11–29 cm. The head is elongated and flattened frontally. The head has a blunt appearance like that of a fish otter. The fur is short and soft and has a silvery gloss. The underfur is very dense. There is one pair of mammae in the groin region. The testes lie within a scrotum. The second and third toes are fused except for the claws. The collar bone is absent. There are 40 teeth: $\frac{3 \cdot 1 \cdot 3 \cdot 3}{3 \cdot 1 \cdot 3 \cdot 3}$. The upper first and second lower incisors are long, pointed, and canine-like. The canines resemble premolars. The main part of the diet consists of crayfish and fish. The animal is found in western and central equatorial Africa. There are two genera with a total of three species.

The GIANT AFRICAN OTTER SHREW (*Potamogale velox;* Color plate p. 178) has an HRL of 29–35 cm and a TL of 24.5–29 cm. The ears and eyes are small. The flattened muzzle has stiff white whiskers. The nostrils are covered by horny flaps that act as valves when the animal is submerged. The hands and feet are short and without webbed skins. The tail is laterally compressed.

The DWARF AFRICAN OTTER SHREW (genus *Micropotamogale*) is only half as large as the giant species. The tail is only slightly compressed laterally. The nose shield is fleshy rather than horn-like as in *Potamogale.* There are two species: 1. The SMALL AFRICAN OTTER SHREW (*Micropotamogale lamottei;* Color plate p. 178) has an HRL of 14.7–15.1 cm and a TL of 10.9–11.1 cm. The feet are not webbed. It is found in West Africa. 2. The RUWENZORI OTTER SHREW (*Micropotamogale ruwenzorii*) is a little larger. The hands and feet are also larger and they are webbed. The tail is longer and stronger. The animal is found in the Ruwenzori Mountains in the Congo.

Giant African otter shrew

The giant African otter shrews look almost like small otters, and like them they are very agile swimmers. They feed on crabs, amphibians, and fish. However, these two groups of animals are not related. The otter shrews are insectivores that have adapted to an aquatic mode of living. The skull and other characteristics are similar to those of the *Tenrecidae.* The otter shrews seem to be most closely related to the *Geogale* but because of their numerous, unique characteristics they have been classified as a distinct family *(Potamogalidae),* separate from the tenrecs. The otter shrew's long head is relatively broad in the front and the muzzle is flattened and blunt. This is more reminiscent of an otter than a tenrec or even a shrew. The long, supple body and the muscular, laterally compressed tail are otter-like adaptations to living in water. This similarity to the otter is most noticeable in the giant African otter shrew *(Potamogale velox),* which is also known as the water shrew or by the native name of Jes. The animal is one of the largest insectivores. The species has been found in the Cameroons, Gabon, Angola, and the Congo. The animals may be found near sluggish, muddy lowland streams or cold, clear mountain creeks, from sea level to about 1800 meters. In some areas otter shrews occupy small forest pools during the rainy season and migrate overland to rivers at the beginning of the dry season. During the day the animals, either alone or in pairs, shelter in holes and tunnels in the banks. The entrances to these burrows are below the water level. They become active in the late afternoon. Otter shrews are extremely agile and rapid swimmers. On land their movements are rather clumsy, but they can move at a considerable speed.

Gerald M. Durrell describes one incident from Cameroon where an African pulled a mature otter shrew out of its burrow by the tail: "In the meantime the otter shrew was rather fed up with dangling by its own tail. One last hiss came from between its bristly beard hairs, and then the animal turned around. It climbed up on itself, very gracefully and lithely, and then sank its teeth into Andraias' thumb." Only with great difficulty could they remove "the little biting devil" from the bleeding and mangled thumb. The animal had bitten right through to the bone. In captivity the otter shrew refused to eat a fish, a frog, and a small water snake. "Then I placed a big crab in his cage. At once he approached, sniffed, and before the crab had time to open its claws it was already flat on its back and was cut in half with a single sharp bite." After this the otter shrew consumed twenty to twenty-five crabs daily. Durrell prepared a mixture of dried freshwater shrimps, raw eggs and scraped meat and put this into an empty crab carapace. This diet was to be fed to the otter shrew on its long journey to England. The otter shrew accepted the substitute diet, and quickly got used to the shrimp pulp which it later even ate out of a bowl. Although it appeared that the animal was doing well on this diet, it was nevertheless found dead one morning. Durrell had the same misfortune with a juvenile female otter shrew. Up to now attempts to keep these interesting insectivores in captivity for longer periods of time have been unsuccessful. It has even been difficult to bring the animals to Europe alive.

The dwarf otter shrew *(Micropotamogale lamottei)* has been found only rarely in the West African states of Guinea, Liberia, and the Ivory Coast. They are nocturnal and aquatic. They seem to prefer swampy areas with a dense growth of vegetation, small pools of water, and shallow creeks. Some dwarf otter shrews have been caught in wire baskets and fish nets where they had killed all the fish. However, the major part of their diet consists of crabs. One specimen was examined whose rectum was filled with the undigested heads of large ants.

Family golden moles: *Chrysochloridae*

The GOLDEN MOLE (Family *Chrysochloridae*) is of the mole type. The HRL is 7.6–23.5 cm. The tail is reduced and is not visible externally. The body is cylindrical. The head is pointed. The vestigial eyes are covered by fur. The ear pinnae are absent. The naked muzzle terminates in a broad horny pad, in which the nostrils terminate. The forelimbs are short. There are four fingers which are heavily clawed (for digging). The tendon between the wrist and the elbow is ossified. The hind limbs are short and weak. The five toes are connected by membranous skin. The ribs and the sternum are bent inwards, and the deep hollow provides space for the thick muscular arms. The sexes cannot be differentiated externally. Females have one pair of abdominal and one pair of inguinal mammae. The skin is loosely attached to the body. The hair is thick and short, with dense woolly underfur. Usually the pelage has a metallic luster. There are 40 high-crowned teeth: $\frac{3 \cdot 1 \cdot 3 \cdot 3}{3 \cdot 1 \cdot 3 \cdot 3}$ (except *Amblysomus,* see p. 201). The second and third incisors and the first premolar are canine-like. The diet consists of invertebrates and lizards. The animals are burrowers, and are found in equatorial and South Africa. There are five genera and a total of fifteen species.

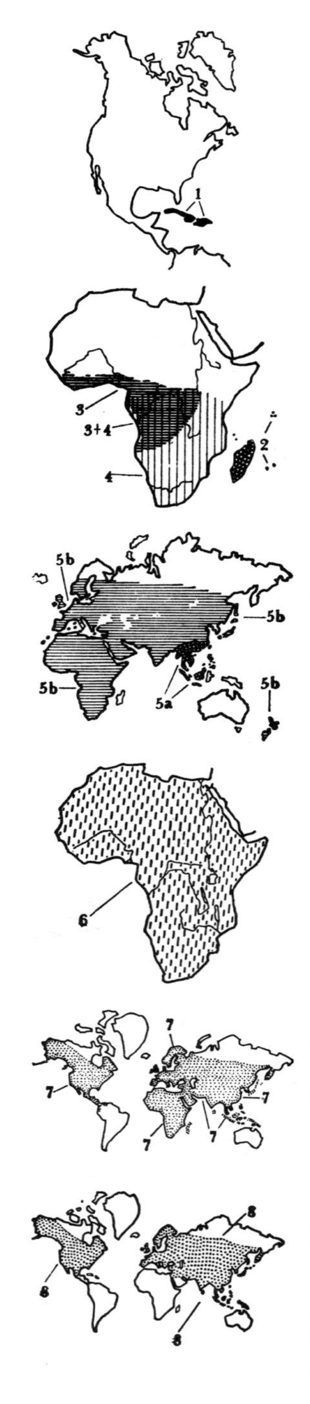

The golden moles *(Chrysochloridae)* or "kruipmollen" (Dutch) are a distinct insectivore family because of their body structure and life habits. They are very old. Miocene records are from Kenya and from the Pleistocene in South Africa. The fossil remains are very similar to the golden moles living today. The dentition and the formation of the tympanic membrane (Bulla tympanica) indicate a phylogenetic link with the tenrecs. On the other hand, the conspicuous similarity between the golden moles and the true moles *(Talpidae)* is not based on a phylogenetic relationship but rather on similar adaptations to a similar way of life (convergence). As ground burrowing animals the golden moles, just like most true moles, have cylindrically shaped bodies that are pointed towards the front. The eyes are vestigial; the ears are concealed and the fur is short-haired and glossy. However, the forelimbs which have been adapted as powerful burrowing tools are very differently constructed in the two groups. The true moles have a broad five-fingered hand, and the golden moles have a relatively narrow hand and only four fingers. The hand is supported by an ossified tendon which is a feature not found in any other mammal. This ossified tendon joins the wrist and the elbow. Golden moles do not burrow with their hands but with the well developed claws of a few fingers. The third finger usually is longest and most powerful and is equipped with the largest claw. In the various species the other three fingers are developed to varying degrees. Other adaptations to living underground include the horny plate at the top of the nose, the absence of the ear pinnae, and the vestigial eyes.

The genus *Chrysochloris* inhabits South Africa *(Chrysochloris asiatica),* East Africa *(Chrysochloris stuhlmanni),* and Central Africa (*Chrysochloris congicus* and *Chrysochloris vermiculus*). The HRL is 9–14 cm. The fur is dense and fine. It is either brown or grey. The fur has an iridescent luster of greenish, violet, or purplish depending on the angle at which the sun hits the animal. The third foreclaw is strongly developed. The first and second foreclaws are shorter. The fourth digit is stump-like.

The Cape golden mole

The best known species, the CAPE GOLDEN MOLE *(Chrysochloris asiatica),* is found in South Africa and particularly in the Cape Province. Here

◁ Fig. 11-7.
1. Solenodons (Family *Solenodontidae*)
2. Tenrecs (Family *Tenrecidae*)
3. Otter shrews (Family *Potamogalidae*)
4. Golden moles (Family *Chrysochloridae*)
5. Hedgehogs (Family *Erinaceidae*) a. Hairy hedgehogs (Subfamily *Echinosoricinae*) b. Spiny hedgehogs (Subfamily *Erinaceinae*)
6. Elephant shrews (Family *Macroscelididae*)
7. Shrews (Family *Soricidae*)
8. Moles (Family *Talpidae*)

it is widely distributed and is very common in certain localities. The Cape golden mole burrows vigorously with its armored nose and scrapes with its powerful foreclaws just below the surface of the ground so that its tunnels can be traced by the broken ground above the surface. Occasionally, however, the animals also burrow at greater depth and they remove soil from the burrow which they push to the surface to form mounds of fresh soil. When it is raining the Cape golden mole sometimes leaves its tunnels and then hunts for insects and small invertebrates on the surface, or roots for earthworms in the damp earth. On occasion the animals surface at night for short excursions. Although the Cape golden mole is very useful because it consumes so many ground insects, it is also considered a pest because of its burrowing activities. The breeding season takes place during the rainy months from April to June. The Cape golden mole makes a round nest of grass in which usually two young are born. The young suckle for two to three months, since the teeth do not cut through the gums until the young are almost fully grown.

Other species of golden moles

The LARGE GOLDEN MOLE (*Chrysospalax trevelyani*; Color plate p. 178) has an HRL up to 23.5 cm. The animal is found in eastern Cape Province. The AFRICAN GOLDEN MOLE (genus *Amblysomus*; Color plate p. 178) has an HRL of 8.5–13 cm. There are only 36 teeth: $\frac{3 \cdot 1 \cdot 3 \cdot 2}{3 \cdot 1 \cdot 3 \cdot 2}$. There are seven species including the HOTTENTOT GOLDEN MOLE (*Amblysomus hottentotus*) from Portuguese East Africa. GRANT'S DESERT GOLDEN MOLE (*Eremitalpa granti*) is the smallest of the golden moles. The HRL is 8 cm and the body weight is 15 g. The temporal bullae are absent. The front feet have three long, leaf-like fore claws. DE WINTON'S GOLDEN MOLE (*Cryptochloris wintoni*) looks very similar to Grant's desert golden mole, but it does have the temporal bullae.

Grant's desert golden mole and De Winton's golden mole inhabit the soft loose sand dunes on the Southwestern coast of Africa. *Eremitalpa* makes shallow, winding tunnels in the sand which it presumably only uses once. Occasionally these surface tunnels connect with deeper excavations which descend several meters below the ground. During the heat of the day the animals dig tunnels up to thirty centimeters deep. Their favorite prey, various sand-burrowing species of legless skinks, also retreat from the heat into deeper ground, where they are easily caught by the Grant's desert golden moles and probably also the De Winton's golden moles. The animals use their long foreclaws to hold the long, moving lizards while they are being killed and consumed. The two species also feed on insects and other small animals.

Family: *Erinaceidae*

The family *Erinaceidae* not only includes the well-known Eurasian hedgehogs and their closest relatives, but also a number of spineless, soft-furred insectivores of the rat type (see p. 176) which are known as the hairy hedgehogs. All members of the *Erinaceidae* walk on the soles of their feet. The head is elongated and blunt. The eyes and ears are well developed. The urogenital opening in the females is well separated from the anus. The penis is directed anteriorly. Hedgehogs prefer to eat animal matter, although occasionally they also feed on vegetation. They are found in Europe, Asia, and Africa. There are two subfamilies and nine genera.

Subfamily hairy hedgehogs: *Echinosoricinae*

Distinguishing characteristics

HAIRY HEDGEHOGS *(Echinosoricinae)* are rat-like in appearance. They have a dense pelage. The long tail is naked or sparsely covered by hair. There are five clawed toes and fingers on each limb. There are 44 teeth: $\frac{3 \cdot 1 \cdot 4 \cdot 3}{3 \cdot 1 \cdot 4 \cdot 3}$. Their diet consists predominantly of invertebrates and small vertebrates. They inhabit the forests in Southeast Asia (see map, p. 200). There are four genera, each with one species.

The MOON RAT (*Echinosorex gymnurus;* Color plate p. 179) is one of the largest insectivores. The HRL is up to 44.5 cm, and the TL is up to 21 cm. The body weight may reach 1.4 kg. The underside of the long mobile nose has a groove which runs from the tip to a point between the upper incisor teeth. The scantily haired tail has scales. The terminal third of the tail is compressed laterally. The pelage is long and rough, with a thick underfur. The black and white coloration varies greatly. There is one pair of mammae in the chest region and one pair in the groin region. Two glands located near the anus emit odors that are described as resembling garlic, sweat, or rotten onions. The species is found in Burma, Thailand, the Malay States, Sumatra, and Borneo. The LESSER GYMNURE (*Hylomys suillus;* Color plate p. 179) has an HRL of 10.5–14.6 cm and a TL of 1.2–3 cm. The species is found in the same areas as the previously-mentioned species, but is also distributed in South China and Java. The MINDANAO GYMNURE (◊*Podogymnurus truei*), with an HRL of approximately 14 cm and a TL of 5–7 cm, has a pelage which is relatively long haired. It is found on Mindanao, in the Philipine Islands, at an elevation of 1600–2300 meters. The SHREW HEDGEHOG (*Neotetracus sinensis;* Color plate p. 179) has an HRL of approximately 13 cm and a TL of approximately 7 cm. The animal inhabits the forests in elevations of 2100–2800 meters in South China, North Burma, and Indochina.

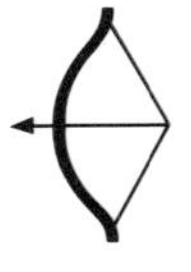

Hairy hedgehogs are very primitive animals

In their external appearance the hairy hedgehogs do not resemble the "real" hedgehogs very much; however, they are very close to the original *Erinaceidae.* Actually the hairy hedgehogs resemble even more the ancestral mammalian type than the short-tailed tenrecs which are partially covered by spines. In their internal structure and particularly the unique features in the dentition, the hairy hedgehogs greatly resemble the spiny hedgehogs. The moon rat primarily inhabits forests and is often found near streams and mangrove swamps. The animals are nocturnal, resting by day in hollow logs, under the roots of trees, or among rocks. At night the moon rat roams about and hunts for invertebrates, fish, and frogs. The smaller species probably feed mainly on invertebrates; however, the stomach content of a shrew hedgehog revealed vegetative remains. The lesser gymnure inhabits humid forests with thick undergrowth. *Hylomys* searches for food during the day and at night. The animal moves on the ground with short leaps, and follows definite pathways. The lesser gymnure has also been observed climbing in the branches of bushes and coffee trees.

Subfamily true or spiny hedgehogs: *Erinaceinae*

All other hedgehogs are covered dorsally with a uniform coat of spines. All these true or spiny hedgehogs belong to the subfamily *Erinaceinae.* The HRL is 13.5–30 cm and the TL is 1–5 cm. The body weight is 400–1100g. The tail is thin, cylindrical, and naked or scantily haired. The sharp, needle-like spines are not grooved. The spines have alternate dark and light color rings. Specialized muscles in the back enable the hedgehog to roll itself into a ball and erect the spines. The animals feed mainly on insects and other small invertebrates, but they also consume small vertebrates, carrion, and occasionally some vegetation. There are 36 teeth: $\frac{3 \cdot 1 \cdot 3 \cdot 3}{2 \cdot 1 \cdot 2 \cdot 3}$. The first incisor is long and sharp. The canine is small. The third premolar and the molars have broad chewing surfaces with round cusps. The animals are widely distributed in Europe, Asia, and Africa (see map, p. 200). There are five genera.

Eurasian hedgehog

The EURASIAN HEDGEHOG (genus *Erinaceus*) possesses broad, roundish ears which barely project beyond the hair and spines. Some scientists include all Eurasian hedgehogs in one species *(Erinaceus europaeus)* which Linné described scientifically on the basis of one specimen found in Sweden. Here, however, we differentiate between the MANCHURIAN HEDGEHOG *(Erinaceus amurensis)*, two Chinese hedgehog species (*Erinaceus dealbatus* and *Erinaceus miodon*), and the KOREAN HEDGEHOG *(Erinaceus koreanus).* We also differentiate between two species and eight subspecies in the spiny hedgehogs found in Europe and the Near East: 1. The WESTERN EUROPEAN HEDGEHOG (*Erinaceus europaeus;* Color plate p. 179) has an HRL of 25–30 cm and a TL of 2.5 cm. The ventral side is brown or gray, and there is almost always a dark spot on the chest. There are four subspecies (see map, p. 204). 2. The EASTERN EUROPEAN HEDGEHOG (*Erinaceus roumanicus;* Color plate p. 179) is the same size as the western European species. There is a glossy white spot in the center of the chest which is in distinct contrast to the brownish-gray or brown background. The light coloration often extends to most of the ventral surface and the flanks. The size relationships in the upper jaw of the western and eastern species differ. There are four subspecies; *Erinaceus roumanicus rhodius* and *Erinaceus roumanicus nesiotes* are probably hybrid forms. In Berlin and Linz (Danube) hybrids of the two species occur.

Fig. 11-8.
Muscles, tendons, and claws on the forefoot of a golden mole.

Prior to the ice age (Pleistocene), the genus *Erinaceus* was already distributed throughout Western and Central Europe but the genus was forced to retreat to South European refuges with the coming of the ice.

Von Wettstein assumes that the spiny hedgehog from the Transcaucasian refuge again invaded the ice-free regions. According to this view the Transcaucasian form is the ancestor of all present European and Asian spiny hedgehogs. However, I think it more probable that there were two completely separated areas to which the spiny hedgehog retreated during the ice ages. The two refuges, one in the Southwest and one in the Southeast, gave rise to two different species of spiny hedgehogs. After the ice ages the western European hedgehog spread from the Southwest to Western and Northern Europe as well as central Russia. The eastern European hedgehog penetrated from the Southeast to Eastern central Europe, the Baltic Sea, and the Dvina.

Fig. 11-9.
□ Western European hedgehogs: 1. Western European hedgehog *(Erinaceus europaeus europaeus)*; 2. North Russian hedgehog *(Erinaceus europaeus centralrossicus)*; 3. Spanish hedgehog *(Erinaceus europaeus hispanicus)*; 4. Italian hedgehog *(Erinaceus europaeus italicus)*; □ Eastern European hedgehogs: 5. East European hedgehog *(Erinaceus roumanicus roumanicus)*; 6. Transcaucasian hedgehog *(Erinaceus roumanicus transcaucasicus)*; 7. Rhodo's hedgehog *(Erinaceus roumanicus rhodius)*; 8. Cretan hedgehog *(Erinaceus roumanicus nesiotes)*; □ Four-toed hedgehog: 9. Algerian hedgehog *(Aethechinus algirus)*; □ Long-eared hedgehog: 10. Long-eared hedgehog *(Hemiechinus auritus)*. Areas of distribution outside of Europe are mentioned in the text.

Occasionally man has displaced hedgehogs and thus has been responsible for blurring the distribution borders of the various species and subspecies. Spiny hedgehogs often shelter in grain stacks, heaps of straw or brushwood, and inside these materials they are easily transported over long distances. Vacationing people sometimes bring hedgehogs home from distant regions, and then release them in their home towns. Hedgehogs have also been introduced to areas where they originally were not found for the purpose of controlling vermin. Man was probably also responsible for introducing the hedgehog from the East Asian mainland to Japan. Between 1870 and 1890 the western European hedgehog was brought from England to the formerly almost mammal-free island of New Zealand. The numerous spiny hedgehogs now found in New Zealand have not changed noticeably in their morphology and behavior in the one hundred years that they have occupied this new habitat.

Farmers and gardeners frequently differentiate between two types of spiny hedgehogs: the "dog hedgehog" which has a blunt snout and high forehead, and the "pig hedgehog" which has a pointed snout and flattened forehead. This could be a mark of differentiation where the two species are found side by side. Yet, in regions where it is certain that only one species is found, one still hears references to the "dog" and the "pig" hedgehogs. A frightened hedgehog which erects the spines on the forehead frontally and pulls in the head slightly resembles a blunt-snouted "dog hedgehog." A spiny hedgehog that is sniffing with a projected nose, and has the spines lying flatly appears to be a "pig hedgehog." The spines, which are thickened at the base, sit loosely in the coarse skin and can be erected when the skin is tightened. Just below the skin on the back lies an oval, thick muscle ring (Musculus orbicularis) which encircles the dome-like tail-back muscle (Musculus caudodorsalis). Both structures form an arched hood which is connected to the spine-bearing dorsal skin, and is separated from the underlying muscles by a layer of fat. When the muscular hood contracts it covers the entire body and head like a sac. The limbs are also pulled in. This results in a tightening of the dorsal skin and an erection of the spines. The hedgehog, in this manner, turns itself into a spiny ball.

Certain species of hedgehogs which originated from warm, dry regions, occasionally shed their spines when kept in Central Europe. A long-eared desert hedgehog *(Hemiechinus auritus persicus)* kept in Stuttgart, Germany, lost one eighth of its spines within three weeks during the summer. The spines were replaced by new ones within a short time. It is not known if spine replacement takes place in nature or if this phenomenon was merely an artifact of captivity. It is possible to estimate the number of spines on a hedgehog's back by counting the spines in one square centimer and multiplying this by the total area of spines. The western European hedgehog has approximately 16,000 spines on his back!

What does a hedgehog hunt?

Hedgehogs prey predominately on insects, earthworms, wood lice, spiders, centipedes, and snails, but also on frogs, toads, lizards, and snakes. Sometimes they capture birds and small mammals up to the size of a young rabbit. Hedgehogs also have the reputation of being good mouse catchers. Often hedgehogs are pictured with a captured mouse in their mouths, and they are recommended for vermin control in stables, barns, and cellars. It is true that hedgehogs dig out mouse nests and eat the helpless young and also occasionally catch a weak mature mouse. Experiments have shown, however, that the hedgehog is normally too awkward to capture healthy, agile mice. The animals will also eat carrion, and most are partial to cooked, fried, smoked, or even pickled fish. Vegetative matter is not popular and is probably only eaten in times of need. Occasionally hedgehogs consume mushrooms, acorns, beechnuts, dropped fruit, or berries.

To what degree is the hedgehog resistant to poisons?

The hedgehog does not even avoid poisonous animals which are shunned by most mammals. He consumes substantial amounts of oil beetles *(Meloe)* and "Spanish flies" *(Lytta)* which belong to the blister beetles. These beetles contain the rubefacient poison cantharidin but this does not seem to affect the hedgehog. One tenth of a gram of this substance is lethal for twenty-five people, but only kills one hedgehog! The animal also eats bees, bumble bees, and wasps, and does not seem to be bothered by their stings. One hedgehog was stung by fifty-two bees, but did not show any discomfort. Generally speaking, the hedgehog is remarkably resistant to various toxins. The hedgehog can tolerate a dose of hydrocyanic acid that would kill five cats in a few minutes, or tetanus toxin in a concentration 7000 times as high as a human can tolerate.

However, the animal is not completely immune to poison. This is demonstrated when it hunts the viper *(Vipera berus).* When attacking this snake, it slowly approaches and then quickly dashes upon its prey with pulled up forehead and erected head spines, and then bites into the snake's body. The snake confronts its enemy with gaping jaws and the forward-pointing poison fangs. Since the hedgehog's spines are longer than the fangs, the poison does not penetrate the hedgehog's body. The snake repeatedly strikes against its foe until finally the hedgehog has bitten through the spinal column. The animal then consumes the entire snake, including the head and the poisonous glands. However, if by chance a hedgehog is bitten by this snake, it will become ill and die within a few hours or days. Nevertheless, a hedgehog is able to tolerate thirty-five to forty times the concentration of poison as an equally heavy guinea pig.

Feeding captive hedgehogs

Hedgehogs are easily kept in captivity on a diet of meat, fish, earthworms, and insects as well as commercial cat and dog food. The animals frequently drink large amounts of fluid. They not only take water but are also very fond of diluted milk. When the hedgehogs are extremely hungry they will eat soaked bread and other vegetative matter. Yet there are some hedgehogs that will eat fruits, nuts, sunflower seeds, hemp, and even custard and chocolate. A steady diet of vegetative matter does not agree with the hedgehog. As a result it will suffer vitamin deficiencies which may be the cause of rickets, muscular weakness, and lameness. It is recommended to supplement the diet regularly with a few drops of multivitamins.

The senses of sight, hearing, and smell

On occasion a hedgehog can consume unbelievable amounts of food. In ten days one ate 1880 grams of meal worm larvae. His body weight increased from 689 to 1155 grams. Another animal of 675 gram body weight consumed in one night 120 grams of bird meat and one sparrow weighing 24 grams. In addition it drank 85 grams of cow's milk. Extremely well fed hedgehogs may reach record weights of 1900 grams. Under proper care a hedgehog may grow to be approximately ten years old.

Although the hedgehog is nocturnal its eye sight seems to be much better than had been previously assumed. Hedgehogs quickly learn to open small, differently colored doors for a reward of food. In this test they were not only able to differentiate between degrees of brightness but also colors. The latter were not perfectly differentiated. One hedgehog preferred a blue door to a yellow one because it had learned that food was hidden behind the former. When the blue door was exchanged for other colors of various brightness, the animal never chose yellow, because this color had been associated with the absence of food.

The hedgehog has a very acute sense of hearing. Even very tame animals are startled by chirping or smacking sounds, and roll into a ball. The animals quickly learn to respond to specific whistles and calls. Outside one hedgehog came from a distance of twenty-five meters if one called its name, "Eri." Yet, like in all insectivores, the olfactory sense is of the greatest importance to the hedgehog. With the aid of its sense of smell the animal finds and tests its food and recognizes its environment, conspecifics, mate, young, and enemies. Free living hedgehogs sometimes are frightened away by human scent at ten paces. Tame hedgehogs can be trained to recognize certain odors, for example the shoes of the warden or the smell of fish. The animals will come forth when they detect the familiar scent. Hedgehogs like warmth. In a heated room one will soon find them resting near the source of heat. Outside they will occasionally bask in the sun. The learning capacity of the hedgehog is well-developed despite the relatively primitive brain. Its behavior is not solely governed by instinct but also through individual experiences. Hedgehogs are solitary, and tend to avoid conspecifics. If one puts several animals into one cage they will fight at first, trying to push the neck spines into each other, bite, and finally roll into a spiny ball. At last one becomes "top hedgehog" and tyranizes all the others. A dominance order is also established among the remaining animals, and it is clearly defined who can bite, box, or push whom. The dominant position is not always determined on the basis of body size or strength. Often a smaller, weaker animal becomes "top hedgehog" because he is the most active, the boldest, and the most aggressive.

When a hedgehog either smells or hears a human it usually walks to the next corner or rolls itself into a ball and hisses. After some time has elapsed, it slowly uncurls, but at the slightest disturbance it rolls itself up again. However, there are less shy hedgehogs which will unroll in one's hand and accept tidbits from one's fingers. Generally, captive hedgehogs do not bite their keeper. Some animals, on the other hand, always snap at the human hand, even if they are tame otherwise.

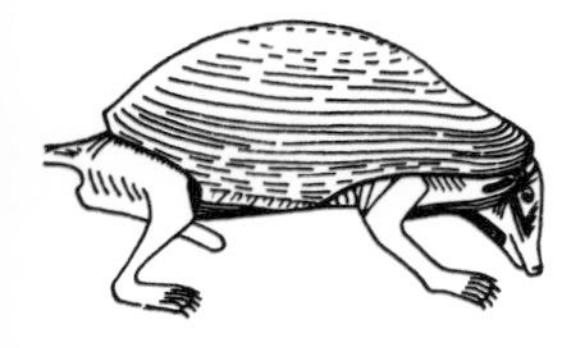

Fig. 11-10.
This is an illustration of a skinned hedgehog to show the thick muscular hood which lies underneath the spine-bearing dorsal skin.

The western European hedgehog is sexually mature at around nine to eleven months. The breeding season begins around April after the animals have left their winter quarters, and lasts to July or August. Initially the female runs from the male. He does not give up, however, and continues to sneak up from behind. Then she turns around quickly and boxes him away. This phase of the courtship continues for hours, until finally they circle each other in ever-increasing excitement. If another conspecific approaches the encircling pair, the male immediately rushes towards him and, snorting and boxing, chases the intruder until he flees. Then the male again circles the female. Finally copulation takes place but it may not occur until the next night. Since Aristotle's times it was assumed that hedgehogs copulate by standing up and facing each other, or that the female lies on her back so that the male would not injure himself on her spines. This assumption was finally refuted by Stieve in 1948, who carefully observed copulation. He observed that the estrous female presses herself to the ground. All her spines are flattened and the hind legs are stretched out backwards.

Fig. 11-11.
When the animal is curled up this hood encloses almost the entire animal.

The pelvic region is raised. The male mounts from behind, supporting his front paws on the female. Intromission lasts only a few seonds. However, copulation may be repeated within a short time. Following these events the male shares the female's shelter until shortly before the birth of the young when she chases him away.

The gestation period is five to six weeks. In Central Europe the offspring are born between May and September. Many females have two litters per year, one in May or June and the other in August or September. While giving birth the female usually lies on her side and pushes the young within the amnion and the placenta to the outside. Then she eats the afterbirth, and licks the baby. In the same manner two to ten more young may be born. The usual litter size is five to seven. As a rule the young are born with the head first, but breech births are not uncommon. When all the young are born the mother picks up each young individually and places it on her abdomen where it gets hold of a teat and starts to suckle. The newly born hedgehog is 5.5–9 cm long and weighs 12–25 grams. The eyes and ears are still closed. The underside of the newborn is pink; the back is grey and completely hairless. The skin is tight and contains a large amount of water. The skin forms a cushion from which the thin, first white spines protrude only about three millimeters. During the birth process the spines are pushed back into the skin and in this way do not injure the mother. Within twenty-four hours most of the water is lost from the skin and it looks wrinkled. At this point the white tips of the spines project about six millimeters. After thirty-six to sixty hours the tips of the light and dark banded second spine sequence follow.

During the third week the young open their eyes and ears. The hair and spines are already fully developed; however, the first spines drop out during the sixth week. The little ones cheep excitely as soon as they crowd around the teats of the mother. Around the end of the third week the young occasionally leave the nest. Later they follow the mother and begin to look for food. If one young loses contact, it emits twittering sounds which quickly bring the mother to the site. She briefly sniffs the young, and then continues on with her search for food. Once the youngsters have become independent the mother chases them away from the communal nest.

The habitat of the hedgehog

The Eurasian hedgehog is found in the forest, heath, plains, and cultivated land, either in the valleys, highlands, or mountains up to the crook-timber line at elevations of 2000 to 3000 meters. They occupy all areas where they can find sufficient food and suitable shelter. They avoid very moist regions or coniferous forests without underbrush because here they do not find dry shelter. Hedgehogs frequently inhabit fences grown over by vegetation or thick hedges, rock crevices, hollow tree trunks, piles of leaves or brushwood, rabbit burrows, areas underneath stables or barn attics. They line their nests with moss, leaves, or hay which they transport in their mouths. Occasionally they also construct their own burrow in the ground. The tunnel frequently has two openings. One is usually plugged depending on the wind direction. Very often hedgehogs are found in the vicinity of human dwellings, even in the center of cities. Animals living on farms, gardens, or parks are occasionally tame to such a degree that they will come to food bowls at specific times and do not object to being touched or being picked up.

The hunting territory of a hedgehog extends up to 200 to 300 meters from its hiding place. On its forays the animal seems to follow definite paths. A hedgehog is very faithful to one particular location, and it may occupy the same shelter for years. A female hedgehog marked with a ring on one foot was released in Berlin-Bush, Germany, in October 1934. A year later the same animal and her four offspring were found 2½ kilometers from the release site. Only rarely do large numbers of hedgehogs migrate long distances. There is one report from the Allgäu, north of the Alps in Germany, of a group of over one hundred hedgehogs that crossed a rural road in a forest.

The daily routine of a hedgehog

During the day the hedgehog sleeps in its hiding place. Rarely does one see a hedgehog outside its nest during the day. However, during the night they are not continuously active. Around 6 P.M. the hedgehog starts to roam about in its territory for two to three hours, and then again for another two hours shortly after midnight, and finally for an hour or so in the early morning hours. The remaining eighteen hours are spent sleeping. The daily activity rhythm of the hedgehog is not greatly influenced by the amount of light. In midsummer, when the sun does not set until 8 P.M. and rises again at 4 A.M., the hedgehog makes its rounds during daylight hours. Obviously its daily rhythm must be regulated by some "internal clock."

After waking up during the day the hedgehog slowly trots through its territory. The animal constantly sniffs the air, smells a piece of wood here, turns over a stone with its nose there, probably also using the front paws to overturn objects, or it may be poking around in moss and leaves, always in search of food. While active the hedgehog's nose is constantly dripping. The fluid keeps the olfactory membranes moist. When the animal detects prey its rate of sniffing increases, or it may snort audibly. It follows the mucous trails of snails or insects, or it burrows with its front claws in likely places in the ground. The hedgehog is attracted to very slight rustling, whistling, or chirping sounds indicative of a prey animal. The hedgehog very accurately locates these sound sources. Dead or sluggish prey is first extensively sniffed at, but quickly moving animals are immediately seized with the teeth. Living lizards, mice, or other larger prey are shaken until dead. Once the hedgehog has seized its prey it starts to eat it even though the victim may still defend itself by kicking. Larger prey is held to the ground with the front paws, and chunks are bitten or torn off. With loud smacking sounds the hedgehog chews and crushes bones, insect carapaces, and snail cases with the aid of the broad crowned molars.

During its diurnal sleep the hedgehog usually rests on its side so that only its head and legs are visible. Its muscles are completely relaxed. If the undercover of the nest is warm, the hedgehog stretches out flatly on the ground. In the sleeping position the spines are usually flattened and pointed backwards. In response to disturbances it twitches, erects the spines, or rolls into a ball. If one touches a sleeping hedgehog in the daytime it will awaken immediately, stretch its limbs, yawn, and scratch itself with the hind foot. With its hind foot it can reach almost all parts of its body.

How a hedgehog moves

The usual gait of the hedgehog is slow with slightly bent legs whereby the belly nearly touches the ground. If the animal moves stiff-leggedly, it is able to run fairly rapidly. The hedgehog is not able to jump nor to sit up on its hind legs and raise the front of its body as is often incorrectly shown on illustrations. For this reason the animal cannot groom its face with the front paws like the Madagascar hedgehogs. Hedgehogs are not good climbers but after repeated attempts they are able to scale the high walls of a box or even walls. Once on top the animal just lets itself drop off the other side all curled up in a ball. The stiff, bendable spines which point in all directions lessen the impact. The musculature of the body can be displaced to an unusual extent, allowing the hedgehog to press himself through very narrow holes and cracks. Hedgehogs avoid water, yet if forced are capable swimmers.

The vocalizations of a hedgehog

When excited the hedgehog's usual loud sniffing and snorting increases to a loud spitting or "puffing" that can be heard over great distances. In darkness it is possible to locate two fighting or courting hedgehogs by these sounds. Frequently one can hear a hedgehog either coughing or sneezing. When in rage or in fright the animals emit loud cackling sounds, and when in great distress they make piercing screams. On the whole a hedgehog is far more vocal than most other animals. It does not attempt to hide its presence. The spiny pelage must obviously provide enough protection to enable the animal to lead such a noisy, conspicuous way of life.

Fluctuations in body temperature

At normal outside temperatures the body temperatures of the hedgehog fluctuate about two degrees during the course of the day. During the second nocturnal rest period around 3 A.M. the temperature is highest, approximately 36.8°C, while during the long diurnal rest the temperature drops to 34.8°C at 3 P.M. The temperature then increases again while the animal is still asleep. Thus the temperature increase is not caused by increased activity. Just the opposite happens; the hedgehog becomes really active after its body temperature has already increased.

Hibernation in the hedgehog

Among the insectivores, the spiny hedgehogs are the only true hibernating species. Tenrecs (see p. 186 ff.) enter into a state of torpidity which shows certain similarities to hibernation, but the internal processes in the former differ from those of animals in true hibernation. The readiness to hibernate in the colder season in the hedgehog is not only dependent on the average outside temperature. The actual causes are not sufficiently known yet. Hibernation may depend on seasonal metabolic fluctuations, or possibly on external stimuli such as temperature, air pressure, light, etc. It is likely that several changes in the body of the animal and in the environment act together. The combined effect of these variations probably acts on the "sleep center" in the diencephalon which produces a "sleeping substance" which in turn influences the hypophysis. The hypophysis or pituitary is one of the most important regulatory centers of the animal or human body. This tissue at the floor of the brain case is responsible for regulating the numerous different hormonal glands distributed throughout the entire body and their smooth interaction which is essential for life. The hormone insulin secreted by the islets of Langerhans (in the submucosa of the pancreas) seems to influence the state of hibernation. Insulin controls the transformation of glucose (blood sugar) to glycogen. When the pituitary of the hedgehog triggers the islets of Langerhans to release insulin into the blood stream, the amount of available glucose required by the musculature decreases, and instead more glycogen is stored in the liver and the muscles for future energy needs. In this way the hedgehog is prepared for hibernation.

In autumn the western European hedgehogs prepare for hibernation when the average air temperature is only 8°-10°C. In the preceding months the hedgehogs consumed copious amounts of food which greatly increased their bodily fat reserves. In the fall increased amounts of nesting materials are carried into the summer shelters. At the time when the temperature inside the nest drops to the critical point of about 15°-17°C, the animal burrows itself into the nest material and rolls itself tightly into a ball. In October or November one usually does not see any hedgehogs around in northern Germany. However, in countries further south, hibernation is of shorter duration or may be absent altogether since the critical temperature within the nest never drops below 15°-17°C. Yet there are species and subspecies where the critical temperature is decidedly higher. Eastern European hedgehogs *(Erinaceus roumanicus nesiotes)* in the Berlin Zoo already hibernated in August or September when the air temperature in their cage was 20°-24°C. In the natural habitat of these animals the summers are very hot and there is hardly any rainfall between May and September. The country becomes very desiccated, and living conditions for the hedgehog become correspondingly very unfavourable. It is reported that one hardly sees any hedgehogs from June

Critical temperature

onward. While it is very hot outside, the animals spend this unfavorable time of year in some cool subterranean den. One of the hedgehogs from Algiers *(Aethechinus algirus)* which I received from Agadir in West Morocco repeatedly went into hibernation during October when the outside temperature was between 11.5° and 21°C. The critical temperature of this animal probably was around 20°. The summers in West Morocco are very dry and there may be frost at night during winter. The animals presumably also hibernate during the summer. Hedgehogs on Gran Canaria are reported to hibernate for four months starting in November. The critical temperature in the Ethiopian hedgehog *(Paraechinus aethiopicus)* also seems to fall around 20°C. Other true hibernators, such as the East European ground squirrel, survive the dry season by entering into a "summer sleep" which is indistinguishable from hibernation on the basis of the metabolic processes.

All life processes are slowed down

All physiological processes decline during hibernation. The rate of breathing in the hedgehog decreases to 5 to 8 times per minute, and the heart rate is only 18 to 20 per minute. Since blood circulates only very slowly through the arteries there is the potential danger of thrombosis. During hibernation, therefore, the blood is of a different composition. Heparin is formed in larger amounts to prevent blood clotting during this time of stasis. A hibernating hedgehog produces less metabolic heat, and progressively cools down to almost the same temperature as the surrounding environment. At this stage the temperature regulatory mechanism is shut off and the hedgehog behaves like a "cold blooded" animal (poikilotherm). However, in contrast to the true poikilotherms like the fishes, amphibians, and reptiles, in which the body temperature can sink below the freezing point, the thermoregulatory mechanism turns on again in the hibernating hedgehog when the animal has cooled down to approximately 6 to 1½°C. This "minimal temperature" serves as a stimulus to increase the metabolic rate until the body temperature reaches about 5°C. This mechanism prevents the hedgehog from freezing to death during the long period of hibernation. During prolonged cold spells, the metabolic rate increases to such an extent that it also accelerates the cardiac and breathing rates and the hedgehog awakens entirely. The occasional active hedgehog that one can see during the winter was probably awakened by extreme cold, excessive warming, or other disturbances. On rare occasions a hedgehog will wake from hibernation if one picks it up.

During hibernation the hedgehog is slightly curled up. In fact, even the application of considerable force will not unroll the animal because the dorsal muscular hood (see p. 204 ff) is in a state of tonic contraction. A hibernating hedgehog does not respond to sounds, and if touched the animal responds by rolling up tighter. The fat reserves supply the energy to maintain the minimal rate of metabolism.

Coming out of hibernation

While insulin plays an important role as a "going-to-sleep hormone," the hormone adrenalin secreted by the adrenal cortex plays an equally important role as a "wake-up hormone." Adrenalin is responsible for the conversion of stored glycogen into glucose which provides the necessary energy to warm up the animal. In Northern Germany the hedgehogs usually awaken from hibernation in March or April when the outside temperature has been higher than the critical temperature for some time. In the laboratory, waking up from hibernation takes from two to five hours. One can observe this process if one brings a hibernating hedgehog into a room in which the temperature is 20°C. At first the animal's body temperature rises slowly and then progressively faster, the heart beat and breathing rates increase, and the rolled-up animal gradually opens up. When the body temperature has reached 20°C the hedgehog opens the eyes. Then the animal shivers strongly and rises to its feet; its nose starts running. However, the hedgehog is only really active when its body temperature has reached about 34°C.

A free living hedgehog that is interrupted during hibernation frequently does not find enough food outside to replenish its used-up energy reserves. The dead hedgehogs that one finds after a severe winter probably did not freeze to death, but starved to death. It is possible to let hedgehogs enter hibernation and to re-awaken them at will any number of times provided the animals are well fed during the waking intervals. Hedgehogs that were kept in warm rooms during the entire winter and which were prevented from hibernating did not seem to be harmed.

The hedgehog exhibits a unique behavior pattern—"self anointing"—which up to now has only been observed in the spiny hedgehogs. If a hedgehog finds an object that has some stimulating odor it will sniff it with growing excitement, lick it, or if the object is small, take it into the mouth and chew it. While chewing the hedgehog produces copius amounts of foaming saliva. After several minutes the hedgehog stiffens the front legs, turns the head to the side or upwards in a cramped manner and flings the foaming mass with the tongue over its spines. The animal repeats this performance three or four times or more. It has been observed to occur up to forty times in succession. Usually the hedgehog distributes this saliva over different parts of its spiny pelage, although never on the hairy fur, in rapid succession. Generally, the hedgehog does not swallow the chewed object but drops it after the end of this proceeding. Spots of dried saliva on the spines show where the animal had distributed the saliva.

Fig. 11-12. This is the manner in which a hedgehog salivates on his own spines when the animal has been stimulated by a conspicuous odor.

Even one-week-old hedgehog young may start self anointing. Self anointing behavior in Eurasian hedgehogs may be released by various objects such as the glued back of a book, newspapers smelling of printer's ink, cigarette stubs, hyacinth blossoms, perfumed soap, medicines, valerian, toad skin, earthworms, rotten meat, or even a strange conspecific. Obviously this unique behavior seems to serve the function of disguising the spines with a strange odor which conceals the animal's personal scent.

The hedgehog has few natural enemies. Large owls and raptors, which have well protected feet, penetrate the body of the hedgehog with their needle sharp claws, and they rip off strips of the spiny pelage with their hard beaks or carve out their victim from the unprotected ventral side. The spiny armor seems to be almost complete protection against carnivorous mammals, although apparently badgers, European polecats, and other martens are able to overpower a rolled-up hedgehog. Sometimes one hears or reads about how the red fox unrolls the hedgehog by pushing the latter into the water or by squirting urine on it. These claims are probably only tall tales since they have not been verified as yet. Most dogs only bark angrily at hedgehogs but do not dare to attack. Others very carefully pick up a hedgehog and bring it to their master, but there are very few dogs which will bite into such a spiny ball with no regard for their own injuries.

It now seems understandable why hedgehogs are less careful and less ready to escape than other small mammals. Unfortunately the spiny armor is no protection against the ever-increasing automobile traffic. In Germany hedgehogs are the most frequently run-over animals on the highways. On a 75-kilometer stretch of highway between Munich and Ingolstadt at least 90 hedgehogs were run over between May and June of 1957. In Denmark approximately 9345 dead hedgehogs were counted for every 1000 kilometers of highway. The greatest enemy of the hedgehog today, then, is human civilization.

Hedgehogs are frequently plagued by parasites. Most hedgehogs captured in the wild are full of fleas *(Archaeopsylla erinacei)* which reach the larval stage inside the host's nest. Other species of fleas are also occasionally found on hedgehogs. Ticks (usually *Ixodes hexagenus)* attach themselves around the ears. The mite *Demodex* infests the hair follicles, and another species of mite, *Caprina,* burrows inside the skin, often causing lethal infections and ulcers. Various species of flukes, tapeworms, and nematodes are found inside the intestine, lung, subcutaneous tissue and other organs. These organisms are often the cause of severe disease. The hedgehog ingests many internal parasites by eating snails and insects which serve as intermediate hosts for many parasites.

In certain areas many people, particularly Gypsies, eat hedgehog meat. The dressed carcass of the hedgehog is fried on the spit, or the dead animal is encased in clay and then roasted over the fire. The spines of the cooked carcass stick to the dried clay when it is done. The Romans used the spiny skins to buff their woolen materials or to comb flax. Farmers and gardeners keep hedgehogs for the control of insects, snails, and mice. Hedgehogs are also very useful in the forest by controlling vermin. However, they also feed on the eggs and chicks of birds that brood on the ground. Occasionally a hedgehog may even take a young hare or a young rabbit. Many hedgehogs can cause considerable damage to quail. Yet this does not mean that the entire hedgehog population of one particular area should be exterminated, as was formerly done. The maintenance of the natural balance in our controlled forests is of far greater importance than any special private interests. According to German conservation law it is rightly prohibited to catch, kill, acquire, or sell hedgehogs. Only between early October and the end of February is it permitted to acquire an individual hedgehog if one wants to keep one. However, this is the time when hedgehogs are very difficult to find because they are hibernating. The conservation boards occasionally permit persons to keep hedgehogs if they have legitimate reasons for doing so, for example, scientists using the animals for research or educational purposes.

The hedgehog has stimulated the imagination of people for ages. Because of its spiny appearance the hedgehog serves as a symbol of unsociability. In German the term "Schweinigel" (swine hedgehog) is a big insult. Nevertheless, hedgehogs are generally well liked because of their unique spiny skin, and certain amusing behavior patterns which even awaken the interest of people that are usually not nature lovers. Of course, because of many of its unique characteristics the hedgehog has been the cause of many legends.

Even Plinius talked about hedgehogs that roll on fruit which lies on the ground, thus driving their spines into it and then carrying it into their shelter. This tale is still believed today, and is constantly elaborated upon. It is even claimed that hedgehogs climb up trees in order to throw down the fruit which the animals either feed to their young or store for the winter. Anyone that knows a little bit about hedgehogs is aware that the animal neither brings food to its young, nor keeps a cache for the winter. The hedgehog is barely able to climb, and it is not capable of rolling on something and then keeping it stuck on the dorsal spines without it falling off. When a hedgehog rolls on his back, the dorsal muscles relax, and the spines do not offer a firm counterforce. Even more fantastic is the story that hedgehogs suckle the milk from the udder of resting cows, and there are even rumors that the hedgehogs spread hoof and mouth disease in this manner. In actual fact the mouth of a hedgehog is far too small to encircle the teat of a cow. Hedgehogs are often accused of being "passionately fond" of chicken or pigeon eggs and "very cunningly suck them out" without spilling a drop. The hedgehogs found in Germany cannot open their mouths wide enough to hold onto a chicken or pigeon egg. It always rolls away from them. However, a hedgehog is capable of biting into small, thin-shelled bird eggs. In this case part of the egg content spills on the ground, so the animal does not suck out the egg content.

One often encounters the hedgehog in fables, poetry, animal stories, and children's books. Often the animal is also shown on more or less corny post cards. In German speaking areas, the story of the race between the hare and the hedgehog is rather well known (in English the equivalent story is about the hare and the tortoise). These anthropomorphized hedgehogs are usually depicted as being good-natured and cooperative, displaying mother-wit and occasionally a degree of cunning, but almost always they are represented as sympathetic, endearing creatures. This characterization stems from the fact that the hedgehogs do not flee at once when one first encounters them, but rather remain in a funny, rolled-up spiny ball, so one can see them close up. The hedgehog also plays a great role in arts and crafts. There are countless graphic and plastic representations of the hedgehog, beginning with clay hedgehogs and ointment jars from Egyptian burial chambers which originated about 2000 B.C., to the manifold hedgehogs made of porcelain, metal, wood, or fabric which are used as room decorations, children's toys, or utensils (i.e., cloth brushes or ash trays). One also finds hedgehogs on stamps or made from marzipan or cake dough.

Additional genera of the Eurasian hedgehog are found in the regions surrounding the Mediterranean, in Africa, Southeastern Europe, and Asia.

Four-toed, long-eared desert, and desert hedgehogs

Distinguishing characteristics

A. The FOUR-TOED HEDGEHOG (genera *Aethechinus* and *Atelerix;* Color plate p. 179) has ears which are larger than those of the western European hedgehog. The legs are longer and thinner. The feet are smaller and the claws are less well developed. The first toe is vestigial or absent. The animals are non-burrowing and inhabit the hard ground of cliffs and plains. They are fast runners and agile hunters—very active animals. 1. *Aethechinus:* ALGERIAN HEDGEHOG *(Aethechinus algirus);* the body size approximates that of a European hedgehog. The head is more distinct from the trunk. There is a bald streak in the center of the head. They are found in North Africa, the Canary Islands, the Isles of Baleares, and Southwestern Europe. There are four subspecies including *Aethechinus algirus vagans.* This subspecies is found in Southwestern Europe and probably came from North Africa rather recently. The CAPE HEDGEHOG *(Aethechinus frontalis)* is found in South Africa and SCLATER'S HEDGEHOG *(Aethechinus sclateri)* is found in Somalia. 2. The CENTRAL AFRICAN HEDGEHOG *(Atelerix)* is distributed from Senegambia to the upper Nile and Abyssinia, and from the Congo to Tanganyika and Kenya. There are six species including the WHITE-BELLIED HEDGEHOG *(Atelerix albiventris)* which is found in Senegambia. PRUNER'S HEDGEHOG *(Atelerix pruneri)* is found in the region of Mount Kilimanjaro up to an elevation of 1800 meters. There is a bald streak on the top of the head, similar to that of the Algerian hedgehog.

Fig. 11-13.
Head of an Ethiopian hedgehog. The "part" of the hair, which in the young of other hedgehogs is found for some time, remains for the rest of its life in Ethiopian hedgehogs.

B. The LONG-EARED DESERT HEDGEHOG (genus *Hemiechinus*) has very prominent and movable external ears. They are nocturnal steppe and desert mammals. There are two species: The LONG-EARED DESERT HEDGEHOG (*Hemiechinus auritus;* Color plate p. 180) has an HRL of only approximately 19 cm and a TL of 1.5 cm. The body weight is approximately 500 g. The ears are 3–4 cm long. If one presses the ears forward they will cover the eyes. The auditory vesicle (bulla tympanica) is present. The snout is pointed and the body is delicate. The legs are long and thin, with large, heavily clawed feet. The spines are the color of the desert, leaving a bare streak on top of the head. The fur is short and soft. There are sixteen subspecies found in North Africa, the Near East, and Central Asia as well as Southeastern Europe. One subspecies is the EUROPEAN LONG-EARED DESERT HEDGEHOG *(Hemiechinus auritus auritus)* which is found in the Southeastern steppe regions of Russia and Cyprus.

C. The DESERT HEDGEHOG (genus *Paraechinus*) has ear pinnae which are very flexible but not quite as large as in the long-eared desert hedgehogs. The tympanic bulla is well developed. The HRL is 14–23 cm and the TL is 1–4 cm. The body weight is 400–700 g. The snout is less pointed, the body is plumper, the legs are shorter with smaller feet and shorter claws than in the long-eared desert hedgehog. There are three species: The ETHIOPIAN HEDGEHOG (*Paraechinus aethiopicus;* Color plate p. 180) has spines which are parted in the middle of the crown. There are several subspecies in the arid regions of North Africa, Arabia, and Iraq. The INDIAN HEDGEHOG *(Paraechinus micropus)* is found in the arid regions of India. BRANDT'S HEDGEHOG *(Paraechinus hypomelas)* inhabits Arabia and Asia Minor up to regions in Panjnad and Russian Turkestan.

Life habits

Captive Algerian hedgehogs are even poorer climbers than the Western European hedgehogs. They prefer to eat insects, particularly grasshoppers. Some Algerian hedgehogs never eat vegetation. In their native habitat grasshoppers probably make up the main part of their diet.

I once looked after a pair of Pruner's hedgehogs which were approximately one month old when I got them. The animals had been caught in the brushwood steppe near Nairobi at an elevation of approximately 1500 meters. When they were full grown the female weighed 675 grams and the male, approximately 900 grams. In December they frequently entered hibernation when the surrounding temperature dropped to 5.5°–6.5°C. Their critical temperature must have been very low indeed. In their native habitat the average temperature during the coldest months is 18°C. There the animals probably never have to make use of their hibernation ability. I was able to observe the courtship behavior of this pair on several occasions in Berlin. The sequence was very similar to that of the western European hedgehog (see p. 207 ff). The female gave birth to three litters a year; however, she always ate the young shortly thereafter. The gestation period is between thirty-seven and thirty-eight days.

The long-eared desert hedgehog inhabits deserts, semi-deserts and steppes. The animal avoids high grass and moist areas. In lower Egypt this hedgehog inhabits the edge of the desert where there are boulders, and the rocky, semi-arid regions of the Nile valley. During the day the animal shelters underneath piles of rocks, in the brush or in the burrows of other small mammals or in holes which it has dug itself. The long-eared desert hedgehog captures insects and other invertebrates, and also occasionally small vertebrates, particularly lizards and snakes. The inhabitants of Cyprus claim that the long-eared desert hedgehogs eat the low-hanging grapes in the vineyards. Yet the hedgehogs which I received from Palestine, Iraq, and Syria and kept in Berlin always refused any kind of vegetable diet. The courtship behavior is similar to that of other hedgehogs. The young, however—at least those from the Eastern steppe of Russia—develop at a faster rate than the Eurasian hedgehogs. The long-eared desert hedgehog young already opened their eyes at one week of age, and were able to look for food independently at two to three weeks. Because the long-eared desert hedgehogs have a pretty appearance, appealing behavior, and are quickly tamed, they are frequently kept as pets in their native lands. In certain areas of Russia they are mainly kept for the eradication of cockroaches.

The Algerian hedgehog inhabits uncultivated areas on the edge of the desert on the Southern border of the Sahara region of the Atlas Mountains in Algeria. During the day the hedgehogs shelter in rock crevices or similar hide-outs. At night they hunt for insects and other invertebrates and also small vertebrates in terrain covered by brush and thorny bushes. Captive Algerian hedgehogs never consumed any vegetable matter.

Soon after its arrival in Stuttgart in April an Algerian hedgehog female gave birth to a litter of four young, which she had in a shallow depression in the sand. The young developed at a slower rate than the Eurasian hedgehog young. On the twenty-second day they finally opened their eyes, and on the forty-fourth day they started to take food independently for the first time. However, the female suckled them until the fifty-ninth day.

Family: shrews

In the insectivora the two families of the shrews and moles are very closely related and are connected with each other by various transitional forms. For this reason these two groups are joined in one superfamily, the SHREW-LIKE *(Soricoidea)*. The SHREWS *(Soricidae)* are the most primitive of this group.

Distinguishing characteristics

These animals are from small to very small in size. The HRL is 3.5–18 cm and the TL is 0.9–12 cm. The body weight is 2–35 g. The animals are shrew-like (see p. 176). The snout is extremely pointed, proboscis-like. The ears protrude from the fur only a little or not at all. Occasionally the ears are vestigial. The eyes are small and in some species are hidden within the fur. The fur is dense, short, and soft. Usually the dorsal side is uniformly dark and the ventral side is lighter or white. Odor glands are found on the flanks, which produce a musky smell. The glands are more developed in the male than in the female. The glands play a great role during the breeding season and probably also in the marking of the territory. The saliva of certain species is poisonous. The elongated skull is small and without a zygomatic arch. The orbital and temporal cavities are not separated. The tympanic bulla is absent. There are twenty-six to thirty-two teeth. The central incisors are often hook-like at the top, long at the bottom, and in a horizontal position. The milk teeth are only found in the embryonic stage, and they do not calcify. Shortly after birth the permanent teeth break through. There are five fingers and five toes. The tibia and the fibula are fused towards the lower end. The pubes are joined by a symphysis which is similar in the moles. Due to this condition parts of the intestine may lie outside of the pelvic girdle (this is not found in any other mammal). The genital and urinary openings are usually covered by a skin fold; this makes it impossible to differentiate the sexes externally. There are three to five pairs of teats. There are 3 subfamilies, 20 genera, and 265 species.

Shrews are the smallest mammals

The shrews include the smallest members of the mammals. The smallest mammal is the SAVI'S PYGMY SHREW *(Suncus etruscus)* which measures only six to eight centimeters, including the tail, and weighs 1.5 to 2 grams. These tiny gnomes are the most voracious and ferocious of all carnivores. The smaller the size of a homeotherm, and particularly if the animal is very active, the larger the energy it uses up and the larger its food requirement. Some shrews consume more food daily than the weight of their body. Nursing mothers require double that amount of food. Shrews prey mainly on insects. These little nimrods truly clean up on these invertebrates. Many insects, particularly if they occur in masses, are very destructive to the forests, fields, and gardens. Shrews are very beneficial to agriculture and forestry by combating these harmful insects. In Germany all species of shrews (except the European water shrew) are therefore protected by law. Aside from insects, spiders, and other invertebrates, certain shrew species also prey on fish, salamanders, reptiles, small birds, and mammals. Shrews also consume carrion. During food shortages shrews may even attack conspecifics. Vegetative matter is only eaten in small amounts, if at all.

How and where they live

Most shrews are terrestrial. They inhabit terrain covered by vegetation which has good hiding places. Shrews usually inhabit moist areas. Very few species are found in the desert or steppe. The desert shrews of the genus *Notiosorex* inhabit the arid regions of the Southwestern

United States and Mexico. The PIEBALD SHREW *(Diplomesodon pulchellum)* is found in the sandy steppes east of the Volga in Russia. The water shrews, which are represented by various closely related genera, live near or in fresh water. They prey on water insects, tadpoles, and fish. The animals are extremely well adapted to swimming and diving. The WATER SHREW *(Neomys fodiens)* and various other forms have stiff hair fringes on the feet to aid in swimming. The SZECHUAN WATER SHREW *(Nectogale elegans)* from Tibet, in addition, has webbed skin between the toes. In most water shrews the tail has two lateral fringes of stiff hair which serve as a rudder. The ASIATIC WATER SHREWS *(Chimmarogale)* are able to seal the ear opening with a flap when submerging.

Shrews shelter under stones, bushes, leaves, self-dug burrows, and some also occupy mole burrows, mouse holes, and similar burrows. Water shrews usually inhabit burrows dug into bushy areas of river banks with entrances above and below the water level. Shrews do not hibernate. In the winter some migrate into basements, barns, and other buildings. Most species are nocturnal but some come into the open during the day for short periods. The daily rhythm of these highly "excited" animals alternates between short rest periods and restless spurts of activity. When active the shrews are constantly in motion, tripping around, testing the air with the highly movable nose and touching all objects with the vibrating vibrissae or the tip of the nose. The tactile and olfactory senses are extremely well developed, and the sense of hearing is also very acute. Occasionally, sudden loud sounds frighten the animal to such an extent that it dies as a result. Death due to shock occurs also in other animals and is usually caused by an excess of certain hormones released into the blood stream by various glands. This condition is known as "stress."

The voice of the shrews

Shrews communicate with their conspecifics in a high, chirping or twittering voice. The high vocal sounds, which are not audible to humans, serve as echo-location, a phenomenon first discovered in bats (Volume XI). At least this echo-location was first described for some of the American species. In contrast, the small eyes, which are often hidden in the fur, are less efficient.

Very often one finds dead shrews. These animals were probably killed by cats, dogs, martens, or other carnivores, and were not eaten by them because of their penetrating musky odor. Some cats consume shrews; others do not touch them at all but let them run again. Some common European white-toothed and bicolor white-toothed shrews have been kept in captivity up to four years, but most free living shrews apparently do not grow older than twelve to eighteen months.

Shrews are usually very unsociable

As a rule, shrews, like most insectivores, are unsociable. Generally they are solitary. A conspecific entering into the territory of another is attacked vigorously. During a fight, the opponents rise up on their hind legs, squeak, bite each other, and hit out with the front paws. In the common shrew the fight is terminated with a submissive posture: one opponent throws itself on its back and shrieks, and then the other combatant instantly runs away. Sometimes both animals lie on their backs until one flees. Trapped shrews occasionally kill and eat the opponent. The LARGE NORTH AMERICAN SHORT-TAILED SHREW *(Blarina brevicauda)* seems to be more gregarious. Although this species preys on other smaller shrew species, captive short-tailed shrews seem to live together peacefully, provided the cage is sufficiently spacious. The tiny North American LEAST SHREW *(Cryptotis parva)* seems to be the most sociable of the more peaceful shrew species. At times several individuals share one nest and together dig the tunnels to the burrow.

Reproductive behavior

In the solitary species, coming together for copulation is often a rather difficult process. The sight of a conspecific automatically releases the aggressive drive and during courtship must somehow be suppressed and overcome. The powerful sexual odors and the rather complicated precopulatory behavior patterns serve to surmount the mutual aversion. The estrous female of the bicolor white-toothed shrew *(Crocidura leucodon)* emits fine peeping sounds almost constantly. A male attracted by this "rutting call" enters into a brief fight with the female. Then the partners sniff each other in the sexual regions and "greet" each other with rapid trilling or twittering. The female walks about with stiff hind legs and the hind end is held up high in a so-called "rutting strut." The male sniffs her, mounts her from behind, and bites into her fur at the nape and, twittering softly, copulates with her. During the actual copulation which lasts approximately ninety seconds the female remains motionless and emits calls which sound like wood rubbed together ("Schnärpsen").

In Germany shrews seem to breed from spring to autumn. Shrews kept at a constant warm temperature in captivity can breed throughout the year. This applies also to the shrew species living in warmer regions. As a rule shrews in Germany bear young several times during the year. For example, the common shrew *(Sorex araneus)* has three to four litters a year with a litter size of five to ten. As far as is known, the gestation period usually lasts three to four weeks. In some species it is shorter, while in others it is somewhat longer. The bicolor white-toothed shrew has a gestation period of thirty-one to thirty-three days. Shortly before parturition the female softly cushions the nest or even builds a new one. The new-born are blind and naked but have well-developed vibrissae. A new-born shrew weighs only one gram. It has a coral pink coloration and the quick snout is purple. From the third day onwards greyish white hairs start to cover the body and these become progressivley darker. The eyes open on the thirteenth day. At sixteen days the offspring look exactly like their parents, but they are slightly smaller. After five or six weeks they are fully grown. Right from the beginning shrew young are extremely active. Already on the

A mole burrow on a meadow

Indigenous Central European Animals: □ Mammals: 1. Field mice *(Microtus arvalis)* and their burrow (1a); 2. A weasel *(Mustela nivalis)* hunting field mice; 3. The common European mole *(Talpa europaea):* a. a fight between rivals; b. in the hunting burrow; c. young inside the nest chamber into which the various tunnels lead (see p. 239); 4. the Western European hedgehog *(Erinaceus europaeus);* 5. The nocturnal bat *(Nyctalus noctula)* in its shelter during the day; 6. Common European white-toothed shrew *(Crocidura russula)*—a co-inhabitant of mole burrows; 7. Bi-color white-toothed shrew *(Crocidura leucodon)* which lives on insects; □ Birds: 8. The meadow lark *(Alauda arvensis)* ascending into the air while singing; 9. Corncrake *(Crex crex);* 10. Whinchat *(Saxicola rubetra);* 11. Tree sparrow *(Passer montanus);* 12. Yellowhammer *(Emberiza citrinella);* 13. Spotted flycatcher *(Muscicapa striata);* 14. White wagtail *(Motacilla alba);* □ Reptiles: 15. Lizard *(Lacerta agilis);* □ Amphibians: 16. Toad *(Bufo bufo)* catching slugs; □ Invertebrates: 17. Spider *(Epeira diademata);* 18. *Anthocaris cardamines;* 19. Small tortoiseshell *(Aglais urticae);* 20. Grubs (larvae of the May beetle *Melolontha*), a favorite prey of the mole. 21. *Carabus auratus* with captured earthworm; 22. Earthworm *(Lumbricus);* 23. Centipede *(Julus);* 24. Snail *(Cepaea nemoralis);* unstriated form; 25. Edible snail *(Helix pomatia);* 26. Cricket *(Gryllus campestris)* first day they crawl all over each other inside the nest making peeping sounds. On the third day they occasionally crawl out of the nest. The little escapees are brought back by the mother who carries them in her mouth. The offspring start to walk around by the fifth to the eighth day. When they are fourteen days old they are almost as fast as the adults. A few days later they begin to feed independently; however, the mother still suckles them until the twenty-sixth day.

The founder of the Münster Zoo in Germany, the humorous zoology professor Hermann Landois, was one of the first to describe a rather unique behavior in the shrews. This behavior, which he described at the turn of the century, he called "Indenschwanzbeissungsgänsemarsch," which translates into "single-file-biting-into-the-tail-march." When Landois first observed this phenomenon he thought he saw a snake in the distance but upon closer inspection the snake "disintegrated" into individual animals. The observation of this old Westfalian humorist was disputed for a long time, but recently this behavior has been observed in several species. Maurice Burton describes such a shrew caravan in the following manner: "As soon as a litter of young shrews leaves the protection of the nest they fall in behind their mother at the slightest disturbance such as a faint noise, a change in temperature, or a light rainfall. One young bites into the mother's fur at the base of the tail, a second bites into the fur of the first and so on until the whole group forms one straight line. Then they all move on as one unit; they invariably fall into step and stay that way. If the mother increases or decreases her speed so will the young. If she jumps over a low obstacle the entire caravan jumps as an entity and still keeps in step. When the mother becomes frightened she freezes motionless. Her young also stop immediately, and stay put without moving a single vibrissa or a nose. Sometimes the animals form a double row or still some other order; however, their movements remain synchronized and they all hold on to each other tightly. If one picks up the mother the young dangle from her as from a string." In older young which already have their eyes open, the first one behind the mother will sometimes let go. In this case the mother usually goes on by herself without paying any further attention to her children. The motherless chain continues on its way under the leadership of the first young, usually at a somewhat slower pace.

If the first juvenile in the lineup meets another shrew it will immediately attach itself to it. On the other hand, if it meets a house mouse the shrew chain will continue on its exploration after brief testing of the mouse's body odor. Initially, such shrew groupings are irregularly organized since each young tries to bite into the closest body part of the other. However, as soon as the clump begins to move all "outsiders" fall into a uniform line. Soon after weaning the family band disintegrates rapidly, and "chain formation" ceases. Some shrews are sexually mature at three months of age. In other species it takes several months longer.

Fig. 11-14.
Common shrew *(Sorex araneus)* in Europe

Shrews are widely distributed throughout the Old and New World, and are absent only in the polar regions, the West Indies, certain islands in the Pacific, Australia, Tasmania, New Zealand, and almost all of South America. Only in the north of South America are five species of the genus *Cryptotis* found. This genus is otherwise indigenous to Central America. During the Tertiary period shrews already occupied these same regions. Fossil records from the lower Oligocene have been found in North America and Europe, from the Miocene in Africa, and from the Pliocene in Asia.

Fig. 11-15.
Lesser shrew *(Sorex minutus)* in Europe

The eight genera with eighty-two species of the RED-TOOTHED SHREWS (subfamily *Soricinae*) are characterized by one conspicuous feature—their teeth have red tips. This group of shrews includes the LONG-TAILED SHREWS (genus *Sorex*, Color plate p. 189) which are also found in Central Europe. They have 32 teeth: $\frac{3 \cdot 1 \cdot 3 \cdot 3}{1 \cdot 1 \cdot 1 \cdot 3}$. The fur is soft. On the dorsal side it is yellowish-brown to black, and ventrally it is usually lighter or white. The uro-genital opening is completely separate from the anus. There are three pairs of mammae. The animals are usually found in moist areas of forests, shrub, and parkland regions, the taiga and the tundra. There are forty species which are mainly distributed in North and Central America. Four species are found in Europe.

Fig. 11-16.
1. Laxmann's shrew *(Sorex caecutiens)* in Europe
2. Alpine shrew *(Sorex alpinus)*

The COMMON SHREW *(Sorex araneus)* has an HRL of 6.2–8.5 cm and a TL of 3–5 cm. The body weight is 6.6–12 g. Its range is from Europe (see map) through Asia and up to Japan. The subspecies *Sorex araneus tetragonurus* is found in the Alps up to an elevation of 2000 meters. The LESSER SHREW *(Sorex minutus)* is the smallest mammal in Germany. The HRL is 4.5–6.6 cm and the TL is 3.1–4.6 cm. The body weight is 3–6 g. It occupies the dry, open regions of Europe (see map) up to Japan. In the mountains it is found at elevations up to 1200 meters. LAXMANN'S SHREW *(Sorex caecutiens)* is somewhat larger in size. It is found in Europe (see map) and in the taiga and tundra regions of North Asia. The ALPINE SHREW *(Sorex alpinus)* has an HRL of 6.2–7.5 cm and a TL of 6–7.5 cm. The body weight is 6–10 g. Its distribution is limited to Europe (see map). The animal predominantly inhabits the upper coniferous forest and crooked timber zone of the mountains. The subspecies *Sorex alpinus hercynicus* is found in the secondary chain of mountains along mountain creeks, and in the cliff regions of old mixed forests at elevations of 300–1000 meters.

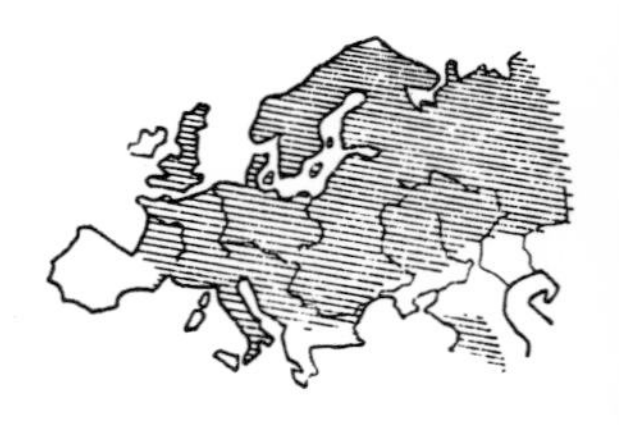

Fig. 11-17.
European water shrew *(Neomys fodiens)* in Europe

There are twenty-eight American species. Some of these are: the MASKED SHREW *(Sorex cinereus)* which is only slightly larger than the lesser shrew. This species is most frequently found in northern North America up to the southern foothills of the Rocky Mountains and the Appalachian Mountains. It prefers moist areas. *Sorex* species that are adapted to the water include the UNALASKA SHREW *(Sorex hydrodromus)* which is found on the Aleutian Islands. The NORTHERN WATER SHREW *(Sorex palustris)* is found near the streams of the north and in the mountains. The PACIFIC WATER SHREW *(Sorex bendirei)* is found near the rivers of the central west coast. The aquatic species possess fringes of stiff hairs around the hind feet which aid in swimming and diving, and trap air bubbles to enable the shrew to run on the surface of the water (*hydrodromus* means water runner).

The COMMON SHREW *(Sorex araneus)* is the most common species of shrew found in Germany. It inhabits predominantly moist forests, alder groves, and swampy meadows. The animal is less frequent in dry terrain, hedges, gardens, fields, and sand dunes. The common shrew builds its nests in vacant mouse burrows or similar subterranean structures. The nest is constructed on the outside of grass and moss, and the inside is lined with leaves. In the winter shrews are often found in house basements and barns. The animals are very unsociable and chase every conspecific from their territory. When one keeps several common shrews in one cage it is quite frequent that one animal will kill the other and eat it. The other species of the genus *Sorex* differ little in their mode of life and behavior from the common shrew.

Fig. 11-18.
Cross section through the den of a European watershrew. At least one entrance leads into the water.

The aquatic red-toothed shrews of Europe and Asia were placed into a separate genus, the WATER SHREWS *(Neomys)*, because of their different dentition. There are 30 teeth: $\frac{3 \cdot 1 \cdot 2 \cdot 3}{2 \cdot 0 \cdot 1 \cdot 3}$. Four pairs of mammae are found. The fur is soft and velvety. The ears hardly protrude from the fur. Dense fringes of stiff hair are found on the lateral sides of the soles and toes. There are two species: 1. The EUROPEAN WATER SHREW (*Neomys fodiens;* Color plate p. 189) has an HRL of 7-11 cm and a TL of 4.7-7.7 cm. The body weight is 10-20 g. The bristles on the feet, which aid in swimming, are particularly long and powerful. There is a keel of stiff hair on the underside of the tail. The ventral side is white or dark grey. The species is distributed in Europe (see map) and Asia north of the steppe girdle up to the Amur region and Sakhalin. 2. The MEDITERRANEAN WATER SHREW *(Neomys anomalus;* Color plate p. 189) has an HRL of 6-9 cm and a TL of 4.4-6.7 cm, and it weighs 16.5 g. The hair fringes on the feet are shorter, and the keel of stiff hair is missing on the tail. In Central Europe (see map) the species is relatively rare. It is found in Asia Minor in mountains up to an elevation of 2000 meters. The animal is not too dependent on water.

The largest indigenous species of shrew in Germany is the European water shrew (*Neomys fodiens*) which is most frequently found in the secondary chain of mountains, but also occurs in the Alps up to an elevation of 2500 meters. This water shrew is usually found in the vicinity of meadow ditches, fish ponds, lakes, mountain creeks or streams, in moist and cool habitats. The animal is rarely found away from water. The European water shrews usually construct their large, subterranean nests in mouse holes or mole burrows; however, they connect these shelters with self-dug tunnels to the nearest source of water where the entrance is located below the water surface. Occasionally the animal constructs its burrow into a river embankment. Water shrews are generally solitary; however, when ice or high water has made their shelters uninhabitable the animals may occasionally share a substitute nest which may be located between stacked ice blocks or even nest boxes for birds. In England it was observed once how many hundreds of these animals were swimming upstream close together in a narrow creek.

In contrast to many indigenous shrews, the European water shrew in Germany seems to be more of a diurnal animal. The animal almost always hunts underwater. While the shrew is submerged air is trapped between the guard hairs of the dense fur. These air bubbles give the animal a silvery outline and give it buoyancy. This buoyant effect is counteracted by the vigorous upward kicking of the hind legs which requires a great deal of energy. The shrew is able to stay submerged from five to twenty seconds. The animal swims about with erected vibrissae and pokes with its pointed snout underneath stones and cracks and in the mud. It locates food via the tactile and olfactory senses. On land the water shrew "smells" out its prey in a similar way.

The European water shrew is the most carnivorous member of the shrews. The animal captures almost all small animals that live in and near the water, such as leeches and other worms, snails, crabs, spiders, water beetles and their larvae, the larvae of the day-fly and stone flies, caddis flies, dragon flies, mosquitoes, the eggs of fish and frogs, and also fully-grown fish, large newts and frogs and even small birds and mammals. As soon as a water shrew has caught a live prey underwater it drags it on land immediately and kills it. Then the shrew shakes the water from its fur, grooms itself with the hind foot, and starts to feed. When it has killed fish and mice the brain is eaten first.

Water shrews occasionally cause some damage in the fisheries industry by destroying spawn and young fish; however, this loss is minimal compared to that caused by industrial waste, insecticides and other by-products of our modern achievements. Nevertheless, European water shrews are able to overpower a carp. Christian Ludwig Brehm, the father of the author of *Brehm's Tierleben,* made this observation at a farm where the farmer kept several carp inside a well box during the winter. One day the farmer, "much to his annoyance, found a dead carp in his well which had its eyes and brain eaten out. After a few days the farmer was angered by a second dead fish which was similarly mutilated, and in this manner the farmer lost additional carp. Finally his wife noticed that a black 'mouse' climbed into the fish tank towards evening, swam around in the water, and then sat on the head of a carp and held on with its front paws. Before the woman was able to open the frozen window and chase the animal away, it had eaten the eyes of the carp. Finally the window opened and the 'mouse' was chased away, only to be caught by a stalking cat. The creature was taken from the cat and was given to me. It was one of our European water shrews all right. I have to mention, however, that the water shrew that was given to me was not the only one that visited that fish tank. One shrew after another followed."

If one considers the extensive poisoning of the water caused by man and the countless edible fish that are killed as a result, then the traditional persecution of fish-eating birds and mammals, which never seriously threaten the population of their prey, seems hardly justified.

In northern Europe the breeding season of the European water shrew lasts from April to September. Wild chasing in the water precedes the actual copulation. The female dives as soon as the male has reached her; for a stretch she runs along on the bottom of the water and emerges again at a different spot as the male resumes his pursuit. This courtship phase can repeat itself every fifteen minutes until the female is finally ready to copulate. After three to four weeks the female gives birth to four to eight young in a well-padded nest. She suckles them for five to six weeks. Often two to three litters are born between May and October. Young born in late spring are usually sexually mature that first summer. Free living European water shrews reach an age of fourteen to nineteen months.

The smallest mammal found in America is the AMERICAN PYGMY SHREW (*Microsorex hoyi;* Color plate p. 189) which is related to the genus *Sorex* and is widely distributed in the north of North America. The SHORT-TAILED SHREW (*Blarina brevicauda;* Color plate p. 189) is much larger than the previously mentioned species, and is one of the most commonly found mammals in North America. The HRL is 7.5-10.5 and the TL is 1.7-3 cm. There are 30 teeth: $\frac{3 \cdot 1 \cdot 2 \cdot 3}{2 \cdot 0 \cdot 1 \cdot 3}$. The salivary glands of these shrews contain neuro toxins which are quickly effective on small animals and can cause a bitten person pain for days. In the vast area between Texas and Quebec the short-tailed shrew occupies a great variety of habitats. The animal may live in hollows under stones, logs, or stumps. The nests, constructed at these sites, are large, about twelve to thirteen cm in diameter, and are lined with dry leaves and grass. With its strong front paws and nose the shrew constructs subterranean runways. The animal hunts for invertebrates mainly in the early morning and late afternoon. These ferocious short-tailed shrews also attack mice *(Peromyscus)* and smaller shrews. The prey is overpowered after heavy fighting and the shrews bite their prey in the throat and face. The victim is dragged into the nest and sometimes the shrew begins to eat while the prey is still alive. The short-tailed shrews cache food, particularly snails and beetles. In captivity the animals store nuts and sunflower seeds and reportedly apples as well.

The SHORT-TAILED MOUPIN SHREW (*Blarinella quadraticauda;* Color plate

p. 189) is somewhat smaller and has a slightly longer tail. The animal has long claws. It is found from Szechuan to North Burma. The LESSER AMERICAN SHORT-TAILED SHREWS (genus *Cryptotis;* Color plate p. 189) are found in North and Central America, northern South America (the only insectivores found there), and the eastern part of the United States. MERRIAM'S DESERT SHREW and CRAWFORD'S DESERT SHREW (*Notiosorex gigas* and *Notiosorex crawfordi;* Color plate p. 189) are found in the desert regions of southwestern North America. **SIKKIM LARGE-CLAWED SHREWS** *(Soriculus nigrescens)* have a somewhat stouter body. The tail is half as long as the body. The front legs have powerful claws, possibly for burrowing. The animal is found in Sikkim. There are five additional *Soriculus* species which have tails longer than their bodies. They are called long-tailed shrews.

Subfamily: white-toothed shrews

The members of this next subfamily have white teeth without red lips. These shrews possess tails that are covered by short close-lying bristles and long projecting hairs. This subfamily, the WHITE-TOOTHED SHREWS *(Crocidurinae),* is only found in the Old World. There are eleven genera and a total of 181 species.

White-toothed shrews (genus *Crocidura;* Color plate p. 189) have an HRL of 4–15 cm and a TL of 4–10 cm. There are 28 teeth: $\frac{3 \cdot 1 \cdot 1 \cdot 3}{1 \cdot 1 \cdot 1 \cdot 3}$. This mammalian genus contains the greatest number of species (144), four of which are European: 1. The BI-COLOR WHITE-TOOTHED SHREW *(Crocidura leucodon)* has an HRL of 6.5–9 cm and a TL of 2.8–4 cm. The body weight is 7–15 gr. There is a sharp color contrast between the light ventral side and the darker dorsal side. The distribution in Europe is shown on the map. This species is found in the alps to an elevation of 1200 meters. The shrew also inhabits Transcaucasia, East Turkestan, and Iran. 2. The COMMON EUROPEAN WHITE-TOOTHED SHREW *(Crocidura russula)* is the same size as the bicolor species. It is found in Europe (see map), Asia Minor and North Africa. In the east it is distributed in Kashmir, China, and Japan. 3. The LESSER-TOOTHED SHREW *(Crocidura suaveolens)* has an HRL of 5.5–7.9 cm and a TL of 2.7–4.3 cm. The body weight is 3–7 g. The species is found in Europe (see map), in central and East Asia and the Near East, and North Africa. 4. The MEDITERRANEAN LONG-TAILED SHREW *(Crocidura caudata)* is found in Sicily, Corsica, and the Balearic Islands. This species closely resembles the common European white-toothed mouse. Some zoologists consider this group a subspecies of *Crocidura russula.*

Fig. 11-19.
Bicolor white-toothed shrew *(Crocidura leucodon)* in Europe

White-toothed shrews inhabit moist and dry forests, parklands, cultivated regions, grass land, steppes, and deserts. During the colder season the shrews may also occupy buildings. These creatures are rather sensitive to cold and moisture. Obligate aquatic forms are not found in this subfamily. Some species are said to live in groups in the winter. The European species of the white-toothed shrew, at least, is more compatible with others when kept in a cage than most other

Fig. 11-20.
Common European white-toothed shrew *(Crocidura russula)*

Fig. 11-21.
Lesser white-toothed shrew *(Crocidura suaveolens)* in Europe

shrews. One bicolor white-toothed shrew lived together with a white male mouse in a cage. The two animals shared the same nest, and the shrew even tolerated the copulatory attempts of the mouse although under "loud cries of protest." The food requirements of the European white-toothed shrews do not seem to be as high as that of most other shrews.

Fifty-one species of the white-toothed shrew are found in Asia and associated islands. In Africa eighty-nine species occur. Examples of the African species are the DESERT MUSK SHREW *(Crocidura smithi)* and the GIANT MUSK SHREW *(Crocidura flavescens)*; this animal is found in the swampy regions of Rhodesia. It has been observed to eat small mammals while in captivity. Two closely related genera in Africa, each with one species, are *Paracrocidura schoutedeni* and the AFRICAN FOREST SHREW (*Praesorex goliath;* Color plate p. 189) which is the largest shrew species. The animal is as large as a rat with an HRL of 15.5–18 cm and a TL of 11 cm.

The bicolor white-toothed shrew *(Crocidura leucodon)* is more easily kept and bred in captivity than other shrews. These animals are also more easily observed in nature. This is the reason why more is known about the life history of *Crocidura leucodon* than any other shrew. The animal prefers dry terrain and is found in fields, gardens, hedges, thickets, underbrush, and forest margins. In the winter they are often observed in buildings. The nest is constructed out of grass and is usually found in brushwood, bushes, or compost heaps. Bicolor white-toothed shrews may be sexually mature in their first year of life. In Germany the breeding season lasts from spring to fall. During this time the males have a strong musky odor. Captive bicolor white-toothed shrews survive well on a diet of raw meat, earthworms, insects, and soft bird food. Shrews caught when very young may live to be three to four years old when kept properly. The animals cache food in their communal sleeping box, sometimes in such large amounts that the animals were lying on a thick layer of bits of meat and chewed-up earthworms. Feces were regularly deposited on specific spots, usually in the corner of the cage. The shrews often attach the fecal material to the wall several centimeters above the ground.

MUSK SHREWS (genus *Suncus*) are characterized by a tail that is swollen at the base. There are large well-developed musky scent glands on the flanks, particularly in the male. There are thirty teeth: $\frac{3 \cdot 1 \cdot 2 \cdot 3}{1 \cdot 1 \cdot 1 \cdot 3}$. SAVI'S PYGMY SHREW (*Suncus etruscus;* Color plate p. 189) has an HRL of 3.5–5 cm and a TL of 2.5–3 cm. The body weight is 1.5–2 g. The ears project well above the fur. The species is found in Southern Europe (see map), Africa, and South Asia to Malaya. The HOUSE SHREW *(Suncus murinus)* has an HRL of 15 cm and a TL of 10 cm. The species is found in South Asia, the Philippines, Guam, the Sunda Islands, and has been introduced into New Guinea, East Africa, and Madagascar. Other species include the MADAGASCAR SHREW *(Suncus madagascariensis)*, which is found in Madagascar. The FOREST SHREW *(Suncus varius)* and the DARK-FOOTED FOREST SHREW *(Suncus cafer)* are found in South Africa. There are twenty-one additional species usually distributed throughout Africa.

Fig. 11-22.
1. Savi's pygmy shrew *(Suncus etruscus)*
2. Piebald shrew *(Diplomesodon pulchellum)* in Europe

The Savi's pygmy shrew *(Suncus etruscus)* is the smallest mammal living today. It is found in terrain heavily overgrown by vegetation in moist, but not wet, regions. The preferred habitat of this tiny creature may consist of cultivated land and gardens, deciduous and coniferous forests, stone hedges, and ruins, piles of rubble, rock debris, and also the banks of small bodies of water. The nest of this shrew is hidden in rock crevices, holes in a wall, between rocks or underneath tree roots. The animals are nocturnal and predominantly prey on spiders, insects, and other invertebrates.

LONG-CLAWED SHREWS (genus *Feroculus* and *Solisorex*; Color plate p. 189) are rare animals found in the mountains of Ceylon. KENYA MOUNTAIN SHREWS (genus *Surdisorex*; Color plate p. 189) are found in the mountains of Kenya at 3000 meters and the Szechuan burrowing shrew (*Anourosorex squamipes*; Color plate p. 189) is found in Southeast Asia and Formosa at elevations of 1500–3100 meters. These genera are more or less adapted to a burrowing, subterranean existence.

The PIEBALD SHREW *(Diplomesodon pulchellum)* has an HRL of 5.4–7.6 cm and a TL of 2.1–3.1 cm. The body weight is 7–13 g. The feet are broad and large and are fringed on both sides with long, stiff hair. The ears and eyes are large. The vibrissae are very long. The shrews inhabit the steppes and savannahs of Eastern Europe (see map), the Kirghiz steppes, Kazakhstan, and Russian Turkestan.

The piebald shrew is named after the white spot on its back. The dark dorsal side is in sharp contrast to the white ventral side. The animal inhabits sandy regions with sparse grass coverage. It shelters in cracks and holes in the ground, piles of grain or human dwellings. The stiff hair fringes on the front and hind legs enable the shrew to run on loose sand. Piebald shrews are able to dig quickly into the sand or to expose a buried prey. The shrews are nocturnal hunters. When large beetles are captured, their legs are first bitten off and eaten, and then the remainder is consumed. A piebald shrew overpowers a lizard by jumping on its back, biting into the neck, and holding on until the victim is motionless. Then the shrew starts eating the lizard's head and finally just leaves the tail and feet. One captive piebald shrew killed eleven lizards in this manner in one day, eating five during the night.

Fig. 11-23.
Front and hind foot of the Piebald shrew *(Diplomesodon pulchellum)* with the bristles which facilitate running in loose sand.

The white-toothed shrews also contain aquatic species which are even better adapted to living in the water than the red-toothed shrew genera *Sorex* and *Neomys*. Like the latter they also possess stiff hair fringes along the sides of the fingers and toes which serve as rudders. The ears and eyes are hidden in the especially soft and waterproof fur. The Asiatic water shrews of the genus *Chimmarogale* are able to close off their ears when diving by means of skin folds formed by the reduction of ear pinnae. The HIMALAYAN WATER SHREW (*Chimmarogale platycephala;* Color plate p. 189) is distributed from the Himalayas to Japan. The BORNEO WATER SHREW *(Chimmarogale phaeura)* is found on Borneo and Sumatra. The animals live near streams and prey on water insects, small crayfish, and fish.

The SZECHUAN WATER SHREW (*Nectogale elegans;* Color plate p. 189) which is found in the high countries of East and Inner Asia has true webbed skin between its toes. Disk-like pads on the underside of the feet, which act as suction cups, assist the shrew in walking over wet stones or holding the prey. The tail consists of four lateral hair fringes with a fifth fringe located on the dorsal side. In swift mountain streams the animals are agile swimmers and expert divers for fish. They construct their nests into burrows in stream beds.

Subfamily: armored shrews

In tropical Africa two very unique shrew species of the subfamily *Scutisoricinae*—the ARMORED SHREWS—are found. These animals are characterized by a spinal column that is unique among the mammals. The spinal column consists of large and strong vertebrae with numerous facets for articulation and lateral interlocking spines, as well as dorsal and ventral spines. These features strengthen the vertebral column enormously and allow it to withstand great pressure. Nevertheless, the animals are able to bend their backs considerably into all directions. The natives of Africa believe that this shrew is magical and that by eating it one becomes brave and immune to wounds.

Armored shrews appear to be almost indestructible. A full grown man can stand on a living armored shrew with his full weight for several minutes, and when he steps off the shrew merely shakes itself and runs off unharmed. The strength and the shape of the vertebral column prevent a crushing of the internal organs. Armored shrews are active both day and night. They search for food under leaves, decayed wood and stones. This subfamily contains only one genus and two species: The ARMORED SHREWS (*Scutisorex;* Color plate p. 189) have an HRL of 12–15 cm and a TL of 6.8–9.5 cm. The fur is long, thick, and coarse. Distribution: the CONGO ARMORED SHREW *(Scutisorex congicus)* is found in the Upper Ituri River region in the Congo; the UGANDA ARMORED SHREW *(Scutisorex somereni)* is known only from near Kampala, Uganda.

Family: moles

Distinguishing characteristics

The MOLES (Family *Talpidae*) have an HRL of 6.3–21.5 cm and a body weight of 9–170 g. The nose is long and tubular and is supported by the prenasal bone or it may be movable like a trunk. The eyes are very small and are occasionally covered by fur. The external ears are reduced or absent. The body is cylindrical. The tail is usually short, although in some species it is medium-long or long. The fur is usually short and soft with the hair lying in all directions. The pelage appears velvety. There are three or four pairs of teats. The limbs are short.

There are claws on all five fingers and five toes. In the burrowing forms the collar bone articulates with the upper arm which is a condition not found in other mammals. The sternum is deeply keeled for the attachment of the burrowing muscles. The hand is used for digging. It is broad and is increased in width by an additional bone, the os falciforme, besides the thumb. As in the shrews the tibia and fibula are partly fused, and the two halves of the pelvic girdle are not joined by the pubis (see p. 219). The skull is elongated, narrow, and somewhat flattened. The zygomatic arch and ossified tympanic bullae are present. There are from thirty-two to forty-four teeth.

There are five subfamilies: 1. SHREW MOLES *(Uropsilinae),* found in Southeast Asia; 2. DESMANS *(Desmaninae),* found in East and Southwestern Europe; 3. OLD WORLD MOLES *(Talpinae),* found in Europe and Asia; 4. AMERICAN-ASIATIC MOLES *(Scalopinae),* found in North America and Eastern Asia; 5. STAR-NOSED MOLES *(Condylurinae),* found in North America. There are a total of twelve genera and nineteen species.

Subfamily: *Uropsilinae*

The SHREW MOLES (subfamily *Uropsilinae*) are shrew-like. There is only one species, the ASIATIC SHREW MOLE (*Uropsilus soricipes;* Color plate p. 190). The HRL is 6.3–8.8 cm and the TL is 5.4–7.5 cm. The long snout is made up of two tubular nostrils which are separated by a groove. The ears protrude somewhat beyond the fur. The tail and the legs are covered by scales and are scantily haired. There are 44 teeth: $\frac{3 \cdot 1 \cdot 4 \cdot 3}{3 \cdot 1 \cdot 4 \cdot 3}$ The species is found in Northern Burma, Szechwan, and Yunnan at elevations of 1250 to 4500 meters.

Subfamily: *Desmaninae*

The DESMANS (subfamily *Desmaninae*) are mole-like. They are characterized by a long trunk which consists of two cartilaginous tubes. It is movable in all directions. The nostrils are located at the tip of these tubes and can be closed by muscular flaps. The eyes are very small. The ear pinnae are absent. The tail, which is as long as the body, is laterally compressed and serves as a rudder. The tail is scantily haired and is covered by scales. At the base of the tail on the underside a large musky scent gland consisting of several diverticulae is located. The legs are short. The forefeet are fringed with stiff hair and are partially webbed. The hind feet are completely webbed. There are forty-four teeth. The diet consists mainly of aquatic animals but some water plants are also consumed. There are two genera, each with one species.

Fig. 11-24.
1. Russian desman *(Desmana moschata)*
2. Pyrenean desman *(Galemys pyrenaicus)*

The RUSSIAN DESMAN (*Desmana moschata;* Color plate p. 190) has an HRL of 18–21.5 cm and a TL of 17–21.5 cm. The fur is glossy, silky, and dense. Originally the animal inhabited the waters near the Volga, Don, and Ural. Fossils have been discovered in Western Europe and the British Isles.

The PYRENEAN DESMAN (◊*Galemys pyrenaicus;* Color plate p. 190) has an HRL of 11–15.6 cm and a TL of 12.6–15.6 cm. The body weight is 50–80 g. The snout is longer than that of the Russian desman. The tail is constricted at the base, and is compressed laterally only near the end.

It is found in mountain streams in the Pyrennees, and in all of Northern Spain and Northern Portugal at elevations of 300–1200 meters.

This species inhabits clear mountain streams, particularly where trout are found.

The Russian desman is limited to bodies of water that have natural banks. Wherever this habitat has been altered by dams or some other form of river regulation, the animals have vacated their original sites. In other regions the species was exterminated by excessive hunting. This valuable fur animal is now protected in the U.S.S.R. In 1920 the government reintroduced approximately 5000 desmans in its former home range, and also introduced the animal into new areas. Today desmans are found at the Ilmen Lake, the river basins of the Volga, Don, Dnieper, and Ob. Limited hunting of this animal is again permitted from time to time.

Desmans inhabit quiet ponds and lower river bays. The animals avoid swift water. They are nocturnal but occasionally in the spring one can observe them during the day. Desmans rarely go on the land; however, they do occasionally sunbathe on the bank. Usually several animals occupy one set of burrows which they construct at relatively low points on the river bank. The burrow entrances are always located below the water level, often as much as forty centimeters below. This enables the animals to enter the burrow even when the water is covered by a sheet of ice. Tunnels, sometimes longer than six meters, lead upwards to the nesting chambers which are above the high water level and are close to the surface of the ground. There are no entrances above ground; however, sufficient fresh air is supplied by the small spaces between the root systems of the plants. It is not uncommon to find several resting chambers which are lined with nesting material, in addition to subterranean food burrows which have one or two entrances below the water level. On steep river banks the burrows are of a simpler structure. The tunnels to the higher chambers are often constructed in a series of steps.

When swimming, desmans frequently stick the tip of the trunk out of the water like a snorkel. As Rickmann describes it, "one sees a whole mass of small swimming objects which seem quite unexplainable." In reality these are the nosetips of desmans. On the shore the animals are quite active. "They chase and play with one another or quietly examine grasses and pieces of plant roots and sniff at them." During the night and at dusk and dawn desmans search for food on the bottom of shallow water and among aquatic plants. The animals very quickly move around under water. Their trunks bend in all directions and skillfully feel out leeches and other worms, snails, and shells, water insects and larvae, and spawn of fish and frogs. Desmans consume tadpoles and occasionally even small fish; however, fast moving prey is beyond their reach. Fish are usually only caught in the winter when they are rather sluggish or half paralyzed in the ice cold water. Stomach contents from desmans caught in the fall occasionally reveal parts of various kinds of water plants. Smaller prey is eaten on the shore, and larger prey is dragged into the burrow. Here one often finds fish bones, shells, or other remains. Desmans are voracious. A captive specimen is able to consume thirty to fifty earthworms or shelled molluscs at one meal.

Reproduction is not limited to a specific time of the year. However, most young are born in the early summer or fall. The gestation period is from forty to fifty days. The usual litter size is three to four, at the most five. The newborn is blind, toothless, and almost naked. The female suckles the young for approximately one month. The desman is sexually mature at six weeks of age and is fully grown by twelve months.

Desmans have few natural enemies. Sometimes they are swallowed by large pikes, or occasionally small mammalian predators kill them but they do not eat the desmans because of the strong musky odor. Desmans forced to migrate on land because of high water or the drying up of their habitat often fall victim to crows, owls, or raptors. However, their worst enemy still is man. He is responsible for the alteration of the animal's aquatic habitat, the straightening and damming of rivers, the reinforcement of the banks and the relentless hunt for the animal's soft, silky, dense, and glossy silver fur which is used for fur trimmings, collars, muffs, and coat lining. The content of the musk glands is used in the production of perfume. Around 1900, 20,000 desman skins were processed yearly. Due to this wasteful exploitation the population went down, so that in 1923/24, only 10,000 to 12,000 furs were traded. Desman furs are insignificant in the fur trade today because of strict protective laws. This situation could change if the protective laws and new introductions would result in a substantial increase in the desman population, making well-regulated hunting once again possible.

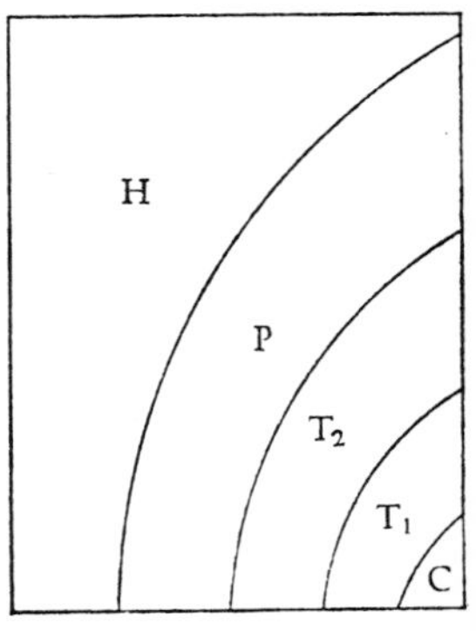

The Evolution of the Primates I

1. Tree shrews: *Tupaia* 2. Adapidae: *Notharctus* 3. Aye-ayes: *Daubentonia* 4. Lemurs: *Microcebus* (Lesser mouse lemur) *Lemur* (Ring-tailed lemur) *Megaladapis* 5. Indris: *Indri* 6. Lorises: *Loris* (Slender loris) 7. Galagos: *Galago* 8. Tarsiers: *Tarsius* 9. Marmosets and tamarins: *Leontideus* (=*Leontocebus*, maned tamarin) 10. Goeldi's monkey: *Callimico* 11. Night and titi monkeys: *Aotes* (night monkey) 12. Sakis: *Pithecia* (white-headed saki) 13. Howler monkeys: *Alouatta* 14. Capuchin monkeys: *Cebus* (apella) (Brown capuchin) 15. Spider monkeys: *Ateles* (C) Cretaceous period, approximately 120–70 million years ago (T_1) Lower Tertiary, approximately 70–25 million years ago (T_2) Upper Tertiary, approximately 25–1 million years ago (P) Pleistocene—"Ice ages"—approximately 1 million to 20,000 years ago (H) Holocene—recent to approximately 20,000 years ago

Subfamily: Old World moles

OLD WORLD MOLES (Subfamily *Talpinae*) are mole-like insectivores adapted for burrowing (see p. 176). There are forty-four teeth. One genus and four species are found. The COMMON EURASIAN MOLE (*Talpa europaea;* Color plate pp. 223–224) has an HRL of 11.5–17 cm and a TL of 2–3.4 cm. The body weight is 65–120 g. The nose is supported by a prenasal bone. The ear openings, hidden in the fur, can be closed by a low skin fold. There are no ear pinnae. The eyes are the size of a poppy seed. When the animal opens the fur-covered eyelids, the fur is parted star-like in all directions. The fur is velvety, very dense, and can be moved in all directions. The nose, tail, and feet are almost naked. The animal possesses broad burrowing hands that are permanently bent outward. For distribution in Europe, the reader is referred to the map. The animal is found in the Alps at elevations up to 2400 meters. It occurs from Central Asia into Mongolia. The ROMAN

Leontocebus
Callimico
Pithecia
Cebus
Ateles
Alouatta
Aotes
Tarsius
Galago
Loris
Indri
Lemur
Megaladapis
Microcebus
Daubentonia
Tupaia
Notharctus
CEBIDAE
PLATYRRHINA
CALLITHRICIDAE
TARSIIDAE
GALAGIDAE
LORISIDAE
ARCHAEOLEMURIDAE
INDRIIDAE
LEMURIDAE
PROSIMIAE
DAUBENTONIIDAE
TUPAIIDAE
ADAPIDAE
PLESIADAPIDAE
NECROLEMURIDAE
ANAPTOMORPHIDAE
OMOMYIDAE

Symphalangus
Pongo
Pan
Homo
sapiens
Hylobates
Gorilla
Papio
Macaca
HYLOBATIDAE
PONGIDAE
Giganto –
pithecus
Homo
neanderthalensis
Cercopithecus
HOMINIDAE
Erythrocebus
Pliopithecus
Colobus
CERCOPITHECIDAE
Oreo –
pithecus
Presbytis
Proconsul
Nasalis
Mesopithecus

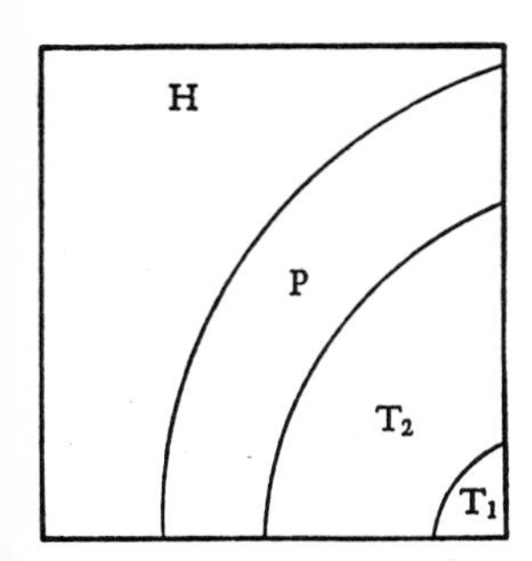

The Phylogeny of the Primates II:

1. Leaf monkeys: *Mesopithecus, Nasalis* (Proboscis monkey), *Presbytis* (Entellus langur), *Colobus* (Guereza).
2. Cercopithecidae: *Erythrocebus* (Red guenon), *Cercopithecus* (Diana monkey), *Macaca* (Grey-armed macaque), *Papio* (Mandrill, according to the classification used in the text it is the genus *Mandrillus*).
3. Gibbons: *Pliopithecus* (Short-armed Gibbon), *Hylobates* (Gibbon), *Symphalangus* (Siamang).
4. Oreopithecidae: *Oreopithecus*
5. Great apes: Proconsul, *Pongo* (Orang-utan), *Gorilla, Pan* (Chimpanzee).
6. Man: *Australopithecus* (Ape man), *Homo neanderthalensis* (Neanderthal man), *Homo sapiens* (modern man).

(T_1)—Lower Tertiary, approximately 70–25 million years ago.
(T_2)—Upper Tertiary, approximately 25–1 million years ago.
(P)—Pleistocene—"ice ages"—approximately 1 million–20,000 years ago.
(H)—Holocene, recent, since approximately 20,000 years ago.

MOLE *(Talpa romana)* is approximately the same size as the common Eurasian mole; however, its skull structure is slightly different. The eyes are always covered by skin. For distribution see the map. The MEDITERRANEAN MOLE *(Talpa caeca)* is somewhat smaller, its lips, tail and feet have lighter hair, and the eyes are usually covered by hair; for its distribution see the map. The EASTERN MOLE (*Talpa micrura;* Color plate p. 190) is more naked. The stump-like tail is completely hidden in the fur. The animal is found from Sikkim across northern India to Mongolia, Manchuria, Korea, Japan, and Formosa.

The body size is highly influenced by environmental conditions. The males are always larger than the females. The mole is able to move equally well backwards or forwards in its burrows because the hair points in all directions. There are occasional color deviations from the common mole-grey fur. Spotted, yellowish, and even totally white moles have been found.

Moles prefer loose, fertile, well-planted ground for digging. Therefore, one finds them most frequently in meadows, grassland, or deciduous forests, and less often in fields and only occasionally in pinewoods. The animals avoid marshy regions as well as very dry sandy areas. The mole bores itself into the ground. When burrowing the animal pulls in its head and rips the ground with the frontally positioned burrowing claws and then throws the soil behind itself. The cylindrical, smooth body of the mole moves through the ground like a turning drill whereby the loosened soil is pushed against the sides of the animal and part of this is firmly pressed against the walls of the tunnel. The mole continuously moves the remaining soil behind itself with its hind legs and occasionally pushes it to the surface at a hand's width off to the side of the burrow, where it forms small mounds. During the summer adult moles burrow 10–40 cm below the surface depending on the condition of the soil, and in the winter they dig to a depth of about 60 cm. Juveniles burrow closer to the surface and are more frequently active above ground.

The completed mole burrow consists of a big mound which contains the nest. Nearby one often finds a series of additional nests which are connected to the main nest by tunnels. One or several connecting tunnels having firm smooth walls lead from the main nest to a highly branched network of hunting-tunnels. The largest mole hills above ground are found in moist regions. These hills are occupied predominantly by moles during the winter, and may be almost 1¾ meters in diameter and 90 cm high. Often the hill contains several dens located one above the other. The top nest is above the ground surface and is driest and serves as living quarters. The lower burrows are more or less moist.

Moles almost exclusively live on animals which they hunt underground. Approximately every three to four hours every mole patrols all its hunting tunnels. Here it captures animals that live in the soil and which have dropped into the tunnel. These are mainly earthworms and insect larvae. Often moles find arrivals that have entered the tunnels from above the ground, such as wood lice, spiders, centipedes, insects, salamanders, reptiles, mice, and shrews. While digging the tunnels the mole preys on grubs and other soil animals which are particularly concentrated around clumps of roots, stumps, and walls. When the upper crust of the ground is frozen or excessively dry, earthworms and other ground insects retreat to greater depth. Under these conditions it is not uncommon to see moles, particularly the juveniles, hunt above ground. The animals run among the vegetation where they find insects, snails, small vertebrates, chicks from ground brooding birds and even dead animals. Larger chunks of prey are dragged into the tunnel and consumed there.

Fig. 11-25.
1. The common European mole *(Talpa europaea)* in Europe
2. The Roman mole *(Talpa romana)*

Fig. 11-26.
Distribution of the Mediterranean mole *(Talpa caeca)*

In the winter moles store food. Subterranean chambers close to the nest contain earthworms and other prey animals. One cache contained 18 grubs and 1280 earthworms which amounted to more than two kilograms. The earthworms were arranged in piles of approximately ten. They had their anterior part bitten off or were otherwise mutilated. Thus they were prevented from crawling away and from burrowing themselves into the ground.

The mole locates prey aided by its active sense of vibration, hearing, and touch. A mole is immediately alarmed by an animal moving on the surface of the ground. The mole searches in the subterranean tunnels until it finds the location of the animal. Here the mole pushes its nose through the ground, excitedly sniffs and feels with its vibrissae until it has discovered the prey. Small creatures are immediately chewed up and eaten. Large earthworms are grasped at the front end and are pulled through the front paws which remove the attached soil and also squeeze out the content of the intestines. Prey that resists is overpowered after a struggle. The mole tears the prey with its front paws and teeth and then eats it. Some moles retreat from a living mouse with loud cries, while others rush right at the largest voles. These are swiftly torn open at the abdomen with the sharp front claws. The mole pushes its snout into the opened animal and feeds on it until only the fur, head, and bones remain.

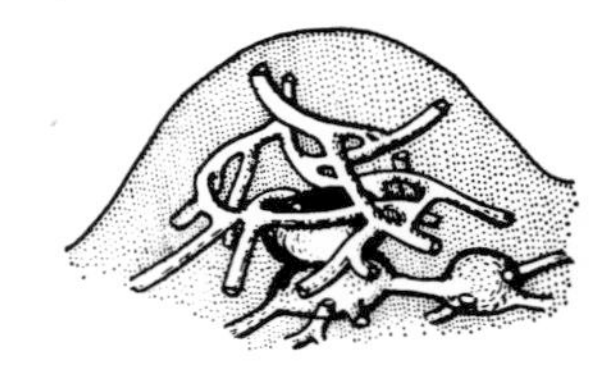

Fig. 11-27.
Mole hills are particularly large when they are located in wet areas. These "swamp castles" often contain several nest chambers, one located above the other (compare also the biotope illustration on p. 223/224).

Repeatedly it was claimed that moles must consume one and one-half their own weight in food daily and that they cannot stay alive without food for more than a few hours. However, captive moles hardly eat more than their own weight in food and can stay alive for at least twelve hours without eating. One animal even survived a fasting period of twenty-seven hours; however, it lost twelve grams of its body weight. The metabolic rate of a mole is very high, and food is digested very rapidly. One cannot keep moles in cages that are closed on the bottom because the soil soon becomes saturated with waste products

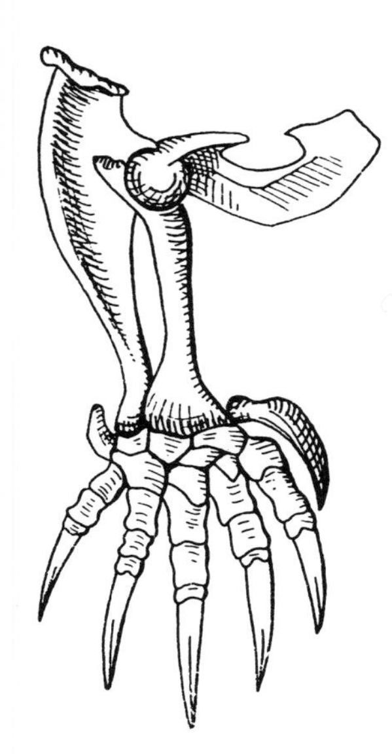

Fig. 11-28.
Skeleton of the front hand of a common Eurasian mole.

and the animals will die. Only if the cage floor is constructed of closely meshed wire and the soil is repeatedly replaced is one able to keep moles alive for several months. Several moles kept in one cage will attack each other viciously. One opponent will kill the other and eat him. Fights among free living moles have been observed. Only occasionally do several moles share the same tunnels. If there are any aggressive interactions in these tunnels the subordinate always has the opportunity to escape in the tunnels. The animals probably sleep alone in their nests which are softly padded with plant material.

Very little is known about the reproduction of the mole. They were probably never raised in captivity. Rutting takes place from March to May or June. During this period the mole enters into strange burrows, usually at night time. If the mole encounters another male in the burrow, violent fights accompanied by loud twittering results. In the end the winner often eats the subdued rival. When the mole meets a female both will initially fight. According to the reports of some investigators the pair hunts together for a period of time until the female constructs a nest chamber with a soft padding. The female probably only bears one litter of four to five young annually, usually in May. However, nine young have been found in one nest. The new-born moles are of a whitish color and are naked. The female is cautious and at the least sign of danger quickly carries the young to another hiding place. The young grow quickly and presumably open their eyes at three weeks of age. They are suckled for four to six weeks. At two months of age they are independent, and are fully grown by twelve months. Moles have a life expectancy of approximately three to four years.

While burrowing, moles often damage the subterranean parts of plants, consume earthworms which are of great importance to the fertility of the soil, and leave heaps of soil on lawns, meadows and vegetable fields. However, this "damage" is more than compensated for by the vermin control of the mole. Moles consume large numbers of mole crickets, grubs, caterpillars, and other insects which are harmful to crops in gardens and fields. In Germany the unauthorized capture of moles on strange property is prohibited. Before the usefulness of the mole was known, farmers or entire municipalities would hire a mole catcher. As late as 1909 such a catcher in a Swiss community caught 4000 moles in eighteen days and collected 800 francs in premiums. Aside from man, major predators are owls, raptors, ravens, and storks. Occasionally martens, foxes, and dogs will kill moles, but they usually do not eat them.

Fig. 11-29.
Moles are rather unsocial animals. If two conspecifics meet they often fight fiercely with each other. Occasionally the winner will eat the loser.

The processing of the fine mole furs is rather difficult. The furs do not last very long and certain spots always reveal the underlying white leather. Nevertheless, mole pelts were already worn in antiquity and the middle ages. In the 1920's, dyed mole furs were high fashion. In 1930 alone, over 20 million furs were sold on the market! The delicate hairs of the fur rub off very easily, and since more women now use cars for transportation, mole fur jackets and coats have become unfashionable, to the great benefit of these small, black, subterranean animals.

Subfamily: American-Asiatic moles

The AMERICAN-ASIATIC MOLES (subfamily *Scalopinae*) are insectivores of the mole type distinguished by burrowing adaptations of varying degrees. The first upper canine-like incisor is enlarged. The canine tooth is small. There are seven genera with a total of twelve species.

The LONG-TAILED MOLE (*Scaptonyx fuscicaudus;* Color plate p. 190) has an HRL of 6.5–9 cm and a TL of approximately 2–3 cm. The ear pinnae are very small. The hands have powerful digging claws; however, they are smaller than in the European mole. The tail is covered with stiff hairs. The species is found in Northern Burma, Szechuan and Yunnan at elevations of 2150–4500 meters.

SHREW MOLES (genus *Urotrichus*) have an HRL of 6.4–10 cm and a TL of 2.4–4.1 cm. These animals look like a combination of shrew and mole. The hands are less broadened. The ear pinnae are small and buried in the fur. The tail is covered by long, thick hair. Often the tail is enlarged by fat deposits. There are two species: TRUE'S SHREW MOLE *(Urotrichus pilirostris)* has only been reported on Hondo Island. The JAPANESE SHREW MOLE (*Urotrichus talpoides;* Color plate p. 190) is found in Japan. The animal inhabits forests, frequently at elevations above 2000 meters. *U. talpoides* burrows under the surface but is also active above ground, and occasionally even climbs low bushes and trees. It feeds on worms, spiders, insects, and other invertebrates.

The AMERICAN SHREW MOLE (*Neurotrichus gibbsi;* Color plate p. 190) has an HRL of 6.9–8.4 cm and a TL of 3.1–4.2 cm. The body weight is 9–11 g. The species resembles the Japanese forms. The ear pinnae are absent. The eyes are minute. The nostrils open at the sides of the snout tip. The hands are longer than they are broad. The claws are very long. The tail is thick and is constricted at the base. It is covered by rows of scales and coarse hair. The fur is soft and dense. It is dark grey to sooty blue black and has a metallic gloss. The fur is used in the fur trade. The species is found on the west coast of North America from British Columbia to central California. It also occurs at elevations up to 2500 meters.

The KANSU MOLE *(Scapanulus oweni)* is only known from about six museum specimens. The dentition and other characteristics are similar to those of the Western and Eastern American moles. It is found in Western China.

The HAIR-TAILED MOLES (*Parascalops breweri;* Color plate p. 190) have an HRL of 11.6–14 cm and a TL of 2.3–3.6 cm. The body weight is 40–55 g. The snout is rather short. The nostrils open dorsally. The ear pinnae are absent. The eyes are almost buried in the fur. The hand surfaces are just as long as they are broad, and are turned backwards in a fashion similar to that of the European moles. The thick, fleshy tail is constricted at the base. The tail is covered by scales and densely spaced long hairs. The dense, soft fur is of a blackish color and is utilized by the furriers. Often white spots are found on the chest and abdomen. With age the snout, tail, and feet turn almost pure white. The animal is found in the eastern regions of North America from Quebec and Ontario to central Ohio and western North Carolina.

The hairy-tailed moles inhabit light, well-drained soils in forests and open terrain. The mole constructs a convoluted tunnel system below the ground surface which extends over a wide area. The burrow system may often extend up to thirty meters in diameter, but it narrows at the deeper levels. The globular breeding nest which is about fifteen cm in diameter is found at an approximate depth of twenty-five to thirty cm. In the fall the hairy-tailed mole extends the lower portion of its tunnel system and constructs a winter nest about forty cm below the surface. This is padded with leaves and grass. The hairy-tailed moles have similar living habits to the European moles. They dig up heaps of soil, hunt in the deeper tunnels when it is cold, but on warm winter days or in the summer they search for prey above ground. At night they frequently come out of their tunnels. The hairy-tailed mole defecates outside its tunnels through openings in the soil. It appears that these moles occupy the same tunnel system for years. In the winter the animals are solitary but in the autumn and winter several may share one burrow system.

During the breeding season from the end of March to the beginning of April several males may often enter the burrow of a female. The gestation period lasts from four to six weeks. The litter size is four to five young. A new-born young is approximately seven cm long, naked, but with vibrissae. The youngsters stay inside the nest for about one month. They are sexually mature at ten months. Presumably hairy-tailed moles have a life expectancy of four years.

The WESTERN AMERICAN MOLE (genus *Scapanus*) has an HRL of 11.1–18.6 cm. The TL is 2.1–5.5 cm. The body weight is 50–170 g. This animal is very similar to the previously mentioned genus but the hand surfaces are broader than they are long. The fur is silky and soft, and is from blackish-brown to almost totally black, but occasionally light grey. The gestation period is about one month. The litter consists of two to five young which are born in March or April; the young are independent by June. Three species are found on the west coast of North America from southern British Columbia to California. They inhabit mountains at elevations up to 2700 meters. The TOWNSEND'S MOLE (*Scapanus townsendi*; Color plate p. 190) is the largest species. It inhabits a variety of habitats in the north coastal regions but avoids deciduous forest. The diet consists of invertebrates and parts of plants. The mole sometimes causes damage to onion cultures. The fur is used commercially. The COAST MOLE *(Scapanus orarius)* is distributed further to the south and east. The animal prefers well-drained soil, and is also found in deciduous forest. During the dry season the mole may burrow down for water to a depth of two meters. The BROAD-FOOTED MOLE *(Scapanus latimanus)* is found in moist regions in the South.

The EASTERN AMERICAN MOLE (*Scalopus aquaticus;* Color plate p. 190) has an HRL of 11–17 cm and a TL of 1.8–3.8 cm. The body weight is 40–140 g. The ears are invisible and the eyes are covered by fur. The palms are turned outwards and are broader than they are long. The fingers and toes are connected by web-like skin tissue. The thin, almost naked tail is faintly annulated with scales. The fur is soft, dense, and velvety. It is made into furs. There are thirty-six teeth. The species is distributed in the eastern United States from the Atlantic coast to the Dakotas and Texas, and from the Canadian border to Florida. It is closely related to the TAMAULIPAN MOLE *(Scalopus inflatus)* and the COAHUILAN MOLE *(Scalopus montanus)*, both of which are found in Mexico.

The "webbed skins" of the eastern American mole prompted Linné to describe the species as *Talpa aquatica*, the aquatic mole. Although the animal, like all moles, is a good swimmer, it does not live at or in the water. It prefers well-drained soil in fields, meadows, gardens, and open forest. The eastern American mole burrows its tunnels close to the ground surface. One can even follow the animal by the trail of raised loose soil ridges. This mole tunnels deeper during the winter. The nesting burrow lies fifteen to twenty cm below the surface and is connected to the hunting tunnels by a series of tubes. Farmers and gardeners persistently pursue this mole, although it is more beneficial than harmful, because of the ridges and mounds it produces in cultivated lands. The eastern American mole mainly preys on earthworms, ground-dwelling insect larvae, and other small invertebrates; however, the mole is also reported to consume plant material. There is only one litter of two to five young a year. The gestation period is about forty-two days. The young are independent after one month. They are sexually mature the following spring.

Subfamily: star-nosed moles

The STAR-NOSED MOLES *(subfamily Condylurinae)* consist of only one genus and one species, the STAR-NOSED MOLE (*Condylura cristata;* Color plate p. 190). The HRL is 10–12.7 cm, the TL is 5.6–8.4 cm, and the body weight is 40–85 g. The palms are turned outward and are as long as they are broad. The tail is constricted at the base. It is annulated with scales and is sparsely covered with hair. In the winter and early spring the tail is enlarged with fat deposits. The animal is found in eastern North America from Labrador and Manitoba to North Carolina and Georgia.

The star-nosed mole is the only mole which has a muzzle ringed by twenty-two naked, finger-like appendages. This feature accounts for the unique appearance of this mole. These fleshy rays or tentacles encircle the nostrils in a star-like fashion. While the mole is searching for food these appendages are constantly in motion, except for the upper two rays which are stiffly erected forwards. When the animal feeds the rays are contracted. Presumably the tentacles are endowed with tactile organs. Up to now it has not been demonstrated if these organs are also sensitive to chemical stimuli.

The star-nosed mole tunnels in moist, swampy ground. The mounds of dug-up soil are approximately twenty-five cm in diameter. Usually some of the tunnels lead directly into the water. The animals are active throughout the year, both in the daytime and at night. They are often observed running on the surface, or on or underneath the snow. The star-nosed mole is an agile swimmer and frequently hunts in the water, even under ice. The animal swims with all four legs. The diet consists of earthworms, crayfish, water insects, small fish, and other small animals. During the winter the animals live in pairs. The gestation period is not known. Between April and June two to seven young are born. The star nose is already prominent at birth. The young are independent after about three weeks, and reach maturity around ten months of age.

Family: elephant shrews

The ELEPHANT SHREWS (*Macroscelididae;* Color plate p. 180) appear like minute kangaroos with small movable elephant-like trunks. This group of animals is rather unique among the insectivores. The resemblance to kangaroos or rather jerboas is only external. These tiny creatures have extremely elongated hind legs which enable them to hop and jump over long distances.

The animal is of the jerboa type (see p. 176). The HRL is 9.5–31.5 cm and the TL is 8–26.5 cm. The snout is tube-like and elongated. It is flexible and movable but not retractile. The nostrils are located at the tip of the snout. The eyes and the ear pinnae are large. The body is short and stout and the tail is long. The tibia and the central part of the foot are greatly elongated. The pelage is dense and soft. The tip of the snout is naked. Numerous long vibrissae grow at the base of the snout. Scent glands are located on the under surface of the tail base. These exude a musky odor, particularly in the males. The penis is divided into three forks. There are two or three pairs of teats. The zygomatic arch is well developed. The hard palate has several openings. There are thirty-six to forty-two teeth. Caeca are present which are not found in other insectivores. The diet consists of insects and other invertebrates. Some species feed mainly on ants and termites. The larger species also prey on small vertebrates, and presumably also eat plant material. There are four genera and a total of twenty-one species.

Elephant shrews

1. The SHORT-EARED ELEPHANT SHREW (genus *Macroscelides*) has an HRL of 9.5–12.4 cm and a TL of 9.7–13.7 cm. The ear pinnae are short and roundish. The tympanic bullae are very well developed. There are 42

Legends to the following color plates

Of the suborder of prosimians, 34 species are represented in the following Color plates by 37 subspecies arranged according to their geographical distribution.

A. Madagascan Prosimians

(Plate I, p. 247)

The aye-aye *(Daubentonia madagascariensis)* uses its extremely attentuated, elongated middle finger to extract food out of narrow holes and cracks.

Dwarf lemurs (Plate II, p. 248)

1. Coquerel's mouse lemur *(Microcebus coquereli)*
2. Lesser mouse lemur *(Microcebus murinus)*, the smallest primate
3. Fat-tailed lemur *(Cheirogaleus medius)*
4. Hairy-eared dwarf lemur *(Cheirogaleus trichotis)*
5. Greater dwarf lemur *(Cheirogaleus major)*
6. Forked-marked mouse lemur *(Phaner furcifer)* The illustrations are not true to scale.

Medium-sized lemurs (Plate III, p. 249)

1. Weasel lemur *(Lepilemur mustelinus)*
2. Sportive lemur *(Lepilemur ruficaudatus)*
3. Broad-nosed gentle lemur *(Hapalemur simus)*
4. Grey gentle lemur *(Hapalemur griseus)*

Medium-sized lemurs (Plate IV, p. 250)

1. Black lemur *(Lemur macaco)*,♂
2. Ring-tailed lemur *(Lemur catta)*
3. Ruffed lemur *(Lemur variegatus)*; occurs in various color combinations and subspecies.
4. Brown lemur *(Lemur fulvus)*, is represented by three subspecies; among them are: a. Rufous lemur *(Lemur fulvus albifrons)* b. Black lemur *(Lemur fulvus rufus)*.
5. Red-bellied lemur *(Lemur rubriventer)*

Aye-ayes (Plate V, p. 251)

1. Aye-aye *(Daubentonia madagascariensis)*

Indris

2. Woolly indris *(Avahi laniger)*
3. Verreaux's sifaka *(Propithecus verreauxi)*
4. Diademed sifaka *(Propithecus diadema)*
5. Indris *(Indri indri)*

B. African Prosimians

Galagos (Plate VI, p. 252)

1. Senegal bushbaby *(Galago senegalensis braccatus)*
2. Allen's bushbaby *(Galago alleni)*
3. Demidoff's bushbaby *(Galago demidovii)*
4. Thick-tailed bushbaby *(Galago crassicaudatus panganensis)*
5. Eastern needle-clawed bushbaby *(Galago inustus)*
6. Western needle-clawed bushbaby *(Galago elegantulus pallidus)*

Lorises

7. Angwantibo *(Arctocebus calabarensis)*
8. Potto *(Perodicticus potto)*

C. Asiatic Prosimians

Tarsiers (Plate VII, p. 253)

1. Western tarsier (from Borneo) *(Tarsius bancanus borneanus)*

Lorises

2. Slow loris *(Nycticebus coucang)*
3. Slender loris *(Loris tardigradus)*

Tree shrews

4. Pen-tailed tree shrew *(Ptilocercus lowii)*
5. Red common tree shrew *(Tupaia glis ferruginea)*

I

II
1
2
3
4
5
6
HELMUT DILLER

1
2
4
3
HELMUT DILLER

IV
1 ♀
1 ♂
2
3a
3b
3c
3d
4a
4c
5
4b ♂
HELMUT DILLER

V
1
2
3
4
5
HELMUT DILLER.

VI
1
2
3
5
6
7
4
8

VII
1
2
3
4
5
Helmut Diller

Großmann

teeth: $\frac{3 \cdot 1 \cdot 4 \cdot 2}{3 \cdot 1 \cdot 4 \cdot 3}$. There are five fingers and five toes. The first toe is small. Three pairs of mammae are present. There is one species *(Macroscelides proboscideus).*

2. The ELEPHANT SHREWS (Genus *Elephantulus*) resemble the previous species; however, the tympanic bullae are less well developed. There are seven species including the SHORT-SNOUTED ELEPHANT SHREW *(Elephantulus brachyrhynchus).* The HRL is 10–13.5 cm and the TL is 8–12.5 cm. The snout is short and tapers off rapidly towards the tip. There are two pairs of mammae. The species is distributed from Kenya to Southwestern Africa, the Transvaal and Mozambique, and the Karooveld. The animal is also found in rocky terrain and open forests with heavy undergrowth. The NORTH AFRICAN ELEPHANT SHREW *(Elephantulus rozeti)* is found in the desert and rocky regions of Morocco to Lybia. The BUSHVELD ELEPHANT SHREW *(Elephantulus intufi)* occurs from Angola to the former Tanganyika and south to the Cape Province. The ROCK ELEPHANT SHREW *(Elephantulus rupestris)* is found in rocky regions of South Africa and Angola.

Forest elephant shrews

3. The FOREST ELEPHANT SHREWS (genus *Petrodromus*) have an HRL of 16.5–22 cm and a TL of 13–18 cm. There are 40 teeth: $\frac{3 \cdot 1 \cdot 4 \cdot 2}{3 \cdot 1 \cdot 4 \cdot 2}$. The first toe is absent. The animals inhabit heavy undergrowth and are active at night and at dusk. There are six species including the FOUR-TOED ELEPHANT SHREW *(Petrodromus tetradactylus)* which is found from the Congo to Kenya to Southwestern Africa and Zululand, and the FOREST ELEPHANT SHREW *(Petrodromus sultan)* which occurs in Kenya, the former Tanganyika, Mozambique, Zanzibar, and Mafia. The species inhabits thick bushland, and shelters in ground burrows or termite hills.

Checkered elephant shrews

4. CHECKERED ELEPHANT SHREWS (genus *Rhynchocyon*), the largest forms, have the longest snouts. The HRL is 24–31.5 cm and the TL is 19–26.5 cm. The hind legs are only slightly elongated. There are only four fingers and four toes. The hairs are stiff. The tail is annulated with scales interspersed with thin, short hair. There are 36 teeth: $\frac{1 \cdot 1 \cdot 4 \cdot 2}{3 \cdot 1 \cdot 4 \cdot 2}$. The upper incisor is vestigial and the upper canines are elongated. The diet consists of insects and probably also snails, small birds, eggs, and small mammals. The animals are mainly active during the day. They inhabit forests and thickets. They are somewhat less agile than other elephant shrews. The animals construct their nests of leaves underneath tree trunks and ground depressions. They live in pairs. There are seven species including the RUFOUS ELEPHANT SHREW *(Rhynchocyon petersi)* which is red to rufous on the anterior dorsal side and black in the back. The species inhabits the bushy grassland in Kenya, all of Tanzania, and Mafia. The BLACK ELEPHANT SHREW *(Rhynchocyon stuhlmanni)* occurs in the Congsur forest. The CHECKERED-BACKED ELEPHANT SHREW *(Rhynchocyon cirnei)* is found in the former Tanganyika, Uganda, the Congo, Rhodesia, and Mozambique. The ZANZIBAR ELEPHANT SHREW *(Rhynchocyon adersi)* inhabits the former Zanzibar.

The Philippine tarsier *(Tarsius syrichta)* and its relatives perform tremendous jumps with their thin, elongated hind legs. The terminal pads on the fingers and toes are soft, flat discs which have adhesive properties.

Elephant shrews are only found in Africa. Some species inhabit the sun-baked regions or rocky areas. Others are found in the thick underbrush of river banks or in the jungle. Most elephant shrews are diurnal, but some are active at night or dusk. Elephant shrews belong to the most highly evolved insectivores. Their body form deviates drastically from that of the other members of this order. In their internal structure they share certain characteristics with the TREE SHREWS *(Tupaiidae)* with which they were formerly classified in one suborder. Later, however the tree shrews were usually classified with the primates (see p. 272). In contrast to other insectivores the female elephant shrew only bears one or two young after a gestation period of two months. The new-born young are completely covered with hair. They are able to see and to run about actively.

The SHORT-EARED ELEPHANT SHREWS inhabit the sandy or gravelly thorn-bush plain of the Karooveld in the Cape Province and Southwestern Africa. Here they dig narrow tunnels under the shrubbery with a shallow entrance and one vertical "emergency exit." Occasionally one pair of shrews occupies the burrow of another small mammal. The animals are diurnal and hunt insects during the hottest hours at noon. "If one is well hidden," writes Ludwig Heck, "one can observe their lively activity; however, the slightest movement will frighten them back into their shelter and then considerable time passes before they reappear. Eventually one will emerge, look around, and listen, and then hop and scurry around hastily. Finally all animals appear again, catching flying insects while jumping and sniffing between rocks, crevices, cracks, and investigating every nook and cranny with their sensitive trunk-like snouts. Often one animal will sit on one of the sunbaked stones, seemingly enjoying the warmth. It is not uncommon to see two animals playing together, possibly a recently formed pair."

The DRYLAND ELEPHANT SHREWS in Southwestern Africa prefer the banks of small dry rivers which are covered with a light growth of bushes. The entrances to their burrows are located in the stone-free ground under single bushes. Well-trodden paths connect one bush to the next or lead to small smooth places where the animals sunbathe. The feces are deposited at the periphery of their territory. They live in loosely organized social groups without any definite social structure. The elephant shrews are active from sunrise to sunset.

These animals do well in captivity and are delightful pets although they often become nocturnal. They do well on a diet that is usually given to insectivorous birds. The Frankfurt Zoo has repeatedly kept North African elephant shrews. One animal lived for three years. In 1908, the first two living forest elephant shrews were brought to Berlin by Vosseler. There they lived for some time on a diet of minced meat, mealworm larvae, and white bread soaked in milk.

Elephant shrews frequently drum with their hind feet. This drumming is probably used to frighten off enemies or in defense of the territory. Ten to twenty beats follow in rapid succession, making it impossible to distinguish the individual beat. One can elicit drumming by clicking of the tongue or similar sounds.

The checkered elephant shrew is a "true diurnal and sun animal" according to Vosseler. He reports that the animals get up early in the morning and move around for the greater part of the day in search of food everywhere. If the animal approaches something edible it first sniffs at it and then grabs it with a hasty jerk. The prey is then eaten while the trunk-like snout is bent down. Larger prey is torn to pieces by holding it down with the front paws. The food is eaten with loud smacking sounds. The animal quenches its thirst by licking dew or rain drops from grasses and leaves. "The delicate, jerky walk changes into lightning-fast jumps at the sign of danger," Vosseler writes. "It becomes clear that the tail is used for balancing. In bushy grassland a Russian greyhound was not able to catch a fleeing animal although the latter did not find cover for some time."

Free living checkered elephant shrews seem to have individual ranges (territories) that contain the nesting place. These animals also make delightful pets. On a diet of milk, raw meat, insects, and fruit, these animals have been kept for more than eighteen months.

The ancient order of the insectivores is represented by many primitive forms which remind us of the appearance of the primeval ancestors of the higher mammals and how they might have lived. Concurrently the insectivores also possess various adaptations to unusual habitats and unique living conditions. Since most insectivores have secretive living habits our knowledge of their behavior under natural conditions is still scant and full of gaps. Yet, what we know already has given us a glimpse into the existence of an ancient order of animals which once gave rise to our more familiar animal forms.

Konrad Herter

12 The Primates

Order: Primates by W. Fiedler

Already in 1758 the great Swedish scholar Carl von Linné, who is the founder of the scientific zoological and botanical systematics, named one order of animals the "Primates." Linné also included man in his system and classified him along with monkeys and apes, prosimians, giant gliders, and bats under this order. In honor of man he united these various animals under the name of primates—"those that are first in rank." One hundred years before Darwin's monumental work, *On the Origin of Species,* appeared, Linné must have been aware of the relationship between the prosimians, the monkeys, the apes, and man without the benefit of the evolutionary insights of Darwin and other researchers of the nineteenth century.

At a later date it was recognized that bats and giant gliders do not belong to the primates but are separate orders. Yet Linné's concept of an order is still valid today as well as his short, concise description of the primates. These are his words translated into English: "Four parallel upper incisors, one canine on each side, paired mammae situated in the chest region, limbs adapted for grasping, the clavicles as important support elements for the functioning of the arms, quadrupedal walk, climbing on trees and the gathering of their fruits." This is a relatively correct definition even though Linné was not aware of any of the species which were of great significance in clarifying the evolutionary relationships.

In Linné's "Systema naturae," the following species in the order primates, besides man, are enumerated: the "Troglodytes" (chimpanzees and the orangutans which Linné could not clearly differentiate), an Indonesian leaf monkey *(Presbytis aygula),* the mandrill and other Old World monkeys (hamadryas baboon, barbary ape, lion-tailed macaque, Diana monkeys, moustached monkeys, and grass monkeys), several species of capuchin monkeys, the squirrel monkey, a black spider monkey, the common marmoset and several other tamarins, the tarsiers, the ring tailed lemur, the slender loris, and several other forms which are not clearly distinguished today. Linné gave man the scientific name of *Homo sapiens,* which is still a valid term today; however, he did not add his customary species or genus description but rather this challenge: "Nosce te ipsum!" (Know yourself!)

Older scientific texts and even the fourth edition of *Brehm's Tierleben* which was published fifty years ago, as a matter of course discussed the primates at the end. For in discussing the various members of the animal kingdom which progress from lower to increasingly complex forms there had to be one mammalian order which also included the "lord of creation" and which constituted the height of development and thereby the crowning conclusion. However, scientific results of the last decade have repeatedly demonstrated the close and multiple relationships of the primates to the insectivores or rather the insectivore-like ancestral mammals. Thus the traditional animal sequence does not comply with the phylogenetic facts. In modern systematics the primates and the bats are grouped immediately following the insectivores.

Distinguishing characteristics

Primates are higher mammals that range in size from that of the mouse to that of the gorilla. They are usually adapted for climbing in trees. The eye sockets are always surrounded by a closed bony ring. The clavicles are present. The first finger or toe (thumb or big toe), and at least one pair of digits, are opposable and able to grasp. Fingers and toes usually have nails, but less frequently claw-like nails. The occipital lobe is well developed (it comprises the visual area), containing the Fissura calcerina. The caecum is well-developed. The penis is suspended. The testes lie within the scrotum. There is usually one pair of mammae in the chest region.

There are two suborders with fifteen presently living families: A. Suborder PROSIMIANS: 1. TREE SHREWS *(Tupaiidae)*; 2. LEMURS *(Lemuridae)*; 3. INDRIS *(Indriidae)*; 4. AYE-AYES *(Daubentoniidae)*; 5. LORISES *(Lorisidae)*; 6. GALAGOS *(Galagidae)*; 7. TARSIERS *(Tarsiidae)*; B. Suborder MONKEYS, APES, and MAN *(Simiae)*; 8. NEW WORLD MONKEYS *(Cebidae)*; 9. GOELDI'S MONKEYS *(Callimiconidae)*; 10. MARMOSETS and TAMARINS *(Callithricidae)*; 11. OLD WORLD MONKEYS *(Cercopithecidae)*; 12. LEAF MONKEYS *(Colobidae)*; 13. GIBBONS *(Hylobatidae)*; 14. GREAT APES *(Pongidae)*; 15. MAN *(Hominidae)*. There are a total of approximately 200 species.

From the tree shrew to man

The mammalian order of the primates comprises many groups of vastly different levels of phylogenetic development. Each individual group demonstrates one or several close common characteristics, but distantly related members of this order differ to such a degree that it is difficult to find general characteristics which apply to all primates. Let us just compare tree shrews, which apparently are close to the ancestral stem of the primate family tree, and which still pose problems with respect to their systematic position, with the great apes or man. If we want to distinguish the primates we must not only consider their characteristics and unique features but above all their phylogenetic history. It reveals to us the origin of the various groups, the evolutionary trends which became important for their development and their relationships towards each other. At one time I (Fiedler) composed such a systematic review in a handbook about primates by Hofer, Schultz, and Starck and established the following "phylogenetic criterion":

Fig. 12-1.
A common marmoset can easily fit onto a human hand; however, a gorilla can weigh up to 350 kilograms.

Many ancestral representatives of the primates and the insectivores share identical characteristics which make it difficult to give a clear-cut definition for many of the extinct forms as well as the tree shrew living today. All forms are arboreal, have enlarged eyes, and an increased brain. The visual centers are well developed, and there is a concurrent development of binocular vision. In the higher primates the brain areas are well-developed which are involved in higher mental processes. Additional characteristics of primates are the opposability of the first digits and a tendency for the formation of flat nails, but this is not always so. Tarsiers, apes, and man have a disk-shaped placenta which can, of course, only be demonstrated for forms living today. The lemurs, on the other hand, still have a villous placenta and a large amniotic cavity.

Our knowledge of primate phylogeny is frequently still incomplete because of insufficient fossil records. This gap can largely be filled by investigating the embryonic development. In his *On the Origin of Species,* Darwin already wrote: "Embryology gains in interest if we imagine that the embryo is more or less the faded blueprint of the basic form common to all members of the same animal class." Embryology, however, is not only a simplified repetition of the phylogeny, as was formally believed. The embryo is an independent organism that has to adapt to its external environment and is influenced by the latter like any other mature organism. The multiple interactions between mother and embryo necessitate certain special adaptations that cannot be interpreted on a phylogenetic basis. Such special adaptations may in turn give rise to new lines of development.

The anatomist D. Starck of Frankfurt followed this line of reasoning when he published his text on embryology in 1965. He outlined the phylogenetic development of the order primates: "Most prosimians possess very primitive features in the development of the embryo and its membranes (yolk sac, chorion, etc.). The tarsier is rather unique and can serve as a model for some of the characteristics found in the higher forms, such as the formation of the primitive connective tissue and the umbilical cord. The placenta of the tarsier deviates greatly in its structure from that of the lemurs but does not lead directly to the apes." There is a higher degree of uniformity in the embryonic membranes and placenta of the New World monkeys, and they show similar but more primitive characteristics than the Old World monkeys.

This does not mean that the present day Old World monkeys originated from the New World monkeys. These two groups probably originated independently from prosimian-like ancestors. The embryonic membranes and placenta are very similar in the great apes and man. On the basis of several distinct characteristics one can easily differentiate them from the other Old World monkeys.

The stages of primate evolution

Starck reached the following conclusion: "It is possible to differentiate between various evolutionary types on the basis of embryological data gathered from primates living today. This data, at the same time, represents the phylogenetic stages: 1. Lemur stage; 2. Tarsier stage; 3. New World monkey stage; 4. The stage of the lower Old World monkeys; 5. The stage of the great apes and man. These stages are significant since they clarify the phylogenetic process of the embryological types."

Besides the insights of comparative morphology and embryology of the living forms and paleontology, which deals with extinct animals, other branches of biology have thrown light on the phylogenetic history of the primates. For instance, the comparative study of chromosomes and blood samples and, above all, comparative ethology, have helped to increase our knowledge of the primates. However, the evaluation of these new results has to be approached with more caution if the basis for comparison is still very small. The unique characteristics of hemoglobin or the number of chromosomes do not explain the relationships between individual primate groups. These research results have to be evaluated against a background of many other criteria as well. Nevertheless, these analyses, as well as the ethological research, have verified the close relationship between the great apes and man, and thereby lend additional support for the long well-established knowledge of the morphologists, embryologists, and paleontologists.

Recently, the American anthropologist DeVore has conducted very intense observations on free living monkeys. This scientist, who is equally familiar with prosimians, monkeys, apes, and man, characterizes this mammalian order which gave rise to and includes man, in the following manner: "The order of primates encompasses nearly two hundred living species. The breadth of this scale can only be judged if one considers that this order extends from such primitive creatures as the insectivorous tree shrews to man. Furthermore, fossil forms and their diverse descendents belong to this order. Therefore it is virtually impossible to have one basic criterion which defines the entire order. For example, the tree shrews seem not to belong to the primates and yet they are included on the basis of their phylogeny and a few features of the skull structure. The one characteristic which the tree shrew shares with all other living and extinct primates is the adaptation to an arboreal existence. Their special adaptations are numerous, stratified, and occasionally quite far reaching. They show up in the structure of the primate brain, the presence of finger and toe nails, and an opposable thumb. They are also seen in the manner in which the primates utilize their olfactory, visual, and tactile senses, and in the way they bear and raise their young. Not all of these characteristics are found in any one species but all species possess rudiments for these characteristics, and these rudiments are characteristic of the entire group."

The ability to climb while grasping

Except for the tree shrews, all species have the ability to climb while grasping a limb. Several other mammals also possess this ability but in the prosimians and monkeys, this adaptation became a basic characteristic of high phylogenetic significance. All primates grasp with the ventral surface of their hands or feet, and often with both. This basic motion is followed by the lesser mouse lemurs as they glide through the branches at night, by the lorises as they dangle from a bough in sloth fashion, by the galagos and tarsiers as they hop and jump, by most primates that live in trees, by the ringtailed lemurs, baboons, and several macaques which are ground-dwellers, by the tail acrobats, the spider monkeys, by the gibbons, by great apes which hang from tree limbs, and by man who is bipedal. Even man cannot climb in any other way but the true primate fashion.

"This talent may seem unimportant," writes DeVore, "but it lies at the root of the entire order of primates. Primates are primarily tree dwellers. They originated, evolved, and did well in trees, and of all the present living primate forms, only man has left this mode of life forever. The primates' ability to encircle a branch with the fingers, instead of putting the claws into it like other mammals, made them the indisputable masters of the trees."

The phylogeny of the primates by E. Thenius

From fossils and the examination of characteristics found in living forms, we have known for some time now that the primates evolved from the insectivores. However, the question of the geologically oldest primate is far more difficult to answer. Zoologists are even divided in their opinions as to where they should classify the living tree shrew. Should it be included in the suborder *Prosimiae* or the *Insectivora* close to the elephant shrews? If we were to examine only the remains of bones and teeth instead of living animals then it becomes impossible to draw a sharp border line between insectivores and prosimians. Such a boundary between these two groups never existed in nature.

The reason for this can be explained by the fact that in the phylogenetic process not all characteristics are changed at the same time. This led to the formation of species that already possess a prosimian-like dentition but have retained a skull and limbs which are still insectivore-like. According to fossil remains the hairy-hedgehog-like insectivores of the upper Cretaceous period were the ancestral forms of the primates. These ancient animals had a complete mammalian den-

▷
The lesser mouse lemur is the smallest primate. It is hardly larger than a house mouse. The large eyes indicate that the animal is nocturnal.

▷▷
Galagos are nocturnal, insectivorous predators. They leap far through the branches. Their long bushy tail serves as a rudder.

tition, a hedgehog-like hearing apparatus, and a non-convoluted brain with a greatly enlarged olfactory center. It seems, then, that the primates originated from ground-dwelling mammals which possessed five digits on their limbs and which oriented mainly with the sense of smell. The lateral position of the small eyes indicates that these upper Cretaceous insectivores did not rely on their sense of vision.

Gradual adaptation to an arboreal mode of life resulted in the enlargement of the eyes. The orbital axes slant frontward which is more advantageous for arboreal animals. This new development eventually resulted in true stereoscopic vision which enabled the climbing primates to see an object in three dimensions. Concurrently the cranium increased in size, and the primitive long snout-like skull as well as the olfactory organ were reduced. Insectivores with a well-developed olfactory sense (Macrosmate) gradually evolved into primates with a poorly developed sense of olfaction (Microsmate). Finally the olfactory lobes became reduced to small bulbs attached to an ever-increasing cerebral brain. The expansion of the neocortex and an initial slight but later more rapid convoluting of the cerebral surface took place and increased in importance. The middle ear became more protected inside a bony bulla (Bulla tympanica). The claws on the fingers and toes were transformed into nails. Concurrently the thumb and first toe could be spread more and more and finally were opposable to the other toes and fingers. This adaptation enabled the primates to climb while grasping a branch and did not require claws to hook into the tree bark. Prosimians still have claws on some toes but these are only used for grooming (grooming claws). The claw-like structures found on the fingers and toes of the marmosets and tamarins and some of the New World monkeys are not true claws but claw-like elongated nails.

These are only the most important trends in the evolutionary process which led from the insectivores to the primates. One can also determine these trends on the basis of fossilized skeletal remains. Numerous other features in the anatomical structure, the composition of the blood serum, the embryological development, and the behavior seem to lend further support to this hypothesis.

Since the phylogenetic process is slow and progresses only in very minute stages, it seems obvious that the boundary between the insectivores and the primates is rather arbitrary. The question comes to mind as to whether the primates, particularly the prosimians, are a natural entity or if they are only an evolutionary stage of various animal groups which have reached approximately the same level of organization. Fossil remains have not been of help in determining if the tree shrews belong to the insectivores, where they were first classified, or if they are the most primitive living primates (subprimates). For a period of time the fossilized *Anagalidae* (*Anagale* and *Anagalopsis*) of the lower

◁◁
The tarsiers (Eastern tarsiers—*Tarsius spectrum*) are even more highly specialized leapers than the galagos. Their tarsus is markedly elongated and greatly facilitates jumping in this primate. The term "tarsier" relates to the characteristically elongated tarsus.

◁
Despite his sloth-like slow movements, this slow loris *(Nycticebus coucang)* is closely related to the agile galagos.

Tertiary in Central and Eastern Asia were believed to be the fossilized relatives of the tree shrew. However, they are insectivores which evolved certain characteristics parallel to those of the tree shrews. Fossilized tree shrews have up to now not been identified with any degree of certainty. It is possible that individual forms from the lower Tertiary (i.e., *Adapisoriculus*) belong to this group. In any case living tree shrews can only be regarded as "model forms" of the primate ancestors but not as direct ancestors.

Fig. 12-2.
Primates are found in most warm countries of the globe (except in Australia). Most forms inhabit forests; however, some are also found in the savannah, steppes, and mountains. A few are found in regions where the winters are cold.

When did the first primates appear? In 1965 the primate-like tooth remains from the upper Cretaceous period of a small mammal of North America *(Purgatorius ceratops)* were described. However, tooth remains alone are rather insufficient evidence to determine whether or not this specimen was truly a primate. Just as we draw an arbitrary line between the extinct reptiles and the mammals on the basis of the lower jaw structure, so we also make an artificial separation between the fossilized insectivores and the primates on the basis of the middle ear development. The criterion for a primate is the bony bulla (Bulla tympanica) and the artery (Arteria promontoria) surrounded by the petrous portion of the temporal bone. Since none of these skull characteristics have been found in *Purgatorius ceratops,* its affiliation with the primates is still questionable.

The geologically oldest primate remains which have been clearly identified originated from the Paleocene of North America and Europe. These remains represent primates of different evolutionary lineages which demonstrates that in the Paleocene, approximately fifty to sixty million years ago, this mammalian order already showed a wide divergence. Besides rather primitive forms that possessed features otherwise unknown in primates, highly specialized members were frequent. Many of these had properties that are characteristic of insectivores (i.e., absence of the postorbitals, claws instead of nails) and yet they have highly developed, specialized structures, particularly in the dentition. The lower Tertiary *Plesiadapidae,* for example, had a reduced number of teeth, a diastema (a gap in the tooth row) and an enlargement of individual incisors. There were attempts to relate these and other lower Tertian primates *(Phenacolemuridae = Paromomyidae)* with the Madagascan aye-aye *(Daubentonia madagascariensis)* because of this specialized dentition. However, this relationship certainly does not hold because of numerous other differences. These remains belong to primate lines which became extinct without leaving any descendants that can be related to any group of prosimians living today.

In contrast, other fossilized forms from the lower Tertiary can be related to modern lemurs or they show similarities to the tarsiers. Up to now the oldest primate fossils have been recovered only from two continents: North America and Europe, which once were connected by a land bridge. Primate fossils of the Paleocene have not been discovered in either Africa or Asia. Nevertheless, it is noteworthy that the primates of the lower Tertiary in North America were either closely related to or had common ancestral forms which gave rise to many divergent groups. According to this view one can assume that this mammalian order, which is of such great significance to man, originated in the European-North American area.

Towards the end of the lower Tertiary and the succeeding epochs these small, inconspicuous primates spread to other regions of the globe. The lemurs flourished on Madagascar. The lorises and galagos maintained themselves in Southern Asia and Africa. The Southeastern Asiatic island complex became the refuge for the tarsiers. In South America the New World monkeys blossomed into a vast variety of forms ranging from the large-eyed night monkeys to the minute marmosets and tamarins, the acrobatic spider monkeys, and the intelligent capuchin monkeys. Finally, besides many arboreal and ground-dwelling monkeys, our closest relatives, the great apes and the ancestors to man, originated in the Old World.

Walter Fiedler
Erich Thenius

13 Tree Shrews and Prosimians

Suborder: Prosimians
by K. Kolar

The PROSIMIANS *(Prosimiae)* are a very diversified suborder among the primates. Some are reminiscent of the insectivores which gave rise to the primates. Others, for example the large lemurs and indris, have monkey-like bodies and limbs. However, their heads with the long pointed snouts are not at all monkey-like but more like dogs and foxes, particularly because the nostrils are located in a hairless moist rhinarium as is common for dogs and most other mammals. However, these are forms that are amazingly similar in their internal structure to the true monkeys.

Distinguishing characteristics

The animals range in size from that of a hamster to larger than foxes. The HRL is thirteen cm (lesser mouse lemur—the smallest primate) to ninety cm (indris). Extinct forms, which were still alive at the beginning of the Pleistocene, reached almost the size of chimpanzees. The extremely soft, broad pads on the fingers and toes and the flat nails are monkey-like features. There is a grooming claw on the second toe. Only the tree shrews have claws on all fingers and toes. The first finger and first toe are opposable except in the tree shrew. The second finger and second toe are often small. The fourth finger and fourth toe are always the longest. The fur is almost always soft and wooly. Tactile hairs are located on the inside of the lower arm above the wrist. These are vestigial only in the lorises and galagos. Certain quite primitive skull characteristics (incus, bony hard palate, ethmoid bone and conchae) link the prosimians, via the tree shrew, to the insectivores.

The animals have conspicuously large eyes, particularly in the nocturnal species where cones are absent from the retina. The nocturnal species cannot perceive color but can recognize the slightest variation of light intensity in the twilight. A slightly light-reflecting layer, the tapetum (Tapetum lucidum) is present in the choroid of nocturnal forms. The tapetum is most noticeable when the eyes are illuminated by light rays at night. The sense of hearing is well developed. Prominent large ear pinnae are frequent. The olfactory lobes of the cerebral hemispheres are larger than in the monkeys but are decidely smaller than in the mammals with a highly developed olfactory sense. In the tarsiers the olfactory lobes are still smaller than in the other prosimians.

In the predominantly jumping forms, the tarsals, tibia, and fibula are elongated, resulting in a particularly long foot. The dental formula is usually as follows: $\frac{2 \cdot 1 \cdot 3 \cdot 3}{2 \cdot 1 \cdot 3 \cdot 3}$, with the exception of the tree shrew (see p. 273). In adults of some genera there is a reduction of teeth. The central incisors in the upper jaw are separated by a gap just as in the insectivores and bats.

The uterus is bicornuate. The placenta is villous with a large amniotic cavity (except in the tarsiers). Usually there are definite breeding seasons. There are one to three young which are cared for by the mother either in the nest or are carried about by her on her abdomen.

There are seven families which we divide into four infraorders: A. TREE SHREWS (Infraorder *Tupaiiformes*). 1. TREE SHREW *(Tupaiidae).* B. LEMURS (Infraorder *Lemuriformes*). 2. LEMURS *(Lemuridae).* 3. INDRIS *(Indriidae).* 4. AYE-AYE *(Daubentoniidae)* C. LORISES (Infraorder *Lorisiformes*). 5. LORISES *(Lorisidae).* 6. GALAGOS *(Galagidae).* D. TARSIERS (Infraorder *Tarsiiformes*). 7. TARSIERS *(Tarsiidae).*

What are prosimians?

When we hear of prosimians we usually think of the Madagascan lemurs, or possibly the galagos or bushbabies of Africa. However, there are many other forms which differ greatly from these two types. The tree shrews are reminiscent of squirrels and move very much like them. The lorises and pottos climb in branches in a slow, sloth-like manner. The unique aye-aye was classified with the rodents by the first discoverers because of its rodent-like incisors. In the large sifakas and indris one can notice the "transformation of the animal face into the physiognomy of the higher monkeys," according to Ludwig Koch-Isenburg. The tarsiers are still more similar to the monkeys because of the structure of the incisors, the formation of the disc-like placenta, and the composition of the blood serum.

The story of their discovery

Because of their very diverse characteristics the discovery of the prosimians is associated with a record of zoological confusion and disputes. The first galagos reached Europe already in the seventeenth century but at that time they were believed to be squirrels. Around 1680 some Dutch merchants found angwantibos on the coast of Guinea, and seventy years later the Dutch captain Bosman discovered the potto. Both species were first described as sloths, just like the lorises. In 1684 the Frenchman Jean de Thévenot saw the loris at the court of an Indian great mogul. The Dutch collector Albertus Seba displayed a picture of such an animal in his museum fifty years later. The French governor Gaston Etienne de Flacourt provided the first report of the Madagascan lemurs in the year 1658. He described white animals that had heads like dogs and bodies like men. He was probably talking about tne sifakas. Philibert de Commerson and Edwards described several more lemurs. On the basis of these accounts Linné coined the term *Prosimiae* ("pre-apes"). Etienne Geoffroy Saint Hilaire classified several species from illustrations prepared by travellers. The first living ring-tailed lemur was brought to Europe by Captain Issac Worth in 1748. Buffon also kept this species already in the menagerie of the Parisian "Jardin des Plantes."

For all that, the zoologists still were not certain which animals belonged to the prosimians and which did not. During his voyages to Madagascar between 1774 and 1781 Pierre Sonnerat was probably the first European to see free living aye-aye, woolly indris, and indris: He was responsible for naming the one species "aye-aye." According to Hill this is the nocturnal call of this animal, but other sources maintain that Sonnerat only misunderstood the exclamation of an astonished native. There was a similar misunderstanding when Sonnerat named the largest living prosimian "indri." This word was merely a native word meaning "There it is!" Buffon referred to the aye-aye as a jerboa, Gmelin referred to it as a squirrel, and other zoologists even classified it with the phalangers. According to Buffon the Southeast Asian tarsiers were also relatives of the jerboa. Linné even placed them with the opossums. Uncertainty as to the classification of prosimians persisted for a long time. Even today not all zoologists are in agreement, for example, on whether the tree shrews and tarsiers belong to the prosimians at all.

The well known English primate researcher W. C. Osman Hill excludes the tree shrews from the order primate because of their insectivore-like characteristics. The presence of the rhinarium in the lemurs, galagos, and lorises prompted Hill to classify these prosimians under the term *Strepsirhini* as opposed to the tarsiers and monkeys which were termed *Haplorhini*—without rhinarium. The view has often been held that the tarsiers, because of their elongated tarsals *(Tarsioidea)* were a suborder in themselves between the prosimians and the monkeys.

Family: *Tupaiidae*

The TREE SHREWS (Family *Tupaiidae)* are the most disputed of all the groups. If one sees these squirrel-like animals for the first time, one would not guess that they belonged to the prosimians. The natives of Malaya have just one name for squirrels and tree shrews and that is "Tupai." Recent investigations, particularly those of Le Gros Clark and Sprankel, have resulted in the idea that these small inconspicuous mammals "were officially raised from the rank of the insectivores to the rank of the prosimians" as a zoologist jokingly formulated it. Although the tree shrews still share certain characteristics with the insectivores, particularly the elephant shrews, they possess the following features of the prosimians: orbits which are completely encircled by bone, the elastic and partially cartilaginous subtongue, and several characteristics in the structure of the reproductive organs, the muscles and the skull. Accordingly, it seems that the tree shrews take up an intermediate position between the insectivores and prosimians whereby the prosimian characteristics predominate. Some researchers have labelled them as "living models" of the oldest primates.

Distinguishing characteristics

The animals may be as large as rats or squirrels. The HRL is 15-20 cm and the TL is just as long as the body but is slightly longer in the feather-tailed tree shrew. The pelage is dense and soft. The tail of the tree shrew is covered by thick, long hair. The tail of the feather-tailed tree shrew is covered by scales and two rows of long white hair at the tail tip. The ear is membranous and large. The centers of vision are well developed while the olfactory lobes are relatively small. There are claws on all fingers and toes. The thumbs are not opposable but the first toe is divergent. The subtongue and lower incisors are used for body care activities.

Fig. 13-1. Indian tree shrew (Genus *Anathana*)

The animals are diurnal and omnivorous. There are 38 teeth: $\frac{2 \cdot 1 \cdot 3 \cdot 3}{3 \cdot 1 \cdot 3 \cdot 3}$. The caecum is present. There are one to three pairs of mammae. This number varies in individual animals. The gestation period lasts 43-56 days. Litter size is from one to three young. They are fully grown at three months of age and are sexually mature by four months. There are two subfamilies and six genera, forty-seven species and approximately a hundred subspecies (a closer examination would probably sharply decrease the number of species and subspecies).

Fig. 13-2. 1. Tree shrew (Genus *Tupaia*)
2. Philippine tree shrew (Genus *Urogale*)
3. Smooth-tailed tree shrew (Genus *Dendrogale*)

A. TREE SHREWS (Subfamily *Tupaiinae*); five genera: 1. TREE SHREWS (genus *Tupaia*); fourteen species including the COMMON TREE SHREW (*Tupaia glis;* Color plate p. 253); Northern India, Southern China. 2. LARGE TREE SHREW (genus *Tana*). The HRL is up to twenty cm; it is found in Northern India and Indonesia. The fur coloration of the LARGE TREE SHREW *(Tana tana)* is highly variable, predominantly dark red to black. The snout is grey and there is a dark dorsal stripe. 3. INDIAN TREE SHREWS (genus *Anathana*) are represented by one species, *Anathana ellioti,* found in peninsular India west of the Ganges. 4. PHILIPPINE TREE SHREWS (genus *Urogale*) inhabit Mindanao Island and neighboring islands. The animals are predominantly found in the brush zone and the thick vegetation along river beds. They feed on the ground. The gestation period is 54-56 days. There is one species, the MINDANAO TREE SHREW *(Urogale everetti)*. 5. The SMALL SMOOTH-TAILED TREE SHREWS (genus *Dendrogale*) are smaller in body size. The ears are larger. The tail is covered by smooth hair. The animals inhabit the mountains of Cambodia, Northern Borneo and Northern Sumatra.

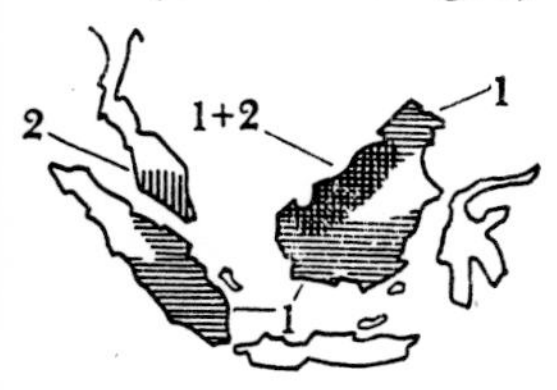

Fig. 13-3. 1. Large tree shrew (Genus *Tana*)
2. Pen-tailed tree shrew (Genus *Ptilocercus*)

B. The FEATHER-TAILED TREE SHREWS (subfamily *Ptilocercinae*) have a long tail which is covered with scales except for the last third which has two rows of long stiff hair on either side. There is only one genus *(Ptilocercus)* with one species *(Ptilocercus lowii)* and two subspecies: LOW'S

FEATHER-TAILED TREE SHREW (*Ptilocercus lowii lowii*; Color plate p. 253) is found on Borneo, and the MAINLAND FEATHER-TAILED TREE SHREW (*Ptilocercus lowii continentis*) is found in Malaysia.

At first they were considered to be squirrels

William Ellis, who accompanied Captain James Cook on his last journey, sketched a tree shrew which one of the ship's crew had shot on the coast of Indochina. Ellis thought that the animal was a squirrel but such cases of mistaken identity were to occur repeatedly. Undoubtedly, there is a similarity in body form, size, and movement between the tree shrews and the squirrels. The tree shrews even sit up on their hind legs in squirrel fashion and hold their food with their hands.

Very little is known about free living feather-tailed shrews. The animals are arboreal and apparently shelter in tree holes during the day where they rest on dried leaves. These tree shrews have never been kept outside of their range of distribution. Their long tails constantly move back and forth slightly, and are bent around their bodies while they are sleeping. The feather-like tail end hangs over their heads like a fan. The slightest touch of the tail hairs causes the animal to jerk back. The "feather hairs" also quiver when an insect approaches even though the tree shrew is asleep. Feather-tailed tree shrews frequently climb around on slim branches and the tail probably serves as a balancing organ.

Tree shrews in human care

Zoologists began to study the tree shrews more intensely when their prosimian characteristics were first observed in 1938. However, the study of these interesting animals has been almost exclusively limited to the most familiar species *Tupaia glis.* In order to supply laboratories with easily obtainable primates, attempts were made in Singapore to raise tree shrews in large numbers. Success was limited because the behavior and requirements of the tree shrews were not known. In the meantime, however, this situation has changed. In Europe tree shrews were first raised in the zoo of Cologne and in the Max-Planck Institute for Brain Research in Frankfurt by H. Sprankel. Later they were also raised in the Max-Planck Institute for Behaviorial Physiology in Seewiesen, in the Dresden Zoo, and in other zoological gardens as well.

Several other *Tupaia* species have been kept in zoos and were also occasionally bred and raised. A pair of great tree shrews, which reached the Breslau Zoo in 1938, were reported to be active from sunrise to dusk. In September the male courted the female, but young were not born.

Territorial behavior and scent marking

Tree shrews live in pairs and can be quite aggressive towards strange conspecifics. Each pair occupies a territory which is delineated by scent markings placed at specific spots. The glandular area on the front of the body secretes an oily, strong, odorous sticky substance. When marking an area the male stands with stiffened legs and rubs the front of his body back and forth on the object. The same areas are repeatedly marked by scent. Tree shrews also mark areas by running in tree branches, depositing drops of urine. Eventually a crust of urine forms on certain branches. Even the hands and feet of the animals may become impregnated with urine because they constantly walk around in the wetted areas. Similar urine impregnation of the hands and feet is found in the galagos (see p. 305).

If one wishes to raise tree shrews one has to provide a cage that is large enough to enable a pair to mark their own territory. Sprankel's tree shrews preferred animal food. Sweet fruit was second choice. The animals were agile in catching flying insects which were caught with the hands or mouth. Stones on the ground were turned over and the tree shrews would collect the exposed insects with their long extended tongue. Only later were the insects chewed and swallowed. Tree shrews also overpower adult mice and juvenile rats which are killed by a skillful bite into the nape. One of Sprankel's tree shrews once captured a newt by flicking it out of an aquarium with one hand.

Reproduction

Three weeks prior to giving birth the parents stop sleeping in the nest box but fill it with a thick layer of dried leaves, moss, pieces of wood and cellulose. The female gives birth in the farthest corner of the box. According to Sprankel, the young are suckled two to three times a day. During suckling the youngsters lie on their backs and kick against the teats with their front legs. When the young are satiated with milk their bellies are distended. On the sixteenth day one can already feel the tips of the teeth. The eyes open on the twentieth day. Towards the end of the fourth or beginning of the fifth week of life the young start to search for the food bowl and feed independently. The young require a lot of warmth and, according to Sprankel, would not survive without an infrared heating lamp placed close to the nest. Six weeks after parturition the female may become pregnant again. At this time the juveniles must be removed from the parent's cage because they become very aggressive towards the new brood.

Tree shrew babies were even raised by hand in the Dresden Zoo. Gotthart Berger reports that in the first week the infants received a mixture of tea and milk enriched with glucose and vitamins which was fed every two hours with a pipette. Robert Martin from the Max-Planck Institute at Seewiesen made the following observations. Into the cage of every pair he placed several nest boxes and observed that the young were raised in a box farthest removed from the sleeping box of the parents. The male was also active in lining the breeding box with beech leaves. He did not enter this nest again until the young left it when about four weeks of age. According to Martin the young hide deep within the beech leaves during the first week and are suckled by the mother only every forty-eight hours. The female too does not visit the young in the meantime.

Martin also observed this unique forty-eight hour rhythm in artificially raised tree shrews. Every second day a young would drink ten to twenty grams of milk which it suckled in the amazingly short period of four to ten minutes. After this the young would snuggle into the dried leaves and go back to sleep. Although the young are left alone, they maintain a constant body temperature of 37°C without heat supplied from the outside. Martin believes that tree shrew young have a greater chance of survival from enemies if the mother only visits them every second day in their shelter. In any case this behavior deviates totally from all other experiences which have been gathered from child care in primates generally. We still do not know if all tree shrew species behave this way or if this phenomenon is only an adaptation in response to specific living conditions.

However, Martin wants to classify the tree shrews separately from the prosimians for several other reasons also: "*Tupaia* shares certain characteristics with the primates only, some only with the marsupials, some only with the rodents, and others only with the lagomorphs." There is certainty (according to Martin) "that the tree shrews are not a link between the insectivores and primates." In this context it is also of interest to mention that the tree shrews in contrast to most other prosimians possess cones in the retina and can perceive colors. Tigges demonstrated in his experiments that tree shrews can distinctly differentiate the colors red, yellow, green, and blue from shades of grey. Red had the greatest stimulating value to the tree shrews.

Thus we are still not certain if the controversial tree shrews are really living models of those early ancestors which took the first steps from the insectivores in the direction of the primates and thereby into our line of ancestors. A final explanation to this important problem is probably only possible if more fossils are found and when we find out more about the juvenile development and behavior of other prosimians.

Lemuriformes

Even young tree shrews enjoy combing each other's fur with the almost horizontally protruding incisors. This mutual grooming is even more frequently observed in the few social species of the Madagascan lemurs. The behavior is similar to that of the monkeys. The three families of the Madagascan prosimians, the lemurs *(Lemuridae)*, indris *(Indriidae)*, and the aye-ayes *(Daubentoniidae)* are already closely related and combined in the infraorder *Lemuriformes*.

The old Romans called the spirits of their dead lemurs. They believed that the spirits sought out the light at night and looked on the living with glowing eyes while lamenting plaintively. Although the Madagascan lemurs do not have any spooky properties, their staring eyes which glow reddish in the dark and their frequent loud voices earned them the name of the Roman ghosts, given to them by French researchers.

Phylogeny by E. Thenius

Relatives of the lemurs have been found from the lower Tertiary strata in Europe and America. Today the lemurs are limited to Madagascar and the neighboring islands of the Comoros. The zoogeography of Madagascar is similar in certain respects to the island continent of Australia. The early separation from neighboring land masses resulted in the preservation and highly specialized development of primitive animal forms. The higher evolved mammalian forms developed at a later date and were only present in small numbers or not at all in these isolated earth masses. The Australian monotremes and marsupials were not subjected to the competition of ungulates and carnivores. Similarly, the higher monkeys could not reach Madagascar and compete for the prosimians' habitat. Two fifths of the Madagascan mammals are therefore prosimians.

Present-day lemurs by K. Kolar

In Madagascar the lemurs play a role similar to that of the monkeys in neighboring Africa. The lemurs have exploited a great variety of ecological niches. As a result of parallel evolution certain species are reminiscent of African prosimians, and others resemble the true monkeys and apes. The lesser mouse lemurs, with their naked ears and heavily padded finger and toe tips, remind one of the galagos. Buettner-Janusch compares the liveliness of the ground dwelling ring-tailed lemur to the social behavior of the baboons and macaques. The sifakas and the almost tailless indris have an erect posture like the gibbons. Even the howling is surprisingly similar to that of the gibbons, according to Andrew.

Family: *Lemuridae*—the true lemurs

Distinguishing characteristics

The lemurs (Family *Lemuridae*) are the most diverse family of the Madagascan prosimians and have the largest number of species. Lemurs vary in size from rat to cat. The HRL is 11–50 cm and the TL is 12–70 cm. The fur is soft and often vividly colored. The tail is covered by dense hair and is often bushy. The mouth region, particularly in the larger species, is elongated into a "fox-like" snout. The ear is of a unique structure: the incus is suspended semicircularly in the middle ear cavity (Bulla auditiva). The hind legs are noticeably longer than the front legs. There are nails on all fingers and toes which are pointed in claw-like fashion in the smaller species. Only the second toe has strongly curved grooming claws.

The diet consists of plant and animal matter. The weasel lemurs are strict leaf eaters. There are 36 teeth: $\frac{2 \cdot 1 \cdot 3 \cdot 3}{2 \cdot 1 \cdot 3 \cdot 3}$, except in the adult weasel lemurs (see page 281). The lower incisors and canines are almost horizontal and form a "comb." The first premolars in the lower jar are shaped like canines and oppose the upper canines.

Subfamily: *Cheirogaleinae*

The DWARF LEMURS (Subfamily *Cheirogaleinae*) are, on the average, smaller than the other lemurs. The ear pinnae are relatively large. The animals have large nocturnal eyes. The tarsals are elongated. The fur is very dense and woolly. The body temperature is only incompletely regulated. During the dry season and in cool temperatures the animals enter into a "lethargic," torpid state. Fat deposits on the hind legs, the surface of the tail base, or the tail supply the body with energy during this torpid state. There are three genera with six species:

A. The MOUSE LEMURS (genus *Microcebus*) are superficially reminiscent of galagos. The animals have large ears which cannot be folded. The hands are less capable of grasping. The thumbs are less well developed and are not as opposable. There are two species: 1. The LESSER MOUSE LEMUR (*Microcebus murinus;* Color plates p. 248, p. 263, and pp. 283-284) is the smallest primate. The HRL is 11-13 cm, and the TL is 12 cm. The body weight is 50 g. This animal is found in moist forests but also in light stands of reed or even matted euphorbia bushes in the arid regions. 2. COQUEREL'S MOUSE LEMUR *(⊕Microcebus coquereli)* is approximately twice as large as the lesser mouse lemur. It is found in Southwestern and Northwestern Madagascar.

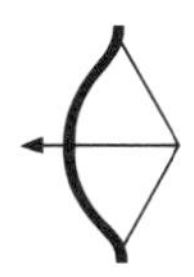

B. The DWARF LEMURS (genus *Cheirogaleus*) are similar to squirrels in body form and movement. There are three species: 1. The HAIRY-EARED DWARF LEMUR *(⊕Cheirogaleus trichotis)* has an HRL of fourteen cm and a TL of approximately sixteen cm. There are pronounced hair tufts on the ear. Up till now only four specimens have been found in mountainous forests of Eastern Madagascar. 2. The FAT-TAILED DWARF LEMUR *(⊕Cheirogaleus medius)* is somewhat larger. A semi-fluid layer of fat up to two cm may be deposited in the tail. 3. THE GREATER DWARF LEMUR *(Cheirogaleus major)* is much larger. The total body length is 55 cm. There are two subspecies found in the forests of Southeastern and Western Madagascar.

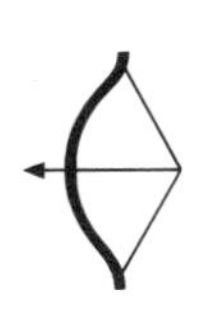

C. The FORK-MARKED MOUSE LEMUR *(⊕Phaner furcifer)* is the largest species of this subfamily. The total body length is 60 cm. The animals are in a subfamily of their own because of the different dentition: the upper incisors are strongly elongated and the upper premolars are canine-like, giving the impression of two canines on each side of the upper jaw. The large ears are naked and rounded.

The lesser mouse lemurs feed mainly on insects. Some of these prosimians kept by humans were observed to hold grasshoppers and beetles with both hands and to eat their soft abdomen first with loud smacking sounds. They also consumed honey and plant juices. It is probable that free living lesser mouse lemurs also take an occasional sleeping bird. When grooming, the animals use both hands at the same time.

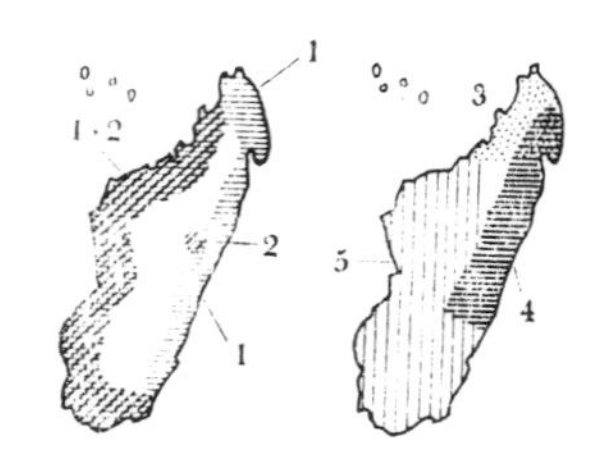

Fig. 13-4.
1. Lesser mouse lemur *(Microcebus murinus)*
2. Dwarf lemurs (Genus *Cheirogaleus)*
3. Fork-marked mouse lemur *(Phaner furcifer)*
4. Broad-nosed gentle lemur *(Hapalemur simus)*
5. Grey gentle lemur *(Hapalemur griseus)*

The lesser mouse lemurs produce calls which are often within our range of hearing and can therefore be perceived by man. One can easily confuse the sounds of free living mouse lemurs with those of insectivores. The animals are nocturnal and are therefore very difficult to observe in their natural habitat. During the day these small lemurs not only sleep in tree holes, but, according to several sources, also construct sleeping nests out of branches and leaves which they line

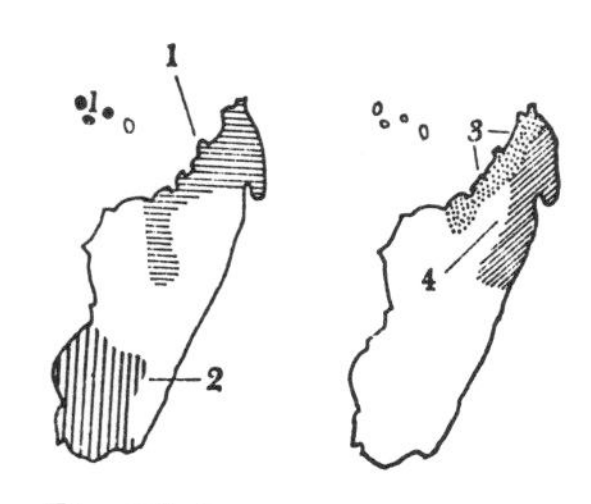

Fig. 13-5.
1. Mongoose lemur *(Lemur mongoz)*
2. Ring-tailed lemur *(Lemur catta)*
3. Black lemur *(Lemur macaco)*
4. Ruffed lemur *(Lemur variegatus)*

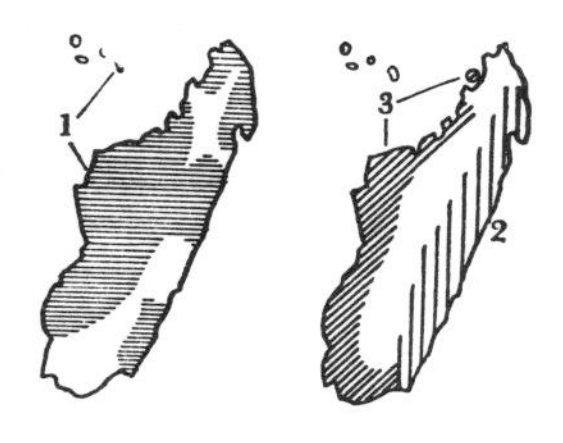

Fig. 13-6.
1. Brown lemur *(Lemur fulvus)*
2. Sportive lemur *(Lepilemur mustelinus;* see p. 281)
3. Lesser sportive lemur *(Lepilemur ruficaudatus;* see p. 281)

with hair. It is difficult to differentiate these nests from those of birds. The possibility exists that the mouse lemur may merely occupy abandoned birds' nests. Starmühlner of the Vienna Institute for Zoology, who kept mouse lemurs for many years, always provided his animals with a plentiful supply of nesting material but he was never able to observe that the animals really constructed a nest. Coquerel's mouse lemurs have repeatedly been kept. Von Haacke kept them in the Frankfurt Zoo. One pair kept in the London Zoo repeatedly had litters. Some of these lived up to fifteen years, which is a surprisingly long life for such a small prosimian.

The much smaller lesser mouse lemur belongs to one of the most frequent lemur species of Madagascar. The animal is even too small for the natives to bother with and to cook in one of their dishes. Attenborough received twenty-two of these dwarfs in one evening. The lemurs were caught in the high trees close to his observation post. Sometimes one finds three or four lesser mouse lemurs huddling together in a tree hollow. Particularly during the dry season it is not too difficult to catch them. During this period they are in a state of torpidity. They are stiff and live off the fat deposits on the hind legs and the base of the tail. According to Petter, these small lemurs fall into a longer state of torpidity only if the outside temperature is below 18°C. When the animal enters a state of torpidity it becomes less and less active and the body temperature drops until the body is "paralyzed." Lesser mouse lemurs kept in the Vienna Institute for Zoology never showed any symptoms of this "hibernating" tendency.

The zoologist Jean Jacques Petter has greatly contributed to our knowledge of the lesser mouse lemurs. He has kept small lemurs for many years and breeds them regularly. The gestation period is fifty-nine to sixty-two days. A female gains six to seven grams during the pregnancy. Two to three young are born in a nest of leaves or in a hollow tree. Right after the birth the mother eats the placenta and bites through the umbilical cord just in front of the young's abdomen. Since lesser mouse lemur adults are small, one can imagine how minute the little ones must be. They weigh between 2.7 and 4.3 g and are only 3.7 to 5 cm long. The tail measures 3 cm. However, the newborn is already covered with greyish brown hair and has canines and incisors. Their eyes open on the fourth day.

During the first hours after birth the mother very eagerly licks the youngsters. She does this while holding them in her hands. Abandoned youngsters emit shrill sounds to which the mother responds immediately with a soft call. In contrast to most prosimians, the lesser mouse lemurs do not hold on to their mother's abdomen but she grabs the little ones in the flank region with her teeth and transports them in her mouth. They are never carried on the back like young of the larger lemurs or monkeys. When three weeks old, the young are already good climbers, and they are beginning to be more skillful in running and jumping. As the little ones grow up they become increasingly more playful. They wrestle with each other, chase their tails, pursue their sisters or brothers and knock each other down. The mother too is often engaged in playing with them. The youngsters climb up on her back and pinch her hands and ears.

At two months of age the juveniles behave like adults. They are probably sexually mature by seven to ten months. However, females kept by Petter became pregnant at a much later date. The youngest female was eighteen months old when she had her first litter. The reproductive process of the greater dwarf lemurs is similar to that of the lesser mouse lemurs except the gestation period lasts seventy days.

The greater dwarf lemurs built elaborate nests of grass and leaves in the crowns of trees. Natives in the eastern region of the island report that these lemurs bury themselves in rotting tree stems during the dry season. During these rest periods the body temperature of all of the dwarf lemur species drops considerably. According to Petter's measurements the body temperature of a fat-tailed dwarf lemur was 17.5°C while the outside temperature was 16°. In the higher mammals this phenomenon is otherwise only known in several insectivores, bats, and rodents.

Subfamily: *Lemurinae*

The TYPICAL LEMURS (Subfamily *Lemurinae*) are much more conspicuous than the small nocturnal lesser mouse lemurs. They are the typical animals of Madagascar. The HRL is 30-50 cm and the TL is 20-70 cm. Usually there are only one pair of teats which are prominant in the breeding season (The gentle lemurs and the ruffed lemurs are exceptions.). Some species are diurnal. There are three genera and ten species.

A. The GENTLE LEMURS (genus *Hapalemur*; Color plate p.249) have round heads and blunt snouts. The small broad ears are covered by dense hair and are almost hidden within the fur. The limbs are short and the tail is long and bushy. Two pairs of mammae are located in the axillary and groin regions. Glands on the underarm are similar to those found in the ring-tailed lemur (for territorial marking). The upper incisors lie somewhat behind each other. The canines are very small. There are two species: 1. The GREY GENTLE LEMUR (♁ *Hapalemur griseus*) has a total body length of approximately seventy cm. It is found in the bamboo thickets of the west coast and inner Madagascar. The animal feeds on bamboo shoots, sugar cane, and grasses. 2. The BROAD-NOSED GENTLE LEMUR (♁ *Hapalemur simus*) has a total body length of approximately ninety cm. It is found in the low reed-covered regions. The animal prefers a reed diet. It even enters drifting reed banks.

B. TRUE LEMURS (Genus *Lemur*; Color plates p. 250, pp. 283-4) have "fox-like" faces. The limbs and the tail are long. The glands on the underside of the arm are only present in the ring-tailed lemur. All species are

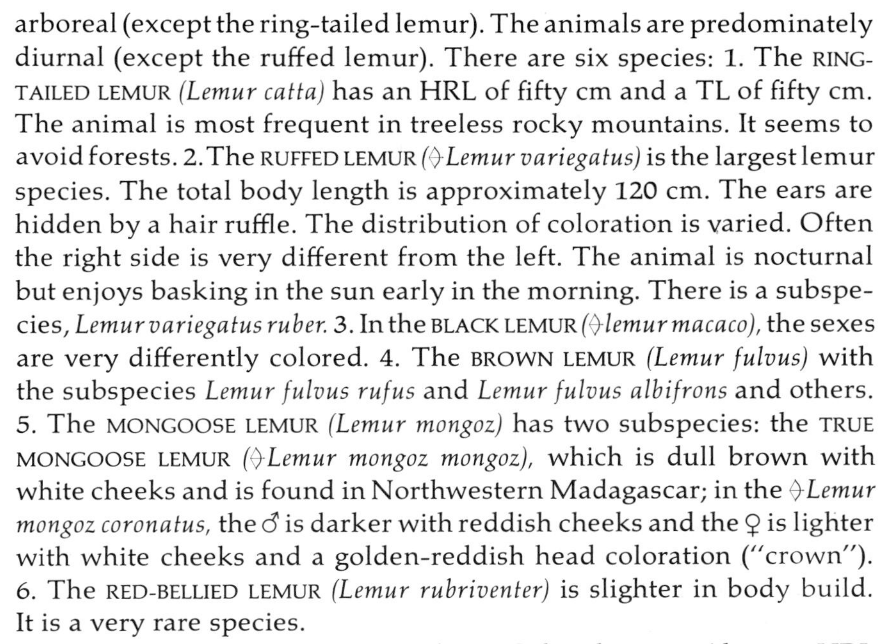

arboreal (except the ring-tailed lemur). The animals are predominately diurnal (except the ruffed lemur). There are six species: 1. The RING-TAILED LEMUR *(Lemur catta)* has an HRL of fifty cm and a TL of fifty cm. The animal is most frequent in treeless rocky mountains. It seems to avoid forests. 2. The RUFFED LEMUR *(Lemur variegatus)* is the largest lemur species. The total body length is approximately 120 cm. The ears are hidden by a hair ruffle. The distribution of coloration is varied. Often the right side is very different from the left. The animal is nocturnal but enjoys basking in the sun early in the morning. There is a subspecies, *Lemur variegatus ruber.* 3. In the BLACK LEMUR *(lemur macaco)*, the sexes are very differently colored. 4. The BROWN LEMUR *(Lemur fulvus)* with the subspecies *Lemur fulvus rufus* and *Lemur fulvus albifrons* and others. 5. The MONGOOSE LEMUR *(Lemur mongoz)* has two subspecies: the TRUE MONGOOSE LEMUR *(Lemur mongoz mongoz)*, which is dull brown with white cheeks and is found in Northwestern Madagascar; in the *Lemur mongoz coronatus,* the ♂ is darker with reddish cheeks and the ♀ is lighter with white cheeks and a golden-reddish head coloration ("crown"). 6. The RED-BELLIED LEMUR *(Lemur rubriventer)* is slighter in body build. It is a very rare species.

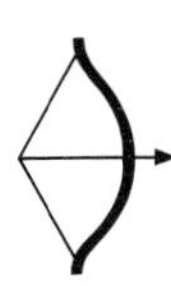

C. The WEASEL LEMURS (genus *Lepilemur;* Color plate p.249) have an HRL of 30–35 cm and a TL of 20–25 cm. The animals are slender and very agile. The head is round and the snout is short and pointed. The eyes are small. The ears are not covered by hair. The limbs are long. The animals feed on leaves. The salivary glands are extremely large. Juveniles lose the upper incisors. The tooth formula for adults is: $\frac{0 \cdot 1 \cdot 3 \cdot 3}{2 \cdot 1 \cdot 3 \cdot 3}$. There are two species: 1. The LESSER SPORTIVE LEMUR *(Lepilemur ruficaudatus)*; and 2. The SPORTIVE LEMUR *(Lepilemur mustelinus).*

The gentle lemurs usually sleep during the day; then they are so sluggish that they can easily be caught. For this reason they are frequently caught by the natives. With the fall of night the animals become very active. They have a reputation as appealing house pets. Under human care they quickly convert to a diurnal rhythm. Although the animals prefer fibrous food, they take without great difficulty to other foods. However, they do show certain food preferences. One prosimian only ate the peels of bananas and left the fruit.

In the zoo of Saarbrücken which has an excellent collection of lemurs, a broad-nosed gentle lemur shares the cage with a mongoose lemur *(Lemur mongoz coronatus).* The two animals get along well and, according to K. H. Winkelsträter, the gentle lemur thrives on a balanced diet of fruit, carrots, chicory, willow, and acacia leaves.

The lemurs (genus *Lemur*) can be clearly differentiated from the gentle lemurs because of their "foxy" faces, long arms and legs, and lively temperament. Aside from the ruffed lemur, the lemurs are primarily diurnal; in fact, they are regular "sun worshippers."

The ring-tailed lemurs possess a hairless glandular region on the

P. BARRUEL

The fauna of Madagascar

It differs greatly from that of the African mainland which is nearby. Instead of monkeys we find a great variety of prosimians. The carnivores are only represented by civet cats, and the insectivores almost exclusively by tenrecs. Even the Madagascan birds are unique. In this case our color plate is exceptional in that it does not demonstrate a specific biotope but illustrates a cross-section of the Madagascan fauna. □ Mammals: 1. Lesser mouse lemur *(Microcebus murinus)*; 2. Verreaux's sifaka *(Propithecus verreauxi)*; 3. Brown lemur *(Lemur fulvus)*; 4. Aye-aye *(Daubentonia madagascariensis)*; 5. Indris *(Indri indri)*; 6. Ring-tailed lemur *(Lemur catta)*; 7. Falanouc *(Eupleres goudoti)*; 8. Tailless tenrec *(Tenrec ecaudatus)*; 9. Madagascar broad-striped mongoose *(Galidictis striata)*; 10. Streaked tenrec *(Hemicentetes semispinosus)*; 11. Fossa *(Cryptoprocta ferox)*; □ Birds: 12. Short-legged ground roller *(Brachypteracias leptosomus)*; 13. Crested coua *(Coua cristata)*; 14. Helmet bird *(Euryceros prevosti)*; 15. *Coua coerulea*; 16. Sicklebill *(Falculea palliata)*; 17. Cuckoo roller *(Leptosomus discolor)*; 18. *Philepitta castanea*; 19. Madagascar weaver *(Foudia madagascariensis)*; 20. Ratvanga *(Schetba rufa)*; 21. Vasa *(Coracopsis nigra)*; 22. Ground roller *(Atelornis pittoides)*; 23. Long-tailed ground roller *(Uratelornis chimaera)*; 24. Brown mesite *(Mesoenas unicolor)*; □ Reptiles: 25. Gecko *(Phelsuma)*; 26. *Uroplates fimbriatus*; 27. Leaf-nosed adder *(Langaha)*; □ Amphibians: 28. Golden frog *(Mantella)*

Fig. 13-7.
This is how the ring-tailed lemur marks a branch with the glands in his anal region.

inside of the lower arm above the wrist. The males have an additional gland on the upper arm which secretes a strong smelling fluid. Every foreign object found in the lemur's territory or cage is marked with this glandular secretion. A typical behavior pattern of the ring-tailed lemur is rubbing with the tail the odorous substances of the underarm glands. The lemur sits upright and alternately pulls the tail through the inside of the left and right arm. This odor marking probably serves to impress (display behavior) other conspecifics. The animal waves the impregnated tail in the direction of a conspecific. Ring-tailed lemurs also mark with glands that surround the anus (anal glands). These lemurs were observed in the Frankfurt Zoo to stand on their hands and rub or press the anal region along the perches, iron bars, and door frames. Winkelsträter of the Saarbrücken Zoo observed similar behavior in the brown lemurs. Gentle lemurs also impregnate their territory with the odorous secretions of their forearm and anal glands.

According to Hill, free living ring-tailed lemurs feed on wild figs, bananas, and above all fig thistles of which the animals remove the hard shell with their incisors. They hold the fruit in their hands and bite off each piece not with their incisors, but with the molars so that the juice runs directly into their mouths and does not wet their fur. On the whole, many observers have remarked on the cleanliness of the animals. They are constantly grooming their fur with the grooming claws of their hind feet. In all likelihood the animals do not drink at all in their rocky habitat but quench their thirst with juicy fruits.

Apparently ring-tailed lemurs jump quite effortlessly up vertical walls up to three meters high. Small ledges are sufficient for the animal to gain a hold and then to propel itself up again. Like all diurnal lemurs the ring-tailed lemur seeks exposure to the sun frequently. When sunbathing the animals sit spreading their legs and arms. One of the best authorities on prosimians, Georges Basilewsky, recommends that one keep ring-tailed lemurs and other lemurs in large cages that are fully exposed to the sun.

Ring-tailed lemurs show themselves to be most contented if they are kept in large groups. During sleep the animal curls up, buries its head and arms between the legs, and swings the tail foreword underneath the body and then back over the shoulder onto the back. When it is cool the lemurs huddle together to keep each other warm. Their mewing and purring mood sounds are reminiscent of cats. The animals also emit these sounds when they are petted or when one gives them a tidbit. The warning sound is a barking call. When the lemurs are frightened they make shrill sounds. The young call their mother in a similar way.

It is interesting that in a group of ring-tailed lemurs the males have to be in the majority to promote a healthy social and reproductive unity. This is also true for their closest relatives. Lemurs kept in pairs are often sluggish and are extremely choosy in their foods, and they breed very poorly. According to Basilewsky, a viable group of lemurs consists of five to six with a majority of males. This male surplus seems to be a stimulating factor during the rutting season. The oldest female is respected by the males and the remaining group members. Groups of fifteen to twenty animals which have a strict rank order are not uncommon.

The young are raised not only by their parents but by the entire group. Frequently mothers exchange their babies. According to Petter the gestation period lasts 120 to 135 days. The young are usually born in March or April. At birth the eyes of the young are blue and only later does the iris turn into a brilliant yellow. Occasionally twins are born. Hill reports that a ring-tailed mother once ate one of the twins but raised the other. According to Déchambre, the ring-tailed lemurs, in contrast to other lemurs, do not carry their young across the abdomen but carry them in line with their body axis like monkeys. The young make their first climbing attempts when they are three weeks old. At one month of age the young separate from the mother but return to her for food and sleeping. According to Basilewsky, ring-tailed lemurs may adopt orphaned youngsters from other groups.

Five month old offspring are still nursed by their mother, which is also true for related species. At six months the young are completely independent and they are sexually mature by eighteen months. The males, however, have few opportunities for copulation. Only when they are 2½ years old are they strong enough to conquer a female for themselves.

Since ring-tailed lemurs, if kept properly, are quite prolific and meet the demands of zoos, they are very popular and are better known than other Madagascan lemurs. In the Pretoria Zoo in South Africa, seventy-two ring-tailed lemurs were born between 1904 and 1964. This included ten sets of twins and one of triplets. The private zoo of Georges Basilewsky in Cros de Cagnes near Nice has supplied many zoos with ring-tailed lemurs that were raised there. In Copenhagen a large group of these lemurs occupy a former outside enclosure for baboons. The animals are able to enter a heated room through a free swinging door any time they wish. In the Cologne Zoo the lemurs spent the greater part of the year on two connected islands located within a large pond. When the warden brings food the lemurs meet him by jumping into his boat. However, lemurs are not able to swim and if one misses its aim and falls into the water it has to be rescued quickly.

All other species of the genus *Lemur* are rarely seen in zoos outside of Madagascar. The Malagasy Republic has issued strict export regulations which make it near to impossible to obtain prosimians; however, these laws have not prevented the gradual dwindling of the larger species. According to recent reports nine tenths of the original

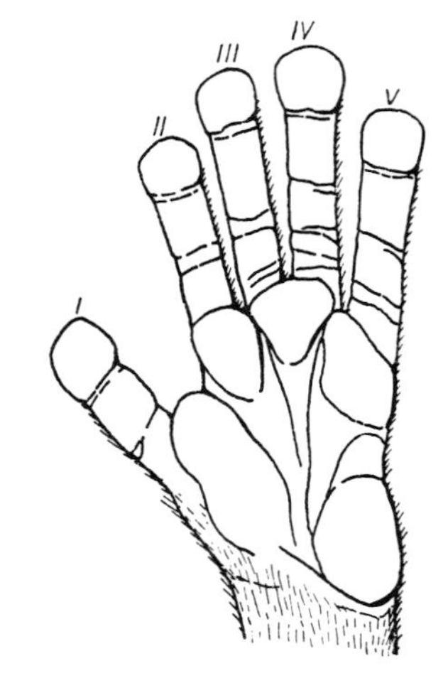

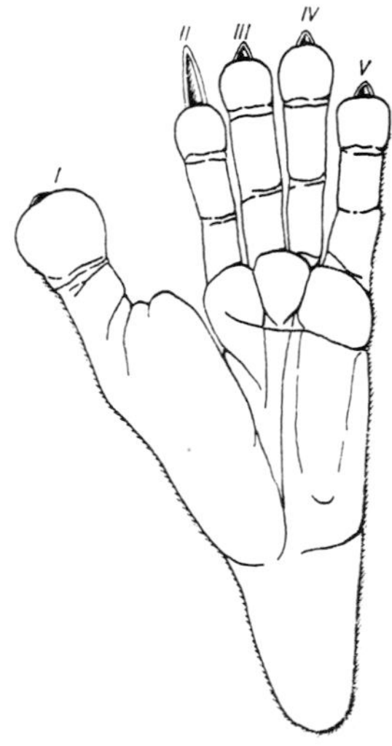

Fig. 13-8.
Left hand (top) and left foot (bottom) of a ring-tailed lemur.

forest in Madagascar has been logged or burned down to provide room for agriculture and grazing land for cattle. Man is constantly reducing the habitat of lemurs and is also killing the animals with modern weapons without heeding the laws. Thus the protective laws do not seem to prevent the extermination of the species but rather only prevent their export into other countries where the animals could be raised and possibly be saved from extinction.

The ruffed lemur belongs to one of the threatened species. In contrast to other lemurs the animal is nocturnal but does sunbathe early in the morning. When basking in the sun during the morning, the ruffed lemur stretches out its legs and turns its face towards the sun. Because of this, the natives formerly believed that the "Varikandanas" were holy animals which worshipped the sun and therefore they hardly hunted them. Unfortunately, civilization has thoroughly destroyed this belief, along with many other legends which proclaimed that the lemurs embodied the souls of the ancestors. With the introduction of modern firearms, the holy respect for the "sun children" vanished as did more and more of the ruffed lemurs.

The ruffed lemur is the only lemur which builds a nest. Before giving birth the female pulls hair out of her flanks and uses it to pad the nursery. New-born ruffed lemurs are able to see and have the same checkered fur pattern as their parents. The female carries the young by the skin of the flanks in a manner similar to the dwarf and weasel lemurs. When suckling, the mother lies directly on her offspring. Up to three pairs of mammae may be found on the chest, abdomen, and the flanks. The young develop slowly. At five weeks they are still poor at climbing. When two months old, they already play with their father, roll about on their backs, and try out their teeth.

Free living ruffed lemurs are conspicuous because of their extremely loud voices. Their howling consists of a sequence of roars which increase in intensity and are followed by various gurgling sounds. When the sudden calls of the ruffed lemur penetrate the nightly silence they have a particularly strange and eerie effect. The extremely thick pelage of the ruffed lemur is an excellent adaptation for living in the rain forest in northern Madagascar. Even heavy downpours cannot penetrate through their fur. When looking at the conspicuously speckled animal inside the cage, it would seem obvious that they are easily seen in the jungle and are therefore endangered. As a matter of fact, the vivid pattern serves as a camouflage. The fur colors seem to blend with the leaves of the trees, making the animal almost invisible. This breaking up of the colors is a phenomenon found in many other species as well. It is known as Somatolysis.

Free living ruffed lemurs primarily feed on leaves and fruits. In zoos the animals also prefer a vegetarian diet. Ruffed lemurs kept in Tananarive, besides bananas and leaves, enjoyed eating stick insects.

Captive black lemurs, on the other hand, consumed fruit, vegetables, ground insects, eggs, soaked dog biscuits, and above all preferred to eat birds. These lemurs very expertly cracked the bird's skull and ate the brain. It seems that most lemurs require much more meat supplement than they were usually given in zoological gardens. Lemurs in the Frankfurt Zoo receive a diet of fruit, vegetables, a cake baked with grain and soybean flour, minced meat supplemented with vitamins and minerals, and freshly killed chicks and mice.

Males and females were once thought to be separate species

Since the males of the black lemurs are black and the females are rufous, they were at first believed to be two different species. The females were described as "white-bearded lemurs *(Lemur leucomystax).*"

Black lemurs are arboreal and inhabit thick jungles. They are particularly active in the morning and evening. The lemurs are able to jump up to eight meters. Some observers have compared their unbelievably quick movements with the flight of a bird. If one of these lemurs is chased for a period of time the animal changes its tactics to deceive the pursuer. It will suddenly drop from the branches and will seek cover in the thicket, then run along the ground for awhile to a distant high tree. There it will climb up to the top with lightning speed. This behavior probably arose from the fact that only birds of prey and snakes are to be feared in the tree tops. The ground is relatively safe since there are only a few species of civet cats on Madagascar.

Although the black lemurs do not possess odor glands on the arms, they mark their territory by rubbing their palms on various objects. Vocalization plays a far greater role in demarcating their territory. This method is also employed by several other prosimians, many New World monkeys, and gibbons; especially in the evenings when the animals gather at their sleeping places, one hears their piercing screams. European travellers have reported that the calls of a large number of Akumbas all at the same time are "fear inducing." Even the natives from Madagascar have a superstitious fear of these lemurs and leave them alone. Modern times, however, have unfortunately done away with these taboos.

Black lemurs, like all lemurs, engage in careful fur grooming. Often they comb their pelage with the horizontal lower incisors, the rough edged underside of the tongue, or the second toe. When a black lemur greets a conspecific or a familiar person he pushes out his lower jaw and goes through the motions of combing and licking the fur of the companion. This is a greeting ceremony which probably originated from the innate behavior patterns of reciprocal fur grooming.

Fig. 13-9.
This is the position of a ring-tailed lemur sunbathing.

Under human care black lemurs soon become tame and make engaging pets. They firmly attach themselves to their keeper and are much more agreeable than the slightly gruff ring-tailed lemurs. They also breed relatively easily in captivity. Following a gestation period of 135 days a single greyish-black offspring with a thin hair coat is born. Just as in the ring-tailed lemurs, the entire group takes an interest in the little one. Other females and even males lick and play with the young. A black lemur young in the Frankfurt Zoo was well known to the visitors because it could squeeze itself through the widely meshed cage wire and then jump around on the roofs of the enclosures. It even dared to enter the cage of the neighboring ring-tailed lemurs and quickly steal their food. The parents of this little fellow were not distressed and the young always returned unharmed.

If kept under favorable conditions lemurs may reach an advanced age. In the London Zoo one animal reached the age of twenty-seven years, and another in Cairo was nearly twenty-two years old. A ruffed lemur in Berlin reached the age of nineteen years. The other lemur species are also not too difficult to keep. Formerly most zoos kept lemurs when they were still available on the market. The animals readily bred in the zoos.

The large lemur collection in the Saarbrücken Zoo, which in 1965 consisted of twenty animals representing six species and ten subspecies, gave rise to hybrids from a cross between a black and a brown lemur and from a cross between a white and a mongoose lemur. Decades ago the Berlin Zoo raised a hybrid from a cross between a black lemur male and a mongoose lemur female. These viable hybrids, which also occured in other zoos, caused a bit of havoc among the taxonomists and resulted in a "new" species or subspecies.

The fact that other lemurs besides the ring-tailed lemur are seen in European zoos is primarily due to Georges Basilewsky's dedicated work. He keeps a record of all lemurs kept outside of Madagascar and he has attempted to prevent extinction of these animals by systematically breeding them. Approximately sixty zoos have been supplied with animals raised by him. The beautiful lemur collections consisting of many species which are found in Cros de Cagnes, Cologne, and Saarbrücken do not obscure the fact that the times have passed when every good zoo could keep a few lemur species in its monkey or small mammal house. Basilewsky has appealed to all zoos which still own lemurs to make an all-out effort to save the unique animals of Madagascar from final extinction.

The agile, almost acrobatic, arboreal weasel lemurs are hardly familiar in name even to persons who are otherwise knowledgeable about animals. Because of their specialized food requirements they are difficult to keep in zoos. They are even hard to keep in Madagascar and they usually die quickly in captivity. Weasel lemurs are less conspicuous than their larger relatives because they are nocturnal. With nightfall the animals come out of their shelters and then congregate in large groups which jump through the tree crowns with gigantic leaps in search of suitable feeding places. The natives of Madagascar

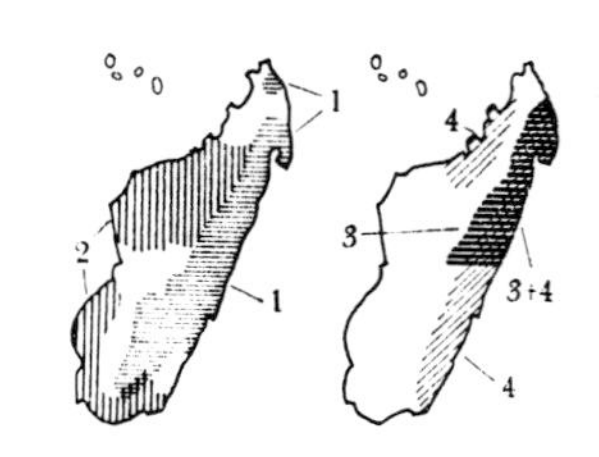

Fig. 13-10.
1. Diademed sifaka *(Propithecus diadema)* 2. Verreaux's sifaka *(Propithecus verreauxi)* 3. Indris *(Indri indri)* 4. Woolly indris *(Avahi laniger)*

consider the weasel lemurs a delicacy because of the animal's vegetarian diet. The Madagascans knock at hollow trees and other hideouts of the small lemurs and then kill them with sticks. However, a greater danger to the weasel lemurs is the destruction of their forest habitat which is constantly shrinking.

The breeding season of the weasel lemurs lasts from the beginning of May to the end of July. Males and females pursue each other and during this time one can frequently hear their calls. The gestation period is about 135 days. Between the middle of September and the end of October a single young is usually born. The young is covered with hair and has its eyes open. The young is ten cm long and the tail is eight cm. In one instance, after one month one young had grown only five cm but had tripled its body weight. During the first days of life the infant stays hidden in the mother's abdominal fur. The young is suckled for four months but at around six weeks it starts to eat from the leaves which the mother is consuming. The youngster still follows its mother when it is one year old and she is pregnant again. Separation between juvenile and mother takes place prior to the birth of the next young. The animal is sexually mature by a year and a half.

When jumping, weasel lemurs hold their bodies vertically and also they always leap from one vertical stem to another.

Family: *Indriidae*

The INDRISOID LEMURS (Family *Indriidae*) include the largest living prosimians. The HRL is 30–90 cm and the TL is 40–55 cm (only the indris is almost tailless). The animals are very slender and the fur is silky, except in the woolly indris. The short, broad snouts give them a more monkey-like facial expression than other lemurs. The hands and feet are very large. The thumbs are small and are only slightly opposable to the remaining fingers. The big toe is a large grasping organ which can be spread laterally for more than 90°. The remaining toes are connected with skin up to the first joint. The foot looks like a grasping claw consisting of two units—the big opposable toe and the joined four toes. The diet is highly specialized and consists of leaves, blossoms, bark, and fruits. Adults have 30 teeth: $\frac{2\cdot 1\cdot 2\cdot 3}{2\cdot 0\cdot 2\cdot 3}$. The digestive organs are adapted to a leaf diet. The salivary glands are large. One pair of mammae is found in the chest region. There are three genera with four species:

A. The SIFAKAS (genus *Propithecus*; Color plates p. 251 and p. 284) have a long tail. There is a gliding membrane (Patagium) between the upper arm and rump. The face is hairless and dark. The fur coloration varies greatly. The dark forms are found in the moist tropical regions and the light forms occur in the cool dry zones. There are two species with twelve subspecies. 1. VERREAUX'S SIFAKA *(◊Propithecus verreauxi)* has an HRL of forty-five cm and a TL of fifty-five cm. 2. The DIADEMED SIFAKA *(◊Propithecus diadema)* has an HRL of fifty-five cm and a TL of fifty cm.

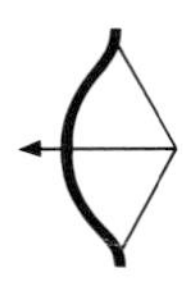

B. The WOOLLY INDRIS *(Avahi laniger*; Color plate p. 251) have an HRL thirty cm and a TL of forty cm. The head is round, almost globular. The large eyes seem to stare. The small ears are hidden in the fur. The thumb and big toe are large while the remaining fingers and toes are short.

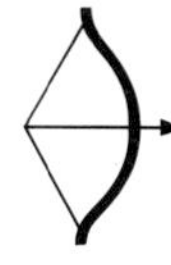

C. The INDRIS (◇ *Indri indri;* Color plates p. 251 and p. 284) are the largest species of lemurs living today. The animals are almost tailless. The HRL is ninety cm and the TL is up to five cm. The long face has a hairless black snout. The ears are large. The hands are six times as long as they are broad. The thumbs and big toes are well developed. The pelage is long and thick. Laterally located throat pouches connect to the larynx between the first and second tracheal cartilage. This probably serves to intensify the voice.

The few indrisoid lemurs which have survived into present times are only a faded splendor of the period when these large prosimians reigned supreme in Madagascar. Prior to human settlement from the Malaysian-Polynesian region the larger lemurs had hardly any enemies on this remote island. Certain Madagascan prosimian families had given rise to gigantic forms. Skulls from the extinct genus *Megaladapis* had almost reached the size of a donkey's head. Probably some of these giant lemurs occupied swamps. Other forms from the Family *Archaeolemuridae (Archaeolemur, Hadropithecus)* could be compared to monkeys and apes. The structure of the skull, dentition, and brain was similar to that of the leaf monkeys. However, these highly evolved promisians that are distantly related to the indris were not "future monkeys," but rather a result of parallel evolution similar to the New World and Old World monkeys. Of all the prosimian families which had a tendency towards "monkey similarity" and gigantism only the indrisoid lemurs remained. Giant indris of almost man size (*Palaeopropithecus, Mesopropithecus,* and *Neopropithecus*) were probably still alive when the first humans invaded Madagascar. The three genera living today are not nearly as large as these giants. Since these indris are highly specialized in their food requirements they cannot adapt themselves to a changing environment. They are only found in a few jungle regions and here their numbers dwindle from year to year.

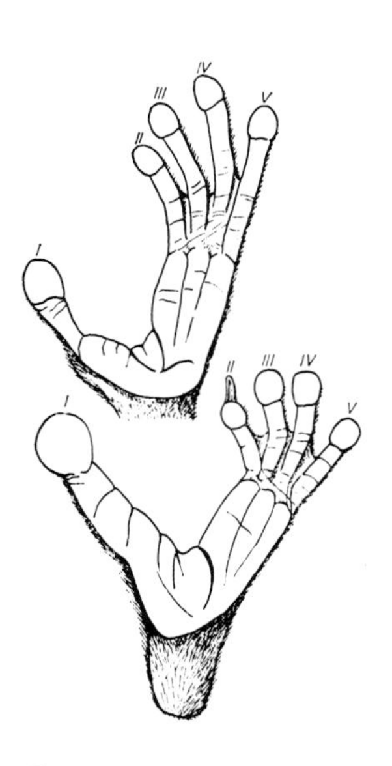

Fig. 13-11.
Left hand (top) and left foot (bottom) of the indri; they are especially suited for grasping branches.

The sifakas reach their peak of activity around the middle of the day when the sun is scorching the foliage of the jungle. During this time the animals can be seen leaping far from tree to tree. The powerful thrust of the legs propels the animal through space over distances of up to ten meters. On the ground the animal not only walks erect but can hop like a kangaroo. A single jump may cover a distance of four meters. In contrast to other lemurs their tail does not function as a balancing organ. When the sifaka is sitting on a branch it rests its hands on its knees and rolls up the tail between the legs into a spiral, very much in the manner of certain New World monkeys. The name "sifaka" was based on a sneezing sound the lemur produces which sounds like "schi-fak." Besides these sneezing sounds one often hears the lemurs produce a series of barking calls which the entire group emits at the same time.

Fig. 13-12.
A sifaka leaps off a branch...

...and after the long jump, extends his arms and hind legs frontwards in order to land vertically on a tree trunk.

Basilewsky reports that sifakas only eat specific leaves, blossoms, fruits, and chunks of bark. They prefer plants which are difficult to substitute outside of Madagascar (*Tamarindus indica, Mangifera indica, Lemuropsium edule* and others). Generally the movements of the sifakas—not counting the long leaps—are more deliberate than those of the true lemurs. A sifaka descends a tree in almost human fashion, backwards and with great caution. During the hot season sifakas climb down to the lower branches, lean against the tree trunk, and let their legs hang down or they even lie on a thick branch with arms and legs hanging down and doze.

In the morning the sifakas climb up on high trees and sit there with raised arms, facing east so that the sun can warm their chest. Because of this behavior the natives formerly believed that the sifakas, like the ruffed lemurs, were "holy sun worshippers." The sifakas were even thought to possess medical knowledge. It was believed that injured sifakas covered their wounds with special leaves which resulted in quick healing. According to the natives, sifaka females pull out the hair from the chest and underarm regions prior to giving birth to build a soft nest; so that the wind will not blow the hair away, the females weigh it down with stones. In reality, however, the young do not require a nest but cling to the mother's abdomen right after birth. Nevertheless, all these legends served to protect the sifakas from hunting for a long time. Today all these ancient cult opinions have disappeared. Our practical-minded modern world has no place for "holy" animals.

A sifaka female is pregnant for five months. In the northwestern part of the island the young are usually born in June, and in the eastern part, at the beginning of August. The new-born young has open eyes. Its short arms and relatively large, domed head are amazingly similar to a human infant's. The sifaka infant hangs across its mother's abdomen, half hiding its head in the fur. During the baby's first days of life the mother licks it frequently and moves very cautiously through the branches. When the young is thirty days old it climbs on its mother's back. At this stage, says Koch-Isenburg, the youngster with its skinny arms and black spindly fingers looks like a goblin. Its sounds are rather like the faint mewing of a cat. The thin tail of the young is always rolled up like the spring of a clock between the thighs. The young is now able to jump in a frog-like fashion and walk upright for a few steps, like a child, with its arms hanging down. "It is a posture which is amazing as well as delightful to watch."

At three months of age the little sifaka meets other conspecifics and plays with them exuberantly. The mother still carries the six to seven month old young which has by now reached two thirds of its adult body size. It is probable that these large lemurs are not sexually mature until they are two and one half years old.

In 1953, Koch-Isenburg artificially raised a young sifaka of approximately four weeks of age. Its mother had been killed by a falling tree. The tiny creature was fed by pipette with a mixture of cow's milk, tea, and sugar. It was given a rabbit fur to cling to. The sifaka quickly became very attached to its human keeper: "He could never get enough of playing with his human friends. One could squeeze and knead him and could pull him on his hind legs and let him dance on the top of the table. He 'joyously' joined in all the fun. He loved to be thrown up into the air and then came down with widely spread hands and feet, much like a parachute. We often tried to test out the limits of his restless energy. When we had already lost our patience he was still ready for more. Finally his eyes closed from exhaustion. He would look at us with glassy eyes and then go to sleep like an exhausted child."

It is impossible to obtain sifakas from Madagascar because of export restrictions. Even Koch-Isenburgh had to leave his pet at the Tananarive Zoo. In any case, sifakas were rarely kept in European zoos because the animals could not accept other foods. In 1908, a diademed sifaka lived for a short time in the London Zoo. The Berlin Zoo kept a Verreaux's sifaka in 1912, but, although the lemur was very tame and frequently sat on its keeper's shoulder, it did not survive for long. Today it is within the realm of possibility to fly in food throughout the year for these fastidious creatures or to plant suitable fruit trees in climatically favorable areas, as for example in San Diego. However, the export restrictions have limited our knowledge of the behavior and food requirements of sifakas in captivity to data which has been collected in Madagascan zoos or on privately kept animals.

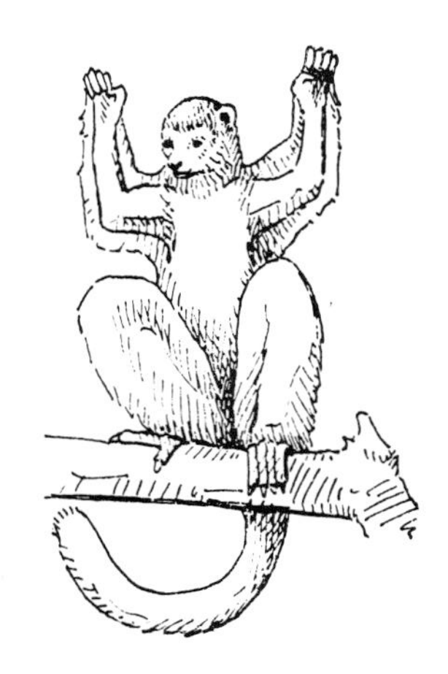

Fig. 13-13.
This is how sifakas sit in the tree crowns early in the morning to sunbathe. It is not surprising that the natives of Madagascar considered them to be sacred sun worshippers!

The sifakas kept in Madagascar usually pick up fruit with their mouths and only rarely with their hands. Hard-shelled fruits are filed open with the lower teeth and the inside is removed piece by piece. According to Webb, who studied sifakas for years, these lemurs require a great deal of room to move around. The animals have a gentle temperament and seldom bite. Webb acquired a diademed sifaka from natives of the central highlands far removed from their natural habitat. The animal was kept completely free and fed on leaves, blossoms, and bark from imported eucalyptus trees because there were no other trees in the highlands. The sifaka consumed only small amounts of bananas. In the evening the animal always returned to Webb's hut to sleep but it also came back periodically during the day. In 1939, Webb brought the sifaka to the Tananarive Zoo where it was still alive in 1946, although the climate there was colder than in its habitat.

In Tananarive the sifakas were fed on a diet of buds and leaves from the white mulberry tree *(Morus alba)*, guava *(Psidium guajava)*, and the leaves and buds of larger grasses *(Panicum maximum, Neyraudia madagascariensis)*. Bamboo, which is a favorite food of the lemur, was left untouched. If one wants to see these beautiful acrobats, one must travel to Tananarive.

The smallest representative of the indrisoid lemurs is the WOOLLY INDRIS OR AVAHI (*Avahi laniger;* Color plate p. 251). The HRL is thirty cm and the TL is forty cm. This animal is rarely seen in Madagascar. The animal is nocturnal and only a few natives are aware of its existence. In its appearance the woolly indris differs greatly from the sifakas and the indris. It has a round, almost globular head, a thick woolly pelage, and small ears which are completely hidden in the fur. The eyes are very large and have a peculiar stare. The thumbs and the big toes appear huge in comparison with the short fingers and toes.

The woolly indris also feeds on leaves, bark, and buds. In trees the animals move in a similar manner to the sifakas, only more slowly and deliberately. During the day the woolly indris sleeps on a forked branch not far from the ground or clings to a vertical stem like a tree frog. The tail is curled up like a watch spring. The animals seldom come down to the ground, but when they do, they walk upright. Today most woolly indris have disappeared from their former home ranges.

The most amazing animal type of the lemur island, Madagascar, is the indris. If one encounters this large black and white animal walking erect on the ground and looking at one from underneath long eyelashes one can easily understand the old Madagascan myth which says that the indris and man had the same ancestors. "A man and a woman walked through the jungle," narrated Attenborough in this delightful, almost Darwinian sounding legend. "After a period of time the woman bore a large number of children. When they had grown up, some, who were industrious by nature, tilled the soil and planted rice. The others continued to feed on roots and leaves of wild plants. In the course of time the members of the first group fought among themselves. These were the ancestors of man. The others sought refuge in the tree tops so that they could continue to live in peace. These were the first indris. Man and indris are related because they had common ancestors."

On the basis of this legend some Madagascan tribes refer to the indris as "babakoto" (father's son). In certain areas the indris is also known as "Amboanala" (forest dog), not only because the eyes and muzzle are dog-like, but because the tale exists which maintains that indris (they are pure vegetarians) can be trained, like dogs, to hunt birds.

These legends demonstrate the great role that the indris plays in the folklore and culture of Madagascar. A few decades ago it would have been unthinkable for a native to hunt a "father's son" like any other game. However, this has not saved the indris from the danger of extinction. The animal is very highly specialized and its potential for further adaptations was exhausted long before the cultivation of Madagascar. The indris is not able to cope with conditions that were created by the settlement of people and the resulting deforestation.

The number of indris decreased rapidly and is still on the decline.

Indris live singly, in pairs, or in small groups of up to five in trees. Today the animals are found in a very small area in the central section of the east coast on forested slopes of a few volcanic mountains. They are diurnal like the sifakas but because of their timidity are heard rather than seen. Their calls are said to be the loudest on Madagascar. A plaintive bark blends with lamenting cries which sound like a mixture of a human crying in pain and the howling of a frightened dog. As in the gibbons, the howling serves to mark the territory. When the animal is disturbed it emits grunting sounds. Attenborough reports that an ear-deafening noise results when several indris call up and down the tonal scale.

Indris become active soon after daybreak. They climb leisurely around in the branches and start their concert close to 5 A.M., right after sunrise. The indris are extremely attached to their home range and within their territory they always use the same trails. Close to 4 P.M. they start to howl again. The animals are very tender with each other. Attenborough observed one young pair which licked each other for hours.

For all practical purposes we do not know anything about the reproduction of the indris since up to now the animals could not be kept in zoos because of their specialized diet of leaves. In 1939, a group of eight to ten indris was introduced to the "Jardin des Plantes" of Paris. There was even a mother which carried a young across her abdomen. However, the entire group died within a month. Even the zoo in the Malagasy capital, Tananarive, has been unable to keep them until now. According to Basilewsky the captive indris were offered all the plants they preferred in nature (for example, *Eugenia sp.* and *Vapaka thouarsii*); however, they refused food altogether. They suffered from diarrhea which doubtlessly was of nervous origin, and they died within a short period.

Attenborough once saw a female carry a young on her back, and Hill mentioned that a male young still had its milk teeth in October. There is a real need for a researcher to make a thorough study of free living indris. This might contribute to the saving of this large, legendary lemur which walks like a man and looks like a dog.

Family: *Daubentoniidae*

The AYE-AYES (Family *Daubentoniidae)* do not share any similarities with other Madagascan prosimians. They are highly specialized in a particular environment and are adapted to a highly specific mode of feeding. It is not surprising that the first scientists did not have any clue as to how to classify this unique creature in the zoological system. At one time or another this animal has been classified as a relative of the squirrels, the jerboas, the phalangers, and temporarily was even classified under a separate order. The German zoologist Schreiber was the first to recognize, around 1775, that this apparent rodent or marsupial was in reality a lemur. However, Schreber's insight was finally confirmed only much later, in the middle of the nineteenth century, when the great English scientist Richard Owen examined the milk dentition of a young lemur. He found that the teeth were completely prosimian-like.

Aside from a gigantic form from southwestern Madagascar which is extinct today, there is only one genus and one species, the AYE-AYE (♁ *Daubentonia madagascariensis;* Color plates p. 247, p. 251, and pp. 283–284). The HRL is forty-five cm and the TL is fifty-five cm. The head is large with a blunt snout. The body is slender and the tail is very bushy. The laterally protruding ears are large and membranous. The fingers and toes are strongly elongated. The central finger is extremely long and spiny and appears withered. Only the thumb and big toe have flat nails. The other digits have sharp claws. The dentition is rodent-like. Each jaw half contains only one large rootless chisel-like incisor and these grow continuously as they wear down. There are only 20 teeth: $\frac{1 \cdot 0 \cdot 1 \cdot 3}{1 \cdot 0 \cdot 1 \cdot 3}$. The milk teeth are similar to those of the lemurs.

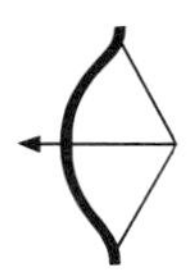

Today the aye-aye is probably limited in its distribution to two separate forest regions (see map). The animal is found primarily in bamboo thickets but it will also climb trees which contain a lot of insect larvae. The aye-aye lives singly or in pairs. During the day it sleeps in hollow trees or in the dense thicket. With the beginning of dusk the animals become active and jump around in the trees like lemurs. Often they dangle from their hind legs like lorises and thereby they have their hands free to eat or to groom themselves. The long middle finger is used for combing and scratching, or cleaning the face, the corners of the eye, the ears, and the nose. During this activity the other fingers are bent back. The long middle finger is primarily used for acquiring food. The aye-aye's main source of food is the beetle larvae which bore deeply into the wood of the jungle trees. The aye-aye knocks along the tree bark until a hollow sound betrays the presence of larvae tunnels. While knocking the aye-aye closely places his large ears near the stem in order to hear even the slightest sounds. With the chisel-like incisors the animal bites a hole into the bark, pokes its thin fingers inside and pulls out the larva. Aye-ayes can even free themselves from a wooden cage with their powerful rodent-like teeth. The incisors can also bite through bamboo and with the middle finger the aye-aye dips out the pith. The animal also chisels open and spoons out the contents of sugar cane, coconuts, and mangoes which grow on plantations close to human dwellings.

How the long middle finger is used

Although free living aye-ayes primarily feed on beetle larvae and bamboo pith, they will also take sugar cane, coconuts, mangoes, bananas, dates, eggs, and cooked rice when in captivity. An egg is opened in the characteristic manner and the content is scratched out with rapid movements of the middle finger until the inside is dry. The

Fig. 13-14.
Aye-aye *(Daubentonia madagascariensis)*

long finger is also used in drinking. Captive aye-ayes repeatedly dip their middle finger into a milk bowl and pull it through to their mouths.

According to Petter, two aye-ayes calling each other sound like the scraping of two pieces of metal being rubbed together. The call of the frightened animal is not "aye-aye," as the discoverer Sonnerat mistakenly assumed, but more a sound like "rron-tsit." These sounds are emitted when the animals are surprised or suddenly struck by a beam of light. If one approaches the aye-aye it hisses and prepares to attack. The animals make grunting sounds when they are feeding.

They build globular nests

The female constructs a globular nest of approximately sixty cm in diameter in a hollow tree or a forked branch. The nest usually consists of rolled leaves from the "tree of travellers" *(Ravenala madagascariensis),* and is lined with twigs and dried branches. During February or March one single young is usually born. The gestation period is not known, although aye-ayes have been born in the zoo of Tsimbazaza near Tananarive. Here one pair lived for five years. Zoo director Ursch steadily employed one native full time to provide the necessary supply of larvae.

However, aye-ayes have been kept in zoos in London, Amsterdam, and Berlin for longer periods of time without the benefit of an expensive larva diet. The animals do well on more accessible food. One aye-aye in Amsterdam lived for twenty-three years. In Berlin an aye-aye was observed to mark its cage with urine. The animal enjoyed cracking nuts and during the night it would jump around actively on the climbing tree, just as a lemur.

In Madagascan folklore this peculiar nocturnal creature plays a great role. The natives say that if a person sleeps in the forest the aye-aye will build a cushion of grass for him. If the person finds the cushion underneath his head he will soon become very rich, but if the cushion is at his feet he will shortly fall victim to the magic of a sorcerer. Anyone who kills an aye-aye, according to folklore, will die within a year. For this reason the animals were left unharmed and if one got accidentally trapped it was quickly released again.

Are aye-ayes in danger of becoming extinct?

Today aye-ayes are extremely rare and their total extinction is feared. During the last years the prosimian researcher J. J. Petter, under the support of the World Wildlife Fund and the I.U.C.N., has attempted to rescue the aye-aye. The government of the Republic of Madagascar made available the small island Nossi-Mangabe in the northwest of the island to provide sanctuary for the aye-ayes. Until 1966 nine aye-ayes were released there. This is, of course, only a first beginning to ensure the survival of the species. Further measures were financed by the "Zoological Association of 1958" in Frankfurt. The habitat of the aye-aye is rapidly dwindling and the aye-aye's days on the Madagascan main island are numbered.

Phylogeny of the remaining Prosimians by E. Thenius

The following infraorders of the lorises and tarsiers differ considerably from the Madagascan lemurs. Hence, the question arises whether the prosimians constitute an evolutionary unit. During the lower Tertiary the prosimians flourished in North America and in Europe. Fossils reveal that there were several divergent lines of evolution. The family *Adapidae* (*Protoadapis* and *Adapis* from Europe, *Smilodectes* and *Notharctus* from North America) seemed to be closely related to today's lemurs. The families *Plesiadapidae, Paromomyidae,* and *Phenacolemuridae* are reminiscent of the Madagascan aye-aye because of their rodent-like front teeth. However, they are not related because the apparent similarity is based on convergent evolution.

These lemur-like prosimians were very different from the small, large-eyed primates which were represented by many species in the beginning of the Tertiary (in the Paleocene and Eocene), particularly in North America and Europe: the *Anaptomorphidae* (among others, the *Anaptomorphus* and the *Tetonius*) and the *Necrolemuridae* (among others, the *Necrolemur* and the *Microchoerus*). Many characteristics indicate that these forms are closely related to today's tarsiers. For a long time zoologists had assumed that the ancestors of the tarsiers gave rise to the monkeys and apes. We shall hear more about this aspect when we discuss the tarsiers (see p. 307 ff.).

Today, however, many scientists assume that monkeys and apes originated from an extremely primitive group of prosimians. The *Omomyidae* (i.e., *Teilhardina* and *Omomys*) which also originated in the lower Tertiary, were small prosimians with long, unshortened jaws and an almost complete dentition from which the monkey dentition could have been derived. This group cannot be classified with any of today's prosimians because the characteristic features of the lemurs, lorises, and tarsiers are lacking. According to this concept the North American *Omomyidae* contained the ancestral forms of the New World monkeys, and the Old World forms gave rise to the Old World monkeys.

Infraorder *Lorisiformes* by K. Kolar

The LORISES (Infraorder *Lorisiformes*) are found in Africa and Asia. They are nocturnal and arboreal. The snout is short. The large nocturnal eyes face forward and are close together. The thumbs and big toes are quite opposable. The hand and foot appear like prehensile pincers, particularly in the loris. The HRL is 14–38 cm. The tail is long and bushy in the galagos and in the loris it is short or vestigial. The incus of the ear forms part of the auditory bulla (Bulla auditiva). They are omnivorous. The tooth formula is as follows: $\frac{2 \cdot 1 \cdot 3 \cdot 3}{2 \cdot 1 \cdot 3 \cdot 3}$. The central upper incisors are separated by a diastema of varying length. There are two families: the LORISES *(Lorisidae)* and the GALAGOS *(Galagidae)*. There are a total of five genera, eleven species, and fifty-nine subspecies.

Family: *Lorisidae*

The LORISES (Family *Lorisidae*) has an HRL of 25-38 cm and a TL of 0–6 cm. The head is short and the back of the head is broad. The snout is short and pointed. The eyes are large to very large. The

Fig. 13-15.
1. Slender loris *(Loris tardigradus)* 2. Slow loris *(Nycticebus coucang)* 3. Potto *(Perodicticus potto)* 4. Angwantibo *(Arctocebus calabarensis)*

medium-sized ears are covered by hair. The arms and legs are of the same length. The grasping hands and feet have strongly opposable thumbs and big toes. The second finger and the second toe are greatly shorted or are vestigial, which results in a more pronounced pincer effect. The fingers and toes are padded at the ends and have nails. Only the second toe is clawed. The fur consists of a short wool and appears plush. Usually there are two pairs of mammae. The gestation period is four months. Newborn young are covered with hair. The animals are found in tropical forests. They are slow hand-over-hand climbers. There are four genera and five species:

A. LORISES i.n.s. (genera *Loris* and *Nycticebus;* Color plates p. 253 and p. 266) is found in South Asia. 1. The SLENDER LORIS *(Loris tardigradus)* has an HRL of twenty-five cm and a TL of approximately one cm. The animal is very slender. The arms and legs are long. The eyes are oval and large. The fur is soft and plush. There are six subspecies. 2. The SLOW LORIS *(Nycticebus coucang)* has an HRL of 32–37 cm and a TL of 1–2 cm. The body is more compact and massive. The limbs, fingers, and toes are shorter and stouter. The ears are covered with short hair and are hidden in the fur. The outer incisors are visibly smaller than the inside ones and may be absent. There are ten subspecies. 3. The LESSER SLOW LORIS *(Nycticebus pygmaeus)* is the same size as the slender loris. The animal is found in Vietnam and Laos.

B. POTTOS (Genera *Arctocebus* and *Perodicticus;* Color plate p. 252) have a still more reduced second finger and second toe. The animals are found in Africa. 1. The ANGWANTIBO *(Arctocebus calabarensis)* has an HRL of 25–30 cm. The tail is a barely discernible stump. The snout is pointed. The eyes and ears are large. The hands and feet are small. The index finger is completely vestigial. The hair is thick, long, and woolly. There are two subspecies. 2. The POTTO *(Perodicticus potto)* has an HRL of approximately thirty-five cm and a TL of approximately six cm. The eyes are medium large. The ears are rather small and membranous. The index finger is only recognizable as a stump. The last neck vertebrae and the first two thoracic vertebrae have long spinal processes. There are five subspecies.

The lorises and galagos can be clearly distinguished. The agile galagos are well adapted for jumping and climbing while the more deliberate loris are not able to jump, but they rather move in the branches almost sloth-like. The loris has an amazingly strong grip with its hands and feet. It is almost impossible to pry a slender loris from the bars of its cage. One has to use both hands to remove a foot or hand of a loris from a cage bar, but as soon as one has loosened one hand the first one has grabbed hold again.

The tendons and muscles of the limbs in the loris function completely mechanically, as in birds. This enables the animal to grasp a limb with the least expenditure of energy. As in the sloths, the rete mirabile in the circulatory system enable a prolonged contraction of the muscles without exhausting them. The loris is therefore able to hang from a branch for hours at a time. In these animals the tail is reduced; it is not utilized as a rudder or balancing organ.

In former times Dutch seafarers frequently brought slender lorises and slow lorises from India or Indonesia to Europe. According to William Baird the word "loris" originated from the Dutch language and means "clown." It was Buffon who first recognized that the lorises were not sloths, as was generally assumed, but were prosimians.

The slender loris

Because of their large nocturnal eyes and small pointed snouts we consider the slender loris very droll and "cute." During the day the animals shelter in hollow trees or firmly press against a tree. They feed on flowers, leaves, young shoots, unripened nuts, insects, and bird eggs. They also hunt geckos and other lizards and prey on small sleeping birds in their nests. Like galagos and capuchin monkeys, they mark their territory with their palms and soles which they wet with their own urine. Each step leaves a trace of odor which also aids the animals in orientation. Slender lorises placed into a cage with unmarked branches move about very insecurely. For instance, they might attempt to cling to the shadow of a branch and try to climb up on it.

Despite their sloth-like slow motion movements, slender lorises can also be rather agile. In many descriptions it was claimed that this animal is solitary and irritable and does not even get along with its mate during the breeding season. Yet Haltenorth observed that several males may court a female without demonstrating any intense jealousy. I kept a pair for six years. The animals lived in harmony until the female died. The male has been with me for more than seven years. His diet consists primarily of mealworm larvae, grasshoppers, and a baby food mixture. Since I fed him the baby food, the slender loris, like the galagos, has refused almost all of the offered fruits. The food is usually held with both hands. If the hands are not used for grasping or holding they are closed into a fist with the thumbs surrounded by the other finger.

My animals never slept in the provided sleeping box, but rather they rolled up somewhere, pressing the head to the body between the thighs and clinging to a branch with the hind legs. The slender lorises rutted twice a year. The gestation period is 122 days. One or two offspring are born at the end of April or beginning of May, and again in November or December. The young are born with open eyes and a thin, silky fur from which protrudes a few long hairs. The mother carries the young for a long period; according to Haltenorth, this period lasts for over a year. If one separates the young from the mother, she frequently does not accept it again. However, Hill observed that a female adopted an orphaned young and fed it. Her older young later displaced the adopted child which then died two days later.

The slow loris

The slow loris moves much slower than its smaller and more slender relatives, seemingly in slow motion. On the ground they trot. When the animal is fleeing it stares at the observer and then climbs step by step by backing up the tree. As Ludwig Heck says, "They climb upwards with their faces directed down." Free living slow lorises feed on various insects, unripened nuts, honey, blossoms, eggs, and small vertebrates.

I fed my slow lorises with freshly killed chicks, mice, and young hamsters. The animals were very fond of large cockchafers. The loris does not kill a defenseless prey before eating it, but rather starts to chew at any spot, completely oblivious to the resistance of the victim. In addition my slow lorises received mealworms, grasshoppers, bananas, and a mixture of beaten egg, baby food, and vitamins. These little fellows were hardy eaters. On the average one animal consumed a banana, a chick, and 1/16 liter of mash daily. A loris first smells the food and then grasps it with its mouth or one hand and finally holds it firmly with one or both hands. The loris frequently clings with its hind legs to the vertical bars and allows the body to hang down, horizontally stretched out or erected. This enables the animal to hold its food with its hands. When eating bananas, the loris occasionally sits up and arches its back. From a similar posture the slow loris will lunge with both hands at a living prey, very much like a praying mantis *(Mantis).*

The slow loris is more sociable than the slender loris. While being transported, three of my animals, two females and a male, were placed in individual boxes. However, when the animals were put together again in one cage, they immediately sought contact. They licked and groomed each other, particularly in the head region. While sleeping, they touched each other. When I introduced an old male into the cage, the three others sniffed at his back and licked his head. Even his appetite was stimulated in company. Suddenly the old male began to eat both living and dead prey, which he had never touched before.

The slow loris uses mainly its mouth and tongue to clean the fur. They scratch with the hind legs, not only in the customary manner "from the back," but occasionally also "from the front" where the hind leg is put between the two arms to the place that is to be scratched on the head. Captive slow lorises sleep in half open boxes. There they will hide their heads between their knees, cross the arms over the head, thereby shutting out all source of light. Their mode of marking a territory is completely different than that of the slender loris. They drag their back sides over all the branches of the cage and urinate copiously. Marking is very pronounced in cages that smell of other lorises. As far as I have observed, males and females mark in the same way. This method is more efficient because it is faster than the rubbing of the hands and feet with the urine, as the slender loris does.

Slow lorises are not difficult to raise in captivity. In Tübingen, Seitz was even able to film the birth of a slow loris. Pregnancy lasts about six months. Most frequently a single young is born during the night. The new-born crawls to the mother's abdomen and clings to the fur. It greatly resembles a slender loris. In the Berlin Zoo, Ludwig Heck observed how a new-born loris mistakenly clung to the father and searched in vain for a teat. The youngster had to be quickly returned to its mother.

Although slow lorises can bite quickly and powerfully, they rapidly become tame and even learn to differentiate different persons and various sounds. To a limited degree, one can train them. Modern accommodations for nocturnal animals are ideal for observing the lively activity that slow lorises display in the dimmed cages.

The angwantibo

Among the African forms is found the rarer and still little known angwantibo. Although the animal already was discovered in 1680 on the coast of Guinea, it was scientifically described only 180 years later. The angwantibo is nocturnal and occupies the high crowns of trees, but because of this inaccessibility very little is known about their behavior. Angwantibos migrate a great deal and are very sensitive to noise. At the slightest noise they pull their heads close to their chests. The angwantibo, which is the native name, can sleep while hanging from its hind legs. If one disturbs such an animal it quickly wakes up. Only the limbs seem to be independent of the central nervous system. The animal does not respond if one pricks it with a needle. The limbs seem cooler than the rest of the body and are completely rigid. Their circulation is probably not as good as it is in the rest of the body.

Until now angwantibos have seldom been kept in zoos. In the years after World War II, Gerald M. Durrell devoted much time to this unique and little known prosimian while conducting his expeditions into the rain forests of Cameroon. He was able to transport several angwantibos to England and did not find them difficult to keep. Right after being caught the animals fed extensively on bananas, grasshoppers, and the breast meat of birds. During the day the animals slept while firmly clinging to a branch with the head between the arms. After sunset the angwantibos awoke, groomed themselves, and yawned, so one could see the brilliantly pink tongue. Then the animals would proceed to climb and dangle from the branches of the cage in search of the food bowl.

Durrell describes one of these nocturnal imps in the following manner: "Sometimes, after it descended out of the branches and stood on the ground with lowered head and bent back, it looked amazingly like a miniature bear, and I was greatly tempted to play with it. Occasionally I put the food bowl directly under the branch to which it was clinging. Immediately it would dangle from its feet and pick up pieces of banana with its pink hands, stuff them into its mouth, smack its lips, and lick the juice from its nose. During the entire time that it spent with me, I never heard it make a sound except for a cat-like purring and a weak hissing sound which it emitted when I tried to touch it. It was not easy to pry it off a branch. Its peculiar hands and feet, with the vestigial second finger, had unbelievable strength. Once it had grasped hold of a branch it hung there as if glued to it. I was forced to hold it around the chest and pull carefully. It responded to my attempts by ducking its head between its arms and then sinking its needle-sharp teeth into my thumbs with lightning speed."

The potto

Over its wide range of distribution the potto occurs frequently. Its discoverer, the Dutch seafarer Bosman, described it in 1699 as "potto." It is not clear if the word originated from a native language. In the pidgin English of the West African coast population, "potto" means "softly-softly" which refers to the slow, noiseless movements of the potto. Just like its relatives the potto is a deliberate climber. It only lets one hand or foot go if the other three limbs are securely anchored.

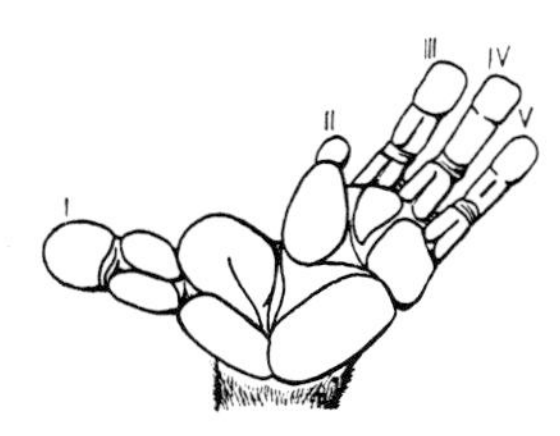

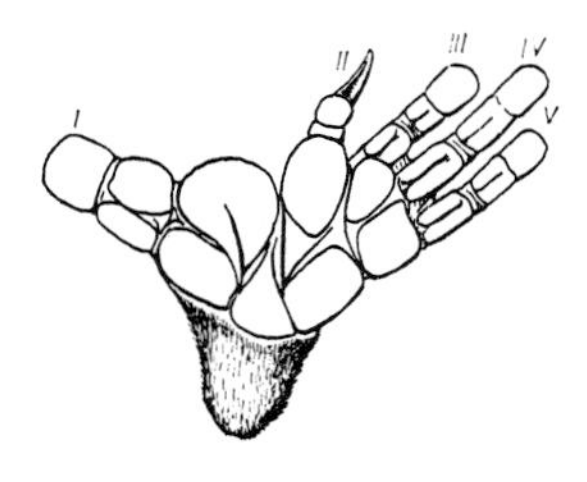

Fig. 13-16.
Left hand (top) and left foot (bottom) of the potto. The shorter digits of the second finger and toe results in a good grip.

The largely elongated spinal processes of the last neck and the first thoracic vertebrae are the most conspicuous characteristics of the potto. These processes lift the skin and protrude like humps out of the neck. The humps are tipped with pointed thorny caps. These structures serve the slow animal as a defense against enemies. If the potto is attacked frontally it raises itself slightly before the attacker has a chance to bite, drops down suddenly, pulls the head to the chest and thrusts the sharp spinal processes into the eyes and nose of the attacker. The main enemy of the potto, the palm civet *(Nandinia binotata),* is, however, apparently very successful in warding off these thrusts.

It is possible that the neck humps were originally not used for defense but served as an anchor when the animal was asleep. The potto bends down and hooks itself into the bark of the tree to prevent it from dropping down. The neck defense probably evolved out of this posture. If one grabs a resting potto the resultant thrust of the neck spines can be rather painful. In addition the potto rapidly rotates its head and bites unexpectedly fast where one least expects it. It is well to know the habits of these small nocturnal wanderers before handling them.

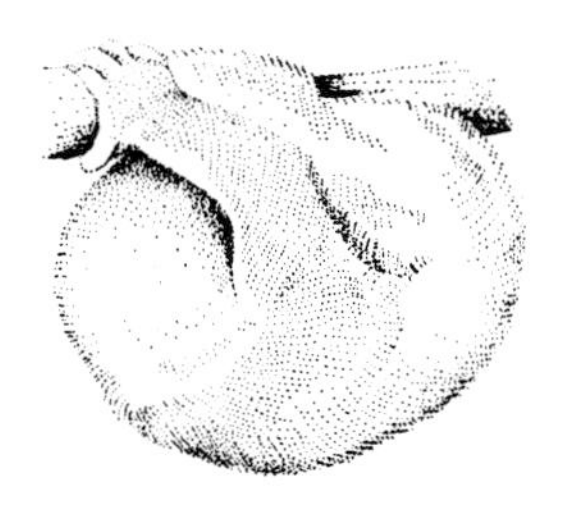

Fig. 13-17
During the day the potto sleeps while clinging horizontally to a branch, rolled into a ball.

During the day the potto shelters in caves and in the crevices of trees. At night they wander through the foliage of the jungle searching for food. Sometimes they come down to the ground and walk slowly, with stiff legs, to the next tree. If the tree has a rough bark it can walk up it for a short distance in this peculiar stance. Aside from various plants, free living pottos also feed on small animals and attack sleeping birds. Captive pottos will consume ripe fruit, cooked rice, and meat. Pottos are not uncommon in zoos. The Frankfurt Zoo has kept them for a considerable time. The animals have bred repeatedly in zoos or in private homes.

Sanderson discovered a potto young only weeks after its birth. The little one had been hidden in its mother's fur so that it was not seen. According to Sanderson, pottos copulate frontally. As far as is known, this occurs only in three other groups of primates: the orangutan, the bonobo, and man. In their natural habitat, most potto births take place at the beginning of the year. According to Hill, one usually sees mothers and their nursing offspring, which have a silvery-white juvenile coloration, during January or February.

Family: *Galagidae*

The GALAGOS OR BUSH BABIES (Family *Galagidae*) reveal at first glance that they are not slow climbers but are very agile leapers. The HRL is 14-33 cm and the TL is 20-35 cm. The tail is long and is partly bushy. The closely set eyes are large. The membranous ears are very large and can be folded like a piece of paper during sleep by means of fine cartilaginous supports and connecting muscle fibers. The foot appears very long because of the elongation of the calcaneum and navicular bones. The legs are much longer than the arms. There are soft pads on the ends of the fingers and toes. The neck is very flexible and the head can be turned around almost 180°, like in owls. The spinal processes of the twelfth and thirteenth thoracic vertebrae point to the front. The upper incisors are small. The first upper premolar is canine-like. There is a pair of mammae in the chest region and another pair in the groin. There are one to two young per litter. There is only one genus (*Galago;* Color plate p. 252) and six species:

Distinguishing characteristics

Fig. 13-18.
1. Senegal bushbaby *(Galago senegalensis)*
2. Western needle-clawed bushbaby *(Galago elegantulus)*
3. Eastern needle-clawed bushbaby *(Galago inustus)*

1. The SENEGAL BUSHBABY *(Galago senegalensis)* has an HRL of 16–20 cm and a TL of 23–25 cm. There are ten subspecies. 2. ALLEN'S BUSHBABY *(Galago alleni)* has an HRL of 20–22 cm and a TL of 24–26 cm. The ears are extremely large. The fingers are greatly elongated. The facial bones are long; there are four subspecies. 3. The THICK-TAILED BUSHBABY *(Galago crassicaudatus)* is the largest species. The HRL is approximately 35 cm and the TL is approximately 35 cm. A bony crest is located on the skull. The tail is thicker and is covered by dense hair, giving a bushy appearance. There are eleven subspecies. 4. DEMIDOFF'S BUSHBABY *(Galago demidovii)* is the smallest species. The HRL is 14 cm and the TL is 18 cm. The head is broad and short, and the snout is short. The ears are not as large. The tail is sparsely covered with hair. In the natural habitat the bushbaby's fur is vividly green on the upper side and yolk yellow on the lower side, but in captivity these colors fade. The animals are forest dwellers. There are seven subspecies. 5. The WESTERN NEEDLE-CLAWED BUSHBABY *(Galago elegantulus)* has a flattened head and ears which are more laterally positioned. The arms and legs are shorter and more powerful. The finger and toenails are pointed and keeled. The fur is dense and woolly. There is a lighter subspecies *(Galago elegantulus pallidus)*, as well as a darker one. 6. The EASTERN NEEDLE-CLAWED BUSHBABY *(Galago inustus)* is very similar. Some zoologists group the needle-clawed galagos under a separate genus of *Euoticus.*

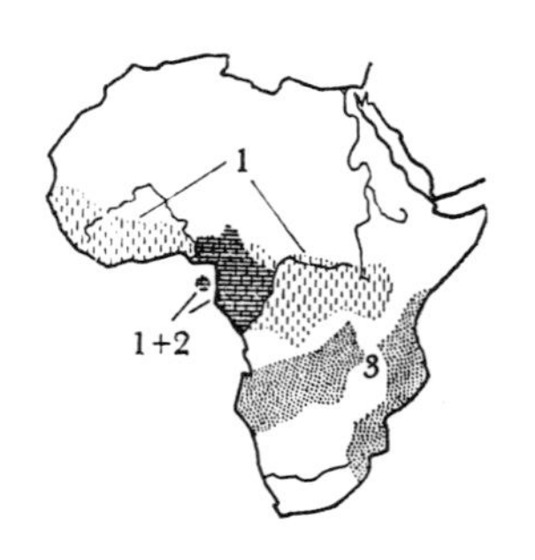

Fig. 13-19.
1. Demidoff's bushbaby *(Galago demidovii)*
2. Allen's bushbaby *(Galago alleni)*
3. Thick-tailed bushbaby *(Galago crassicaudatus)*

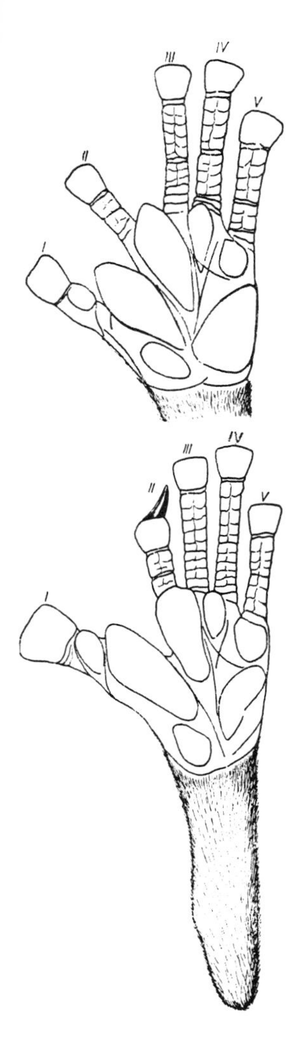

Fig. 13-20.
Left hand (top) and left foot (bottom) of the Senegal bushbaby. The claw-like nail of the second toe is characteristic of the prosimians.

In Africa these tiny prosimians occupy a great variety of ecological niches. They are found in tropical rain forests, in the fringes of forests, tree savannahs, and bush steppes. At dusk, the animals wake up and emerge from their shelters. They unfold their ear pinnae and direct them towards the sources of sound. Even the soft humming of a fly produces a turning of the ear. The olfactory sense and sight are also well developed. Vösseler describes the pupil "as a small vertical ellipse during the day which becomes completely circular at night. The eyes shine splendidly. At dusk the eyes look like dark, yellow, glowing coals when hit by a beam of light from the proper direction. However, when the light comes from a certain angle (as when the light source is located between the eye of the observer and the animal) the eye looks like a greenish blue opal which is actually dazzling in its brilliance."

It is expected that these sharp-eyed jumping acrobats are excellent hunters. Even the small Demidoff's bushbaby does not limit itself to cracking unripened nuts, but also collects snails, insects, and tree frogs. The larger species prey on lizards, bird eggs, and fledglings, and they can also skillfully grasp a mouse or small bird. In addition, they feed on various fruits, nuts, and other plants. Only Allen's bushbaby, the best jumper of the family which is able to "fly" from tree top to tree top in giant leaps, is predominantly dependent on a vegetarian diet. The loud cries of the galagos are reminiscent of the yelling of children. For this reason, the English have called them "bushbabies."

The first bushbabies known to science came from Senegal. Adanson discovered the animals and in 1796 Etienne Geoffroy Saint Hilaire described them. This species is still known as the Senegal galago although the animals are widely distributed in Southwestern Africa and particularly in the Eastern part of the continent. The Senegal bushbaby inhabits dry forests, the savannah and bush steppes. It is the most frequently kept bushbaby. If the animals are kept properly, it is not uncommon for them to bear young.

Under human care juvenile galagos become extremely tame and affectionate pets. They seek contact with their owner, lick his face, and climb into his jacket pocket. However, one has to consider that even the tamest and most pleasant bushbaby wets its palms and soles with urine and that each footstep leaves behind an odor. If one scrubs the climbing bars of the cage of such a bushbaby, it will mark even more intensively. The bushbaby's jumping abilities are remarkable. This small animal is capable of jumping upwards at a slant of up to three meters. On the ground the bushbaby hops like a jumping mouse.

Even tame galagos begin to shriek when they see unfamiliar objects. At the sight of cats, dogs, and snakes my animals emitted staccato alarm calls. It always took a long time before the bushbabies settled down again after an unfamiliar sighting. On the whole, bushbabies are much more sociable than lorises or pottos. Although I provided several sleeping boxes for my galagos, frequently six of them would try to crowd into one. This did not only include mothers and their babies or juveniles, but also mature males which usually did not get along too well. In their natural habitat, one usually finds several bushbabies in one shelter. These tiny Demidoff's bushbabies prefer to seek cover in abandoned squirrel nests.

The female gives birth twice a year. Of nine births which occurred in my animals, four were in April, two in June, two at the end of August, and one at the end of October. According to Hill, Senegal bushbabies in the London Zoo were born in May, June, and September. Pregnant females of Allen's bushbaby, Demidoff's bushbaby, and needle-clawed bushbabies were sighted at the end of the year, and in May and July. At the beginning of October a Senegal bushbaby female with a week old infant arrived in Vienna.

Late one afternoon at the end of April, 1961, I was able to observe the last phase of a birth which only took thirty seconds. The head of the young was already visible, yet it appeared that the female was not experiencing any discomfort. In this position she still leapt about and then sat on a wall board and licked the infant's head. Shortly afterwards the entire body of the young slipped out. The mother picked up the young with her mouth and jumped to the sleeping box. A pair of twins were also born on April 25, 1962, during mid-morning.

A newborn Senegal bushbaby weighs about fifteen grams and measures around fifty-eight millimeters without the tail. The young is sparsely covered with hair and the abdomen is naked. There the hair begins to grow around the seventh to ninth day. When disturbed by conspecifics or humans, the mother frequently picks up her young at the middle of its back and removes it from the sleeping box to the wall board. The little one firmly presses its legs against her abdomen and curls his tail around the neck of the mother. While the mother is carrying the young, its body remains motionless; only its head moves occasionally in response to some outside stimulus. When two weeks old, the galago young is able to leave the sleeping box without the aid of the mother, but it always stays close to her. It has doubled in length by the time it is four weeks old; however, the mother still picks it up in her mouth. At this stage the offspring is able to climb well and is able to jump up to one half a meter. Around four months juveniles have nearly attained adult size.

Although the animals are sociable the males do not tolerate each other. During the breeding season older males pursue younger ones and attack them. If there are no escape possibilities, one male may even kill the other. One of my four month old males was fatally injured by the bites of an opponent. Particularly intense fights broke out when I had removed four of my ten galagos. As a result of the smaller number of individuals, the stronger males did not limit their attacks to all subordinates but picked out single individuals which were pursued with concentrated effort. Many socially living species of monkeys and numerous other animals fight more intensely in smaller than in larger groups.

The thick-tailed galago inhabits light forests. Its fur coloration varies greatly. Even black animals exist. In captivity this species is more difficult to handle than their smaller cousins. My thick-tailed bushbabies, which, however, were adults when I received them, always raised up on their hind legs and lifted their arms in a threatening posture whenever a person approached their cage. Earlier reports always refer to the fact that thick-tailed galagos enjoy licking palm wine, and when they are intoxicated they are easily caught.

Infraorder *Tarsiiformes* by K. Kolar

The TARSIERS (Infraorder *Tarsiiformes*) differ vastly from all the other prosimians. Experts have repeatedly discussed classifying the tarsiers closer to the monkeys. According to the opinion of certain scientists, the immediate ancestors of monkeys are to be found among the numerous extinct ancestors of the tarsiers in the Tertiary period. Others have even suggested that the tarsiers gave rise to the apes and thereby to man. Still others regard the tarsiers as a highly specialized group within the prosimians. During the Tertiary the tarsiers were represented by an amazing number of forms in Europe and North America. Today only the last remains of this once flourishing group are found on the island complex of Southeastern Asia.

Analyses of blood serum have shown that the tarsiers are in fact more closely related to the monkeys than to the prosimians. Even the structure of the ear of these interesting animals is surprisingly similar to the ape ear. The accommodation center in the brain, which is responsible for adjusting the eye to a specific distance, is highly developed. All this data suggests that the tarsiers originated from primitive forms which were closely related to the ancestors of the monkeys in the lower Tertiary. However, it is still doubtful as to whether the tarsiers are part of our direct line of ancestors, because the similarities which were discussed could be based on convergent evolution, in particular the erect body posture which is common to the tarsiers as well as human-like apes. "One should not forget," writes Georg Steinbacher, "that the apes and man are more or less upright 'striders' whereas tarsiers are erect 'jumpers' like the kangaroos." In any case, these tiny nocturnal creatures are of particular significance to scientists who are studying our own phylogenetic history.

Family: *Tarsiidae*

There is only one family, the TARSIERS (*Tarsiidae*) and one genus (*Tarsius*; Color plates p. 253, p. 254, and p. 265). The HRL is approximately 15–18 cm and the TL is approximately 22–25 cm. The skull is greatly rounded with a reduced muzzle. The eyes are extremely large and are directed forward. They seem to take up almost the entire face (the

Distinguishing characteristics

diameter is approximately seventeen mm, and the volume of a single eye is only slightly less than the brain mass). The ears are large and membraneous. The incus forms part of the external ear passage and part of the auditory bulla. The brain is not convoluted. The cerebellum is not covered by the cerebral hemispheres. The upper lip is covered by hair but is not divided by a subnasal groove. The nostrils do not end in a moist naked muzzle like in other prosimians. The calcaneum and navicular bones are greatly elongated and have the appearance of a tubular bone. The tibia and fibula are fused. The fingers and toes are tipped with disc-like soft pads. There are flat nails on all digits except the second and third toe, which have short upright claws. The thumb and big toe are opposable. Rete mirabile in the arteries are similar to those in the loris (see p. 299). Their diet is primarily carnivorous. The dental formula is as follows: $\frac{2 \cdot 1 \cdot 3 \cdot 3}{2 \cdot 1 \cdot 3 \cdot 3}$. There is no space between the upper incisors. The lower incisors are almost vertical. The lower canines are not like incisors as in many prosimians, but are true canines as in the monkeys. The placenta is disc-like as in the monkeys. The uterus is bicornuate as in the prosimians. There is a pair of mammae in the chest region and another pair in the groin (*Tarsius syrichta* occasionally has three pairs of mammae). There is only one young.

Fig. 13-21.
1. Philippine tarsier (*Tarsius syrichta*) 2. Eastern tarsier (*Tarsius spectrum*) 3. Western tarsier (*Tarsius bancanus*)

Today there are still three living species and twelve subspecies. The WESTERN TARSIER (*Tarsius bancanus*) has feet which are not covered by hair. The underside of the tail is naked and smooth. The skull is usually shorter than that of the PHILIPPINE TARSIER (*Tarsius syrichta*), whose feet are not covered by hair. The underside of the tail is smooth and naked. This is the largest species with the longest skull. The feet of the EASTERN TARSIER (*Tarsius spectrum*) are covered with hair. The underside of the tail is covered by coarse hair arranged in rows of three.

Monkey, opossum, or jerboa?

Towards the end of the seventeenth century the Jesuit father J. G. Camel or Camelli described a "small long-tailed monkey from Luzon." This was the Philippine tarsier. The Englishman Petiver sketched this species in 1702. Linné knew the eastern tarsier but he grouped it with the monkeys at first and later with the opossums. Buffon considered the tarsier a type of jerboa. In 1777, Erxleben classified the big-eyed "spooky animal" under the lemurs. Although various tarsiers were caught and kept by Cuming, Jagor, and other researchers during the nineteenth century, very little is known about their behavior in their natural habitat. All three species have become very rare. For many millions of years this ancient animal form was able to maintain itself on the southeastern Asiatic island complex despite all geological changes. Today, however, the tarsier's habitat is threatened by plantations and cultivated land which seriously interfere with the reproduction of the species.

Originally the western tarsiers were strict jungle dwellers. They are most commonly found in the coastal forests and in the vicinity of rivers and creeks. Harrisson, however, also observed these tarsiers at 1200 meters above sea level in Northern Borneo. To a certain degree the western and the Philippine tarsiers are able to adapt to an environment changed by man. Recently the animals have been observed close to human settlements, such as in corn fields, hemp plantations, and even gardens. Here they usually do not climb more than two meters above the ground. Tarsiers prefer to sleep on thin branches which are in the shade. They cling vertically to these stems. Of one hundred Philippine tarsiers caught on Mindanao only three were found in hollow trees.

On the ground tarsiers sometimes jump on two legs like jerboas but usually they leap like frogs. Occasionally they walk on all fours. According to Hans von Boetticher they flit from branch to branch "by the sudden stretching of the thighs, shooting like a bullet through the air." While jumping the tarsier pulls its arms and legs close to the body and uses its long tail for steering. They are able to leap one to two meters and over half a meter high. Shortly before landing they hold up their tails vertically, stretch out their arms and legs and the discs on the finger and toes, and then adhere to the surface of the branch (Color plate p. 254).

During the day their vision is probably quite poor. The pupil is only a tiny dot. In the dark, however, the pupil dilates widely and the iris shrinks to a small ring. A tarsier can very rapidly dilate and contract the pupil. The olfactory sense is not too well developed, but the sense of hearing is just as acute as in the galagos. In a state of excitement the ears are in constant motion. Simultaneously one ear pinnae can be turned forwards while the other is turned back. Tarsiers disturbed during the day slowly turn their ears in the direction of the sound. Then they open their eyes. The head can be turned in any direction up to 180°.

The Dayaks and other native tribes of Northern Borneo who formerly were head hunters were afraid of the tarsiers because they could turn their heads to such a degree. When the natives were on the war path and encountered a tarsier this was considered to be a bad omen. It meant, so the belief went, the loss of one's own head. Even today the forest tribes of Borneo are afraid of tarsiers. They either try to avoid the animals or kill them.

The tarsiers do not only feed on insects and lizards but also catch small fish and crabs out of creeks. Captive tarsiers were observed to grasp fish and crabs directly in the water or to flick them out of the container and then to eat them. One subspecies *(Tarsius syrichta carbonarius)* on Mindanao seemed to prefer aquatic animals which were to be found in the vicinity of creeks.

With the onset of darkness the tarsiers emit high twittering sounds. Males chirp and snarl in the presence of an estrous female. According to Jakobs tarsiers are most active at the time of sunset, between 5 and 7 P.M., and again at 9 A.M. It is believed that the tarsiers find an abundant amount of food during this period. At sunset swarms of insects settle in the jungle and then fly out again in the morning.

Only since 1938 when air transport became possible were tarsiers kept and studied outside of their home territory. After 1947 several animals were distributed to American and English institutes and zoos. One female lived in the Philadelphia Zoo for almost twelve years. For several years they have been successfully kept in the Frankfurt Zoo and in the Max-Planck Institute for Brain Research at Frankfurt.

According to observations made up to now, western tarsiers live singly or in pairs. The Philippine tarsiers are somewhat more sociable and are therefore much tamer in captivity. The tarsier grooms its fur by licking or scratching it with the claws of the second and third toe. Like certain monkeys, the tarsier does not wash its face but rubs it along small branches or similar objects. During the breeding season both mates groom each other. The male sniffs around the genital area of the female. Copulation takes place at night. The gestation period is not known. At birth the young has open eyes and is completely covered by hair. It holds on to the fur of the mother's abdomen with both hands and feet. Right after birth the young is already able to climb in the branches. It is not known for how long the young are nursed.

Reproduction of the tarsier

According to Ulmer, tarsiers were born twice in the Philadelphia Zoo. The newborn young was half the length of the mother and weighed one fourth of her body weight. Right after birth the young clings to its mother on its own. The young clings lengthwise on its mother's abdomen like an Old World monkey. Later, the young clings at an angle and then finally right across the mother's belly like most prosimian young. During a hurried escape the mother picks the young up with her mouth. Lost offspring emit a high multi-syllabic squeaking call. The parents become alarmed immediately and look in the direction of the cries. About fifty percent of the other mothers also responded and picked up their own young. Individual females were seen to carry up to five young. It seems that the female is not able to distinguish her young from the others. She just picks up any squeaking baby. Young that are quiet are ignored.

Fig. 13-22.
A tarsier in defensive posture.

Hill reported extensively on three Philippine tarsiers, a male and two females, which arrived in the London Zoo from Mindanao on March 19, 1948. They preferred warmth, and were cold at 18°C. In the beginning the bottom of their cage was covered with wood shavings but this did not work out. The pieces of wood were eaten along with the food and caused constipation. In London the animals regularly woke up at 6 P.M. Two of the animals formed a pair and soon adjusted to their new surroundings. The single female, however, remained nervous and went into a defense posture as soon as someone approached. She would rear up partially or entirely on her legs with the aid of her tail, lift her arms, and open her mouth widely. If one approached her very closely, she would widen her mouth even more, close her eyes, and bite with lightning speed.

What do tarsiers eat in the zoo?

In London the tarsiers showed interest in small animals. They were not afraid of snakes but were frightened by dogs and other lively animals. Before the animals arrived in the zoo they were fed young mice and strips of raw meat. In the zoo, at least in the beginning, they were given a very diverse diet. The animals were particularly fond of lizards *(Lacerta dugesii* and *Lacerta vivipara),* and also locusts and mealworm larvae. In one night an animal was able to eat up to one hundred mealworm larvae. A tarsier would first bite the head and thorax of a grasshopper into small pieces and then eat the pieces. Then the abdomen was sucked out and the hard chitinous parts and digestive organs were spit out. It was never observed that two tarsiers would fight over food.

Their jumping ability

When jumping the tarsier is able to land on a vertical plate of glass, provided he can get a hold with one finger. One of our animals repeatedly demonstrated this feat. Although the tarsier is nocturnal it was insensitive to the glare of spot lights which we required for filming and television recording. According to Sprankel, who observed tarsiers at the Max-Planck Institute for Brain Research, the animals did not jump around in search of prey but rather they lay in ambush. They locate prey with their ears—a lizard, for example—and then sight it and approach it. A tarsier which is hunting always holds the tail slightly curved to the side in excitement. When the tarsier grabs the prey it closes its eyes tightly to prevent possible injury from the defending prey. The tarsier rapidly and repeatedly bites into the back of the lizard, close to the spinal column, and then begins to feed on the lizard's head.

If a tarsier clings to thin branches it does not let its tail hang down vertically but presses it closely to the bark where it serves as a support organ. The Eastern tarsier has small coarse hair on the underside of the tail while the tails of the other two species are hairless and have a skin structure which prevents gliding down. A sleeping tarsier sits directly on its tail.

Although today's tarsiers are specifically adapted to their environment, their behavior nevertheless gives us a glimpse into the life history of those ancient primates which fifty million years ago roamed through the nightly forests of the lower Tertiary. From these extinct, hardly differentiated forms developed the kaleidoscope of today's prosimians, monkeys, and man.

Kurt Kolar

14 Monkeys and Apes

Distinguishing characteristics

Suborder *Simiae:* In size the animals range from the squirrel-sized to the gorilla-sized; the body weight ranges from 70 grams to 250 kilograms. The general appearance varies greatly. The head is often round. In many species the jaw protrudes slightly or not at all. In others it is strongly elongated. The eyes are always directed frontwards. There is binocular vision. Ears are usually human-like. The face is more or less naked, although there is often a beard-like structure. The naked, moist rhinarium is absent. Hands and feet are almost always capable of grasping, usually with opposable thumbs and first toes. The body fur is usually less dense than in other fur-bearing mammals. As a rule the coloration is inconspicuous but occasionally it is cryptic. There are naked skin areas in the face, the seat, and the genital region which is occasionally brilliantly colored (i.e., the mandrill). There are true sweat glands. Odoriferous glands are not well developed. Two mammae are always found in the chest regions. Buttock pads (Ischial callosities) are only present in the Old World monkeys and the gibbons.

The front and hind limbs vary in length depending on the main mode of locomotion. The tail may be very long or may be absent. The tail is prehensile in only a few New World monkeys. The facial musculature in conjunction with facial expressions is more or less highly developed. There are from 32 to 36 teeth: $\frac{2 \cdot 1 \cdot 2\text{-}3 \cdot 2\text{-}3}{2 \cdot 1 \cdot 2\text{-}3 \cdot 2\text{-}3}$. The canine tooth is often large and pointed.

There are two infraorders with a total of eight families: A. NEW WORLD MONKEYS (*Platyrrhina;* see p. 326); 1. TYPICAL SOUTH AMERICAN MONKEYS *(Cebidae);* 2. GOELDI'S MONKEY *(Callimiconidae);* 3. MARMOSETS and TAMARINS *(Callithricidae).* B. OLD WORLD MONKEYS, APES, and MAN (*Catarrhina;* see p. 396); 4. OLD WORLD MONKEYS *(Cercopithecidae);* 5. LEAF MONKEYS *(Colobidae);* 6. GIBBONS *(Hylobatidae);* 7. GREAT APES *(Pongidae);* 8. MAN *(Hominidae).*

Suborder: *Simiae* by W. Fiedler

The group of the monkeys, apes, and man *(Simiae)* are the second and last suborder of the primates. Man, the species *Homo sapiens,* is also of this suborder. Biologically he belongs to the monkeys and apes. If one does not like this there is the alternative of describing the suborder *Simiae* as the "higher primates." Actually this means the same thing but it is a bit more cumbersome and the word "monkey" would be void as a zoological concept. It would be a name without meaning. Those qualities that separate man from the animal kingdom and thereby from his monkey relations may, to a large part, not be in the realm of natural science; however, to a very considerable extent this uniqueness of man has its precursors in the higher primates. One cannot understand man at all without having a knowledge of monkeys and apes and their behavior.

Two morphological groups of characteristics are the distinguishing features for the development of monkeys. These features occur in various modifications in the small New World monkeys, the Old World Monkeys, the great apes, and man. Man was greatly impressed with these characteristics because they revealed the similarity between the monkeys and ourselves. First of all, there is the head with the eyes directed forward, the almost human ears, and, in many monkeys, the roughly man-like facial features. Also one often finds hands and feet of monkeys which are reminiscent of those of man. Certain groups deviate from this norm, for example the marmosets and tamarins with their claw-like nails and the large Old World monkeys with their long snouts and powerful canines. However, these are specialized developments and adaptations. These groups are further removed from the original monkey characteristics than man himself. This is borne out by numerous comparative studies about the anatomy, embryology, and paleontology of the monkeys, apes, and man.

The lucky combination of these two sets of characteristics were also decisive for our development. It has made man what he is today. Along with the development of those important grasping hands occured the highly specialized development of our brain and our erect posture which ultimately freed our hands for other tasks. A tendency to the bipedal stance is already present in many monkeys and even in a few prosimians (indris and sifakas) is this tendency more or less expressed.

Fig. 14-1.
A juvenile hamadryas baboon stands on the look-out. Occasionally many monkey species rise up on two legs. Some can walk reasonably well on two legs. However, only in man has the erect bipedal mode become the main mode of locomotion.

As a rule the body of the monkey is covered by hair; however, the density and type of hair varies greatly. Normally the back and the outside of the limbs are more densely covered than the abdomen and the inside of the limbs which are sometimes very sparsely covered. Often one can see the skin in such barely covered spots. The night monkey and the woolly monkeys of the New World have the densest fur, and of the Old World monkeys, it is the guenons and gibbons. If one compares the spider monkeys and the woolly monkeys, the macaques and guenons, and finally the siamangs and the gibbons, it becomes apparent that closely related species or groups of species are represented by sparsely-haired and densely-haired types. The Swiss anthropologist A. H. Schultz counted the hair thickness per square centimeter in many species of monkeys and he found extraordinary differences. He came to the conclusion that the slight hair cover of man is only the final development of a noticeable trend towards a loss of hair among the primates.

The totally bare inner surfaces of the hands and feet are made up of a fine pattern of ridges and grooves which remain constant for the entire life of the animal. Therefore it is possible to recognize every monkey and man by the finger prints. The pattern of ridges is closely connected to the tactile sense and is particularly well-developed in the finger pads. The howler monkeys, woolly monkeys, and spider monkeys have a similar pattern on the naked prehensile area of the tip of the tail.

Buttock pads are only found in the Old World monkeys. These callouses are of very different shapes and are located on the ischial bone. In the Old World monkeys these ischial callosities are well-developed and in the gibbons only slightly. The structures are absent in the great apes but occasionally there are indications of pads in very old animals. Man, who has developed a powerful set of muscles in the seat region in connection with this erect posture, does not show any sign of these callouses.

Monkeys either run, climb, or dangle from branch to branch. Most are arboreal and some are terrestrial. The length of the limbs reflects the locomotor habit. The guenons and many other forms possess legs that are longer than their arms. Baboons and large macaques have arms and legs of equal length. Both groups climb well but they also take the opportunity to run along horizontal branches. The hands of these groups always have well developed thumbs. The hand of man most closely resembles these monkeys' forms. In contrast, the South American spider monkeys and the Old World leaf monkeys have very small thumbs or none at all. Their arms and legs are of approximately the same length. These monkeys swing through the trees by alternately hanging on to branches; however, they still do not possess the extremely long arms of the gibbons and orang-utans. In these forms the hands are very long, powerful, and narrow while the thumbs are small and weak. Similar length relationships but less well developed are found in the chimpanzees and the gorilla. Mature gorillas are too heavy to dangle and swing in the branches. In man, finally, the bipedal mode of locomotion was most highly developed. He almost exclusively uses his legs for locomotion.

The development of the monkey tail is in close conjunction with the various modes of movement. Many monkeys have long tails which serve as a rudder and a balancing organ during climbing and jumping. True prehensile tails are only found in the New World monkeys. In some species the tail is only prehensile during early childhood. At a later period this ability is lost. In others the tail serves as a "fifth hand." The conspicuously colored and furred tails of certain Old World monkeys, for example the white "horse tails" of the guerezas, also function as signals. Terrestrial macaques and baboons almost always have medium-long, or greatly reduced, vestigial tails. The tail of the barbary ape is the most reduced and is externally not visible. Gibbons, the great apes, and man are tailless.

When looking at the skeletons of monkeys the spacious skull is particularly conspicuous. Even in groups which are only remotely related to man, we consider the skull "human" if it is large. In some monkey forms (howler monkeys, great macaques, baboons, and also the great apes) the males have powerfully developed upper jaws which deviate from the ancestral norm. Superficially the skull of the smaller New World monkeys, gibbons, and other forms, where the muzzle formation is less pronounced, seem to be more "human-like" than those of our closest relatives. The chest cavity of the quadrupedal monkeys is narrow and short while it is broad and tent-shaped in the great apes. Only in the gibbon do we find the barrel-shaped form which is familiar to us from the human skeleton. The pelvis is even more altered than the chest cavity in connection with the bipedal stance. This morphological characteristic is less human-like in the other forms of monkeys. The great apes possess a wider pelvic girdle than man. It has broad powerful ilia.

The naked and sparsely haired monkey face facilitates more versatile and more expressive facial expressions than do the furred faces of other mammals. There is an ever increasing development of the facial muscles. Ever more complex and versatile facial effects were evolved. However, the high degree of development which is found in the human facial muscles and the associated nerves is not present in any of the monkeys and apes. In monkeys, and to a lesser degree in man, the entire body participates in gestures and expressive movements.

Dentition, intestine, and larynx

In monkeys the dentition is much more uniform than in the prosimians. The canines are often very large and dagger-like. In the male baboons and great apes the canines are frequently used as weapons in interactions with conspecifics and in fights with other species. A baboon that opens his mouth wide and shows his dangerous teeth is displaying an innate threat behavior pattern (Color plate p. 439). As with several other characteristics, man is more primitive in certain aspects of his dentition than many of today's monkeys.

The oral cavity of the monkeys is small. Only the baboons possess cheek pouches for the temporary storing of food. The stomach is relatively simple. Only in the leaf monkeys, which are primarily vegetarians, is the stomach large, divided into several chambers, and provided with powerful muscular bands. The small intestine is relatively long. In lower monkeys the large intestine is more compartmentalized than in the higher forms. In the great apes and in man the caecum is present only as a small sac-like pouch. The kidneys, if compared to the lobed structures of certain ungulates and carnivores, are more primitive. In several groups glands are found in the anal region which serve to mark the animal's territory or its mate.

The squirrel monkey (*Saimiri sciureus*) rubs his urine into the fur with its hands and feet. The tail is particularly well penetrated with the fluid (see p. 343). These "odor markings" serve to mark the territory of this small monkey.

The larynx is particularly well-developed because it does not only serve to regulate the body's air supply but also performs an important function in connection with the vocal cords in a wide repertoire of vocalizations. In certain monkeys there are additional structures for the modulation and magnification of the voice. The hollow bulb in the hyoid bone of the howler monkeys serves as a resonance organ. The air sac systems in many of the New World and Old World monkeys probably fulfill a similar function. Well-developed air sac structures are present in individual leaf monkeys, the siamangs, and the larger great apes. Various vocalizations play a great role in the social behavior of the orangutans.

Sensory organs and their capacities

In the monkeys vision is of primary importance; it is, in fact, of paramount significance. The fundamental structure of the eye is the same in all monkeys, and is essentially the same as man's. Nocturnal forms like the night monkey have neither cones associated with color vision nor a fovea in the retina which is the locus of the most acute vision. All monkeys have binocular vision and good accommodation (the shape of the lens is modified). These two properties are very significant for orientation in space. Several Old World monkeys have an even more highly developed retina than man. The tactile and auditory senses are also well developed. The New World monkeys with prehensile tails, like all other monkeys, not only possess tactile skin on palms and soles but also on the underside of the tail. In comparison the olfactory sense is of small importance. Yet not all monkeys have such "poor noses" as man. Many monkeys right up to the great apes very often test objects or persons by sniffing them.

Brain and intelligence

Still more important than all these characteristics of the monkeys is the ever increasing development of the brain and the nervous system. Here lie the roots of our own intelligence, and here is the decisive factor which enables one of the descendants, man, to become "master of the earth." In the monkeys one finds a great range of brain formations. In the simplest forms the cerebral cortex has no convolutions. We observe an increase in brain size and weight in the more highly developed species until the circumference and surface area are vastly increased by folding. These developments, according to scientists Noback and Moskowitz, are the "dramatic characteristics of phylogeny." Originally the cerebral hemispheres were primarily an olfactory brain. As the olfactory sense regressed, the cerebral cortex gained in importance and became the locus for the centers of all complicated behavior patterns.

Großmann

The night monkey *(Aotes trivirgatus)* is the only nocturnal monkey. Therefore its eyes are especially large.

In order to demonstrate how much, in this respect, the lower monkeys differ from the prosimians we have compared the brain size of the Demidoff's bushbaby and the marmosets. Both animals have approximately the same body size, but yet the brain of the marmoset is approximately three times the size as that of the small prosimian. Aside from the areas for vision and the tactile and grasping ability of the hands, association centers which serve the learning and memory abilities have greatly increased in size in the monkeys. Similarly, the cerebellum, which regulates the maintenance of equilibrium and coordination of motor activities, increased in complexity.

From the phylogenetic point of view, it is interesting that at the time of birth the brains of monkeys are substantially more developed than the jaw region which is not of great importance during the infant stage. In man with his highly developed brain and his relatively little-used jaw, this external condition is maintained throughout his life. This is one reason why monkey infants appear "human" to us. The hypothesis of the Dutch anatomist, Bolk, that man has remained physically and, in respect to his "curiosity," at the level of an infant monkey, and that it was due to this that he became man, has been refuted by E. Slijper, D. Starck, and other scientists during the last years.

Sexual differences and reproduction

Totally different from other mammals, a very important hormonally-directed process influences the reproductive cycle of many prosimians but especially that of the great apes and man. Generally mammals have one or several breeding seasons during the year. Only during these periods does copulation take place, while the remainder of the time the hormonally-controlled sexual organs are inactive. Monkeys and apes, on the other hand, can breed at any time, just like man. However, there is a monthly cycle in the females which revolves around the maturation of the ovum (in man it is usually every four weeks; in some groups the period is a little shorter or longer). Therefore females are only fertile on specific days of this menstrual cycle.

In many female Old World monkeys certain skin areas around the genital region swell and change color during the estrous phase. This configuration serves as a "releaser" for the male and stimulates his sexual interest in the direction of these females. Female baboons have very conspicuous swellings, and many concerned zoo visitors have mistaken these for pathological tumors. Mangabeys and short-tailed macaques also have swellings but these are less conspicuous in the long-tailed macaques and the African forest monkeys. This phenomenon is present in certain colobus monkeys also, but is absent in the New World monkeys, gibbons, and man. Interestingly enough, our closest relatives, the chimpanzees, have very conspicuous swellings. They are less noticeable in the gorilla and seem to be absent in the orang-utan. Up to now similar swellings have been observed in pregnant orang-utan females, but of course in these cases there is no connection with the estrous phase.

Occasionally a concentration of the births during certain months has been recorded in certain monkey species. This could be related to the hormonal control of living conditions in general or to the rhythmic alternation of favorable and unfavorable seasons.

As in many other mammals, a number of monkeys and apes possess secondary sex characteristics by which one can distinguish the two sexes at a glance. Mature males are often distinctly larger than the females. Some males have almost carnivore-like dentitions, and a different color of fur in the face or genital region. Sexual dimorphism is very conspicuous in the howler monkeys and certain baboons. The beard in man is such a secondary sexual characteristic. The gigantic cucumber-shaped nose of the male proboscis monkey is an extreme example of a secondary sex characteristic. The large nose could also function as a resonance organ in vocalization. Males and females of other similar species can, however, only be differentiated on the basis of "primary sexual characteristics," the genital organs.

In monkeys, apes, and man, and already in prosimians the penis hangs freely from the body wall. The foreskin and the tip of the penis vary greatly in the various species. Primates normally have a bony structure in the penis (os penis). This structure may more or less be vestigial or may be absent, as in the tree shrews, tarsiers, spider and woolly monkeys, and man. Generally the scrotum is also suspended. In certain species or groups of species, such as certain marmosets and tamarins, the howler monkeys and the African forest monkeys, the penis is brightly colored and serves as a display organ. Dominant males display the penis towards conspecifics or utilize it as a "taillight" which aids in keeping the group together.

The external female genitalia differ in certain New World monkeys from those of the Old World monkeys. Frequently the labia majora are large and appear swollen. These swellings are probably not associated with the female cycle. The clitoris is often arched forward and in the spider monkeys may look like a penis at first glance. Unlike in prosimians, the uterus is not bicornuate, but in all monkeys, including man, it is a uniform, pear-shaped structure. Most monkeys possess an additional, smaller disc-like placenta. This structure is absent only in the great apes and man.

The smaller New World monkeys have a gestation period of barely five months. In the lower Old World monkeys the duration is 165–240 days, and in the great apes and man it is from 230 to 290 days. Unlike in many other mammals, the nutritive substances for the fetus do not have to penetrate through several maternal and fetal tissue layers. Rather, numerous villi supplied with minute blood vessels from the embryo come in touch with the maternal blood system within the placenta. This means that at the time of birth of monkeys and man, a wound is produced in the uterus. There is a certain amount of blood loss even in the smallest species of monkeys. Aside from certain New World monkeys, usually only a single young is born. However, twin births occur in many species, including the great apes. During the first weeks of life monkey infants are fed exclusively by their mother's milk, but during the nursing phase they gradually begin to eat other food as well. In the great apes the young are weaned well after one year. It seems that nursing on the mother's breasts is also associated with a feeling of security in the growing monkey infants. Some still suckle long after they can feed independently and are no longer infants.

Phylogeny of *Simiae*: by E. Thenius

Originally all monkeys and apes probably inhabited forest regions in the warmer zones. Only a very few species—the barbary ape, the Japanese macaque, and the stump-tailed macaque—were able to invade the cooler regions. Many *Cercopithecidae* are found in open terrain. Certain macaques, the baboons, and the red guenon spent the major part of their lives on the ground. The hamadryas baboon and the gelada baboon are best adapted to a terrestrial existence. Of the great apes, the orangutan is arboreal while the gorillas and chimpanzees spend a considerable amount of time on the ground. Man is the only member of the suborder *Simiae* who was capable of conquering all corners of the earth.

The phylogeny of the monkeys and apes

According to recent investigations and fossil discoveries, the New World monkeys (Infraorder *Platyrrhina*) and the Old World monkeys (Infraorder *Catarrhina*) did not originate from two separate prosimian branches as zoologists once believed. It is probable that all monkeys and apes arose from a single ancestral group. These were the *Omomyidae,* a prosimian family which was distributed throughout North America and Europe during the early Tertiary (see p. 268). Therefore, the place of origin of the monkeys and apes seems to have been the North American-European area. From these regions the ancestral forms of the *Platyrrhina* spread to South America and the ancestors of the *Catarrhina* to Africa. In these continents the two groups flourished and evolved but followed different lines of development.

Although there are fossil remains to explain the basic trends of the phylogeny of the Old World monkeys, only meager fossil discoveries of New World monkeys have been found until now. The geologically oldest *Platyrrhina* are of the more recent upper Tertiary of Argentina. These early forms must have reached South America via "island hopping," since there was no land bridge between South and North America during the upper and lower Tertiary. During their life time, in the Oligocene, the primates became extinct in North America. The oldest South American discovery, *Dolichocebus gaimanensis,* appears to be a rather specialized form because it has the same dental formula as the very highly specialized and probably much more recent marmosets and tamarins (see p. 361). This similarity may only be based on the smallness of the jaw because in the skull structure, *Dolichocebus* seems to resemble more the night and titi monkeys (see p. 329).

Most South American fossil remains originated from the following epoch, the Miocene, and thus undoubtedly belong to the New World monkeys (*Cebidae;* see p. 329). At that point night monkey-like and howler monkey-like forms *(Homunculus* and *Pitheculus)* and also saki-like and capuchin-like forms *(Cebupithecia* and *Neosaimiri)* were present. The genus *Xenothrix* from the Quaternary of Jamaica seems to represent a totally unique group. The branching out of the New World monkeys into today's subfamilies must already have taken place during the middle Tertiary.

Fossil remains have not shed any light on the origin of the marmosets and tamarins *(Callithricidae).* However, today it is assumed that they are not primitive monkeys but rather are a side line which has evolved separately (compare, p. 358 ff.).

Fossil history of the Old World monkeys

The fossil history of the Old World monkeys is much better known. However, the direct ancestral forms which gave rise to both of today's superfamilies, the Old World monkeys *(Cercopithecoidea)* and the apes and man *(Hominoidea)* are still missing. As in North America, the primates disappeared from Europe during the recent upper Tertiary. These extinctions may have been a result of climatic changes. Up to this epoch, Europe had a tropical and sub-tropical climate, but since the middle Eocene the temperatures have been colder. Fossil records from the next epoch, the Oligocene, are more abundant from Africa. New discoveries, particularly from the Fayum zone of Egypt, have demonstrated that the *Cercopithecoidea,* the Old World monkeys, were separated from the great apes and man at this period. This discovery also shows the phylogenetic unity of the Old World monkeys.

Besides members from various extinct branches *(Parapithecus* and *Moeripithecus)* from the upper and lower strata of the African Oligocene, not only cercopithecid-like *(Oligopithecus)* fossil remains were found but also forms related to the gibbons *(Aeolopithecus)* and the great apes *(Propliopithecus, Aegyptopithecus).* The forerunners of the Oreopithecids *(Apidium)* were also discovered there. This extinct family of monkeys was temporarily classified close to the pre-hominids because of their apparent human-like characteristics; however, it was only an extinct side branch distantly related to the great apes.

The fossil remains also settled one controversial issue which had occupied zoologists for a long time. It was demonstrated that the great apes (including man) possessed a more primitive dentition than the Old World monkeys with their bilophodont molars, powerful canines, and the knife-like lower premolars which oppose the upper canines. Some researchers concluded from the differentiation of this dentition that the ancestors of the great apes are not to be found among the lower Old World monkeys. However, the oldest known Old World monkey,

Oligopithecus, also possessed a primitive dentition in which the specialization of the later Old World monkeys was already indicated. It thus appears that there is no doubt that the Old World monkeys and great apes had a common origin.

The early history of the apes and man *(Hominoidea:* gibbons, great apes, and man) is discussed in a later chapter (see Volume XI). Fossil remains of the Old World monkeys from the Miocene of Africa have now been found. At first these remains were mistakenly considered to be the ancestors of the gibbons and were termed *Prohylobates* ("pregibbon"). In the Pliocene the two present families of the Old World monkeys, the Cercopithecids *(Cercopithecidae)* and the leaf monkeys *(Colobidae)* were already clearly separated. During this epoch the genus *Mesopithecus* was widely distributed throughout Eurasia and Africa. Their first fossil records were discovered near Pikermi, Greece, and judging by the skull and dentition, this group belonged to the leaf monkeys just as *Libypithecus* from North Africa. However, in *Dolichopithecus* the frontal portion of the skull is more elongated and the proportions of the limbs are baboon-like. This form is therefore considered to be a primitive baboon and not a leaf monkey.

Pleistocene deposits have revealed leaf monkeys, macaques, cercopithecids, and baboons that could essentially not be differentiated from their present-day relatives. During that period certain baboon species (i.e., *Dinopithecus)* had developed into gigantic forms. Even macaques were present in central Europe during the warm interglacial periods of the middle Pleistocene (e.g., Swabian Barbary ape [*Macaca sylvana suevica*]). On the basis of their Pleistocene and present distribution it is assumed that these forms originated in Eurasia and from there spread to North Africa. The Gelada baboon *(Theropithecus gelada)* which is found in the Abyssinian highlands today, is usually classified with the baboons; however, according to Remane it is a representative of the macaques and phylogenetically is considered an isolated branch of this monkey group. In recent times the cercopithecids are only found in Africa south of the Sahara. They are a uniform group except for the swamp guenon *(Allenopithecus)* perhaps, which began to diversify into various forms only in the upper Tertiary. The phylogenetic relationship of the mangabeys *(Cercocebus)* has not been clarified beyond doubt. The number of their chromosomes coincides with that of the macaques; however, they share the same habitat with the *Cercopithecidae* and are probably closer related to them. During the ice ages the baboons, the most highly specialized of the *Cercopithecidae,* also occurred in Asia, but today they are only found in Africa except for a small number of Arabian hamadryas baboons.

The leaf monkeys *(Colobidae),* the second family of the Old World monkeys, are a special case. In their dentition and skull structure they appear primitive; on the other hand they are also quite specialized in that their stomachs are adapted to a leaf diet and by their brachiating means of locomotion, found in some species. Although the area in which they mainly developed lies in Asia, where they are today still represented by five genera, it is assumed that they originated in Africa if judged by *Libypithecus* as representing the African short-tailed monkeys *(Colobus, Procolobus).*

Monkeys in zoos by W. Fiedler

Monkeys kept in zoos today have all benefited from progress made in modern zookeeping. Monkeys and apes survive much longer than in former times. A total ban on feeding by zoo visitors has been a most effective measure. The Frankfurt Zoo pioneered in bringing this issue to the public and the press in a hard, long struggle after World War II. Meanwhile, most scientifically managed zoos in Europe have adopted this ban. In order to maintain a healthy monkey and ape population it is very important to separate the animals from the visitors by indoor glass walls. This way one prevents the spread of human infections such as influenza, colds, etc., to the very sensitive monkeys and apes. Furthermore, monkeys should have equal access to the outside enclosure and the inside shelter during any season. The monkey enclosures of the Frankfurt Zoo have soft, heavy plastic flaps which the animals can open at will and which are kept closed by their own weight pressing them against the slanted door frame so that draft and a cooling of the interior is prevented. The old metal shutters that zoo director Professor Brandes of Dresden recommended for this purpose decades ago have the big disadvantage of being very noisy and of pinching the monkeys' tails. Visual barriers in the interior rooms and rock piles in the center of the outside enclosure provide escape possibilities for the subordinates from the "boss." These provisions serve to lessen and to "deactivate" the excessive dominance fights which are often increased in captivity, and thus they aid in the general well-being of the animals.

Particularly significant in keeping healthy monkeys is the control of parasites. Since most monkeys and apes are arboreal in the free living state, the feces drop to the ground immediately and there is no further contact. Reinfection with eggs from intestinal parasites which are almost always present in small numbers is rare. Thus, the frequency of host-parasite contact occurs in enough instances to prevent the extinction of the parasite, but there are not enough for the monkeys to incur harmful effects. A continuous significant shift in the balance would have resulted in the extinction of either the parasite or of the host. However, in cages or enclosures the danger of reinfection with the eggs of the parasite is almost always greater than in the free living state. The balance shifts in favor of the parasite. Worm eggs are not killed by the usual disinfectants and persist in wood cracks. In the Frankfurt Zoo wood is banned from the monkey house and from many other animal shelters. It is replaced by tiles, plastic walls, nonrusting steel structures for climbing, fiberglass, and similar materials.

Peril and protection of monkeys by W. Fiedler

Many monkeys and apes are also seriously threatened by extinction today. In the Red Data Book of the International Union for Conservation of Nature, which lists the most endangered animal species, we find a whole series of our relatives: Goeldi's monkey, the white-nosed saki, the woolly spider monkeys, the agile mangabey, the snub-nosed monkey from Tibet, the red colobus monkey from Zanzibar, and two related forms, the dwarf chimpanzee, the mountain gorilla, and the orang-utan. Other species as well have alarmingly decreased in numbers, such as the lowland gorilla and the maned tamarin. Today the Zanzibar colobus monkey and the orang-utan are in the greatest danger of extinction. Man, who so highly values his uniqueness as compared to the other members of his suborder, forgets very easily that he also has special obligations and duties which include the protection of species from whose ranks he arose and with whom he forms a biological unit.

The unity of all living things

The idea of the unity of all living things has more and more gained acceptance since the appearance of Darwin's work. No one doubts any longer that man and the great apes are far closer in many characteristics than, for example, the gorilla and the baboon. Yet no biologist denies the uniqueness of man. Man is perhaps, as the Paleontologist Kuhn-Schnyder from Zurich once said, "only a special case in the long chain of becoming on earth." The development of his brain and intellect has given him the power to change the earth more than any other creature previously. On the other hand, this thinking creature ("Psychozoon"), man, has been the cause of many problems and dangers which can annihilate him if he is not able to offset the lost animal inheritance with reason and responsible morality. Konrad Lorenz in *On Aggression* pointed out the loss of the inhibition to kill conspecifics. This innate inhibition generally prevents the higher animals from killing or seriously maiming their conspecifics.

Recent observations on free living baboons, chimpanzees, and other monkeys have greatly helped to increase our understanding of these forms. Above all, these studies have shown that our distant and close relatives do not live by the "rule of the jungle," which formerly was wrongly interpreted from the concept of the struggle for existence. Rather, they follow a meaningful and orderly group existence which suggests comparisons to our own group living. Some of our noblest emotions had their origins long before our rise to humanity. Konrad Lorenz comments, "We do not know what a monkey feels during his social defense actions, but certainly he fights just as selflessly and gallantly as an inspired person. There is no doubt about the evolutionary concurrence of the group defense in chimpanzees and human enthusiasm. One can easily conceive of the one originating from the other."

Walter Fiedler
Erich Thenius

15 Capuchin-like Monkeys *(Cebidae)*

"Historical" films which deal with Cleopatra, the Queen of Sheba, or Emperor Nero's persecution of the Christians occasionally show a spoiled society lady with a marmoset or a ring-tailed capuchin. Neither the film producer nor the majority of the public is aware that these droll jungle gnomes come from an entirely different continent, South and Central America, which was discovered fifteen hundred years after Nero. In caricatures monkeys generally are shown with a prehensile tail. This characteristic is found in only a few groups of New World monkeys, and is absent in all Old World monkeys. Most people today cannot tell one monkey from the next, but someone who cannot tell the difference between a Picasso and a Kokoschka is considered uneducated.

Infraorder: New World monkeys *(Platyrrhina)* by D. Heinemann

All prehensile-tailed monkeys and marmosets and tamarins are limited in their distribution to the American tropics. On the basis of many morphological characteristics, the monkeys of the New World and various other types of monkeys have been divided into the infraorder New World monkeys or *Platyrrhina,* and the Old World monkeys are classified under the infraorder of Old World monkeys or *Catarrhina.* Externally the New World monkeys are distinguished from the Old World monkeys by nostrils which are widely separated and more or less open to the side. The body and the limbs are usually slender or very slender. As a rule the tail is long and well-developed. In many species the tail is prehensile and in only a few species is the tail shortened.

Saki monkeys

1. Bald uakari *(Cacajao calvus)* 2. Red uakari *(Cacajao rubicundus)* 3. Hairy saki *(Pithecia monacha)* 4. Pale-headed saki *(Pithecia pithecia pithecia—♂)* 5. Black saki *(Chiropotes satanas)* 6. Red-backed saki *(Chiropotes chiropotes)*

The ancestors of today's New World monkeys separated from the other monkeys at a very early period, for in the lower Tertiary the land bridge between South and North America, which at that time was still connected to the Old World, had disappeared. For many millions of years South America was an island continent. Consequently, in many aspects the New World monkeys have retained a more primitive body structure than the Old World monkeys. The *Platyrrhina,* for instance,

1
2
3
4
5
6

1♀
1♂
2
3
4
5

Howler monkeys
1. Black howler monkey *(Alouatta caraya*—♀ and juvenile♂) 2. Rufus-handed howler monkey *(Alouatta belzebul)* 3. Red howler monkey *(Alouatta seniculus)* 4. Mantled howler monkey *(Alouatta palliata)* 5. Guatemalan howler monkey *(Alouatta villosa)*

cannot oppose their thumbs to the remaining fingers to such a degree as can the majority of the *Catarrhina.* Only a small number of species are capable of grasping with their thumbs; however, the first digit of their feet is just as well-developed and opposable as those of the Old World monkeys. The prehensile tail of the capuchin monkeys, howler monkeys, woolly and spider monkeys, on the other hand, is a highly developed adaptation which is unique among the primates and is only present in the New World monkeys. Most *Platyrrhina* possess a brain which is less complex and capable than that of the *Catarrhina;* however, there are exceptions. The capuchin monkeys do not lag behind most Old World monkeys either in brain development or intelligence. In fact, the capuchins surpass many of the Old World monkeys, and because of this they have been used for decades for "intelligence tests."

There are no terrestrial forms among the New World monkeys. All are arboreal and inhabit the jungle forests. In the mountains they are found at elevations up to 1000–1500 meters. The many broad and deep rivers in the jungle regions of South America serve as insurmountable boundaries. The members of one genus or species often look very different on either side of a river. For this reason, many subspecies were frequently described as species. Often these subspecies differ so greatly that one wonders whether they should not be classified as a separate species. Even within one home range or within the same group the individuals may vary considerably. Since large sections of the South American jungle belong to the least explored regions on this earth, the classification into the various genera, species, and subspecies has been very difficult and many questions are still unsolved.

Distinguishing characteristics

Of the infraorder NEW WORLD MONKEYS *(Platyrrhina),* there is one superfamily *(Ceboidea)* with three families: 1. CEBUS MONKEYS *(Cebidae);* 2. GOELDI'S MONKEY *(Callimiconidae);* 3. MARMOSETS and TAMARINS *(Callithricidae).*

CEBUS MONKEYS (Family *Cebidae*) may range in size from that of a squirrel to that of a cat. There are 36 teeth: $\frac{2 \cdot 1 \cdot 3 \cdot 3}{2 \cdot 1 \cdot 3 \cdot 3}$. There are five subfamilies: 1. NIGHT and TITI MONKEYS *(Aotinae);* 2. SAKIS *(Pitheciinae);* 3. CAPUCHIN MONKEYS *(Cebinae);* 4. HOWLER MONKEYS *(Alouattinae);* 5. SPIDER MONKEYS *(Atelinae);* there are ten genera and thirty-four species.

Subfamily: night and titi monkeys

Among the many species of monkeys in the Old and New World, the NIGHT MONKEY *(Aots trivirgatus;* Color plate p. 138) is the only nocturnal form. Despite its many special adaptations many scientists consider the night monkey to be the most primitive living New World monkey. The HRL is approximately thirty-five cm and the TL is fifty cm. The head is round without a protruding muzzle. The very large nocturnal eyes have a strongly convex retina. The nostrils are closer together than in most New World monkeys. The body is slender and is covered by dense fine wool. The tail is also densely furred and is not prehensile. The hands are shorter and broader than in other New World monkeys. The feet are long and narrow. The fingertips are padded. The nails are not elongated into claws. Body coloration and the stripe pattern on the head vary greatly. The mountainous forms are covered by particularly dense and long hair, and have bushy tails. These animals are widely distributed throughout the moist and dry South American tropical forests.

The immense nocturnal eyes enable the douroucoulis to see well enough even in the darkest nights when other nocturnal animals are only able to orient with the senses of hearing and touch. The animals live in pairs. During the day they sleep in tree hollows. The animals seem to see poorly during the day because if they are disturbed in their sleep, they jump up blindly and bite in the direction of the intruder. At dusk the animals become active and stay in motion throughout the night. Night monkeys seem to have peaks of activity at dusk and dawn. Every pair occupies a relatively small territory and always returns to the same tree hollow like the marmosets and tamarins. Most other monkeys usually wander over large areas in search of food and often they change their place of shelter.

In the fertile South American tropics, food sources do not seem to be exhausted even in small areas. Fruits, berries, soft unripened nuts, insects, spiders, tree snails, and other small animals are present in all seasons. Douroucoulis are able to capture even flying insects with large jumps. They also capture tree frogs, small lizards, and plunder birds' nests, occasionally even licking honey from blossoms.

When hunting, the douroucoulis behave in a manner similar to that of the larger marmosets and tamarins, and kill with the same type of bite. "If one permits them to run freely," writes Sanderson, "they run through the entire apartment. If they catch a large insect they turn it around with both hands until they can bite off the wings or legs. Then they take the insect in one hand and pull its head off with the other, and eat it leisurely. Birds are seized with a sudden jump. The skull is bitten and the prey is plucked and eaten." Crandall recommends the following diet for the night monkey: bananas, other fruits, a small amount of salt, cabbage, raw carrots, canned dog food, orange juice, milk, and vitamin supplements. The animals are also fond of boiled eggs, nuts, insects, and lizards.

Very young douroucoulis make delightful house pets; however, one has to accept the fact that the animals are asleep throughout the day and consequently are very lively during the night. In the early morning, the night monkeys conduct loud, resounding howling concerts. The voices of these small and delicately built creatures are much louder and more powerful than one would expect. The animals can alternate between quiet melodic twittering and chirping sounds, and deep resonant calls. Torchlights at night do not seem to disturb the activity of cap-

Fig. 15-1.
Night monkey *(Aotes trivirgatus)*

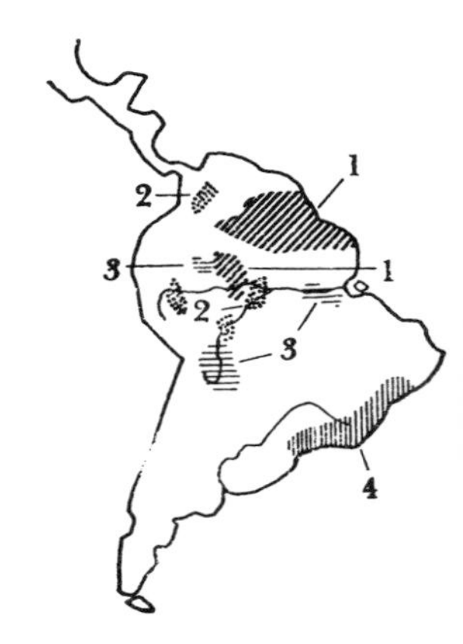

Fig. 15-2.
1. Collared titi *(Callicebus torquatus)* 2. Red titi *(Callicebus cupreus)* 3. Orabussu titi *(Callicebus moloch)*
4. Masked titi *(Callicebus personatus)*

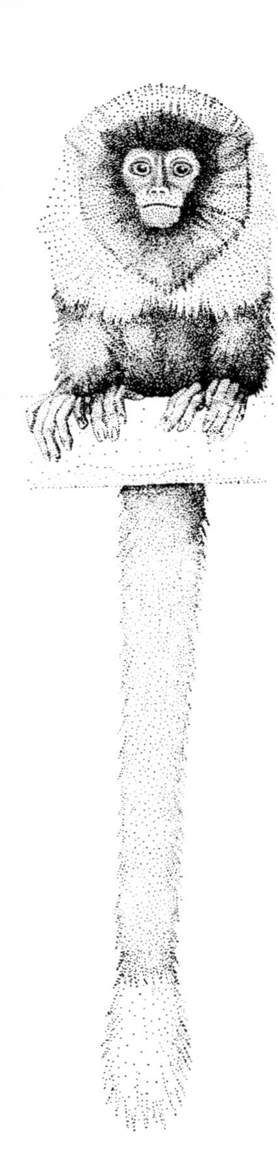

Fig. 15-3.
Resting titi monkeys sit hunched across a branch with bent legs, with tails hanging down. Their body weight is mainly supported by the feet. The hands grasp the branch right beside the feet. The animal is able to stretch its legs suddenly from this hunched position and quickly jump long distances either to escape or to catch prey.

tive night monkeys. Herbert Wendt told of some douroucoulis kept in Lima which arose around 5 P.M. and then chased about in their outside enclosure with wide jumps although it was still light outside.

Occasionally douroucoulis have bred in zoos. They have been raised in the zoos of Frankfurt, San Diego, and Washington. Twin births are frequent. The female lines her sleeping box with soft materials prior to giving birth. During the first weeks of life the young clings to the back of either the mother or the father. The young stays with the parents until it is nearly grown.

The TITI MONKEYS (Genus *Callicebus;* Color plate p. 337) are about the same size as the night monkey. Both forms are classified under the same subfamily. As is the case for most New World monkeys in the insufficiently explored tropical regions of South America, a distinction of the numerous forms and their range of distribution is very difficult. There are four species with twenty-four subspecies: 1. The COLLARED TITI *(Callicebus torquatus)* has an HRL of 40–45 cm and a TL of 40–50 cm. The fur coloration varies with the regions. There are several subspecies. 2. The RED TITI *(Callicebus cupreus)* is somewhat smaller, not quite as colorful, and does not have the white chest spot. All subspecies have red shades on the ventral side. 3. The ORABUSSU TITI *(Callicebus moloch)* includes, besides other subspecies, HOFFMANN'S TITI *(Callicebus moloch hoffmanni)* which possesses a white to ochre hair fringe surrounding the face and covering the neck. 4. The MASKED TITI *(Callicebus personatus)* has a conspicuously dark or black front on the head and chest. In all titis the nails are elongated into claw-like structures similar to the marmosets and tamarins.

Although the silky fur and the claw-like structures on the fingers and toes of the titi monkeys are somewhat reminiscent of the marmosets and tamarins, the former group is the diurnal counterpart in appearance and behavior of the night monkeys. We know very little about free living titi monkeys. Hans Krieg saw the titis mainly in the thick underbrush of forests, even when hollow trees were not present. The titis were usually seen in pairs, or occasionally with two young of different ages. The titi monkeys capture insects, lizards, and small birds. They also feed on soft fruits, leaves, blossoms, and other plant material.

All titi monkeys are very difficult to keep, and they rarely survive in captivity. This may be due to the fact that sufficiently large enclosures for these very agile jumpers are not provided. In the Delta Regional Primate Research Center at Covington, Louisiana, which is climatically located very favorably, the Orabussu titi *Callicebus moloch)* is doing very well in conditions similar to the free living state. There the comparative psychologist E. A. Mason is studying their fascinating territorial behavior. Usually captive titis were not provided with sufficient animal food. The animals are primarily meat eaters but must have roughage for proper digestion. Captive titis greedily consumed crabs, small fish, chicks, and young mice. Too much plant material in the diet leads to chronic intestinal disorders which may cause death.

Around 1930 an adult masked titi lived in the Hamburg Zoo for two years. Bungartz reported that this animal preferred eating bananas, figs, dates, and other sweet fruits, but was very greedy for mealworm larvae and cockroaches. This titi even used a straw to push into cracks and crevices to extract the cockroaches, and removed its sleeping box from the wall in order to search for those insects there. This search for cockroaches, which also has been observed in other small monkeys in captivity, is not always without inherent dangers. Cockroaches are intermediate hosts for a variety of intestinal parasites *(Acanthocephala)* which are difficult to combat. The masked titi from Hamburg also consumed plucked sparrows and drank the contents of bird eggs. In the spring it would feed on twigs, buds, and young leaves. Uta Hick described a Hoffman's titi in the Cologne Zoo as "distinctly tame and good natured." It is also a "demanding pet which requires a lot of movement and a very diversified diet."

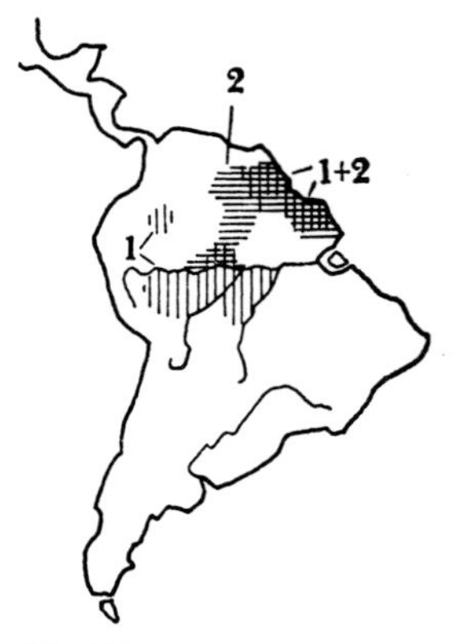

Fig. 15-4.
1. Hairy saki *(Pithecia monacha)* 2. Pale-headed saki *(Pithecia pithecia)*

These small titi monkeys have extraordinarily loud voices which almost remind one of the howling of howler monkeys. On summer days one can hear them very early in the morning. Not much is known about reproduction in titis, although it is known that the females usually bear only one young.

Subfamily: *Pitheciinae*

The SAKIS and UAKARIS (Genera *Pithecia, Chiropotes,* and *Cacajao;* Color plate p. 327) closely resemble each other. They are combined as the saki monkeys (Subfamily *Pitheciinae*) and are grouped with night and titi monkeys. These animals are very slender with delicate limbs; however, they appear stout because they usually have long thick fur. The nostrils are particularly widely spaced and open to the side. The nostrils cannot be seen from the front. The thumb and index finger oppose the other fingers in grasping. There are three genera and a total of nine species:

A. The SAKIS' (Genus *Pithecia*) fur is coarse and consists of very long hairs. The tail is long and bushy. The frontal thyroid cartilage of the larynx is enlarged as a resonating chamber. There are two species and several subspecies: 1. The HAIRY SAKI *(Pithecia monacha)* has an HRL of approximately forty cm and a TL of approximately forty-six cm. It often has a wig-like forehead fringe. 2. The PALE-HEADED SAKI *(Pithecia pithecia)* has an HRL of approximately thirty-seven cm and a TL of approximately thirty-six cm. The wig-like forehead fringe is absent and the central part of the face around the eyes, nose, and mouth is black and almost naked. The cheeks and forehead are covered by a broad face mask of stiff hair which is yellowish-white to ochre-yellow in the ♂. In the ♀ it is shorter and black with a white stripe on each side of the nose. The two subspecies are the WHITE-HEADED SAKI *(Pithecia pithecia*

Fig. 15-5.
1. Black saki *(Chiropotes satanas)* 2. White-nosed saki *(Chiropotes albinasa)* 3. Red-backed saki *(Chiropotes chiropotes)*

pithecia), whose ♂ has a face mask of yellowish-white, and the GOLDEN-HEADED SAKI *(Pithecia pithecia chrysocephala)* with a face mask of ochre-yellow in the ♀.

B. The BEARDED SAKIS (Genus *Chiropotes*) have an HRL of approximately forty cm and a TL of approximately thirty-eight cm. When skinned the animal is very similar to the sakis but is different in skull characteristics and dentition. The head hair is long and drops down from the center part to all sides. In two of the three species the ♂♂ and ♀♀ have well developed beards on the chin. The body hair looks "well groomed," particularly the hair on the head, the beard, and the tail. 1. The BLACK SAKI *(Chiropotes satanas)* is dark brown to blackish-brown with a parted "wig" and a well developed beard. 2. The RED-BACKED SAKI *(Chiropotes chiropotes)* is yellowish-grey to dark reddish-brown; the back is almost black and the chin beard is even more developed. 3. The WHITE-NOSED SAKI *(Chiropotes albinasa)* is black, with a conspicuously red nose and upper lip region covered with dense white hair. All three species probably represent each other ecologically and probably can be regarded as subspecies.

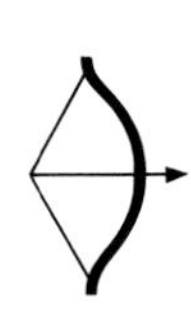

C. The UAKARIS (Genus *Cacajao*) are probably the only New World monkeys with a reduced tail. The HRL is 40–45 cm, but the TL varies. The body is somewhat more compact than that of the saki. The face, the sides of the head, and the frontal end of the parted hair are totally or almost totally naked. There are four species: 1. The BALD UAKARI (*Cacajao calvus;* Color plate p. 327) has a red face and bald pate; the fur is whitish. 2. The RED UAKARI (*Cacajao rubicundus;* Color plate p. 327) has a red face and bald pate; the fur is reddish-brown. 3. The BLACK-HEADED UAKARI *(Cacajao melanocephalus)* has a black face. The head is normally covered by hair, and the fur is black and chestnut brown. 4. The BLACK UAKARI *(Cacajao roosevelti)* is the largest uakari species with a relatively long tail; the HRL is approximately forty-five cm and the TL is approximately thirty-nine cm. The face and fur are black. See the maps for the range of distribution.

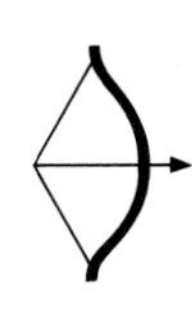

Fig. 15-6.
1. Bald uakari *(Cacajao calvus)* 2. Red uakari *(Cacajao rubicundus)*
3. Black-headed uakari *(Cacajao melanocephalus)*
4. Black uakari *(Cacajao roosevelti)*

In some Indian languages the words "sakiwonka," "sakiwinki," and "sakimiri" simply mean "small monkey." "Saimiri" was derived from this and is the name of the squirrel monkey (Genus *Saimiri*) which belongs to the capuchin monkeys. The name "saki" is not a good choice because in South America this may lead to confusion with other species of monkeys. Even more confusing is the fur coloration of the two saki species. Usually the fur is dusky grey or blackish brown. The individual hairs often have silvery, yellow, or ochre colored tips. The back of the hand is either black or white. There is an unbelievable variation in the types with all sorts of intermediate gradations. In certain regions even the sexes are of conspicuously different coloration.

Sakis are typical occupants of the hot and humid rain forests of central South America. The respiratory organs of the sakis are well adapted to the sultry climate where the air is saturated with moisture. The long coarse fur protects the animals against the heavy downpours. The sakis move around as a family unit. Occasionally they form larger groups, particularly in the dense foliage at the edge of the forest where berry bushes are in abundance. The animals not only feed on berries, nuts, leaves, and other plant material, but also on small birds and other mammals. Sanderson told of an Indian hunter who once showed him a hollow tree into which the sakis kept entering: "They had discovered the shelter of bats inside the giant jungle tree. They caught the sleeping animals, pulled off their fur, and ate them in the branches."

During their migrations to suitable feeding places sakis leap in tremendous jumps over wide abysses from tree crown to tree crown. This ability is also very adaptive when the jungle forests are flooded by high water. Their jumping capabilities far exceed those of most New World monkeys. However, they are very cautious climbers. When climbing down they carefully put down their feet and test each branch to see if it will support them before they let go with their hands. In this aspect they seem to be like people. The sakis walk upright on the ground or thick branches; like gibbons, however, they do not hold their arms to the side but rather they balance themselves with raised arms.

In South America sakis are often kept as pets because they are very cute and affectionate. Yet even in South America these animals are considered to be very fastidious. Formerly, sakis did not survive long in European and North American zoos. They usually sat listlessly in their cages and made a boring exhibit. Soon they died. The dry air inside the animals houses did not agree with them. There was not enough room for the sakis to exercise and very little was known about their food habits. Most New World monkeys and, in particular, the smaller and more sensitive species, require a very diversified diet which contains sufficient amounts of animal protein, many vitamins, and trace elements. Since these facts have become known, sakis in some zoos have stayed in excellent health. One female hairy saki lived in the Philadelphia Zoo for over seven years. This animal became very tame and shared a cage with a uakari female in great harmony. A magnificent pair of hairy sakis, who were always ready to play, lived in the small mammal house in the Frankfurt Zoo and they are still (1967) active and healthy. In the San Diego Zoo golden-headed sakis have been raised for four years in succession. One golden-headed saki lived there from November 29, 1946, to August 8, 1960. This is almost fourteen years and is the longest life span attained by a saki in captivity up to now.

In recent years the Cologne Zoo has been very successful in raising sakis. The director, Dr. W. Windecker, has hired biologically trained assistants to care for these sensitive animals and this measure has been of great value. Uta Hick, who looked after white-headed sakis, red-backed sakis, and white-nosed sakis in 1966, described the saki

"Sascha" as a particularly lively pet: "He has tremendous jumping abilities of which he makes extensive use. Often he crosses the large room with gigantic leaps. He prefers to jump into the branches of the tree placed in his cage, hang from his hind legs, and swing his body back and forth. Except for small resting periods, he is active all day long and is always ready to play."

He enjoyed eating nuts, sunflower seeds, melon seeds, egg yolk, and buds and leaves. He refused insects. The vitamins and trace elements were mixed into milk and tea with added honey, egg yolk, soja flour, vitamins and calcium, of which he seemed very fond.

The white-headed saki which has repeatedly been kept in the Cologne Zoo lives primarily on the maanys nuts when in the wild. These nuts are filled with a milk-like fluid. It also eats marmadosa, which are small red berries. In Cologne the animals adapted to a diet of nuts, chestnuts, sunflower seeds, cucumbers, tomatoes, grapes, grasshoppers, and mealworm larvae. Formerly these monkeys were reported to be quiet and "melancholy," but this type of behavior was only due to improper care. The last saki male which arrived in Cologne (1966) is very temperamental and is constantly on the move. Often one can also hear his bird-like voice. Of course these creatures of the humid and hot tropical forests not only need a large space for movement and a carefully chosen diet but also a high degree of humidity in the air. A humidifier is of necessity in a saki enclosure.

In Cologne a white-headed saki shared a cage with a white-nosed saki. It was interesting how these two different animals reached a particular destination. The white-headed saki covered the distance in long jumps while his companion cautiously climbed there. The bearded sakis, which include the white-nosed saki, are not such agile jumpers as the sakis. They move more cautiously through the branches and carry their tails upright, shaped like a question mark. Their diet consists primarily of fruits and nuts, and probably also includes insects but not mammals. While searching for food, the animals split up into small groups but unite again when continuing their migration or in case of danger.

Fig. 15-7.
When the red-backed saki *(Chiropotes chiropotes)* walks on all four legs, it often holds up its bushy tail.

The San Diego Zoo, in California, which is famous for being able to keep many difficult to keep animal species, was able to keep a black saki alive for fifteen years. Formerly this species was frequently seen in zoological gardens but the animals never lived very long. More recently black sakis have only been kept in the Amsterdam Zoo. The diet of these sakis is very diverse and consists of apples, bananas, oranges, carrots, and other plant materials in addition to baby biscuits, meat, boiled eggs, and mealworms. This mixture is also supplemented with glucose and minerals. The red-backed sakis in the Cologne Zoo are kept primarily on a diet of fruits with some grasshoppers and mealworms.

All bearded sakis groom their hair and beard very carefully. A scraggly saki is not healthy. These monkeys are not as sensitive to climatic conditions as was originally assumed. In Germany the animals stayed outside during the warm seasons. Red-backed saki males, which the Cologne Zoo has kept in an enclosure with the pheasants since 1961, have the opportunity to choose inside or outside cages. According to Humboldt, red-backed sakis drink by dipping their palms into the water and then sucking the fluid from the cupped hand. This behavior pattern was not observed on the first sakis that arrived in Cologne, but was seen in the last arrival, as well as in the white-nosed saki. An excited or angry red-backed saki stands upright, flashes his teeth and shakes his beard.

The white-nosed saki is a greater rarity than other bearded sakis kept in zoos. Currently the Cologne Zoo is keeping two of these monkeys. The animals prefer a diet of nuts, sunflower seeds, and grasshoppers. Occasionally they also eat young mice from the nest. The white-nosed saki is able to crack nuts with its well-developed teeth. They are very greedy eaters. They take as much food as possible into their hands and toes and eat very rapidly, already on the look-out for the next piece. They like to rub lemon peel into their fur. When drinking, they occasionally dip their hands into the liquid and lick it off. The white-nosed saki produces a weak chirping sound. In an exited state the animal's voice sounds rather peculiar, almost like the whistling of a bird.

In spite of their protruding muzzles and broad noses, the bald-headed uakari species appear more like humans than do the sakis. The naked faces and particularly the large bald pate, which mimics a high human forehead, are responsible for this apparent human resemblance. There are good reasons why we are repulsed by the facial features and particularly by the red-faced forms. The more facial characteristics an animal shares with man, the greater the tendency in us to attribute human emotions and values to this animal. This perception seems to be a deeply-rooted innate feature which probably aided our prehominid ancestors to determine the intentions and emotional state of other conspecifics. If, on the basis of this innate prejudice, we have mistaken the animal face for a human face, we automatically take offense at any features which deviate from the normal human type. The uakari's face is reminiscent of all too many unpleasant human facial expressions.

The bald pate of the uakari is not "handsome" in our eyes, either. Bald headed people do not fit our image of a beautiful person. The broad nose of the uakari looks as if it was knocked flat. The shape of the eyes and nose and the missing eyebrows only seem to increase our natural repulsion. The naked facial and forehead skin of the bald and red uakaris looks "unhealthily" red, as if the areas were inflamed or infected with some dangerous communicable disease. In this respect

▷

Squirrel monkeys

1. Red-backed squirrel monkey *(Saimiri oerstedi)*
2. Common squirrel monkey *(Saimiri sciureus)*

Titi monkeys

3. Collared titi *(Callicebus torquatus)* 4a. Orabussu titi *(Callicebus moloch moloch)* 4b. Hoffman's titi *(Callicebus moloch hoffmannsi)* 5. Red titi *(Callicebus cupreus)*

Goeldi's monkey

6. Goeldi's monkey *(Callimico goeldii)*

▷▷

Capuchins

1. Weeper capuchin *(Cebus nigrivittatus)*
2. White-throated capuchin *(Cebus capucinus)*
3. White-fronted capuchin *(Cebus albifrons)*
4. Various subspecies of the black-capped capuchin *(Cebus apella)*

The illustrated capuchins represent an example of the great variability among these monkeys. The classification of individual animals into one of the various species and subspecies is sometimes impossible.

▷▷▷

1. Woolly spider monkey *(Brachyteles arachnoides)*
2. Humboldt's woolly monkey *(Lagothrix lagothricha)*

Subspecies: a. Grey woolly monkey *(Lagothrix lagothricha cana)* b. Brown woolly monkey *(Lagothrix lagothricha poeppigii)* c. Colombian mountain woolly monkey *(Lagothrix lagotricha lugens)*

1
2
4a
3
6
4b
5

1
2
4a
3
4b
4c

1
2b
2a
2c

1a
2a
2b
1b
3

Spider monkeys

1. Long-haired spider monkey *(Ateles belzebuth)* Subspecies: a. *Ateles belzebuth marginatus* b. *Ateles belzebuth belzebuth*
2. Black spider monkey *(Ateles paniscus)* Subspecies: a. *Ateles paniscus chamek* b. *Ateles paniscus paniscus*
3. Central American spider monkey *(Ateles geoffroyi panamensis)*

a second human innate reaction finds expression. An innate reaction causes us to avoid anyone who has an obvious skin infection. We treat such an infected person as a "leper" and expel him from our community in order to protect other members from the infection. However, when the red-faced uakaris become ill their faces do not increase in redness but turn pale like corpses. Ironically, the completely healthy "scarlet faces" evoke in us, who are so proud of our intellect, the ancient, cruel primitive innate "expulsion reaction." This demonstrates how deeply our feelings are still rooted in our animal past.

Like other sakis, the uakaris occasionally walk upright. They hold up their arms to maintain balance. The uakaris are very active and agile in their jungle habitat. Since the proper care and diet of these sensitive monkeys has become known, the animals do not sit around their cages listlessly and sluggishly, as in former times. The Frankfurt Zoo displays its red uakaris very well. The monkeys are housed in a hygienically clean "play room" constructed out of a synthetic material and metal climbing bars which are separated from the public by a huge picture window. During the warm season the animals have access to a similar outside enclosure. One red uakari lived in the Philadelphia Zoo for nearly nine years, while another in the New York Zoo survived for at least five years. J. Schmitt diagnosed filaria in the uakaris kept in Frankfurt. These blood parasites, which had not previously been described for uakaris, can often be fatal. The parasites cannot be detected in the feces and only with the greatest difficulty from blood smears. Since the uakaris in Frankfurt have been treated against filaria their life expectancy has been greatly increased. One male has been living in the zoo for over seven years. One young was even born in June, 1967, the first one born in the Temperate Zone.

Under human care uakaris develop into regular gourmets which like great variation in their diet. The uakari in New York was fed commercial monkey chow which is used in many modern zoos today, as well as fruits, green vegetables, a little milk, and hard-boiled eggs. It cracked nuts with its teeth. In contrast to other uakaris this animal also ate some meat. It always enjoyed mealworms and *Anolis* lizards.

Subfamily: capuchin monkeys

The New World monkeys discussed up to now give the impression that they did not reach the same developmental stage as did the Old World monkeys in body structure and behavior. However, their apparent primitive characteristics may possibly be connected to their small size. The small delicate squirrel monkeys (Genus *Saimiri*) also have such features and therefore Sanderson and Steinbacher place them close to the relatives of the sakis. However, squirrel monkeys already look like "real" monkeys. Their saki-like dentition is probably associated with their small size and the concomitant restricted method of obtaining food. For a number of important reasons, the genus *Saimiri* and the genus *Cebus*, the capuchins, are combined in the subfamily CAPUCHIN MONKEYS *(Cebinae)*.

There are great variations in the squirrel monkeys which depend on the geographic locality; they have been classified into four species: 1. The COMMON SQUIRREL MONKEY *(Saimiri sciureus)*; 2. The RED-BACKED SQUIRREL MONKEY *(Saimiri oerstedi)*; 3. The BLACK-CROWNED SQUIRREL MONKEY *(Saimiri boliviensis)*; 4. The MADEIRA RIVER SQUIRREL MONKEY *(Saimiri madeirae)*. The species are primarily differentiated on the basis of their fur color (Color plate p. 337). The normal HRL is approximately thirty cm and the TL is approximately forty cm. Certain individuals reach almost twice the normal size. In grasping, the thumbs oppose the other fingers. The tail is very long but is not prehensile.

At a quick glance the unique facial coloration of the squirrel monkey looks like a bony skull. The German term for this monkey "Totenkopfäffchen" (skull monkey) is not very suitable; rather, the animal appears as if it had dipped its mouth too deeply into a jar of blueberry jam. The faces of squirrel monkeys are very appealing to us and we consider them to be "cute." In German the syllable "chen" (diminutive) is always added to a name if the object or animal is appealing. In describing the squirrel monkey the suffix "chen" has been added; it has not, however, been added to the description of the equally small but "ugly" bald-headed uakaris. Konrad Lorenz has shown that the diminutive "chen," at least in German, tends to be added when describing small animal species that possess a round head with large eyes which are relatively low in the face and have round, soft cheeks. The animal also moves a little clumsily. All these characteristics appeal to our innate "baby schema" and awaken our maternal or paternal feelings.

Squirrel monkeys are widely distributed in South and Central American tropical forest regions, and in some areas they are very frequent. In Guayana Sanderson observed a group of approximately 550 animals which jumped, one after another, in continuous succession across a forest clearing. In tropical Bolivia, Herbert Wendt could detect the presence of squirrel monkeys only by the fact that suddenly leaves would drop from the tree crowns without there being the slightest breeze. Upon closer inspection he recognized a large group of squirrel monkeys which had moved through the foliage almost noiselessly. The monkeys were hardly discernible in the greenish twilight and were excellently camouflaged by the leaves on which the sunlight was glittering. Within a few minutes the animals had disappeared. In the South American tropical forest huge trees from twenty-five to fifty meters in height reach up to the sky. There is hardly any underbrush on the dark forest floor but the foliage of the tree crowns forms a thick canopy. Here live the squirrel monkeys, under the cupola of a green dome.

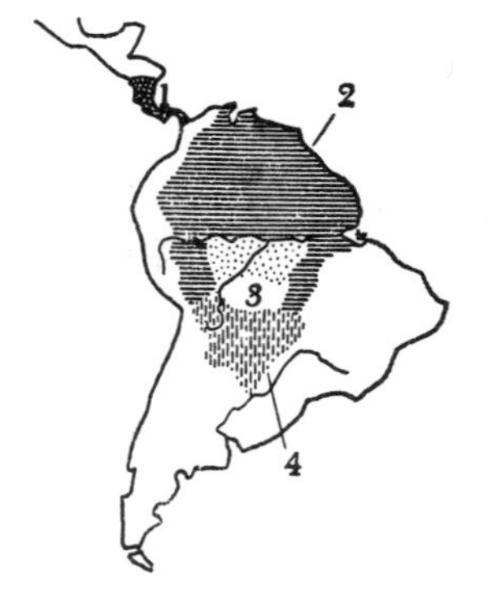

Fig. 15-8.
1. Red-backed squirrel monkey *(Saimiri oerstedi)*
2. Common squirrel monkey *(Saimiri sciureus)*
3. Madeira River squirrel monkey *(Saimiri madeirae)*
4. Black-headed squirrel monkey *(Saimiri boliviensis)*

Along the forest edge and along rivers this canopy descends to the ground like a heavy curtain. The canopy is interwoven with countless vines at these points and is fully exposed to the sun. This is also the place where many fruits, berries, and nuts grow. Like many other New World monkeys, squirrel monkeys frequent these leafy curtains and they even use specific paths which lead through this leafy jungle. Squirrel monkeys climb through the brushes and vines with unbelievable speed. They might grab a few berries here, capture a snail, spider, tree frog, or terrestrial crab at another place. High in the crowns of the trees, they hunt, leaping far, for beetles, butterflies, and even flies and mosquitoes. Their jumping ability is closely linked to the size and shape of their brain. The long, oval head of the squirrel monkey appears more human than any other monkey head. In relation to body weight the squirrel monkeys have the largest brain of all primates. The brain mass of these tiny monkeys is 1/17 of its body weight. In man it is only 1/35 of the total weight. However, the brain is not convoluted to such a degree as it is in the more highly evolved primates. In the squirrel monkey the brain regions which control difficult movement maneuvers with the eye are highly developed.

These true arboreal animals are rarely seen on the ground. Individual animals or small groups probably do not dare to come down at all; however, occasionally large herds of monkeys descend to the ground in search of food.

Squirrel monkeys are quite difficult to keep in captivity. They require a greatly varied vegetarian diet in addition to animal protein which they take in the form of insects, bird eggs, and poultry. They are very susceptible to infections of intestinal worms and other parasites. Usually squirrel monkeys are captured in large numbers and are kept under crowded and not very sanitary conditions. Even before the monkeys are bought and are delivered to European and American zoos they are heavily infested with parasites and only very few survive. However, if one is successful in curing the squirrel monkey of its parasites with the available medicines and by regularly removing fecal material and also adapting the animal to the new diet, then they will often live for many years.

In recent years, the behavior of squirrel monkeys has been extensively studied by several scientists. In certain behavioral characteristics these monkeys are similar to the capuchin monkeys. The squirrel monkeys, for instance, also impregnate themselves with urine. While the capuchin monkeys rub mainly the palms and soles, and also, to a small degree, the fur, with urine, the saimiris liberally distribute urine throughout their fur, particularly on the tail hairs which are often dripping wet. The color photograph on p. 317 shows a squirrel monkey in the process of impregnating its tail with urine. This behavior, which appears rather repulsive to us, of course has nothing to do with uncleanliness. The converse is true; squirrel monkeys are extremely clean animals. They constantly groom their fur by combing and brushing the ventral part of the body, legs, and tail with their fingers. The head, ears, and shoulders are cleaned with the toes. Up to now it has not been observed that squirrel monkeys groom each other as is the practice in the Old World monkeys, capuchin monkeys, marmosets, and tamarins.

Squirrel monkeys display to their conspecifics by standing up on their hind legs, conspicuously spreading the legs, and exposing the genitalia. Even young juveniles determine their rank position within a group in this manner.

In 1578, the Frenchman Jean de Léry published his travel report about distant Brazil in La Rochelle. Here he mentioned a "small black guenon which the natives called 'cay'." These lines written by the French nobleman are the first report of the capuchins (Genus *Cebus*). The French refer to these monkeys as Sajou or Sapajou, which are derivations of the Indian terms "Sai-hui" or "Cai-hui."

Of all the New World monkeys, the capuchins most resemble the arboreal cercopithecids of the Old World in their appearance and behavior. The capuchins lack the conspicuous specializations, such as the large nocturnal eyes of the night monkeys, the shaggy fur of the sakis, the claw-like nails of the tamarins, the extremely long limbs and naked prehensile pad on the underside of the tail tip of the spider monkeys, or the powerful voice apparatus of the howler monkeys. In a way the capuchins represent the "completely normal, average monkey." It is therefore not surprising that for a long time zoologists considered them to be close relatives of Old World cercopithecids and macaques. The great Swedish scientist Carl von Linné gave the name *Simia capucina* to one type of capuchin monkey which possessed a black cap on its head. Today the monkeys are called capuchins in English, Kapuziner in German, and Mono Capuchino in Spanish.

These monkeys are of medium size; the HRL is 32–56 cm, the TL is 38–56 cm, and the body weight is approximately 1100–3300 g. The legs and arms are of about the same length. The thumbs are opposable to the other fingers. The big toe is also prehensile. The tail is fully covered by hair and lacks the naked "prehensile sole." The tail is only partially prehensile. It serves as a prop but does not serve as a "fifth hand" as in the following genera. The hair coat is very variable. The head fur is usually dark and is sharply separated from the naked forehead or is set off by white forehead hairs ("cap"). Hair on the head is often long. The skull is round and the brain is more highly developed than in other New World monkeys. There are four species and many subspecies:

1. The WHITE-THROATED CAPUCHIN *(Cebus capucinus)* is relatively small and slender with fine limbs. The dark head pate is covered by short hair but only on the back of the head.

2. The WHITE-FRONTED CAPUCHIN *(Cebus albifrons)* is also small and

Fig. 15-9.
1. White-throated capuchin *(Cebus capucinus)*
2. Weeper capuchin *(Cebus nigrivittatus)*

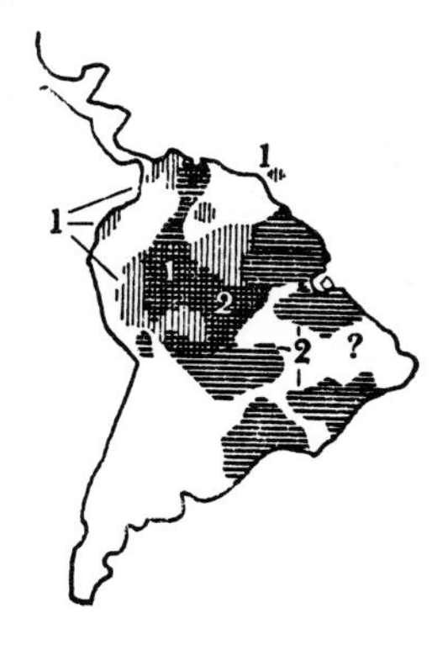

Fig. 15-10.
1. White-fronted capuchin *(Cebus albifrons)*
2. Brown capuchin *(Cebus apella)*

slender, with delicate limbs. The coloration is lighter; the dark head pate is only present on the back of the head.

3. The WEEPER CAPUCHIN *(Cebus nigrivittatus)* is somewhat larger; the limbs are longer and the fur is of a looser and coarser texture. The dark, short-haired head pate is found around the central head part.

4. The BROWN CAPUCHIN *(Cebus apella)* has a more robust body form. The limbs are more compact and shorter, and the tail is stronger and covered by dense hair. The hair is longer and coarser. The dark head pate extends over the entire head and often hair tufts of varying lengths and shapes are present.

Capuchins are widely distributed. They are found in the hot, humid forests of the Amazon Delta as well as in mountainous forests of the Andes up to elevations of over 1600 meters. Within this vast geographical area, which is interrupted by rivers, chains of mountains, steppes, and cultivated land, the appearance of the capuchins varies immensely. The classification into species and subspecies is even more difficult than in other genera of New World monkeys. Nevertheless, we know that in various regions several forms of capuchins live side by side without interbreeding. This seems to be a certain indication that one is dealing with various species and not subspecies. However, our two maps of distribution and our color plate (see p. 337) are only an approximation of the occurrence and the appearance of the capuchins.

All capuchins are wholly arboreal and generally only descend to the ground to drink. They prefer dense forests with little undergrowth. Capuchins may occur in orange, corn, or cocoa plantations. They pluck the fruits and carry them off into the tree tops of the forest where they can eat them in safety. Apparently individual groups do not roam over large distances but always use the same routes through the forest canopy. Capuchins always seem to move in the same order. Kühlhorn observed that juveniles lead the procession and are followed by mature males and females. Mothers with babies hanging under their abdomens or on their backs bring up the tail end of the group. Causey and his co-workers did not find any evidence indicating that one group entered into the territory of another, when they observed black-capped capuchins in Brazil. The actual feeding territories in Brazil were usually only a few hundred meters in circumference. Only occasionally did the monkeys visit other nearby regions of the forest. Capuchins regularly urinate on their hands and then rub their feet and fur with the fluid. This is probably the manner in which a troupe of capuchins lays down its odor-marked trails in the forest canopy, which they can easily follow with their rather poorly developed sense of smell.

In the years after 1930, F. M. Chapman studied the behavior of the white-throated capuchins on the Barro-Colorado Island in the artificially created Lake Gatun in the Panama Canal. He found that individual capuchin troops are not as cohesive, for instance, as are the societies of the howler monkeys. The animals are often scattered over a wide area in the forest. A group of approximately thirty capuchins may be spread over an area of approximately 400 meters. However, the monkeys keep in constant contact via occasional loud, hoarse calls.

More than other New World monkeys, capuchins are frequently kept in European and North American zoos, research institutes, and as pets by private persons. The monkeys are not particularly hard to keep. They adapt very easily to our climate and are not too finicky in their food requirements. However, one has to feed them more than the required amount of food because the monkeys waste and smear great quantities of food around in their cages. The basic monkey diet of bananas, oranges, apples, carrots, raw and cooked potatoes, small amounts of lettuce or cabbage, pieces of bread, peanuts, and sunflower seeds has proved to be successful with the capuchins. Yet, as in all New World monkeys, one should supplement the diet with meat, eggs, or some other animal protein.

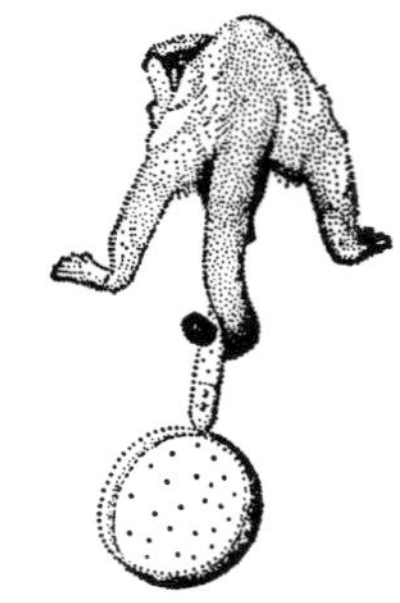

Fig. 15-11.
This is how the brown capuchin "Pablo" *(Cebus apella)* holds his food pan with his prehensile tail.

The zoologist Angela Nolte of Münster over a period of time offered a great variety of insects to her brown capuchin "Pablo." He ate most insects but avoided cockroaches, earwigs, cockchafers, bumblebees, and mealworm beetles. He very quickly grabbed locusts and ate them greedily. "When Pablo received an ample number of grasshoppers he used to play a 'prey capture game' with the last insects," reports Mrs. Nolte. "He would let them fly away and then would rush after them and grab them with both hands. He repeated this maneuver several times. Pablo would often rear up on his hind legs, circle around the grasshoppers, and then touch them with his index finger if they did not move." Brown capuchins also catch flies out of the air quite expertly.

Fig. 15-12.
Pablo has made a "table." He eats the seeds from his underarms which are closely pressed together.

Capuchins like to fill their mouths with seeds and then sit upright and hold out their arms with their palms held closely together. In this manner they make a little platform on which they spit the entire mouthful of seeds. They then pick up the seeds individually and eat them. This "table" also catches all the food remains that drop out of the mouth during their hasty eating. Mrs. Nolte observed what she called "table-making behavior" in all her brown capuchins and also in other capuchin species. In the Amsterdam Zoo she observed this behavior in a squirrel monkey, which seems to resemble the capuchins in this innate behavior pattern.

Fig. 15-13.
When Mrs. Nolte gave "Pablo," her capuchin monkey, orange peels, he would roll up and at the same time try to rub the strong odorous substances into his head hair with both hands and into his neck with his foot.

Since capuchins are frequently kept and are easy to care for, they occasionally have been bred in captivity. Oddly enough, no young have been born to capuchin families in the best zoos and under excellent care while young are often born in small zoos or in private homes. Why this is so, no one knows. The gestation period lasts approximately six months. Angela Nolte and Gerti Dücker were able to observe the development of a brown capuchin *(Cebus apella),*

Fig. 15-14.
This is the manner in which a capuchin male courts his female.

which was born in the Münster Zoo and raised in the Zoological Institute, in 1957. The mother, Suse, received the regular diet plus milk and calcium and vitamin additives and live grasshoppers. Her cage was held at an constant temperature with an infrared heater. For the first five weeks the young, called "Bambino," constantly clung to his mother's body, but gradually he jumped clumsily around close to his mother. When Bambino first tried to climb, Suse held his tail with her foot to prevent him from straying too far. Occasionally Bambino tried to tear himself loose but Suse always pulled him back. If she pulled too hard, Bambino immediately climbed on Suse's back. After eight weeks he moved more freely and more independently, but still seemed to enjoy being carried by his mother.

For decades capuchins have been the "display animals" of animal psychologists and, later, ethologists. In relationship to their body weight the size of their brain is amazingly large. It is only somewhat smaller than that of the squirrel monkey, but it is much better developed. The manner in which captive capuchins use tools to open nuts has often been observed. They crack the shells with either stones or wooden blocks. Sometimes they hit the nuts vigorously with one or both hands on the ground, their sitting bar, or the wall. Once I observed a capuchin in the Halle Zoo tossing the nuts that were offered to him into the air with the expectation that they would crack once they hit the floor. Other capuchins in the same cage cracked their nuts with a stone or by banging them on the floor. Yet this one particular monkey did not seem capable of imitating the successful method of his conspecifics. A chimpanzee would quickly be able to imitate the specific use of a tool from a conspecific. It is also able to solve problems which require thinking, or, in other words, by insight learning. The learned behavior of the capuchins, however, is still far removed from the matter-of-course way with which the great apes are able to understand and solve quite difficult problems.

Like other monkeys the capuchins groom their own fur and that of their conspecifics with great care and persistence. They hastily brush the hair with their fingers and remove dandruff or bits of dirt with the fingers, lips, or tongue, and swallow it. Only very rarely can ectoparasites maintain themselves under such intensive body care. During these grooming sessions the monkeys often grunt contentedly, indicating a sense of well being. Capuchins like to rub strong-smelling materials like onions, cologne, or orange peel into their fur with their hands and feet. Pablo used to curl himself up in order to rub the top of his head with both hands and to rub his nape with one foot, all at the same time. Mrs. Nolte got the impression that "this rubbing caused a very pleasant sensation in the monkey."

Subfamily: howler monkeys

In the following genera (howler monkeys, woolly monkeys, woolly spider monkeys, and spider monkeys) the tail has literally become a "fifth hand." The tail is highly muscular and the underside of the tail-tip possesses a naked "prehensile sole." The animals can hang by their tails, utilize it during climbing, or even hold objects with it. There are two subfamilies; HOWLER MONKEYS (*Alouattinae*) and the SPIDER MONKEYS (*Atelinae*).

The howler monkeys, together with the spider monkeys, woolly spider monkeys, and woolly monkeys, are the largest New World monkeys. The HRL is approximately fifty-seven cm and the TL is approximately sixty cm. The head is flat with a protruding muzzle. The facial profile is slanted. The angle of the lower jaw is very high, particularly in the back portion. The hyoid bone is enlarged to a sac-like structure. The thyroid cartilage of the larynx is greatly enlarged. The body is robust; the arms and legs are long but compact. The thumb is fully developed and together with the index finger is opposed to the other fingers. The fur is long and silky. The specialized larynx is covered by longer hair, giving the impression of a beard. The fur coloration varies greatly in the species and subspecies. In certain species color varies with age and sex. There are six species and many subspecies: 1. The GUATEMALAN HOWLER MONKEY (*Alouatta villosa*) has fur which is soft, silky, and always black, regardless of sex or age. 2. The MANTLED HOWLER MONKEY (*Alouatta palliata*) has blackish-brown fur. On the sides the hair is longer and is usually of a lighter color. The hyoid is less enlarged than in the other species. 3. The RED HOWLER MONKEY (*Alouatta seniculus*) is larger than the mantled howler and is more colorful. Coloration and pattern vary greatly in the different subspecies, and also within one region. The coloration is usually red or reddish-brown. 4. The BROWN HOWLER MONKEY (*Alouatta fusca*) has almost uniformly brown or reddish-brown fur with a golden gleam; the face is framed by a heavy beard. 5. The RUFOUS-HANDED HOWLER MONKEY (*Alouatta belzebul*) has black or blackish-brown fur. The hands, feet, and tail tip are reddish-brown but depending on the subspecies may blend into the red, golden, or yellow color. 6. The BLACK HOWLER MONKEY (*Alouatta caraya*) is of large body size; the ♂♂ are black, while the ♀♀ and juveniles are yellowish olive-brown. The hair is long and stiff, and the beard is well-developed.

Fig. 15-15. 1. Mantled howler monkey (*Alouatta palliata*)
2. Rufous-handed howler monkey (*Allouatta belzebul*)
3. Black howler monkey (*Alouatta caraya*)

As in many song birds, the howler monkeys stake out their territories with the aid of their voices. Their calls carry over a long distance and are considered to be the loudest sounds that animals are capable of producing. Alexander von Humboldt once estimated that their calls can be heard 2½ kilometers through the jungle. In London one could hear the calls of three howler monkeys in a pet shop five kilometers down wind. Hans Krieg believes that these calls not only mark a territory but are also an expression of well being.

Before a living emperor tamarin (*Saguinus imperator*, see p. 374 ff.) had been seen, taxidermists in museums always liked to twist up the beard hairs of this squirrel-sized monkey in the fashion of Emperor Wilhelm II. The Swiss zoologist Goeldi, therefore, irreverently called it the "emperor tamarin."

In a howler monkey troop, the strongest male often calls. Occasionally several males howl together or the entire group cries in unison. The females have a slightly less well-developed larynx, and their calls

Fig. 15-16. 1. Guatemalan howler monkey *(Alouatta villosa)*
2. Red howler monkey *(Alouatta seniculus)*
3. Brown howler monkey *(Alouatta fusca)*

are not as loud as those of the males. When Darwin was formulating his theory on sexual natural selection he thought that the males cried in order to impress the females. The loudest male would have the best chances with the "weaker sex." It seems more probable that the major job of marking off the territory fell to the male and therefore his larynx became more highly developed. A similar situation is found in many other animals.

When calling the howler monkeys open the jaws and purse the lips in a funnel shape. Krieg describes their sounds as a rapid repetition of "a-hü, a-hü, a-hü," whereby "a" is emitted during inhalation and "hü" during exhalation. In an excited state the sounds produced during exhalation are rougher and louder. The calls sound like "a-hö, a-hö, a-hö."

The movements of the howler monkeys are slower and more deliberate than those of other monkeys. They jump very rarely and when climbing usually hold on with four of their "five" hands. Nevertheless, they are able to move through the branches faster than a man walking on the forest bottom. Howler monkeys are difficult to keep. Even the South American Indians, who have tamed all other New World monkeys, very rarely keep a captive howler monkey. They have seldom reached zoos because they are highly specialized in their food requirements. Their diet consists primarily of leaves, buds, and blossoms in addition to nuts and fruits. Some of the larger North American Zoos, such as those in New York and San Diego, or South American zoos occasionally display howlers. A red howler monkey which primarily ate *Wisteria* leaves, and grapefruit lived in the Bronx Zoo in New York for over two years.

Subfamily: spider monkeys

The prehensile tail is most highly developed in the spider monkeys (Subfamily *Atelinae*). In a way this fifth hand has moved up to the position of first hand. These monkeys are large and have long arms and legs. The prehensile tail is the longest and most powerful tail found among the primates. The spider monkeys are adapted for brachiation in a manner similar to the Old World gibbons. Brachiation is only weakly developed in the slower woolly monkeys but is fully developed in the agile woolly spider monkey and spider monkeys. Their bodies show the same adaptations to this arboreal mode of movement as do the gibbons and great apes. The limbs are elongated, the hands are extended and thumbs reduced because they are not used in the "arm stride." The vertebral column is reduced and strengthened. The thorax is enlarged and the pelvis is transformed.

Cotton-head tamarins *(Oedipomidas oedipus)* for a long time were considered extremely hard to keep. On page 381, Herbert Wendt reports how one can care for these tiny animals properly.

Nevertheless, there are still considerable differences between the brachiators of the New and Old World. In the gibbons and great apes, brachiation resulted in a final loss of the tail which was a hindrance in this mode of movement. However, in the spider monkeys the prehensile tail developed into the most significant "brachiating arm." It is probable that this "double insurance" of the prehensile tail in spider monkeys takes from them this last element of weightless elegance which we so admire in the gibbons.

The prehensile tail is not only a locomotor organ; spider monkeys also grasp objects with their tail, particularly if they are not able to reach them with their shorter arms. Winkelsträter observed a woolly monkey which begged by holding its tail through the bars in front of the zoo visitors, just as other monkeys and man hold out their hands. A male spider monkey which Grzimek kept in his apartment was able to open door handles with its tail. I was present once when Philipp, the monkey, tried to leave the room which he was not allowed to do. Philipp walked through the room on all fours and approached the door as if by chance. Then he abruptly seemed to turn back and glance "innocently" at his mistress. While doing this he edged backwards almost imperceptibly towards the door and then with lightning speed opened the door handle with the tail and whisked down the hallway.

Two species of woolly monkeys

In woolly monkeys (Genus *Lagothrix*) brachiation is only slightly developed. The arms and legs are barely longer than those of the capuchins and howler monkeys. The thumbs are still well developed. There are two species: 1. HUMBOLDT'S WOOLLY MONKEY (*Lagothrix lagotricha* Color plate p. 339) has an HRL of approximately 50–60 cm; in some old males the HRL is up to 70 cm. The TL is approximately 60–70 cm. The fur is soft, woolly, and very dense. The face is almost naked and is uniformly dark brown to blackish brown. The predominant color of the fur is ashy-brown or yellowish-gray to blackish; there are four subspecies. 2. The PERUVIAN MOUNTAIN WOOLLY MONKEY (*Lagothrix flavicauda*) is deep mahogany colored and darker in the front. A dense yellowish genital tuft is found in the male. The underside of the tail is yellow from the middle to the tip and is sharply set off against the dark upper side. The brown face has a clearly defined brownish-yellow triangular spot from the base of the nose to the upper tip.

Woolly monkeys inhabit the rain forest of the central and upper Amazon basin. On the slopes of the Andes they are found in the higher regions. At higher elevations the monkeys are darker and have longer hair than the lowland forms which prefer the swampy and occasionally flooded forests. There is great individual and geographical variation in fur coloration, just as in other New World monkeys, which makes the classification into subspecies very difficult. The grey woolly monkey is the most familiar and most frequent form seen in zoos. This name, however, refers to two different, yet similarly colored, subspecies, *Lagothrix lagothricha lagothricha* and *Lagothrix lagothricha cana.* Formerly, the brown woolly monkey (*Lagothrix lagothricha poeppigii*) was considered to be a separate species, but there are many transitional forms between these types. Some monkeys have dark heads, light or almost black fur. Brown woolly monkeys may be of an ochre color, brownish-yellow with reddish sides, or dark chestnut brown.

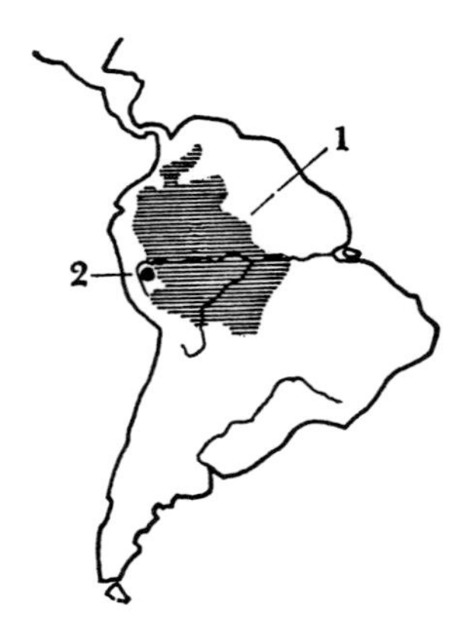

Fig. 15-17.
1. Humboldt's woolly monkey (*Lagothrix lagotricha*) 2. Peruvian Mountain woolly monkey (*Lagothrix flavicauda*)

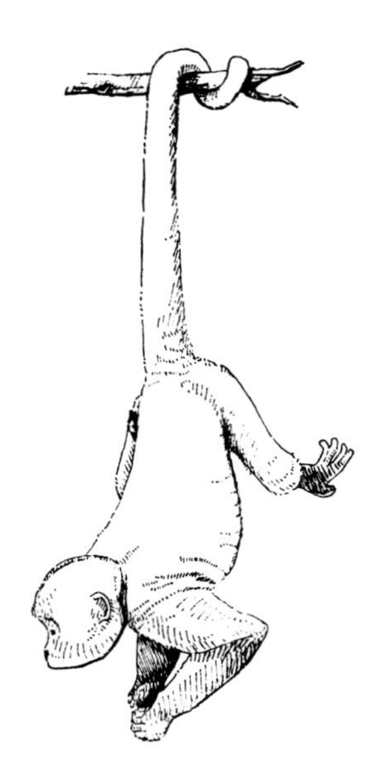

Fig. 15-18.
Woolly monkeys often suspend themselves from their strong tails which enables them to have their hands and legs free for manipulation.

The occasional old woolly monkey male may be solitary, but as a rule there are small or larger troops consisting of males, females, and young. Occasionally the woolly monkeys unite with groups of capuchins, spider or howler monkeys. The woolly monkey group usually migrates through the branches at a slow pace. The monkey cautiously places one hand in front of the other, and foot in front of foot. Additional support is given by the prehensile tail. Larger spaces between trees are crossed by brachiating to the outside branches and then dropping vertically into the brush. According to Miller they may drop for more than ten meters. However, these large monkeys can also be very agile and are able to "arm stride" great distances like spider monkeys. When the animals are resting or settling down to eat they usually sit upright, which frees their hands and tail to investigate the nearby surroundings or fish for food and to put it in their mouths. Woolly monkeys are seldom seen on the ground. However, they are able to walk bipedally with or without using the tail as support.

Their facial expressions are very lively and occasionally appear rather human. Often they hold their hands in front of their faces like frightened or grieving people. The lips are extraordinarily movable and can be extended more than in all other New World monkeys. When H. H. Keays shot one woolly monkey out of a troop of fifty, most members were not disturbed by the shot but two individuals came down to the dead animal. They lifted him up "and propped him against a tree as if expecting him to climb up. Then they gave the impression that they understood that he was dead. They dropped him and started to scream. The entire group joined in the cry, and rushed away through the tree tops."

"The natural diet of the woolly monkeys consists primarily of fruits and leaves," writes Jack Fooden who has intensively studied woolly monkeys. "The extremely worn incisors and canines of examined animals are indicative of a diet of hard-shelled fruits." Captive woolly monkeys also eat cooked and raw meat, as well as fruit and vegetables. It seems that animal protein is part of their natural diet. Woolly monkeys seem to have an insatiable appetite. They can consume large amounts of food at one time, enough to make their bellies look distended.

Like most New World monkeys the woolly monkeys are difficult to keep in captivity. They are considered to be the most pleasant, placid, and friendly pets among the larger monkeys. Around the 1930's a five year life expectancy in the zoo was considered a record. However, recent nutritional research and advancements in zookeeping have prolonged the zoo life expectancy of woolly monkeys from seven to nine years. In Berlin a zoologically trained private person was repeatedly successful in breeding woolly monkeys. These monkeys, when kept in the home, are not as destructive as capuchins. Even Alfred Russel Wallace already praised their "very mild character."

Woolly monkeys raised as youngsters usually remain very tame and are very amusing in their deliberate manner. Only older males which are kept in crowded cages and under improper care will occasionally inflict painful bites.

Woolly spider monkeys and spider monkeys are very closely related. Yet the few zoologists who were fortunate enough to see a living woolly spider monkey point out that this large, densely furred monkey with its claw-like elongated nails gives a completely different impression than the spider monkeys. Therefore this monkey was placed into the separate genus *Brachyteles.*

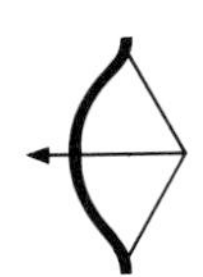

The WOOLLY SPIDER MONKEY (⊕ *Brachyteles arachnoides;* Color plate p. 339) has an HRL of 46–63 cm, a TL of 65–80 cm, and a body weight of 10 kg and more. The flat, naked face is flesh-colored or dark. There is a black hair fringe above the eyebrows. The fur is yellowish grey brown in color, dense, with long guard hairs and very thick underwool. The thumbs are vestigial with nails absent or present. Thumbs may be absent or are reduced to different degrees on both hands. The second and fifth fingers possess strongly laterally compressed, elongated, almost claw-like nails. There are possibly two subspecies.

Woolly spider monkeys belong to the rare New World monkeys. Their distribution is limited to the Tupi forests of Southeast Brazil. Only rarely have these animals been shipped to Europe or North America. They are very shy in the free living state and Indian hunters almost never saw them. Since large tracts of their habitat have been cultivated it seems that this species is on the way to extinction. Brazilian zoos consider the woolly monkey to be a great treasure. Herbert Wendt was able to observe one pair of woolly spider monkeys in a Brazilian private zoo. He found that these monkeys were reminiscent in their movements of the woolly monkeys and they were very different from the much more agile spider monkeys. Their naked, occasionally brilliant red faces, almost similar to those of the uakaris, contrast sharply to the greyish-brown fur. In 1937, a woolly spider monkey reached the Leipzig Zoo. Zoos in Breslau, Düsseldorf, and Wuppertal have displayed woolly spider monkeys for brief periods. In 1959, the first living woolly spider monkey reached the Bronx Zoo in New York, where it was observed by Lee Crandall. Almost nothing is known about free living woolly spider monkeys.

Four species of spider monkeys

SPIDER MONKEYS (Genus *Ateles;* Color plate p. 340) have an HRL of 34–59 cm and a TL of 61–92 cm. The face is naked and less flattened. The pelage is coarse and without underwool. The thumbs are vestigial or absent. The second and fifth fingers possess flat nails. There are four species: 1. CENTRAL AMERICAN SPIDER MONKEY *(Ateles geoffroyi);* 2. BROWN-HEADED SPIDER MONKEY *(Ateles fusciceps);* 3. LONG-HAIRED SPIDER MONKEY *(Ateles belzebuth);* 4. BLACK SPIDER MONKEY *(Ateles paniscus).*

The spider monkeys are the "artists" among the New World

Fig. 15-19. 1. Central American spider monkey *(Ateles geoffroyi)*
2. Long-haired spider monkey *(Ateles belzebuth)*
3. Woolly spider monkey *(Brachyteles arachnoides)*

Fig. 15-20. 1. Brown-headed spider monkey *(Ateles fusciceps)*
2. Black spider monkey *(Ateles paniscus)*

monkeys. Sanderson compares them to the trapeze performers in a circus. They are able to swing by means of arms and tail and arm-stride from tree to tree for up to ten meters. When they are in a hurry they run on all fours. They accomplish the quadrupedal mode better than the gibbons which they resemble, but which have much longer arms. Like other New World monkeys, they occasionally stand up or walk bipedally, especially when they are on the look out. If at all possible they try to cling to a support with their tail when up on two feet. In the free living state they rarely come down to the jungle floor.

Spider monkeys usually congregate in large groups. According to Humboldt the various species keep strictly to themselves. Each troop has its own territory which it patrols daily on specific pathways. The neighboring territory is usually not entered. The American zoologist Clarence Ray Carpenter has made an intensive study of the black-handed spider monkey from Panama in the free living state. He observed, however, that the territorial boundaries occasionally overlap, and that these can slowly be readjusted. At times a troop of spider monkeys would attack Carpenter. As soon as the monkeys discovered the intruder they would emit rough barking sounds and descend to the lower limbs. Their barks came in faster succession the closer they approached the stranger. Finally the shouts sounded like unified metallic chattering. Occasionally some of the stronger males and possibly a few females would shake the branches, accompanying this with threatening growls. Several times the monkeys approached Carpenter as close as 12–13 meters; then they broke off dead branches with their hands, feet, or tail, and let these drop on him. This maneuver was not without peril because some of the wood chunks weighed from eight to ten pounds. During these attacks the monkeys would nervously scratch their fur with their fingers and toe nails, just like we scratch our heads when we are ill at ease.

Actually spider monkeys are not very courageous. Their extremely wild behavior is mainly bluff, a feint attack, which is supposed to frighten and chase away the enemy. If this behavior is of no avail the troop subsequently breaks up into several subgroups which slowly move off into different directions. Spider monkeys are only daring towards man if they have not encountered him previously. Once the monkeys have unpleasant experiences with the "lord of creation" they tend to be cautious and try to get away from his presence silently and unnoticeably.

Indians have hunted spider monkeys with poisoned arrows. The hunters, however, lost many of the shot prey because the monkeys would often anchor themselves in the branches with hands, feet, or tail. Often they did not drop down to the ground even after they had died.

Since the thumb is absent in the spider monkeys they are less skillful in grooming themselves than are other monkeys. They scratch themselves a great deal with their hands and feet. The thorough reciprocal grooming is not highly developed in them as it is in their relatives. Carpenter mostly observed mothers grooming the fur of their young.

The long armed and tailed acrobats prefer more of a fruit diet than most other New World monkeys. The monkeys which Carpenter observed in Panama ate a diet which consisted 90% of nuts and fruits. Carpenter also saw these spider monkeys search under the bark of a tree or between leaves, presumably for insects and their larvae. According to observations made by H. O. Wagner, they also plunder bird nests, eat young shoots, and pick *Dendrophytus* blossoms which they rip open in order to get at the honey. Feeding spider monkeys in zoos is therefore less difficult than feeding most other New World monkeys. Spider monkeys eat about the same diet as the capuchins. However, they have to be housed in spacious enclosures with many climbing opportunities, and have to be kept as a group. If at all possible, one adult male should be kept with several females. It is difficult to distinguish the sexes when such a group is assembled. The females possess a long dangling clitoris while the penis in the male is short and inconspicuous.

Spider monkeys rarely give birth in captivity in Europe but it is not rare in the United States. According to current findings the gestation period lasts approximately 139 days. Normally only one young is born but the Myrick Park Zoo in La Crosse, Wisconsin, U.S.A., reported the birth of twins in 1960. In one of the zoos in New York a black spider monkey female lived for over twenty years. Generally a zoo life expectancy of four to six years is considered to be a good record. Although these monkeys are unassuming in their food requirements they do not adapt too well to the climate change.

Dietrich Heinemann

16 Goeldi's Monkeys, Marmosets, and Tamarins

In the year 1904 the Brazilian State Museum for Natural Science and Ethnology in the port city of Para (Belem, today) received a small, shiny black monkey which almost looked like a maned tamarin and which was unknown to zoologists. The director of the museum, the Swiss zoologist E. A. Goeldi, accommodated the monkey in the zoo which belonged to the museum. Unfortunately, it only stayed alive for a short period. Goeldi contributed greatly to the investigation of the South American fauna and the museum of Belem carries his name today. Goeldi was unable to determine the origin of this unknown monkey. It had probably been brought in by one of the river steamers which at that time not only transported caoutchouc and other natural products but also freshly captured animals from the western Amazon region. Goeldi sent the skin of the monkey to the British Museum of London for closer classification. However, the skull was missing and an exact identification was not possible. The English mammalogist Oldfield Thomas considered it a new marmoset on the basis of the claw-like nails and silky fur coat. He named the monkey after Goeldi and classified it close to the tamarins, the hairy-faced tamarins and moustached tamarins, which also inhabited the tropical rain forests of the Amazon.

Between 1911 and 1914 the next two Goeldi's tamarins arrived in the zoo of Para, probably also by route of the river steamer. These monkeys also did not live very long; however, the Brazilian zoologist Miranda Ribeiro was able to make an intensive study of their bodies. These studies were very surprising. Mrs. Ribeiro's results showed that in body structure Goeldi's monkey resembled the marmosets and tamarins but in skull structure and dentition was related to the Cebidae. For these reasons one can consider Goeldi's monkey as a living link between these two New World monkey families. Mrs. Ribeiro gave this new genus the beautiful name *Callimico* which means "beautiful little monkey."

Family: Goeldi's monkey by H. Wendt

Ever since its discovery this little monkey has remained a matter of dispute to taxonomists. Some classify the animal with the Cebidae, others with the marmosets and tamarins *(Callithricidae).* Actually this monkey possesses characteristics which fit both families. Yet none of the characteristics predominate in either family. Therefore it was decided to classify this monkey in its own family, *Callimiconidae,* which is between the *Cebidae* and marmosets and tamarins. There is only one species (Color plate p. 337).

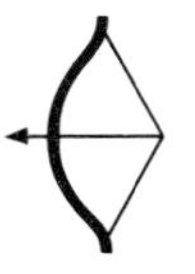

GOELDI'S MONKEY *(◊Callimico goeldii)* is slightly larger than a squirrel. The HRL is approximately 27 cm, the TL is 30 cm, and the body weight is 420–500 g. The head is roundish. The muzzle is slightly more developed than in the marmosets and tamarins. The large ears are round and hairless and are totally or partially hidden when in the mane. The hair crown is puffed out. The fur coat is silky, deep black with a brownish, golden, or yellowish shimmer which results from the light hair tips. There is a hair crown on the head, and a mane in the nape. The tail is slightly bushy. The front limbs are shorter than the hind limbs. The fingers and toes have claw-like elongated nails. Only the big toe has a flat nail. The thumb is short but slightly opposable. The diet is omnivorous and consists of fruits, insects, bird eggs, and small vertebrates. There are 36 teeth; $\frac{2 \cdot 1 \cdot 3 \cdot 3}{2 \cdot 1 \cdot 3 \cdot 3}$. After nine months the monkey grows its permanent teeth. It inhabits the tropical forests of the upper Amazon. The gestation period is 4½–5 months. There is only one young per litter. In the only observed case of twin births the young were stillborn. The young opens the eyes on the first day of life. The nursing period is from 60–70 days.

Distinguishing characteristics

Up to the date of the discovery and closer study of the Goeldi's marmoset, the marmosets and tamarins were regarded as the most primitive, physically and mentally least developed, of the true monkeys. Phylogenetically they were believed to be not highly developed "not-yet-monkeys" or a "rough approximation of monkeys." Some zoologists even wanted to classify them in a separate category way below the Old World Catarrhina and the New World Platyrrhina. However, it has been shown that marmosets and tamarins are closely related to the other Platyrrhina via Goeldi's marmoset. Is it therefore still possible to support the thesis that the Cebidae originated from Callithricid-like ancestors? In 1915 when the first Goeldi's marmoset reached Europe and lived in the London Zoo for one week the first objections to this view were heard. The English zoologist Pocock was first to assume that the Callithricidae were probably a dwarfed side branch of the New World monkeys which could be traced from the Cebidae via Goeldi's marmoset. Since then this view has found acceptance. The Callithricidae are no longer considered the most phylogenetically primitive living New World monkeys, but rather a single specialized, probably still very young and still highly evolving branch

Fig. 16-1. Goeldi's marmoset *(Callimico goeldii)*

of the Platyrrhina stem. Goeldi's marmoset, which was the cause of this change of view, is still a matter of great dispute. The four previously mentioned Goeldi's marmosets were the only specimens that were known to have come from their habitat in thirty years. This sample was not large enough to clarify the taxonomic position of this "beautiful little monkey" within the zoological system once and for all. Again in 1945 Goeldi's marmoset was kept in the Belém Zoo (Pará) and was described by Da Cruz Lima. Still nothing was known about the behavior and reproduction of Goeldi's marmoset. The little monkey is considered to be such a rarity that it prompted Ernest P. Walker to write in *The Monkey Book* in 1954: "Should someone have the great fortune of seeing this extremely rare little monkey, he should do his utmost to keep this treasure alive, photograph it, and make careful observations about its habits."

Keeping and raising Goeldi's marmoset

In 1954 and 1955 several American zoologists had this good fortune. Several Goeldi's marmosets reached North and South American zoos, universities, and private persons via the Peruvian harbor of Iquitos. Unfortunately this event was short-lived because, like their predecessors in London and Belém, all the monkeys died quickly except for one animal in the Bronx Zoo. Since 1957 living Goeldi's marmosets were seen again in Europe, in the Museum for Natural Science in Bern and five German zoos. Most monkeys only survived for a few weeks or even days, with a few encouraging exceptions. One male stayed alive for eight months in Cologne, another for five and one half years which is the longest period up to now which a Goeldi's marmoset has survived in captivity. Two Goeldi's marmosets lived for more than two and one half years in the Frankfurt Zoo.

Some animal-loving private persons, the Heinemann family, were successful in breeding a pair of Goeldi's marmosets which they brought from the United States to Germany. The female had had a previous abortion in America. In Wiesbaden the pair was housed in a spacious climbing and jumping cage. The front wall of the cage could easily be removed so that the monkeys were also able to move around in the indoor room. These lively animals, which were always ready to jump around, adjusted well to their new surroundings. They copulated and raised four young between 1964 and 1965.

The Goeldi's marmosets in Wiesbaden were thoroughly studied and described by Mrs. Heike Heinemann and Rainer Lorenz. According to their observations, the Goeldi's marmoset is not only an intermediate form in its skull and anatomy, but also in certain behavioral characteristics, particularly their sounds, where they are truly transitional between the larger New World monkeys and the Callithricidae. The thumb is of particular interest. It appears like a normal marmoset thumb but is more divergent and is opposable to the other fingers during climbing. It seems then that the Goeldi's marmoset, along with the capuchins and squirrel monkeys, belongs to the few New World monkeys that possess a true grasping hand.

The diet of Goeldi's marmoset

Just like captive marmosets and tamarins, captive Goeldi's marmosets can become very tame. They express their affection or dislike towards specific persons in a rather temperamental manner. The Goeldi's marmoset which lived in the Cologne Zoo for five and one half years received a diet of apples, oranges, bananas, mealworm larvae, grasshoppers, bread, zwieback, and milk. "Of all the small monkeys which shared his cage, he was by far the friendliest and most active," reported Uta Hick. "When I opened the door to his cage he would immediately jump into my arms and greatly enjoyed being petted on the head and behind the ears. He preferred a stretched-out body position, putting his head in the hollow of his palms. However, he did not maintain this resting position for long since he always reacted intensively to any change in the environment. He typically greeted familiar persons with an opened mouth and soft twittering. He expressed anger by erecting his body, arching the back and shoulders, and erecting the mane in the nape and on the head. At the same time he emitted shrieks which consisted of a quick succession of shrill sounds."

Birth and raising of the young

Mrs. Heinemann was able to observe the final events of the birth process in her group of Goeldi's marmosets: "Towards 8 P.M. all animals had disappeared into the sleeping box. Seventy-five minutes later I heard excited sounds. I saw the female and a juvenile male outside the sleeping box. The female was hanging on a branch and was fending off the young male with loud cries. As the female was climbing up the branch I noticed a newborn young between her thighs. The young crawled on to the right thigh and clung to it. The other monkeys in the cage were not interested in these events. During the next twenty-five minutes the female remained in a hunched position and continued to fend off the curious juvenile with loud cries. Then she retreated to the sleeping box with the afterbirth in her mouth."

According to Mrs. Heinemann and Rainer Lorenz, all visible parts of the newborn Goeldi's marmoset were covered with short black fur. The ventral side, on the other hand, was still hairless, but within two months hair grew in. The hair on the forehead is still flat, and is only erected in the fifth month into the "crown." The newborn infant clings to the back of its mother and in the beginning is almost entirely covered by the mane of her nape. The young puts its arms around the neck and sides of the mother while it anchors its feet into the fur of the belly and flanks. The tail is curled underneath the mother's abdomen, and whenever she moves the tail is more firmly pressed against her.

Fiedler and Starck also observed this tail-curling reflex in the marmosets and tamarins. In the first weeks of life young marmosets and tamarins always press the tail underneath the abdomen of the carrying adult and curl it up on the other side of its body. If one picks up a newborn marmoset, tamarin, or Goeldi's marmoset, it not only holds on with its hands and feet, but occasionally also wraps the tail around the finger of the person. Adult marmosets and tamarins coil up their tails between their hind legs. Hence, tail-coiling is not a specific characteristic of just the capuchins and the spider monkeys. The Goeldi's marmoset in particular, even more so than the tamarins and marmosets, is able to use the extended or coiled-up tail vigorously for bracing itself. Goeldi's marmoset also uses its tail to touch its caretaker but it never uses it to actually hold on tightly.

In the third week of life the young Goeldi's marmoset climbs on its father for the first time. From then on the mother only carries the infant for nursing. During the fourth month the juvenile is rejected by its mother when it begs to be fed. The father cares for and feeds the young for another few weeks. At the time of its first birthday the Goeldi's marmoset is fully grown; it is probably sexually mature by eighteen months.

Jumping abilities of Goeldi's marmoset

Goeldi's marmoset has better jumping abilities than most marmosets and tamarins, and it is only slightly less expert at it than the titi monkeys. In its horizontal jumps in a room this little monkey literally "flies" from cupboard to cupboard, extending his arms, slightly flexing his legs, and using the long tail for balancing. It can effortlessly cover distances of up to 3½ meters. In the wild it can jump 5 meters from tree to tree if it jumps downward even a little. In addition to these gliding jumps, Rainer Lorenz has also observed the typical galloping jumps, quadrupedal walking, and occasional bipedal walking similar to that of the larger New World monkeys. However, marmosets and tamarins occasionally also walk on two legs.

Family: marmosets and tamarins by H. Wendt

Distinguishing characteristics

In the MARMOSETS and TAMARINS (Family *Callithricidae;* Color plates p. 375 and p. 376) the face is flatter than in Goeldi's marmoset. In size they range from that of the rat to that of the squirrel or slightly larger. The HRL is 16–31 cm and the TL is 18–42 cm. The body weight is 85–560 g. The head is roundish. The ears are large, naked, or only slightly covered by hair or are adorned with hair tufts or stiff hair fringes. The fur is silky, and often conspicuously colored. In certain species manes of hair are found on the head; in others the hair on the shoulders or back is elongated in a mane-like fashion. The very long tail is covered by bushy hair at the distal end. The forelimbs are shorter than the hindlimbs. The short-limbed species are primarily climbers while the long-limbed forms are excellent jumpers. There are small, claw-like elongated nails on fingers and toes. Only the big toe has a broad flat nail. Embryological studies have shown that these claws are not a primitive characteristic but that they originated from transformed nails. The thumbs are short, not opposable or only very slightly. The big toe, on the other hand, is opposable to the other toes. These monkeys are omnivorous and their diet consists of fruits, nuts, buds, and other plant material, as well as insects, spiders, tree frogs, small lizards, bird eggs, and fledglings. There are 32 teeth: $\frac{2 \cdot 1 \cdot 3 \cdot 2}{2 \cdot 1 \cdot 3 \cdot 2}$. The last molar (the wisdom tooth) does not break through because of the smallness of the animal and the shortness of the jaw. In tamarins the canines are well developed. Permanent teeth are exchanged for milk teeth between the seventh and ninth months. The gestation period is 140–160 days. Usually twins are born, less often single young or triplets. The eyes are already open on the first day. During the first months of life the father carries the youngsters; the mother or other group members carry them occasionally. The nursing period is approximately 6–12 weeks.

Inhabitants of the "forest above the forest"

Only a few primates are so totally adapted to an arboreal existence as are the marmosets and tamarins. They are the typical inhabitants of the upper strata of the forest canopy. Alexander von Humboldt has referred to this strata as the "forest above the forest." This part of the South American rain forest is matted with epiphytes and the marmosets and tamarins hardly ever leave it to come down to the ground. Only Miller has observed silver marmosets in the high steppe grass of Mato Grosso. Geographical landmarks like mountains, dry forests, savannahs, barren lands, streams, larger rivers and treeless cultivated areas serve as insurmountable boundaries to these monkeys. The South American tropical forest is far from being a uniform region. Mountains, bodies of water, steppes, and semi-deserts divided the tropical forest into numerous regions. Each is inhabited by its own kind of marmosets and tamarins.

Exuberant growth of vegetation is prevalent in the Tupi forests of the Brazilian coastal mountains, and particularly in the wide river valleys of the large tributaries of the Amazon. At the edge of the forest one usually finds a dense curtain, often several meters in thickness, which consists of lianes, bromeliaceous plants, orchids, berry bushes, giant ferns, and climbing grasses. This dense growth makes it extremely difficult for man and larger animals to penetrate. However, this web of climbing plants and epiphytes provides ideal living conditions for the marmosets and tamarins. Here they find fruit, berries and nuts, smaller and larger insects, tree frogs which spawn in the water-filled funnels of the bromeliaceous plants, and small birds which construct their nests in this jungle curtain. The small monkeys, therefore, have a wide variety of foods available.

Rivers serve as distribution boundaries

These little monkeys are incapable of crossing the large rivers or the steppe regions between the various lowlands. This has often resulted in one particular species or subspecies being on one side of the river while on the opposite side a completely different type is found. Since many South American forest rivers have often changed their courses during the ages and have created new valleys, it becomes understandable why the marmosets and tamarins have split into countless different forms in the vast Amazon basin.

Even the original Indian inhabitants considered the marmosets and tamarins favorite house pets. After the discovery of America, European conquerors, colonists, and explorers soon took a liking to these delicate, charming, and often very pretty and attractively colored monkeys. In 1551 Konrad Gesner illustrated the golden lion marmoset *(Leontideus rosalia)* and the common marmoset *(Callithrix jacchus)* in his comprehensive work, *Historia animalium.*

Since then countless marmosets and tamarins have been transported to Europe. They were not only displayed in zoos but were kept by many private people. During the Baroque period in France it was the fashion among nobility to give one's wife or mistress such a tiny monkey as a special gift. Madame Pompadour's maned tamarin has been introduced into literature by Buffon and Brisson as "le petit singe lion." The first crested bare-faced tamarin which the South American explorer La Condamine, the discoverer of the caoutchouc tree, presented to the Count of Buffon enjoyed no less a reputation. Buffon is responsible for giving this monkey the name "le pinché" which is still used today. However, this is a native name which probably refers to a different monkey species living in Northeast Peru.

Origin of the names "marmoset" and "tamarin"

The names "tamarin" and "marmoset" were also coined by a French scientist during the Baroque period. Sanderson and Steinbacher describe the origin of the word "marmoset" in the following manner: "In France marmosets and tamarins were extremely popular, and were called "Marmouset." Originally this referred to a small boy or dwarf. Later the meaning changed and it referred to a little man carved into stone which held up the baptismal bowl in churches. Finally in the sixteenth century, very grotesque small figurines, particularly those stone caricatures of man, were called 'marmouset.' To people of that period marmosets appeared like miniature caricatures of themselves."

The origin of the term "tamarin" is uncertain. Gesner already used the word and it could possibly have referred to a tropical tree genus, the *Tamarinde.* Tamarins, therefore, were those "monkeys that lived on the tamarinds." In any case, this is the explanation that Le Barrère gave in reference to Gesner's remark in 1749. A short period thereafter the French explorer Antoine Binet referred to the red-handed tamarin *(Saguinus midas)* from Guayane as "tamarin." He mistakenly assumed that an Indian tribe from Cayenne called this monkey by that term. Buffon adopted this name and used it to refer to other tamarins and marmosets as well. He was responsible for bringing the term "tamarin" into general usage.

These terms which originated in the French Baroque period are still in use today. On the basis of dentition the *Callithricidae* are divided into two large groups: 1. Marmosets, their lower incisors being only slightly or not at all extended beyond the adjacent incisors; and 2. Tamarins, their lower canines being longer than the incisors.

The marmosets

MARMOSETS (Genus *Callithrix*) usually possess bushy, brush-like or fan-like ear tufts which often sharply contrast with the rest of the fur. The HRL is 20–25 cm and the TL is 29–35 cm. Only the pygmy marmosets are smaller. The hair color and fur pattern varies greatly, often within the species and subspecies. The geographical distribution of most forms is not well known. The classification of the various species and subspecies is difficult and is still disputed.

Distinguishing characteristics

A. BÜSCHELÄFFCHEN (SHORT-TUSKED MARMOSETS) have bushy ear tufts which consist of erect white or light yellow hair. There are two species, which possibly are only subspecies. 1. COMMON MARMOSET *(Callithrix jacchus).* 2. BUFF-HEADED MARMOSET *(Callithrix flaviceps).* The head is yellowish-brown, the muzzle is whitish; the ear tufts are yellowish-white and are somewhat shorter and fan-shaped. This monkey is very rare.

Fig. 16-2. Pinseläffchen *(Callithrix penicillata; C. aurita; C. leucocephala; C. humeralifer; C. albicollis)* 2. Pygmy marmoset *(Callithrix pygmaea)*

B. SEIDENÄFFCHEN (SILKY MARMOSETS) usually have long, pointed ear tufts which are bent upwards. The body is slight and the limbs are shorter. There are at least five different forms, some of which are closely related to the type species, the black-eared marmoset, and there is probably only one subspecies. 1. BLACK-EARED MARMOSET *(Callithrix penicillata).* This is a lowland dweller which occupies the largest area in the geographical distribution of this group. The ear tufts are black. 2. The WHITE-EARED MARMOSET *(Callithrix aurita)* is found on the east slopes of the South Brazilian coastal range. The ear tufts are white or grey, the abdomen is ochre, and the hands and feet are yellowish. In other respects it is like the black-eared marmoset. 3. The WHITE-FRONTED MARMOSET *(Callithrix leucocephala)* is found in the forest regions in Espirito Santo and Eastern Minas Gerais. The face is covered by short white hair. The throat and chest are white, the mane in the neck is black, the back is ochre with black stripes and yellowish-white spots, and the underside is blackish brown. 4. The WHITE-SHOULDERED MARMOSET *(Callithrix humeralifer)* is found in the forest around the All Saints Bay in Baia. The ear tufts are black, the shoulder, throat, abdomen, and arms are white, and the remainder of the body is blackish-brown with a greyish tinge in spots. 5. The WHITE-NECKED MARMOSET *(Callithrix albicollis)* is also found in the All Saints Bay in Baia. It resembles the white-fronted marmoset but the ear tufts are short and yellowish-white; the nape is whitish-yellow.

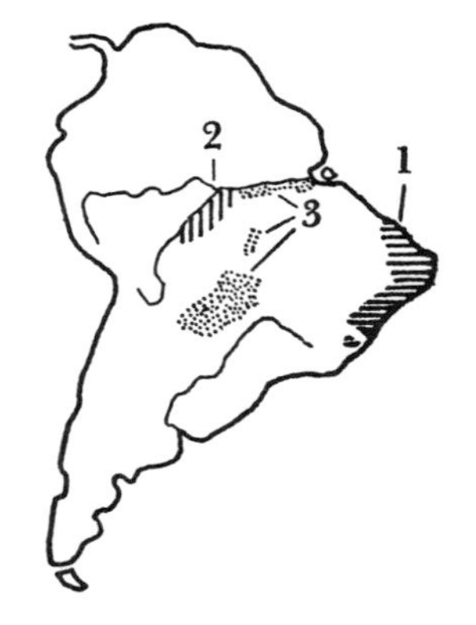

Fig. 16-3. 1. Büscheläffchen *(Callithrix jacchus, C. flaviceps)* 2. Amazon marmosets *(Callithrix santaremensis; C. chrysoleucos)* 3. Silvery marmoset *(Callithrix argentata)*

C. AMAZONIAN MARMOSETS: 1. The SANTAREM MARMOSET *(Callithrix santaremensis)* has relatively long legs. It is a good jumper. 2. The YELLOW-LEGGED MARMOSET *(Callithrix chrysoleucos)* is silky white. The head crown and back are occasionally yellow to ochre, while the hands, feet, and tail are always yellow to golden ochre.

D. The SILVERY MARMOSET (some zoologists classify it under the separate genus *Mico*) has one species and several subspecies: 1. The BLACK-TAILED SILVERY MARMOSET *(Callithrix argentata argentata)* is found on the South shore of the lower Amazon. 2. *Callithrix argentata melanura* is brown or yellowish-brown. The underside is yellowish-white and the tail is black. It is found in Southern Mato Grosso. 3. *Callithrix argentata emiliae* is silvery grey to light brownish grey. The head hair is a dark brownish grey and the hands and feet are brown. It is found in Southern Para.

E. The PYGMY MARMOSET (Subgenus *Cebuella;* see p. 369) has only one species, the PYGMY MARMOSET (*Callithrix pygmaea).*

Common marmosets

The common marmoset is the best known marmoset and is most frequently kept by man. Every year thousands of common marmosets are captured and are shipped to Europe, North America and Asia, and yet there are no indications that their population has decreased. These marmosets have been introduced into other parts of tropical America, for example, near the mouth of the Amazon, the region around Buenos Aires, and even Central America. In addition this marmoset belongs to the few *Callithricidae* which have followed man into plantations, gardens, and city parks. Konrad Guenther discovered them even in the center of the city of Recife: "After I had eaten on the open and airy back veranda of one of the local hotels I liked to sit by myself at a table under palm trees and enjoy these engaging little guests that regularly showed up for desert. Like a small kitten, the little monkey would emerge from the feather-like palm leaves. Its woolly fur was grey and its thick tail was covered by black and white stripes. The amazing feature about this creature was its gnome-like face and the round eyes which attentively watched the observer. Suddenly the little monkey jumped on the veranda support, and with a second leap it had reached another dining table where it started to clean up the left over food. Eventually the marmoset mother [probably the father; Ed.] also made its way to the table by climbing along the electrical wire underneath the veranda ceiling. This monkey carried a little one on its back which clung to it timidly. The baby was the size of a mouse and was simply delightful. As soon as these two reached the table the infant's fur was thoroughly groomed. The adult monkey was particularly concerned about the little one's hair-do. The father had his hair parted in the middle with a white tuft on either side but the infant's hair was brushed back smoothly and was hanging loosely on the nape in the manner that young people wear their hair today. At once the youngster felt at ease on the table and started to move about. It put sugar in its mouth and then blew it in all directions which seemed to greatly delight it."

Common marmosets in city gardens

Common marmosets as house pets

The common marmoset is by far the most popular pet among all the monkeys. Yet the number of private persons which keep these marmosets over longer periods of time or regularly raise them is not that large. This also explains why up to recent times there were conflicting reports about the social behavior, reproduction, and length of gestation period of this well-known marmoset. According to Sanderson, Fitzgerald, Stellar, and other authors these marmosets are monogamous and are rather aggressive towards their sexually mature offspring. More recent investigations have cast doubts on these previous findings. The American, F. Pratt, has cared for and raised several generations of common marmosets for over twenty years right in the city of New York. According to Pratt, marmosets aggregate in larger groups which are organized in a strict linear hierarchy, and are usually dominated by an older male. He observed that the mature males of mated pairs fed first and this occurred in the order of age. After the adult males the females of mated pairs ate, then the young unmated males, which were then followed by the young unmated females, and finally the juveniles.

Social life and rank order

Gisela Epple of the Zoological Institute of the University of Frankfurt kept 40 common marmosets in three large rooms with connecting outside enclosures. In four years 15 young were born there. In contrast to Pratt's observations Gisela Epple saw that a dominant male and a dominant female shared the leadership of the group. Both of these animals tolerated the copulation of the partner with a less dominant animal. However, dominant males and, particularly, dominant females frequently become aggressive if their partner courts a conspecific of lower rank. Often intense fights may result. According to Epple's observation, biting matches settle the rank order within the group. Apparently only members of the same sex fight. Males tolerate females and vice versa. In Pratt's marmoset group, on the other hand, very few biting matches occurred.

Display behavior

Ear-tufted marmosets and other marmosets which live in large family groups demonstrate their social rank by a variety of display patterns: threat mimicry, ear flapping, arching of the back, and marking of the territory with secretions from the "genital glands," which are odorous glands in the vicinity of the genital organs. In 1967 Gisela Epple described a rather characteristic display pattern. She observed that dominant animals turn their backs to subordinate animals of the same sex. They lift up their tails and show off their genitalia. This behavior pattern is not the submissive gesture of the baboons and other Old World monkeys, but is a definite threat display. "Whimpering, the threatened opponent approaches the displaying animal in the typical crouching submissive posture with its fur lying flatly," as Mrs. Epple reported. "It very thoroughly sniffs at the displayed genitals of the dominant animal. During this sniffing the displaying animal usually stands still with its tail raised up. If the dominant animal moves off from the subordinate, it will follow the former, uttering submissive

Fig. 16-4. A dominant common marmoset *(Callithrix jacchus)* points his posterior and raised tail towards a subordinate. This behavior pattern is shown by dominant animals.

sounds. If the subordinate is greatly frightened it often only dares to sniff the dominant animal's tail tip." The odorous glands in the genital region obviously fulfill an important signaling function in the marmosets. The glands help to identify the conspecifics of various ranks. The male marmosets often possess a conspicuously colored scrotum which they seem to display as visual signals. They even threaten an opponent on the other side of a glass plate in this manner.

Grooming plays an important role in the social life of the ear-tufted marmosets and other marmosets. The monkeys comb through their hair with the claw-like nails, pick out dandruff and remove foreign material that has been trapped in the hair. An animal that wants to be groomed approaches another, looks directly at it, and stretches out in front of the other in an inviting position. "The animal then relaxes completely," observed Mrs. A. Fitzgerald. "Its body moves slightly in order to expose all parts to the 'groomer.' An experienced groomer is anchored firmly in the branches and has both arms free. Only a beginner has to lean against the partner, and is therefore able to groom with only one arm. The grooming skill seems to improve with age. A good groomer seems to be in high demand by the other group members."

While courting, the male also combs the hair of the female. He nibbles her fur with the incisors. "During courtship male and female walk behind each other unhurriedly, with arched backs, stiff limbs, and puffed-up fur," reports Gisela Epple. "This arched back is almost exclusively displayed by dominant animals and represents a directed display. Both partners interrupt their display, walk repeatedly, and rub the genital region, which is endowed with numerous odorous glands, against some object. Then the male approaches the female with rhythmic smacking sounds. The lip-smacking is interspersed with tongue flicking whereby the tongue is rhythmically stuck out five to seven times. If the faces of both partners approach closely, the tongues sometimes touch."

When cared for properly, marmosets also breed in captivity. Common marmosets, of course, have bred most frequently because they are owned by many zoos and private persons. More recent findings by Lucas, Hume, and G. Epple indicate that the gestation period of marmosets is almost five months. Newborn marmosets and tamarins are able to cling to their parents' fur, without any aid from them, from the first day of life. Shortly after birth the father takes over the care and carrying of the young. Other group members, as well as the young from the previous litter, also participate as babysitters. According to G. Epple, the males carry the young the longest and most frequently. However, I have also observed "aunts," older females, which are just as interested in the babies as is the male. Nursing mothers generally only take the infants at nursing time. Yet there are also mothers which carry their young outside the nursing period.

Black-eared marmosets occur near Salvador da Bahia and close to the city of Rio de Janeiro; however, their original home seems to be the impenetrable forests of the coastal mountains and the interior. This may be one reason why they are less frequently seen at animal supply houses than the common marmoset. Today the white-eared marmoset also belongs to the less frequent species, and is only occasionally seen in zoos. Formerly this species was exported in great numbers as the "common marmoset" from the densely populated and easily accessible coastal regions between Rio de Janeiro and Santa Catarina to Europe and North America. However, in the meantime the trees of these southeast Brazilian coastal forests have been extensively cut down and these areas have become cultivated. This has robbed the marmosets of their habitat, and there are hardly any there anymore.

A pair of black-eared marmosets, which I owned, adopted three hand-raised white-eared marmosets of approximately two months of age. They fed and carried the little ones and were mainly responsible for the successful upbringing of the orphans. In the marmosets the male usually carries the young; however, in this case the female always carried the adopted young. Young white-eared marmosets, which are still without the ear tufts, can hardly be distinguished from black-eared marmosets of the same age. Recently the closely related white-fronted marmoset and black-eared marmoset were crossed in the Frankfurt Zoo. In the Cologne Zoo Santarém marmosets live in good harmony with other marmosets.

Black-tailed marmosets *(Callithrix argentata argentata)* markedly differ in their behavior from other marmosets. In groups of 12 or more animals they roam through the forest, and seem to prefer the shrub layer and the less dense bush forest. More than other marmosets they are skillful hunters of birds and small mammals.

Many people do not feel the black-tailed marmoset is as "pretty" as the other marmosets because of his naked flesh-colored head and the less pronounced ear tufts. However, this first impression is quickly compensated for by this monkey's gentle and trusting nature. This little monkey is feared in its home country because of certain superstitious beliefs of half-breed Indian settlers in the Amazon, who assume, mistakenly, that its bite is poisonous. On the other hand, though, it is a favorite house pet.

Black-tailed marmosets as "babysitters"

Two male black-tailed marmosets, which Gisela Epple kept with the common marmosets and maned tamarins, enthusiastically participated in the carrying of the common marmoset young. At the beginning, however, there were regular fights between the two black-tailed marmosets and the common marmoset male about who was to care for the young. In one of these disputes one of the young lost a piece of its tail. In the course of time, however, the common marmosets became accustomed to their new "babysitters."

One can find black-tailed marmosets in the marmoset and tamarin collection of most large zoos. Occasionally they have reproduced in captivity but usually these breeding groups cease to exist after a certain time period. Real breeding success over a number of years and over several generations is extremely rare. The Frankfurt Zoo once attempted to raise a black-tailed infant that had been abandoned by its parents. The tiny creature only weighed 27 grams. Every hour it received a bottle of warm baby food. After a few days it took three grams at every feeding. At the end of six weeks it already weighed 62 grams and had its complete set of milk teeth.

Pygmy marmosets—the smallest species of monkeys

The last species of marmosets to be discussed is the PYGMY MARMOSET *(Callithrix pygmaea).* It is the smallest of all the monkeys. The HRL is 16 cm, the TL is 18 cm and the body weight is 85 grams. The head is round. The eyes have a "mongoloid" appearance. The ears are not tufted and are hidden within the head mane. Since the number of chromosomes is less than in the other species and its appearance and behavior is also different, Fiedler has classified this group as the subgenus *Cebuella* (Gray, Thomas, and Hill even group them as a separate genus).

In 1823 the South American explorer Spix discovered this tiny monkey in the forests near Tabatinga, north of Rio Solimões where Peru, Brazil, and Columbia join borders. In subsequent years pygmy marmosets were also discovered in many other regions. For some time, some zoologists considered these smallest of monkeys to be juveniles of the common marmoset or some related species. This confusion came to an end when the French zoologists Deville and Count Francis de Castelnau displayed a large number of pygmy marmosets of various ages in 1847.

These tiny gnomes are not quite as good in jumping as are other marmosets, but they are very agile climbers and runners among the tangled branches. They are very difficult to observe in the wild state because their rapid movements can barely be followed with the eyes. When in danger they spiral up a tree in the manner of squirrels or woodpeckers and then hide behind the branches. They are even able to take small leaps backwards. Their twittering and trilling calls are partially reminiscent of soft bird songs, or the sounds of grasshoppers and crickets. Some tones are very high and are beyond the human range of hearing. When threatened these marmosets erect the hair mane and shriek loudly.

More recently pygmy marmosets are not too uncommon in zoos or even in private hands. They have occasionally reproduced in captivity. The newborn young are only the size of a large bean. One has to use a strong magnifying glass in order to see their tiny fingers.

Tamarins

While the marmosets, with the exception of the pygmy marmoset, are relatively uniform in body structure and behavior, the tamarins are a very diversified group. They are not only distinguished from marmosets by their dentition but also by their longer limbs and better jumping ability. As a rule, the tamarins are somewhat or considerably larger than the marmosets (the HRL is 22–31 cm, the TL is 30–42 cm, and the body weight is 300–560 g). The maned tamarin is farthest removed systematically from the other tamarins. It is not quite certain whether they are closely related to the tamarins at all. Yet other subgenera are not equivalent to each other. The crested bare-faced tamarins, for instance, are quite different from the majority of other tamarins.

The tamarins are divided into three genera: 1. MANED TAMARINS *(Leontideus)*; 2. TAMARINS *(Saguinus)* with subgenera TAMARINS *(Saguinus)*, MOUSTACHED TAMARINS *(Tamarinus)* and PIED TAMARINS *(Marikina)*; 3. CRESTED BARE-FACED TAMARINS *(Oedipomidas)*.

The maned tamarin

The MANED TAMARINS (Genus *Leontideus*) are the largest and by human standards the most beautiful or at least the most unusual tamarins. Their silky golden-yellow or yellowish-red fur, which shines like gold dust in the light, produces amazement and delight in any lover of animals. The maned tamarins possess characteristically long and narrow hands and feet. The extremely elongated fingers are just as long as the forearms. A skin fold connects the three middle fingers up to the central joint. The face is almost hairless and is of a pale purple-violet pigmentation. A long metallic-shining head mane cloaks the neck and shoulder regions like a cape.

There are three species (or subspecies): 1. The GOLDEN LION MARMOSET (⊕*Leontideus rosalia)* has golden fur almost without dark areas. It is found in the coastal mountains southwest of Rio de Janeiro, and inhabits forests at elevations of 500–1000 meters. 2. The GOLDEN-HEADED TAMARIN *(Leontideus chrysomelas)* is black to reddish-black; only the head, nape, upper arms, lumbar region, and upper side of the tail are golden. It is found in a very small area of the coastal mountains in the south of the state of Bahia. 3. The GOLDEN-RUMPED TAMARIN *(Leontideus chrysopygus)* has an extremely long mane which falls down over the elbows. Only the scalp is golden; the rump, hands, and feet are black, the lower back, upper thighs, and upper part of the tail are a deep rusty red. Up to now it has only been recorded in a few places in the state of São Paulo.

The insular, widely separated home ranges of the maned tamarins are probably the last refuge of a formerly large range of distribution which up to recent times extended from Bahia to São Paulo and to the Rio Paraná. It is possible that man is mainly responsible for forcing the maned tamarin into small isolated areas. Their former habitat has been extensively cultivated and turned into plantations since the early colonization of Brazil. In contrast to the common and black-eared marmosets and the hairy-faced tamarin, maned tamarins do not invade cultivated land. The most recent reports indicate that this species is also threatened with extinction. Their already small home territory is constantly reduced by slash-burn deforestation and cultivation. It seems high time to provide strict protection for these delightful monkeys.

Maned tamarins seem to inhabit still higher regions in the tree crowns than do the marmosets. They leap or jump with unbelievable speed from branch to branch. They can jump over the same distances as the Goeldi's marmoset. The elongated and partially webbed fingers allow these monkeys to securely hold on to a branch after a long jump and prevent them from falling into the lower tree strata. Even captive maned tamarins prefer to stay up as high as possible in their cages. Often they are immediately below the ceiling. Up to recent times the golden-yellow species have been kept more frequently than all other tamarins in zoos and by individuals. Even at the time of Count Buffon the maned tamarin had the reputation of being more tenacious and cold resistant than the majority of tamarins and marmosets. However, like all marmosets and tamarins, it is very sensitive to direct sunlight. In the free living state the maned tamarins retreat into the dense foliage during the hottest hours of the day.

Although many observers have praised their gentleness and easy tamability, maned tamarins may be very aggressive towards strange people or animals. They threaten by erecting their manes, showing their teeth, and emitting shrill and very high shrieking sounds. They are able to inflict painful bites. It is best to keep them in pairs because in larger groups, intense biting matches between two adults of the same sex may occasionally break out. These fights usually end with the death of the subordinate. A pair or two animals of the same sex which are familiar with each other, on the other hand, keep close together. A single maned tamarin usually attaches itself to its keeper. This monkey becomes very fond of its substitute partner and shows signs of jealousy if the former gives his attention to another animal.

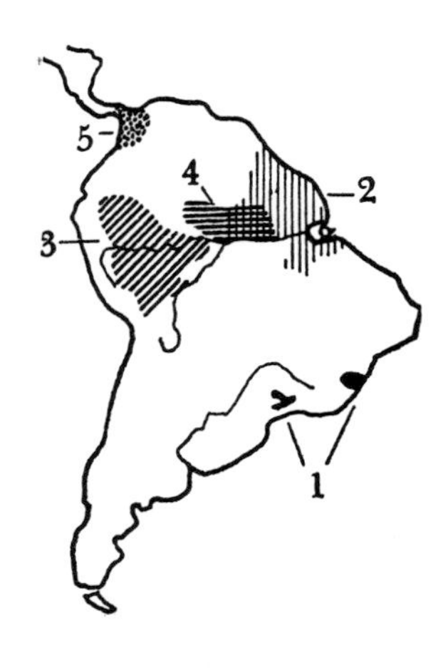

Fig. 16-5. 1. Maned tamarin (Genus *Leontideus*) 2. Tamarins (Subgenus *Saguinus*) 3. Moustached tamarins (Subgenus *Tamarinus*) 4. Pied tamarin (Subgenus *Marikina*) 5. Crested bare-faced tamarin (Genus *Oedipomidas*)

Sanderson praises the intelligence and good memory of the maned tamarins. Steinbacher reports the following episode about a maned tamarin that lived with Sanderson for many years: "As soon as its master came home it emitted a high twittering sound—it did not matter where it was at the time, in his country house, town apartment, or in the trees in the garden. The maned tamarin always called loudly before Sanderson had even left his car. It still remains a mystery how the animal recognized its human friend even though it could neither see, smell, nor hear him. It probably recognized the sound of the car from a long distance and could distinguish it from other cars. Sanderson was able to demonstrate this amazing feat to many of his visitors."

When maned tamarins were still quite common they were frequently on display in the larger zoos, including Frankfurt, Berlin, Antwerp, and New York. Often one could even view breeding groups of these golden dwarfs. Small young look somewhat like lighter replicas of their parents. Right after birth they climb up on their mother and cling to her hindquarters. Every two hours they climb to her chest in order to nurse. J. Frantz observed in the Frankfurt Zoo in 1957 how a two-day-old newborn maned tamarin was carried about by all group members. After four to ten days the young are taken over by the father, and the mother only nurses them from then on. As the nursing period approaches the female moves close to the male and stretches out her arms. The male responds by handing the young to her. The usual duration of nursing is 15 minutes. According to Ditmars, the young are nursed for more than three months. He also observed how the father mashed pieces of banana in his hand and then fed it to the young.

Long before weaning, the little ones leave their father for short periods of time during which they practice climbing or try a little of the food from the bowl. However, at the slightest sign of danger, alarmed by their parents' warning calls, they scurry back into the paternal fur. They become independent around three to five months of age and then they only return to their parents' fur in case of danger or to seek a warm place to sleep. If the juveniles are too demanding, both parents may occasionally fend them off with the bite-threats or with slaps. Nevertheless, young maned tamarins up to the age of four and one half months still beg food from their parents and take it away from them.

Tamarins

Among the tamarins (Genus *Saguinus*) are two species from the lower Amazon and from Guiana which are very conspicuous because of their hairless black faces, their extremely large ears, and their black fur. These are the TAMARINS (Subgenus *Saguinus* i.n.s.) which are distinguished from the MOUSTACHED TAMARINS (Subgenus *Tamarinus*, p. 373) by their longer skulls and the absence of the white hair on the face.

1. The fur of the NEGRO TAMARIN *(Saguinus tamarin)* is a deep black and, depending on the subspecies, there is a more or less distinct grey or ochre colored saddle; this species is found south of the mouth of the Amazon. 2. The RED-HANDED TAMARIN *(Saguinus midas)* has similar fur; depending on the subspecies the hands and feet may be red, orange, ochre-yellow, or chestnut-brown and are sharply contrasted with the black fur. It is found north of the Amazon.

Since negro tamarins are found at the mouth of the easily accessible Amazon, they were already known to Gesner, Linné, Buffon, and other classics of zoology. Great explorers like Alexander von Humboldt and Henry Walter Bates already provided detailed descriptions of this black forest gnome in the first half of the nineteenth century. Even today one can see the negro tamarin or "Pará-monkey" in the suburbs of the large harbor of Belém (formerly Pará). The inhabitants often capture and tame negro tamarins. Some of these pets have escaped and have occupied gardens and city parks. Otherwise these tamarins prefer secondary forest, which is forest that has grown on formerly cultivated land, and tree plantations. The monkeys prefer the tree crowns. They run spirally around thick branches like squirrels.

The authorities are not in agreement about the temperament of the negro and red-handed tamarins. Bates and other older authors described this monkey as good natured, clever, and easily tameable. According to Bates, the negro tamarin is even "more friendly to man than any other monkey." Da Cruz Lima also emphasized their trust in humans but also observed that they are nervous, irritable, and wild shortly after capture. Observations on free living negro tamarins have shown that they do not flee quickly into the impenetrable thicket when pursued but that they climb into the highest branches of the tree crowns with loud outcries. They also aggregate peacefully with capuchins. Negro tamarins and capuchins tolerate each other and rarely engage in fights.

Sanderson, in contrast, got a completely different impression from the negro tamarins which he kept in his enclosures. He called them "irascible little black devils that wildly bit with their long canines." All of Sanderson's efforts to establish a good relationship with them were to no avail. I personally met negro and red-handed tamarins in zoos that were gentle and tame and could be turned loose in the keeper's room. As with other tamarins and marmosets, negro and red-handed tamarins are probably greatly influenced by previous experiences with people and this is greatly reflected in the subsequent behavior with humans. If juveniles come into good hands right away they do not develop into "little devils" but into affectionate and delightful pets. However, one should never attempt to tame them, or for that matter other marmosets and tamarins, by forcefully grabbing or holding them. If, on the other hand, the animals are caught as adults and are torn away from their families and then confined in small boxes or tied on strings, as unfortunately is the custom in South American harbors where the animals are traded, then it is not too surprising that the monkeys will not forget these frightening experiences for a long time, if at all.

MOUSTACHED TAMARINS (Subgenus *Tamarinus*) are usually somewhat smaller and more delicate than the negro tamarins. The skull is shorter and the fur coloration varies greatly, from a dark blackish-brown to almost white or multi-colored. The beard hairs are always white. There are 12 species with 23 subspecies.

A. Those species with only an indication of a moustache include: 1. The BLACK-AND-RED TAMARIN *(Saguinus nigricollis)* is dark in the front and light in the back, which is a black to leathery brown. The hindquarters are a marbled reddish brown, and the mouth is encircled by white hair, "as if dipped in milk." 2. The BROWN-HEADED TAMARIN *(Sa-*

guinus fuscicollis) has brown shoulder fur; the middle of the body is darker than the front or the back. The fur on the back is black and a marbled reddish-yellow, occasionally a deep rusty red. The hindquarters are ashy brown, although the color boundaries are often faded. There are many subspecies and color variations. 3. WEDDELL'S TAMARIN *(Saguinus weddelli).* 4. GOLDEN-MANTLED TAMARIN *(Saguinus tripartitus).* 5. RED-MANTLED TAMARIN *(Saguinus illigeri).* 6. WHITE TAMARIN *(Saguinus melanoleucus)*; despite its different coloration, it is closely related to the brown-headed tamarin.

B. Those species with regular white mustaches include: 7. The MOUSTACHED TAMARIN *(Saguinus mystax)* has a white face and white hairs around the face, "as if it was holding a flake of snow-white cotton wool in its mouth" (Bates); the dorsal side is a marbled black-yellow-brown-grey and the ventral side is blackish-brown. The head mane, throat, limbs, and tail are a glistening black. It is widely distributed from the slopes of the Andes up into the interior of Western Brazil. 8. LÖNNBERG'S TAMARIN *(Saguinus pluto)* and 9. The RIO NAPO TAMARIN *(Saguinus graellsi)* are very similar. 10. The RED-CAPPED TAMARIN *(Saguinus pileatus)* and 11. The RED-BELLIED WHITE-LIPPED TAMARIN *(Saguinus labiatus)* is larger. The L is up to 66 cm, and the moustache is small. There are conspicuous red areas in the fur. The head crown in the red-capped tamarin is a reddish-cinnamon. In the red-bellied white-lipped tamarin the ventral side is a deep red. 12. The EMPEROR TAMARIN (*Saguinus imperator*; Color plates p. 349 and p. 376) has an extremely long white moustache which hangs down. The behavior and voice are reminiscent of silky marmosets.

Moustached tamarins inhabit the vast, partially unexplored, Amazonian tropical forest regions which extend from inner Brazil to the eastern slopes of the Andes and from Ecuador to the lowlands of northern Bolivia. Since their habitats were located in such inaccessible places, most of them were discovered and described relatively late like the majority of the other marmosets and tamarins. Five forms were scientifically described only in this century. Moustached, black-and-red, and brown-headed tamarins still reach zoological gardens relatively frequently. The conspicuous white tamarin, on the other hand, reached Europe alive only once. It lived in the Cologne Zoo from 1965 to 1967.

The Ruhr Zoo of Gelsenkirchen received three moustached tamarins in 1953 that could not be classified in any of the twelve species. The two males resembled a red-mouthed tamarin and the female looked somewhat like Weddell's tamarin. Despite the large external differences, the animals obviously belonged to the same species or subspecies. The female infants that were born resembled their mother exactly. A thorough examination of sexually determined color differences will probably result in a considerable reduction in the number

Marmosets and tamarins

1. Black-eared marmoset *(Callithrix penicillata)* 2. Common marmoset *(Callithrix jacchus)* 3. Pied tamarin *(Saguinus bicolor martinsi)* 4. Cotton-head tamarin *(Oedipomidas oedipus)* 5. Geoffroy's tamarin *(Oedipomidas geoffroyi)* 6. Golden lion marmoset *(Leontideus rosalia)* 7. Black and red tamarin *(Saguinus nigricollis)* 8. Pygmy marmoset *(Callithrix pygmaea)*

1
2
3
5
4
7
6
8

1
8
7
2
3
6
4
5

of species and subspecies in the moustached tamarins.

The most conspicuous member of the subgenus is the emperor tamarin *(Saguinus imperator)* which is represented by two subspecies and occurs on the rivers Acre, Purús, and Juruá in the Southwest of the Brazilian state Amazon and possible also in the East Peruvian border regions. In these tamarins the white hairs on the upper lip have developed into an almost monstrous moustache which in the older animals hangs down to the chest in two frazzled strands.

The emperor tamarin

When the first specimens of the rare emperor tamarin, which was only discovered in 1907, were prepared in the museums, the taxidermists mistakenly twirled the moustache up in the fashion of Emperor Wilhelm of Germany. For this reason Goeldi gave it the proud, humorous species name *imperator.* In reality, however, the droll giant moustache of the emperor tamarin does not in the least resemble the beards of Wilhelm's era, but rather the wild drooping moustache of an old czarist cossack.

Emperor tamarins are graceful, friendly, and playful monkeys. Da Cruz Lima emphasized their liveliness and need for tenderness. He described one of his animals in the Belém Zoo as follows: "With a very expressive face, it begged for caresses and would lie on its back so that one could gently scratch its belly." This female emperor tamarin was crossed with a moustached tamarin but she died shortly before giving birth. An autopsy revealed that she carried twins. The emperor tamarin is not seen frequently in our zoos although the monkey is in great demand as a zoo display. It has been kept in the New York Bronx Zoo, however. One male has been living in the Frankfurt Zoo for five years, which is a considerable time span.

The PIED TAMARINS (Subgenus *Marikina*), with their naked, bald, deep-black or brown-spotted pigmented heads and their jug-handle "bat ears" do not fit into our "baby schema" as do the otherwise pretty tamarins and marmosets. These tamarins look strange to us and we even consider them "ugly." In reality, however, they are no less delightful or friendly than other marmosets and tamarins. For instance, we consider a freshly hatched, downy gosling as "darling," while many people find a freshly hatched, naked song bird "not beautiful." We have to overcome these innate prejudices if we are going to be as fair to the bald-headed pied tamarin as to the others.

Marmosets and tamarins

1. Pied tamarin *(Saguinus bicolor bicolor)* 2. Moustached tamarin *(Saguinus mystax)* 3. Red-handed tamarin *(Saguinus midas)* 4. Brown-headed tamarin *(Saguinus fuscicollis)* 5. Emperor tamarin *(Saguinus imperator)* 6. White tamarin *(Saguinus melanoleucus)* 7. Silvery marmoset *(Callithrix argentata)* 8. Santarém marmoset *(Callithrix santaremensis)*

There is one species, the PIED TAMARIN *(Saguinus bicolor)*, with three subspecies: 1. The PIED TAMARIN'S *(Saguinus bicolor bicolor)* frontal body is almost white to the middle of the abdomen; the upper back and the hindquarters are a dark brownish-yellow. The two colors are sharply separated. 2. MARTIN'S PIED TAMARIN *(Saguinus bicolor martinsi)* is a solid dark brown. 3. The OCHRE-COLORED PIED TAMARIN *(Saguinus bicolor ochraceus)* is a solid, ashy, yellowish-brown.

The pied tamarins have very vivid facial expressions and in this respect they resemble the nice, but occasionally quite aggressive, crested bare-faced tamarins. In the zoos they have repeatedly borne and raised young. It is best to keep only one pair in the smaller cages because in larger groups bitter social interactions about rank order often break out in these temperamental animals. Individual pied tamarins generally get along with other tamarins if they are gradually introduced to each other. Occasionally they will even mate with different but related species.

In a zoo a male pied tamarin peacefully shared a cage with a male moustached tamarin and a female red-handed tamarin. The surprise was great when the red-handed tamarin bore two young which were obviously sired by the pied tamarin. Unfortunately the mother died right after the birth of the second infant, which was stillborn. The male pied tamarin cared for the surviving offspring and carried it on its back like any good marmoset or tamarin father.

Artificial raising in the Cologne Zoo

Because of the mother's death, Margaret Immendorf attempted to artificially raise the hybrid young. During the first days of its life, Miss Immendorf fed it a little thinned cow milk out of a doll's bottle every two hours. A few days later the infant also took oatmeal fluid and mashed bananas. It clung to a piece of silver fox fur and during the day rested in a cushioned pail. At night Miss Immendorf let the monkey infant rest against her body. During the second week, "Stups," the monkey's name, already climbed around a little and greeted its adopted mother with tongue flicking and an occasional vigorous bite. When it was one month old it jumped actively around the room, followed its substitute mother, and started to attack strange persons. This artificial raising was successful because of the knowing, loving care and the correct transition to a diet of fruit, vegetables, and meat.

Crested bare-faced tamarins

The CRESTED BARE-FACED TAMARIN (Genus *Oedipomidas;* Color plate p. 350) is the largest tamarin except for the maned tamarin. Many zoologists regard them as relatives of the pied tamarins. The ears are small and there is a hair adornment on the head. They are the only tamarin west of the Andes. They are a distinct group in relation to the other tamarins. There are three species: 1. The COTTON-HEAD TAMARIN *(Oedipomidas oedipus)*; 2. GEOFFROY'S TAMARIN *(Oedipomidas geoffroyi)*; and 3. The WHITE-FOOTED TAMARIN *(Oedipomidas leucopus)*, which has a silky fur and is of a light greyish-brown color with a rusty-red ventral side. The hands and feet are white, and the small head crest is not developed into a mane. The crest is white on the top of the head and then blends into a brownish-grey.

They can jump more than three meters

The long limbs and the extremely long tail indicate that the crested bare-faced tamarins are excellent jumpers. When jumping from tree to tree, they can effortlessly cover distances of over three meters. They almost equal the abilities of the common marmoset, and "fly" through the air in a similar manner. They sail through the air with laterally stretched-out arms, spread-out legs, expanded flank hairs, and puffed-out tail rudder. Unfortunately, the feats of these quick, agile forest ghosts are not often seen in the zoo because the animals are usually housed in small cages. This is also probably one of the contributing facts to their high rate of mortality in captivity.

The behavior of the "cotton-heads" or "cotton-tops" as the crested bare-faced tamarins are commonly known in North America, shows certain similarities to the larger New World monkeys, particularly the capuchins. In the free state they aggregate in larger family groups like the capuchins. The group is led by a powerful male. Their facial expressions are very expressive and often appear amazingly human. While in most tamarins and marmosets the facial expressions only appear in conjunction with other threat display behavior (arched back, etc.), the crested bare-faced tamarin turns its "threat face" towards the opponent usually without emphasis of other expressive movements. This behavior pattern consists of lowering the forehead until it forms a bulge which almost covers the eyes; the lips are pushed forward and the head and neck manes are erected. Clearly the facial expression serves as a signal as it does in the capuchins and New World monkeys.

The "threat face"

The threat display of the raised tail and exposed genitalia, which is so characteristic of the marmosets, is not prevalent in adult cotton-heads. They frequently roll up their tails between their legs in front of the lower abdomen and groom them in a manner similar to that of the squirrel monkeys. In courting the tail is curled like a big loop and is impregnated with odorous substances from the genital glands. The female encourages the male to copulate by flicking her tongue and licking his face. The male picks up the "perfumed" moist tail of the female in both hands and sniffs at it.

Crested bare-faced tamarins as predators

It is likely that the cotton-head is more of a meat eater than other tamarins. Their well-developed dentition with the powerful canines is indicative of a predator existence. They can quickly kill mice and birds up to the size of a grass parakeet. One can keep captive cotton-heads healthy by giving them fruit, oatmeal liquid, and insects in addition to young mice, lizards, and frogs or as a substitute, liver and raw egg, white chicken breast and fresh sea fish cut into small deboned pieces. If one animal out of a larger group has taken a liking to fish, a food preference that is passed on by tradition is soon established. Other group members taste the fish and eat it. They pass this preference on to their young. Thus all juveniles of such a group become fish eaters without great difficulty.

Much has been written about the characteristic killing-bite of the marmosets and tamarins. Otmar Schäuffelen describes this in a crested bare-faced tamarin as follows: "If one puts a bird into their cage, usually only one monkey hunts at a time, while the others watch attentively. When, after much chasing, the bird is finally caught with the hands, the monkey holds it tightly around the neck and quickly bites into the skull. The bird almost always dies immediately. Right after the killing-bite, the monkey turns the bird's head around and hastily bites off the beak. All bitten-off parts are thrown away with a fling of the head. If one offers dead birds to the tamarins the killing-bite is omitted."

Living mice and frogs are also killed quickly with the skillful head-bite. This killing-bite does not seem to be innate. Rather, the adult family members teach the younger ones how to kill the prey. I often observed the helpless manner in which young cotton-heads approached a young mouse or cockchafer. They would bite anywhere and were obviously frightened by the struggling victim until their father or another older group member showed them how to overpower the prey in the quickest possible way. In a pair of cotton-heads that I kept together in one cage with tree shrews at Seewiesen, the four-year-old female had mastered the killing-bite, but the somewhat older male was not very skilled and the tree shrews frequently stole his young mice. It seems that not only form, size, and movement pattern of a prey are releasers for the killing-bite, but experience also plays a decisive role. This is not necessarily to be expected because many other animals also "kill" dead prey instinctively.

The crested bare-faced tamarins emit very loud, ringing, song-bird-like trills which are unique in the realm of the mammals and probably also among other marmosets and tamarins as well. Their song starts with a high "dididi" and is followed by long drawn-out flute-like tones which are interrupted by trills. The song increases in pitch and frequency and becomes a kind of "nightingale song." In an old biology text from the beginning of the nineteenth century, I discovered the description "nightingale monkey" which probably referred to the crested bare-faced tamarin. In fact the song of the cotton-head serves the same function as that of the song bird. It functions as an acoustical marking of territory. However, the cotton-heads also "sing" when highly excited. Their "zick-zick-zick" or "ga-ga-zack" does not develop into a long drawn-out shrieking but into a loud belting sound. When a group of six to ten animals engages in such a "song," windows literally vibrate.

Cotton-head tamarins

The best known species, the cotton-head tamarin, is characterized by a large snow-white head mane. In German this species is known as "Lisztäffchen." Ludwig Heck humorously compared the monkey's hair adornment with the white shock of hair of the composer Franz Liszt. To me the impressive white wig is more reminiscent of the festive headdress of an Indian chief. This is particularly true if the mane is erected when the animal is in a state of excitement or display.

Cotton-head tamarins inhabit a relatively small territory in the

Tierra Caliente which is located in the warm humid regions of the Caribbean coast of Colombia. West and northwest of this area one finds Geoffroy's tamarin. The white-footed tamarin occupies the river valleys of mountainous Central Colombia. This species is not well known and was formerly classified with the moustached tamarins. Since the white-footed tamarin's habitat borders next to the cotton-head tamarin's, and both species look almost alike when skinned, it is more probable that this somewhat doubtful species belongs to the crested bare-faced tamarins rather than to the moustached tamarins.

The home range of the crested bare-faced tamarin is constantly shrinking, just as the maned tamarin's, due to land cultivation in recent times. Only on Barro Colorado, an island in the Gatun Lake in the Panama Canal, does the Geoffroy's tamarin enjoy total protection. Since Geoffroy's tamarin and the crested bare-faced tamarins like to visit fruit plantations and often get caught in traps or snares, they are still sold in large numbers in the animal trade in the harbors despite their diminishing numbers. They are often sold to North America. In the southern states of the United States, where the climate is favorable to these monkeys, they do very well and often raise young, especially if they are kept in outdoor enclosures. In certain institutes they have become favorite experimental animals for ethologists. Despite excellent care, it is difficult to keep these monkeys in Central Europe. Except for a few isolated cases, most crested bare-faced tamarins just sit in their cages and not uncommonly they will die overnight without any apparent cause.

The high mortality rate of captured cotton-head tamarins is often due to a lack of room for exercise or a deficiency of vertebrate meat and undigestible roughage. Newly imported crested bare-faced tamarins or those that were bought from a merchant often greedily pounce upon grasshoppers and cockchafers in order to eat the chitin. They also eat the skin which they have removed from the mice. They will even consume fish scales and bird feathers. On a pure diet of fruits, plants, and mealworm larvae these tamarins easily fall victim to stomach and intestinal infections and suffer from chronic diarrhea which results in a gradual wasting away, with eventual death. Cotton-head tamarins fed on a balanced diet which contains sufficient amounts of large insects, fish, and mammal meat ignore insect wings, bird feathers, and mammalian hair because their food contains the proper amount of roughage.

It has been shown that crested bare-faced tamarins are not at all sensitive but surprisingly hardy if they are fed a proper diet and are kept in large cages, so that they will have opportunities to jump, preferably with outside enclosures in sunny locations planted with living bushes. Even when there is frost and snow, my cotton-head tamarins would spend at least a short period in their outside enclosure and would jump around in the branches before returning to the heated inside cage. They are not subject to many colds and thrive at normal room temperatures provided their cage is large enough. A tamarin which hunches in a corner, has a swollen neck, disorderly fur, coughs, sneezes, and has a colorless discharge running out of the nose and eyes does not have a cold, as we would assume. It is suffering from a dangerous intestinal parasite infection, *Prosthenorchis elegans,* which belongs to the Trichostrongylids. This parasite migrates from the intestine into the lymph nodes of the neck and will cause eventual death of the host if the disease is not treated in time with one of the commercial worm medicines. Ear-tufted and silvery marmosets and maned tarmarins often fall victim to this same parasite.

Captive crested bare-faced tamarins, like all tamarins and marmosets, do not only suffer from Trichostrongylids, scratches, and other intestinal parasites, but also from an illness known as "cage paralysis." This disease of the spinal cord, to which probably thousands of captive marmosets and tamarins fall victim, can be prevented or relieved by natural sunlight or ultraviolet light. The view that little monkeys only need little cages is widely held, but it is wrong. If one keeps these active, climbing and jumping enthusiastic monkey dwarfs in large cages with adjacent outside enclosures which they can enter at will, they will thrive much better than in an inside cage. Even monkeys which were formerly regarded as extremely difficult to keep will breed under these conditions and form family groups that give us an opportunity to study their social behavior. In the Frankfurt Zoo one crested bare-faced tamarin has lived for almost eight years. My oldest male, kept in Baden-Baden, is now nine years old and does not show any signs of old age.

Geoffroy's tamarins seem to be more sensitive to the climate than the cotton-head tamarins. In Europe, as far as is known, these tamarins have only bred under the care of Gisela Epple in the Zoological Institute of Frankfurt and my daughter Sabine Wendt at the Zoological Institute of Giessen. Sabine could demonstrate that her Geoffroy's tamarins responded to colored pictures and were able to distinguish the pictured animals and objects: "When shown pictures of insects they would smack their lips and try to grasp the picture. They also became very excited about pictures of fledgling birds. Illustrations of fish, frogs, turtles, large birds, and elephants were ignored. They were interested in snakes but not as much as insects. When I showed them colored pictures of predators, particularly one of a leopard that was in the process of killing a monkey, they screamed with fright and could not be calmed."

Crested bare-faced tamarins have been bred surprisingly little in European zoos. However, in my own experience, they are relatively easy to breed under proper care and can even be crossed with Geoffroy's tamarins. The gestation period is considerably longer than in the marmosets. The American scientists Hampton and Landwehr give 140 days. However, I observed three females and a total of 13 births, and it must have been at least 160 to 170 days. I kept my cotton-head tamarins in larger groups which consisted of one pair of parents and their adult or juvenile progeny. In large enclosures the parents will tolerate younger group members even though they have reached sexual maturity, particularly if the latter will aid in the carrying and feeding of the younger siblings.

A type of "monogamy" exists in such a natural group which is made up of eight to twelve animals. Monogamy is probably also prevalent in other tamarins, because only the dominant female bears young. If she dies a younger female takes her place and soon becomes a mother. Young crested bare-faced tamarins look most appealing as they curiously peer out of the fur of the adult with their little round heads. Just as in the maned tamarins, I observed a busy passing around of the infant from adult to adult. When a cotton-head child indicates its hunger by a squalling call, the mother approaches with stretched-out hands in order to receive her little one from the adult that has been carrying it.

I rarely observed fierce social interactions in such natural groups. The "boss" or the "old man" rules his members not by biting but by his facial expressions, calls, or an occasional erection of the hair mane. The highest ranking female marks her territory with the odorous substances secreted by her genital glands. She rubs this substance on twigs and branches. During the estrous phase, i.e. release of ovum, this marking is very pronounced.

Crested bare-faced tamarins in the garden

The Cologne Zoo and the Zoological Institute in Frankfurt left their marmosets and tamarins outside for the entire summer. This worked surprisingly well, and refuted the oft-repeated view that these monkeys are very sensitive to weather and temperature, and could only be kept in well-heated rooms. It is even possible to let the animals live in the garden. I kept a pair of cotton-head tamarins in the garden during the entire summer right into late fall. The animals were released every morning and had the opportunity to roam around the trees and bushes. They took the opportunity to search for spiders, beetles, caterpillars, and tree-bugs. The tamarins greedily pounced on these strong-smelling insects, picked them up with their mouths and tried to avoid smudging their hands with the stinking insect secretion.

In the garden the two animals never strayed far from each other. They kept in contact by constantly making soft, bird-like calls. With the approaching dusk, they arrived punctually at a specific window, jumped into the room, and entered their sleeping cage. When young were born to this pair I decided to terminate this arrangement. A garden at the edge of the city is too dangerous a place for a child-rearing pair of dwarf monkeys.

The number of animal lovers that have the desire to keep marmosets

Explanation of following color plates:

The brilliant skin colors in the face of the mandrill male *(Mandrillus sphinx)* are a type of "war paint." The colors play a role in the threat display in interactions with conspecifics (compare p. 424 ff).

On the 9 color plates (I-IX) following the mandrill (opposite page), thirty-six species of the forty species that comprise the guenon-like monkeys are represented by forty-seven subspecies.

Macaques: (Plate I, p. 386)

1. Pig-tailed macaque *(Macaca nemestrina)* Subspecies:
a. Burmese pig-tailed macaque *(Macaca nemestrina leonina)*
b. *Macaca nemestrina nemestrina*
2. Rhesus macaque *(Macaca mulatta)*
3. Crab-eating macaque *(Macaca irus)*
4. Toque monkey *(Macaca sinica)*
5. Lion-tailed macaque *(Macaca silenus)*

Macaques: (Plate II, p. 387)

1. Japanese macaque *(Macaca fuscata)*
2. Stump-tailed macaque *(Macaca arctoides)*
3. Moor macaque *(Macaca maura)* Subspecies: a. *Macaca maura maura* b. *Macaca maura ochreata*
4. Barbary ape *(Macaca sylvana)*
5. Celebes crested macaque *(Cynopithecus niger)*

Baboons: (Plate III, p. 388)

1. Guinea baboon *(Papio papio)*
2. Yellow baboon *(Papio cynocephalus)* Subspecies: a. *Papio cyncephalus ruhei* b. *Papio cynocephalus ochraceus*
3. Chacma baboon *(Papio ursinus)*
4. Anubis baboon *(Papio anubis)*

Drills and mandrills: (Plate IV, p. 389)

1. Mandrill *(Mandrillus sphinx)*
2. Drill *(Mandrillus leucophaeus)*

Gelada baboons:

3. Gelada baboon *(Theropithecus gelada)*; young male

Baboons:

4. Hamadryas baboon *(Papio hamadryas)* a. juvenile

Mangabeys: (Plate V, p. 390)

1. Grey-cheeked mangabey *(Cercocebus albigena)*
2. Black mangabey *(Cercocebus aterrimus)*
3. Sooty mangabey *(Cercocebus torquatus)* Subspecies: a. *Cercocebus torquatus torquatus* b. *Cercocebus torquatus atys*
4. Agile mangabey *(Cercocebus galeritus)* Subspecies: a. *Cercocebus galeritus agilis*

Guenons: (Plate VI, p. 391)

1. Dwarf guenon *(Cercopithecus talapoin)*
2. Moustached guenon *(Cercopithecus cephus)* Subspecies:
a. Red-eared guenon *(Cercopithecus cephus erythrotis)*
b. Moustached guenon i.n.s. *(Cercopithecus cephus cephus)*
3. Lesser white-nosed guenon group:
a. *Cercopithecus petaurista fantiensis*
b. Red-bellied guenon *(Cercopithecus erythrogaster)*
c. Schmidt's white-nosed guenon *(Cercopithecus petaurista schmidti)*

Guenons: (Plate VII, p. 392)

1. Diademed guenon *(Cercopithecus mitis)* Subspecies:
a. White-throated guenon *(Cercopithecus mitis albogularis)*
b. Diademed guenon *(Cercopithecus mitis mitis)*
2. Owl-faced guenon *(Cercopithecus hamlyni)*
3. L'Hoest's monkey *(Cercopithecus lhoesti lhoesti)*
4. DeBrazza's monkey *(Cercopithecus neglectus)*

Guenons: (Plate VIII, p. 393)

1. Diana monkeys *(Cercopithecus diana)* Subspecies:
a. *Cercopithecus diana diana* b. Roloway *(Cercopithecus diana roloway)*
2. Mona group:
a. Mona monkey *(Cercopithecus mona mona)*
b. Campbell's monkey *(Cercopithecus mona campbelli)*
c. Gray's crowned guenon *(Cercopithecus pogonias grayi)*

Guenons: (Plate IX, p. 394)

1. Grass monkey *(Cercopithecus aethiops)* Subspecies:
a. Grey grass monkey *(Cercopithecus aethiops aethiops)*
b. *Cercopithecus aethiops sabaeus*
2. Swamp guenon *(Cercopithecus nigroviridis)*
3. Red guenon *(Erythrocebus patas)* Subspecies: a. Nisnas monkey *(Erythrocebus patas pyrrhonotus)* b. Patas monkey *(Erythrocebus patas patas)*

I
1
a
b
2
3
4
5

II

1
2a
3
4
2b

1♂
2♂
2♀
1♀
♀
4
♂
♀
♂
3
4a

1
2
3 a
3b
4

1
2
a
b
a
b
3
c

VII

1a
1b
2a
2b
2c

IX

1a

2

1b

3a

3b

and tamarins in their apartments is continually increasing. Yet many do not have the opportunity to offer open air cages with natural sunlight. In this case, large movable cages are recommended which can easily be pushed to an open window or onto the balcony. It is important to avoid drafts and intense sunlight. Shady, draft-free corners are just as essential as climbing bars and wooden sleeping boxes inside the tamarin cage. During the cold season one should expose the monkey twice weekly for ten minutes to an ultraviolet lamp to avoid possible deficiencies which might occur, particularly during the winter. All marmosets and tamarins should receive a sufficient and varied diet. Even the largest and most beautiful aviary is a very empty environment when compared to the lush vegetation of the tropical rain forests, teeming with life. The many stimuli that are missing in the artificial setting can partially be replaced by a colorful and varied menu which should keep the animal alert and active.

Herbert Wendt

17 Guenons and Their Relatives

Infraorder *Catarrhina* or Old World monkeys by W. Fiedler

All other primates, including man, are combined in the infraorder of the OLD WORLD SIMIAN PRIMATES *(Catarrhina)*. The space between the nostrils is narrow. The external auditory meatus is long. The tail is not prehensile, with the exception of young guenons. There are 32 teeth: $\frac{2 \cdot 1 \cdot 2 \cdot 3}{2 \cdot 1 \cdot 2 \cdot 3}$. There are two superfamilies with a total of five families: 1. OLD WORLD MONKEYS *(Cercopithecoidea)*: Cercopithecids, leaf monkeys; 2. APES and MAN *(Hominoidea)*: gibbons, great apes, and man. Three additional families are extinct.

Superfamily *Cercopithecoidea*

OLD WORLD MONKEYS *(Cercopithecoidea)* have an HRL of 32.5–110 cm, and the TL is 0–103 cm. The body weight may reach 50 kg (mandrill). The nails on all fingers and toes are flattened. They are diurnal animals which are either terrestrial or arboreal, and they are quadrupeds. Certain forms are capable of walking bipedally for a short distance. They are usually social animals. Cheek pouches are present. The gestation period ranges from approximately 165 to over 240 days. There is usually only one young.

There are two families (some zoologists classify these as two subfamilies under one family): 1. The *Cercopithecidae* possess relatively broad hands and feet, and the thumb is well developed. They walk with flat soles or with slightly lifted wrists and ankles. The cheek pouches are well developed. The stomach is simple and they are usually omnivorous. There are eight genera: MACAQUES, CELEBES CRESTED MACAQUES, BABOONS, MANDRILLS and DRILLS, GELADA BABOONS, MANGABEYS, CERCOPITHECIDS, and RED GUENONS. 2. The LEAF MONKEYS *(Colobidae)* include six genera: LANGURS, DOUC LANGUR, SNUB-NOSED MONKEYS, PIG-TAILED LANGURS, PROBOSCIS MONKEYS, and COLOBUS MONKEYS.

Family: *Cercopithecidae*

The CERCOPITHECIDS (Family *Cercopithecidae)* are the typical monkeys in the popular sense. They represent that restless, lively crowd which Kipling so masterfully described in his *Jungle Book* in a highly anthropomorphic way. These monkeys were familiar to civilizations of antiquity. They are represented on many pictures, sculptures, and temple friezes from Egypt to India and eastern Asia. At a later date the cercopithecids are also found on works of art in Greece and the Roman Empire. The Egyptians considered the hamadryas baboon and the anubis baboon to be sacred animals. The anubis baboon today still bears the name of the Egyptian god Anubis. The great Greek doctor Galen dissected Barbary apes and other monkey species in order to reach a better understanding of human anatomy. Medical doctors in the Middle Ages dissected the bodies of monkeys for similar reasons, since performing autopsies on human cadavers was strictly prohibited. Since then the cercopithecids have become increasingly familiar to us. Today even the smallest zoo possesses some cercopithecid representatives. These monkeys became the experimental objects of medical science, and man is greatly indebted to them for many important discoveries, including the Rh factor in our blood. In most recent times they have even become "pioneers" in man's probes into outer space.

Their well-known restless "monkey temperament" is partly due to their way of feeding. The cercopithecids enjoy a very diverse diet which includes fruits, blossoms, buds, stems of plants, leaves, nuts, roots, onions and tubers, insects and other small animals. They also eat bird eggs and fledglings, and in certain cases even feed on the meat of larger vertebrates. Thus these monkey troops or smaller groups are almost constantly roaming around and searching for food during the day. They are "constant eaters," always discovering something new or trying to reach a better feeding place via an unfamiliar route.

During the last decades especially, cercopithecids in the free living state have been thoroughly studied. As a result, many old prejudices had to be discarded. Formerly macaques and baboons, in particular, were considered to be very aggressive towards conspecifics. It was believed that the old powerful "pashas" of a monkey troop exerted "despotic powers," and that they suppressed weaker group members or even tried to kill them during an attempted "rebellion." The primate researcher De Vore referred to this view as "the myth of unsociableness" which was based on incorrect observations and is not in the least related to the real facts. The groups of monkeys kept in zoological gardens often consisted of an un-biological ratio of males to females. Of course, the combined effect of a narrow enclosure and this unfavorable sex ratio had to result in severe social interactions. In 1938 a group of approximately four hundred rhesus monkeys was released on the island of Cayo Santiago near Puerto Rico. The monkeys did not know each other, and as a consequence many serious fights broke out.

Today, however, we know that free living monkeys behave differently. Generally, various troops avoid each other where their territories join. Within a group there is a definite social order, and no arbitrariness and suppression. Nevertheless, the baboons and ground-dwelling macaques inhabit a far more dangerous environment than the arboreal forms. This is one of the reasons which necessitates a more disciplined social order where each group member is certain of its specific rank position. Mature males, which are "responsible" for the group, have to be physically strong and have to possess "self-confidence." Consequently the other members treat these leaders respectfully. If one keeps this fact in mind, the baboons and larger macaques are not really that much more aggressive than the arboreal monkey, for instance, or the howler or leaf monkeys (in ethology the words "aggressive" and "aggression" are only used to describe the antagonistic behavior towards conspecifics. Attacks against predators or foreign species are not considered aggressive.). Monkeys living out in the open areas are in continuous danger; hence, in order to survive in these situations, there has been a selection pressure for physical strength, large canines, and a well-ordered society.

The macaques

MACAQUES (Genus *Macaca*) are represented by a great variety of forms. The HRL is 38–76 cm, the TL is 0–61 cm, and the body weight is up to 13 kg. The macaques are heavily built with compact, robust limbs. Distinct low brow ridges often give the older males a "sinister" appearance. Fur color is usually a yellowish or olive brown on the dorsal side and a lighter color on the ventral side. After puberty, conspicuous red, frequently large, ischial callosities develop. The monkeys may be arboreal, terrestrial, or rock dwellers. They are found mainly in Asia. Only the Barbary ape is found in North Africa and Gibraltar. There are seven subgenera and a total of twelve species.

Barbary ape

A. Subgenus *Macaca* i.n.s. has only one species, the BARBARY APE (*Macaca sylvana*; Color plate p. 387). The HRL is up to 75 cm, and the shoulder height is approximately 50 cm. The animal is quite slender but appears more compact because of the thick fur and the absence of a tail. This is the only monkey living in Europe (Gibraltar). In North Africa it is often found at high elevations.

In central Europe these hardy monkeys can easily be kept outdoors. In 1763, Count Schlieffen imported a troop of Barbary apes to Germany and released them in the forest of his country estate near Kassel. They adapted well to the climate, multiplied, and thrived until 1784; however, they were rather unpopular because of their raids. Unfortunately a rabid dog bit several monkeys, and the count, with a heavy heart, had to shoot them all. Today one can still view the commemorative stone in the park of the estate Wildhausen which was erected in memory of the sixty Barbary apes that are buried beneath it and which at one time made up the "Hessian monkey colony."

The Barbary apes on Gibraltar are another colony which is very famous. Their fate has undergone many fluctuations. Unfortunately it is not possible to prove without a doubt that the Barbary apes have always inhabited Gibraltar's rocky terrain, which has always been a part of Europe, or if man introduced them. Otherwise the only other living Barbary apes are found in North Africa today. However, fossil remains of Barbary ape-like monkeys have been discovered in various locations in Europe. If the original Gibraltar monkeys are the last progeny of these "European monkeys," their position has, however, undergone great changes. The English have repeatedly reintroduced Barbary apes from North Africa to the fortress of Gibraltar.

Consequently the Gibraltar monkeys have played a role in political history. According to tradition, the English will lose their strategic fortress at the entrance to the Mediterranean when the last monkey has disappeared. England respects such old traditions. Whenever there was a crisis concerning Gibraltar, Barbary apes were quickly imported from North Africa. In the summer of 1942, Great Britain's Prime Minister Winston Churchill telegraphed the British High Command in North Africa as follows: "Catch some monkeys for Gibraltar at once!" The General promptly ordered a group of soldiers to catch African Barbary apes in order to increase the dwindling population on the rock of Gibraltar.

It is possible that the Phoenicians, Carthaginians, or Romans brought the ancestors of the monkeys to Gibraltar and set them free at "the European pillar of Hercules." In 711 A.D., when the Arabian conqueror Tarik Ibn Sijad, whose name was later given to the Rock of Gibraltar (Djebel al Tarik = Hill of Tarik), first entered Gibraltar, the monkeys were already present. England officially mentioned them for the first time in 1856. At that time approximately 130 Barbary apes were present on the rock. The British governor ordered total protection of the monkeys in a special decree. However, two years later an epidemic broke out, and only three Barbary apes survived. By order of the governor, additional Barbary apes were brought to Gibraltar from North Africa.

Under British protection the animals did well, and soon took over the whole town. They descended in groups from the rocks and plundered houses and stores, destroyed gardens, and wrung the necks of chickens. Reputedly they even hit women and children. When one of the monkeys stole the decorative feathered helmet of the governor, which he was wearing on the occasion of a celebration, this caricatured his excellency on the pinnacles of the fortress in front of the gathered public. Patience had run out. The monkeys were banned from the vicinity of the town and chased to the most desolate part of the rock. The protective decree of the Barbary apes, however, remained in force. After World War I, a proposal reached the House of Commons which requested the total ban of Barbary apes on Gibraltar because the animals raised too much havoc. Fortunately, the Barbary apes had more friends than enemies among the representatives and the majority rejected the proposal.

However, since 1913 the British have governed the lives of these subjects by stringent restrictions. The number of Barbary apes, which around 1910 was more than two hundred, was limited to thirty to forty animals. An "officer in charge of apes" and an enlisted man were made responsible for the well-being of the animals. The monkeys are under the jurisdiction of the war ministry. Records of the animals are kept, and there is money for a food allowance—four pence per day per monkey.

Like macaques, Barbary apes always live in groups. On Gibraltar there are two troops. One of these is almost wild and occupies the highest and most inaccessible parts of the rock. The other group is located halfway between the top of the rock and the harbor, along Queen's Road. These are the monkeys that are viewed daily by tourists. Because of their constant association with humans, these Barbary apes have lost all natural fear of people. Even the young ones are already clever pickpockets. They plunder cars with open windows. They pull money purses out of pockets, or snatch cameras and handbags from the tourists. After they have taken an object, they rapidly climb to an inaccessible rock ledge. They thoroughly investigate the stolen article, tearing to pieces whatever they cannot eat and dropping it off the rock.

In the African National Parks the monkeys also learned quickly that man no longer hunts them, and they act like the Gibraltar monkeys. Although the Barbary apes on Gibraltar appear very tame and even climb up on the visitors in order to reach their pockets better, they object to being held. Signs have been posted which warn tourists not to touch the monkeys. If one, nevertheless, attempts to hold an animal it will immediately emit a hoarse cry which will summon the entire horde to its aid. Only a hasty retreat will save one from rough treatment.

Barbary apes, like all macaques, have a life span exceeding twenty years. "Old Bess" of Gibraltar is still very lively despite her twenty-four years. The tourists particularly enjoy the young monkeys which are usually born in July or August. The infants are small black-haired fellows which have a grouchy, "old-man's" face. The mothers spoil their children with typical "monkey love." In case of danger, all young are defended by the entire group.

Barbary apes in the mountainous countries of North Africa are often very bold. Frequently they descend from the mountains in troops and plunder fields and gardens. For this reason the natives pursue them, so in various places Barbary apes have become quite rare. It is therefore very encouraging to know that on Gibraltar the British and the monkeys have come to a type of "gentlemen's agreement." The monkeys receive their pay in the form of fruits, and today are rarely found inside the city. If the population increases too rapidly, some monkeys are caught and exported to zoos. The remaining ones enjoy a beautiful and secure existence as "protectors of the empire."

Stump-tailed and Japanese macaques

B. Subgenus *Lyssodes* (Color plate p. 387); these monkeys are also very hardy animals. The northern forms especially have a dense, long-haired pelage on the back. The ventral side is only sparsely covered by hair. They are heavily built and massive. The stump-like tail, which is often hairless, flattened, and bent to the right, is almost unmovable; in a way it forms part of the ischial callosities. There are two species: 1. The STUMP-TAILED MACAQUE *(Macaca arctoides)* has a well developed shoulder mane. The face is red and occasionally spotted; during excitement or heat it becomes redder, whereas when cold or in a state of weakness it becomes blueish. The young have pale faces. 2. The JAPANESE MACAQUE *(Macaca fuscata)* has a uniformly light red face. It is the most northerly species of monkeys. Two subspecies are found on the large South islands of Japan (they are very common near Kyoto) and on Yakushima. The Japanese macaques are under strict protection, and of course, are regularly kept in Japanese zoos. One macaque in the Kumamato Zoo was still alive after thirty-five years. These monkeys also play a special role in the mythology, folklore, and art of the Japanese people. Most familiar are the three monkeys which represent the wisdom of Buddha: "See no evil, hear no evil, speak no evil." These are Japanese macaques; one covers his eyes, the second holds his ears, and the third covers his mouth.

In recent times the Japanese macaques have become even more well known because of their astonishingly man-like behavior patterns. Japanese scientists have studied the social behavior of these macaques for years. Professor Junichiro Itani of the University of Kyoto and his associates have observed Japanese macaques at the most northerly point of the island Honshu for twelve years. This is also the most northerly region on earth where monkeys are found. Weather conditions are rather harsh there. In the winter, temperatures often fall below −5°C and snow may be as high as one and a half meters. When the trees are bare on the forested mountain slopes, the macaques are forced to feed on hardy winter buds and shoots and even on the bark of trees.

Fig. 17-1. 1. Stump-tailed macaque *(Macaca arctoides)* 2. Japanese macaque *(Macaca fuscata)* 3. Hat monkeys: a. Bonnet monkey *(Macaca radiata)* b. Toque monkey *(Macaca sinica)*

"When we began our observations in 1948," said Professor Itani in the magazine *Das Tier*, "almost nothing was known about the behavior and ecology of the Japanese macaques. They inhabited the thick forest of steep mountain slopes at high elevations, and would not let people approach them. However, in the summer of 1952 we were finally successful in persuading one group of wild monkeys on the island of Koshima to accept food from us. By accepting sweet potatoes and wheat grains that we offered them they overcame their shyness which until then had separated us from them. The monkeys became used to us and we were soon able to recognize individual animals, to recognize their social interactions, and to gather information about their social structure within the group."

Social behavior of the Japanese macaques by J. Itani

Currently there are approximately thirty groups with a total of 4300 macaques which are fed and observed by man. These observations are part of a long-range program. Comparative studies of the various monkey societies have already resulted in a variety of interesting problems. For instance, we not only found differences in the manner in which the animals fed, but also in how individual societies differed in the degree of sociability, in the frequency of their encounters, and many other factors. In order to study the basic characteristic of their social order, it is not sufficient to observe just one society. It is necessary to observe many groups and then compare them.

The animals of a group know each other. Each group has its own territory which ranges in size from approximately 2-15 square kilometers. Within this territory the troop wanders in search of leaves and ripe fruits. The animals spend the night in thick forest or on a steep rock where the monkeys sleep in the trees. A society may consist of four or five members but also of a group of six hundred. Societies of from 30 to 150 members are most frequent.

Group composition

When roaming through an area the group is led by several young males. These are followed by a larger number of females, infants, and juveniles interspersed with several mature adult males. The rear is again made up of a group of young adult males. When the macaques settled down to feed we noticed that they formed two circles, an inner and an outer one. The central circle was made up of all the females, juveniles, young, and several large mature males. The outer circle consisted of the young adult males.

A distinct dominance order is recognizable among adult males. In the Japanese macaques this order is generally not established by fights. Rather, male animals of the same age group gradually get to know one another through play, such as tag and wrestling. In the beginning the dominance order among the young males is still very flexible and changes frequently. Only after the macaques have reached their fifth or sixth birthday does the dominance position become more permanent. The three dominant males in the Koshima monkeys have maintained their dominance position since 1952. In 1954-1955 we observed an amazingly clear-cut and easily recognizable rank order (dominance hierarchy) in the Takasakijama society. There was a straight linear relationship from the first place to the fortieth.

Dominance order in the males

There are various classes among the adult males. In Takasakijama I was able to determine three classes: leader, subleader, and males of the outer circle. The latter were again subdivided into three classes—

upper, middle, and lower class. The males of the upper leading class used to squat in the center with the females, juveniles, and children, while the subleaders were grouped around this inner circle. The remaining males make up the outer circle. The classes develop from the various age groups which grew up together as playmates. Gradually the class is formed. Rank order within one class is much more clear-cut than among members of different classes.

The leaders guide and watch the group. They settle quarrels, and watch out for stronger enemies which they will confront if necessary. During the time that the female is looking after her newborn infant the males take over the care of the yearlings. The responsibilities of the subleaders are more or less the same. It seems that they aid the leaders with their tasks; however, they are not allowed to squat in the inner circle. The males of the outer circle act as scouts and leaders when the group is migrating. They watch out for danger and attack enemies.

All activities of the above species are called group directed behavior patterns in contrast to self-directed activities such as eating, sleeping, or walking. The more willingly an individual animal performs these group-directed activities, the more does it establish its own position within its class.

Social status of the female

In the males a social class system and rank order persists while in the females the social position is determined by kinship. Nevertheless, a type of rank order can also be observed among them but it is not as well established as in the males. This situation results in constant quarrels. A female's social position frequently changes, and is also subject to personal influences. Age does not play a role. The daughters of a dominant female enjoy a privileged position because they are under the influence of their mother and are dependent on her. The sons, on the other hand, leave their mother when they are one and a half to two years old. They enter the outer circle where they live together with members of their own age group. Of course, even in this case there may be exceptions.

An old macaque grandmother, all her daughters and granddaughters, are emotionally dependent on each other as members of one family group which has to assert itself as a unit against other family groups. In the Kashima society, which consisted of seventy members and where the social dominance order was clear-cut, it was noticed that the social status of the family group increased with the number of female offspring present.

Reproduction and group composition

Usually the breeding season of Japanese macaques begins in November or December and lasts until February or March. It has often been claimed that the social organization in a monkey society is centered around sexual relationships. If so, then how can one explain the fact that this organization is also maintained from April to October when the monkeys do not engage in any sexual relations? During the breeding season the established social order is not disrupted by the sexual activity of the group members. As soon as this period begins, pairs form within the general circle or a little removed from it. If the leader is busy with his female there is always another male that will substitute for him and take on his responsibilities.

Naturally, leading males enjoy certain prerequisites in the choice of their female and in the uninterrupted pleasure of their "connubial bliss." However, there are only a few leaders and they do not demand access to all females. The males of the outer circle are allowed to choose a female and copulate with her. Yet if one of the dominant males of the inner circle finds such a pair, he will abduct the female. However, this abduction is not a serious matter. Usually when a female sees the dominant male approaching, she runs away and later returns to her mate who has remained undisturbed in his place. As a rule a mating relationship lasts from four to seven days but occasionally can end on the same day it has started. Since the gestation period lasts for approximately five months, the births fall between April and August.

Solitary animals

Some monkeys, in contrast to most members of their species, are solitary. These are always males. We observed this transition to the solitary state in a great number of males from various societies. At first we assumed that these animals were perhaps inclined to be unsociable and different and could not adapt to the social life. Since then, however, we have realized that this explanation is not so clear-cut. In some cases the reason for leaving the society may have been based on a misunderstanding between age and social rank of the specific monkey or a drastic change which placed him in a much lower rank. Generally, though, the males were large powerful animals which were fully aware of their social duties, and which suddenly left the group.

Additional observations and studies will probably lead us to the explanation of this behavior. However, it is likely that transfers from one society to another do occur, although until now Japanese macaques groups were considered to be closed societies. We actually know of a few cases where males left their group and joined another.

Another Japanese macaque researcher, Dr. Masao Kawai, was able to observe the origin of new behavior patterns and the subsequent spreading of such "inventions" in macaques on the small island of Koshima. Such passing on of personal experiences from one individual to his companions is regarded by scientists studying human culture and prehistory as an important prerequisite for the origin of a culture, or can even be regarded as the beginning of culture.

Culture in monkeys by M. Kawai

"In the fall of 1953," writes Dr. Kawai in the magazine *Das Tier,* "a one and a half year old female, which we named Imo, one day picked up a sweet potato which was covered with sand. She dipped the potato, probably by pure accident, into water and rubbed off the sand with

One female macaque started to wash sweet potatoes

her hands. By this inconspicuous act, Imo thus introduced monkey culture to Koshima, which later made it famous. One month afterwards, one of Imo's companions also started to wash sweet potatoes, and after four months Imo's mother did the same. This behavior gradually spread through daily interactions between mothers and young, sisters and brothers, members of the same age and play companions. In 1957, fifteen monkeys were showing this behavior. Usually the one- to three-year-olds were learning it. None of the males which were older than four years (i.e., sub-adults and adults) at that time adopted the new behavior pattern. Among the females, three five- to seven-year-olds and two adults learned it.

Can one conclude from this example that females are more adaptable than males? Probably not; the difference seems to be due to the different social positions of males and females. As we discussed previously, males from their fourth birthday have to live in the outer circle. Here they hardly ever came into contact with the young female which had discovered washing sweet potatoes. Thus the females of the inner circle had much more opportunity to imitate new habits from one another.

In the beginning the new behavior spread via the association of playmates and members of the same family, particularly from the young to the mothers, and from younger sisters and brothers to older ones. Later, when potato washing was more widespread, mothers passed it on to their children. In 1957–1958 and thereafter, newborn youngsters adopted this pattern as a normal sequence in eating behavior since they spent the entire day with their mothers.

Finally almost everyone learned washing

By 1962, forty-two of the fifty-nine members of this macaque society had adopted the habit of washing their sweet potatoes. The few monkeys that did not do it were all mature males and females which were already adults in 1953. Mature monkeys apparently seem to be too conservative to incorporate new patterns into their behavior, while the young are still very flexible.

After ten years, the washing of sweet potatoes was part of the normal eating routine of this monkey society; it had been invented by a young female. Each generation passed it on to the following. Can one regard this as "culture"? Strictly speaking, it is not equivalent to what we consider to be human culture. Hence, we called this behavior 'monkey-or pre-culture'."

Later these monkeys became used to washing the sweet potatoes in salt water, and not in fresh water as previously. Perhaps they tasted better with some salt on them. Kawai also observed the beginning of several other traditions. In part he purposely influenced the monkeys by his experiments. On the other hand, the macaques learned certain things the investigator had not intended. Thus, Kawai enticed several monkeys to enter the water by giving them peanuts. After three years all youngsters and juveniles had adopted the habit of bathing regularly in the ocean. They swam and even dived. The macaques also started to wash grains of wheat which had been scattered on the sand for them. At first they laboriously picked each grain out of the sand. Later they just threw a handful of sand and grain into the water. The sand sank to the bottom and the light wheat grains floated on top. All the monkeys had to do now was to gather the grains from the surface of the water in order to eat them. It was Imo, who had first discovered sweet potato washing, who also invented wheat washing. It seems that talents are also differently distributed among monkeys. Among Imo's closest relatives almost all learned the new behavior patterns, while of the macaque mother Nami's children, only few.

The most amazing factor was the fact that during the course of time, more and more macaques started to walk on two legs. "This seemed to be closely related to the described pre-cultural behavioral patterns," says Kawai, "because the monkeys often walked upright for almost thirty meters with their hands full of sweet potatoes or wheat." The Dutch zoologist Kortlandt observed that chimpanzees walk bipedally if they have to carry something. Grzimek observed a similar phenomenon in his colony of released chimpanzees on the island of Rubondo in Lake Victoria. These observations are of great interest since in human phylogeny there also must have been close connections between the development of culture and the erection of the body for a bipedal stance. If we humans are finally to understand ourselves and our behavior, it is of the utmost importance that we study thoroughly, over many decades, the behavior of higher primates in their natural surroundings.

Rhesus macaques by W. Fiedler

C. Subgenus *Rhesus.* The tail is medium long and the hair cover is not conspicuous. The ♂♂ are decidedly larger than the ♀♀. There are Three species: 1. RHESUS MACAQUE (*Macaca mulatta;* Color plate p. 386); 2. The ASSAMESE MACAQUE (*Macaca assamensis);* and 3. The FORMOSA MACAQUE (*Macaca cyclopis).*

Medical experiments with monkeys

The rhesus macaque is that species of monkey that is most frequently kept in zoos, even in the smallest zoological gardens. For many years the animal has played a paramount role in medical research. The Rh factor was detected in the rhesus monkey after which it was named. The Rh(esus) factor is a genetically determined substance present in the blood protein of which the person is usually unaware. However, under certain circumstances it can be fatal to unborn and newly born human infants. The first group of rhesus macaques which were moved into the bomb-damaged Frankfurt Zoo shortly after World War II was at the disposal of the Paul-Ehrlich Institute for obtaining serum. This was the only way it was possible to obtain foreign currency in order to import the monkeys. Every few weeks the caretakers caught the macaques for members of the institute, who removed a small amount

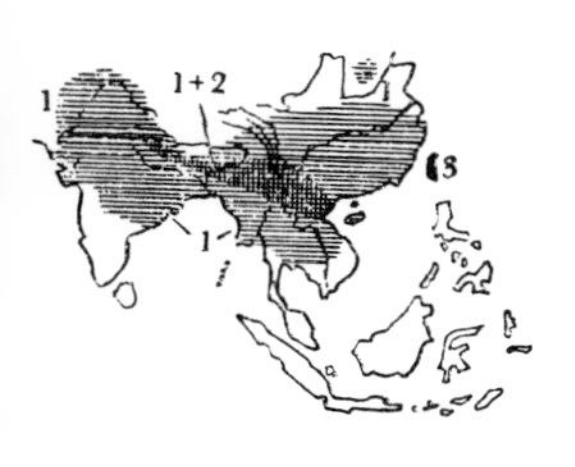

Fig. 17-2. 1. Rhesus macaque *(Macaca mulatta)* 2. Assamese macaque *(Macaca assamensis)* 3. Formosa macaque *(Macaca cyclopis)* The geological range of the Barbary ape *(Macaca sylvana)* is shown in Fig. 17-6.

of blood from the vein in the monkey's hollow of the knee. The monkeys did not suffer unduly from this. In those times, just like human blood donors, they received extra food rations, and generally they were healthy and active. With the aid of the blood from the rhesus macaque, the doctors were able to determine in suspected cases whether or not the life of an unborn child was in danger because of the Rh factor. The serum helped to save the life of many children. Today, thanks to the knowledge that was gained from experiments with rhesus macaques, we are able to save most children which are endangered by the Rh factor and help them to grow into healthy adults. The rhesus group in the Frankfurt Zoo has thrived and multiplied for over two decades. This example shows the beneficial effects of animal experiments, including those done with monkeys. If these tests are done within certain limits, without undue exploitation of the numbers of free living monkeys, and with the welfare of suffering humanity and empathy with our fellow creatures at heart, the tests are justifiable. Scientific experimentation with animals is important and necessary in order to provide help for men and animals. However, the extent to which captured wild animals, and in particular, monkeys, are used should not be dependent only on convenience and availability of research funds. Neither should this decision be solely determined by the needs of science. In no case should the need for experimental animals ever exterminate or even threaten a monkey species. Certain cheap species of monkeys, like the rhesus macaque, are often carelessly transported and so half of them die on the way and others are cruelly tormented. In addition, many research institutes are guilty of keeping animals in solitary confinement or in cages which are too small. Often the specialists are not familiar with the psychic and physical needs of these highly developed animals. Scientists can easily fall victim to a certain habituation and job blindness. They should always ask themselves if a certain series of tests could not just as well be carried out on less highly developed experimental animals, and if certain operations on experimental animals are really essential and justifiable.

In India, rhesus macaques inhabit forests and mountainous regions but also cultivated land with trees and river banks which are covered with dense vegetation. These macaques like to enter water and are agile swimmers. The main part of their diet consists of grain, fruits, seed, and onions and tubers, but also all kinds of small animals. In India where love of animals is part of the religion, these monkeys have very successfully adapted to man's civilization. According to a recent census conducted by Southwick in the North Indian state of Uttar Pradesh, 48% of all local monkeys occupied the villages, 30% the cities, and others were found along roads or in temples. Only 12% inhabited the forest. The rhesus macaques also enter houses and steal all edible things, and often they cause a lot of damage. In 1962 the Indian government was even concerned about the monkeys because officials of the war ministry complained that the monkeys entered their offices. There they ripped or stole official documents which they would only hand back for a piece of fruit or candy.

Nevertheless, the Indians tried to get along with the monkeys and all attempts of the government to reduce the number of rhesus macaques in the over-populated regions met with the embittered opposition of the pious residents. In certain regions, for example in Bengal, the religious attitude towards the monkeys is losing ground. Mass captures and exports have already greatly decreased the numbers of rhesus macaques.

Like the Japanese macaques, rhesus monkeys also aggregate in large troops. In the wild state the territories of the individual troops overlap considerably. Generally, however, one group tends to avoid the other when they meet.

Rhesus macaques under human care have demonstrated that they can be remarkably intelligent. One rhesus female at the Institute of Zoology in Münster learned to recognize various color rings as representing different values, much the same way we distinguish between the values of various coins. This female was able to exchange these rings immediately or after a certain time interval for various numbers of peanuts or a piece of banana at an automatic machine. She was able to distinguish, almost without errors, six different values. During the tests she always took the ring with the highest reward value first. Among the least valued rings she could distinguish, in human terms, between "dimes" and "pennies." These abilities were otherwise only known in chimpanzees. Such experiments conducted with the great apes will be discussed in Volume XI.

Rhesus macaques have also been instrumental in gaining important insights into child psychology. The American psychologist Harlow placed rhesus infants into cages which contained "artificial mothers" made of meshed wire with wooden heads and artificial "breasts" which contained nipples filled with milk. One of the "mother" dummies was covered by terry cloth, and the other was not. Harlow observed that rhesus infants would drink milk from the meshed wire doll but definitely preferred the cloth doll as a "mother substitute," since they could snuggle into the terry cloth as they would into the fur of a living animal.

Harlow placed a wound-up toy teddy bear with a drum into the cage of the young monkeys. Of course, the youngsters were badly frightened but they immediately felt secure if they could rush to their cloth mother. From the secure perch on "mother," they gathered sufficient courage to view this horrible apparition, and later they even played with it. On the basis of these experiments with the rhesus infants, Harlow could demonstrate the great significance of security during early childhood and its effect on the continued psychic development of the child. Rhesus infants with cloth mothers did not develop as atypically as those infants that only knew the "meshed-wire" mother.

Lion-tailed macaque and pig-tailed macaque

D. Subgenus *Silenus* (Color plate p. 386), includes two externally very different species which are, however, similar in beard growth, shape of the tail, and the conspicuous swelling and reddening of the ischial callosities: 1. The LION-TAILED MACAQUE *(Macaca silenus)* is medium large, and has abundant gray side whiskers which frame the entire face. The tail is medium-long with a conspicuous tuft. This macaque inhabits dense mountainous forests. 2. The PIG-TAILED MACAQUE *(Macaca nemestrina)* is large and robust. The side whiskers are short and form only a type of "mariner's fringe." The tail is short and thin and is almost always held in a bent position. The Burmese subspecies *Macaca nemestrina leonina* has a distinct tail tuft.

The lion-tailed macaque is also known as "wanderoo." This term is based on a misunderstanding because it refers to a leaf monkey species living on Ceylon. However, such misunderstandings are quite common in animal terminology.

The pig-tailed macaques are among the largest of the macaques. The males with their long bulky muzzles are especially reminiscent of baboons. A "friendly greeting" in the pig-tailed macaque consists of a rather peculiar gesture which is different from the usual mumbling and chattering of other monkeys. The macaque jerks his head up, raises his eyebrows and squints rather oddly over his nose towards the partner. He also pushes his lips forward considerably and "pouts" in an exaggerated fashion with his lips. It is quite easy to imitate this, although human lips are not nearly so large. In any case pig-tailed macaques comprehend at once when a person greets them in "broken pig-tailed macaque language," and if they feel inclined they will respond with the friendly greeting.

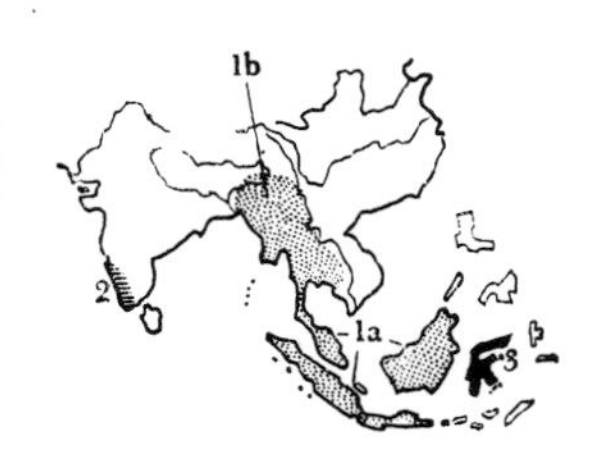

Fig. 17-3. 1. Pig-tailed macaque *(Macaca nemestrina)* Subspecies: a. *Macaca nemestrina nemestrina* b. *Macaca nemestrina leonina* 2. Lion-tailed macaque *(Macaca silenus)* 3. Moor macaque *(Macaca maura)* In the same region in Celebes the Celebes crested macaque *(Cynopithecus niger)* is found.

A pair of pig-tailed macaques which had lived for years in the Frankfurt Zoo regularly exchanged friendly greetings with D. Heinemann. "Each of the two monkeys would greet me from a distance with an extended mouth, but only if its mate did not notice it. When both were sitting together I could 'pout' in front of their cage as much as I wanted. Each looked at the other and if the mate even seemed to look in my direction the greeting was not given. On such occasions the female sometimes threatened menacingly at me with widely opened jaws. It appeared as if neither of the two mates could admit to the other that it was friendly towards me."

In certain areas of their homeland, pig-tailed macaques are trained as helpers during the harvest. They learn to climb up on coconut palms, pick the ripe nuts, and drop them. However, one can only use females and young adults; the old males are too powerful and too dangerous.

Baboons in the Amboseli National Park (Kenya).

Anubis baboons *(Papio anubis)* (1) inhabit the East African steppe and savannah. They often search for food among herds of zebras and antelopes (Grant's zebras, *Equus quagga boehmi*). (2), Grant's gazelle, *Gazella granti.* (3). They dig for roots, tubers, and onions, gather fruits, nibble young shoots and also feed on insects, lizards, and other small animals. Occasionally they also capture larger animals. Washburn and De Vore were able to witness the capture of a newborn gazelle fawn (4) and the eating of it. Aside from man and the lion, the leopard *(Panthera pardus)* (5) is the most important and dangerous predator of these aggressive monkeys. Even this big cat usually avoids the powerful male baboons and attacks careless females and young baboons that have strayed away from the group.

Hat monkeys and crab-eating macaques

E. Subgenus HAT MONKEYS *(Zati)* are small to medium-large, slender, and long-tailed. Long hairs on the crown radiate outwards in all directions. There are two species: 1. The BONNET MACAQUE *(Macaca radiata)* has a high, bare forehead; the hair on the crown is arranged in an "orderly fashion." 2. The TOQUE MACAQUE (*Macaca sinica;* Color plate p. 386) has hair on the crown which stands up "as if it were not cut properly" (Sanderson).

F. Subgenus *Cynomolgus.* There is only one species, the CRAB-EATING MACAQUE (*Macaca irus;* Color plate p. 386). It is widely distributed, even beyond Java. There are many subspecies differing in size and coloration including the PHILLIPPINE MACAQUE *(Macaca irus philippinensis).*

Crab-eating macaques prefer to live in the vicinity of rivers and lakes, and also on the ocean beaches. They inhabit mangrove forests or other coastal forests. They fish in creeks and rivers, and also swim and dive expertly after crabs and other crayfish.

These macaques were formerly considered sacred on Bali, like the entellus langurs in India. The people would bring them rice and other foods, placing these gifts in the forests near the outskirts of the villages. Because of these practices, many monkeys, such as the entellus langurs and rhesus monkeys in India, have become regular "temple monkeys." Hindus on Bali still protect and feed some of these monkeys in specific "monkey forests."

Fig. 17-4. Crab-eating macaque *(Macaca irus)*

For many decades Arnold Spiegel thoroughly studied the reproductive behavior of crab-eating macaques in the Zoological Institute at Jena. The animals are sexually mature (i.e., fertile) when they are three and one half to five and one half years old. The average menstrual cycle of the females was twenty-eight days. The menstrual flow lasted from three to six days at the most. The female with her conspicuous sexual swellings stays in the immediate vicinity of the male. The pair copulates several times during the day. Fertilization takes place between the sixth and twentieth day after the beginning of the last menstrual period. Between 1927 and 1951, ninety females became pregnant. There were two miscarriages and seventeen infants were stillborn. The gestation period lasts 159 to 178 days. After five to thirty-seven contractions the baby is born. The mother licks it off immediately and greedily eats up the afterbirth, even in the case of stillborn infants. The young nurse for fifteen to eighteen months. One female even gave birth to a young at the age of seventeen years. She still menstruated when she was nineteen years old; however, by then she was so senile that she had to be euthanized.

Moor macaques and Celebes crested macaques

G. Subgenus CELEBES MACAQUES *(Gymnopyga)* has only one species, the MOOR MACAQUE (*Macaca maura;* Color plate p. 387). It is black with a button-like vestigial tail. In size and appearance it greatly resembles the Barbary ape. There are three subspecies on each of the three southern peninsulas of Celebes, including the *Macaca maura ochreata* which is found on the southeastern peninsula and the bordering small islands.

The CELEBES CRESTED MACAQUE (*Cynopithecus niger;* Color plate p. 387) is closely related to the macaques. The HRL of the ♂ is over 60 cm; the ♀ is smaller. The tail is vestigial and only a few centimeters long. The tail is vestigial and only a few centimeters long. There are bony ridges on either side of the nose, similar to the mandrills. They live in pairs or small groups. It is likely that they are pure fruit eaters. The gestation period is seven months (San Diego Zoo).

All species of macaques require little care, and even the smallest zoos have macaques on display. Formerly macaques, like all monkeys, were usually kept in overheated, humid animal houses. They were overfed and took unlimited amounts of food from zoo visitors. This type of life did not agree with many of them. However, since the monkey houses have been less heated and better ventilated, macaques have belonged to the longest living monkeys in zoos. Zuckermann reported about 117 monkeys that were born in the London Zoo between 1928 and 1937. Twelve were New World monkeys and 105 were Old World monkeys. Among the Old World monkeys were four chimpanzees, four leaf monkeys, nine guenons, seven mangabeys, twenty-three baboons and a total of fifty-eight macaques alone. One Moor macaque in the Philadelphia Zoo reached an age of over twenty-eight years. Macaques are not very suitable as pets in private apartments. The young can be delightful and friendly but as they grow older they often become unpredictable and bite. Really, a monkey that cannot be with his own kind is indeed a creature to be pitied.

The baboons by W. Fiedler

Distinguishing characteristics

The BABOONS (Genus *Papio*) are on the average larger and heavier than the macaques. The HRL is 51–114 cm, the TL is 5–71 cm, and the body weight is 14–54 kg. The head is long and heavy, and the muzzle is greatly elongated, looking bulky and angular with the pointed nose. The small eyes are deeply set under prominent brow ridges, and the ears are also small. The body is compact; the limbs are of equal length and are robust. There is marked sexual dimorphism with respect to size; the ♂♂ often have manes or beards, while the fur in other places is less dense and usually insignificantly colored. Ischial callosities are strongly developed and are often vividly colored. In estrous ♀♀ the sexual and anal regions swell greatly and often appear to us as a "sickly" red. Teeth are well developed; the canines are exceptionally long, pointed, and powerful, with sharp ventral edges, in the ♂♂. There are gaps between the teeth in front or in the back of the canines to provide room for the upper canine when the mouth is closed. The cheek pouches are well developed. The diet consists of onions, tubers, grasses, herbs, fruits, and other plant material, as well as eggs, small animals, and meat of captured vertebrates up to the size of gazelles. During the ice ages the baboon was also present in India and China. Today five species are found in Africa and southern

Arabia. A. The STEPPE BABOONS (Color plate p. 388) are considered by some zoologists to be a widely distributed and diverse species; however, there are only four basic types which are considered to be separate species. 1. The CHACKMA BABOON (*Papio ursinus;* see Color plate p. 388) is very large. 2. The YELLOW BABOON *(Papio cynocephalus)* is more slender. 3. The ANUBIS BABOON (*Papio anubis;* see Color plate p. 388 and p. 411/412) is heavier. 4. The GUINEA BABOON *(Papio papio)* is the smallest species. All species have several subspecies. B. The HAMADRYAS BABOON (*Papio hamadryas;* Color plates p. 389 and p. 439) is considered by some zoologists to be a separate subgenus *(Comopithecus).* There is conspicuous sexual dimorphism; while the ♀♀ are reminiscent of steppe baboons, the ♂♂ have a prominent silvery-gray mane.

Baboons are terrestrial monkeys

In contrast to the macaques, which are represented by many different forms that inhabit trees, the ground and rock terrain, the baboons are a much more uniform group. They are primarily ground dwellers and only climb into the trees in order to sleep or to escape danger. Aside from the great apes and man, we find the largest primates among the baboons. During the ice age a baboon *(Dinopithecus)* existed in South Africa which almost reached the size of a gorilla. In contrast to other ground-dwelling forms, such as the chimpanzees and gorillas, the baboons are decidedly quadrupedal. They walk more on all fours than any other monkey. When walking they often lift the wrists and ankles (some ground-dwelling macaques also do this). Strictly speaking, they are not plantigrade as are most primates. Their equally long limbs and the stubby hands and feet are special adaptations to their terrestrial existence. The long bulky muzzles with the powerful canines also deviate greatly from the facial, skull, and dental formation of other monkeys.

Fig. 17-5. 1. Chachma baboon *(Papio ursinus)* 2. Yellow baboon *(Papio cynocephalus)* 3. Anubis baboon *(Papio anubis)* 4. Guinea baboon *(Papio papio)*

Despite these features one should not regard the baboons as the most animal-like monkeys. Those properties which particularly impress us as being animal-like are highly developed adaptations and specializations. However, these highly developed adaptations have not evolved in the direction of man, but away from him. In order to appreciate baboons we should not call them "the ugliest, rudest, most ill-behaved and therefore the most unpleasant members of the entire order," as Brehm described them in the anthropomorphic style of his times. In reality the baboons, with their extensive physical adaptations to very specialized living conditions and also their highly developed social organization within their groups, have reached a second high point in the evolution of the primates. The other high point is the lineage of the *Hominoidea* leading to man.

Fig 17-6. 1. Hamadryas baboon *(Papio hamadryas)* 2. Gelada baboon *(Theropithecus gelada)* 3. Barbary ape *(Macaca sylvana)*

Today baboons are still very numerous in most steppe and savannah regions of Africa. They are also found in forests, rocky terrain, and mountainous plateaus. Baboons belong to the few large wild animals which do not retreat from human settlements. Quite the contrary, they utilize fields and plantations as easily accessible food sources, and are not too disturbed about the presence of man. They cleverly evade any persecution. If they have been shot at, they do not let a person with a gun approach within range on subsequent occasions. In South Africa a premium for baboons was offered in 1925. Within two years, 200,000 animals were turned in which also included baboons killed by poison; however, the number of baboons has not decreased. In certain areas baboons were regular pests. Yet man is partially to blame for this. He decreased the number of leopards because these spotted cats occasionally stole a chicken, sheep, or calf, and also because leopard fur has unfortunately become "high fashion." As a result, leopards are shot or cruelly captured in traps in ever-increasing numbers. Leopards, on the other hand, are the greatest enemy of the baboons. The former were responsible for a healthy balance in the habitats of the baboons, and overpopulation did not occur. Wherever the number of leopards has greatly dwindled, the baboons increase without limit and can cause great agricultural damage to wide areas. As on so many other occasions where man acts without sense and understanding and does not pay attention to the biological balance, just to fulfill the special interests of a few people, eventually many people have to suffer the consequences.

Field studies

During the last years the American professors S. L. Washburn and I. DeVore have extensively studied free living baboons and the organizations of their societies in East Africa. Both scientists are anthropologists who have realized, as so many others have, that one first has to gain a knowledge about the behavior of monkeys and the great apes before one can interpret pre-human behavior and thus gain a correct insight into the development of human society. Washburn and DeVore studied steppe baboons because these baboons inhabit the same steppe and savannah biotope where our ancestors presumably completed their transition from animal to man. The omnivorous baboons of today are, in addition, subject to similar environmental influences as were our ancestors millions of years ago.

The two scientists observed more than thirty baboon troops. First, they studied baboons in the small Nairobi National Park near the capital of Kenya and later in the more distant and far less visited Amboseli Park where the behavior of the baboons had hardly been influenced by people. Each troop consisted of forty to sixty individuals which occasionally formed into larger groups of several hundred head near water holes or other places. These groups did not fight one another. The individual groups remain together and do not mix with the others.

"Our observations showed," wrote Washburn and DeVore in the magazine *Das Tier,* "that the social bond of the baboons is one of the most important prerequisites for the survival of the species. Almost the entire life of an individual is spent in the immediate vicinity of his conspecifics. A baboon is only secure within his troop and only as a group do baboons roam through their territory. A troop of baboons is not merely an aggregation of many individuals but an association of very different animal personalities. Our observations did not support the old view that a group of monkeys is mainly held together by its sexual relationships. A group is held together by the extremely social nature of these monkeys, which is expressed in a great variety of ways between individual animals. These different social bonds still have to be studied more closely. Today we do know that the close knit social life is highly significant and that a baboon can only survive within the group."

The region of Amboseli Park that the researchers studied contained fifteen troops with a total of approximately 1200 baboons. The smallest group was made up of thirteen members, and the largest of 185 monkeys. The troops in Nairobi Park were smaller.

DeVore and Washburn further reported: "A baboon group can roam through an area of 5–15 km but it only utilizes small parts of these regions. If food and water are found over a wide area, the troops rarely meet. Nevertheless, the home ranges of neighboring groups overlap extensively. Towards the end of the dry season this could be observed in Amboseli Park easily. Water was available in specific spots and different groups often used the same water hole to drink or to feed on the succulent plants which surrounded it. Once we counted over 400 baboons which were gathered around a single water hole at the same time. A superficial observer might have considered them to be one troop, whereas actually there were three large groups. They came and left without mixing with the others although they were sitting in close proximity to each other and were feeding. Near the waterhole we did not observe any fights between the groups, although we did notice that the smaller troops slowly retreated from the larger ones. Groups which rarely met showed great interest in each other.

The marching order within a group of baboons

"If one sees a group of baboons for the first time, one gets the impression that there is no order at all. The basic organization of the troop may be recognized most clearly when a large group descends from the safety of the trees into the open plain. When the animals start to move the less dominant males and occasionally some strong subadult males make up the vanguard. Juveniles and older females follow. The center of the herd consists of females and young, smaller juveniles, and a large percentage of dominant males. The rear guard is identical to the vanguard, but with less dominant adult males. The group is loosely organized yet in a way that provides protection for the females and young in the center. From whatever direction an enemy would approach, he would thus first encounter the mature males.

Males defend the group

"One day we observed how two barking dogs rushed towards a

group of baboons. Females and young hurried along but the males, on the other hand, calmly continued on. Quickly a group of over twenty adult males came between the dogs and the rest of the group. When one of these males approached the dogs, they fled. We even observed baboons in the close vicinity of hyenas, cheetahs, and jackals. Usually the baboons did not pay any attention to the predators. However, they fled from lions and climbed into trees. From their perch they scolded the big cats, but on the ground they could not resist them. Baboons may often be seen among topis, gazelles and other antelopes, zebras, giraffes, and buffalos. Baboons step aside for elephants and rhinoceros only at the last moment. We observed warthogs and a rhinoceros run right through a troop of baboons. The baboons only moved aside a few steps.

Relationship to other animal species

"The peaceful coexistence of several species of animals is often advantageous for all concerned. On the open plain baboons often stay close to impalas. In forest regions baboons seek out bushbucks as partners. The ungulates have an acute sense of smell while the baboons rely on their eyes. While eating, these monkeys frequently look up in all directions. When they have spied an enemy they emit warning calls which not only alarm other baboons but also all other animals close by. Similarly a warning call of a bushbuck or impala can cause the flight of a group of baboons. A group of impalas and baboons is very difficult to surprise. Once we observed a group of impalas and baboons feeding together. Three cheetahs approached this mixed group, but the impalas showed no signs of flight. They only looked on as a mature baboon male emitted a challenging call while approaching the cheetahs and chasing them away.

"The mutual dependence of the various species on each other is best seen near a waterhole where the bush is dense and visibility is poor. Baboons in the Wankie Reservation approached water with great caution. They often paused and played in the bush for awhile before they rushed in to drink. Apparently many animals know the behavior of others, and so give alarm signals.

"Guenons frequently visited the same water holes as the baboons. They usually moved closely with the baboons, or even among them, without any friction between them. Only once did we observe a baboon killing a small grass monkey and eating it. In our observation area the diet of baboons consisted almost entirely of vegetative matter; rarely did they eat meat. We observed two baboon males killing and eating two newborn Thomson gazelles (color plate p. 411). Apparently they like to eat chicks and bird eggs, and it has even been reported that they dig out crocodile eggs. They also feed on insects. However, the main part of their diet consists of grass, fruits, buds, and plant shoots of various species."

A diet of plants and meat

When baboons are fed by tourists in the national parks they often

Accidents with baboons

become very obtrusive. The wardens had to kill a baboon on an island in Lake George (Uganda) because this animal had the habit of sneaking up on fishermen and stealing their food, and also on occasion hurting them badly. In 1965 one baboon stole a baby out of a carriage in the mother's presence in one of the immigration camps set up in the South African city of Brakpan and killed the child by repeatedly biting it in the head. In the last century in Uganda, Cuningham saw two village women that had just been bitten to death by baboons, as well as several seriously injured children.

Defense against dogs and leopards

During a fight with a dog, wrote Blaney Percival, the baboon flings his opponent around by one leg, sinks his teeth into the dog's throat and then pushes it away with arms and legs and tears out a large piece of flesh. A leopard which had attacked baboons and was driven away by them was shortly afterwards shot by Lord Delamere. The leopard was "bitten terribly and undoubtedly would have died of his wounds." "A baboon caught a hare," reported the South African L. MacWilliam in the magazine *Das Tier,* "and he sat down and looked at the screaming and struggling animal in his hands. He then bit off an ear and ate it. Then he held up the still struggling hare, bit into it and tore out a larger piece of skin and meat. After a few additional mouthfuls he followed his troop." A fight between a single baboon male and a leopard was described by the South African J. Bester in *Das Tier.* "When he returned with his oxcart, he found the dead baboon sitting erect against a tree with an injury in the neck. A little way off lay the leopard with its intestines torn out (compare color plate p. 31)."

Sleeping places in the trees

While some baboons, particularly hamadryas baboons, seek out rocky areas as a safe retreat or for sleeping places, the troops in the steppes and savannahs always climb up trees to sleep. In border areas, they even penetrate the rain forest. They not only climb trees in order to get a better view, but they can also escape through the trees with great speed. "In the canopy of the trees they are most secure," write Washburn and DeVore. "Trees are therefore just as vital to them as food and water. In a slightly swampy region of the Amboseli Park where food and water were plentiful, the baboons were absent because lions were present and the region was treeless. Nearby was an area with trees and many lions, and many baboons lived there. Three large groups regularly visited this area."

At night when predators and snakes are most active, the baboons sleep in the tops of high trees. Because of their ischial callosities the monkeys can even sit and sleep on very thin branches. One whole group of baboons can be accommodated on only a few trees. It is a known fact that colobus monkeys take turns sleeping and watching during the entire night. The same is probably true of the baboons as well. In any case, baboons are afraid of the dark. They approach the trees before nightfall and stay in them until it is light.

Internal organization of a group of baboons

The cohesion within the society is established through the internal organization and the manifold interrelationships among individual baboons. This social organization can best be observed when the monkeys are eating or resting. On these occasions the majority of the troop is split up into small groups. Some members are grooming each other while others are just quietly sitting. Such a small group within the large troop often consists of two females and their young of different ages or a mature male with one or several females and juveniles which are constantly grooming him. Frequently such groups stay together when the troop is moving.

Reciprocal grooming

The focal point of such a grooming session is usually a dominant male or a mother with an infant. Dominant males attract other animals, and the latter deliberately seek out the big males. In sharp contrast to this is the behavior of the males of many social ungulates which constantly herd their groups. Baboon males do not have to coerce the other troop members to stay together. Just the opposite is true; their mere presence is sufficient to keep the troop together.

Initial grooming stabilizes the social group. The mother grooms her newborn on the day it is born. Most older baboons engage in grooming but it is most pronounced in adult females. They groom infants, juveniles, mature males, and other females. Baboons regularly visit each other in order to be groomed. These activities not only serve to maintain group cohesion, but, in addition, contribute to the cleanliness of the animals. Many ticks are found in the habitat of baboons. Lions and many other wild animals are infected with them. Because of their grooming activities the skin of baboons is free of these parasites."

Newborn baboons cling to the fur of their mother's chests. If they would have to learn this they would be lost. This behavior pattern, which they already exhibit on their first day of life, is innate, as in most primates. A newborn baby is the center of attention in the group. Dominant males sit close to the young mothers or walk closely beside them. Adult females and juveniles groom the mother and also attempt to groom the baby. In this manner the infant gets acquainted with the members of his group at an early age, and learns the social position of each individual, thus becoming integrated into the society. Baboon young raised in the laboratory without their mother or even a human substitute mother are later incapable of interacting socially with others. They are like those young rhesus monkeys which were only acquainted with a wire dummy which served as a "mother substitute" (see p. 408). When adulthood was reached, they were incapable of mating.

Washburn and DeVore were able to closely observe the development and social activities of baboon infants raised in the wild. They were also able to gain new information about the social dominance order and the sexual behavior of baboons: "As the young grows up it learns to ride on its mother's back. At first it still clings tightly to the fur but later it sits upright. It starts to take solid food and to leave the mother for ever-increasing periods of time in order to play with other children. In the course of time it plays with other children for many hours every day. Gradually playmates gain in importance over the mother. Play group activities prepare the youngsters for the interaction of the adult world. The habits of the adult, such as climbing, are frequently imitated. Usually, however, games are a mixture of reciprocal chasing, tail pulling, and wrestling. Joy Adamson observed how a small baboon fell off a tree into the river and was promptly rescued by an old baboon. When a youngster is injured and cries, adults rush to his side and stop the games. The presence of an adult male inhibits tormenting of small youngsters. Thus in the protected atmosphere of the play group, the social ties of the youngster are expanded and gradually stabilized.

"There is a clear dominance hierarchy among the adults of a baboon troop. Dominant males are more frequently groomed, and they can eat and rest at their pleasure. If a dominant animal approaches a subordinate the latter will walk away. An observer can determine the social rank on the basis of how a baboon behaves when it passes another. One can verify these findings in a tame troop by feeding them. If one throws food between two baboons the dominant one will pick it up. The subordinate often does not even dare to look at the food.

"The rank position of a baboon male within the dominance hierarchy is not only dependent on his physical condition and his ability to fight but also on his relationship to the other males. Certain mature males in every large troop band together for a major portion of the time. If one member is threatened all the others will come to his aid. Against such a male 'club,' any single animal is subordinate even if he were able to dominate each member of this 'club' individually. The social dominance order within the entire troop is quite stable. This is largely determined by the fact that the rank order is dependent on the male groups rather than on the fighting ability of the individual. In groups where the dominance hierarchy has been firmly established few fights occur. Once the relationships between the males have been settled the rank order usually decreases the probability that the group will split. As soon as a quarrel breaks out within the group the dominant males rush to the scene and separate the two opponents. In this way the weak ones are protected and peace and security are restored. Females and juveniles seek out the presence of the males in order to groom them or just to sit close to them. Although the dominant animals rule the group they are not despots but rather are responsible for the prevention of intraspecific fighting and the general well-being of the group.

"The sexual behavior of baboons has often been called 'licentious.

The dominant baboon males have also been compared to 'pashas' which keep a 'harem' of subordinate females. In reality the situation looks different. An adult female enters the estrous phase for approximately one week every month, provided she is not pregnant or suckling. During this phase the females leave their small group and weaned young, and join the males. At first they may mate with subordinate males or sub-adults. At the peak of estrous, however, they consort with dominant males. If a male is not sexually interested, the female will groom him. However, during the last days of estrous the dominant males become very interested. One male and one female will then form a pair. This consort-pairing may sometimes only last for an hour, or sometimes for several days. Oogenesis and ovulation in the ovary disrupts all other social ties. Usually pairs are found at the periphery of a troop. At this time fights may occur if the rank order among the males is not definitely established. Usually, however, there are no fights about a female, and although a male may have a high social rank he will never keep a female for himself for too long. A male does not court more than one estrous female at a time. In baboons we did not observe anything that was reminiscent of a family or a harem. The sexual drive is obviously not an important factor in keeping the troop together. Baboon females are only estrous for a very small part of their life. For the majority of the time they are either immature, pregnant, or nursing an infant. In smaller troops months may go by in which none of the females is in estrous. Yet none of the animals leaves the troop. The highly developed social relationships are continued without disruptions."

Pair formation and reproduction

Group behavior of the hamadryas baboons

The rock-inhabiting hamadryas baboons diverge somewhat in their social behavior from that of the plains baboons. According to the observations of Kummer and Kurt, these baboons do not form closed societies but live in sub-groups which consist of very constant "one-man groups." Such a group is made up of a mature male and one to nine females with their young. Various one-man groups sleep together on a rock ledge, and altogether they may form troops ranging in numbers from 12 to 750 members. During the day the troop splits up into individual groups which then meet again at night time in the mountainous terrain. This organization is probably due to the scarcity of food and sleeping places in the habitat of the hamadryas baboons. The result has been small family units which form loose aggregations with other units during the night.

The most important result of such studies is perhaps the observation that the social behavior of these monkeys is obviously less determined by rigid innate behavior patterns than is the case in most other herd animals. All scientifically trained observers of free living primates were repeatedly impressed by the amazing variability of their behavior. This was not only observed in the steppe baboons by DeVore and Washburn, but also in the Japanese macaques by Itani and Kawai (see

p. 402). Goodall, Korlandt, Reynolds, and Schaller observed the same phenomenon in the great apes, but with one difference— their behavior, or at least that of the chimpanzee, is still more flexible than that of the macaques and baboons.

Baboons as goatherders

The late zoologist Dr. Hoesch from Southwest Africa reported an interesting case where innate behavior patterns and learned behavior interlocked. Mrs. Aston, the owner of a farm, got the idea of using chacma baboons as goatherds. For a short period she locked a young baboon female into the goat corral, so that she would get used to the goats. After a few days she accompanied the goats when they were grazing. The baboon had become completely adapted to her alien "troop companions." The baboon kept the goat herd together and prevented the straying of single goats. In the evening the baboon always returned to the corral with the whole goat herd. Mrs. Aston did not have to go to any lengths to train the baboon; rather, she followed her innate social inclinations.

One morning the baboon female "Ahla III" returned to the corral from the goat herd, scolding loudly. The Ovambo shepherd who milked the goats in the morning had forgotten to release two older kids from the enclosure. Ahla had noticed that they were missing from a herd of eighty animals. This baboon female enticed the kids back to the herd with constant calls. Even if one assumes that Ahla was alarmed by the bleating of the mother goats out in the field and that this prompted her to return to the corral, it took a remarkable amount of insight and persistence to accomplish this task.

The smallest kids did not go out into the field. In the evening when the goat herd had returned Ahla often picked up each individual kid out of a group of twenty and very gently brought it to its bleating mother, putting it down by her udder. Ahla knew exactly which kid belonged to which female. She never made a mistake. Kids which were still weak were very patiently supported by Ahla and held to the udder. In female goats that had only one kid, she tested the milk pressure in the udder. When the kid was satiated, she would reduce the still-considerable pressure by sucking the milk from the goat. There was one thing which the goat-herding Ahla did not tolerate: If one removed one kid from a female that had three offspring and placed it with a different female, Ahla would always bring the third kid back to its rightful mother.

One baboon female even carried kids which had been born on the field back to the corral. She carried the newborn kid under her arm. This behavior is of great interest from a psychological point of view because here one is dealing with a "learning process," i.e., the meaningful application of the maternal drive to a situation in a different species. This behavior is not normally found in baboons because young baboons are not carried but cling independently to the mother's fur.

Baboons in zoos

In zoos baboons are often kept on spacious rock enclosures. Here small groups are formed as in the wild state. They also breed regularly, and several times even twins were born. Hediger reported one such case: "In the Zürich Zoo we observed that the mother of the twins was obviously confused at first as to whether she should carry the twins in one arm, or one in each arm. The latter situation greatly interfered with her climbing and feeding. Finally she settled on usually carrying both young under her left arm while engaged in these activities."

In such naturally arranged enclosures one can frequently observe male baboons displaying their characteristic "threat yawning." Yet a group with a biologically appropriate composition is usually subject to much less aggression between members than was formerly assumed because of improper care and observation of the animals. During the 1920s a group of mature hamadryas baboons was released in an outside enclosure in the London Zoo. Of course, in this situation fierce fighting broke out. At that period people were not aware of the strict social order within the various groups. Today we are better informed and such incidents do not occur in scientifically managed zoos. The "troop boss" of the large breeding population of hamadryas baboons, which has existed in the Frankfurt Zoo for decades, had become senile in due course, as reported by Grzimek. The old male began to lose his beautiful shoulder mane and his fur became thin. A younger male displaced him and took over the dominant position without the occurrence of bloody interactions. The old male has continued to exist peacefully within the troop and is still accepted by its members. Estrous females and juveniles still looked after him and groomed him. In another case, in the Münster Zoo, an old senile hamadryas baboon male with totally worn and stump-like canines maintained his dominant position because the small group did not have another mature male. Heinemann had the idea to suddenly flash a life-sized picture of a yawning hamadryas baboon male with powerful canines (the photo is on p. 439) at this oldster from the outside of the cage. The old baboon, which had stood at the bars and threatened Heinemann, let out a cry of terror when the threatening picture suddenly appeared less than a meter away from him. He fled into the farthest corner of his cage. Later Heinemann repeated this experiment in the Frankfurt Zoo. The male baboon, in that case, paid hardly any attention to the picture. This male was at the height of his physical strength and possessed powerful canines. That this was the reason for his ignoring the picture cannot, of course, be stated without conducting further experiments.

The mandrills

The MANDRILLS (Genus *Mandrillus*) are terrestrial monkeys which inhabit the jungle. The HRL of the ♂♂ is over 80 cm, the TL is 5–8 cm, the shoulder height is over 50 cm, and the body weight is over 50 kg. The ♀♀ are considerably smaller and lighter in weight. The head is very large, and the facial part of the skull is greatly elongated. There are prominent ridges on either side of the nasal bones. The ridges are covered by skin which is grooved lengthwise. The fleshy nose terminates in large, forward-directed nostrils. Naked skin areas are conspicuously colored or are, in part, a glossy black color. The body form is compact, with long, powerful limbs. The stubby tail is held upright. There are two species: 1. The MANDRILL (*Mandrillus sphinx;* Color plates p. 385 and p. 389); its young initially have light, and later uniformly dark, faces with ridges just beginning to show; 2. DRILL (*Mandrillus leucophaeus;* Color plate p. 389).

The mandrill and drill inhabit the forest

Although mandrills and drills are found in the forest, they usually stay on the ground. In groups numbering from only a few individuals to over fifty, they roam among the trees and search for edible plants, dig for roots, turn over stones and eat the small animals they find there. At night they probably climb up into the trees to sleep. During their movements, the animals also venture out into open terrain. Often the strong males are found at a little distance from the groups; however, they are immediately present if there is any sign of danger. A mandrill or drill threatens an opponent by spreading his arms, displaying his lowered, impressive head, and flashing his powerful teeth towards the enemy in the yawning gesture.

The mandrill is more conspicuous than the drill because of its bright coloration. The colors on the bare facial skin and the buttocks are the most flashy among the monkeys, and probably among all mammals. With increasing excitement the colors increase in brilliance, and even the chest turns blue with red dots appearing on the surface of the wrists and ankles. Undoubtedly these pigmented skin areas play a role in the threat behavior of the mandrill. Probably the change in skin coloration serves more as a visual signal in interactions with other mandrills in rank order disputes than as defensive or fighting behavior against other species. The advantage of an effective threat behavior is obvious, because in many cases the opponent is intimidated without an actual fight having taken place. Thus more serious interactions can be avoided which would otherwise cause more harm than benefit to the species. The brilliant colors in the face, on the chest, and on the joints of the mandrill are probably conspicuous threat signals just as the black face and red chin region are in the drill.

Fig. 17-7. 1. Mandrill (*Mandrillus sphinx*)
2. Drill (*Mandrillus leucophaeus*)

The vivid skin coloration on the buttocks, on the other hand, seems to have a different meaning. The Swiss scientist Konrad Gesner (1516–1565) was already aware of the submissive posture of the mandrill: "If one threatened with a finger or pointed it at him, he would present his behind." This submissive posture, whereby the buttocks are displayed, is a phenomenon widely distributed among the Old World monkeys. Originally this behavior was probably derived from the manner in which an estrous female offers her posterior to the male.

During the course of evolution this submissive posture took on a different meaning which is not sexual in nature. Subordinate males present their posteriors to dominant males in the same manner as the females. In many monkey species this behavior pattern is part of "proper conduct." It is the perquisite of a dominant to briefly mount a subordinate animal which approaches him in this manner. However, this behavior only serves to reconfirm group cohesion and the existing rank order.

Fig. 17-8. The showing of teeth in the mandrill is a gesture of friendliness but in most monkeys it means a threat.

Formerly, this behavior, which "mimics" sexual behavior, was usually misinterpreted. The monkeys, and in particular the baboons, had the reputation of being extremely lustful. It was assumed that their social groups were merely kept together on the basis of sexual relationships. In reality, however, the male "mimicked" the conspicuously brilliant colors of the sexual parts of the female, as well as the concurrent female behavior which was adopted in subordinate situations, thus keeping social stability within the group (this change in function, of course, is not a true imitation but is rather a genetically determined factor). The frequently erected penis observed in baboons, according to Wickler, is not a sexually motivated act. The males mark the area occupied by their group against others with these visual signals.

The mandrills and drills are omnivorous. Besides roots, tubers, fruit which has dropped from trees, and other vegetative matter, they also eat snails and worms, insects, frogs, lizards, and even snakes. They also may consume mice and other small mammals. In the dense forest, where visibility is often poor, the members of one group keep in contact by constantly grunting or making other sounds. Gerald M. Durrell, when he was in the Cameroon, put a living chameleon among his tame drills. They avoided this animal while chattering and making faces at it, but were otherwise not disturbed. However, when Durrell put a large water snake in their vicinity the drills fled at once, and shrieked until the "monster" was removed.

Although mandrills and drills are always unpredictable, they occasionally make nice and friendly pets even when caught as adults. I (Fiedler) observed how the warden in the Zürich Zoo could enter the cage of an adult male drill without special caution, although he always had a few tidbits in his pockets. The drill immediately got up on his hind legs and very thoroughly searched the pockets of his human friend. On the other hand, mandrills and drills have the reputation of being especially skillful in taking various articles from visitors. Zoo director Hediger reported: "In the Zürich Zoo a young mandrill had specialized in snatching glasses and pipes. The insurance company finally requested that a glass partition be put in front of his cage bars."

Helga Lindroth studied the significance of the various gestures and postures of the mandrills in the Frankfurt Zoo. If the mandrill male shakes his head and shoulders he is issuing an invitation to be groomed. The playful exposure of the teeth where the lips are only slightly lifted and where the teeth occasionally chatter is not only an indication of friendliness and a greeting but also a general expression of well being. An angry mandrill will vigorously slap the ground with one hand. His entire upper body will drop for a fraction of a second and the hair in the nape becomes erected. Such an excited animal will constantly gaze at the observer. After this threat posture, which can be repeated several times, the animal often sits down immediately and vigorously scratches his forearm or thigh, yet never letting the intruder out of his eyes. Scratching under these circumstances is called "displacement activity" by ethologists. It can often develop into a regular session of "self-grooming," which is intensive care of one's own fur.

In these conflict situations, a mandrill male may also show the characteristic yawning. He yawns when he cannot carry out his intended actions. A conflict situation may occur, for instance, when a male wants to groom a female and he touches her with his fingertips on the buttock pads. Normally the female will pause and wait until he begins to groom her on some part of her fur. If she does not respond in the expected manner and continues to move away, the male will react with a threat yawn and then will groom his own fur. The partner being groomed sits or lies comfortably, sometimes with the eyes half closed. While engaged in grooming, the male presses the bared teeth closely together and emits vigorous smacking sounds which are similar to the sounds he makes during copulation.

In the Baltimore Zoo twins were once born to mandrills. Several mandrills and drills in American and English zoos have had life spans of twenty-six to twenty-eight years. One monkey in the Kumamoto Zoo (Japan) even reached the age of thirty-two. These jungle residents live just as long in captivity as do the baboons and hamadryas baboons. American zoos in 1966 had a total of seventy-seven mandrills and forty drills.

Gelada baboons—baboons or macaques?

The GELADA BABOON (*Theropithecus gelada;* Color plate p. 389) is rather unique. This monkey resembles a baboon but, according to some scientists, it represents a separate branch of the macaques. Other scientists think that the gelada baboon shares certain characteristics with the guenons. The HRL is approximately 70 cm and the TL is approximately 50 cm. The skull is short and the muzzle is round. The body form is slender; the fur is dark brown and on the chest is a naked red skin region. The ♂♂ have long whiskers and a broad brown shoulder mane. The light eyebrows are conspicuous. The nostrils open to the side. The well developed fingernails are sharp. The tail is tufted at the tip. The gelada baboon inhabits the plateau of Ethiopia close to the boundary of Eritrea up to the region of Addis Ababa.

Gelada baboons are well adapted to living in the mountains. They are found at elevations up to 3000–4000 meters. Their thick fur protects them against low temperatures at night. With their powerful fingernails they dig in the hard ground of the mountain plateaus or in the meadows and fields in search of roots and onions. They also eat large amounts of grass and other greens. Occasionally they also dig out small animals. They sleep in the rock cliffs above the deep valleys. Before they venture out in the morning, they spend much time sunbathing.

According to Sanderson, these mountain monkeys are not jumpers but rather "clingers"; they cannot jump very well. Sanderson reported: "A gelada baboon will usually injure himself fatally if he falls only a few meters. Instead, he clings to the rocks with his sharp fingernails and rarely slips. I repeatedly observed how these monkeys, in captivity, climbed five to six meters on walls that were covered with smooth tiles. Although the cracks between the tiles had been filled in carefully it nevertheless provided sufficient hold for the gelada baboons."

Like the baboons and macaques, the gelada baboons form smaller or larger troops. The gelada baboons have a very characteristic and impressive threat display which is totally different from that of the baboons. While threatening, they curl up the upper lip and display the light, brilliant pink inner skin surface and the teeth with the huge canines. At the same time, they lift the eyebrows and flash the bright eyelids which are in very striking contrast to the black face. During intense excitement the naked reddish skin regions on the chest turn a deep red. A threatening gelada baboon makes a frightful sight indeed. Yet, in reality, gelada baboons are by far not as aggressive as baboons. Usually the entire group will escape at the slightest approach of a potential enemy. Little is known about the social behavior of these interesting and taxonomically problematic monkeys. In our zoological gardens, gelada baboons do as well as steppe and hamadryas baboons, mandrills, and drills. Gelada baboons inhabit high elevations and are exposed to intense sunlight and an almost dust free atmosphere. In captivity, they should be provided with ultraviolet light, particularly during the winter, to prevent the occurrence of stiff joints or other deficiency diseases. Until the bombs destroyed it during World War II, the Berlin Zoo kept a healthy breeding troop of gelada baboons. Today one can view these Ethiopian mountain monkeys in outside enclosures in the zoos in Zürich, Duisburg and various larger zoos in North America. One gelada baboon has been living for fifteen years in Philadelphia.

Most species in the various genera of the *Cercopithecidae* discussed up to now are primarily ground-dwelling forms, or at least they spend most of their time on the ground. Completely arboreal forms are not found among them. Thus they have relatively broad hands and feet. In many species the tail is reduced or is absent. Only a few species of macaques have truly long tails. Most members of the following genera are, almost without exception, arboreal forms (only the red guenon is a true ground-dwelling form). The arboreal forms have narrower hands and feet, and they always have a long or very long tail. The long tail in the genera *Cercocebus* (mangabeys) and *Cercopithecus* (guenons) was responsible for their scientific name. Both Greek-Latin words mean "tail monkey."

The mangabeys by W. Fiedler

Distinguishing characteristics

Among the long-tailed, arboreal cercopithecids the MANGABEYS (Genus *Cercocebus*) are most closely related to the macaques. Their external appearance is reminiscent of the large guenons. The HRL is 38–70 cm, the TL is 43–76 cm, and the body weight is 3–15 kg. They are very slender and are excellent jumpers. The facial part of the skull is long. The supraorbitals are well developed and are emphasized by the pale eyelids. The cheek pouches are large. The air sacs in the laryngeal region are poorly developed. The voice is usually not very loud. The third lower molar has a tribercus, as in the macaques and baboons. There is an indication that the tail is used as a prop. Menstrual swellings are not as pronounced as in the macaques and baboons, and sexual dimorphism is less well developed. There are four species with many subspecies: 1. The SOOTY MANGABEY (*Cercocebus torquatus;* Color plate p. 390) has an HRL up to 70 cm and a TL somewhat longer. The body weight goes almost up to 15 kg (largest species). There is great color variation in the various subspecies and even in individuals. The dorsal side is gray to brown, and the ventral side is paler. The hair on the crown is olive to brownish-red and is framed by lighter or darker hair; the juveniles are somewhat darker. 2. The GRAY-CHEEKED MANGABEY (*Cercocebus albigena;* Color plate p. 390) is somewhat smaller. It is gray to black; a crest of short hair is on the back of the head and a long shoulder mane is present. The laryngeal air sacs are relatively well developed. 3. The BLACK MANGABEY (*Cercocebus aterrimus;* Color plate p. 390) is black with grayish-brown whiskers. 4. The AGILE MANGABEY (◊ *Cercocebus galeritus;* Color plate p. 390) has subspecies differentiated on the basis of hair adornments and details in the fur coloration; however, all have a brownish base color which is more or less sprinkled with yellow spots. The ventral side is lighter.

The name "mangabey" is due, like the names of so many foreign animals, to a misunderstanding. These monkeys were named after the port city of Mangabe on Madagascar. Reportedly the first mangabeys were shipped from there to Europe. Yet, as we know, there are no monkeys on Madagascar, only prosimians.

Long-tailed tree dwellers

Mangabeys are primarily arboreal. This is particularly true of the gray-cheeked mangabey, for the other species come down to the ground to search for food. All mangabeys seem to prefer moist terrain. Often they inhabit swampy forest, and according to Tappen, sooty mangabeys also inhabit the aerial root systems in the mangrove forest near the mouths of rivers. Like all monkeys, when they descend to the ground they can be a plague for local farmers. A subspecies of the agile mangabey, which lives in the Congo, plunders the wet rice paddies, and the most westerly form of the sooty mangabey has the reputation of being "the most destructive monkey of Sierra Leone—a serious threat to the cocoa harvest." This is a rather dubious "honor" for an animal because such generalizations can easily lead to extermination programs, regardless of whether the damage is actually large or is virtually non-existent.

Fig. 17-9. 1. Sooty mangabey *(Cercocebus torquatus)* 2. Black mangabey *(Cercocebus aterrimus)*

The mangabeys do not have loud voices. This may be due to the poorly developed air sacs on the larynx. The gray-cheeked mangabey is, however, an exception because it possesses larger air sacs. Pechuel-Loesche, the editor of "Brehm's" third edition, said of the loud voice of the gray-cheeked mangabey: "The Mbukumbuku, as the natives on the coast of Loange call it, is found in the vast forests. It does not occur frequently and never in troops but only in two or threes; some old males occur singly. The name Mbukumbuku is based on the way the animal calls, but only males can call loudly. The calls are rumbling and rolling and are emitted with grimacing, erection of the hair crest, arching of the back, and often with the long tail held in a vertical position." The other species, which are less vocal, are not by any means less "communicative." They communicate less with sounds but more with their extremely expressive facial expressions. Sanderson writes in his monkey book: "If one mangabey wants to communicate with another it gazes into the other's face and begins to flutter the eyelids quickly and irregularly. Furthermore, the mangabey chatters with its lips or opens its mouth, and quickly wags the extended tongue up and down or left and right." The white coloration exaggerates the movements and greatly facilitates the communication of emotions and moods.

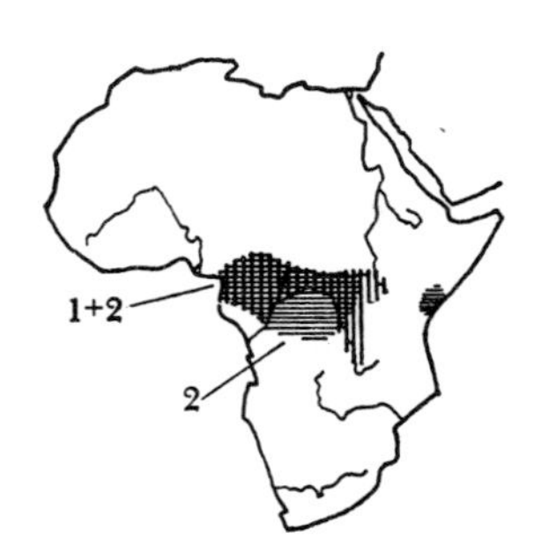

Fig. 17-10. 1. Gray-cheeked mangabey *(Cercocebus albigena)* 2. Agile mangabey *(Cercocebus galeritus)*

The guenons by W. Fiedler

The GUENONS (Genus *Cercopithecus*) are small to medium-sized Old World monkeys with very long tails. The HRL is 32.5-70 cm, the TL is 50-87.5 cm, and the body weight is in excess of 10 kg. The head is roundish and the supraorbitals are only slightly developed. The jaws are short; the cheek pouches are large. The overall body form is slender, and the limbs are medium-long. While the young are in the clinging stage the tail is still slightly prehensile. Ischial callosities are small. In Africa there are approximately fifteen species with a total of over seventy subspecies.

All guenons are characterized by colorful fur, or at least a few colorful spots. In some species the naked or sparsely haired body regions are also very colorful. They have colorful designs in the face, ischial callosities, and the males frequently have prominently colored scrotums. Juvenile coloration is occasionally different from that of the adults. It has been assumed that the different coloration in the young evolved in those species where the males are particularly aggressive, so that the color variation served to distinguish and thus protect the young from the attacks of the males. However, such simple and plausible explanations should be approached with great caution. The interrelationships in nature are usually much more complex than appears at first. Similar, but often more extreme color differences than in the guenons, are also found in the gentle leaf monkeys (see p. 443). The very defensive and aggressive steppe baboon males "love children," so to speak, although the young in the "play age" are of the same coloration as the adults.

Guenons inhabit forests or the open savannah in Africa south of the Sahara. The species which occur more in the open spaces, e.g., the grass monkey, avoid waterless arid areas, and prefer the vicinity of rivers and particularly the gallery forest or the river banks. In these vast forest and savannah regions, a seemingly infinite variety of guenons is found. An attempt to describe the diversity of these various, yet very closely related, tribes is a "hopeless undertaking," according to the experienced African primate authority, Sanderson. It actually would take an entire volume by itself to describe all the guenons. In one small area of only 90 kilometers in diameter, Sanderson counted not less than eight different species and each of these had several subspecies; "yet 150 kilometers to the east another half dozen different guenon forms were found, and to the west of the river the situation was not much different." DeVore has stated: "The number of guenon species that exist depends on the specialist who counts."

During the early stages of the exploration of Africa very slightly different local forms were immediately hailed as new species or even genera. Today the concept of a species is more comprehensive. Locally, closely related forms which exclude each other geographically, and occupy the same ecological niche, and in a way represent each other, are considered subspecies of one species. Since vast regions of Africa have not been sufficiently explored, this has contributed to the uncertainty of classifying the guenons and has been the cause of dispute among zoologists. The geographical range of the individual guenon species is not even certain. Therefore our maps of distribution are only approximations.

Since every large zoo displays several species and subspecies of these pretty and colorful monkeys, and their great diversity catches the interest of the visitors, we shall give a summary of the individual species using the key of the English primate authority O. Hill.

I. Predominant color is light green to pale gray: 1. The GRASS MONKEY *(Cercopithecus aethiops;* Color plate p. 394) has an HRL of 41-62 cm and a TL of 53-72 cm. The fur is slick and is of different shades of green. The ventral side is light. There are approximately twenty subspecies which are distinguished from each other on the basis of the shade of green and hair adornment, light bands on the forehead and whiskers. The ♂♂ can also be distinguished by the color of their scrotum. They inhabit the open savannah.

II. The predominant color is not green; the limbs are black below the elbow and knee.

A. The abdomen is blackish and the whiskers are white: 2. L'HOEST'S MONKEY (*Cercopithecus lhoesti*; Color plate p.392); the HRL is 45–70 cm and the TL is 55–80 cm. The fur is dense; the whiskers are white to pale gray. Otherwise the animal is dark with a light saddle on the reddish brown back. Two geographically widely separated subspecies are found in Cameroon and the East Congo and Uganda. They inhabit mountain forests, and are primarily berry and leaf eaters.

B. The abdomen is whitish and the whiskers are dark: 3. The DIADEMED GUENON (*Cercopithecus mitis;* Color plate p. 392) has an HRL of over 60 cm and a TL of over 75 cm; the ♀♀ are considerably smaller than the ♂♂. There are hair tufts on the ears. There are approximately twenty subspecies which have either a dark head, a light band on the forehead, a green or reddish tinged back ("true" diademed guenon) or a light head and paler throat (white-throated guenons). The animals are quiet and shy. They are found in very diverse habitats, i.e., mountain forests, forests near riverbanks, bamboo growth, and they occasionally penetrate a little distance into the savannah.

III. Predominant color is not green; the limbs below elbow and knee are not completely black.

A. The face is black with a white supraorbital band and long white beard. 4. The DIANA MONKEY *(Cercopithecus diana)* has very brightly colored fur, which is predominantly gray with orange-reddish lines on the sides of the thighs, with white patches (Color plate p. 393) and beards which are shaped differently in the various subspecies. They are very active and curious monkeys and they are more arboreal than the other species. They feed mainly on fruits. Three species are found in the coastal forests from Sierra Leone to Ghana and a small region in the Congo delta.

B. The face is black with chestnut-brownish headband and short white beard: 5. DeBRAZZA'S MONKEY (*Cercopithecus neglectus;* Color plate p. 392) has an HRL of 43–60 cm and a TL of 58–70 cm; they may weigh up to 8 kg. Despite their vast geographical range (from Cameroon to Kenya and Katanga), these guenons are uniform in external appearance. They inhabit mountainous and flat terrain, and are frequently active on the ground. They are more insectivorous than other guenons.

C. Face with fleshy colored mouth region: 6. The MONA MONKEY (*Cercopithecus mona;* Color plate p. 393; "Mona" is a Moorish term for long-tailed monkeys) has an HRL of up to 55 cm and a TL of up to 80 cm. They are strong animals with long, thin tails; they live in the

Tropical rain forest
Savannah
Steppe
Desert

Fig. 17-11. The different species of guenons prefer different habitats. The following maps show that certain species inhabit the tropical rain forest, while others live in the savannah girdle.

Characteristics of the guenon species

Fig. 17-12. The grass monkey *(Cercopithecus aethiops)* is a savannah-dweller (compare with Fig. 17-11). The very discontinuous distribution areas are largely determined by environmental conditions (biotope).

Fig. 17-13. The diademed monkey *(Cercopithecus mitis)* is both a savannah and forest dweller (compare with Fig. 17-11). 1. Diademed guenon i.n.s.; 2. White-throated guenon and related subspecies.

upper strata of the trees. There are seven subspecies (Senegambia to the Congo delta). 7. The CROWNED GUENON (*Cercopithecus pogonias;* Color plate p. 393) is very similar. There are four subspecies (Cameroon, Gabun, Congo).

D. Face with a heart-shaped white, yellow, or red spot on the nose: 8. The LESSER WHITE-NOSED GUENON (*Cercopithecus petaurista:* Color plate p. 391); *Cercopithecus petaurista petaurista,* a subspecies, has an HRL of up to 45 cm and a TL of 60 cm and more. It is found from Gambia to Ghana; additional subspecies in the Congo delta are occasionally classified as a separate species (Redtail, *Cercopithecus ascanius*). To these belong, among others, SCHMIDT'S WHITE-NOSED GUENON (*Cercopithecus petaurista schmidti* or *Cercopithecus ascanius schmidti;* Color plate p. 391). Probably closely related to the lesser white-nosed guenon is: 9. The RED-BELLIED GUENON (*Cercopithecus erythrogaster;* Color plate p. 391), which is found in South Nigeria.

E. White oval nose spot: 10. The GREATER WHITE-NOSED GUENON (*Cercopithecus nictitans*) has an HRL of 40–64 cm and a TL of up to almost 1 meter. It is found from Liberia to Chad. The fur coloration is fairly uniformly dark; the nose spot is very prominent. They are forest dwellers that rarely descend to the ground, although they are said to be more territorial than most guenons. There are three subspecies.

F. Pale blue horizontal stripe on the upper lip, yellow whiskers: 11. The MOUSTACHED GUENON (*Cercopithecus cephus;* Color plate p. 391) has an HRL of 38–58 cm and a TL of 57–84 cm. Their behavior is reminiscent of the lesser white-nosed guenons. They do not descend to the ground very often. Their diet consists mainly of fruits and seeds. They are very fond of palm nut marrow. Five species are found to the west of the lower Congo; three of these are occasionally classified as a species (RED-EARED GUENON, *Cercopithecus erythrotis;* Color plate p. 391).

G. Vertical white stripe on the bridge of the nose: 12. The OWL-FACED GUENON (*Cercopithecus hamlyni;* Color plate p. 392) has an HRL of approximately 55 cm and a TL of approximately 58 cm.

This key does not include two species of guenons because these differ in certain aspects from the usual guenon type. Hill and other zoologists have classified them in a separate genus. One of these, the TALAPOIN MONKEY (*Cercopithecus talapoin;* Color plate p. 391) is much smaller than all other guenons (the HRL is up to 25 cm and the TL is up to 40 cm. They weigh approximately 1.4 kg); most other differences in their body structure are without doubt related to their small size. The head seems large in relation to the size of the body and the large cranium and small jaws give the impression of roundness. However, there is one unique feature in the dwarf guenons that is not related to their smallness: the females have conspicuous menstrual swellings.

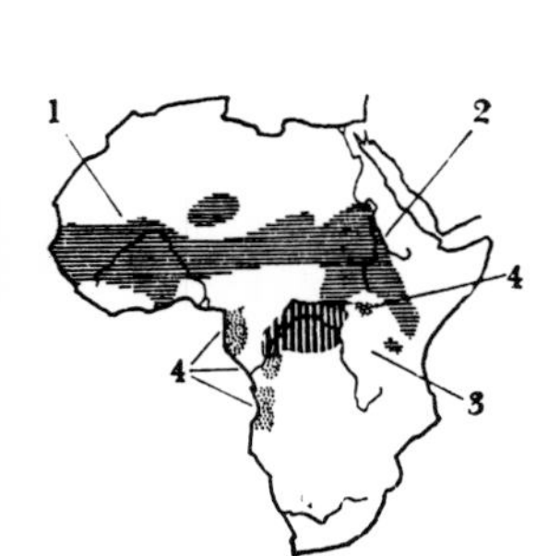

Fig. 17-14. 1. Patas monkey *(Erythrocebus patas patas)* 2. Nisnas monkey *(Erythrocebus patas pyrrhonotus)* 3. Swamp guenon *(Cercopithecus nigroviridis)* 4. Dwarf guenon *(Cercopithecus talapoin)*.

If one wants to separate the dwarf guenons from all other guenons, it seems sufficient to classify them in a subgenus *(Miopithecus)*. We know of four subspecies which are found in West Africa, and far removed from there in Ruwenzori. The animals are primarily found in the swampy forests and mangrove swamps near the coast. They are mainly plant eaters, and are very fond of palm nuts. They also capture insects and other small animals.

The second "outsider" among the guenons is the SWAMP GUENON (*Cercopithecus nigroviridis;* Color plate p. 394; the HRL is approximately 46 cm and the TL is approximately 50 cm). The skull and dentition and a few other anatomical characteristics are somewhat reminiscent of mangabeys and baboons. The behavior and voice of swamp guenons is also different from other guenons. Some experts classify them in a separate genus *(Allenopithecus)*, which could suggest a link between the mangabeys and guenons. However, one should not underrate the numerous guenon characteristics of this species. Therefore, we have classified the swamp guenon in the subgenus *Allenopithecus* within the genus *Cercopithecus.*

Pocock first discovered the swamp guenon in 1907, not in its native habitat but in London. A darkly colored guenon had lived in the zoo for some time and had died in 1894. During its lifetime no one had recognized this monkey as a new species or genus. Yet the remains of the animal had been carefully preserved for science. Pocock finally recognized that he had found a new species. Later, however, additional swamp guenons reached zoological gardens. They were extremely lively and agile animals which tirelessly romped through their cages. In the National Zoological Park in Washington, one could always observe them splashing in the shallow water. It is possible that they also do this in the flooded swamp forests of their native habitat. Very little is known about free living swamp guenons. According to Sanderson and Steinbacher, they live in small groups and feed on fruits, leaves, and unripe nuts. During 1959, 1960, and 1961, three young were born in the San Diego (California) Zoo, always during the months of June and July.

Grass monkey—savannah-dweller

Among the true guenons the GRASS MONKEY *(Cercopithecus aethiops)* is primarily a savannah dweller. The anatomist and mammalogist Dietrich Starck and Hans Frick of Frankfurt have described them as "typical animals of the bush steppe, particularly at the boundary regions of the rain forest (Hylaea). They occur only near running water, and avoid the open, arid steppes, as well as the interior of continuous forests. We (Starck, Frick, and Fiedler) observed grass monkeys in the highlands of Shoa north of Addis Ababa at elevations of almost 3000 meters, as well as in the hot valleys and lowlands. Where living conditions are favorable, one usually finds groups of twenty to fifty animals. Occasionally we also saw smaller groups and single animals which, however, may have only separated from their group for short

Fig. 17-15. Most guenons are jungle inhabitants (compare with Fig. 17-11): 1. L'Hoest's monkey *(Cercopithecus lhoesti);* Diana monkey *(Cercopithecus diana);* DeBrazza's monkey *(Cercopithecus neglectus);* Mona monkey *(Cercopithecus mona);* Crowned guenon *(Cercopithecus pogonias);* Lesser white-nosed guenon *(Cercopithecus petaurista);* Red-bellied guenon *(Cercopithecus erythrogaster);* Greater white-nosed guenon *(Cercopithecus nictitans);* Moustached guenon *(Cercopithecus cephus);* Owl-faced guenon *(Cercopithecus hamlyni).*
The geographical range of the various species has only been insufficiently investigated up to now. The map, therefore, only provides a rough outline.

periods. We observed that guenons preferred the gallery forests near the river banks, where each troop has a specific territory. From there they would regularly migrate into the open steppe and cultivated fields. Baboons are known to do this even more frequently. In the forest they prefer certain trees where one can always see them. The grass monkeys which we observed were very fond of the fruits of the wild fig tree. Usually one could approach the monkeys as close as thirty to fifty meters before they fled. This flight distance, as in all arboreal monkeys, depends on visibility. When chased, guenons often press themselves against tree branches, an excellent camouflage."

"It has frequently been maintained that feeding monkey troops have sentries. We cannot verify these statements," reported Starck and Frick. In any case, one should be cautious about using such human concepts as "sentry" or "lookout" when talking of animals. This inevitably leads to anthropomorphism, and thus tends to obstruct our understanding of the nature of the animals. Concepts like "sentry," "leader," or "guard" lead one to assume a human-like social order, as Hall emphasized in his work about chacma baboons. Only a clear-cut separation of concepts can lead to a clear understanding of the various animals, and also that special case, man. This, of course, does not mean that we cannot relate personal experience to something comparable in the realm of spontaneous and insightful behavior with terminology referring to our own experiences.

During his travels through various African national parks, Hediger observed grass monkeys and told how these lively creatures occasionally played tricks on other animals: "I observed how a grass monkey purposefully dropped on the back of an impala antelope which of course got greatly frightened. On another occasion a troop descended to the ground to observe some elephants from close up. The most daring of them approached up to within reach of the trunks of the 'patient' elephants. Grass monkeys often seem unconcerned in the presence of tourists, and at opportune moments help themselves to food from picnic baskets that contained one's own meal."

Forest-dwelling guenons

In contrast to the savannah-dwelling grass monkeys, the remaining species are more or less forest dwellers; some of them are typical creatures of the tropical rain forest. Individual species prefer specific types of forest. Some live high up in the canopy and rarely descend to the lower strata of the forest. Others, again, are more active near the ground, and descend from trees more frequently. Swamp, dwarf, and white-nosed guenons are most frequently found in swampy forests, and DeBrazza's monkey also occurs here very frequently. Diademed guenons, various forms of the mona group, and crowned guenons inhabit the highest strata of their native forests. Schmidt's white-nosed guenon is very adaptable and occurs in very diverse habitats, but it is primarily a forest dweller. In the lower regions it inhabits all forms of rain forest, including swampy forests. The remaining white-nosed forms, L'Hoest's monkey, and diademed guenons also inhabit very diverse forest regions. However, the distribution of the species in a particular habitat is influenced by many different conditions. The distribution can change if the population of a species decreases, perhaps due to man's interference. Then one species may migrate into an area where previously it was not found. Grass monkeys avoid thickets at the edge of the forest which are occupied by lesser white-nosed guenons. If the latter, however, disappear, the grass monkeys will inhabit this region. Similarly, Congo white-nosed guenons will occupy jungle regions which have been deserted by other guenon species.

Habitat and mode of feeding are closely interrelated. The diet of Schmidt's white-nosed guenon has been studied very thoroughly. These guenons primarily feed on leaves, young shoots, and tree fruits. They will also come down to the ground and will plunder fields and plantations. L'Hoest's monkeys apparently do this even more frequently. It has been repeatedly maintained that free living guenons also eat bird eggs but no one has been able to verify this beyond doubt. Captive guenons are frightened if one gives them an egg or another round, smooth object. It usually takes a long time before they overcome their fear and learn to eat eggs. Later on they become very fond of eggs. A male grass monkey in the Münster Zoo greedily grasped locusts and ate them with obvious enjoyment. A moustached guenon which Pechuel-Loesche had running free in his house liked to look at illustrations in books: "At first it quickly reached out for pictures of grasshoppers and spiders, but soon learned that these were not edible."

The size of a guenon troop varies greatly. In some species the troops roam through a certain region which obviously does not have any definite boundaries. Others, like the greater white-nosed guenons and particularly the moustached guenons, maintain more or less strict territories. The moustached guenons have a relatively well-defined social order. Bourliere mentioned this species as an example of small family bands with one dominant male. The well-developed rank order of baboons and macaques and hence the well-established social order is only little developed in the other guenon species.

As most other monkeys, guenons are most active during the forenoon and afternoon. During the hot noon hours they rest. The noon break is often utilized for extensive mutual grooming which in the guenons seems to be the only significant form of social contact. While searching for food, the members stay in contact by "contented," peaceful calls. If one guenon discovers something potentially dangerous, it emits sharp barks and is joined by all others immediately. Each monkey attempts to gain a perch from where to better see the cause

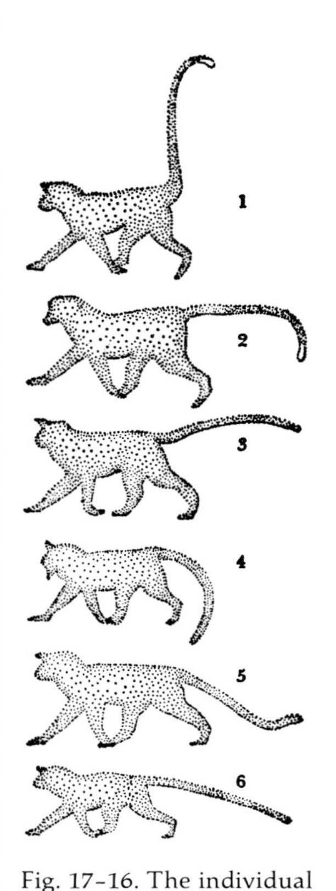

Fig. 17-16. The individual guenon species can occasionally be recognized by the manner in which they hold their tails: 1. Grass monkey when walking through tall grass; 2. Grass monkey running along on branches; 3. Diademed guenon; 4. De-Brazz's monkey; 5. L'Hoest's monkey; 6. Schmidt's white-nosed guenon, all running along on branches.

of the disturbance. Finally the entire troop flees. Running and jumping through the crowns of trees, the monkeys hurry along. They leap tremendous distances from the outermost branches of one tree to the next, landing in a vertical position and extending all four limbs in order to gain a hold. Occasionally they miss and drop down but usually they manage to hold on to one of the lower branches. Sometimes, however, they may fall all the way to the ground. Sanderson observed how one entire troop of mona monkeys dropped one by one into a creek which they intended to cross. "All mona monkeys screamed loudly, but they quickly and expertly swam towards land, paddling like dogs, and quickly climbed back into the branches." The grass monkeys also seem to be good swimmers.

Without doubt, the long tails serve as balancing organs during climbing and jumping, but to a certain extent they are also important in steering. However, as in many long-tailed animals, the tail is also important as a signal and as a display organ. In Ethiopia we observed that grass monkeys held their tails up when running through high grass. In forms that have a white tip on their tail, this probably is an effective signal. When running on branches, the grass monkey carries its tail in a more horizontal manner. Such differences in the tail carriage are superimposed on tail positions which characterize each subspecies. Haddow was able to distinguish the various species by the manner in which they held their tails in comparable situations.

As all monkeys, guenons breed throughout the year. In some species, however, the births of the young seem to be concentrated during a specific part of the year. The gestation period is approximately seven months. Usually one single young is born but occasionally twins occur. The newborn young clings to the fur on the mother's abdomen. It is also able to hold on a little with its tail, but this slight prehensility soon disappears. Often the mother supports the young with one hand. According to Haddow, small Schmidt's white-nosed guenons are able to walk after only a few days, at about the time when the navel has healed. Around one or two months the young are already very active. At two months they take their first solid food in addition to the mother's milk. When they are four months old one can remove them from their mothers, and if need be they can look after themselves. Shortly after six months, they possess their complete set of milk teeth. They are sexually mature at about four years of age.

In zoos guenons are just as great favorites as macaques or baboons. Formerly various species were kept together in one cage. Today they are usually kept in family groups or pairs. Like macaques, they received a mixed diet of fruits, carrots, potatoes, bread, lettuce or cabbage, celery, peanuts, and similar vegetable foods. In addition, they receive milk and occasionally a small amount of raw egg and meat as well as vitamin additives and minerals. Many are very fond of grasshoppers and other insects. One can also feed them with commercial monkey pellets, and supplement this basic diet with fresh fruit and vegetables. Guenons in captivity do not breed as readily as do macaques or baboons, yet breeding them is not too difficult. A number of zoos were able to cross various guenon species and to obtain viable young.

Since a shortage of rhesus monkeys has developed, guenons have increasingly been used for medical and pharmaceutical experiments. In 1962, 25,000 guenons were exported from Kenya alone for this purpose. The only other monkey where such a population decrease would probably do little harm would be the baboons. However, these are too large and powerful, and would require too much space for keeping in a laboratory; severe bites could also be expected.

When properly kept, guenons live long in captivity. In zoos, individual animals have lived longer than twenty years. One grass monkey *(Cercopithecus aethiops aethiops)* in the zoo of Gizeh near Cairo lived a record twenty-four years. A mona monkey lived in an American zoo for twenty-six years. Free living guenons probably do not reach such an advanced age. While young guenons make very nice, trusting, and gentle pets, they often become unpredictable and aggressive with increasing age. The males in particular can inflict dangerous bites with their dagger-like canines. Therefore, it is not advisable to keep them as house pets.

The RED GUENON *(Erythrocebus patas;* Color plate p. 394) is the true ground-living member among the guenons. Ground living requires a series of special adaptations which distinguish the red guenons from all other guenons. Therefore this species is classified as a separate genus *(Erythrocebus):* The HRL is 60–87.5 cm, the TL is 50–75 cm, and the weight of the ♂♂ is 12–25 kg; the ♀ is much smaller. The arms and legs are long and slender, and the hands and feet are shorter than in the guenons. The thumbs are also shorter. When walking or standing, the patas lifts up the wrists and ankles (walks on knuckles). The fur is coarse; old ♂♂ have shoulder manes, whiskers, and moustaches. A troop usually consists of seven to fifteen animals. There are two subspecies: the PATAS MONKEY *(Erythrocebus patas patas)* and the NISNAS MONKEY *(Erythrocebus patas pyrrhonotus).*

Red guenons are regular inhabitants of the steppe, preferring grass plains with few trees and avoiding the forest completely. They do not even enter the forest islands found in the steppe. Like all monkeys, they are able to climb trees, but they rarely do so. Even in the case of danger where any other monkey would run up a tree, the red guenon almost always escapes on the ground in a steady galloping gait.

Hall, the previously mentioned primate authority, has observed free living red guenons. Some of his results have been published in the journal *Tier*: The red guenon is the greyhound among the primates.

A hamadryas baboon male (*Papio hamadryas;* see p. 422 and p. 424) threatens with his dangerous teeth. Zoo visitors often mistake this "rage yawning" as an expression of boredom. When this photograph was suddenly flashed in front of an old, almost toothless baboon male in the Münster Zoo, he ran away.

A Northern black-and-white colobus mother *(Colobus abyssinicus)* with her infant which was born in the Berlin Zoo.

It has been clocked at 50 km per hour. On bad roads it effortlessly passes any car. Otherwise it does not make use of these capabilities. Only young animals chase each other through the steppe, and they playfully wrestle with each other. As Hall observed, the animals are able to execute jumps over bushes at a full run while changing direction at the same time.

While baboon troops survive, aided by the extraordinary defensive and aggressive behavior of the large males, the patas survive because they remain silent. At night they distribute themselves over various trees, and during the day they hide in the tall grass and vocalize only with almost-closed mouths. One can barely hear their voices. Each red guenon troop contains only one male. He is twice as large as the largest female. He is an impressive figure, and yet the appearance is misleading. He is not the ruler but the watchdog of his troop, and is constantly on the lookout for possible enemies. In case of need, he distracts them by his conspicuous escape. There is a strict linear hierarchy among the female members of the troop but it is rarely upset by dominance fights. The social rank is recognizable by the distance females sit from the male. Even the young of a dominant female enjoy a more favorable position. The "pasha" avoids these young because otherwise he might risk a dangerous encounter with its mother or all the females of the group. The young males become sexually mature at about three and a half years. At this point the "pasha" attacks them viciously and chases them from the troop.

Aside from plants, red guenons also feed on insects and other small animals. Red guenons keep well in captivity but have bred relatively rarely up to now; we were successful only once in the Vienna Zoo, for example. Some reached a life span of more than twenty years in zoos. Young red guenons are delightful house pets, and even adults are less difficult to handle than baboons and many macaques. One can even train them to defecate in a specific place. This is a habit which an arboreal monkey can never learn, because its feces drop out of sight from the tree tops. Old red guenon males have dangerous, dagger-like canines, and even females can inflict dangerous wounds. Monkeys, in particular, are clever and mentally active creatures; they are far more flexible in their social relationships than other animals whose behavior is innately determined. In monkeys, traditions and personal experiences play a far greater role. Thus, a monkey which has grown up with a human family and has adapted to their social arrangements is really no longer a "true monkey." If such an animal is placed in a zoo with a troop of conspecifics, it is not fully accepted and usually remains a pitiful outsider. However, to keep a monkey in a cage by itself is sheer torture. Every animal lover who wishes to keep a young monkey as a house pet should seriously consider the consequences of this act.

Walter Fiedler

18 Leaf Monkeys and Colobus Monkeys

Distinguishing characteristics

The second family of the OLD WORLD MONKEYS is the LEAF MONKEYS *(Colobidae)*; some zoologists consider these a subfamily *(Colobinae)* of the *Cercopithecidae,* as is seen, for example, in the phylogenetic table on p. 238 which was designed by Erich Thenius. The HRL is 43–83 cm, the TL is 15–107 cm, and the body weight is 7–22.5 kg. The body form is slender and light; only the colobus monkeys are more heavily built. The legs are usually longer than the arms; the thumbs are reduced and are vestigial in the colobus monkeys. The head is rounder than in the cercopithecids. The face is naked or sparsely covered by hair. The muzzle is short, and the cheek pouches are absent. The tail is very long except in the pig-tailed langur. The molars have cross ridges (high cusps) for grinding leaves. The stomach is sacculated and is three times as large as in other monkeys. The intestines are large. There are five genera in South Asia: LANGURS, DOUC LANGURS, SNUB-NOSED MONKEYS, PIG-TAILED LANGURS,. and PROBOSCIS MONKEYS. There is one genus in Africa: the COLOBUS MONKEYS.

Leaf monkeys by W. Fiedler and H. Wendt

Because of their leaf diet, these monkeys have been called "leaf monkeys." Their large, sacculated stomach is reminiscent of an ungulate stomach and fulfills a similar function, as H. J. Kuhn was able to demonstrate in the colobus monkeys. The first two compartments are fermentation chambers where the food content is broken down by bacterial action in a fashion similar to the ruminants, camels, and kangaroos. The same is probably also true for the South Asian species. The Asiatic genera are slender while the African genus is more compact. In South Asia the leaf monkeys occupy a great variety of habitats. Many inhabit the tropical rain forests, some prefer the mangrove swamps and swampy forests, and others are found in the cold, high altitude zone within the tropical longitudes. They even occur in mountainous forests which are often deeply covered by snow during the winter in the south and east of the Himalayas of Tibet.

Since the leaf monkeys are more or less highly specialized leaf eaters, they are by far not as active and restless as the cercopithecids. The large amounts of leaves of poor nutritional value that these monkeys must consume are within easy reach, but, on the other hand, they are more difficult to digest. The restless temperament which characterizes the cercopithecids would not be very advantageous for the leaf monkeys, but would only disrupt the rest required for their digestion. Thus they often sit in the crowns of trees for the greater part of the day, actually in the midst of their food, looking "reflective and dignified" while they are digesting. When the entellus langurs quietly bask in the sun at sunrise and sunset they appear, anthropomorphically speaking, as if they were deeply absorbed in prayer. This may be the reason why the Hindus deemed them sacred. Other leaf monkeys which demonstrated the same behavior were similarly revered. Tribes in the southern region of the Himalayas formerly worshipped the native snub-nosed monkeys. The Bataks of Sumatra have incorporated the island langur into their cult. The African colobus monkeys also were regarded as God's messengers by certain West African tribes.

Perhaps the first leaf monkey to be pictured in a book in Europe was in a volume by Breydenbach called, in translation, *Trip into the Promised Land,* which appeared in 1486. Hill is of the opinion that it may be either a douc langur or a lion-tailed macaque.

The Langurs

The LANGURS (Genus *Presbytis*) are the best known and at the same time the most diverse of the leaf monkeys of Asia. The HRL is 43–79 cm, the TL is 54–107 cm, and the body weight is 7–18 kg. They are the most primitive representatives of the family. The body is very slender and the limbs are long and slim. The thumbs are greatly reduced while the other fingers are long and well developed. The tail is very long. The larynx is large with laryngeal saccules for amplification of the voice. The fur is usually uniformly colored; dorsally it is brownish, gray, or black, while ventrally it is pale. Often there are color patches on the head or other varied "adornments" (crests, crowns, tufts, whiskers, etc.). There are four subgenera which are distinguished by the hair adornments on the head, but mainly by the very conspicuous color differentiation found in the newborn which deviates greatly from the coloration of the adults. It is probable that at least in colder regions reproduction depends on the season.

A. The young of the ENTELLUS LANGURS *(Semnopithecus)* are brown to black; the adults usually have coarse gray fur, long bristle-like eyebrows, and well-developed supraorbitals on the skull of mature males.

B. Except for one form, the young of the PURPLE-FACED LEAF MONKEYS *(Kasi)* are usually gray with white cheeks (the juveniles of the exception are black); the adults are silvery gray to black with whiskers and reddish faces.

C. The young of the CAPPED LANGURS *(Trachypithecus)* are usually brownish-yellow to rusty-red; the adults have fine fur, often with hair tufts on the head. The eyebrows and supraorbitals are less conspicuous.

D. The young of the ISLAND LEAF MONKEYS *(Presbytis)* are white with a dark cross mark over the back and arms; the adults are very different in color, possessing a variety of head adornments and often with stripes on the legs.

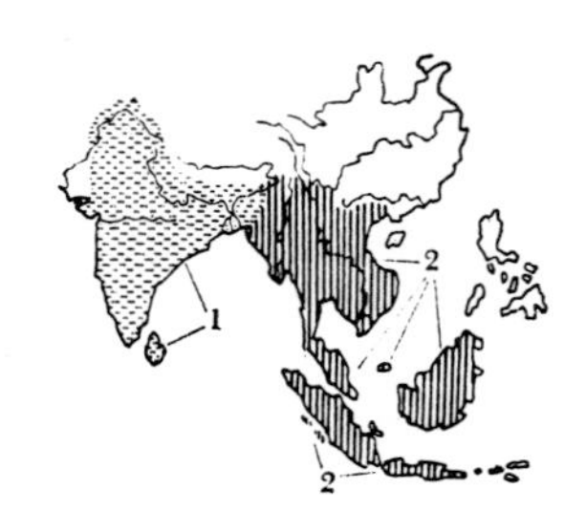

Fig. 18–1. 1 Entellus Langurs (Subgenus *Semnopithecus*) 2. Capped Langurs (Subgenus *Trachypithecus*)

Fig. 18–2. 1. Purple-faced Langurs (Subgenus *Kasi*) 2. Island Langurs (Subgenus *Presbytis*)

In this connection a report in Brehm's animal books is of interest: "In April 1912 a capped langur *(Presbytis lar pileatus)* was born in the London Zoo which had abnormal juvenile coloration: The face, ears, hands, and feet were a pale flesh color which seems to be true for all newborn monkeys. The forehead was bare, the pelage was a golden brown, and it had a thin line of stiff, black eyebrow hairs. After three months the darker skin pigmentation began to appear, and after six months all bare regions were black. In the fall the infant coat was replaced by longer, grey-brown hair, but when the animal was eighteen months old it still did not have the adult hair color, nor did it have the long, white beard and the erect hair-crest on its head. The second young again looked quite different; it had a sandy-yellow pelage, a less bare forehead, and smaller eyes and eyebrows that were not as black. This raised the suspicion that the father of the first young was an entellus langur who at the time lived in the same cage." When we consider the usual coat coloration of young entellus and capped langurs, this hypothesis could be correct. This would be an indication of the close relationships between members of the various subspecies.

Entellus Langurs

The ENTELLUS LANGURS (Subgenus *Semnopithecus*) are the largest langurs. According to the latest opinion there is only one species, the HANUMAN LANGUR or ENTELLUS LANGUR (*Presbytis entellus*; Color plate p. 453) with approximately fifteen subspecies, including: the ENTELLUS LANGUR *(Presbytis entellus entellus)*, which is found in northern South India.

Presbytis entellus schistaceus is particularly large: the HRL is up to 79 cm, the TL is up to 107 cm. It has broad whiskers. It inhabits conifer and rhododendron forests in the South Himalayas, in the summer up to elevations of 4000 meters.

Presbytis entellus hypoleucos has black feet and tail. It is found on the west coast of South India. The CEYLON ENTELUS LANGUR *(Presbytis entellus priam)* has a tuft-like black hair crest on the head; it is found in Ceylon and Southeast India.

The Red Colobus *(Colobus badius kirkii)* is almost extinct. Unfortunately it has not been possible to keep this beautiful monkey species in the zoo for longer periods, or even to breed it.

The large Indian national epic Ramayana, which tells the life history of the king's son Rama and his wife Sita, reveals why entellus langurs are sacred in India. When Sita was abducted by the wicked giant Ravana and was brought over the sea to the island Lanka (probably Ceylon), Rama, the reincarnation of the god Vishnu, received help from the monkey tribe. Sugriva, the king of the monkeys, and his

clever minister Hanuman rescued Sita and brought her back to her husband. Hanuman also removed the precious mango fruit from the garden of the giant to the great joy of the people. However, the giant caught him in the garden and Hanuman was to be burned at the stake. He managed to extinguish the flames, although he burnt his hands and face. Hanuman's children, the hanuman langurs, still have black faces and hands today.

The entellus langur is the true sacred monkey of India. It is a symbol of self sacrifice for a friend. No Indian who still believes in Vishnu would harm one of the langurs. In many regions of India, entellus langurs obviously seem to know that the people are harmless. The langurs wander back and forth between forests and villages; they even visit cities, jump around on roofs, and are allowed to pick garden fruits or take food from tables without being stopped. Often they are regularly fed by believers in the temples. If foreigners do not show tolerance for this monkey cult or attempt to chase away begging langurs, Vishnu believers will consider this a grave insult to their religious feelings.

In European circles, this Indian religious tradition was often ridiculed. One often heard the view that an overpopulated and partially underdeveloped country could not "afford" to tolerate and to protect sacred monkeys. In reality, however, entellus langurs do very little damage since they feed primarily on leaves, blossoms, and tree fruits. Surely the langur's visits into gardens and fruit plantations have not contributed to the poverty and starvation in India. In any case, we Europeans should concede to the Hindus that the worship of living animals deserves the same respect as the sanctification of any other religious symbol.

In recent times free living entellus langurs have been extensively studied by the anthropologist Phyllis Jay. Entellus langurs are extremely agile and when pursued can effortlessly jump distances up to ten meters. When leaping from a tree top to somewhat lower branches of another tree, they land on the branch so that the impact of the body on the branch propels them up again on the next jump. Occasionally they even change direction in mid-air in order to grasp a more suitable branch. Because of their adaptibility to man, they are more active on the ground than other langurs; however, they always are only a few seconds out of the sight of trees. In more densely populated regions of Northern India, the hanuman langurs feed mainly on plants which they can find in plantations and fields, according to Phyllis Jay. In sparsely populated regions they feed exclusively on food found in the forests.

The females and young males of the Proboscis Monkey *(Nasalis larvatus)* have turned-up noses. In the old males the nose grows into a monstrous "cucumber" as is illustrated on p. 455.

The social behavior of the entellus langurs has not evolved to protect the individual as much as in the macaques and baboons. In case of danger a single langur does not join the group but rather escapes into the trees alone. Nevertheless, the group does provide security for the individual animal because the chances of spotting danger in time are always greater in a group. The main function of a langur group is the "socialization" of the young, which consists of integrating the young into the social order. When a young is born, the mother and child become, above all others, the center of attention within the group regardless of rank. All females take an active interest in the youngster, but the males are not as concerned about the young as are, for example, the baboon males.

The development of young langurs clearly falls into several stages: the nursing stage, in which the infant is distinguished by the conspicuous natal coloration, leads into the playing and learning stage which occurs around twelve to fifteen months of age, when the youngster is weaned. The pre-adult stage begins in the third year for the females and in the fourth for the male. During this period the young adults become part of the social hierarchy of the adults. The young males stay more on the periphery of the group. Young females, however, seek the company of their mothers and are very attentive to the youngsters. Similar to the guenons, mating behavior does not play as great a role in adult langurs as it does in certain baboons and macaques.

The langur territories which Phyllis Jay visited measured approximately eight square kilometers each. The troop was usually made up of twenty to forty animals and was headed by one large male. Where entellus langurs are free living, independent of man, they migrate from the higher elevations to the lower during the fall, and in the spring they return to the higher and colder regions. One often finds them in the vicinity of water or on rocky or stony terrain. According to some observations the langurs feed on buds, leaves, blossoms, and fruits of all kinds, but seem to prefer red hibiscus blossoms. They will also eat paradise figs and bananas. When the sun rises the entellus langurs get up and migrate to the trees to feed. At noon they take a long rest, but start to feed again in the afternoon, and finally in the evening they climb to their sleeping trees.

Formerly langurs were only kept in zoos which were climatically favorable, such as in their native India or in California. It is rather difficult to properly and sufficiently feed these leaf eaters during the winter months. In American zoos, good success has been obtained by feeding lettuce and other greens, as G. S. Cansdale reported, and also with hardy species of bamboo. After an investigation into the causes of the deaths of leaf monkeys in zoos, Hill came to the conclusion that at least half of the langurs died as a result of stomach-intestinal disorders.

"Although langurs are normally quiet and almost phlegmatic, they are at certain times extremely active," wrote L. S. Crandall. "During such occasions they can leap seven to ten meters, according to Beddard's data. Therefore large enclosures with the usual branches and perches are as essential as they are for all arboreal monkeys. Very strong fences are not necessary. Although some langur species are very large and strong, they lack the aggressive nature of other monkeys. Some zoos, San Diego for example, had the opportunity to build up large family groups without many aggressive interactions occurring." Occasionally captive langurs have given birth to young. One entellus langur in the San Diego Zoo lived for over twenty years.

Since 1959 the Frankfurt Zoo has kept entellus langurs, but reliable breeding success has only occurred since 1962 when the langurs were transferred to the new monkey house which was designed with hygienic and psychological considerations in mind. Until 1967, eight entellus infants were born there. The newborn young have a thin black fur which gradually turns into the color of the parents by the third month. In the first few days the young is entirely preoccupied with clinging to the mother and nursing. They do not respond to the larger environment at this point. During the second week the young occasionally cling to other group members. When the mother is not present the infant occasionally sucks its thumb. This probably serves as a substitute for suckling on the mother, as it does in human infants.

Lutz Ehmler intensively studied the development of young entellus langurs in the Frankfurt Zoo. The young make their first crawling and walking attempts during the second and third week. At the beginning of the fourth week they already begin to climb, but they first practice this on their mother. Soon thereafter the entellus infant is able to jump quite well. With a skipping jump it may suddenly land on the back of an older family member. It may sit there, or decide to immediately jump somewhere else. After two months the young tirelessly gallops through the cage with rhythmically bouncing fore and hind quarters.

Lutz Ehmler reports about the wild chases and wrestling matches with other youngsters: "Both partners grab each other by the head or on the shoulder, and for a few seconds roll on the floor, tightly clinging to each other. Then they suddenly release each other, and the wild chase resumes once more. The pursued youngster never attempts to escape outside the enclosure. These play-fights are a part of the youngster's daily activities and are kept up into the early adult age." All not-fully-mature entellus langurs participate in this play-fighting. The young langurs like to chase a strong male which "pretends" to be "chased." These fighting bouts may occasionally appear quite rough and tough, and entellus langurs often emit shrill calls during these games. Generally, however, serious injuries do not occur during this "necessary preparation for life."

Healthy and well-adjusted entellus langurs do not appear "unmonkey-like" in zoos, as was formerly said of all leaf monkeys. On the contrary, they are very active and often surprisingly curious. In Frankfurt the entellus langurs were temporarily housed in an enclosure which bordered on the outdoor baboon enclosure. All entellus langurs played "zoo visitors." They were "glued" to the bars and stared with great interest at the baboons. Older females seem to be "persons of respect" within an entellus family. At feeding time they always reach for the food first, and successfully repulse all others.

Purple-faced Langurs

The PURPLE-FACED LANGURS (Subgenus *Kasi*) are still more slender and lighter in body form. The face is reddish to purplish-red; there is beard growth on the cheeks. There are two species: 1. JOHN'S LANGUR (*Presbytis johni;* Color plate p. 453) is found in the mountains of southwestern India; 2. The PURPLE-FACED LANGUR (*Presbytis senex;* Color plate p. 453) has several subspecies which are found on Ceylon.

Although purple-faced langurs rarely descend to the ground, when need be they can develop considerable speed there. Walker has timed a purple-faced langur at 35 kilometers per hour. In its South Indian mountain habitat John's langurs are often hunted for their fur and meat by tribes which do not believe in the Hindu religion. Apparently Hindus also take part in these monkey hunts under the pretext "that such good-tasting langurs do not possess any god-like qualities," as Sanderson reported. Purple-faced langurs are only rarely seen in zoos outside of India. Among the zoos, they have been kept in London and San Diego. W. C. Osman Hill kept them in Ceylon but the langurs did not last long. The Duisburg Zoo is presently keeping both species. One purple-faced langur lived for eight years in the zoo in Colombo, Ceylon.

The Capped Langurs

The CAPPED LANGURS (Subgenus *Trachypithecus*) are characterized by their fine fur. They have hair tufts on the head, and the supraorbitals are less well developed. They are widely distributed in northern India and the large Sunda Isles. There are approximately seven species: 1. The SILVERED LEAF MONKEY (*Presbytis cristatus;* Color plate p. 453) has several subspecies which are found in northern India, south of Tenasserim, and in Sumatra and Borneo. The subspecies *Presbytis cristatus pyrrhus* is found on Java and Bali. 2. The CAPPED LANGUR (*Presbytis pileatus;* Color plate p. 453) is found in Assam, Burma, and Yünnan; the northern subspecies are very large and the southern ones are smaller (the HRL is 50–65 cm). 3. PHAYRE'S LEAF MONKEY (*Presbytis phayrei*) is very similar to the capped langur but has pale spots, rather like glasses, around the eyes like the dusky leaf monkey; several subspecies are found in East Bengal, Assam, Burma, Thailand, and South Vietnam. 4. The DUSKY LEAF MONKEY (*Presbytis obscurus;* Color plate p. 453) is found in Tenasserim, South Thailand, Malacca, and bordering islands. 5. FRANÇOIS'S MONKEY (*Presbytis francoisi*) has a white stripe between the corners of the mouth and the ear; some forms are also characterized by a pale or even a golden-yellowish color on the crown and neck. 6.

The WHITE-HEADED LANGUR *(Presbytis francoisi leucocephalus)* is probably included in the former group. *P. f. leucocephalus* was only discovered in 1966 by Chinese zoologists in the mountains of Kwangsi. The head, shoulders, and feet are white; the rest of the body is black. 7. The MENTAWI LEAF MONKEY *(Presbytis potenzani)* has white cheeks and chin. The ventral side is yellow to golden orange, and the rest is black. This leaf monkey is found on the west coast of Sumatra, which is far removed from the geographical range of the other species.

"Bezoars" used as a wonder drug

The alimentary canal, and in particular the gall bladder and the intestine, of certain capped langurs occasionally contain bezoars which are similar to the ones found in the ibex and other wild goats. These are peculiar hard concretions which consist of swallowed rolled-up hairs or calcium carbonate. The superstitious tribes in Bengal attribute special healing powers to the bezoars found in a subspecies of the Phayre's leaf monkey. These calciferous formations are occasionally the size of a chicken egg and in a Chinese drug store one would have to pay several hundred dollars for these "wonder drugs." It is therefore not surprising that the langurs in Bengal have been mercilessly hunted. Today these langurs have become exceedingly rare and their bezoars rarely reach the market. For these very same reasons, the ibex and wild goats were hunted in Europe not too long ago and were exterminated because of their supposedly curative bezoars.

Capped Langurs in human care

Formerly Javanese silvered leaf monkeys were occasionally seen in zoos. Brehm already praised the beauty of this subspecies *(P. cristatus pyrrhus)* and emphasized the fact that this langur "was velvety in the face and on the hands, and silky on the back." It used to be customary to keep these langurs together with macaques or guenons. The langurs suffered greatly from these impudent cage mates, and they usually did not live long. Capped langurs were occasionally displayed in zoos also. It is surprising that under these conditions one langur lived for eight years in London. Years later a capped langur in San Diego held the longevity record with twenty-three years, which is the longest period that any leaf monkey has survived in captivity.

Today François's monkeys are seen in zoos, the Duisburg Zoo being one, in addition to the pretty dusky leaf monkeys which are not uncommon in captivity. However, the latter still remain difficult and frail pets although some dusky leaf monkeys in American zoos grew to be fourteen years old. The new accommodations and feeding methods in the Frankfurt Zoo have demonstrated that it is possible to keep these animals in good condition if they are fed sufficient amounts of fresh leaves and greens when the season allows. The Frankfurt Zoo has kept several dusky leaf monkeys since 1960 and, up to 1967, three young were born and raised. Newborn langurs have a conspicuous orange-yellowish plush-like fur. As the youngsters grow up they gradually become gray with a black back. However, they still have brilliant golden-brownish back sides when they are ten weeks old. The "eye glasses" in the face appear in the third week of life.

Dusky Leaf monkeys in the Frankfurt Zoo

Lutz Ehmler of the Frankfurt Zoo observed that the development of young dusky leaf monkeys was similar to that of the entellus langur young. Other group members and particularly the juveniles show great interest in the smaller young, and often try to pull it towards them. In such a situation, the baby emits shrill calls; however, it seems reassured as soon as it can press itself closely to the mother. Even during the longest leaps the young clings securely on its mother's fur. Other females occasionally pick up the infant but will immediately return it to the mother if she should want it. Even in the second month, the young did not show fear of humans. "At one time a youngster pushed his head and arm through the bars and watched me curiously and finally stretched out its hand towards me," reported Ehmler. "When I held out my hand, it did not show any fear. However, the mother immediately came, and hastily retrieved her young and fled to the wall board." Such incidents demonstrate that monkeys also act with reason concerning child care. The mother might not interfere when the youngster emits shrill calls while wrestling with its playmates because she knows the child is not really in danger. However, she immediately recognized a truly dangerous situation although the youngster had not called and seemed quite at ease.

Small dusky leaf monkeys are extremely lively while they grow up. They jump far and high, and permit themselves to be caught in the air by other troop members, forever inventing new climbing games which they repeat over and over with great patience and persistence. As they become older, rough-house play and play-fighting become wild chases throughout the cage, and they end when two playing animals try to push each other away, either on the ground or on the perches. In the Frankfurt Zoo a younger male often chased the older male without apparent reason; shortly after that he would permit the older one to groom his fur. When it appeared as if the older male did not get enough rest he was removed from the group. A strict rank order did not seem to exist, nor did the animals sleep in a particular order and place. Frequently all the animals huddled together when sleeping, with the mothers holding their young in their arms; however, it also occurred that one of the other animals slept away from the group by itself.

▷

1. John's Langur *(Presbytis johni)*
2. Purple-faced Langur *(Presbytis senex nestor)*
3. Silvered Leaf Monkey *(Presbytis cristatus)* Subspecies: a. *Presbytis cristatus germaini;* b. *Presbytis cristatus cristatus*
4. Dusky Leaf Monkey *(Presbytis obscurus flavicauda)*
5. Entellus Langur *(Presbytis entellus).* Subspecies: a. *Presbytis entellus entellus;* b. *Presbytis entellus schistaceus*

▷▷

1. Douc Langur *(Pygathrix nemaeus)*
2. Thomas's Sundra Island Leaf Monkey *(Presbytis aygula thomasi)*
3. Banded Leaf Monkey *(Presbytis melalophus)* Subspecies: *Presbytis melalophus chrysomelas;* a. color phase "crucigera"; b. darker form
4. Maroon Leaf Monkey *(Presbytis rubicundus)* Subspecies: a. *Presbytis rubicundus rubicundus;* b. *Presbytis rubicundus carimatae*
5. Capped Langur *(Presbytis pileatus)* Subspecies: a. *Presbytis pileatus gei;* b. *Presbytis pileatus durga*

The Island Langurs

The ISLAND LANGURS (Subgenus *Presbytis* i.n.s.) are smaller than the capped langurs. The hair on the head is often shaped into caps, hoods, or other hair styles. The fur coloration varies greatly. It is colorful or has conspicuous designs. There are many species and subspecies, including 1. The BANDED LEAF MONKEY (*Presbytis melalophus;* Color plate p. 454), which is found on Sumatra. Certain subspecies have a brown to blackish dorsal side. 2. The MAROON LEAF MONKEY (*Presbytis rubi-*

1
2
4
3 a
5 b
5 a
3 b

1
2
4a
4b
3a
5a
3b
5b

1
2
3
♀
♂
a

1b
2a
1a
2b
3a
3c
3b

◁◁
1. Brown Snub-nosed Langur *(Rhinopithecus bieti)*
2. Snub-nosed Monkey *(Rhinopithecus roxellanae)*
3. Proboscis Monkey *(Nasalis larvatus)* a. young animal

◁
Colobus Monkeys:
1. Southern Black and White Colobus *(Colobus polykomos)* Subspecies:
a. *Colobus polykomos polykomos* b. Angola Guereza *(Colobus polykomos angolensis)*
2. Northern Black and White Colobus *(Colobus abyssinicus)* Subspecies:
a. Abyssinian Guereza *(Colobus abyssinicus abyssinicus)* b. Kilimanjaro Guereza *(Colobus abyssinicus caudatus)*
3. Red Colobus *(Colobus badius)* Subspecies:
a. Rust-red Colobus *(Colobus badius ferrugineus)* b. Elliot Colobus *(Colobus badius ellioti)*
c. Zanzibar Colobus *(Colobus badius kirkii)*

cundus; Color plate p. 454) is found in Borneo. 3. The WHITE-FRONTED LEAF MONKEY *(Presbytis frontatus)* is grayish brown with black limbs; the forehead is hairless. There are two hair whirls on the crown with a cap behind these. 4. The SUNDA ISLAND LEAF MONKEY (*Presbytis aygula;* Color plate p. 454) is found on Borneo, Java, and Sumatra. Subspecies are distinguished by the shape and size of the white spots on the head.

Although island langurs are very rarely seen in zoos and usually do not survive for very long, the length of the gestation period has finally been determined: It is 140 days. The Sunda Island leaf monkey *(Presbytis aygula)* is the only leaf monkey that Linné mentioned in his "Systema naturae." The Bataks of Sumatra still revere the "Lutongs," the native name for the Sunda Island leaf monkey, in a manner similar to the way in which the Hindus worship the entellus langur.

Wherever island langurs are not pursued, they fearlessly come out of the forests, like the entellus langurs, and help themselves to food from plantations and gardens, even in the presence of people. Very little is known about free living island langurs. They probably also feed occasionally on animal matter. Sanderson observed that dusky leaf monkeys were feeding on large tree snails. Decades ago the former Berlin Zoo and the Antwerp Zoo kept a banded leaf monkey for a brief period. The London Zoo has also kept Sunda Island leaf monkeys and other species. The great difficulty of keeping these small leaf monkeys is indicated in a report by Hill. Even on Ceylon where climatic and dietary conditions were very favorable, Hill's banded leaf monkeys only survived for two years and then died because of a stomach-intestinal infection.

The Douc Langurs and Proboscis monkeys

The DOUC LANGUR (◊ *Pygathrix nemaeus;* Color plate p. 454) is closely related to the langurs but because of certain skull characteristics and a different color pattern, it is classified in a separate genus *(Pygathrix)*. It is heavier than the langurs; the HRL is 61–76 cm and the TL is 56–76 cm. The limbs are of approximately equal length. The eyes are slanted and appear "mongoloid." The coloration gives the effect of the monkey wearing a jacket, vest, knickerbockers, stockings, and shoes. Juveniles are whitish-red. About the life history little is known, and the monkey is very rare. Attempts to keep douc langurs in zoos have been unsuccessful. The zoo record (London) is four months. The population may be endangered by the Vietnam war.

Proboscis monkeys are quite distinct from the other leaf monkeys. They are more robust than the langurs. The fleshy parts of the nose protrude. The nose is human-like in structure. There are three genera; however, the skull and foot characteristics of the various genera often overlap (possibly they are only subgenera).

A. SNUB-NOSED LANGURS *(Rhinopithecus)*; B. PIG-TAILED LANGURS *(Simias)*; C. PROBOSCIS OR LONG-NOSED MONKEYS *(Nasalis)*.

The snub-nosed langurs *(Rhinopithecus)* are characterized by a relatively massive skull, a robust figure, and rather long upper arms. The HRL is 52–83 cm and the TL is 61–97 cm. The turned-up snub nose is present in both sexes. The hairs on the fur are long, and in some forms measure 15–18 cm. The stomach is very large and is sacculated. As in the langurs, the cheek pouches are absent. The snub-nosed langurs occupy high mountain forests (they are resistant to cold). Depending on how one looks at it, there are two to four species: 1. The SNUB-NOSED MONKEY (♁ *Rhinopithecus roxellanae;* Color plate p. 455) is found in Szechuan, Sikang and up to Upper Burma where it is represented by related species or subspecies: a. The BROWN SNUB-NOSED LANGUR (*Rhinopithecus bieti;* Color plate p. 455) is somewhat larger, and is found in Yünnan; and b. BRELICH'S MONKEY *(Rhinopithecus brelichi)* is still larger. The back is gray and a white shoulder mane is suggested. The arms are yellowish and the crown is yellow. It is found in Kweichow (Southwest China). 2. The TONKIN SNUB-NOSED MONKEY *(Rhinopithecus avunculus)* is somewhat smaller: the HRL is 52 cm and the TL is 66 cm. The feet and hands are short and broad with relatively long fingers and toes. The back and limbs are black, while the forehead and cheeks are cream-colored and the sides of the neck are orange to leathery brown. The ventral side is pale yellow. This species is found in North Vietnam. Formerly it was regarded as a separate genus or subgenus *(Presbytiscus).*

Old Chinese vases from 2200 B.C. were decorated with golden, human-like creatures with long tasseled tails and snub noses that were turned up almost to the eyes, and which appeared almost bald. Up to 1871 no one in Europe believed that such a creature really existed. It had been assumed that the illustrations on the vases were creatures of the imagination that represented some fabled figure. Then the French Jesuit priest Armand David returned to Paris with a monkey skin from the west Chinese-Tibetan border region. The fur was grayish black with golden yellow head hair and a golden underside. The illustrations on the vases and the fur were identical. When the French zoologist Alphonse Milne-Edwards was describing the snub-nosed monkey he must have remembered another picture from the sixteenth century which represented the famous Russian courtesan Roxellane. Since Roxellane had just as reddish-golden hair and a similar delightful turned-up nose, Milne-Edwards gave her name to the new Chinese monkey: *Rhinopithecus roxellanae.*

It is very difficult to observe the snub-nosed monkeys in the high timber and bamboo jungle. They apparently form large groups, and, according to Walker, feed on fruits, buds, and young bamboo shoots. Snub-nosed monkeys are remarkably insensitive to low temperatures. These monkeys have been seen in forests which were covered by a thick layer of snow. The long-haired dense fur protects these animals against the cold climate. During the winter the monkeys come down from the mountain forests into lower regions. These weather-hardened food specialists are very rare and difficult to approach and consequently they have not been seen in European or North American zoos. It is therefore very commendable that the large zoos in Shanghai and Peking were successful in keeping two groups of snub-nosed monkeys in recent times. The monkeys even have reproduced several times.

In the magazine *Das Tier* the Canadian zoo director Al Oeming wrote about his encounter with snub-nosed monkeys in Shanghai: "The first exciting experience which I had here was to see no less than ten snub-nosed monkeys which come from the cold high mountain regions of Szechuan in Southwest China. I saw four young which still clung to the breasts of their mothers. One very beautiful, large male already had its thick winter coat. They have a gleaming golden fur and pale violet faces with conspicuous snub noses. Everything possible is done to give the best of care to these rare animals. Inside and outside they are always kept behind protective glass walls, and they have high, sunny terraces in their cages. My Chinese companions told me that it had taken them years to overcome all the problems and to maintain the animals in good health and to get them to reproduce regularly. The tricky problems with respect to proper diet were also solved. I envisioned how these longhaired leaf monkeys would look in a large enclosure in my home in Alberta, Canada. They could be kept outside the year round, even at —36 degrees Centigrade."

The PIG-TAILED LANGUR (⊕ *Simias concolor)* is an intermediate form between the snub-nosed monkeys and the true proboscis monkeys. The HRL is approximately 55 cm and the TL is approximately 15 cm. The shape of the nose and the proportions of the limbs are similar to the snub-nosed monkeys. The structure of the skull is proboscis monkey-like. The tail is short and naked except for the hair tuft at the tip. The hairs of the crown point ventrally, and there are hair tufts on the ears. The fur is brown with paler whiskers. The face, hands, and feet are black. The distribution is limited to the islands of South Pagai, Sipora, and Siberut in the Mentawi group west of Sumatra.

The pig-tailed langur with its strong arms and body, and short downward curled tail is superficially reminiscent of a pig-tailed macaque. Yet, unlike the pig-tailed macaque, it is not a ground-dwelling form but is adapted to the thick, nearly impenetrable swampy forests of its native islands. Walker thinks it is possible that these monkeys also eat crabs and other small aquatic creatures, in addition to leaves and shoots. If this is true, their diet would greatly deviate from that of all other leaf monkeys.

The PROBOSCIS MONKEY (*Nasalis larvatus;* Color plates p. 446 and p. 455) is one of the strangest looking monkeys. It is large and robust; the HRL of the ♂♂ is up to 76 cm, while the ♀♀ are approximately 60 cm. The TL is 56–76 cm, and the body weight of the ♂♂ is 16–22.5 kg and of the ♀♀ is 7-11 kg. The nose of the ♂♂ is long and bulbous and is up to 10 cm long, but in the ♀♀ it is considerably smaller, although longer than in the genera *Rhinopithecus* and *Simias.*

The delightful snub nose which characterizes the snub-nosed monkeys is only present in the young and in young female proboscis monkeys. After the males become sexually mature, which is around seven years of age, the nose becomes monstrous and increases in size as the animal grows older. In old males the nose droops down over the mouth. When eating, they have to push aside their bulbous noses with their hands so that they can put food in their mouths. The Freiburg anatomist Wiedersheim studied the development of this peculiar nose and found that the embryological phase does not differ from that of man. The question then arises as to what purpose is fulfilled by such a nose. It probably functions as a voice amplifier. The males emit long, drawn-out calls which sound like "honk" or "kihonk." The American zoo director William T. Hornaday has compared the sound with that of a bass violin. The softer calls of the female, according to Walker, sound more like the contact calls of geese.

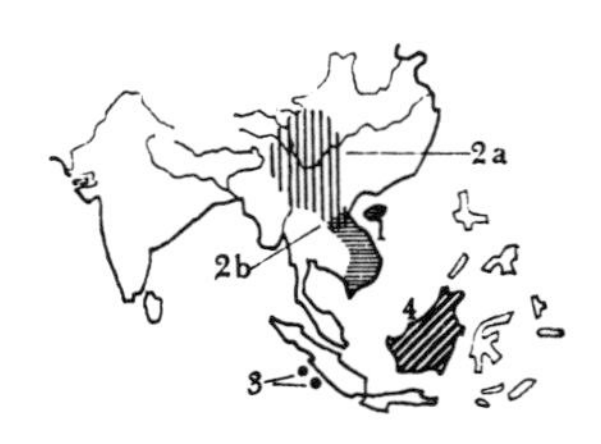

Fig. 18-3. 1. Douc Langur *(Pygathrix nemaeus;* p. 457) 2. Snub-nosed Monkeys (Genus *Rhinopithecus);* a. Snub-nosed Monkeys *(Rhinopithecus roxellanae);* and related forms: b. Tonkin Snub-nosed Monkey *(Rhinopithecus avunculus)* 3. Pig-tailed Langur *(Simias concolor)* 4. Proboscis Monkey *(Nasalis larvatus)*

Behavior of free living proboscis monkeys has scarcely been investigated. It has been observed that the groups move very quickly through the branches of the mangrove forests. William Beebe once saw a swimming proboscis monkey in the middle of the wide Rejang River. As the boat approached, the monkey dived underwater for twenty-eight seconds and then continued to swim vigorously. Other explorers have repeatedly described the swimming and diving capabilities of these monkeys at the mouths of rivers.

Proboscis monkeys primarily feed on buds and young leaves from tree species that are common in the swamp or on riverbanks, and according to older reports they also eat various swamp plants and small bitter forest fruits. It is therefore not surprising that until recently it had been impossible to keep these unique monkeys outside of their geographical range.

The first proboscis monkeys which were still alive when they reached their destination on another continent were shipped from Borneo to Egypt in 1899 and were destined for the Gizeh Zoo in Cairo. Only three of the five animals were alive when they reached the Suez Canal. One died during unloading procedures, and another one was near death by the time it reached the zoo. It was said that it "did not even revive under the influence of a warm fire and a shot of gin." The last survivor, a male, died one month after its arrival. The circus owner Wilhelm Hagenbeck, a brother to the famous animal merchant and zoo founder Carl Hagenbeck, kept a young proboscis monkey for a short period in 1901 which, however, suffered the same fate as the other monkeys. A year later a young male proboscis monkey stayed alive in the London Zoo for sixty-nine days. At the same time the

Calcutta Zoo in India acquired a proboscis monkey which lived for two and one half years. This was the longevity record for half a century.

Fig. 18-4. This is the manner in which a Proboscis Monkey jumps from the high branch of one tree into the lower ones of another. The movement sequence was taken from motion picture frames.

These failures were probably not only due to faulty feeding methods. In their native land proboscis monkeys are frequently kept as "domestic animals," and they most certainly are not fed with food that is difficult to obtain. At the beginning of this century, the zoo of Surabaja on Java very successfully kept several proboscis monkeys on a different leaf diet, and even on rice. Hill reported an almost unbelievable case in 1965: An English official had brought a young female proboscis monkey from North Borneo to Singapore. He kept it as a house pet and since he was not aware of the animal's dietary needs, he fed it with his own food. The unfortunate creature was fed with fruit, eggs, bread, barbecued chicken, lamb and pork chops, bacon, sausages, fried fish, and even chocolate. In addition, the monkey received insects and freshly killed birds. After four months the monkey still looked very healthy, and ate the unusual diet with great appetite. It is not known how long the animal survived but it is close to a miracle that it stayed alive at all with such demands on its digestive system.

The San Diego Zoo finally has had its first true success with keeping proboscis monkeys in captivity. In January of 1956 they received a young proboscis monkey pair named Cyrano and Roxanne. The female, Roxanne, died in November of that year but the male Cyrano lived until January 18, 1960. He died as a result of skull fractures obtained in a fall.

In general, leaf monkeys can easily sustain damage when they fall from great heights. Uta Hick reports on two proboscis monkeys which arrived at the Cologne Zoo in the middle of August of 1966: "For the first few minutes in their new home it would have been better to have had wire mesh on the ceiling as well. The two proboscis monkeys jumped up at the ceiling, tried to hold on, and then dropped onto the ground, luckily on a prepared layer of straw.

After Cyrano's accidental death, the San Diego Zoo acquired another pair of young proboscis monkeys in September, 1961. They were Pinocchio and Penelope. After their arrival these fastidious monkeys were fed with a mixture of fresh hibiscus blossoms, acacia leaves, and *Eugenia* berries. After several weeks they were also fed bananas, oranges, apples, yams, and lettuce. Later the monkeys also accepted bread containing vitamin supplements, boiled potatoes, corn on the cob, peanuts, celery, and a great variety of other vegetable matter. This careful readjustment met with success. In the fall of 1965 the first proboscis monkey was born outside the Southeast Asian islands in the San Diego Zoo.

Reproduction of Proboscis monkeys in zoos.

George H. Pournelle, the curator for mammals in San Diego, describes this happy occasion and the further development: "During the first weeks the face of the young was deep blue, very different from the flesh-colored faces of its parents. At around three months the bluish coloration gradually faded into dark slate-gray. Another distinguishing feature between parents and young was the nose which, in contrast to the large, drooping nose of the father, was small and turned up. In due course the young became increasingly more active and spent more and more time investigating all the corners of the enclosure, always under the watchful eyes of Penelope. Occasionally the baby got into trouble with its otherwise very patient father if it suddenly pinched him in the nose or jolted him unexpectedly out of his sleep. Such 'disrespectful' behavior resulted in an immediate stern reprimand from the father and sent the child scurrying off into the protective arms of its mother."

Proboscis monkeys in the zoo of Surabaja on Java had given birth to young on previous occasions. These successes encouraged the zoos of New York (the Bronx), Frankfurt, Duisburg, and Cologne to import proboscis monkeys. On March 8, 1965, two pairs of sub-adults born in the zoo of Surabaja arrived by plane in Frankfurt. They had bad colds and were heavily infected with intestinal parasites. The zoo people in Frankfurt were prepared for the worst or at least foresaw a difficult and busy period of readjustment.

Acclimatization to the Frankfurt Zoo

Yet, as so often happens in zoo keeping, the situation turned out differently than anticipated. The four monkeys ate practically all types of fruit and vegetables that were fed to the other leaf monkeys in Frankfurt right from the beginning. They behaved as if they had lived in the Frankfurt monkey house for years. Of course, attempts were made to offer a sufficient and diverse diet. When in season, fresh twigs and plentiful supplies of leaves were fed to all the leaf monkeys. Immediate measures had to be taken to treat the dangerous infection of intestinal worms. Within six months the parasites were brought under control by applying hygienic measures and giving suitable worm medicines. Although the scientific staff still enters the proboscis monkey enclosure with great apprehension, after a few months the caretaker said that he would like to see such monkeys more often because they were easy to keep, and that he has had far more difficult animals to look after.

In the meantime (1967) the animals have almost doubled in size. Their healthy appetite has remained. The first signs of sexual maturity became apparent, and several mounting attempts have already been observed. On November 1, 1966, the Frankfurt Zoo ventured to import two additional sub-adult females. Frankfurt now possesses six proboscis monkeys which is the largest group of these rare and conspicuous monkeys outside of Indonesia. Thus it would seem that the zoo-keeping problem of proboscis monkeys has been solved which had troubled zoo keepers for nearly seventy years. In contrast to earlier times, proboscis monkeys thrive in other zoos as well. The pair in Cologne, which also were born in Surabaja, is kept on a diet of leaves from the acacia, birch, acorn, willows, hazelnut bushes, roses, blackberries, mulberry-tree, other vegetative matter, and also mealworm larvae and grasshoppers. The proboscis monkeys, in particular, have demonstrated that it is possible to keep "unkeepable" species thanks to the developments of modern zookeeping and a certain amount of patience, persistence, and the necessary scientific knowledge. The studies conducted in zoos may be beneficial to the animals because the knowledge gained about their behavior and living conditions may ultimately be applied in case their native range is subjected to increased cultivation, and may help to save the species from extinction.

It is probable that the leaf monkeys evolved in Asia. Their African representatives, the colobus monkeys, probably migrated to their present range via South Africa, as did many antelopes. The African leaf monkeys are clearly distinguished from the Asiatic forms by their greatly reduced thumbs, frequently well-developed shoulder and flank manes, and a trend towards brachiation.

The Colobus monkeys

Distinguishing characteristics

The COLOBUS MONKEYS (Genus *Colobus*) have an HRL of 50–80 cm, and TL of 62–100 cm, and a body weight in the large ♂♂ of up to 12 kg. The stomach is enlarged and sacculated; the first and second compartments are fermentation chambers. The cheek pouches are only indicated. The body and limbs are stouter than in the Asiatic leaf monkeys. The thumb is hardly visible and is present only in the embryonic stage. Conspicuous hair adornments are often present, such as head crests, manes, and bushy "horsetails." The monkeys inhabit tropical rain and mountain forests in Africa. They are found from Senegal to Ethiopia and in the South down to Angola. There three very different forms, or rather groups of forms, with four species and many subspecies, are found:

A. GREEN and RED COLOBUS MONKEYS (considered by some authors as a separate subgenus *Procolobus*; Color plate p. 445 and p. 456) have conspicuous genital swellings, and no different juvenile coloration. 1. The GREEN COLOBUS MONKEY (*⍚ Colobus verus*) is the smallest and probably the most primitive species. The HRL is 50 cm and the TL is 65 cm; the head is small and the face is naked or sparsely covered by hair. Females have distinct genital swellings even before sexual maturity has been reached. This monkey inhabits various forest types from the swampy, tropical rain forest to the arid bush country. Newborn young are carried in the mouth (this is unusual for monkeys). 2. The RED COLOBUS MONKEY (*⍚ Colobus badius*) has an HRL up to 75 cm and a TL up to 95 cm. There are many (up to 19?) very diverse subspecies. The tail is covered by long or short hair and never by a "horsetail." It inhabits tropical rain forest and occasionally also dry forests. The habitat of these monkeys is constantly shrinking due to

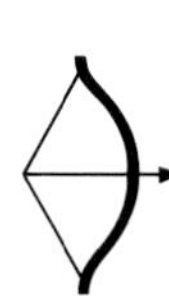

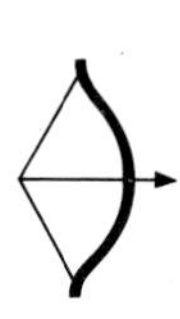

the pressures of human settlements and deforestation. Three East African subspecies are threatened as a result: a. ⊕ *Colobus badius gordonorum;* b. ⊕ *Colobus badius rufomitratus;* and c. KIRK'S COLOBUS MONKEY (⊕ *Colobus badius kirkii).*

B. BLACK-AND-WHITE COLOBUS MONKEYS or GUEREZAS (Color plate p. 440 and p. 456) do not have conspicuous genital swellings. Newborn young have white natal coloration. There are many subspecies with greatly different color designs and "fur adornments"; Ernst Schwarz considers them as one species *(Colobus polykomos).* In certain areas two subspecies occur side by side and have therefore been classified into at least two species: 1. The NORTHERN BLACK-AND-WHITE COLOBUS *(Colobus abyssinicus)* has heavy white fringes on the flanks; it inhabits mountain forests. 2. In the SOUTHERN BLACK-AND-WHITE COLOBUS *(Colobus polykomos)* the fingers on the flanks are not continuous; the subspecies are variously colored, e.g., the BLACK COLOBUS *(Colobus polykomos satanas)* is pure black.

They have no thumbs

The first zoologists who saw the guerezas were surprised by the two missing thumbs, and believed that some one must have cut them off. Therefore the monkeys received the generic name *Colobus* which means the "mutilated one." Later it became obvious that all colobus monkeys lacked thumbs. The reduction of the thumbs is related to the monkey's brachiating locomotion. The colobus monkeys are not true brachiators, as are the gibbons, which swing freely with their arms. Colobus monkeys usually extend their hands and feet frontally when landing after a jump. The Dresden Zoo director, Wolfgang Ullrich, has observed that guerezas occasionally engage in true brachiation which means that they swing only by their arms. Before jumping off, the monkeys often jump on the branches like on a trampoline. When jumping off, they extend their arms and legs but immediately pull them in again, and land in this position. At first the body is in a horizontal position but then the head and chest sink down and the arms are slightly bent. When dropping downward, the animals fall with stretched-out arms and legs. It looks like a "gliding flight." In this body position, the monkey lands on the branch which can be up to 15 meters below the jumping-off point.

Formerly guerezas were excessively hunted because of their beautiful furs. The mountain forms were particularly pursued because of their white flank fringes and "horsetails." Until World War I, "monkey fur" was very fashionable with European women. For this reason, guerezas were shot in masses, and in some regions were almost exterminated. In 1892 alone, 170,000 guereza furs were sold on the market. In the meantime the fickle fashion has concentrated on other fur animals and the guereza populations have had a chance to recuperate.

Guerezas in the wild are barely discernible in the tree crowns despite their conspicuous fur. Christian Scherpner from the Frankfurt Zoo saw

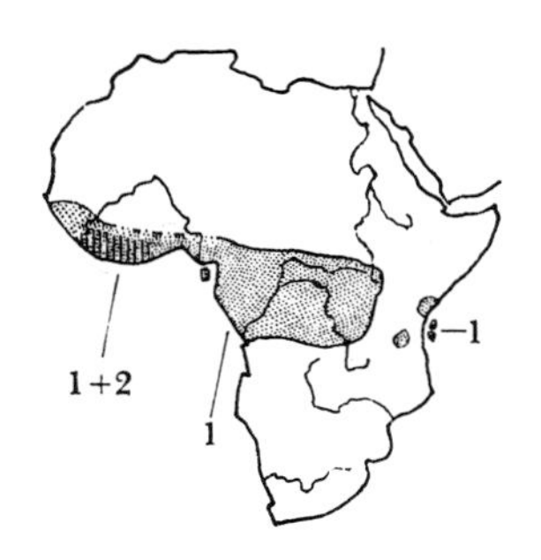

Fig. 18-5. 1. Red Colobus *(Colobus badius)* 2. Green Colobus *(Colobus verus)*

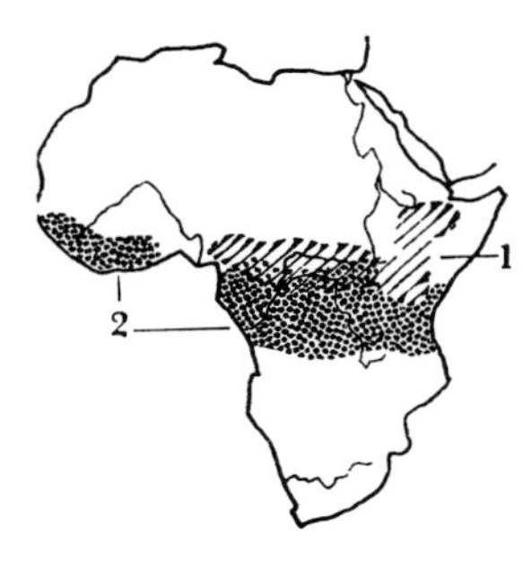

Fig. 18-6. 1. Northern Black and White Colobus *(Colobus abyssinicus)* 2. Southern Black and White Colobus *(Colobus polykomos)*

one troop of five guerezas and another of twelve animals in the Southern Sudan in 1955. The branches of one tree on which they had spent the night were chewed bare at approximately the 35-meter level. One warning call which Scherpner heard set the entire troop in flight. The guerezas jumped up, then dropped down into an oblique "gliding flight." A guereza troop of ten to twelve members seen by Bernhard and Michael Grzimek on Mount Hoyo in the Congo always moved along the same paths, which were approximately 40 meters up in the tree canopy, at the same time of day.

The Dresden Zoo director Wolfgang Ullrich also observed in guerezas this high degree of adherence to a place and to pathways. Ullrich intensively studies Kilimanjaro guerezas *(Colobus abyssinicus caudatus)* in the free living state. This subspecies is characterized by a particularly long white "horse tail," and occurs sporadically in the tropical and temperate rain forests on the eastern slope of the Kilimanjaro complex.

According to Ullrich, the home ranges of the individual troops are clearly defined. The size of a home range is influenced by the number of troop members, food supply, and the compatibility with other troops. Ullrich often heard their unique creaking and rattling vocalizations which served to mark their territory accoustically. Ullrich also tried to determine the significance of the brilliant black-white coloration. A possible function would be that it camouflages the monkey at rest by breaking up the animal's outline. When the animal is in motion the white flank fringes look as if the animals "wave" to each other, and thus they serve as a clearly visible signal. When two troops meet at the territorial boundaries the members of both groups jump up and down excitedly up to four minutes so that the white flank fringes wave. In this manner the guerezas indicate that foreign intruders are not permitted inside their home territory.

A true rank order does not exist within the group. However, a strong male acts as lead animal. He threatens enemies, and if necessary covers the retreat of the troop. If a guereza has lost sight of the group it emits a call like "orr-orr-orr" to which the other monkeys respond immediately. When the young emit squeaking sounds the mother comes at once. Ullrich differentiates three calls: the song, the warning call, and the mating call. As in many monkeys, mutual grooming is also found in the colobus monkeys. There is a certain hierarchy among the various species with which the guerezas share their biotope: the baboons are dominant, then the hornbills, followed by the guerezas and finally the white-throated guenons. However, the rank order between the guenons and guerezas is less well defined. The hierarchy among various animal species is known as the biological rank order in contrast to the dominance hierarchy within a group of one species.

The major predators of the guerezas, next to man, are only the leop-

Predators of the Guerezas

ards and primarily the large eagles. Generally, the flight distance in response to a threat is 20 to 30 meters, but it increases if the lead monkey is absent. In Ethiopia, Fiedler, Frick, and Starck observed that guerezas often hid instead of escaping. Starck was able to approach within ten meters of a pair of feeding guerezas. The observer stood on the crest of a small mountain and was at the same height as the two monkeys which were sitting in the crown of a tree which had the trunk leaning against the slope: "It was not difficult to observe these monkeys for approximately 15 minutes from this close distance. When I purposefully disturbed them they escaped down the stem of the tree."

The diet of the Guerezas

When feeding, the guerezas quietly sit on a branch and pull twigs towards them with their hands, and then finally rip off bunches of leaves and eat them. They also pluck off individual leaves and put them in their mouths with their hands. According to some reports, guerezas prefer the leaves and shoots of *Rauwolfia.* Fiedler, Starck, and Frick, on the other hand, observed that these monkeys are only found where sufficient yew trees *(Podocarpus gracillior)* occur. The Frankfurt scientists, when examining the stomach contents of some shot guerezas, were only able to detect finely chewed leaf pulp but no traces of fruit. The stomach was always well filled. The first stomach compartment contained a green mash while the contents of the next compartment were paste-like. Usually the monkeys stay on one or two feeding trees for several days. They prefer trees with young leaves. After they have eaten all the tender leaves, they change feeding places.

Reproduction and care of the young

During breeding, fighting rarely occurs. Males and estrous females withdraw from the troop and engage in a type of "communal marriage" which is of short duration. The various males copulate with the females in succession. A few days before parturition the pregnant female withdraws from the group. A male accompanies her. One day after the birth, at the latest, the female returns to the troop.

"The infant is carried on the abdomen," described Ullrich. "The head of the young lies on the chest of the mother. The little one clings with hands and feet onto the mother's fur at her sides. The tail is extended horizontally and is pushed between the legs of the mother. During flight the mother holds the young against her abdomen with one hand. However, after several days it can cling to the mother without help during flight or a jump. After one week the infant already starts to move about freely on its mother's lap. If, during the course of the next week, the young strays from its mother, she still holds it back and pulls it towards herself. I saw a five-week-old colobus young climbing around on other adult conspecifics who tolerated this. When doing this, the young made frog-like jumps." According to Ullrich the young learn most about their environment during the third and twentieth months of their lives.

It can occasionally be rather cold in the mountainous habitat of the

Guerezas also tolerate cooler temperatures

guerezas. Ullrich's observation area, which was located in the forests of Mount Meru and was at an elevation of 1500 meters, was characterized by a moist climate with extraordinary temperature fluctuations of 40°C at noon and 3°C at night. For this reason, these monkeys survive well in Central European zoos, and even on cold days they can spend a short period in the outside enclosures. In 1962 a troop of guerezas escaped from a monkey island in the Hannover Zoo. The animals roamed about in the trees, and fed extensively from the leaves and fruits. They even withstood the cool temperatures at night which were sometimes below freezing. Zoo Director Dittrich wrote of the incident: "The branches of our trees do not seem to have the necessary firmness and suitable strength to support the weight and the gait and leaping of guerezas. Therefore it is not recommended to keep guerezas on islands with high trees, as one does with gibbons or spider monkeys."

Keeping Guerezas in zoos

As other leaf monkeys, guerezas formerly had the reputation of being extremely difficult to keep in zoos. A group of the subspecies of the Northern black-and-white colobus *(Colobus abyssinicus kikuyuensis)* which has lived in the Frankfurt Zoo in 1956, received the following diet after an adjustment period of several months, according to Rosl Kirchshofer: leaves from willows, birches, roses, green twigs, and ever increasing amounts of germinating barley, green lettuce, and fruit. Additionally, the monkeys received a cake consisting of a pressed mixture of vegetative and animal proteins and carbohydrates, yeast, minerals, and vitamins. Gradually the guerezas adjusted to a diet that was more readily available and still predominantly consisted of raw, pure vegetative food, and they remained in good health.

On July 9, 1959, the first guereza young was born in Frankfurt, which is a first for Europe. The youngster had a typically white and slightly ruffed natal fur. It developed normally. Since then several other young have been born in Frankfurt. Approximately four weeks after birth the young gradually change color, but not until about three months of age does the adult coloration appear, except for the shoulder mane. "The mother carries her young for eight months, and thereafter mainly in case of danger," writes Rosl Kirchshofer. "At night the young sleeps near its mother, firmly holding on between her thighs. The father, too, takes an active interest in the infant. He frequently sniffed the young, and several times tried to snatch it from its mother. Later, when the young occasionally left the mother, it often sought out the father. The young snatched away his food, pulled his hair or slapped his face with both its hands. The father was extremely tolerant of these antics. He was never seen chasing his child, screaming at it, or mistreating it in any way."

Play in Guereza young

Young guerezas are very playful. During the first months of life they exercise their bodies by various "games." They seem to delight in running a little distance from their mothers and then jumping at them. They also explore the cage and the immediate environment of the mother. Rosl Kirchshofer was able to observe a peculiar escape-playing: "The young went away from its mother, showed a startle reaction for no apparent reason and fled back to her, but without crying. However, barely back at its mother's side, the youngster turned away, ran in the same direction, stopped, jerked, and quickly ran back to its mother. This manoeuver was repeated several times. The youngster never ran beyond a certain distance, and did not really need its mother's protection. As an onlooker, one got the impression that the youngster was enjoying this game."

Another "game" of guereza young seems to mean "startle" or "danger." "The youngster runs away from the mother, climbs up on the supporting bar of the sitting board, and then carefully and slowly pushes the head over the edge of the board and looks at its father. He has heard the climbing and unhurriedly turns around. Thereupon the young very quickly disappears down below and "escapes" to mother, immediately climbs up again, waits until the father looks up, "gets frightened," and again flees to its mother. This "game" can be repeated for approximately ten minutes. The intention of the young "to become frightened" was clearly recognizable. In reality the situation was totally harmless, for the father never harms his young even when they pull his hair or slap his face. The "fun" of the game is not only the running away, but begins with the seeking out of "danger," from which one naturally has to escape." Lorenz in starlings and Leyhausen in bats have observed that in the absence of the biologically appropriate stimulus object (for prey in their examples), entire behavior sequences run off as so-called vacuum activities. These examples seem to fit this category of behavior (Ed.).

Human children also take great pleasure in games which involve tension and fright. They even enter totally dark rooms in order to enjoy the "creepy" feeling. In many of our children's games, for example in "Who's afraid of the boogey man?", children treat something that is nonexistent as reality. Even little guerezas engage in such hallucinatory games. The father becomes the "boogey man" of whom they pretend to be afraid. It is a game which the little ones, anthropomorphically speaking, seems to greatly enjoy.

Towards the end of the seventh month, the small guereza also begins to play with the juveniles in its enclosure. It shows a behavior similar to that of human children: All of a sudden the parents are no longer good enough as playmates, and it seeks the companionship of younger group members. This is the first time the youngster leaves the family and joins groups of its own age. The "games" of the juveniles gradually develop into sham fights. At the beginning of such a fight, the partners stand erect, opposite each other with laterally lifted arms, opened mouth, and bared teeth. This is strongly reminiscent of true threat behavior as is demonstrated by adults at the beginning of a fight. The subsequent chasing "games" occasionally appear very dangerous but are almost always harmless. The wrestling and "cops and robbers" games of our children appear very much like "fighting behavior" of adults. In a way these games are a kind of dress rehearsal for real life. "One cannot help but draw on a comparison with human children whose games, independent of time, culture, or race, are similar, and which obviously develop in a specific sequence out of a genetic base," states Mrs. Kirchshofer.

Other zoos have also been successful in breeding these magnificent monkeys. Kilimanjaro guerezas have recently been born in Berlin, Dresden, and Cleveland (Ohio) among other zoos. Kikuyu guerezas have been born in Basel and San Diego. Guerezas of an unspecified subspecies have been born in Colorado. It is always very encouraging if monkeys which are threatened in their home ranges live and breed in zoos.

The Zanzibar Colobus is a threatened species

Unfortunately red colobus monkeys seem to be less adaptable and do not do well in zoos. This species is much more threatened by civilization, particularly in East Africa. The red colobus form on Zanzibar is greatly endangered. At most, there are only a few hundred of these animals left on this widely cultivated island. An attempt has been made to introduce a small group of Zanzibar colobus monkeys, which had been caught during logging operations, into Kenya. There these monkeys were to live in a protected area. However, this attempt to save this subspecies from extinction did not succeed. Only two males survived, and they were transported to the Frankfurt Zoo. A few females were also imported from Zanzibar with the hope of establishing a breeding population so as to preserve this subspecies, at least in zoos. Yet the experiences of the Frankfurt Zoo personnel indicate that red colobus monkeys are vastly more difficult to keep than other leaf monkeys. Unfortunately the animals could be kept alive for only two years. Additional attempts are impossible because of export restrictions. One can only hope that the relocation of the last Zanzibar colobus monkeys to a sanctuary on the mainland will meet with success.

Walter Fiedler
Herbert Wendt

19 Gibbons

Superfamily: *Hominoidea*

A hundred years ago when the German biologist Ernst Haeckel was reflecting on the origin of man, he supposed that our closest primate relatives might be the elegant, long-armed gibbons. As a matter of fact, the gibbons are the only monkeys which always move upright on two legs while either walking or running. In addition, the gibbon embryo looks surprisingly human. The elongated arms develop late in the embryo. Gibbon-like monkeys from the middle Tertiary, which were also found in Europe and Africa during that time, had less elongated arms and possibly still had tails. Certain features in their skull and skeletal structure led to the view held for a long time that such primitive "pre-gibbons" were the link between the lower Old World monkeys and the great apes.

Today's gibbons are not related to the human ancestral lineage. The extremely elongated "super arms" and the narrow, long-fingered hands with the deep cleft between index and almost non-opposable thumb indicate that the gibbons followed a different evolutionary path. They are highly specialized brachiators. They are the accomplished aerial acrobats among the primates.

Within the group of man and great apes (Superfamily *Hominoidea*) which includes the gibbons, great apes, and man, zoologists have classified the gibbons in a separate family.

Family: Gibbons, by H. Wendt

The GIBBONS (*Hylobatidae*; Color plates p. 473-474 and p. 476) have an HRL of 46-90 cm. They are tailless and the head is small and round. Sagittal crests are lacking. The snout is not protruding; the nostrils are more widely spaced and are more lateral than in other Old World monkeys; and they have small, short, non-protruding jaws. The brain is relatively small. The body form is slender. The chest cavity is short and broad. The vertebral column is not S-shaped as in the great apes and man. The fur is very dense and silky. The arms are greatly elongated, with the forearm and hand being the most elongated. The thumbs are small, and there is a deep cleft between the index finger

Distinguishing characteristics

and the thumb. The pelvis is long and narrow with externally everted ischial protuberances; there are distinct ischial callosities. The diet consists of vegetation and animal matter. The canines are strongly elongated. Sexual maturity begins at seven years. The gestation period is approximately 210 days. There is only one young which clings to the mother's abdomen for the first few months. The animal is found in Southeast Asia and the Malay archipelago.

Two genera are differentiated: 1. The SIAMANGS *(Symphalangus)* have an HRL of 75-90 cm. There is webbing of the second and third toe. The face is naked except for a few pale hairs, the forehead is low, and the nose is broad and flat. The nostrils are very large. The chin is reduced; a naked throat sac is present which is inflated during calling. There are two species. 2. The GIBBONS *(Hylobates)* have an HRL of 46-64 cm. The second and third fingers are not webbed. The face is often framed by a hair fringe. The forehead is flat, and the chin is more highly developed. The eyes are not as deeply sunk into the orbits. The monkey is smaller than the siamang but the arms are relatively longer. The throat sac is absent except in the black gibbon. There are five species.

Siamangs and Dwarf Siamangs

The SIAMANG *(Symphalangus syndactylus)* is the largest and most robust gibbon; the HRL is approximately 90 cm, and when the animal is standing it may attain a height of up to 100 cm. The arms measure up to 180 cm, and the body weight (♂♂) is up to 22.5 kg. The fur is long, soft, and shining, but is less dense than in the true gibbons. It is always black; only the eyebrows are reddish brown. Extremely long hair covers the scrotum. The animal inhabits the jungle and mountain forests up to elevations of 1500 meters. The DWARF SIAMANG *(Symphalangus klossi)* is decidedly smaller. The fur is particularly soft and silky. The species is rare and is only found on the southern Pakang islands. The status of the species is controversial; it may belong to the genus *Hylobates*.

Compared to the aerial acrobatic artistry of the smaller gibbons, the movements of the siamangs appear more sedate and a little clumsy. When running on the ground or on thick branches they move in a short gallop with arms extended laterally or lifted up. They are not quite as agile in climbing trees as are their fast cousins. Even Alfred Russel Wallace, who one hundred years ago was one of the first scientists to study gibbons in the wild, emphasized the fact that siamangs move much slower than other gibbons and rarely "fly" from tree to tree. In fact, the animals have no need for fast movements in the dense, rainy, and foggy forest of their native land where the tree crowns form an almost uninterrupted green canopy.

With their muscular arms they swing from branch to branch, one hand grasping in front of the other, and with a powerful vault they propel themselves to the next tree top. In flight this movement sequence accelerates so that more time is spent gliding through the air than dangling from the branches. If their escape route through the tree tops has been obstructed they usually climb to the highest point and pause before they dare to bound over to the next tree top over the larger distance. Yet when they are able to gather sufficient momentum by bounding on elastic branches, they freely jump into the air in the manner of flying acrobats, and unerringly grasp a hold with their hands on a tree several meters away.

The songs of Siamangs

Their loud and resonant songs have earned gibbons the reputation of being the "howler monkeys of the Old World." However, the songs of the siamangs by far excel those of all other species because of the sound-amplifying throat sac. At sunrise and sunset the valleys of the siamang's home range regularly boom with their bellowing and fading and rising sounds which drown out all other noises. The individual family groups thereby mark their home territories. In the early morning they howl before they move to the feeding trees. The group making a lesser noise usually gives precedence to the louder group without any fights occurring. In the evening they sing again at their sleeping places to indicate ownership. The song is also probably an expression of a feeling of well-being. The siamangs are silent when the weather is bad.

The zoologist Emil Selenka, who had intensively studied the embryology of the gibbons, together with his wife Lenore visited the mountain forests of southwestern Sumatra. "The morning concert of the siamangs increased to a jubilant shouting chorus," reported the scientist couple. "Every morning the troops filled forest and valley with their song. Usually the song only began around nine o'clock, as soon as the sun had broken through the regular morning fog cover. A few old males started the song by emitting very deep, bell-like sounds and then females and younger animals joined in with a resounding, high, jubilant shout which was followed by a deafening high laughter. This forest music, which is audible for the distance of an hour's walk, slowly fades into increasingly softer tones. These sounds accompanied us almost daily in the morning hours."

B. Schirmeyer was so impressed during her observations of the four siamangs in the Frankfurt Zoo by their "songs" that she paid special attention to them. The siamangs "sing" two to three times a day: in the morning shortly after awakening, at noon between eleven and twelve o'clock, and in the afternoon around three o'clock. Before "singing," they brachiate quickly and excitedly about in their cage. Their throat sacs are inflated and initially they utter only single calls. A general restlessness is noticeable. Bothy, the oldest in the ground, occupies the highest perch. All other animals (a younger male and two females) sit on the other perches in the room. The male carries on the song with dark, long drawn out and then shortened "ooo" calls. His

▷

Gibbons bound from springy branches of high trees and are able to "fly" in this manner over distances of up to 12 meters to the next tree.

▷▷

In this sequential picture one can see how a gibbon swings (brachiates) along a rope by its arms.

1
2
3
4
5

Gibbons:
1. Siamang *(Symphalangus syndactylus)*
2. Black gibbon *(Hylobates concolor)*
3. White-handed gibbon *(Hylobates lar)*
4. Grey gibbon *(Hylobates moloch)*
5. Hoolock gibbon *(Hylobates hoolock)*

partners then join with staccato-like, more high-pitched screams and accompany them by swinging to and fro. When "barking" they throw their heads far back or to the side, and the mouth opens abruptly. Short rest periods end by the old male's "inciting" calls. The others then follow with piercing guttural calls. When the vocalizations are piercing and guttural, then the greatest intensity of the "song" has been reached. The siamangs modulate the calls they produce by gently slapping their mouths with a cupped hand. After several such high points the strength of the vocalizations diminishes. They swing about less and the pauses become longer. This daily "performance" lasts twenty to twenty-five minutes.

In their native land, the siamangs are occasionally kept by the natives and settlers as semi-tame "domestic animals" but not as frequently as other gibbons. According to older reports these monkeys appeared quiet, depressed, and sad under human care. This was probably due to the fact that these creatures of the humid jungle and mountain forests were suffering from the heat, were poorly cared for, and usually could move around only on long chains. If they are kept in a semi-free state, as the Malaysians keep the smaller species of gibbons, they escape into the forest, never to be seen again. Tame dark-handed gibbons and gray gibbons, on the other hand, usually stay in the gardens and villages.

Until recently, siamangs were considered to be rare, fastidious, and sensitive animals by zoological gardens. Thanks to modern zookeeping methods and the nutritional sciences this situation has been changed in the last years. The Frankfurt Zoo has kept a nice family group of four siamangs since 1962. The American zoos of San Diego, San Francisco, and Milwaukee have even bred siamangs. One specimen in Washington lived for more than sixteen years. These black giant gibbons are reproducing very well in the zoo of Surabaja on Java where they live freely in the trees of several islands. In contrast to the smaller gibbons, siamangs are good swimmers. They use the breast stroke, and very easily can escape from the modern gibbon islands by swimming across the pond of water. In certain zoos, for example in Berlin, the siamangs live in outside enclosures which are isolated by a dry ditch.

True Gibbons

The smaller gibbons of the genus *Hylobates* appear much more elegant and graceful than the large siamangs. The gibbon, with its much longer arms and very narrow grasping hands where the small thumbs do not get hooked around a branch, seems to have been "born" for swinging through the tree crowns. They are probably the fastest of all the primates. Their mode of locomotion closely approaches flight. It is amazing how gracefully and elegantly these monkeys brachiate from branch to branch either horizontally or vertically. Their movements appear effortless. With the dare-devil assurance of a masterful circus acrobat they seem to barely hold on to the smallest twig or thinnest vine and then pull in the feet and suddenly sit quietly on the tree as if they had not been in motion at all.

Distinguishing characteristics

It is not easy to differentiate between the various species of the gibbons since the fur coloration differs greatly in the various home ranges and even in individuals of the same family group. The young are often of a totally different color than the adults. Occasionally the females retain the juvenile coloration for life, and others may change color several times. There are five species: 1. The BLACK GIBBON *(Hylobates concolor)* is characterized by a throat sac similar to the siamang. For this reason some scientists classify them in a separate genus or subgenus *(Nomascus).* There is a lateral crest of long hair on the crown in the ♂♂; there are smaller hair crests on the sides of the head of the ♀♀. There are three different color types, which may possibly be subspecies: (a) on Hainan Island and on the coast of North Vietnam the ♂♂ and the ♀♀ are usually uniformly black or dark brown; (b) in central North Vietnam the ♂♂ are black and the ♀♀ are dark to yellowish-brown with a black crown; (c) from western North Vietnam to Laos to eastern Thailand and South Vietnam, the ♂♂ are black with white cheeks, and 40% of the ♀♀ are pale yellow (the WHITE-CHEEKED GIBBON, *Hylobates concolor leucogenys*). 2. The HOOLOCK GIBBON *(Hylobates hoolock)* has very long and dense fur; the ischial callosities are covered by the long fur. The ♂♂ are always deep black, while the coloration of the ♀♀ varies; they may be black to brownish or gray. There is always a white forehead band. The last three species are similar; the fur is not as long and dense, the ischial callosities are clearly visible: 3. The WHITE-HANDED GIBBON *(Hylobates lar)* has black to pale brown or yellowish-gray fur; the face is black and naked and is framed by white hairs. The upper side of the hands and feet is always white. There are two subspecies: the eastern one (The CAPPED GIBBON, *Hylobates lar pileatus)* is characterized by a black crown and black spot on the chest. Some scientists classify them as a separate species. 4. The DARK-HANDED GIBBON *(Hylobates agilis)* has a dark upper surface on the hands and feet, but otherwise the animal varies from dark brown to pale yellow or is multicolored. The face is sometimes framed by light hair. Otherwise it is very similar. 5. The GRAY GIBBON *(Hylobates moloch)* has fur which is usually uniformly gray or silvery gray; the face is black.

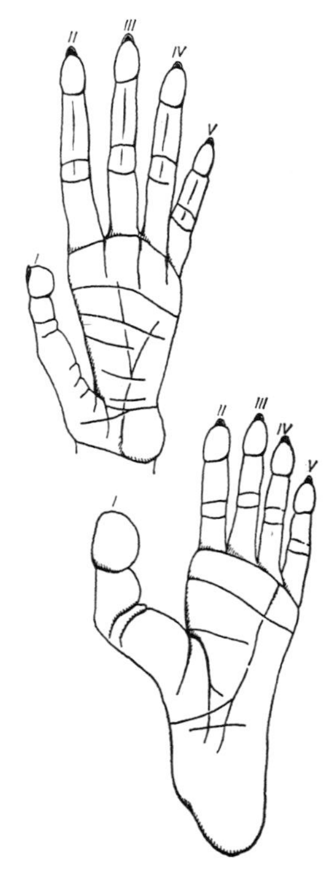

Fig. 19-1. Left hand (top) and left foot (bottom) of a Gibbon *(Hylobates).* Gibbons are the most skillful brachiators among the apes; their hands and feet are accordingly long and narrow

Just as in the tropical rain forests of South America and Africa, the trees in the forests of southeast Asia grow into immense giants. Many are supported by buttress roots, others by board roots. The tree crowns tower above each other, and the branches of one layer make contact with the branches of another layer of tree crowns. This results in several horizontal strata. Climbing plants push through the gaps towards the sunlight. These vines cling to the tree stems, and string from tree to tree or wind down in contortions like ropes. These lianas

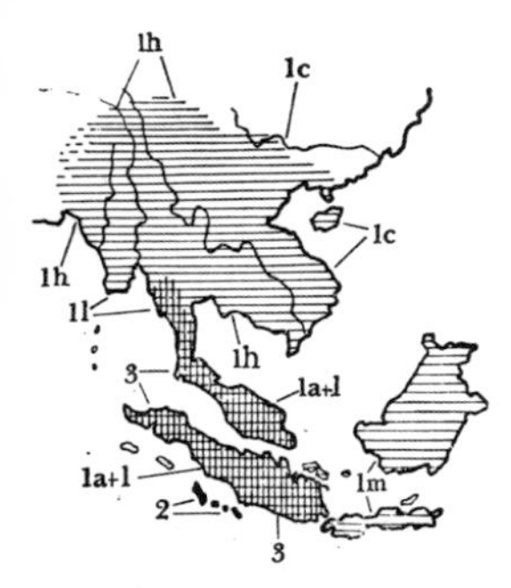

Fig. 19-2. 1. Gibbons (genus *Hylobates*): a. Dark-handed gibbon *(Hylobates agilis)* b. Black gibbon *(Hylobates concolor)* h. Hoolock gibbon *(Hylobates hoolock)* l. White-handed gibbon *(Hylobates lar)* m. Grey gibbon *(Hylobates moloch)*
2. Dwarf siamang *(Symphalangus klossi)*
3. Siamang *(Symphalangus syndactylus)*

seem to invite a primate to dangle from them, to set them in motion and then to swing into free space like an aerial acrobat. It is therefore characteristic of monkeys in southeast Asia as well as in South America and Africa that they have evolved into brachiators.

In the South American spider monkeys, the tail acts as a "fifth" or rather "first" hand in swinging through the trees. Brachiation is only rudimentary in the African colobus monkeys. The tailless gibbons of southeast Asia are dependent on their "superarms" for arboreal locomotion. The evolution of these elongated arms enables these monkeys to "fly" through the tree crowns as long as they want to and without a deliberate running start. In the zoo, one has the opportunity to observe these aerial feats that the animals perform on climbing ropes and other equipment. They are able to change direction even during the fastest bounding flight by slightly touching the branch. While swinging through the air, gibbons are able to catch fruits and other objects which one tosses to them. In the free living state they can even catch flying birds out of the air and eat them after having landed. The gibbon propels itself from the bouncing branches of high trees, and merely by swinging its arms it gains enough momentum to fly through the air up to twelve meters. The gibbon is often able to cross wide gaps in the forest, or over rivers in this manner, and it always lands securely in another group of trees.

Even their bipedal walk was more "designed" for running on horizontal branches rather than for walking on the ground. When they walk or run on the ground, they move in a waving manner. Like rope dancers, they hold up their arms to the sides in order to balance. It seems as if they were reaching a hold in the air. The long arms, which reach down to the ankle bones, are also used as "crutches." A gibbon in a hurry uses the back of the hand to push itself off the ground. In this way, it can move at high speeds, and a man is barely able to keep up with it. Young gibbons often slide along on the palms of their hands, which are pressed against the ground forward and out, while the legs are providing the forward thrust.

These arms, specialized for brachiating, are amazingly strong. A gibbon in the zoo can reach through the bars and pull an adult towards itself over about 30 cm on smooth ground. This equals an expenditure of energy which would enable us to move fifteen times the weight of our own bodies. The arms also protect the gibbon from falling from great heights. Gibbons are not very adept at falling. They would badly crash on the forest floor if they could not correct a mistaken grasp in the tangle of branches of the jungle giants by very quickly finding a hold with the other hand.

When drinking, they dip the back of the hand into the water, and then suck the drops from the hairs and knuckles. They also do this on the river banks since they do not descend to the ground. Rather, they hang from an overhanging branch and quickly dip their hands into the water. Here they can even less afford a fall than in the forest. A fall into the water would not only mean danger from crocodiles but also from drowning. Their dense fur would quickly absorb water. Unlike siamangs, gibbons would probably never attempt to swim on their own. In the Whipsnade Zoo near London, however, there are some white-handed gibbons which swim voluntarily and then return to their tree-covered islands. However, gibbons that fall into the water usually seem to drown, in captivity as well as in the wild.

Gibbons live in family groups which contain one male, one to two females, and several youngsters of various ages. In such a group, only one mature male and female is usually found; they exclude all competitors of the same sex from the group by biting them. However, old males, juveniles, and even younger females which have given birth already but are not "considered fully adult yet" by the leading female are tolerated within the family. Since free living gibbon females probably give birth only every second year and juveniles only reach sexual maturity at seven years, the family stays together for a long time. Such a troop occasionally consists of eight to fifteen members. Family members are very peaceful among themselves and are a closely knit group. They obviously "know" neighboring families which they often meet on the feeding places. Here one can see groups of more than thirty gibbons.

The American zoologist Clarence Ray Carpenter has spent a long time studying the behavior of the white-handed gibbon in the jungles of Thailand. According to Carpenter the family groups follow a strict daily rhythm. They wake up in their sleeping trees early in the morning, around six o'clock. For more than an hour they sing in long drawn-out howling sounds. The song differs in the various species. The song of the gray gibbon sounds like "Ooa-ooa-o-oooh" followed by a quavering click. The black gibbon starts with lower tones which rise to a high-pitched trill. The high trills are repeated in quick succession in three to six long, drawn-out sounds which Wolfgang Fischer described as "Gjoo gjoo gjoooo gjooooo." The hoolock gibbon alternates between two different pitches which sound like "hahoo-hahoo-hooha huhuhuhu hahahaha."

Each tribe marks its territory which is approximately 12–20 hectares large by its song. Since the territory of gibbons is three dimensional, the volume of the territory is more significant than its surface area. The various territories are separated by a strip of "no man's land." Here the various groups gather to perform their morning concert. In this "no man's land," various troops will also feed together on tree fruits, leaves, buds, and blossoms; they will gather tree ants and other insects, snails and small vertebrates, plunder bird nests and very expertly catch small birds. In his zoological park in Cleres in Normandy, Delacour was able to observe black gibbons which were living in the trees sucking out bird eggs: The monkey bit into an egg, lifted it up and let the content trickle into its mouth, and finally it very gingerly licked its soiled hands clean.

After the gibbons have finished their meal they wander back into their own territories. There the adults will take an afternoon nap, while the young and juveniles merrily play around and test their strength in wrestling and biting matches. When engaged in such games, they sharply gaze at each other, wrinkle their forehead and squint their eyes to narrow slits so as to protect them. Even more serious bites do not hurt the animal because of the denseness of the fur. In the afternoon around three o'clock the group again seeks out their feeding trees, and returns to their sleeping places with the onset of darkness. The animals always sleep in their own territory. Gibbons do not construct sleeping nests but show a preference for specific "sleeping trees" where no other family group is tolerated. Each family defends its "inherited" corner of the forest by song and threat displays against encroachment from neighboring groups.

Gibbons on cultivated lands

Cultivation of land in the homeland of the gibbons has opened up additional food sources for them. However, here they can also very easily fall victim to man, particularly since they do not enjoy any appreciable protection. Otherwise the major enemies of the gibbon are leopards, clouded leopards, a mustelid, large raptors, and giant snakes. Gibbons are extremely alert and even at the slightest creaking of a branch or the rustling of leaves they emit their barking warning calls and very quickly brachiate away. Thanks to their keen senses, agility, and adaptability to the occasional plantation, man's progressive settling of their habitat has not caused them as extensive a damage as is experienced by the great apes. However, the number of siamangs has greatly decreased during the last decades. Gray gibbons on the densely populated island of Java are now only limited to a few mountain forests on the slopes of the eastern volcanos.

Gibbons as house pets

The gibbon is also favored by the fact that in certain regions of its home range it is well liked. Natives will often sport captive gibbons like little children. However, during the capture of these gibbon youngsters, entire family groups are frequently chased into isolated groups of trees and the adults are killed. The captured young gibbons, on the other hand, enjoy extensive freedom as semi-tame "house pets." They swing around in the trees, appear at the houses for their meals, sleep in a forked branch of some tree, and usually do not stray too far from the villages.

A female gray gibbon, "Cherry," which a friend of mine kept on Java, impressed me with her gentle nature, affection, and impeccable manners. Cherry was able to move about freely in the garden where she searched for insects and in a very short time ate all the flowers and fruits of the Papaya trees. At mealtime she came into the house and sat down in a polite manner at the table. She ate cautiously without wasting any food. Using only the thumb or two fingers, but never the whole hand, she ate the same food as her human companions: meat, eggs, rice, and all different kinds of vegetables and fruits. She very carefully sucked on pieces of mandarin. When drinking, she dipped her knuckles into the cup and then licked off the fluid from the hairs. All these activities were performed in a "sophisticated" manner without any of the typical haste and greed demonstrated by many New and Old World monkeys. Even when several gibbons are feeding together, they generally do not fight over the best tid-bits. Two gibbons may even take a bite off of the same fruit. This sharing is otherwise only observed in the leaf monkeys.

Within a short time Cherry had befriended a boxer dog, and took over the dominant role in this unequal companionship. She cleaned and groomed the dog's fur, warned him against strangers, protected him, and was much more watchful than her companion. It is generally known that tame gibbons often befriend other animals. Wolfgang Fischer, the inspector of the Berlin Zoo Friedrichsfelde, described two young black gibbons that were "crazy" about golden hamsters. They carried the hamsters about, squeezed and licked them, yet never hurt the little rodents. Although tame gibbons can never be house broken, they are among the "cleanest" monkeys. They are constantly cleaning their fur, and very painstakingly they remove all dirt particles and crumbs from the hairs. Captive gibbons are also busy "groomers" and choose man as a partner in mutual grooming. Cherry was passionately fond of being combed. As soon as her human friends appeared with a comb, she was all set to be groomed. She would bend her back, expose the throat and with an almost machine-like rhythm, would first hold out one arm and then the other in order to have everything thoroughly combed. If one stopped combing, she would reach for the comb or impatiently drum with her hands on the floor.

They are very clean

The extreme affection that a young gibbon raised by humans displays towards people is probably based on the fact that gibbons do not live in large troops but in families. The pet gibbon regards the person that is looking after him as a substitute mother at first and later as a family companion. "This attachment to a human companion can last for the lifetime of the gibbon," writes Wolfgang Fischer, who is very familiar with the behavior of gibbons under human care. "That means that it persists even after the animal has reached sexual maturity. One can grow just as fond of a gibbon in the house as of a member of the family. The animal is not like a 'monkey' at all and in many characteristics is politely reserved like a well-brought-up person."

They require a lot of exercise

But these "born cuddle babies," as the zoologist Karl Vosseler once termed the gibbons, belong only in the hands of an experienced animal keeper. Quite rightly Wolfgang Fischer stated: "There are several factors which make the keeping of gibbons such a fascinating experience that one does not want to part with the animals—there is the lightness and elegance of their movements, their trusting and affectionate nature, and the unfathomable, mysterious look of their dark eyes. The first basic principle of keeping gibbons is the provision of enough space for sufficient exercise. For this reason it is not always very enjoyable to keep a gibbon in one's house. If a gibbon has upset or pulled something down it would be a great mistake to punish him because the monkey was merely satisfying his enormous urge for movement. A slap or smack on the face would only train the gibbon to do the opposite; hence, on a future occasion, it may show the sharpness of its teeth. Repetition of such disciplinary measures will only result in bites, and older gibbons possess considerable strength and respectable teeth. There is the additional factor: older gibbons kept in the house will only accept familiar persons and recognize them only as 'family members.' Strange visitors to the house will be attacked immediately."

Biting matches and rank order

Such attacks usually take the person completely by surprise. The gibbon will cross the room with playful-appearing bounds in a few seconds. It will cling to the stranger, grasp his hair, bite him, and then brachiate away as quickly as it came. The resultant wounds may be quite serious. However, gibbons are not aware that the human skin is far more easily injured than their own densely covered skin. The playful "biting match" (which can be quite painful to man) seems to determine the rank order of the maturing juveniles within the family group. Wolfgang Fischer observed this in the male black gibbon "Mohrle" who was cared for by his son Wilfried: "Since Mohrle had tried to bite during the first few days, his caretaker put on some heavy clothing to protect himself against Mohrle's sharp canine teeth. However, the gibbon continued to attempt to improve his social rank through severe biting into hands, arms, shoulders, and chest. Since his caretaker was well protected, Mohrle soon accepted Wilfried's higher rank. However, the arms, shoulders, and chest of Wilfried were green, yellow, and blue from the gibbon's many bites. If the gibbon had been able to attain superiority by his bites, we would never have been able to give him one whole summer of freedom."

Gibbons living freely within the zoo

Thus Wolfgang Fischer released two black gibbons which he had cared for in the Berlin Zoo of Friedrichsfelde for one whole summer. In this free state the gibbons ate leaves, flowers, dandelions (which they liked very much), and insects. They plundered nests and gradually drove away every song bird from the trees of their "territory." They even attacked guinea fowl, peacocks, and other large birds. Usually their attacks were unsuccessful. Only once "Mohrle" caught a guinea-fowl but immediately released it when it started to squawk.

One white-handed gibbon on an island in the zoo pond in Cologne not only captured sparrows and magpies but also a duck—"which was the reason he was put into the cage again, since we did not feel inclined to see our famous bird collection disappear as a dietary supplement for a gibbon," reported zoo director Windecker.

The gibbons can tolerate quite low temperatures because of their dense fur. In their native habitat gibbons are also subjected to frequent damp weather because of the constant downpours during the monsoon, and the foggy clouds which hang over the jungle in the winter. In the higher elevations it can get severely cold during the night. Since gibbons are not kept in overheated monkey houses as in former times, but rather in large brachiating cages or island-like outside enclosures surrounded by water, they seem to survive the winter astonishingly well in Central Europe. However, they require a dry, warm inside room in which they can retreat during the rain or severe cold. The gibbons on the island in the Hannover Zoo spent many hours outside even at −15°C. These monkeys can demonstrate their aerial feats in the trees on the monkey islands, or in special cages equipped with hanging ropes and other exercise apparatus. Here the animals can enjoy a proper "work out" which is essential for their well being.

Since 1960, the Berlin Zoo has kept a pair of white-handed gibbons on two islands in a larger pond. Zoo director Dr. Heinz George Klös described the situation thus: "The islands contain climbing trees which are connected with ropes for brachiating. Since the gibbons have left a large bush completely alone, the dead branches will in time be replaced by living trees. Depending on weather conditions, the gibbons are released on the islands either in the beginning or middle of May and are taken to the monkey house in November, when it freezes at night and the possibility exists that the animals might escape across the frozen pond. Until now, keeping gibbons on these islands has proven very successful. The gibbons are always in excellent condition and have a dense, shiny fur. For the major part of the day they are actively climbing and brachiating. They spend the night and the unfavorable hours during the day inside a barrel which was installed on top of one of the trees. Occasionally the gibbons have caught blackbirds, wild ducks, and other birds which they ripped to pieces and presumably partially ate. When the animals were caught for the winter, one occasionally jumped into the water. It was observed that the gibbons never made any swimming attempts, and always had to be fished quickly out of the water with a net."

Breeding gibbons in zoos

One hoolock gibbon in the zoo of Lucknow (India) reached the astonishingly old age of more than thirty-two years. White-handed and gray gibbons kept in the zoos of Philadelphia and New York held longevity records of twenty-two to thirty-one years. Nevertheless, gibbon births in zoos are still rare occurrences. The first gibbon to be born outside its Asiatic range happened in 1936 in a small zoo in Aarhus in Denmark. The first "American" gibbon was born four years later in Philadelphia. In the meantime, however, many gibbons have been born in the United States, including a cross breed between a capped gibbon and a dark-handed gibbon which was born in Washington in 1944. This famous gibbon, "Barbara," later bore several young sired by white-handed and capped gibbons. All her young, irrespective of sex, had a uniformly grayish fur which they retained for life. Los Angeles, San Francisco, Zürich, and Berlin also bred such hybrids.

The Zürich Zoo is particularly proud of its gibbon groups. The two families of white-handed gibbons which they kept did not tolerate each other. Their inside cages had to be divided so that they could not see each other anymore. Otherwise they were constantly fighting over their "territory" through the bars of their cages. Several gibbon young that were born there have since grown up and can now be seen in the large outside enclosure which has been equipped with bouncing bamboo trestles, all to the fancy of the gibbons.

In Germany the first gibbon was born in Berlin in 1961; however, the mother, a white-handed gibbon female named "Yellow," later gave birth to several other young. Since then several other German zoos have also been able to record this rare and happy occurrence.

Reproduction and care of the young

It is not very easy to get these favorite and much admired zoo dwellers to breed in captivity. Many modern zoos which have provided beautiful enclosures for their gibbons have not been successful up to now. Gibbons are self-willed "personalities" in their mating preferences as well. In the wild state, young gibbons which have reached sexual maturity meet members of their age group from other families at the territorial boundaries. They get acquainted and form new families. Two gibbons which meet in this manner and form a pair demonstrate a far more gentle and "cuddly" courtship than most other monkeys. According to Carpenter, the male embraces the female prior to copulation. During copulation, which appears rather acrobatic—partially standing up and partially hanging from a higher branch—the female may occasionally extend her arm towards the back and pull the male closer to her. The large great apes *(Pongidae)* are generally much rougher with their sexual partners.

The development of young gibbons

Newborn gibbon babies with their round, human-like baby heads, strongly appeal to our "baby schema." During the first months of life the young clings to its mother's abdomen night and day. The mother is only slightly handicapped while brachiating since the child holds on to her fur by itself. Holding on is of the utmost significance for a young gibbon, since even a brief disruption of this behavior pattern can result in a fatal fall, according to L. Ibscher. Mature gibbons have astonishingly many accidents. If a decayed branch breaks under the weight of the animal, even acrobatic agility will not save it. No less than 33% of the adult gibbons examined by Schultz had healed bone fractures, yet only one young had one. The young only hold on to the mother's hairs, which are three times the width of the newborn's hand at the sides and back of her body. "It is not infrequent that the security of the grip is increased by the child which intertwines the mother's hairs between its fingers, and not by simply encircling the hair with its hands," observed Ibscher. "In the first four months of life, all four limbs are never released from the mother's fur at the same time. One arm, at least, always stays in contact with the mother."

Brachiation is the first mode of locomotion for the young. At first this movement consists only of a hurried, hand over hand forward motion. The momentum of arm-swinging seen in the adults is not yet utilized by young gibbons. On the ground they move in short jumps at first. Walking is learned at a much later date. Around six months they start to walk on two legs with their knees bent. In the beginning the youngster is cautious and unsure, just like a human baby when it learns to walk. By seven to eight months the youngsters have mastered most locomotory patterns of the adults. However, at this age they are not too successful with the long distance jumps.

The games of young gibbons

The mother does not teach her youngster much, but to the little one, according to Monika Meyer-Holzapfel, she means "Home, a resting and sleeping place, cover in case of danger, the main source of food, and a playmate." Among the gibbons in the Zürich Zoo, it has been observed that the mother occasionally lies on her back and lets the young play on her. The mothers also play "catching," "biting," "pushing," and "teasing" with their youngsters. When mother and child were rolling on the floor in a tight bundle, the mother emitted "sounds that resembled real laughter." As a rule, gibbons are very playful in nature. Occasionally they dangle from two feet with the head down, perform somersaults on bamboo bars, or do backwards somersaults on the ground. Monika Meyer-Holzapfel observed once how a female gibbon placed a bell on her nose and juggled it for a moment; "this serves as a good example that above and beyond testing small objects for palatability, handling develops into true play."

Young gibbons always pretend to be "blind" when they play, just as young langurs do. They close their eyes, walk through their cages, bump themselves or fall down, laugh and continue without opening their eyes. This is quite reminiscent of the playing of blind man's buff by human children.

"Play culture" in juvenile groups

It is probable that the young do not learn the actual swinging and jumping techniques from the adults but learn it in play with other gibbon youngsters. Observations in the wild have revealed that young gibbons of various ages aggregate. A unique form of "play culture" develops in these open "juvenile groups" which contributes to mutual training and up-bringing. Young gibbons under human care seek contact with small children with whom they eagerly play. "Because of their small size, children are obviously regarded as 'socially equal,'" commented Wolfgang Fischer. "Children can handle young gibbons much better than we can. For example, they may tightly squeeze a gibbon young, or carry it about in a variety of postures."

In the free living state, older gibbon youngsters do just this with the younger ones, and the small ones learn from them. When one of the older gibbon children swings through the branches, the little one would like to imitate it. Thus, for the first few times it fumbles with its arms and misses the jumping distance before it attempts the first modest jumps through the air. In this way the youngster prepares itself in play, but soon thereafter the jump can be performed without undue preliminaries. Our own children also develop a similar "play culture" in the home, playground, and street. The younger children learn from the older ones how to make mud pies, play with marbles or balls, the various games of running and catching, skipping and jumping. Finally the little ones have learned how to control their body movements and how to master the various sports and playground equipment.

During play, young gibbons develop the same enthusiasm as small human children. Many times they might repeat the same jump, play catch, hide, or throw and catch the same object. All these activities are accompanied by a great variety of gestures and facial expressions just as in human children on the playground. Maybe this is the reason why gibbons appear so attractive to us.

Herbert Wendt

20 The Great Apes

Family: Anthropoid Apes, by D. Heinemann

"They are not people—but they are actually not animals either!" This is the impression that the Dutch ethologist Dr. Adriaan Kortlandt has gained from the chimpanzee. Dr. Kortlandt is one of the foremost authorities on free living chimpanzees. The great apes (Family *Pongidae*) are not only anatomically the most human-like of all animals, but they are also more closely related to man's mental capacities than any other animal. The English scientist A. Keith compared the anatomical characteristics of man with the various species of great apes and monkeys. Of 1065 individual characteristics the chimpanzees shared 369 with man, the gorilla 385, the orang-utan 359, the gibbon 117, and common monkeys, 113. Only 312 characteristics are exclusive to man.

Distinguishing characteristics

The largest primate besides man has a body weight of 45–290 kg. It has no tail. The arms are longer than the legs. The back slopes downwards when the animal stands on four legs. The muzzle is well developed in the adults. The fur is medium long to long, and is most developed on the shoulders and arms. The chest and abdomen are less densely covered by hair. The face, as in man, is only covered by fine hairs. Beard growth on the upper lip, chin, and cheeks occurs. As in man, the hair line on the upper arm is downwards and upwards on the lower arm. The hands and feet are adapted for grasping. The thumbs and first toes are opposable. The teeth are well developed and strong. The canines in the ♂♂ are long and pointed, which has resulted in a gap in front of the upper canine and behind the lower one. The jaws are well developed with powerful chewing muscles. Prominent sagittal crests are found in male orang-utans and gorillas. Cheek pads are present in ♂♂ orang-utans.

There are three genera with a total of four species: 1. The ORANG-UTANS *(Pongo)* have reddish-brown fur. They are found in Sumatra and Borneo. 2. The GORILLAS *(Gorilla)* have predominantly black fur. They are found in tropical Africa. 3. CHIMPANZEES *(Pan)* have black fur, and they are also found in tropical Africa.

Fossil remains of great apes have been found in the various strata from the Tertiary and Pleistocene. The phylogeny of the great apes is, however, very closely tied to that of man. This aspect is discussed in a subsequent chapter (Volume 11).

Body form and movements

The great apes living today are predominantly inhabitants of the tropical forests and the most common mode of locomotion is brachiation; that means that the animals hang more by the arms than stand on the legs. The hand is used like a hook in climbing. It is used less for grasping. For this reason the thumbs are small, and the remaining fingers are long and strong. However, only the orang-utan resembles this brachiator type in all aspects. Chimpanzees also occur in the savannah and spend considerable time on the ground just as gorillas. The massive gorilla males very rarely climb trees. The great apes not only utilize their hands as a hook, but can also use them for a variety of manipulations. Even in the other non-human primates, the hand is not solely used as a locomotory organ but is also used in grasping and manipulation. This ability is more highly developed in the great apes. Their small thumbs are not atrophied to any degree. The animals use them very skillfully and in the same manner as man. It is this particular feature which makes them appear so extremely human-like.

The cranium of the great apes is larger than those of any other monkeys, but it is smaller than man's. In mature gorillas and chimpanzees the flat forehead is emphasized by prominent supraorbital crests. The strongly developed teeth necessitate powerful protruding jaws and highly developed chewing muscles which require large areas for attachment on the cranium. These muscular attachments have constantly moved up higher on the skull until they met in the center of the crown. In the gorilla and orang-utan males, a sagittal crest developed which considerably increased the area of attachment for the temporal muscles. Some extinct proto hominids also had such bony crests.

The diet of the great apes

The great apes are more adapted to a vegetable diet than the baboons and other omnivorous monkeys. However, they do not dislike animal food. Some groups of chimpanzees very expertly hunt small vertebrates, even dwarf antelopes. On the whole there are great differences in the food habits of the great apes. The differences do not exist so much between individual species as between the various groups within a species. Great apes are less strictly tied to innate behavior patterns than most other animals. For this reason the great apes are able to adapt to the particular foods available in their territory, similar to our early human ancestors.

Our closest relatives

The shape, size, internal position, and even the microscopic structure of the internal organs of the great apes are very similar to those of man. One hundred and fifty years before Darwin, dissected chimpanzees were considered as "man" or very close relatives to man.

Once, when the anatomist Fick was opening an orang-utan cadaver he posed the question, what was missing in the internal structure of this *Homo satyrus* which would not make a *Homo sapiens.* Even the blood of the great apes greatly resembles ours. They have the same blood groups as does man, which, however, is also true for other species of monkeys. Until very recently it was impossible to distinguish chimpanzee blood from human blood, while on the other hand one can very well differentiate between deer and goat blood or cat and dog blood. Only in the last few years were refined methods developed by which one can determine the definite identity of the blood of the higher primates.

Blood analysis

On the basis of these blood analyses, we can describe the African great ape species as our "blood relatives" in the truest sense of the word. The protein differs in the various animal species, and the structure of the blood protein along with the similarities and differences in the anatomical structure and the innate behavior patterns serve as the most important indicators in determining the interrelationships of the various animal species. Jakob Schmitt of the Frankfurt Zoo was able to determine by modern methods, for example electrophoresis, that the blood serum protein of the gorilla and chimpanzee is more similar to that of man than to that of the orang-utan. There is practically no difference between the blood of the chimpanzee and the dwarf chimpanzee *(Pan paniscus)*; however, the serum protein of the dwarf chimpanzee is more closely related to man than to the chimpanzee. When trying to determine if one could transfuse human blood to the great apes, it was found that only the blood serum of the dwarf chimpanzee did not have antibodies against the human blood corpuscles. The dwarf chimpanzee, therefore, would best survive a transfusion of human blood of his own blood group. The other great apes would at best only tolerate a transfusion of serum without the blood corpuscles. This analysis only verified the view which had previously been assumed for a variety of other reasons: that the orang-utan branched off the main stem of the great apes and man at a far earlier time than the gorilla, chimpanzee, and man, which separated much later. These serological analyses conducted in Frankfurt also showed that chimpanzees and man are closely related and that the dwarf chimpanzees still possess several more human characteristics. These same similarities had previously been demonstrated on the basis of anatomical comparisons.

Sense organs and the brain

The sense organs and the sensory capacities of the great apes are essentially not different from man's. The basic capacities of the nervous system in the great apes and man do not differ. However, there are great differences in the highest brain functions. Although the brain of the great apes and man is very similar, and is also the largest animal brain in relation to body size (except for the much more simply constructed brain of some New World monkeys), in the mighty gorilla the brain size is only approximately 685 cm^3, while in modern man it is 1350–1500 cm^3. Our brain is much larger than that of the great apes. In relation to body weight, the brain of an adult man is relatively four times larger than that of the great apes. However, the skulls of many of our pre-hominid and early human ancestors only had a brain volume of 450–900 cm^3. The increase in the brain mass of man only occurred several hundred thousand years ago and applies especially to the parts of the cerebral cortex which are regarded as the seat of higher brain functions. These capacities alone justify us in elevating man above the entire animal kingdom despite his close relationship to the great apes.

The highest mental capacities

Aside from man, the great apes undoubtedly possess the highest mental abilities of all animals. They may not have a spoken language, but their mental abilities are of the same nature as man's. More than other animals, they have the capacity to learn from experience, to act with reason, to make and to utilize simple tools, and thereby to become independent of innate behavior patterns.

Our knowledge of the highest achievements and capacities of the animal brain has increased greatly during the past decades due to the thorough and comprehensive work of many scientists. As a result any attempt at an understanding of man must be made in the context of these findings. No responsible scientist would use this new knowledge to deny the unique position which man occupies in the world of living things. On the other hand, all that we know about the achievements of the brain in animals and man leads to the conclusion that not only the physical, but also the mental position of man, unique as it is, is the result of an historical development. Man must have passed through earlier stages in the development of his mental capacities such as we can still see as "models" in the behavior of presently living animals. Hence it is no accident that our nearest relatives, the anthropoid apes, are closer to us in the development of the cortex of the brain as well as in its functions than are any other animals. It is true that capuchins and baboons can show considerable achievements (see p. 330 and p. 408), but they are clearly inferior to those of the anthropoid apes.

From the wealth of studies that are available we can refer to only a few. Those experiments in which monkeys and apes were presented with pictures are especially revealing. "Vicky," a chimpanzee raised like a human child by the American psychologists Hayes and Hayes, held a picture of a wristwatch to her ear so as to hear its ticking. She recognized reduced and in part simplified black-and-white pictures of cars, pocket watches, binoculars, shoes, dogs, or roses in sixty-eight percent of the trials; she recognized color pictures for the same objects in ninety-five percent of the cases.

All of the more highly evolved animals can comprehend spatiotemporal relationships and can hence expect the occurrence of future events. Training that is based on reward or punishment depends on this ability of animals. "A true understanding of relationships is probably not associated with this capacity as is the case for mentally retarded people," says Bernhard Rensch. According to Rensch, the ability to comprehend relationships and a concomitant ability to foresee consequences are more highly developed in the monkeys, especially the anthropoid apes, than in predatory animals, porpoises, and other higher mammals. He says: "Already the spontaneous imitation of human actions presupposes that the perception of a sequence of events forms a conceptual entity which becomes linked with the concept of a self in the ape, which then enables the animal to perform a sequence of motor patterns in the sense of an appropriate sequence of actions. This is true, for example, for the smoking of cigarettes which was demonstrated to me in the zoological gardens of St. Louis and Duisburg. It is especially true for the many imitations about which C. Hayes (1952) reports: 'to wash herself, to brush the hair, to powder the face, to pound wooden nails with a hammer into place, to pick up a telephone receiver when a phone rang in the next room, or to dial various numbers and to call out 'booh' when a connection was made by chance and then to replace the receiver.'"

Anthropoid apes recognize pictures

Goal-directed behavior

Vicky could, as Rensch further reports, act appropriately in response to specific instructions: She would close the door when told, turn off the light, enter the bathroom, or give her foster parents specific toys. Once, when asked to get her toy dog, she failed to find it for five minutes. She looked into boxes, under the couch, and finally found it behind her bed.

"Understanding" words

When I was working in the Munich Animal Park at Hellabrunn in 1945 I told the adult female chimpanzee "Lissy" to bring me her mug from which she had just finished drinking. She understood at once that she was to bring me something, but she did not know what it was. One after another she brought me her blanket, various wooden blocks, and other toys as well as all objects that were in the cage. I kept repeating my demand, "Bring me the mug," and in the end she seemed to be quite frustrated since everything she brought me was the wrong thing. Finally she brought the cup to the bars and seemed quite relieved when I praised her lavishly. Later I discovered that the Hellabrunn chimps had been taught the word "cup" instead of "mug," and hence I had used the wrong word.

The large orang-utan male "Marius," who also lived in the Munich Animal Park used to take his time before he followed the order, "Marius, come out," which meant leaving his sleeping room and entering the dayroom via a remote-control door that was operated by the caretaker. We had to be very patient before he decided to come out. A blue macaw parrot who had never spoken had his place near Marius' cage and heard our daily calling for several weeks. One morning when the bird heard the opening of the door, he said, before the keeper had a chance to speak, "Marius, come out." This time Marius bolted out of the door at once and closed the sliding door behind himself. I had the impression that the orang-utan became frightened by the command coming from the bird. Perhaps this is too anthropomorphic an interpretation, but I would not rule out the possibility that the orang-utan was "surprised" about the talking parrot.

During an air raid the glass roof above the orang-utan enclosure was broken, and it took quite some time before the caretakers had collected all the glass splinters that were in the cage and on the wire roof. Marius kept bringing glass splinters out of hiding places and playing with them without ever hurting himself. We became quite concerned, since he preferred to keep them in his mouth and we were unable to entice him to give them up. Only when we brought his food were we able to trade with him. Marius used to be fed at noon with a spoon through the cage bars. However, before he was given the first spoonful of food, he had to hand over the pieces of glass. He understood at once what was wanted, and he passed one piece of glass after another through the bars with his lips, which he shaped into a point. However, after having completed his meal he brought forth yet another piece of glass from his mouth, showed it to me, but kept out of my reach. Hence he had been able to hide the piece of glass in his large mouth while I inspected it.

Apes trade objects with each other

Anthropoid apes also exchange things among each other. Of three half-grown chimpanzees which had survived World War II in the monkey house in Hellabrunn, "Michael," the strongest, had diarrhea. He swallowed his medicine without protest, as do most anthropoid apes, but he also had to eat a special diet, unlike his cage mates "Vroni" and "Susi," who received their usual head lettuce and fruit. For this reason he was separated from them for his dinner in a separate cage which had a fine wire mesh. The two healthy chimpanzees then received their food, while Michael received only dry, toasted bread. When I checked again after a few minutes, they had exchanged the food; Vroni and Susi ate toasted bread, and Michael ate the lettuce.

The observation that anthropoid apes exchange objects was elaborated on by two American scientists, J. B. Wolfe and J. T. Cowles, in a series of experiments designed to further the study of the mental capacities of the apes. They had taught their chimpanzees to use tokens of various colors in exchange for food from a dispenser. The chimps received two grapes for one blue chip, but for a white chip they received only one grape. They received nothing for a brass token. The animals learned very rapidly that the blue chips were the most valuable ones. When given the opportunity to choose, they took the blue chips and ignored the white and brass ones. When they were not hungry or when the dispenser was not available to them, they took the blue tokens anyway and saved them. One animal once collected thirty tokens before he cashed them in. Similar "value concepts" can also be formed by other primates, as was described for rhesus monkeys.

Curiosity behavior

An important part of higher brain functions in vertebrates is "curiosity behavior," which is often expressed by a tendency to explore. This is not only found in anthropoid apes, but is most highly developed in them. A. Wünschmann tested it in various species of varying stages of evolution. He found that carps showed very little curiosity behavior; in birds it is quite recognizable, and in rats it is still more developed. It is especially pronounced in chimpanzees. This is particularly important in light of the frequently made statement that only man is such a curious being and that he alone retains this essentially juvenile behavior into adulthood. Young animals and children seem to be more curious than adult species members, and man is perhaps the most curious of all animals. However, these differences seem to be more a matter of degree than of basic differences. Undoubtedly, the highly developed curiosity behavior of the most highly developed non-human primates was one of the most important prerequisites for the evolution of man. Even our passion for scientific discovery, which brought us much of the technological progress, has been inherited from our anthropoid ancestors which were also the ancestors of presently living apes. This curiosity behavior is based, according to Rensch, on the fact that "new stimuli have a stronger arousal effect than familiar stimulus situations." Carl Friedrich von Weizsäcker, the Nobel Laureate, comments on this very old, and in man especially strongly developed drive as follows: "There is this peculiar fascination with technology which brings us to the point where we think that we must carry out anything which is technically feasible. This is, in my opinion, not progressive but childish. It is the typical behavior of a first generation which explores all possibilities merely because they are new, in the manner of a small child or a young ape."

Playful handling of objects

The many forms of play and games, which are based on deep-seated drives in anthropoid apes, are, especially in early childhood, quite varied. Frequently they are pure experimentation. An unknown object is playfully examined for all kinds of possible uses, be it to take it apart into its basic components or for its edibility, or to try out all kinds of practical uses. Often the line between play and "serious" actions is crossed. In the animal park of Munich an adult female chimpanzee named "Marietta" had the opportunity to watch workmen for months at their work, and this stimulated her to similar activities. She tried to use a strip of sheet metal as a saw and worked on a particular place in her cage for hours. With her large incisors she cut off large chips of wood from her perches, cut them to size and placed them into the square openings of the key holes in the door to her cage. When these tools were square enough and of the right thickness she was able to open the doors with these "keys." When she was given a pencil and paper, she would lay down on her stomach, place the paper in front of her and make lines on the paper. One day Marietta escaped unnoticed from her cage. I found her in the caretaker's room sitting at the table; she had opened the drawer, taken out my notebook, opened it to a new page, and with a pencil she was diligently drawing her lines. Marietta always drew parallel lines, and when the paper slipped, the new lines crossed the old ones.

Comparative studies of a human child and a young chimpanzee

The first scientific report about an anthropoid ape who painted came from Dr. W. N. Kellog and his wife who, in 1932, raised a chimpanzee female named "Gua" and compared her mental development with that of their own son Donald. For this they used Gesell's tests, a number of carefully chosen trials, which were repeated monthly and with increasing complexity and which allow evaluations about normal or retarded development in children. In the fifth trial a "writing test" was included. Gua was twelve months old at that time and Donald was fourteen months. The Kellogs gave each a pencil and paper and watched what they did with them. They reported that Gua at once followed their instructions to draw once she had been shown the use of these objects. Donald did not have to be shown how to use the pencil, he began at once to scribble away.

One month later Gua also began to draw spontaneously. Two months later, however, the superiority of the human child was clearly evident: Donald was able to draw a straight line as he was shown, Gua continued to scribble.

The Russian investigator Nadja Kohts had begun similar experiments in 1913, but her report was not published until 1935, two years after the Kellog report. Mrs. Kohts reported about the drawings of the chimpanzee child "Joni" and her own son Rudi: "Although Joni scribbles away continuously with his pencil, he does not get beyond drawing lines that cross one another. Rudi, on the other hand, at the age of two to three years, has already drawn something which clearly resembles some objects in his environment."

D. Morris, a London zoologist, writes in his book, *The Biology of Art*, about these and all later attempts at painting by anthropoid apes. Morris calls attention to Mrs. Kohts' important discovery that the scribbles of a chimpanzee do change and improve to a certain degree, but that they do not reach the stage of representing basic figures which can be detected in the drawings of two to three year old children.

It was not until after World War II that a more systematic analysis of the ability to draw and paint in anthropoid apes was begun. Chimpanzees, orang-utans, and gorillas were given pencils, crayons, brushes, water, and oil paints, or they were allowed to finger-paint.

B. Rensch in Münster, West Germany, and Nadja Kohts in Moscow made similar experiments with capuchin monkeys. The most comprehensive experiments were carried out by Paul Schiller of the Yerkes Primate Laboratory in Florida, by Hermann Goja at the Vienna Zoo in Schönbrunn, and by Desmond Morris of the London Zoo.

Painting by the chimpanzee "Congo"

Morris used six chimpanzees and one orang-utan in his experiments in London. The star performer among his animals was the young male chimpanzee "Congo," who came to the zoo when he was about one year old after being caught in the wild, and who produced his first drawing at the age of one and a half years. Morris writes about the creation of this first drawing: "I held a pencil up in front of him. His curiosity made him approach this new object. I carefully placed his fingers around the pencil and led the tip over the paper, then I let go. His arm moved a little and then stopped. Congo stared at the paper. Something strange seemed to have come out of the tip of that thing—Congo's first line. It had moved a little and had again stopped. Would it work again? It did, and again and again. Congo stared at the paper and drew one line after another, and as I watched I noticed that all his lines converged toward one place on the paper where I could see a small spot. This could mean that even on his first attempt he not merely scribbled away, but that he had, like Alpha, some kind of control over what he was doing, primitive as it may be."

"From this moment on," Morris writes, "Congo produced one drawing after another. Even based on his earliest scribbling it soon became clear to me that his drawings would provide an excellent basis for analysis in systematic experiments." Morris did not, however, begin at once with specific tasks for his experimental subject. It seemed important to him to maintain as close a personal relationship between himself and the chimp and to achieve a good rapport. He writes: "Unfortunately this important preliminary step was not considered in earlier experiments with chimpanzees. The result was that test results were adversely affected by the temperament, personal idiosyncracies and disturbing influences in the immediate environment and hence lost their value. Such disruptions occur when these highly sensitive animals are treated as any other, less advanced experimental animal. One zoologist complained that he had tried for three days to carry out an experiment with a chimpanzee without success. If his experiments had been carried out over three years instead of three days, I would have been able to understand his problem." Altogether, Congo produced 384 drawings and paintings over a period of two years. Then sexual maturity began to emerge and his motivation to paint was supplanted by other concerns.

From the results of these many drawing and painting experiments involving anthropoid apes in England, the United States, Russia, Germany, and other countries, it became obvious that great differences

Style differences among paintings of chimpanzees

in style exist between individual animals. Unfortunately the testing procedures were not uniform; hence it is not possible to say with certainty whether the differences are due solely to the differences between the apes. For this reason, D. Morris began a test with six chimpanzees in 1959, all at the same time. "Each animal," he reports, "was tested individually and sent to a quiet place until all animals had had their turn. In this manner several test series were carried out over a period of time. It was soon possible to distinguish each individual animal clearly from all others based on its style of drawing. The youngest, a one year old female, could not be induced to participate. It was impossible to get her to place the pencil on the paper."

None of these chimpanzees had ever drawn before, and Morris assumed that he would have to show each one how to hold the pencil as he had done with Congo. "The first three chimps, 'Josie,' 'Beebee' and 'Charlie' had actually to be shown, but after they had drawn the first line they needed no further assistance. When the fourth chimp, a female named 'Fifi,' was given a pencil, she at once set to work. I assumed at once that this was an example of imitation. I turned around and saw the other animals cling to the wire mesh of their cage, one next to the other, at a place from which they could best observe the drawing which Fifi was making. I had been so concerned with watching the individual drawings being made that I had not noticed the unusual silence in the chimp enclosure. I was later told how the entire group of animals had watched breathlessly behind my back how the other chimps had made drawings. The animals had followed each movement with interest, as if their very life depended on it." Fifi was the leader of the group, and Morris thought at first that this was the reason for her spontaneous painting. But even the small, delicate "Jubi," the most independent of them all, at once began to draw, although she did not take the pencil out of Morris' hand.

They paint with great enthusiasm

Morris was impressed again and again by the unbelievable enthusiasm with which the chimps made drawings and painted. Congo always flew into a rage and screamed loudly when he was interrupted in his painting. Kortland reports the same about the female chimpanzee "Bella," and Lilo Hess had the same experience with her chimpanzee "Christine." The six chimpanzee youngsters tested by Morris were "completely fascinated, lost in an activity that was new to them and for which they could not expect a reward of the kind that was necessary to get them to perform one of the more standard intelligence tests. It appears as if painting has the same effect that it has on people. One is always amazed, perhaps unjustifiably so, that this activity has to be initiated by man and is not begun spontaneously by the apes themselves."

I would doubt, however, whether it is necessary in each case that man initiate this activity. The baby chimpanzees which I cared for over a period of half a year concerned themselves, not surprising in zoo animals, quite intensively with their feces. They spread soft feces with their fingers or lips over the walls in the same manner as the finger-painting chimpanzees "Christine," "Beth," and "Dr. Tom" did in the United States. In this case people certainly had not given them the idea. On the contrary we tried very hard to prevent this kind of activity. It is true, however, that in the final analysis man was the cause, since preoccupation with feces is surely an activity induced by the conditions of captivity.

All these experiments would be without further significance if these drawings and paintings by the apes were nothing more than aimless scribbling. However, the changes and improvements in the drawings, already discovered by Mrs. Kohts, and the differences in style point to the fact that there is more to the pictures produced by the apes than was thought at first.

Relationships between drawings and format

Paul Schiller was the first to notice that his experimental animal, the eighteen-year-old female chimpanzee "Alpha," concentrated her drawings in the center of the paper and not all over. She did not paint over the edge and used to first mark the corners of the paper before filling in the center. This gave Schiller the idea to give the animal some papers which already contained some simple figures, lines, and patterns. If a sheet contained a large figure or area, then Alpha almost always drew in the center of this figure and left the rest of the paper empty. If the test-sheet contained a small, filled-in pattern off to one side, or if the examples were large and non-symmetric, then Alpha drew her lines in such a way as to complement that which was given and to achieve some kind of balance. She completed a circle with a missing segment in two instances, and in four additional cases she covered the incomplete figure with lines, just as she did with completed figures. In one test series Alpha was given sheets which contained the outline of a triangle with a circle or some other figure inside. "In five out of seven trials," reports Morris, "Alpha marked not only the inside of these triangles as the experimenter had expected, but she was also inclined to draw a number of scribbled figures of her own near the sides of the triangles in a symmetric arrangement." This rather amazing feat clearly indicates the presence of a certain sense of symmetry and supports this idea in a manner different from the "balance tests" reported earlier. In both cases the animal selected areas for drawing whose position reveals an aesthetic organization, gross as it may be. The experimenter Schiller writes: "Within the series of trials the number of symmetrically distributed scribble figures is so large that Alpha's feeling for a balance of pictorial elements can be clearly demonstrated."

"Congo's" fan pattern

Congo also did not draw over the edge in a single one of his drawings in any real sense; he followed predominantly the format dictated by the size of the sheet. In twenty-five out of forty cases he covered the entire area with his scribbled figures, but in twenty of these cases there is a concentration of lines in the center of the sheet; in eight cases Congo drew only in the center of the paper. Congo's personal trademark was a fan-shaped pattern in which he would begin at the top of the page and pull the pencil toward himself. Congo also made fan-shaped patterns when he painted with paint or chalk. Similar fan-shaped figures were produced by Kortlandt's chimpanzee female Bella in Amsterdam and Rensch's capuchin "Pablo." Morris observed indications of this also in the six chimpanzees which were described earlier. In all other chimpanzees, orang-utans, and gorillas these fan-shaped patterns were not observed.

Marking and completion of given figures

In the extensive experiments with already given forms, carried out by Morris with Congo, this animal behaved not much differently than Alpha. He also marked existing figures, filled in empty spaces, and balanced unsymmetrical figures, but he completed an incomplete figure only once. The adult gorilla "Sophie" at the Rotterdam zoo, who was tested with the same procedure as Congo, behaved hardly any differently; however, she was not very concerned with balancing unsymmetric forms.

These same tests were also given to a two-year-old human child. His scribblings were much less balanced than those of the chimpanzees, but instead another trait characteristic of man became evident: The child mastered lines much better than Congo and Sophie. He was so engrossed in this invention that he neglected the details in favor of the total picture. "At times the child would place only a small figure on the paper and go on to the next variation found on the next free space. He only rarely paid any attention to the surroundings or the location of the figure," says Morris, and he emphasizes that this control over lines and their variation leads eventually to the representing schema that is characteristic of children, a level not reached by anthropoid apes.

However, even the further development of lines exists in anthropoid apes in at least rudimentary form. The first discovery of a spiral that he had drawn himself occupied Congo to such a degree "that he concentrated all of his attention upon this figure so that the position of the figure on the paper and its relationship to the open spaces on it became completely unimportant. Perhaps," Morris continues, "this is an explanation for the fact that many modern painters have returned to very simple elements in their pictures, especially when they have tried to develop new ideas with respect to composition."

The differences between shorter and longer test series are quite conspicuous. The basic behavior patterns are the same in both cases, but Congo and Alpha, who had been tested over long periods of time, only gradually developed the ability to produce balanced pictures.

Animals that were tested for only a short time, such as the gorilla Sophie and the six young chimps in London, never reached this stage. "If our experiments had been solely with short test runs we would undoubtedly have come to the conclusion that only man is able to produce balance in the composition of pictures," says Morris. He thinks that it is highly unlikely that later, more thorough, investigations will uncover still higher capacities in anthropoid apes, but he does not rule it out in principle.

The paintings of apes and its relationship to art

Of course the question must be asked whether this ability to draw and paint has anything to do with the origin of our own art. However, it has been shown that there is a relationship between the early stages of drawing in human children, before they begin to copy something, and similar performance in apes. Only if a person were to deny that there is any relationship between the early scribbling of children and the basis of art later on, would he be able to deny also a connection between the drawings and paintings of anthropoid apes and human art.

The fact that the aesthetic appreciation of man belongs to those inborn capacities which he inherited from his pre-human ancestors, and which he shares with his close anthropoid relatives, is an idea that is unacceptable to many people. The conclusion that our own art is rooted ultimately in this age-old capacity which we share with other animals, and that the first stages can be attained by the anthropoid apes, is considered by many nothing less than a scandal. However, we must ask whether the rejection of such a conclusion can also be supported by the facts. Is it not true that in other areas of art the same laws of harmony and rhythm are the basis of similar or related activities in animals? It seems that a sensitive, aesthetically aware person, who objects to the drawings of chimpanzees and its implications, should also be upset when he hears that the song of a nightingale follows the same laws of tonal harmony that are also a part of human music. Why should not the same yardstick apply to the paintings of chimpanzees as to the singing of the nightingale?

Prejudices against anthropoid apes

Here we are undoubtedly dealing with a deep-seated rejection of man against "the apes," a rejection whose biological basis is perhaps to be found in an earlier species barrier between the quite ape-like human ancestors and the still more ape-like ancestors of present-day anthropoid apes. This highly emotionally charged rejection of the idea of considering anthropoid apes as our nearest relatives unfortunately presents quite a barrier for our understanding of ourselves. Very few persons are able to overcome this biological inhibition and to recognize, against their own feelings, that an interpretation of human behavior patterns is impossible today without a knowledge and consideration of their phylogenetic roots. The anthropoid apes remain for most people rather "embarrassing" relatives.

Can anthropoid apes learn to talk?

One of the most important differences between man and all other animals is human language. However, the ability to speak is already present in animals. Otto Koehler and his students have investigated the pre-lingual abilities of birds and mammals for decades. We report on some of these experiments in Volume 8. However, even here the anthropoid apes (and perhaps the porpoises, see Volume 11), come closest to man in this ability. Chimpanzees, gorillas, and orang-utans neither have a special inclination to imitate sounds nor do they succeed in imitating the human voice as the parrots or other talking birds are able to do. However, while birds are merely able to imitate the sound of the human voice, and are at most able to reproduce the words in the appropriate situation, anthropoid apes are able to use words which they have learned appropriately. The young chimpanzee female "Vicky," who was raised by the Hayes, had learned to say three words; she was able to use two of these correctly, "Mama" for her human foster mother, and "cup" for the object as well as for drinking. As B. Rensch reports: "At breakfast Vicky would hold the empty cup to the coffee pot and say 'Mama' and then 'cup.' Another time Vicky opened a magazine, pointed to the picture of a cup, said 'cup,' and led her foster father to the kitchen where she wanted to drink."

Rensch continues: "We can conclude then that in many anthropoid apes prerequisites for the possible development of a material culture are given. They have a great capacity to learn, can form abstract concepts, can generalize, can recognize casual relationships and symbolic representations, and, because of their ability to manipulate tools, they can improve them or even invent them." In man, however, these abilities are infinitely more developed, and have become the biological basis which has elevated him to a special category within the realm of living things. Man alone has a language which permits him to express insights. Independent of individual memory, he can acquire and pass on experiences so that future generations can elaborate on the experiences of their ancestors. C. W. Corner once expressed it in these words: "If man is a monkey, then he is the only monkey which discusses what type of a monkey he is."

Anthropoid apes are in danger of becoming extinct

Wherever free living great apes have met with man, they frequently prefer the same food plants that man has cultivated. For this reason the population of great apes is threatened today wherever no protective measures have been taken. Great apes are also endangered by medical research which requires ever increasing numbers of these animals because they react to disease-inducing agents and medicines just as man does. Their reproductive rate, however, is too small to balance their losses.

As in most higher primates, the great apes do not have a specific breeding season. The males are always ready to copulate. The females usually are ready between two menstrual periods, at ovulation. Just as in human females, the menstrual cycle occurs approximately every four weeks. Menstruation does not occur during pregnancy, and usually not during the nursing period, which may last years. Depending on the species, the gestation period may last from eight to nine months. It is just as long as or slightly shorter than in man. The newborn great ape is just as helpless as a human infant. The former is able to cling to its mother's fur with hands and feet, but has to be supported with one hand in the beginning.

Readiness to mate

If there were no great apes today, many questions of our own phylogeny would have remained unanswered. It is therefore an obligation of mankind to protect and to preserve our closest relatives for the future. Perhaps the consensus of future generations may share the opinion of those who today consider the murder of these human-like creatures as inhuman.

Dietrich Heinemann

21 The Orang-utan

Distinguishing characteristics

Of all the great apes, the ORANG-UTAN (⊕*Pongo pygmaeus*; Color plates p. 510, 511, and 512) is the most distantly related to man. The HRL is approximately 1.25–1.5 meters (exceptions may reach 1.8 meters); the body weight of the ♀ is approximately 40 kg, and of the ♂ it is approximately 75–100 kg (obese zoo orang-utans are occasionally considerably heavier; "Jiggs" in Detroit produced three young despite a body weight of 188 kg; "Andy" in the Bronx Zoo, New York, weighed 150 kg when he accidentally died at the age of 19). The arm spread is approximately 2.25 meters. The supraorbitals are only slightly indicated which results in the very "human" appearance of the upper face.

The protuberance of the jaw region is very prominent. Most ♂♂ have large cheek flanges of fibrous tissue which are at the side of the face. These skin structures often develop after sexual maturity has already been reached. In mature ♂♂ the cheek flange may measure 10 cm in width and 20 cm in length. The fur is reddish, dark, or pale brown, and very long and dense, particularly on the shoulders and the arms. The shoulder hair measures 50 cm, and hair on the fingers 10 cm. Hair in front of the whirl points to the front. The ♂♂ have reddish blond whiskers and beards reminiscent of humans. The nails on the fingers and toes are strongly curved. Orangs of certain regions lack the nails on the first toes. The ♂♂ have sagittal crests for the attachment of the chewing musculature. The throat sac is extremely developed. The ♂♂ can take in several liters of air (the throat sac is equivalent to the "Morgagnitic pouches" in man which are usually quite small but can be very sizable in trumpeters, bass singers, and Muslim prayer callers). The dentition occasionally contains an extra molar (this also occurs in man). There are probably two different subspecies: 1. SUMATRAN ORANG-UTAN *(Pongo pygmaeus abeli)*; 2. BORNEAN ORANG-UTAN *(Pongo pygmaeus pygmaeus)*.

The orang-utan is the only true arboreal member of the great apes.

The Orang-utan, by B. Harrisson

It is highly adapted for an arboreal mode of life. Movement through the jungle forest is by brachiation and swinging. While moving, orangs have to reach out far with their arms, and in order to support the major part of the body weight, their arms have developed longer and more strongly than in the other great apes. Even the fingers of the orang are longer and the thumbs shorter than in the gorilla and chimpanzee. The grasping feet lend extra support on the lower branches. However, the legs of orang-utans are shorter and weaker than those of other apes.

Orang males have never been observed fighting over a female in the free living state; however, there are sufficient indications that such fights take place. Many males which have been caught as adults had injuries such as a torn nasal septum, deep bites, and scars on the back and neck, bitten-off fingers, and scars on the arms and upper lip.

Reproduction of Orang-utans

Pre-copulatory behavior begins with the "singing" of the males which is a quiet, vibrating hum, gradually increasing in loudness and becoming a deep howling. Then the song subsides into a low hum again. Finally the male moves closer to the female and excitedly starts to play with her. Both accompany this lively play with deep grunting sounds. During copulation itself, both are silent. Orangs usually copulate in a dangling body posture. The former zoo director of Dresden, Brandes, observed that occasionally the female lies on her back, and the male sits in front of her with his body bent back. To facilitate penetration of the penis, the male occasionally helps with his large toes. Brandes once observed how after copulation the female bent over to her partner and nestled her head on his thigh.

After a gestation period of almost eight months (255–275 days) a totally helpless infant is born which is carefully attended by its mother. The infant only weighs 1.1–1.5 kg. It is able to cling to its mother's fur but does require some support from her. Newborn human infants also possess this grip reflex but this is lost after a short period.

In the wild state, the nursing period probably lasts from three to four years. However, at an early date the mother feeds her child with food which she has thoroughly pre-chewed. She pushes this mash into the child's mouth with her lips. After the first year, the small orang-utan starts to eat solid food on its own. He gradually becomes independent by four years of age, and at this point he is also strong enough to hold his own in a group of juveniles and to look after himself.

Orangs have a slow growth rate. They are sexually mature at approximately ten years. Their life span in the free living state is probably thirty years. Many probably die at a much earlier age. Orang mothers do not conceive during the nursing period. Thus if the young are nursed for four years a free living female is only able to bear four to five young. If one also takes into consideration the infant mortality rate, which is 40% in the wild state, then the average number of young per female is two to three.

Fig. 21-1. Orang-utan *(Pongo pygmaeus).* Today the orang is only found in a few areas of this region.

As in so many other animals, orang-utans in zoos are sexually mature at an earlier age than in the wild. In the zoo the females bore their first young at eight years of age. The average birth weight was 1700 grams. By six months the infant has doubled its birth weight. At their first birthday, their average weight is seven kilograms. The physical development of an orang infant is hardly different from that of a human baby during its first year. Even the eruption of the milk teeth is similar. The orang infant is also very dependent on its mother's protection, and its behavior compares to that of the human baby. It also cries when it is uncomfortable and wants food. It requires warmth, food, and tender care from its mother, and only very slowly learns to control its visual sense and its movements. The only distinguishing feature from its human counterpart is the manner in which it clings to its mother's fur with hands and feet. The orang young is helpless without its mother because its arms and legs do not have their natural support yet. Orphaned infants have to be supplied with a substitute to which they can cling. Clinging onto something with its four limbs provides the infant with a stimulus for movement and climbing, for learning and healthy development.

The orang mother not only cuddles and nurses her baby but she also keeps it clean. She cleans its fur with her teeth and shortens its fingernails. She also washes the infant with rain water, and holds it off her body when it has to urinate or defecate. At a very early stage, the mother attempts to motivate her child to cling to branches and other objects on its own. In the beginning the young is only able to cling to its mother's fur. Climbing develops only very gradually.

Fig. 21-2. The orang "Jocki" is the only true arboreal inhabitant, as well as the most skillful among the great apes. This orang male of the Berlin Zoo hangs from a branch with his left hand and left foot.

All orang-utan infants separated from their mothers show their affection to the person that is feeding and taking care of them. They adopt the person as a mother substitute. Depending on the treatment such infants receive, they develop specific characteristics. If they are ill treated they become intimidated. If they are allowed total freedom, they become naughty and insolent. If they are kept in a small confinement without objects to stimulate their interest, they become dull and disinterested. All these conditions can lead to neuroses and can even result in death. Aside from mother-child contact, which is essential for the infant's survival, an orang child is also able to express affection, rejection, or indifference to specific individuals. Such an attachment is often developed at the first glance and becomes more marked the older the animal is when such new encounters take place. A young orang will quietly look a person over from head to foot, then look back at him directly and take his scent. The animal then touches the person and then smells the finger with which it touched, and finally it will bring its nose close to his arm or hand. Sometimes the orang may reject the person immediately on the basis of his appearance. In such a case the animal does not even bother to smell the person.

Orang-utans remain today only in a few areas of Sumatra and

Borneo. Even in these areas, however, they were found in larger numbers and in a wilder area in the recent past than they are now. In fairly recent geological times they also lived on the Asiatic mainland. At the time when the early ancestors of man, *Pithecanthropus* on Java and *Sinanthropus* in China (*Homo erectus;* see Volume 11) began to expand in East and Southeast Asia, there were many orang-utans. Their remains were found in stone age excavations from Peking to Celebes. Dubois, the discoverer of *Pithecanthropus,* found large numbers of orang-utan teeth from more recent times in the limestone caves in the highlands of Padan in the interior of Sumatra. In the last five years similar discoveries of teeth were made during excavations by members of the Museum of Sarawak in the Niah caves and other areas in western Borneo, although there are no more living orang-utans within three hundred kilometers of this region. In the large caves at Niah, during our excavations down to the strata of the Lower Stone Age, which represents an age of thirty-seven thousand years, we discovered repeatedly charred remains of orang-utan bones. These early remains show, as a rule, larger and more powerful sets of teeth than those which are known from historical times. From this we can infer that orang-utans have become smaller during the last few thousand years. On the Asiatic mainland there were even orangs the size of gorillas.

Orang-utan prehistoric fossils

In the caves of Niah the remains of young and female orang-utans predominate; this can be determined readily since the bones and teeth show differences according to sex. On the basis of other evidence as well it was already surmised that the inhabitants of the caves at Niah were hardly able to catch large animals which lived high in the trees. They were able to do this only, it seems, when they could separate a mother orang-utan and her young from the other animals, a procedure which even today is practiced with such devastating success by professional animal catchers.

There is also evidence in the floors of the caves of Niah which indicates that orang-utans were kept in the dwellings of these stone age people. The inhabitants of Asia have been familiar with orang-utans for thousands of years. It is most likely that they were killed by man and kept in captivity long before European explorers heard of them.

However, the early people, especially those of tropical Asia, have rarely destroyed their food base through excessive exploitation. This is because only a measured hunting and proper utilization of nature's gifts make it possible to survive in the jungle.

The history of their discovery

Today orang-utans are only found in a few regions of Sumatra and Borneo, although not too long ago orang-utans were also found on the Asiatic mainland. The word "orang-utan" originated from the Malayan language and means "forest man." Originally this was the term which the coastal Malayan tribes designated to the tribes in the interior of the Indonesian islands. Later this term was also applied to this large great ape, probably because many tribes which shared the region with great apes did not consider them as animals but as different looking "wild" people. The first European explorers gained a similar impression. A Dutch medical doctor, Jakob Bontius, in 1630 described that with "great astonishment" he had seen several orangs "walking upright on two legs." We do not know if Bontius had really seen orang-utans or members of an Indonesian jungle tribe. Bontius had added a drawing to his report of a "female" which "seemed to possess modesty and which covered herself with a hand in the presence of unknown men, which cried and groaned and could do all other actions known of man except speak." This sounds almost as if the Dutch doctor had not seen a great ape but a deformed or even a feeble-minded native. Nevertheless, the doctor mentioned the widely held belief which referred to the orangs. "The Javanese maintained that these animals could speak but did not do so because they did not want to be forced to work."

A European, the Dutchman Nicholas Tulp, was the first to have used the name orang-utan, in 1641. However, he applied it to a chimpanzee that came from Angola. The Englishman Edward Typson, a London physician and a member of the Royal Asiatic Society, also applied the name orang-utan to a chimpanzee. He had purchased a dead chimpanzee baby which had died on the trip from Angola to London. In 1712 the English captain Daniel Beeckman arrived in South Borneo. In his travel report he writes about orang-utans: "There are many kinds of monkeys, apes, and baboons which all diverge greatly from one another in form. The most remarkable among them are the orang-utans. They are up to six feet tall, walk upright on their feet, have longer arms than man, and possess a tolerable face in appearance. They seemed more attractive to me than many a Hottentot that I have seen. Furthermore, they have large teeth, no tails, and have hair only in those places where man also has hair. They move easily and have tremendous strength. They throw stones, sticks, and pieces of wood at people when they feel that they are being attacked. The natives are firmly convinced that they were once men, but that they were transformed into animals because they had mocked God."

The first Orang-utans in Europe

During the second half of the eighteenth century, several orang-utans which were mainly from Sumatra reached Holland and other European countries. One orang female from Borneo was kept in the private zoo of the Prince of Orange in Het Loo in 1776. The animal did not live long in captivity, and after its death it was dissected by the Dutch anatomist Peter Camper. However, it took another eighty years before the English evolution theorist Alfred Russel Wallace and other travelers to Indonesia finally observed living orangs in the wild state. The controversy about evolutionary theory increased the public's interest in all of the great apes. Consequently, towards the end of the nineteenth century and the beginning of the twentieth century, more and more orang-utans reached European and later also North American zoos. However, the necessary experience in keeping these great apes probably was still lacking. In those times, many orangs died in zoos, many by contracting human diseases and, in particular, tuberculosis. Usually infants whose mothers had been shot were imported. In 1928 the first infants were born in captivity, and this occurred almost at the same time in the zoos in Berlin, Nürnberg, and Philadelphia. All three babies died soon after birth because their mothers had too little milk which was also vitamin deficient because of a faulty diet. However, with time, feeding and care of these great apes improved. In the following years more orang infants were born in the zoos of Düsseldorf, Moscow, Dresden, Rome, St. Louis, and Havana. Some of these young were raised with the bottle. The first orang to grow up in a zoo had been born at an earlier period, in 1927. The mother, "Suma," was pregnant when she was caught, and during the transport through the Red Sea she gave birth to her young. The new home for this family was the Dresden Zoo. Since the mammary glands of the mother seemed small, the zoo director of Dresden, Brandes, was debating whether the young should be fed by bottle. Suma, however, proved to be an exemplary mother for she nursed her child, "Buschi," for more than six and one half years. Both animals developed well. There were no indications of rickets or undernourishment, and this was probably due to the fact that Professor Brandes was the first zoo director to have the idea of feeding a daily supply of fresh leaves and branches to these sensitive great apes. Orangs require a very diverse diet. They primarily feed on various types of fruit, leaves, buds, and young shoots, insects, bird eggs, lizards and other animal matter. They often debark trees or chew on rotten wood and mushrooms.

▷
The young orang "Jocki" of the Berlin Zoo already has an impressive beard. His jowls, however, are not fully developed as yet.

▷▷
The orang-utan female "Eva" which Mrs. Barbara Harrisson raised in her home on Borneo and of which she writes in this chapter. Today Eva lives in the Berlin Zoo. This photo shows her as a "teenager."

▷▷▷
A baby orang-utan in the Frankfurt Zoo. Breeding these threatened great apes has not been met with consistent success. Thus, their preservation in zoos is not assured.

Much to their distress, many zoo directors discovered that not every orang male will court any orang female, and vice versa. Thus the birth of an orang-utan in the zoo still remains an extraordinary event. Certain orang females regularly feed their men piece by piece. Suma often refused the most delicious tidbits that Brandes tried to push into her mouth. Only after he had finished his meal did she eat herself. Yet the converse occurred in the Hellabrunn Zoo in Munich. The small frail orang female, Juliette, did not give a single bite of food to her powerful mate, "Marius," after they had been separated for a long period of time. She not only very energetically demanded her share but regularly demanded the morsel that Marius was about to eat. In this respect, orangs are also very independent creatures.

Exploitation of the Orang-utan

Many adult zoo orangs, including those in the Dresden Zoo, were captured with the aid of new capture methods. The Dutch animal catcher Van Goens encircled a large forest region with a group of native helpers. Then he had all the trees cut except a few to which the fright-

Great Apes: Orang-utan *(Pongo pygmaeus)*; the male is in the foreground, the mother and young are behind him.

ened orangs had retreated. Thus he had robbed the animals of their escape route through the tree tops. Finally the animals were driven out of their hide-out by hunger. At this point entire families of males, females, and young were trapped in the nets put up by the catchers, and there the former were overpowered. Many orang-utans were killed in these trapping expeditions, or they died miserably during the transport. Nevertheless, between 1927 and 1928 three transports with a total of 102 Sumatran orangs reached Europe. The German animal dealer Ruhe then resold the animals to zoos, circuses, and private persons. One American circus alone bought 33 orangs.

Finally the Dutch Colonial Administration put an end to this exploitation by issuing an order against trapping and exporting these animals. However, the illegal shooting of orang mothers for the illicit trade with their young has continued in Sumatra and South Borneo despite the ban. The new state, Indonesia, took over the Dutch protective laws but is unable to enforce them because of a lack of suitable personnel. In order to discourage the capture of young orangs by killing the entire family, associations of scientifically managed zoos in Germany, England, the United States, and Japan, as well as the International Association for Zoo Directors, have decided not to import orangs without the export permission of the original country. Unfortunately, there are still enough "fanciers" who continue to buy these illegally captured and smuggled orangs.

The raising of orphaned Orang babies

The story of my orang-utans began in Sarawak when my husband brought me an orang baby on Christmas day of 1956. Forestry officials had delivered the young to the Museum of Sarawak. The provincial forestry superintendent of Sarawak was greatly interested in conserving the last orang populations. Whenever he heard of an illegally kept orang in the hands of natives, Europeans, or Chinese he would confiscate the animal in accordance with the game preservation act and bring it to us for raising. This way he saved the babies from almost certain death because someone who does not have the necessary knowledge and experience cannot keep small orangs alive. Thus we got Bob, our first orang baby, who was soon followed by Eva who came from the back country of Lundu where orang-utans still occur in the jungle. After the mother of Eva had been shot, the infant had been kept in a Dayak house for several weeks. Then she was sold to a Chinese dealer near the coast who kept the small orang chained like a dog beneath his house. The animal was fed bananas and biscuits. When this story circulated, the forestry authorities confiscated the animal and brought it to us.

The Orang children Bob and Eva

Eva was much smaller than Bob and very thin. Her neck was sore and infected from the rubbing chain. Although she had her full set of milk teeth, she weighed three kilograms. At first she refused all food except bananas. Bidai, our young Dayak assistant, made her a little nest inside a cage into which she retreated at once. The next morning I forced milk into her mouth with a glass pipette. After a fruitless attempt to escape, she did take the milk, but hesitatingly at first. Later I fed her with a baby bottle out of which she soon learned to drink properly.

Every afternoon Bob and Eva were allowed to roam around in the garden for several hours. As soon as Bob was released from his cage, he would run wild and somersault on the lawn. Usually he moved on all fours. He supported himself with his fists, closely spaced together underneath his body, on the ground; his feet were slightly spread apart. We had to teach Eva how to walk. Bidai or I would put her on the lawn and move away from her for approximately three meters. At first she cried dreadfully because she felt deserted, but then she would crawl towards us, still screaming. When we slowly moved on, she was totally engrossed with the problem of locomotion and the correct use of her limbs that she forgot to cry. After a period of time she was able to walk quite well, and now started independent exploratory trips. Nevertheless, she never moved off too far, and if she lost sight of us she would immediately start to cry. During her first months with us, Eva was spoiled like a small child. Bidai and I cuddled her a lot so that she would feel secure with us. Gradually Eva developed into a healthy, active orang child. In the meantime Bob had grown up and we decided to give him to the zoo in San Diego, California. However, before we had made all the arrangements for his transport we received three new male orang babies. Now it was high time to provide more space. Later, we sent Eva to Germany where she found a new home in the Berlin Zoo (see color plate, p. 510). Eva's successors finally traveled to Amsterdam and Hamburg. "Nigel," whom we received as a two-year-old orphan in 1958, became a father for the first time in 1966 in the Hagenbeck Zoo.

Under my care Nigel was very adventurous and independent, and I felt very sorry to condemn him to a life within a cage. Orangs do adapt quite well to an existence behind bars, provided they have been cared for well and have enjoyed an upbringing which was suited to their personalities. The readjustment works best if the animal is brought to the zoo within the first three years of life. Nevertheless, an existence within the vastness of the jungle provides by far a richer opportunity for developing the natural aptitudes of the animal; however, the juvenile has to be with its mother and within the family group for several years of learning.

A young Orang has much to learn

It is wrong to assume that the young orang-utans act mainly on the basis of "instincts." In my opinion, the opposite is true. Most things are learned by seeing and doing. A newborn orang-utan is totally dependent on its mother which feeds and protects it, and keeps it warm. Later the orang mother supervises the first climbing attempts of her child, and one gains the impression that she encourages and properly instructs the child. Infants which have been raised on the ground by humans, for instance, do not climb up trees on their own. Some of them are even afraid of any high places. During the first four to five years of life the orang young learn mainly from their mother; however, by approximately eighteen months the little ones learn quite a bit from their playmates as well.

The director of the Düsseldorf, Germany, Zoo, Aulmann, observed how a mother orang-utan tried to teach her baby to climb: "To this date, at the age of three months, the young baby did not undertake independent excursions, although the mother tried to teach it to climb and hold on to the cage bars. She does this by holding the baby with one hand around its torso and then bringing its hands and feet into contact with the cage bars, which the baby even today merely touches in a clumsy way without actually grasping. Thus the young at three months of age is unable to grip an object other than the hair of its mother. The mother tries in other ways as well to teach her young to move independently. She places it on its stomach on the ground and watches it from a branch with seeming interest while the baby tries to move about, screaming all the while. When the little one fails to get anywhere, she climbs down and gives it one of her fingers, which it holds on to. Then she pulls it across the floor carefully."

Similar observations were also made by Brandes in the Dresden Zoo. When the orang baby "Buschi" was two months old, the mother began to feed it with pre-chewed food. When it was three and a half months old the mother placed it on the floor by itself. At this age the young was able to hold on to branches, but its legs were still quite weak.

Training for cleanliness in the zoo

Training the young to be clean presents a problem for the mother, at least in zoos. It is only with great difficulty that an orang-utan can be taught to go to one particular place for urination and defecation. Its innate behavior is, of course, adapted to living in trees. Their excrements simply drop to the ground and are not seen, so odor is not a problem. Still, orangs try to keep their sleeping places clean. The male orang-utan Marius at the Munich Animal Park in Hellabrunn preferred to use an old helmet as its toilet, as Dietrich Heinemann observed in 1945. He either sat on or held the helmet under himself, and urinated into it. Then he carefully climbed down to the ground with his helmet without spilling anything, waddled to the cage bars, and emptied the helmet into the drain. Marius did not always do this; as a rule he simply urinated or defecated and it fell to the ground and he paid no further attention to it. On occasion Marius swept the perches with the long hairs of his underarm, sweeping food remains and feces to the ground. Then he cleaned the rest of his cage in the same way and swept everything through the cage bars to the ground outside.

Sometimes he was so thorough in doing this that the keepers did not need to sweep the cage. The orang baby "Ossy" which I hand raised never soiled its bed. Not until it was picked up did it show the need to empty its bladder and defecate. Independently of this probably inborn behavior, the mothers too seemed to care for their young in this respect. Thus G. Caulfield observed the following: "An additional kind of 'acrobatic' behavior is undertaken with the baby once a day. It consists of the mother freeing herself from the hold of the baby, grasping one arm or leg and holding it far away from herself into the air, and climbing through the cage in this manner. The young dangles in the air and screams." Finally it was observed that "shortly after or during this procedure the young would defecate. Perhaps the movements of the young plus the massage of the abdominal muscles in crying are a substitute for the lack of activity."

Gradually the young orang-utan also learns how to orient within the tangled mass of vegetation in the rain forest. The young animal climbs into the highest tree tops, mountain ridges, and rocks, and thereby learns about its surroundings. Finally it has learned where to go to obtain specific ripened fruits. It also learns how to react to the sounds and other signs of its conspecifics. For example, it learns the meaning of the smell of urine which adheres to some plants, or to recognize a conspecific moving away by the manner in which the branches are slowly swaying. It knows that loud snapping of twigs means danger and a quick departure, while the quiet rhythmic breaking of branches indicates the construction of a nest which means a timely invitation for a rest period of the entire family group.

Nest-building: innate or learned?

Aside from his great learning capacity, the orang-utan naturally also has a set of innate behavior patterns which do not have to be learned. There are many indications that nest-building is an innate behavior pattern, although formerly this ability was often hailed as proof of the high intelligence of great apes. Even today, certain scientists who do not consider nest-building in the great apes as a genetically determined behavior pattern consider it to be a tradition which is passed on from the parents to their offspring. However, at least in the orang we must assume that an innate nest-building behavior plays a decisive role.

Phillip Street reported an unusual event that took place in the mid-1950s in the London Zoo's Regent's Park and which shows how deeply rooted nest-building is in orang-utans. The orang "Jacob" escaped one night from his cage. He was able to get out of the primate house through a window onto the roof and from there into a tree where he was found the next morning after his escape had been detected. He was lying curled up in a nest which he had built in the top of the tree. The important point is that Jacob had already been living for years in the zoo and had not seen a tree nest or built one for himself. However, he had not lost his ability to build a sleeping nest and used the first opportunity to build one.

In 1965 Dietrich Heinemann observed a mature orang male in the Munich Zoo building a "nest in vacuo," which means without nesting material. The animal was sitting in the corner of the cage on a slightly elevated sitting board and was lifting imaginary "branches" all around itself. Then he bent these towards himself and carefully pushed them into position with the back of his hands. Such "vacuum activity" is a sure sign that the basic elements of this particular behavior sequence are not learned, but are innate.

The selection pressure for healthy and strong orang-utans is great because of the humid atmosphere in the rain forest and the manner of feeding in the orang, which forces him to continuously migrate through the trees. During heavy downpours the orangs sit in a hunched position and let the raindrops run off their necks and backs. Usually, however, they crouch close to the main tree trunk protected by the canopy. The nests, however, are more exposed to the wind and rain, and during bad weather the orangs may become quite chilled, so even growing young which have left their mothers huddle together for warmth.

Orangs range over wide areas

During their search for food the orangs migrate over a large area of the jungle. It is not uncommon for food to become scarce, and so the orangs are forced to migrate to areas where there is more. This often brings the orangs into contact with the rapidly spreading human population in Borneo. The fertile regions which provide sufficient food for the orangs are particularly desirable to the land-hungry Dayaks. There are new settlements everywhere. The orang-utans which do not have any other enemies besides man are constantly forced into smaller and more remote regions, and are separated into increasingly smaller and more widely scattered troops. Today there are many single orangs which will never have the chance to meet a conspecific, to mate, and to reproduce.

In the mountains the orangs only climb to an elevation of 2000 meters at most. High mountain chains are just as insurmountable boundaries to their migrations as are jungle rivers. The orang-utans only cross water over which vegetation has provided a natural bridge.

It is obvious that man, who is dependent on so many things and is quite unable to climb, cannot replace the mother for an orphaned orang in its natural surroundings. Nevertheless, I thought to myself that it should be possible to raise and to look after orangs in the jungle, and to get them to the point where they gradually gathered sufficient experiences from their outside environment to readapt to their natural habitat. In 1961 this idea received support from the forestry agencies in Sarawak and Sabah, the World Wildlife Fund (IUCN-Switzerland) and the Wenner-Gren Foundation for Anthropological Research (U.S.A.). If this experiment was successful, it would mean that the orang orphans which constantly come into human hands would in the future be saved for their home ranges in Borneo and might restock the threatened wild populations. In addition, such an experiment might reveal new knowledge about free living orang-utans. Of all the great apes, the orang-utans are the most difficult to observe because of their arboreal existence, and because of their social organization, which consists of widely scattered small groups.

I supervised the first experiment of this kind which took place in the spring of 1961 in Sarawak's first National Park, Bako. This park could be reached within just a few hours, in any kind of weather, from the capital city. Within the forest and one kilometer from the camp of the supervisory personnel on the beach, I constructed an enclosure of meshed wire. There I placed three young orangs, one pair from Sabah and a young male from Sarawak. Here the animals were to familiarize themselves with the jungle, but still within an enclosure to which they were accustomed. After one week, we opened the enclosure for a few hours at a time. Later it was always left opened. The animals were very reluctant to move from this center where we still fed them. We became aware of the fact that we had to entice our charges to come out and explore the jungle. Therefore we often changed the feeding locations to ever increasing distances. We also encouraged the animals to follow us on our excursions through the forest. We constantly called them and allowed them lots of time to climb in the trees parallel to us. This provided them with the opportunity to gather first-hand experiences and to try out natural foods. Our orangs, who were younger than three years, obviously needed the presence of a human guardian who was always within their sight and hearing. If these young lost contact with us they cried loudly and returned to the vicinity of the enclosure or our beach camp as quickly as they could. As the animals grew older they eventually were content in the company of a young conspecific while conducting short exploratory trips. Arthur, our oldest male, developed a great interest in rock caves where he used to construct nests out of dry grasses and twigs. Often he would also spend the night there.

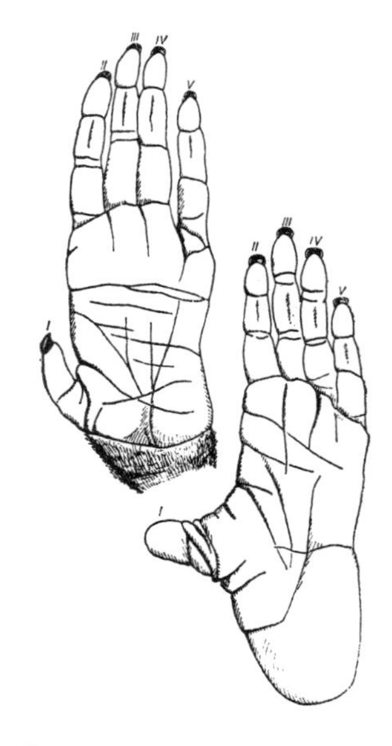

Fig. 21-3. Left hand (top) and left foot (bottom) of an orang-utan. The orang is the most highly developed brachiator among the anthropoid apes. His hands are adapted for this mode of locomotion by having evolved into long "hooks" with only a short thumb that is well back on his hand. The feet are also long and narrow.

After having spent three years in the jungle, the five-year-old Arthur weighed 25 kg. He was slim and very strong. He was not afraid of people, and he still recognized his caretakers as being of a higher rank. However, he was dangerous to strangers who walked into his territory without us. Arthur was not aware of his own strength, and did not have the faintest notion that people are so easily injured. Yet this was not the only reason why we transported our charges to another region. More importantly, wild orangs were absent in Bako, and thus Arthur did not have a chance to interact with conspecifics and thereby to become part of a natural orang-utan society. Towards the end of 1964, the forestry service of Sabah was willing to continue the experiment on their own under the supervision of warden G. S. deSilva. In April, 1965, we transported our orangs from Bako into a protected jungle region in the vicinity of Sandakan (Sepilok) which was inaccessible to visitors. This region also had the advantage of being regularly frequented by wild orang-utans. We forced our older orangs to range more widely into the jungle by giving them very little food. For weeks at a time the animals fended for themselves when the wild fruits were ripe. In the spring of 1967, the oldest fully mature female, Joan, returned with her first infant to the vicinity of the observation site after having been absent for several months. During one estrous period she had paired off with a wild orang male. Thus she had provided the needed proof that re-introduction into the wild state from captivity is possible.

Only a few thousand Orangs are left

In 1961 my husband and I estimated that the total number of orang-utans was 5000 at most. Today their numbers are probably only half that. It is probable that in the first centuries after Christ, they numbered at least one half million. One thousand years ago there were far more great apes than men on Borneo. Today the few thousand orangs on the islands are outnumbered by approximately three million people.

Of the total number, perhaps less than one thousand orangs live in the forests of Sarawak. It was not possible to determine the number of orang-utans living in Sabah, the former British Borneo. Certainly there are no more than in Sarawak, although this is often claimed. In Brunei, the area between Sabah and Sarawak, there are no more orangs. What little we know about the large areas of Borneo, which belong to Indonesia, indicates that they are persecuted by man without restraint, and this is in spite of all good intentions to the contrary and official decrees to protect them. In Sumatra the situation is even more serious, since the proximity of the mainland favors the smuggling of orangs.

Orangs in the zoo for 35 years

Presently there are about 250 orang-utans in the zoos of the world. One pair in Philadelphia has lived there for thirty-five years. Marvin L. Jones estimates the age of the animals to be forty-seven years. The daughter of this pair was born in 1937. She is also still alive. In the zoos of New York, Birmingham (Alabama), London, and others, orangs have lived for over twenty years. The zoo of Rotterdam had a beautiful and famous group, but unfortunately this group was almost totally wiped out by a monkey smallpox epidemic in 1965. In order to aid the zoo of Rotterdam to establish a new breeding group, I sent them a pair of young orangs in 1966. Only Rotterdam and Philadelphia have up to now (1967) been successful in breeding second generation orangs. In the last years the zoos in Frankfurt and Berlin have repeatedly reported successful births of orang-utans.

Despite these great successes, the total number of orang babies born and raised in zoos is only about four or five per year. This number is far too low for maintaining the species under human care in case the animals become extinct in the wild. The orang-utan will only then survive into the next century if its protection is left not only to the few states where the animal still occurs. These developing nations have so many other pressing problems that little effort is expended for the effective protection of animals. A general law against the import and trade of orang-utans in all countries of the world would be the only help. Above all, it is extremely important to provide sufficiently large areas for the orangs where they are protected under international law. Equally important are studies on free living orang-utans, because only thorough scientific investigations will provide insights to help solve problems, which if solved can still contribute to the survival of this threatened species of great apes.

Is the Orang-utan becoming extinct?

The basic question is: Do we still have time, or is it too late already? If we are not successful in keeping our cousin, the orang-utan, alive outside of cage bars, mankind will have become a threat to nature, and a higher authority may deny mankind the right to continue to "rule" the world.

Evolution will not stand still. It may bypass the exterminated orang-utan and its exterminator, *Homo sapiens,* and may give rise to a creature which practices tolerance and self-conquest and which is better able to solve the difficult equations of the universe.

Barbara Harrisson

▷
A half-grown gorilla in the Frankfurt Zoo.

▷▷
The groups are always led by a "silver-backed" male (right). In the Virunga region gorillas eat especially the pith of bamboo stalks which grow there in great abundance (left). They build new sleeping nests each night—usually on the ground (right foreground).

22 The Gorilla

The Gorilla by
B. Grzimek

The American palaeontologist and zoologist G. G. Simpson combines the gorillas and chimpanzees into a single genus *(Pan).* Yet these two African great apes are quite different and we have maintained the familiar separation of the two genera. The gorillas and chimpanzees are more closely related to each other than to the orang-utan. According to Thenius, both genera originated from a common root. However, during the Upper Tertiary, or at the latest in the beginning of the Pleistocene, when the ice ages started in Europe, this stem branched into two evolutionary lines which gave rise to the genera living today. The ancestral form is not known but must have been arboreal. In contrast to chimpanzees, the exclusively arboreal period of the direct ancestors of the gorilla must not have lasted for too long. Today gorillas are predominantly active on the ground, and are not as highly adapted to brachiating as the chimpanzees. Gorillas have short broad hands and feet which are not hook-like. The genus *Gorilla* consists of one species.

Distinguishing characteristics

The GORILLA (*Gorilla gorilla;* Color plate p. 521) is the largest of the great apes. In fact, it is the largest primate living today; the height is 1.25–1.75 meters when the body is erect (unnatural position), and when the knees are straight the animal measures over 2.30 meters. The span of the outstretched arms is 2.00–2.75 meters. The circumference of the chest of the ♂ is up to 1.75 meters; the body weight of the ♀ is 70–140 kg, and of the ♂ it is 135–275 kg. In zoos it can occasionally reach 350 kg. The body form is compact. The fur is dense, black or grayish-black, occasionally with a reddish crown. The back of old males is silvery-gray, and the beard is absent. The face, hands, and feet are hairless and black, and so is the chest of old males. The ear pinnae are small; the nostrils are surrounded by large bulges. The eyes are small and lie beneath well-developed supraorbitals. There are two subspecies in tropical Africa (see map): 1. The LOWLAND GORILLA *(Gorilla gorilla gorilla)* is found in the lowlands in the west of Equatorial Africa; 2. The MOUN-

"Max" the first gorilla baby born in Germany

TAIN GORILLA (♂*Gorilla gorilla beringei*) is found in the lowlands and mountains up to elevations of 3500 meters in Central Africa.

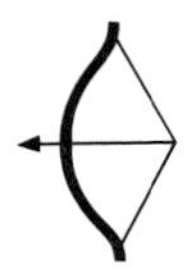

How Gorillas were discovered

In 460 B.C., which is close to 2500 years ago, sixty large ships set out to sea from Carthage. On board were 30,000 seamen, oarsmen, and soldiers. Under the command of Hanno, who was one of the highest ranking statesmen of that large trading city, this fleet sailed through the straits of Gibraltar and continued southwards along the coast of Africa up to Cameroon. Never before had anyone sailed this far. The fleet attempted to go up the Niger but had to turn back at the rapids. Presumably it was in Gabon where the expedition encountered some gigantic black people that were covered by hair. These creatures gnashed their teeth, grunted, and threatened. When the Carthaginians, in line with the usual slave-making methods, tried to catch some of them, even the females inflicted such dreadful bites on their captors that they had to be killed.

Hanno ordered these peculiar hairy human creatures to be skinned. The skins were taken to Carthage where they were hung up inside a temple. There they apparently remained for several centuries. According to Plinius, who perished during the eruption of Vesuvius in A.D. 79, the Roman conquerors had seen these skins 500 years after the expedition. The interpreters had told Hanno that these strange wild creatures were called gorillas. It probably will never be known if these were really members of the large species of great apes. Some scientists assume that it may have only been baboons which Hanno's people caught. However, this does not seem probable because the Phoenicians must have been familiar with baboons because of their extensive trading journeys in North Africa. It seems unlikely that they would bring baboon skins from Gabon and display them as a great rarity.

It took a considerable period of time before additional news of this gigantic animal-man reached the North. Towards the end of the sixteenth century, the English sailor Andrew Battel was captured by the Portuguese and detained for years in West Africa. He described two species of great apes which are not too difficult to recognize as gorillas and chimpanzees. He called the gorillas "Pongo," and he maintained that these great apes came to the camp fire after the people had left in order to warm themselves but that they were not clever enough to put new wood on the fire. Many other things which he related about these black giants later proved to be right.

Fig. 22-1. 1. The lowland gorilla (*Gorilla gorilla gorilla*) has only a spotty distribution in this region. 2. Mountain gorilla (*Gorilla gorilla beringei*).

The first living chimpanzees reached Europe as early as 1640, but almost nothing was known of the other great apes. The scientific texts of that period constantly confused the gorillas, orang-utans, and chimpanzees, but also mixed up the great apes with the pygmy tribes and various other primitive races of man. Only by the year 1860 did the missionary Savage and the tourist DuChaillu bring more detailed descriptions, skulls, and skins back to Europe. According to DuChaillu, gorillas were ferocious monsters of the jungle which attacked any person and tore him apart. He described the horrible face, the gigantic body, the terrifying roars, and the drumming of the fists on the chest prior to an attack. For half a century his descriptions served to imprint the European mind with the image of the gorilla as a hulking forest devil. Just a few years ago a German journalist maintained that he had "to execute" a gorilla male in West Africa because the latter "assaulted every gorilla female," broke into a native's hut, killed a gorilla female, and abducted her child. Later Grzimek was able to demonstrate that this "murdering" gorilla had been a female. Generally speaking, until very recently little was known about the life history and behavior of our closely related powerful African cousins.

The first zoo Gorillas

Only towards the end of the last century did living gorillas reach Europe. Major Dominik summoned 800–1000 natives to finally catch three gorillas with nets. One of these lived for thirteen and the other for seventeen days. The old zoo director Hagenbeck said that they died because of "home sickness." This may not have been without foundation as far as older animals were concerned, but the former "animal storage buildings" (early zoos) hardly contained any of the equipment which is taken for granted today in modern great ape houses. A gorilla which lived in the Berlin Aquarium for some time received "a breakfast of a pair of wieners, frankfurters, sausages, Hamburg smoked meat, Berlin cow's cheese, or some other type of sandwich. He preferred a glass of white beer with a fruit flavor." Lunch consisted of a bottle of light beer, plus rice or vegetables, potatoes, carrots, or broccoli and meat—all foods that were totally new to the gorilla. Today we know more about the food requirements of wild animals in the zoo, and consequently zoos have been more successful in keeping the animals.

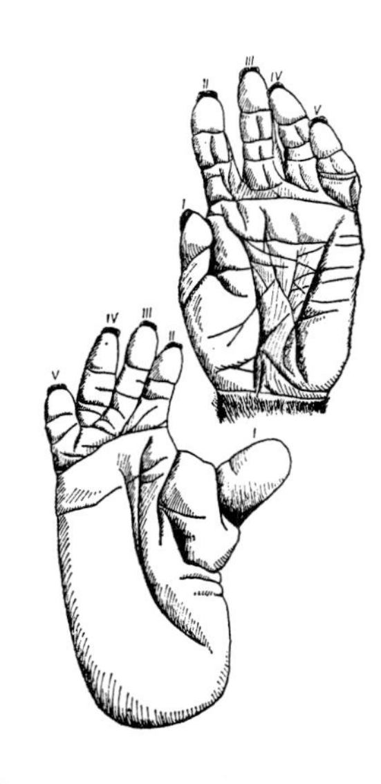

Fig. 22-2.
Left hand (top) and right foot (bottom) of a gorilla. Gorillas live more on the ground than do all other anthropoid apes. Their hands and feet are therefore not as well developed for brachiating and climbing as is the case in orang-utans (see fig. 21-3) and in gibbons (see fig. 19-1).

When seeing a mature or nearly mature gorilla male in the zoo for the first time, one has to pause for a moment to reflect on the impression he makes. Such an animal can reach a height of 2.30 meters, although the legs are relatively short. Some have a shoulder width of more than one meter. These male gorillas appear monstrous and superhuman, and one can well understand the horror which the first whites must have experienced when they first confronted these human-like animals without sufficient weapons. Such giants were rarely seen in zoos during the first decades of the twentieth century. In 1961 "Bamboo" died at the age of thirty-four years in the Philadelphia Zoo. Today (1967) the gorilla male "Mossa," which was born in the summer of 1931 in Africa, is still alive in the same zoo. Thus he is already thirty-six years old. A female gorilla, "Toto," was brought from Africa as a two-month-old infant by Kenneth Hoyt. When Toto was fully grown she was placed in a custom-built trailer in a circus in Florida. In 1967 she was still living there. The zoos of New York have three female gorillas which they have cared for since 1941. These six gorillas in America hold the longevity record up to now.

Captive gorillas should be kept as family units so that maturing females have the opportunity to observe the birth process and small offspring. It is sheer cruelty to keep great apes in solitary confinement. In the Frankfurt Zoo even mature gorilla males are not kept behind bars but behind glass plates measuring 1.8 × 3 meters and 40–42 mm thickness. An innovation introduced by the Frankfurt Zoo allows all monkeys access to the outside through doors covered by soft, heavy plastic flaps; these also proved successful for the gorillas. The sliding doors, which are hydraulically operated and lead to sleeping bunks, can be adjusted to any desirable position. Thus it is possible that only young or females can slip through, providing them with security by barring the large males.

Today there are approximately 15,000–20,000 gorillas; others estimate the number to be 50,000–60,000. In any case this number seems small if compared to over three billion people. The majority of gorillas are found in the jungles of West Africa around the Gulf of Guinea. The mountain gorillas inhabit Central Africa, more to the east, in the Virunga Volcanoes between Lakes Edward and Kivi. In 1925 the Albert National Park was established in this region. This large preserve of 315,000 hectares was primarily set aside for the protection of the gorilla. Some of the gorillas there live at elevations of three to four thousand meters. Since the other subspecies has hardly been studied in the free living state, we have limited our discussion mainly to the mountain gorilla. The gorilla females are barely half as large as the males. A full grown gorilla is almost impossible to capture, and if captured, difficult to haul away. Females are often shot by the natives and eaten, in spite of all laws. Their young are seized and then usually sold to so-called "catchers." This is the way most gorillas eventually get into zoos.

In February, 1959, Dr. George and Kay Schaller, young American scientists, went into the same gorilla range that I had visited four years earlier. They lived among the gorillas for twenty months, and were the first to observe the behavior of these giant apes in their natural environment. Schaller spent 466 hours in direct observation of these animals and met them 314 different times. Six of the gorilla groups became completely accustomed to his presence.

Observations on free living Gorillas, by G. B. Schaller

"We established a base camp in a hut in the Kabara district in the center of the gorilla home range. Kabara is situated in a charming park-like area with a good view into the distance, at three thousand five hundred meters elevation, high on the saddle between the two volcanoes Mikeno and Karisimbi. There is only little bush cover and the trees are far apart. Usually they are *Hagenia abyssinica,* whose long, feathered leaves form umbrella-like crowns.

"It is not so easy to observe these apes during their daily activities;

the brush and grasses are quite dense. The Karaba region was favorable for our research since the gorillas living there had had little or no contact with human beings before we came. When an observer approached them alone they rarely fled.

"During the first two days in Kabara we found no gorillas, only abandoned sleeping nests. On the third day I went again to explore with our African assistant N'sekanabo. We walked through dense groups of trees; the trunks were bare for two meters up and were topped by a bundle of large leaves. Off to the side and above us we suddenly heard a shrill cry. I signalled N'sekanabo to wait, crept forward and reached a depression in the ground which I observed from the cover of a tree. A gorilla female emerged from the brush and climbed slowly onto a tree stump, holding a stick of wild celery in her mouth as if it were a cigar. She sat down, grabbed the stick of celery with two hands, ripped off the tough outer layer and ate the juicy center. A second gorilla female with a young on her back appeared and grasped a stick of celery close to the roots and pulled it out of the ground with a jerky movement. Then she pressed down the grass with her hand, squatted down, and ate the plant. Wild celery is the second most important food plant for gorillas in the Kabara region. In order to be able to see the other animals of the group I edged forward a little carelessly and one of the females saw me. She uttered a short cry and disappeared among the trees. A large juvenile animal weighing about eighty pounds climbed up on a felled tree, looked in my direction, and climbed back down again. Suddenly seven animals walked past, single file, at a distance of about thirty meters; a large silver-backed male brought up the rear. He stopped briefly and observed me, hidden behind some leaves so that only the top of his head was visible. Then he uttered a series of grunting calls which seemed intended to warn both me and his group, and then he too moved away. Closely behind him followed three females and four young, two of which were carried piggyback by two of the females.

"The next day Kay and I saw the gorillas only from a distance, but on the third day I was able to stalk close enough to the group. I was hunched over, covered only by a few branches in a fork of a large Kusso tree and was able to observe the animals as they were feeding and resting between the high bushes without them being aware of my presence. The silver-backed male approached a small tree, grabbed the trunk about 1.8 meters above the ground, and pulled it out with a sudden jerk. Just as slowly and deliberately he grasped a branch, ripped it apart with his teeth and nibbled on the soft, white core as if he were eating corn on the cob. A female was leaning on a slanting tree with her back. She had pulled the crown of a nearby tree towards herself and ate the dark-red blossoms which she plucked one by one and then put into her mouth.

"These first gorillas which I had encountered in the Kabara region

I designated as Group I; however, they were not the only group in this region. On August 22 I heard a male beating his chest about one hundred meters from the troop which I was observing. The next morning when I checked I found that no less than three different groups had built their sleeping nests very close together. One of these groups was quite large; a later count revealed that it consisted of nineteen members. The third group, on the other hand, consisted of only five animals, a male, two females, and two young. These three groups (I, II, and III) remained for five days in the same forest region. Later on I met many more groups.

"When I was looking for a group I usually went to the place where I had seen them the day before. I carefully followed their tracks, not knowing whether the animals were two hundred meters or two kilometers ahead of me, or if they had circled back and were now behind me. Sometimes I came upon a stuffy odor similar to that one finds on farms. This told me that I had reached the place where they had spent the night. The sleeping nests and their immediate surroundings were covered with feces; numerous small flies buzzed around, depositing minute white eggs on the still moist excrements. I often needed half an hour to find all the nests since one gorilla occasionally slept twenty or more meters away from its neighbor.

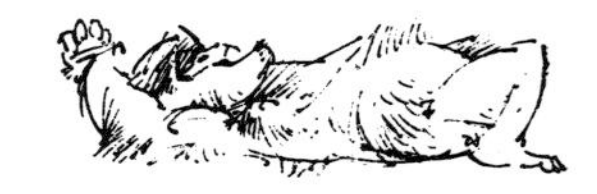

"Gorillas go to sleep wherever they are overcome by nightfall; the only requirement is the presence of plants that are suitable for building nests. These are constructed either on the ground or in trees. In the *Hagenia* forests around Kabara 97 percent of the nests were on the ground since the brittle branches of these trees are not suitable for nestbuilding. In the Utu region, however, only twenty out of a hundred nests are on the ground. Juveniles built their nests in trees twice as often as females and black-backed males. Silver-backed males always slept on the ground, but in nearby Kisoro they also built an occasional nest two and a half meters high in the brush. The view is widely held that females and juveniles sleep on trees for reasons of safety while the males sit on the ground as lookouts. However, this is not the case. Black-backed males sleep twenty or more meters away from the group, while juveniles often sleep near the females or in the same nest with them. A small gorilla baby remains with its mother; only occasionally does a larger young build its own nest near her.

"When constructing a nest on the ground gorillas bend the stems of all grasses or bushes towards a common center so that a springy platform results. They build similar platforms in the canopies of trees or in the forks of branches. Such a nest is, of course, also capable of supporting the weight of a person. Chimpanzee nests look similar, but they are usually much higher in the trees. A gorilla builds a nest each night, and almost always chooses new locations. The nests are only a few meters apart; there seem to be no hard and fast rules where,

Fig. 22-3. During the day gorillas often rest on the ground for extended periods of time...

within the general area, females, young, or the leader may bed down. In the Utu and Kisoro region the gorillas sometimes slept a mere thirty to seventy meters away from human settlements. Incidentally, gorillas do not snore.

"It is sometimes asserted that the ability to build nests must be learned; however, I doubt this (see p. 516). Undoubtedly the young learn how to build a sturdy nest by utilizing the various plants appropriately; however, I believe that gorillas instinctively build so as to have something under themselves before going to sleep.

"Usually gorillas build their nests at the beginning of dusk and they rise in the morning after sleeping about thirteen hours, generally, at the latest, about an hour after the sun has risen.

"In the morning gorillas need to feed about two hours before they are satiated. In the Kabara region their basic food is *Galium*, a vine, wild celery, thistles, and nettles. Depending on the season they also eat bamboo shoots and blue *Pygeum* berries. They eat primarily leaves, shoots, and the hearts of stalks and branches. Sometimes they pull off the bark and eat it. When eating they almost always use their hands and they rarely bite off the plants directly. Here they differ from chimpanzees, which primarily eat fruits and less frequently shoots and leaves. I identified about one hundred different kinds of plants that are eaten by gorillas. Grzimek followed the tracks of gorillas in the Kisoro region and tasted all the plants that they had eaten. Most of them tasted bitter. Since leaves and plants are not very nutritious gorillas must eat large amounts of them. The dry matter alone in the daily production of feces weighs one to one and a half kilograms; thus large amounts of waste are excreted. It seems that the plants also supply the needed water. Gorillas have never been observed to drink in the wild.

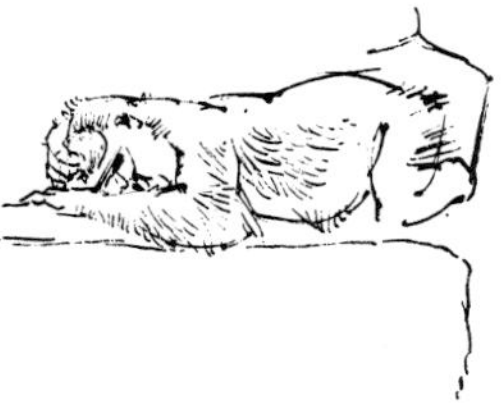

"Gorillas which are free living have never been observed eating animals. I examined several thousand dung heaps from gorillas and have never found a single instance of hair, chitin casings of insects, bones, skin, or other signs of animal food. Under human care, however, gorillas readily eat meat. This is probably due to the complete change of diet and the lack of protein.

"In the wild they pay no attention to freshly killed animals. Once, a gorilla group rested three meters from an incubating dove which they could not have overlooked. However, they did not disturb the bird. Natives have often told me that gorillas often plunder the nests of bees, but in the Kabara region I was never able to observe this.

Fig. 22-4. ...or on surprisingly thin branches.

"They usually stop eating between nine and ten o'clock in the morning and then they rest until the afternoon. Gorillas present a picture of complete contentment when they surround the silver-backed male, all lying on the ground, especially when the sun shines on them. Many stretch out on the ground, lie on their backs, on their sides, or on their stomachs, with their arms and legs spread out; others may lean against a tree. Occasionally a gorilla will build himself a nest to rest in. However, during this rest period they do not just sleep, doze, or sit around. Some gorillas groom themselves by scratching and cleaning themselves. Some also clean each other, especially the young. Among adults, mutual grooming does not play as large as role as with other apes (see p. 540). Gorilla young use the rest period for playing and exploring. It is just as hard for them to sit still as it is for human children, especially when the mother wants to rest in peace. They move about between the resting adults, playing alone or with others. However, young gorillas do not always play. Sometimes I did not see a playing animal for days, especially when the clouds were low and the plants were wet. Enjoyment of play stops in gorillas at about six years of age, at the time when they become sexually mature. Favorite games consist of playing catch, or defense of a tree stump or a hill against the assault of others; in other words, these are the games which, with various names, are also played by human children.

"Gorillas rest for about one to three hours at noon. Sometimes the leader decides to get up and move about fifty to one hundred meters and then lie down again, without any apparent reason. Of course, the entire group follows him, since the highest-ranking animal determines by his actions in general the course of the day's activities: when and where the troop will rest, how far and in what direction they will move. With simple gestures and vocalizations he directs the behavior of the troup. When he suddenly rises and stiff-leggedly faces in one direction they know not only that he is ready to leave, but the direction as well.

"The gorillas spend the afternoon sleeping, eating, and sleeping, after which they eat once more continuously until dusk. They search very deliberately for food, sit around frequently, and move with a speed of about three to five kilometers per hour. When it gets darker in the forest they begin to slow down and gather about the highest-ranking male. They sit around undecidedly, as if each was waiting for the other to begin building a nest. At around six o'clock, or five if the sky is covered with clouds, the highest-ranking male begins to break branches for his sleeping nest, and all others follow his example.

"In my first encounters with gorillas I had tried to approach them unseen. However, I soon discovered that I was unable to make many important observations if I hid too well. As soon as I tried to reach a better position for observation the animals would see me and become excited. Therefore I walked slowly and openly towards the gorillas, climbed a tree, and settled down comfortably, pretending not to be very interested in the gorillas. When I climbed such an observation post I not only could see the gorillas, but they too could always keep an eye on me. I was convinced that the gorillas would soon discover that I was harmless when I approached them alone in this manner. In some animals, for example in dogs, rhesus monkeys, and man, it is

Fig. 22-5. Young gorillas climb more frequently into trees than do adults.

a kind of threat to look directly into the eyes of another. Even when I observed gorillas from a distance I had to remember always not to look directly at them for too long and to move my head away from time to time so that they would not become upset. They also regarded it as a threat when I pointed the camera or my binoculars at them; hence I could use both only sparingly.

"While observing gorilla Group IV I had frequently noticed that the young, black-backed male which I called 'Junior,' or sometimes another animal, would come up to me to a distance of about twenty meters and would then shake his head. This gesture seemed to mean: 'I mean no harm.' In order to see what the gorillas would do if I shook my own head, I waited until one day Junior was about ten steps away from me and was observing me closely while I was putting new film into my camera. As soon as I shook my head he looked away; perhaps he thought that I took his gazing at me as a threat. When I looked directly into his eyes, he shook his head. We continued this for ten minutes. After that, whenever I met gorillas I shook my head and they seemed to understand this gesture, which I used to signal my peaceful intentions.

"Sometimes curiosity overcame the gorillas, although there seemed to be no particular reason for it. One group which I had met and observed seventy-six times approached me one day until they were ten meters away from me. They looked me over with great curiosity. One female with a three month old baby clinging to her reached up and sharply bumped the branch on which I sat. Then she looked up to see what I would do. A juvenile then did the same and a female even climbed upon the branch for a few seconds. Still more frequently the males, especially the leader, made sham attacks to see what I would do, no doubt, or to chase me away. Once a large gorilla male ran seven meters towards me but stopped at a distance of about twenty-five meters.

"It is possible to recognize individual gorillas, like people, not only by their body build, but also by their entire demeanor and especially their faces. The members of the group which I observed most frequently I had given names; this facilitated my observations and the writing of my notes. I discovered that the groups are very tightly organized and show much cohesion. This is quite unlike the smaller groups of orang-utans and the larger troups of chimpanzees where individuals always seem to be coming and going.

The Outsider

"With Group IV, which I knew especially well and which was led by 'Big Daddy,' was another large adult male. He moved with great deliberation among the group members, but otherwise he ignored the other animals. This is why I called him 'The Outsider.' He was a giant and in full possession of his strength, clearly larger than Big Daddy and by far the largest animal in the entire region. During the first few weeks of my presence in the Kabara region he left the troop at least twice and returned again. Some gorilla males live alone, but join up with a group from time to time for a period of time. In the Kabara region I discovered seven such lone males, four silver-backed and three black-backed; possibly there were even more. These males joined up exclusively with Groups IV and VI. It seems as if some groups are willing to accept loners, and these discover eventually where they are welcome and where they are not.

"Among the loners I knew best was an animal which I dubbed 'The Lone Stranger,' a silver-backed male in the prime of life who had a thoughtful gaze and a somewhat arrogant expression around the corners of his mouth. He did not like to see me and usually took off screaming when he saw me. I first observed him on November 18 near Group VI. 'Mr. Dillon,' the highest-ranking male in this group, rested among his females and young, he seemed not to notice The Lone Stranger. When the animals of this group moved past him, he remained behind and stared at the visitor. It appeared as if he wanted to show him that it was time to leave. The Lone Stranger disappeared into the forest, but in the following weeks I could observe him on several occasions as he followed these animals. When Group IV moved into the slopes of Mount Mikeno, The Lone Stranger was with them for at least one week. Soon afterwards, however, he was off by himself again in the forest.

The social rank order makes peaceful coexistence possible

"When more than one silver-backed male belongs to a group there is always a clearly defined social hierarchy. A higher-ranking animal has the right to pass first on a narrow path and can drive off a lower-ranking animal from its place at any time. We know of such rank orders in many other social vertebrates. Contrary to widely held beliefs, this does not lead to fighting and friction, but makes a peaceful association possible because each individual has his definite place in the group; each animal knows exactly which are his rights and obligations towards other group members. All others follow their silver-backed male leader, although he gives no orders as far as I could see. He is the most alert and most excitable, still screaming and threatening when the others have already calmed down. At the same time he is also the shyest and often hides behind branches.

"Silver-backed males rank above all other animals of the group, since to an extent the social rank of an animal seems to depend on its size. Likewise, the females rank above the juveniles, which in turn rank above youngsters that no longer live with their mothers. Once, when it began to rain, a juvenile sought shelter under a sloping tree, pressed against the tree, and looked at the dense falling rain from his dry place. However, when a female ran towards the tree the young animal rose at once and ran off into the rain. No sooner had the female sat down in the dry spot than an adult male appeared from the bushes.

Fig. 22-6. They like to place leaves on their heads for "decorations."

He sat next to her, but pushed her slowly but firmly away until she sat in the rain and he occupied her former place.

Among females there is no definite rank order

"Among the females of a group there seems to be no clear-cut rank order. It is interesting to note that primarily the females squabble, while the males do not actively participate in these altercations. I am not certain of this, but I believe there is a "sliding" rank order: mothers with smaller young hold a higher rank than those with older ones. Of course, individual temperament also places a role in the establishment of rank; more excitable group members are usually avoided.

"The highest-ranking silver-backed males are undisputed rulers who, because of their size and social position, always have their way. However, they are also very tolerant and peaceful. The females and young of a group seem to like the silver-backed male. Sometimes a female rested her head on his silvery back or leaned on him. At times as many as five youngsters went to such a male, sitting around him and in his lap, climbing all over him, and making quite a nuisance of themselves. However, the males reacted only when they became too wild. Then a single stare was enough to calm the youngsters down."

I (Grzimek) sent a co-worker, the camera man Alan Root and his young wife Joan, into the same region of the Congo where I had stayed previously from September to December, 1963.

Gorillas, and in particular mountain gorillas, are usually extremely difficult to film. They are shy animals and are usually covered by brush and plants; in addition, they are black. The few scenes of free living gorillas which have been shown in two or three films at cinemas actually came from animals that had been caught with the help of hundreds of beaters. The gorillas were released again in fenced areas and filmed. Our observation area was located between the Congo and Ruanda in a mountain chain of six Virunga Volcanoes which were no longer active. These volcanoes protruded out of very fertile land which is one of the most densely populated areas of Africa. Although the new African Congo administration protected the gorillas more than the Belgians, in spite of the civil war the mountain gorillas are nevertheless constantly threatened by the expansion of agriculture which means the clearing of forests by hunters and Watussi herders who try to drive their large cattle herds, that are of little economic value, into the area occupied by the gorilla.

The misunderstood Gorilla, by B. Grzimek

In the heart of Africa, the only place where gorillas occur, probably a population of not less than 5000 but not more than 15,000 mountain gorillas exists. They are not unwilling to live close to man. They occur in the vicinity of villages and plantations, streets and mines. In these areas the gigantic jungle trees have been logged and the plants, bushes, and young trees which subsequently have grown up supply the gorillas with more abundant food than they can find in the jungle, not to mention the fruits cultivated by man. Consequently the number of gorillas had always been overestimated, because it was assumed that gorillas were just as numerous in the vast rain forest as they were in the vicinity of streets and settlements.

Of the approximately four to five thousand mountain gorillas in the region around the Virunga volcanoes, about two hundred live in ten different groups in the Kabara region where the Schallers and the Roots have worked.

Rarely has a creature on earth been so unjustly misunderstood as the gorilla. One comes to this conclusion when one reads the descriptions which Schaller has entered into his diaries during his long hours of observation.

They are more peaceful than most monkeys

Most animals, including man, take possession of specific land areas which they defend angrily. Man even kills other men for this reason. A gorilla group also has its home region which consists of a territory of 25–40 square kilometers. The gorillas, which are almost always active on the ground, are continually on the move in search of food. Sometimes they only move 100 meters in one day, but at other times they move up to five kilometers. The same area may be shared by up to six other gorilla groups. If two different groups meet, they do not fight but the two dominant males may stare at and display before each other briefly. Most often, however, even this does not happen. The members of two groups may even meet and then part again. Schaller never observed quarrels or fights among the gorillas.

There are also hardly any disputes about females. Sexual activity in the gorillas, in contrast to most other monkeys, plays a rather subordinate role. The group leader may tolerate another male courting one of his females only three meters away from him, and copulation may take place between them.

Mating in Gorillas

Schaller was twice able to observe copulation in gorillas, and in both cases the male partner was not the dominant silver-backed male. In one case the female was kneeling and supported herself on the ground on the elbows, while "D.J.," the second most dominant male of the troop, stood behind her and grasped her hips. Since the slope was rather steep the animals slowly slipped downwards. The other time the sitting male was holding the female on his lap. The gorillas, which are usually silent, can be rather noisy during copulation. Schaller noticed a very characteristic mating call emitted by the males. Other members of the troop do not pay any attention to the mating pair.

Pregnancy

Gorilla females bear a young every three and one half to four and one half years. Almost 50% of the offspring die as infants or juveniles. Within troops there are two females to every male. If one counts the singles or the gorilla males living in pairs, then the ratio of the total population is probably 1.5 females to each male. It is probable that the gorilla males have a higher mortality rate.

Pregnancy is barely discernible in gorilla females since their abdomens are normally well developed. Judging by the eleven births which have occurred in zoos up to the spring of 1967, the gestation period is barely eight and one half months, as opposed to nearly eight months in the chimpanzees, eight months in the orang-utans, and nine months in man. During pregnancy some gorilla females have swollen ankles at times. During birth, which only lasts a few minutes, the female lies down. Subsequently the female severs the umbilical cord and holds the infant tightly to her chest. In contrast to young of many lower monkeys, the gorilla baby cannot hold on to the mother on its own. Youngsters are born throughout the year; there is no specific breeding season.

When we received our gorilla female "Makula" who in the meantime has become a mother in the Frankfurt Zoo, she was so small that she could not even lift her head by herself. She was raised just like a human child in my family. Her bed stood beside ours in the bedroom. Gorilla young develop approximately twice as fast as human infants. Within six months they have become active, cheerful youngsters. In the Kabara region the infants started to feed on plant material around two and one half months of age, and it seemed obvious that this was the main source of food at around six to seven months. Nevertheless, some still nursed occasionally from the mother at one and one half years of age. The young are able to crawl by three months. By four and one half months they walk very well on all fours, and by six to seven months they are able to climb.

Gorilla mothers which have young of their own are not unfriendly towards others. It can happen that a strange young will come, sit next to the young on the mother's lap, and press itself against her chest without being pushed away. Even the strong males tolerate youngsters playing around them, or permit them to climb up on themselves. One gorilla mother had an eight-month-old young with a large wound on its rump. This infant did not ride on her back as was customary of other youngsters because it was obviously too weak to hold on by itself. The mother always carried it very carefully in her arm so that no part of the wound touched her body. The infant was carried so that its abdomen faced downwards. Periodically the mother examined the wound thoroughly, and briefly picked at it with her hand. Once another female which had lost her own child two months previously joined them and bent over the injured child and touched its face with her pursed lips.

Fig. 22-7. The games of gorilla young are strikingly similar to those of human children.

It was observed in zoos that gorilla females which bore young for the first time did not know what to do with it, just like women inexperienced with infants do not know how to handle them. The females are afraid of the infant and are also upset by the birth process. In the end they may show hostility towards the newborn infants.

Since the gorilla mother "Achilla" of Basel did not nurse her infant

"Goma," which was born in 1955, the baby was removed from the cage and was raised by Ernst Lang, the zoo director, in his home. He and his co-workers provided the first detailed description of the development of a gorilla from birth.

"Max," the first gorilla born in the Frankfurt Zoo on June 22, 1965, according to R. Kirchshofer's notes weighed 2100 grams at birth; after four weeks he weighed 2780 grams, at sixteen weeks 4800 grams, at twenty-eight weeks 8680 grams, at forty-four weeks 14,060 grams, and at fifty-two weeks 16,600 grams. Max cut his first milk incisors in the sixth week, in the ninth and tenth week the two outer upper incisors, in the seventeenth week the first molars, in the fortieth week the lower milk canines, and in the forty-second week the upper canines. In the second week he could lift his head when lying on his abdomen. He followed moving objects with his eyes, and he laughed when one tickled him. In the tenth week he could distinguish his two foster parents, and could turn from his stomach onto his back. In the nineteenth week he walked on all fours, and in the twenty-sixth week he could stand from a walk and drum against the walls. In week twenty-seven the hairs on arms and shoulders would stand erect when he was excited, and in week thirty-one he still did not recognize precipices; he would simply let himself drop, but he could hold on by one hand only. In the thirty-fourth week he could walk a few steps bipedally, and in the thirty-first he hit towards strangers with a clenched fist with a downward motion. At the end of his first year he weighed eight times (16.6 kg) his birth weight. The gorilla mother "Makula," which had not accepted her first young, bore female twins twenty-two months after Max. These also had to be raised artificially.

Alice and Ellen, the first Gorilla twins

"Alice" and "Ellen" are the first known gorilla twins. As with human twins their individual birth weight was lower than that for a single baby. Alice weighed seventeen hundred grams and Ellen weighed nineteen hundred grams. According to the notes of Mrs. Kirchshofer this weight difference was still evident ten months later. Alice weighed eight thousand grams and Ellen, ninety-five hundred grams. From the beginning the question was whether the twins were monozygous or heterozygous. Unfortunately the mother had already eaten large parts of the amniotic membranes when the birth was discovered; hence the question could not be answered at once. Kirchshofer, Weisse, and Berenz observed their behavior development and physical growth in great detail and concluded ten weeks later that the twins must be heterozygous. They were quite unlike. Ellen had an oval face from birth, and she was paler than Alice and had dark brown hair. From the very beginning she was easily frightened and tended to scream and cry often. During the first few months she was by far the more active of the two. She practiced new behavior patterns very persistently, but she also avoided contact as much as possible and kept to herself. This has not changed to this date. Her only desire for contact is directed to her sister, and she is calm and satisfied only as long as she is near her sister. When she is afraid, she clasps her sister, and when she is "fleeing" she pulls her along or pushes her in front. Alice, on the other hand, had a heart-shaped face, and her skin and hair were jet black. To this day she is, in contrast to her sister, outgoing and very interested in people and objects in her environment. She completely lacks the reserve and caution which Ellen shows towards new things; on the other hand, she is less persistent. From the time she was born she had a more friendly expression on her face; the corners of her mouth were drawn up slightly and her lower lip drooped a little (such anthropomorphizing is permissible perhaps with our nearest relatives, since they, in contrast to all other animals, have almost the same innate expressive behaviors as man). Familiar people, such as the foster mother Mrs. Podolczak or the observer R. Kirchshofer, can induce smiling or loud laughing in her at any time. Her sister, on the other hand, always pressed her lips together somewhat, which not only looks unfriendly but also indicates rejection and withdrawal. During her first weeks of life Alice, in contrast to Ellen who was withdrawn but physically quite active otherwise, was less active even though she was friendly and outgoing. This has changed more recently, however, and Alice is now much more active and tireless in all kinds of activities, while her sister has become less active.

These two animals are, then, quite different from each other. These behavior differences are also substantiated by a comparison of foot and palm prints, the shape of the ears, and other physical characteristics. Hence there can be no doubt that they are heterozygotic twins. The behavioral characteristics have shown especially how different young anthropoid apes can be. In addition to the innate movement patterns and postures, such as lying down, sitting, walking on all fours, standing and walking upright, they do exhibit quite different temperaments and differential needs for contact and openness toward the world around them. This is especially significant since they grew up in identical environments, are of the same age, and are sisters. It shows us that one ape is not like another and gives us a glimpse of the range of "personality" development in one species of ape.

The gorilla female "Achilla" of Basel, on the other hand, raised her second and third young by herself. On April 17, 1961, the caretaker had looked in on Achilla around 7 A.M. but noticed nothing unusual. When he looked in one half hour later Achilla was carrying her newborn infant in her arms. She was carefully examining it and licking its hands and feet, and was by far not as excited as she was after the birth of her first child. After the first day the mother nursed the little one, and in the course of a month the youngster developed beautifully. "Jambo," as the youngster was named, and "Migger," his brother, were the first gorilla youngsters which grew up in their mother's care in captivity. All other infants born in the zoos had to be raised artificially. Other exceptions are the female gorilla "Porta" in the Zoo of Toledo who raised her first-born young that was born on November 14, 1967, and cared for it well, after having had a miscarriage earlier. The same was true for "Colo," a female gorilla in the Columbus Zoo who, although totally inexperienced, accepted her young which was born on February 1, 1968. However, the baby was removed from her mother for hygienic reasons and was raised by hand. This precaution is understandable since the Columbus Zoo achieved two records at once with Colo and the newborn baby: First, Colo, who was born on December 22, 1956, was the first gorilla born and raised successfully in a zoo; and second, she is the first zoo-born female with her own young. Thus, third generation gorillas now live in the Columbus Zoo.

Gorillas born in zoos are usually raised artificially

The records of R. Kirchshofer, who keeps the International Gorilla Breeding records, show that until March, 1968, a total of fifteen gorillas were born and survived in captivity. All these animals are lowland gorillas. Three each were born in Basel, Switzerland, Frankfurt, Germany, and Washington, D.C.; two each were born in Dallas, Texas, and Columbus, Ohio; and finally, one each was born in the San Diego Zoo, California, and in Toledo, Ohio.

Infants are frequently held in both arms when the mother is sitting. One wild gorilla mother carried her dead child around for four days. Thumb sucking in gorillas was never observed. The youngsters stay with their mothers for three years during which period they also sleep in her nest. Only by four or five years does the mother-child bond dissolve, but this is not an abrupt process. The older young may still sleep with the mother frequently, although she already has a new baby. Later the mother may occasionally gently remove the hand of the juvenile as it still tries to hold on to her hair while they are on the move.

Female gorillas are sexually mature by six to seven years, and the males by nine to ten years. Nothing much is known about the life span of free living gorillas but in the zoo they became thirty-six years old. Around ten years of age the central part of the back of the male turns silvery gray. Only these "silver-backed males" are leaders of a family group. Yet even without their silvery saddle, one cannot mistake them because of their enormous size and dignified bearing. Their dominance is established, in contrast to baboons without quarrels, beatings, or bites. There is also no quarreling about food.

Gorillas very rarely show mutual grooming. It has been observed between mothers and their young, and occasionally in the strong silver-backed males, but never in the black-backed males. B. Schirmeyer never observed any grooming in the lowland gorillas housed in the Frankfurt Zoo. However, these gorillas frequently picked out hairs, scratched, and cleaned the nose, eyes, and ears with their index

Fig. 22-8. Mutual grooming is rare in gorillas in comparison to other primates.

or middle finger. They only cleaned their fingernails with the mouth, but not the fur, as was observed in chimpanzees. Heavy dirt is shaken off the fur, or the fur is rubbed against solid objects. While baboon and chimpanzee young cannot get enough grooming, and frequently solicit it, gorilla infants not uncommonly seem to dislike it. They struggle against being groomed like human children do when their faces are being washed.

Gorillas yawn, huff, cough, burp, hiccough, and scratch themselves like people do. They make twenty-two different sounds with which to communicate, although they have no language; this is developed only in man. Something like friendship between members of one group, as found in chimpanzees, does not seem to exist among gorillas as agreeable as these large apes are among each other.

The intimidation display of the silver-backed male

The silver-backed male keeps his group as well as intruders and people in line by performing a type of furious dance. He starts this with a series of screams which are interrupted when he puts a leaf between his lips. This is followed by a more rapid succession of screams. The male stands up on his hind legs and throws leaves and twigs into the air, and as a climax this giant beats his hands against his chest several times. Then he runs sideways on all fours, and rips down plants. Finally he hits the ground with his flat palm.

Even four-month-old infants beat their chests and pull out plants, but only large males scream during this intimidation display. The male dances of the Indians, African natives, and primitive tribes seem to fulfill a similar function. Today's version of intimidation may be parades, protest marches, and maneuvers. The purpose is to make an impression or to intimidate without actually resorting immediately to beatings and killings.

Looking directly at them is a threat

Gorillas who really intend to attack behave differently, and not unlike man, in a similar situation. If a gorilla looks hard at an approaching person or another conspecific for a long period of time it means a threat. Conversely, he too considers our direct glance as a threat. Therefore, it is very important for an observer or a cameraman not to look directly at gorillas. Sometimes the threatening male gorilla wrinkles the skin on his forehead into a scowl and contracts his eyebrows. He emits short staccato-like grunting sounds at the same time. Finally, he points the head into the direction of the opponent, and often the entire body, as if he were going to rush towards the opponent. Eventually he will actually do this but usually he only steps a few meters in front of the target or runs past it off to the side. A hunter with a gun in such a situation usually shoots the animal in "self-defense." All reports of the past decades are full of such killings.

The situation can be dangerous if one becomes frightened of such a gigantic male as it approaches and starts to run away. The animal then usually keeps on following on all fours. He may bite him in the legs, the seat, the arms or back and still continue running without inflicting further injuries or even killing the victim.

Staring directly at someone is also considered a provocation in man. Just a few decades ago, members of certain student fraternities in Europe were offended when someone in a restaurant stared at them. The person was then asked by the fraternity man to step out into the corridor or washroom. Here slaps in the face were exchanged as well as calling cards. The opponents sent for seconds and later they duelled. Thus we too possess a lot of behavior which we share with the great apes that may have an innate basis. Although the death rate in these disputes and duels was probably not very high, it must have been even lower in the "wild" gorillas.

Looking away is appeasement

There are built-in "inhibitions" so that gorillas do not cripple or kill each other. A female or subordinate male usually turns its head away when the "boss" looks at them directly. This merely indicates that the subordinate animal has no desire to start a quarrel. In antithesis, during courting the gorilla male in the Columbus Zoo looked away from the female so as not to intimidate her. In the course of time it developed that the Frankfurt male "Abraham" and I (Grzimek) always observed each other out of the corners of our eyes every morning. If one looked at the other directly, the other one had to look away immediately, seemingly "uninterested." In the New York Zoo two mature gorillas, which had been acquainted for a long period but only through bars, were put together. When they approached each other they usually averted their faces and heads. One day the female hit at the male and received a long gash on the skin of her lower arm because she had inadvertently hit his canine tooth. All day long she was preoccupied with this wound. The gorilla male was visibly upset by this, and from then on he completely avoided the female, and he almost always left the room shortly after she entered.

Submissive posture

If a gorilla wants to avoid the show-down of the furious dance display, or if he wants to clearly demonstrate his subordination, he will crouch. The animal will lie on his belly, push his head down, and pull his arms and legs underneath his body so that only the broad back is shown. Juveniles will occasionally place a hand over the back of their heads in this posture. In this manner all vulnerable parts of the body are covered, and above all it inhibits all aggressive tendencies of the dominant animal. Man has an amazingly similar submissive posture. The deep bow or the curtsy are really nothing more than an expression of loyalty and submission. When the native "Friday" first met Robinson Crusoe he prostrated himself on the ground and even placed Robinson's foot on his neck. In many native tribes it is customary for subjects to crawl towards the ruler on all fours, or they must lie prostrate, as was the custom in former periods of Europe. It could well be that head-nodding, which is practiced as a greeting in Western countries, is an expression of good will just as it is in the gorilla.

Two Gorilla groups meet, by G. B. Schaller

Once Schaller was able to observe the meeting of two gorilla groups. "When I encountered the animals, 'Climber,' the silver-backed leader of one group, was sitting in a crouched position and stared straight ahead, seemingly 'lost in thought.' The other members of his group were closely gathered around him. The silver-backed male of the other troop was crouching down underneath a tree barely six meters away. The fifteen members of his troop were scattered throughout the underbrush behind him. This male was excited and emitted muted sounds in quick succession which changed into threatening growls. He beat his chest, turned around, and climbed up a tree trunk, then let himself drop to the ground again with much noise. At the end he resoundingly hit the ground with his flat hand. Thereupon 'Climber' quickly approached the other male until their faces were only thirty centimeters apart and the two looked hard into each other's eyes. These giants of the forest, who each had the strength of several men, did not fight in order to settle the question of priority, but a settlement would be reached as soon as one opponent averted his glance. The two confronted each other threateningly for twenty to thirty seconds, but since neither gave in they parted again. 'Climber' returned to the spot where he had previously sat, and within the next several minutes he made two more attempts to settle the dispute but without success. I could not help thinking of two belligerent school boys that were attempting to intimidate each other without actually using force. In a last attempt to win this war of the nerves, 'Climber' threw up a handful of grass and stood up very erect to demonstrate all his strength. Then he ran towards the other male, and stared into his eyes from a distance of only a few centimeters. His opponent, however, was just as determined as he was, and this maneuver also settled nothing. The other members of the two groups paid very little attention to their leaders, and they pretended not to be in the least interested in the outcome of this dispute. After 'Climber' had been unsuccessful in forcing the opponent to yield, he started to feed and the females and juveniles of his group followed his example. 'Climber,' however, was not ready to give up yet. He approached the opponent for the last time, stared at him briefly, and then he averted his glance and rushed away. His troop followed him."

Gorillas in the wild and in the zoo, by B. Grzimek

My co-worker Jakob Schmitt in the Frankfurt Zoo conducted electrophoretic and other comparative blood tests on great apes, men, and other monkeys during the past few years. According to this data the gorilla is second after the chimpanzee in being related to man, but closer than the Asiatic orang-utan. This also seems to be supported by the behavior and way of life of gorillas. Despite this fact, we are more attracted to the gorilla than to the behavior of the chimpanzee, which seems to reflect the more unfavorable qualities of man as well in a rather embarrassing manner for us.

There are no observations of free living gorillas ever drinking water, although they waded through streams which were thirty to sixty centimeters deep. The animals do not cross deeper or broader rivers unless there is a tree that forms a bridge across the water. Gorillas, just like man and the other two species of great apes, are not natural swimmers, in contrast to almost all the lower monkeys and most mammals. In the New York Zoo the gorilla male "Makobo" drowned in the ditch of water which surrounded his new outside enclosure in 1951. The animal had bent over the water, then tumbled in and sank like a stone. "Makoba" made no attempts to swim or to use the ropes and ladders which had been provided for this purpose. The mature "Abraham" of the Frankfurt Zoo died in a similar manner in a water ditch in 1967.

In contrast to chimpanzees, free living gorillas have never been observed to utilize sticks or other objects as tools. They also do not show any intense curiosity for strange objects. On one occasion, Schaller's backpack was lying in full view of a black-backed male, no more than five meters away from him. The animal looked at it once but then did not pay any more attention to it. A piece of paper which was a gleaming white against the green of the forest was used to mark a path; however, the gorillas which walked past it paid no attention to it. They behaved similarly in other situations. Gorillas are "introverts" in the wild, as well as in the zoo. Man would probably be a lot friendlier and more peace-loving if he were more closely related to the gorilla instead of the chimpanzee.

Gorillas see, hear, and smell just about as well as people. They get up between six and eight o'clock in the morning out of beds which they constructed themselves. Then they feed rather intensively for about two hours. Between 10 A.M. and 2 P.M. they rest, but occasionally may get up to eat a little more. Around 6 P.M., just prior to nightfall, they build their padded nests. There are local variations in nest construction, just as in food preferences. Certain plants, which constitute the major part of the diet of the gorillas in one mountain chain, may not be touched by others living in a different location although the plants grow there. Gorillas which were already quite large or fully grown when captured would not touch bread, bananas, or plants they did not know for a long time. Gorilla infants are more willing to try new foods.

Gorillas like to sun themselves

A gorilla is not a creature who feels well only within the shadowy darkness of the forest. Although they live in the forest, and particularly the mountain gorilla in regions that are both foggy and rainy, they do not avoid the sun. Just the opposite is true. They seem to be delighted when the sun comes out. Some animals lie on their backs for more than two hours to bask in the sun. Drops of sweat form on their upper lips, and sweat runs down their chests. Under no circumstances do they avoid the sun. They even get up from their resting places in order to be in an area where there is sun.

They are used to rain, but are not very enthusiastic about it. Quite frequently gorillas will get up at the beginning of a rain shower and stand beneath a tree or sit closely pressed against a tree trunk where they will stay dry. Just as often, however, they remain sitting in the rain. They do not mind light showers. Usually they stop searching for food. If they have already been resting, they turn from their backs to their stomachs or else sit up. At the beginning of a heavy downpour, the gorillas which were up in the trees climbed down, and the infants returned to their mothers. If the animals are sitting in the open they duck their heads so that the chin touches the chest. They cross their arms over their chests or put the left hand on the right shoulder. Mothers often pick up their infants in their arms, and lean forwards in such a way as to keep the infants dry. In this way the gorillas sit motionless and silently and let the water run off their shoulders and eyebrows. They are then a rather sorry sight. In this state they are hardly disturbed by anything. Schaller once walked right through such a crouching gorilla group without any previous warning. Only one of the animals lifted its head. On another occasion he approached the gorillas in full sight and sat down at a distance of three to ten meters beneath an overhanging tree trunk. Although the animals saw him they did not go away. It has to rain hard for more than two hours before the animals will get up to look for food despite the downpour. They never leave their sleeping nests because of rain, and during heavy hail they behave similarly.

Interactions with other animals

These self-confident giants pay little attention to elephants, buffalos, Duiker antelopes, and other animals. Groups of chimpanzees occasionally occupy the same region as the gorillas without fighting taking place. Sometimes chimpanzee and gorilla groups may sleep in close proximity.

The literature often mentioned the fact that leopards prey on young gorillas, especially at night. In Kabara there was no evidence of this. During the entire observation period only one infant disappeared from the group; the cause remained unknown. The feces of leopards in this region contained almost only hairs and remains of duikers and tree hyrax. In the region of Kisoro, Ruanda, a gorilla male was found which had been killed in a fight with a leopard. On another occasion a killed gorilla female was found. The leopard had surprised the sleeping animal in its nest, and both had rolled down the slope.

Encounters with man

When gorilla groups encountered people, they usually retreated peacefully and disappeared. This is their usual behavior. Injuries inflicted by gorillas occur only if one interferes with the animals, for example, if one is out to kill the gorillas, or to encircle a group, or to catch individuals, or, rarely, when chasing them out of cultivated fields and plantations. After native hunters have killed the silver-backed male they encircle the females and kill them with clubs. The animals do not even make attempts to escape. It is pitiful to see how they only hold their arms over their heads to fend off the blows and how they remain in this submissive posture.

Gorillas are not dangerous

Between 1950 and 1959, the mission hospital at Kitsombiro near Lubera treated nine cases of injuries inflicted by gorillas. Only six of these required a long period of hospitalization. Three Africans were bitten because they had encircled a lone silver-backed male with the intent to kill it. The injuries were located in the thigh, the calf, and the hand. On another occasion a single black-backed male was surrounded. One of the hunters fled and slipped. The gorilla grabbed him by the knee and the heel, and bit into the calf, ripping out a piece of muscle eighteen inches long. Several years ago a Bantu in Kayonza was bitten in the seat. The Vativa are still laughing about this incident today because the injured man could not sit down for a long time. It is usually the gorilla males that attack when defending the group, and only occasionally females. In the Cameroon it is a disgrace to be wounded by a gorilla because everyone knows that the animal would not have attacked if the man had not become frightened and run away.

The gorilla is not a malicious forest devil. Despite his superior strength and power, he is peaceful and sociable. The gorilla's reputation which hunters have spread is plainly wrong. This may have been due to their own guilty consciences. When a hunter shoots a large great ape, he may experience emotions which are similar to those he would feel if he had shot a man. The facial expression, the behavior, and the look of a gorilla are so unbearably human. Before our own conscience and in front of other people, however, we best rationalize a killing by slandering the opponent as a criminal, a monster, and a murderer, or at least as an attacker who had to be shot in self defense.

Growth of human populations endangers the Gorillas

All over Africa, as in the rest of the world, the human population is increasing rapidly and requires additional room to live. Where a few years ago gorillas roamed in dense forests, today cattle graze on green meadows. In the gorilla range of the eastern Congo region, the ground is covered with a thick, fertile layer of humus which makes agricultural exploitation possible. The original plant growth has been largely displaced; only high on the mountains and in the forest preserves does one find relatively untouched strips of land. Experience has shown that rare animals like the mountain gorilla are in great danger. Continuous vigilance is required if this animal is not to become extinct one day. The lowland gorilla's condition is apparently not much better.

Julian Huxley has written, with justification: "Conservation must become a central concern in all decisions. The young African nations must understand the bitter fact that without proper conservation of the soil, water resources, and natural plant growth their lands will loose their value. They also need to recognize that their natural riches, including the wild animals and the beauty of their land, are positive assets."

Basically, the fate of animals depends on how man manages his environment. Is he going to continue to rob nature, or is he going to take his role as her custodian seriously? Man is conquering the diseases which once kept human populations in check, and he has increased his range by exterminating animals and depleting the soil. The same attitudes which once enabled him to overcome lions and bears are now helping him to subdue nature, and he sacrifices that which is lasting for the satisfaction of short-lived, immediate needs. He has the power to destroy the earth, but he seems unable to recognize that he is dependent upon plants, animals, soil, and water, and that he is one with them. Man depends on these things just as do protozoans, tse-tse flies, and gorillas, but man has removed himself from this ecological community. He has become the tyrant, but a tyrant whose fate is sealed if he should succeed in displacing all those with whom he has been in such murderous competition until now.

Prognosis for the future, by E. Klinghammer

Since the first German edition of this volume was written, much has happened with respect to the public's awareness of environmental problems, man's place in nature, and his interdependence with other animals. While admittedly much needs to be done, and while indeed conservation and preservation efforts may fail in the long run, it is true that conservation groups, concerned citizens, and young people are exerting an increasing influence on governmental policies with respect to these problems, especially in the United States. It is hoped that these concerns will become worldwide, that they will continue with increasing effectiveness, and that they will eventually permeate all human societies. Our hope especially lies with the young people of the world who are turning away, especially in the technologically most advanced countries, from the attitudes of progress for its own sake without regard to the long term effects. The defeat of a bill in the United States Congress in 1971 to support the development of the American Supersonic Transport is a first and hopeful symbol that technology must not dominate our lives but be our servant. This means that we must consider the implications of man's activities on the total environment and its effect on the quality of life.

Needless to say, the human population growth must come to a stop. Not just zero population growth, but a gradual decline for many years must be our goal. The real challenge of the future lies in our ability to accomplish these necessary tasks. The threat to various species discussed in this volume cannot be averted in the long run by rules and regulations about hunting them, or by establishing preserves for them. As long as human numbers increase unchecked, it is inevitable that the habitats of many presently existing species will be destroyed and their inhabitants along with them. The sad but inescapable fact is that the disappearing wild species are but heralds of our own demise. Yet, tragically, while vanishing, these species have shown us the basic requirements for a habitat in balance, and we thus have a model for what we must do to preserve ourselves. If we can learn a lesson from their passing, perhaps we can save ourselves.

For this to come about we must bring to the public—worldwide—an awareness of the nature of living things, and a recognition that we too, as men, are subject to the same laws as are the animals discussed in these pages. Should this awareness become strong and influential enough in time, we might yet preserve most of the world as we know it and pass it on to our descendants.

The pictures of the earth suspended in space that were brought back by the astronauts should bring home the point that our habitat, like that of any other species, is limited and must be managed wisely. Man must use his reason to accomplish what the blindly acting forces of instinct and physical realities impose on the animals, and which they will ultimately impose on us should we fail to use our brain for our ultimate survival. Perhaps this goal could unite mankind in a way that nothing so far has been able to do.

Bernhard Grzimek
George B. Schaller

Systematic Classification

Fossil species have not been included. The page numbers refer to the main chapter; page numbers in parentheses refer to illustrations of species or subspecies which are, however, not mentioned in the text.

Class Mammals (Mammalia)

Subclass Egglaying Mammals (Prototheria)

1. Order Monotremes (Monotremata)

Family: Echidnas (Tachyglossidae) 40
- *Tachyglossus* (Short-nosed spiny anteaters, or echidnas) 40
  - Australian echidna, *T. aculeatus* (Shaw & Nodder, 1792) 43
  - Tasmanian echidna, *T. setosus* (Geoffroy, 1803) 43
- *Zaglossus* (Long-nosed spiny anteaters or echidnas) 43
  - Barton's echidna, *Z. bartoni* (Thomas, 1907) 43
  - Bruijn's echidna, *Z. bruijni* (Peters & Doria, 1876) 43
  - Bubu echidna, *Z. bubuensis* Laurie, 1952 43

Family: Duck-billed platypus (Ornithorhynchidae) 45
- *Ornithorhynchus* 45
  - Duck-billed platypus, *O. anatinus* (Shaw & Nodder, 1799) 45

Subclass Marsupials (Metatheria)

2. Order Marsupials (Marsupialia)

Family: Opossums (Didelphidae) 57
- *Caluromys* (Woolly opossums) 57
  - Black-shouldered woolly opossum, *C. irrupta* (Sanborn, 1951) 58
  - Philander opossum, *C. philander* (Linné, 1758) 58
  - Woolly opossum, *C. laniger* (Desmarest, 1820) 58
- *Monodelphis* (Short bare-tailed opossum) 58
  - House short bare-tailed opossum, *M. domestica* (Wagner, 1842) 58
  - Striped short bare-tailed opossum *M. americana* (Müller, 1776) 58
- *Glironia* (Bushy-tailed opossums) 58
  - *G. venusta* Thomas, 1912 58
  - *G. aequatorialis* Antony, 1926 58
- *Marmosa* (South American mouse opossums) 58
  - Mexican mouse opossum, *M. mexicana* Merriam, 1897 58
  - Murine opossum, *M. murina* (Linné, 1758) 58
  - Ashy opossum, *M. cinerea* (Temminck, 1834) 58
- *Metachirops* 58
  - Four-eyed opossum, *M. opossum* (Linné, 1758) 58
- *Metachirus* 58
  - Rat-tailed opossum, *M. nudicaudatus* (Geoffroy, 1803) 58
- *Lutreolina* 58
  - Little water opossum, *L. crassicaudata* (Desmarest, 1804) 59
- *Chironectes* (Water opossums) 59
  - Yapok, *Ch. minimus* Zimmermann, 1780 59
- *Didelphis* (Common opossum) 59
  - Common opossum, *D. marsupialis* Linné, 1758 59
  - North American opossum, *D. marsupialis virginiana* Kerr, 1792 (62)
  - Azara's opossum, *D. paraguayensis* Oken, 1816 59

Family: Dasyures (Dasyuridae) 70

Subfamily: Marsupial mice (Phascogalinae) 73
- *Antechinus* (Broad-footed marsupial mice) 73
  - Yellow-footed marsupial mouse, *A. flavipes* (Waterhouse, 1831) 73
  - Pygmy marsupial mouse, *A. maculatus* Gould, 1851 73
  - Subgenus *Pseudantechinus:*
  - Fat-tailed marsupial mouse, *A. (Pseudantechinus) macdonnellensis* (Spencer, 1896)
  - Subgenus *Parantechinus:* 74
  - Speckled marsupial mouse, *A. (Parantechinus)*

1818) 113
Bear phalanger, *Ph. ursinus* (Temminck, 1824) 113
Dactylopsila (Striped phalanger) 113
Striped phalanger, *D. trivirgata* Gray, 1858 114
Long-fingered striped phalanger, *D. palpator* (Milne-Edwards, 1888) 114
Eudromicia (Pygmy possums) 114
New Guinean pygmy possum, *E. caudata* (Milne-Edwards, 1877) 114
Cercaërtus (Dormouse possum) 114
Thick-tailed dormouse possum *C. nanus* (Desmarest, 1818) 114
South-western pygmy phalanger, *C. concinnus* (Gould, 1845) 114
Burramys 114
B. parvus Broom, 1895 114
Gymnobelideus 114
Leadbeater's phalanger, *G. leadbeateri* McCoy, 1867 ◊ 114
Petaurus (Lesser gliding possum) 115
Honey glider, *P. breviceps* (Waterhouse, 1839) 115
Papua honey glider, *P. breviceps papuanus* Thomas, 1888 117
Lesser gliding possum, *P. norfolcensis* (Kerr, 1792) 115
Yellow-bellied glider, *P. australis* Shaw & Nodder, 1791 115
Distoechurus 117
Pen-tailed phalanger, *D. pennatus* (Peters, 1874) 117
Acrobates (Pygmy flying phalanger) 117
Pygmy flying phalanger, *A. pygmaeus* (Shaw, 1793) 117
New Guinea pygmy flying phalanger, *A. pulchellus* Rothschild, 1893 117

Subfamily: Long-snouted phalanger (Tarsipedinae) 118
Tarsipes 118
Honey possum, *T. spenserae* Gray, 1842 118

Subfamily: Koalas (Phascolarctinae) 119
Pseudocheirus (Ring-tailed possums) 119
Subgenus *Pseudocheirus* (i.n.s.): 119
Queensland ring tail, *P. (Pseudocheirus) peregrinus* (Boddaert, 1785) 119
Subgenus Pseudocheirops: 119
Striped ring tail, *P. (Pseudocheirops) archeri* (Collett, 1884) 119
Subgenus *Petropseudes:* 119
Rock-haunting ring tail, *P. (Petropseudes) dahli* (Collett, 1895) 119
Subgenus *Hemibelideus:* 119
Brush-tipped ring tail, *P. (Hemibelideus) lemuroides* (Collett, 1884) 119
Schoinobates 119
Greater gliding possum, *Sch. volans* (Kerr, 1792) 119
Phascolarctos 119
Koala, *Ph. cinereus* (Goldfuss, 1817) 119

Family: Wombats (Vombatidae) 129
Vombatus 129
Common wombat, *V. ursinus* (Shaw, 1800) ◊ 139
Australian common wombat, *V. ursinus platyrrhinus* (Owen, 1853) ◊ (101)
Tasmanian common wombat, *V. ursinus tasmaniensis* (Spencer & Kershaw, 1910) (101)
Lasiorhinus
Hairy-nosed wombat, *L. latifrons* (Owen, 1845) ◊ 139

Family: Kangaroos (Macropodidae) 147

Subfamily: Musky rat kangaroos (Hypsiprymnodontinae) 159
Hypsiprymnodon 159
Musky rat kangaroo, *H. moschatus* Ramsay, 1876 ◊ 159

Subfamily: Rat kangaroo (Potoroinae) 160
Potoroops 160
Broad-faced rat kangaroo, *P. platyops* (Gould, 1844) 160
Potorous (Long-nosed rat kangaroo) 160
Gilbert's rat kangaroo, *P. gilberti* (Gould, 1841) 160
Long-nosed rat kangaroo, *P. tridactylus* (Kerr, 1792) 160
Caloprymnus 160
Desert rat kangaroo, *C. campestris* (Gould, 1843) ◊ 160
Bettongia (Short-nosed rat kangaroos) 160
Gaimard's rat kangaroo, *B. gaimardi* (Desmarest, 1822) 161
Lesueur's rat kangaroo, *B. lesueur* (Quoy & Gaimard, 1824) ◊ 161
Aepyprymnus 161
Rufous rat kangaroo, *Ae. rufescens* (Gray, 1837) 161

Subfamily: True kangaroos (Macropodinae) 162
Lagorchestes (Hare wallabies) 162
Brown hare wallaby, *L. leporoides* (Gould, 1841) 162
Western hare wallaby, *L. hirsutus* Gould, 1844 ◊ 162
Bernier's hare wallaby, *L. hirsutus bernieri* Thomas, 1907 162
Dorre hare wallaby, *L. hirsutus dorreae* Thomas, 1907 162
Spectacled hare wallaby, *L. conspicillatus* Gould, 1841 162
Leichhardt's spectacled hare wallaby, *L. conspicillatus leichardti* Gould, 1853 162
Lagostrophus 163
Banded hare wallaby, *L. fasciatus* (Péron & Lesueur, 1887) ◊ 163
Petrogale (Rock wallabies) 163
Brush-tailed rock wallaby, *P. penicillata* (Griffith, 1827) 163

P. penicillata penicillata (Griffith, 1827) ◊ 163
Ring-tailed rock wallaby, *P. xanthopus* Gray, 1885 163
Palin rock wallaby, *P. inornata* Gould, 1842 (133)
Short-eared rock wallaby, *P. brachyotis* Gould, 1841 (133)
Peradorcas 164
Little rock wallaby, *P. concinna* (Gould, 1842) 164
Onychogalea (Nail-tail wallabies) 164
Northern nail-tail wallaby, *O. unguifer* (Gould, 1840) 164
Bridled nail-tail wallaby, *O. fraenata,* (Gould, 1841) ◊ 164
Crescent nail-tail wallaby, *O. lunata* (Gould, 1840) ◊ 164
Dendrolagus (Tree kangaroos) 164
Black tree kangaroo, *D. ursinus* Temminck, 1836 164
Goodfellow's tree kangaroo, *D. goodfellowi* Thomas, 1908 164
Matschie's tree kangaroo, *D. matschiei* Förster & Rothschild, 1907 164
Doria's tree kangaroos, *D. dorianus* Ramsay, 1883 164
Bennett's tree kangaroo, *D. dorianus bennettianus* De Vis, 1887 164
Lumholtz's tree kangaroo, *D. lumholtzi* Collett, 1884 (135)
Dorcopsis (New Guinean wallabies) 165
Northern New Guinean wallaby, *D. hageni* Heller, 1897 166
Goodenough's New Guinean wallaby, *D. atrata* Van Deusen, 1957 166
Subgenus *Dorcopsulus:* 166
New Guinean mountain wallaby, *D. (Dorcopsulus) macleayi* Miklouho-Macleay, 1885 166
Thylogale (Pademelons) 166
Red-necked pademelon, *Th. thetis* (Lesson, 1828) 166
Red-legged pademelon, *Th. stigmatica* Gould, 1860 166
Bruijn's pademelon, *Th. brunii* (Schreber, 1778) 166
Rufous-bellied pademelon, *Th. billardierii* (Desmarest, 1822) 166
Setonix (Quokkas) 169
Quokka, *S. brachyurus* (Quoy & Gaimard, 1830) 169
Wallabia (Wallabies) 169
Tammar, *W. eugenii* (Desmarest, 1817) 169
Parma wallaby, *W. parma* (Waterhouse, 1846) ◊ (137)
Red-necked wallaby, *W. rufogrisea* (Desmarest, 1817) 169
Tasmanian red-necked wallaby, *W. rufogrisea frutica* (Ogilby, 1838) 169
Mainland red-necked wallaby, *W. rufogrisea banksiana* (Quoy & Gaimard, 1825) 169
Island red-necked wallaby, *W. rufogrisea rufogrisea* (Desmarest, 1817) 169
Black-striped wallaby, *W. dorsalis* (Gray, 1837) 170
Pretty-face wallaby, *W. canguru* (Müller, 1776) 170
Black-tailed wallaby, *W. bicolor* (Desmarest, 1804) 170
Sandy wallaby, *W. agilis* (Gould, 1842) 170
Black-gloved wallaby, *W. irma* (Jourdan, 1837) 170
Eastern black-gloved wallaby, *W. irma greyi* (Waterhouse, 1846) 170
Macropus (Kangaroos) 171
Subgenus *Megaleia:* 171
Red kangaroo, *M. (Megaleia) rufus* (Desmarest, 1822) 171
Western red kangaroo, *M. (Megaleia) rufus dissimulatus* Rothschild, 1905 172
Northern red kangaroo, *M. (Megaleia) rufus pallidus* Schwarz, 1910 172
Subgenus *Macropus* (i.n.s.) 171
Grey kangaroo, *M. (Macropus) giganteus* (Zimmermann, 1777) 171
Tasmanian grey kangaroo, *M. (Macropus) giganteus tasmaniensis* Le Souef, 1923 172
Subgenus *Osphranter:* 171
Wallaroo, *M. (Osphranter) robustus* Gould, 1841 171
South-eastern wallaroo, *M. (Osphranter) robustus robustus* Gould, 1841 (138)
Antelope wallaroo, *M. (Osphranter) robustus antilopinus* (Gould, 1842) 173
Deer wallaroo, *M. (Osphranter) robustus cervinus* Thomas, 1900 173

Subclass Higher Mammals (Eutheria)

3. Order Insectivora

Superfamily: Tenrecs (Tenrecoidea)
Family: Solenodons (Solenodontidae) 182
Solenodon 182
Haitian solenodon, *S. paradoxus* Brandt, 1833 182
Atopogale 182
Cuban solenodon, *A. cubana* (Peters, 1861) ◊ 182
Family: Tenrecs (Tenrecidae) 185

4. Order Primates

Suborder Prosimiae

Infraorder tree shrews (Tupaiiformes)

Infraorder Lemurs (Lemuriformes)

Family: Lemurs (Lemuridae) 277
Subfamily: Dwarf lemurs (Cheirogaleinae) 277
Microcebus (Mouse lemurs) 278
Lesser mouse lemur, *M. murinus* (J. F. Miller, 1777) 278
Coquerel's mouse lemur, *M. coquereli* (Grandidier, 1867) ◊ 278
Cheirogaleus (Dwarf lemurs) 278
Hairy-eared dwarf lemur, *Ch. trichotis* Günther, 1875 ◊ 278
Fat-tailed dwarf lemur, *Ch. medius* E. Geoffroy, 1812 ◊ 278
Greater dwarf lemur, *Ch. major* E. Geoffroy, 1812 278
Phaner 278
Fork-marked mouse lemur, *Ph. furcifer* (Blainville, 1839) ◊ 278
Subfamily: Typical lemurs (Lemurinae) 280
Hapalemur (Gentle lemurs) 280
Grey gentle lemur, *H. griseus* (Link, 1795) ◊ 280
Broad-nosed gentle lemur, *H. simus* Gray, 1870 ◊ 280
Lemur (Lemurs) 280
Ring-tailed lemur, *L. catta* Linné, 1758 281
Ruffed lemur, *L. variegatus* Kerr, 1792 281
L. variegatus ruber E. Geoffroy, 1812 281
Black lemur, *L. macaco* Linné, 1766 281
Brown lemur, *L. fulvus* E. Geoffroy, 1812 281
L. fulvus rufus Audebert, 1799 ◊ 281
L. fulvus albifrons E. Geoffroy, 1776 281
Mongoose lemur, *L. mongoz* Linné, 1766 281
True mongoose lemur, *L. mongoz mongoz* Linné, 1766 281
L. mongoz coronatus Gray, 1842 281
Red-bellied lemur, *L. rubriventer* I. Geoffroy, 1850 281
Lepilemur (Weasel lemurs) 281
Sportive lemur, *L. ruficaudatus* Grandidier, 1867 ◊ 281
Weasel lemur, *L. mustelinus* I. Geoffroy, 1851 281
Family: Indrisoid lemurs (Indriidae) 290
Propithecus (Sifakas) 290
Verreaux's sifaka, *P. verreauxi* Grandidier, 1867 290
Diademed sifaka, *P. diadema* Bennett, 1832 290
Avahi 290
Woolly indri, *A. laniger* (Gmelin, 1788)
Indri 291
Indri, *I. indri* (Gmelin, 1788) ◊ 291
Family: Aye-ayes (Daubentoniidae) 295
Daubentonia 296
Aye-aye, *D. madagascariensis* (Gmelin, 1788) ◊ 296

Infraorder Lorises (Lorisiformes)

Family: Lorises (Lorisidae) 298
Loris 299
Slender loris, *L. tardigradus* (Linné, 1758) 299
Nycticebus 299
Slow loris, *N. coucang* (Boddaert, 1758) 299
Lesser slow loris, *N. pygmaeus* Bonhote, 1907 299
Arctocebus 299
Angwantibo, *A. calabarensis* (Smith, 1860) 299
Perodicticus 299
Potto, *P. potto* (P. L. S. Müller, 1766) 299
Family: Galagos (Galagidae) 304
Galago 304
Senegal bushbaby, *G. senegalensis* E. Geoffroy, 1796 304
G. senegalensis braccatus Elliot, 1907 (252)
Allen's bushbaby, *G. alleni* Waterhouse, 1837 304
Thick-tailed bushbaby, *G. crassicaudatus* E. Geoffroy, 1812 304
G. crassicaudatus panganensis Matschie, 1906 (252)
Demidoff's bushbaby, *G. demidovii* Fischer, 1808 304
Western needle-clawed bushbaby, *G. elegantulus* (Le Conte, 1857) 304
G. elegantulus pallidus Gray, 1863 304
Eastern needle-clawed bushbaby, *G. inustus* Schwarz, 1930 304

Infraorder Tarsiers (Tarsiiformes)

Family: Tarsiers (Tarsiidae) 307
Tarsius 307
Western tarsier, *T. bancanus* Horsfield, 1821 308
Bornean tarsier, *T. bancanus borneanus* Elliot, 1910 (253)
Philippine tarsier, *T. syrichta* (Linné, 1758) 308
Mindanao tarsier, *T. syrichta carbonarius* Heude, 1898 309
Eastern tarsier, *T. spectrum* Pallas, 1778 308

Suborder Monkeys, Apes and Man (Simiae)

Infraorder New World Monkeys (Platyrrhina)

Infraorder Old World Simian Primates (Catarrhina)

(Pang-Chieh, 1957) 451
Mentawi leaf monkey, *P. (Trachypithecus) potenzani* (Bonaparte, 1856) 451
Subgenus *Presbytis* (i.n.s.) (Island langurs): 452
Banded leaf monkey, *P. (Presbytis) melalophus* (Raffles, 1821) 452
Maroon leaf monkey, *P. (Presbytis) rubicundus* (Müller, 1838) 452
P. (Presbytis) melalophus chrysomelas (Müller, 1838) (454)
P. (Presbytis) rubicunda carimatae Miller, 1906 (454)
White-fronted leaf monkey, *P. (Presbytis) frontatus* (Müller, 1838) 457
Sunda Island leaf monkey, *P. (Presbytis) aygula* (Linné, 1758) 457
P. (Presbytis) aygula thomasi (Collett, 1892) (454)

Pygathrix 457
Douc langur, *P. nemaeus* (Linné, 1771) ◊ 457

Rhinopithecus (Snub-nosed monkeys) 457
Snub-nosed monkey, *Rh. roxellanae* Milne-Edwards, 1870 ◊ 458
Brown snub-nosed monkey, *Rh. bieti* Milne-Edwards, 1897 458
Brelich's monkey, *Rh.* Thomas, 1903 458
Tonkin snub-nosed monkey, *Rh. avunculus* Dollman, 1912 458

Simias 459
Pig-tailed langur, *S. concolor* Miller, 1903 ◊ 459

Nasalis 459
Proboscis monkey, *N. larvatus* (Wurmb, 1781) 459

Colobus (Colobus monkeys) 463
Green colobus, *C. verus* (Van Beneden, 1838) ◊ 463
Red colobus, *C. badius* (Kerr, 1792) 463
C. badius gordonorum (Matschie, 1900) ◊ 464
Red-headed colobus, *C. badius rufomitratus* Peters, 1879 ◊ 464
C. badius kirkii Gray, 1868 ◊ 464
C. badius ferrugineus (Shaw, 1800) (456)
C. badius ellioti, Dollman, 1909 (456)
Northern black-and-white colobus, *C. abyssinicus* (Oken, 1816) 464
Kilimanjaro colobus, *C. abyssinicus caudatus* Thomas, 1885 465
Kikuyu colobus, *C. abyssinicus kikuyuensis* Lönnberg, 1912
Southern black-and-white colobus, *C. polykomos* (Zimmermann, 1780) 464
Black colobus, *C. polykomos satanas* Waterhouse, 1838 464
C. polykomos angolensis Sclater, 1860 (456)

Superfamily: Apes and Men (Hominoidea)

Family: Gibbons (Hylobatidae) 470

Symphalangus (Siamangs) 471
Siamang, *S. syndactylus*, Raffles, 1821 471
Dwarf siamang, *S. klossi*, (Miller, 1903) 471

Hylobates (Typical gibbons) 478
Black gibbon, *H. concolor* (Harlan, 1826) 478
White-cheeked gibbon, *H. concolor leucogenys* Ogilby, 1840 478
Hoolock gibbon, *H. hoolock* (Harlan, 1834) 478
White-handed gibbon, *H. lar* (Linné, 1771) 478
H. lar pileatus Gray, 1861 478
Dark-handed gibbon, *H. agilis* Cuvier, 1821 478
Grey gibbon, *H. moloch*, (Audebert, 1797) 478

Family: Great apes (Pongidae) 488

Pongo 503
Orang-utan, *P. pygmaeus* (Linné, 1760) ◊ 503
Bornean orang-utan, *P. pygmaeus pygmaeus* (Linné, 1760) 503
Sumatran orang-utan, *P. pygmaeus abelii* Lesson, 1827 503

Gorilla 525
Gorilla, *G. gorilla* (Savage & Wyman, 1847) 525
Lowland gorilla, *G. gorilla gorilla* (Savage & Wyman, 1847) 525
Mountain gorilla, *G. gorilla beringei* Matschie, 1903 ◊ 526

The chimpanzees (Genus *Pan*) which together with the gibbons and great apes belong to the superfamily Hominoidea (which also includes man [Family Hominidae]), are discussed in Volume XI.

On the Zoological Classification and Names

For many years, zoologists and botanists have tried to classify animals and plants into a system which would be a survey of the abundance of forms in fauna and flora. Such a system, of course, may be established under very different aspects. Since Charles Darwin, his predecessors, and his successors have found that all creatures have evolved out of common ancestors, species of animals and plants have been classified according to their natural relationships. Our knowledge about the phylogeny, and thus the relationship of each living being to the other, is augmented every year by new discoveries and insights. Old ideas are replaced with more recent and more appropriate ones. Therefore, the natural classification of the animal kingdom (and the plant kingdom) is subject to changes. Furthermore, the opinions of zoologists, who are working on the classification of animals into the various groups, are anything but uniform. These differences and changes are usually insignificant. The classification of vertebrates into the classes of fish, amphibians, reptiles, birds, and mammals has been fixed for many decades. Only the Cyclostomata were recently separated from the fish and all other classes of vertebrates as the "jawless" Agnatha (comp. Vol. 4).

The animal kingdom has been split into several subkingdoms and these were again divided into further sections, subsections, and so on. The scale of the most important systematic categories follows in a descending rank order:

Kingdom
Subkingdom
Phylum
Subphylum
Class
Subclass
Superorder
Order
Suborder
Infraorder
Family
Subfamily
Tribe
Genus
Subgenus
Species
Subspecies

The scientific names of the animals and their spelling follow the international rules for the zoological nomenclature as agreed upon by the XV International Congress for Zoology and are obligatory for all zoological publications. The name of the genus, which is a Latin or Latinized noun, is singular and capitalized. After the name of the genus follows the name of the species and of the subspecies. The names of the species and subspecies may be nouns or adjectives, and they are spelled in the lower case. The name of a subgenus, which is formed in the same manner as a genus, may be added in brackets following the name of the genus. The names of the tribes, subfamilies, families, and superfamilies are plural capitalized nouns. They are formed from the name of a given genus by adding to the principal word the endings -ini for the tribe, -inae for the subfamily, -idae for the family, and -oidea for the superfamily. The names of the authors who were the first to describe and to name a species, subspecies, or group of animals should be cited with the year of this naming at least once in each scientific publication. The name of the author and year are not enclosed in brackets when the species or subspecies is classified as belonging to the same genus with which the author had originally classified it. They are in brackets when another genus name is used in the present publication. The scientific names of the genus, subgenus, species, and subspecies are supposed to be printed with different letters, usually italics.

ANIMAL DICTIONARY

I. English—German—French—Russian

For scientific names consult the German—English—French—Russian section of the dictionary, or the index.

In most cases names of subspecies are formed by putting an adjective or geographical specification before the name of species. These English names of subspecies will, as a rule, not appear in this part of the zoological dictionary.

ENGLISH NAME	GERMAN NAME	FRENCH NAME	RUSSIAN NAME
African Forest Shrew	Afrikanische Riesenspitzmaus	Musaraigne géante	гигантская белозубка
Agile Mangabey	Haubenmangabe	Mangabey à ventre doré	чубастый мангабей
Algerian Hedgehog	Algerischer Igel	Hérisson d'Algérie	алжирский еж
Allen's Bushbaby	Buschwaldgalago	Galago d'Allen	
Alpine Shrew	Alpenspitzmaus	Musaraigne des montagnes	альпийская бурозубка
American and Asian Moles	Amerikanisch-Asiatische Maulwürfe	Taupes d'Asie et d'Amérique du Nord	американско-азиатские кроты
– Pigmy Shrew	Amerikanische Zwergspitzmaus	Musaraigne pygmée d'Amérique	североамериканская карликовая бурозубка
– Shrew Mole	Amerikanischer Spitzmaus-maulwurf	Taupe de Gibbs	американский землеройковый крот
Amur Hedgehog	Amurigel	Hérisson de l'Amour	амурский еж
Angwantibo	Bärenmaki	Angwantibo	медвежий маки
Anubis Baboon	Grüner Pavian	Papion anubis	анубис
Apes	Menschenaffen	Pongidés	человекообразные обезьяны
Armoured Shrews	Panzerspitzmäuse		белозубки-броненоски
Ashy Opossum	Aschgraue Zwergbeutelratte		пепельная сумчатая крыса
Asian Water Shrews	Asiatische Wasserspitzmäuse	Chimarrogales	
Assamese Macaque	Bergrhesus	Macaque d'Assam	горный резус
Australian Echidna	Australien-Kurzschnabeligel	Échidné de l'Australie	австралийская ехидна
Aye-Aye	Fingertier	Aye-Aye	мадагаскарская руконожка
Azara's Opossum	Paraguayanisches Opossum		казака
Baboons	Paviane	Papions	павианы
Bald Uakari	Scharlachgesicht	Ouakari chauve	лысый уакори
Banded Hare Wallaby	Bänderkänguruh	Wallaby rayé	поперечнополосатый кенгуру
Banded Leaf Monkey	Roter Langur	Semnopithèque melalophe	рыжий тонкотел
Bandicoots	Nasenbeutler	Bandicoots	сумчатые барсуки
Barbary Ape	Magot	Magot	магот
Bare-faced Tamarins	Nacktgesichttamarins	Marikinas	
Barton's Echidna	Barton-Langschnabeligel	Échidné de Barton	проехидна Бартона
Bearded Sakis	Bartsakis		
Bear Phalanger	Bärenkuskus	Couscous de Célèbes	черный кускус
Bennett's Tree Kangaroo	Bennett-Baumkänguruh	Kangourou arboricole de Bennett	древесный кенгуру Беннетта
Bicolor White-toothed Shrew	Feldspitzmaus	Crocidure leucode	белобрюхая белозубка
Black-and-red Tamarin	Schwarzrückentamarin	Tamarin rouge et noir	черноспинный тамарин
Black-and-white Colobuses	Guerezas	Guérézas	
Black Gibbon	Schopfgibbon	Gibbon noir	одноцветный гиббон
Black-gloved Wallaby	Irmawallaby	Wallaby d'Irma	кенгуру ирма
Black-headed Squirrel Monkey	Schwarzköpfiges Totenköpfchen	Sapajou à tête noire	черноголовая саймири
Black-headed Uakari	Schwarzkopfuakari	Ouakari à tête noire	черноголовый уакори
Black Howler Monkey	Schwarzer Brüllaffe	Hurleur noir	черный ревун
Black Lemur	Mohrenmaki	Lémur macaco	черный маки
Black Mangabey	Schopfmangabe	Mangabey noir	бородатый мангабей
Black-pencilled Marmoset	Schwarzpinseläffchen	Ouistiti à pinceau noir	кисточковая игрунка
Black Saki	Satansaffe	Saki noir	чертов саки
– Spider Monkey	Schwarzer Klammeraffe	Singe-Araignée noir	черная коата
Black-striped Wallaby	Rückenstreifkänguruh		полосатый кустовый кенгуру
Black-tailed Phascogale	Großer Pinselschwanzbeutler		кистехвостая мышевидка
– Wallaby	Sumpfwallaby	Wallaby bicolore	чернохвостый кустовый кенгуру
Black Tree Kangaroo	Bären-Baumkänguruh		кенгуру-медведь
– Uakari	Schwarzer Uakari	Ouakari de Roosevelt	черный уакори
Bonnet Monkey	Indischer Hutaffe	Macaque bonnet chinois	индийский макак
Brandt's Hedgehog	Brandts Igel	Hérisson de Brandt	длинноиглый еж
Bridled Nail-tail Wallaby	Kurznagelkänguruh		уздечковый когтехвостый кенгуру
Brindled Bandicoot	Großer Kurznasenbeutler		большой курносый бандикут
Broad-faced Rat Kangaroo	Breitkopfkänguruh		широколицый потору
Broad-footed Marsupial Mice	Breitfußbeutelmäuse	Rats marsupiaux	
– Mole	Kalifornischer Maulwurf		калифорнийский крот
Broad-nosed Gentle Lemur	Breitschnauzenhalbmaki	Hapalémur à nez large	широконосый полумаки
Brown Capuchin	Apella	Sapajou apelle	капуцин-фавн
– Hare Wallaby	Langohr-Hasenkänguruh		обыкновенный заячий кенгуру
Brown-headed Spider Monkey	Braunkopfklammeraffe	Singe-Araignée à tête brune	буроголовая коата
– Tamarin	Braunrückentamarin	Tamarin à tête brune	буроспинный тамарин
Brown Howler Monkey	Brauner Brüllaffe	Hurleur brun	бурый ревун

ENGLISH NAME	GERMAN NAME	FRENCH NAME	RUSSIAN NAME
– Lemur	Schwarzkopfmaki	Lémur brun	черноголовый маки
– Snub-nosed Langur	Braune Stumpfnase	Rhinopithèque brun	биэтовский ринопитек
– Woolly Monkey	Brauner Wollaffe	Lagotriche de Castelnau	бурая шерстистая обезьяна
Bruijn's Echidna	Bruijn-Langschnabeligel	Échidné de Bruijn	проехидна Бруйна
– Pademelon	Neuguineafilander	Wallaby de Bruijn	аруанский кустовый кенгуру
Brush-tailed Opossums	Buschschwanzbeutelratten		кистехвостые сумчатые крысы
– Phalanger	Fuchskusu	Phalanger-Renard	лисий кузу
– Rock Wallaby	Bürsten-Felskänguruh		кистехвостый каменный кенгуру
– Tree Shrews	Buschschwanztupaias	Tupaiinés	собственно тупайи
Brush-tipped Ring Tail	Lemuren-Ringelschwanzbeutler		лемуровый кускус
Brush Wallabies	Wallabys	Wallabies	кустовые кенгуру
Bubu-Echidna	Bubu-Langschnabeligel		проехидна острова Бубу
Budeng	Budeng	Budeng	
Buff-headed Marmoset	Gelbkopfbüscheläffchen	Ouistiti à tête jaune	желтоголовая игрунка
Burramys Pigmy Opossum		Souris-Opossum de Burramy	
Bushveld Elephant Shrew	Trockenland-Elefantenspitzmaus		
Campbell's Guenon	Campbells Meerkatze	Mone de Campbell	мартышка Кампбелла
Cape Golden Mole	Kapgoldmull	Taupe dorée du Cap	капский златокрот
– Hedgehog	Kapigel		капский еж
Capped Langur	Schopflangur		хохлатый тонкотел
Capuchin Monkeys	Kapuzineraffen	Cébinés	цебусовые обезьяны
Capuchins	Kapuziner	Sapajous	капуцины
Celebes crested Macaque	Schopfmakak	Cynopithèque nègre	хохлатый павиан
Central African Hedgehogs	Mittelafrikanische Igel	Hérissons d'Afrique centrale	среднеафриканские ежи
– American Spider Monkey	Geoffroy-Klammeraffe	Singe-Araignée aux Mains noirs	коата Жоффруа
– Jerboa Marsupial	Inneraustralische Springbeutelmaus		среднеавстралийский сумчатый тушканчик
Ceram Long-nosed Bandicoot	Ceramnasenbeutler		
Chacma Baboon	Bärenpavian	Chacma	чакма
Checkered Elephant Shrew	Geflecktes Rüsselhündchen		пятнистая хоботковая собачка
Chilean Rat Opossum	Chile-Opossummaus		чилийская первокрыса
Coahuilan Mole	Coahuilamaulwurf	Taupe de Coahuila	
Coast Mole	Pazifischer Maulwurf	Taupe de côte	тихоокеанский крот
Collared Titi	Witwenaffe	Callicèbe à fraise	
Colobus Monkeys	Stummelaffen	Colobes	толстотелы
Common Eurasian Mole	Europäischer Maulwurf	Taupe commune	обыкновенный крот
– European White-toothed Shrew	Hausspitzmaus	Musaraigne musette	бурая белозубка
– Hedgehogs	Kleinohrigel	Hérissons communs	обыкновенные ежи
– Marmoset	Weißbüscheläffchen	Ouistiti	обыкновенная игрунка
– Opossum	Nordopossum	Opossum commun	обыкновенный опоссум
– Shrew	Waldspitzmaus	Musaraigne carrelet	обыкновенная бурозубка
– Squirrel Monkey	Totenkopfäffchen	Sapajou jaune	саймири-белка
– Tree-shrew	Gewöhnliches Spitzhörnchen	Toupaïe	обыкновенная тупайя
– Wombat	Nacktnasenwombat	Wombat à narines dénudées	медвежий вомбат
Congo Armoured Shrew	Kongopanzerspitzmaus		конголезская броненоска
Coquerel's Mouse Lemur	Coquerels Zwergmaki	Mirza de Coquerel	
Cotton-head Tamarin	Lisztäffchen	Pinché	эдипова игрунка
Crab-eating Macaque	Javaneraffe	Macaque de Buffon	яванский макак
Crawford's Desert Shrew	Graue Wüstenspitzmaus	Musaraigne du désert	серая пустынная бурозубка
Crescent Nail-tail Wallaby	Mondnagelkänguruh		полулунный когтехвостый кенгуру
Crested Bare-faced Tamarins	Perückenäffchen	Pinchés	пинче
Crest-tailed Marsupial Mice	Kammschwanzbeutelmäuse		
– Marsupial Rat	Doppelkammbeutelmaus		гребешковая мышевидка
Crowned Guenon	Kronenmeerkatze	Cercopithèque pogonias	чубатая мартышка
– Lemur	Kronenmaki	Lémur couronné	хохлатый монго
Cuban Solenodon	Kuba-Schlitzrüßler	Almiqui	кубинский щелезуб
Cuscuses	Kuskuse	Couscous	кускусы
Dark-footed Forest Shrew	Dunkelfüßige Waldspitzmaus		темнопалая белозубка
Dark-handed Gibbon	Ungka	Gibbon agile	быстрый гиббон
Dasyures	Beutelmarder	Dasyurinés	собственно сумчатые куницы
De Brazza's Monkey	Brazzameerkatze	Cercopithèque de Brazza	бразовская мартышка
Demidoff's Bushbaby	Zwerggalago	Galago de Demidoff	галаго Демидова
Desert Musk Shrew	Wüstenwimperspitzmaus	Crocidure du désert	пустынная белозубка
– Rat Kangaroo	Nacktbrustkänguruh	Rat-Kangourou du désert	степная кенгуровая крыса
Desmans	Desmane	Desmaninés	выхухоли
De Winton's Golden Mole	Winton-Goldmull	Taupe dorée de Winton	златокрот
Diademed Guenon	Diademmeerkatze	Cercopithèque diadème	
– Sifaka	Diademsifaka	Propithèque diadème	белолобый сифака
Diana Monkey	Dianameerkatze	Cercopithèque diane	диана
Doria's Tree Kangaroo	Doria-Baumkänguruh	Kangourou arboricole de Doria	древесный кенгуру Дория
Dormouse Possum	Dickschwanz-Schlafbeutler		обыкновенная сумчатая соня
Douc Langur	Kleideraffe	Douc	немейский тонкотел
Douroucouli	Nachtaffe	Singe de nuit	мирикина
Drill	Drill	Drill	дрил
Duck-billed Platypus	Schnabeltier	Ornithorhynque	утконос
Dusky Leaf Monkey	Brillenlangur	Semnopithèque obscur	очковый тонкотел

ENGLISH NAME	GERMAN NAME	FRENCH NAME	RUSSIAN NAME
Dwarf Guenon	Zwergmeerkatze	Talapoin	крошечная мартышка
– Lemurs	Katzenmakis	Chirogales	крысиные маки
– Siamang	Zwergsiamang	Siamang de Kloss	карликовый сиаманг
Eared Hedgehog	Ohrenigel		ушастые ежи
Eastern American Mole	Ostamerikanischer Maulwurf	Taupe à queue glabre	восточноамериканский крот
– Barred Bandicoot	Bänder-Langnasenbeutler		полосатый бандикут
– Dasyure	Tüpfelbeutelmarder		крапчатая сумчатая куница
– European Hedgehog	Weißbrustigel	Hérisson d'Europe de l'Est	белогрудый еж
– Jerboa Marsupial	Östliche Springbeutelmaus		восточноавстралийский сумчатый тушканчик
– Mole	Ostmaulwurf	Taupe d'Europe de l'Est	короткохвостый крот
– Needle-clawed Bushbaby	Östlicher Kielnagelgalago	Galago du Congo	
– Tarsier	Celebeskoboldmaki	Tarsier spectre	долгопят-привидение
Echidnas	Ameisenigel	Échidnés	ехидны
Ecuador Rat Opossum	Eierlegende Säugetiere	Protothériens	клоачные
Egg-laying Mammals	Ekuador-Opossummaus		бурая первокрыса
Elephant Shrews	Rüsselspringer	Macroscélidés	прыгунчики
Emperor Tamarin	Kaiserschnurrbarttamarin	Tamarin empereur	императорский усатый тамарин
Entellus Langur	Hulman	Houleman	гульман
Ethiopian Hedgehog	Äthiopischer Igel	Hérisson du désert	абиссинский еж
European Water Shrew	Wasserspitzmaus	Musaraigne aquatique	обыкновенная кутора
Fat-tailed Dwarf Lemur	Mittlerer Katzenmaki		
– Marsupial Mouse	Fettschwänzige Breitfußbeutelmaus		толстохвостая мышевидка
– Sminthopsis	Dickschwänzige Schmalfußbeutelmaus		толстохвостая сумчатая землеройка
Flat-skulled Marsupials	Flachkopfbeutelmäuse		плоскоголовые мышевидки
Flesh-eating Marsupials	Raubbeutler	Dasyuridés	хищные сумчатые
Fontoynont's Hedgehog Tenrec	Fontoynonts Igeltanrek		ежевый тенрек Фонтуанона
Forest Elephant Shrew	Waldrüsselratte		лесная хоботковая крыса
– Shrew	Afrikanische Waldspitzmaus		африканская лесная белозубка
Fork-marked Dwarf Lemur	Gabelstreifiger Katzenmaki	Phaner à fourche	вильчатый маки
Formosa Macaque	Formosamakak	Macaque de Formose	тайванский резус
Four-eyed Opossum	Vieraugenbeutelratte	Quatre-oeil	опоссум квика
Four-toed Elephant Shrew	Vierzehenrüsselratte		четырехпалая хоботковая крыса
François' Monkey	Tonkinlangur	Semnopithèque de François	тонкинский тонкотел
Gaimard's Rat Kangaroo	Festland-Bürstenkänguruh	Rat-Kangourou de Gaimard	австралийский кистехвостый кенгуру
Galago	Galago	Galago	галаго
Gelada Baboon	Dschelada	Gelada	джелада
Gentle Lemurs	Halbmakis	Hapalémur	полумаки
Geoffroy's Tamarin	Panamaperückenäffchen	Pinché de Geoffroy	игрунка Жоффруа
Giant Golden Mole	Riesengoldmull	Grande Taupe dorée	исполинский златокрот
– Musk Shrew	Riesenwimperspitzmaus		большая белозубка
Gibbons	Gibbons	Gibbons	гиббоны
Gilbert's Rat Kangaroo	Gilbert-Kaninchenkänguruh	Rat-Kangourou de Gilbert	потору Джильберта
Goeldi's Monkey	Springtamarin	Tamarin de Goeldi	
Golden-headed Saki	Goldkopfsaki	Saki à tête dorée	золотистый саки
– Tamarin	Goldkopflöwenäffchen	Singe-lion à tête dorée	желтоголовая львиная игрунка
Golden Lion Marmoset	Goldgelbes Löwenäffchen	Petit singe-lion	розалия
Golden-mantled Tamarin	Goldmanteltamarin	Tamarin à manteau doré	
Golden Moles	Goldmulle	Taupes dorées	златокроты
Golden-rumped Tamarin	Rotsteißlöwenäffchen	Singe-lion à queue jaune	краснозадая львиная игрунка
Goodfellow's Tree Kangaroo	Goodfellow-Baumkänguruh	Kangourou arboricole de Goodfellow	
Gorilla	Gorilla	Gorille	горилла
Grant's Desert Golden Mole	Wüstengoldmull	Taupe dorée de Grant	пустынный крот
Grass Monkey	Grüne Meerkatze	Singe vert	зеленая мартышка
Gray's Guenon	Grays Kronenmeerkatze	Cercopithèque de Gray	мартышка Грея
Great Apes	Menschenaffen	Pongidés	собственно человекообразные обезьяны
Greater Dwarf Lemur	Großer Katzenmaki	Chirogale de Milius	
– Gliding Phalanger	Riesengleitbeutler	Grand Phalanger volant	исполинская сумчатая летяга
– Hedgehog Tenrec	Großer Igeltanrek		обыкновенный ежевый тенрек
– Marsupial Mole	– Beutelmull	Grande Taupe marsupiale	сумчатый крот
– Sportive Lemur	– Wieselmaki	Lépilémur mustélin	большой куний маки
– White-nosed Guenon	Große Weißnasenmeerkatze	Cercopithèque hocheur	большая белоносая мартышка
Green Colobus	Grüner Stummelaffe	Colobe vrai	ванбенеденовский толстотел
Grey-cheeked Mangabey	Mantelmangabe	Mangabey à gorge blanche	гривистый мангабей
Grey Cuscus	Wollkuskus		серый кускус
– Gentle Lemur	Grauer Halbmaki	Hapalémur	серый полумаки
– Gibbon	Silbergibbon	Gibbon cendré	серебристый гиббон
– Kangaroo	Graues Riesenkänguruh	Kangourou géant	серый исполинский кенгуру
Grey's Wallaby	Östliches Irmawallaby	Wallaby de Grey	восточный кенгуру ирма
Grey Woolly Monkey	Grauer Wollaffe	Lagotriche grison	серая шестистая обезьяна
Grivet	Graugrüne Meerkatze	Grivet	серозеленая мартышка
Guatemalan Howler Monkey	Guatemalabrüllaffe	Hurleur de Guatemala	гватемальский ревун

ENGLISH NAME	GERMAN NAME	FRENCH NAME	RUSSIAN NAME
Guenon-like Monkeys	Meerkatzenartige	Cercopithecidés	мартышки
Guenons	Meerkatzen	Cercopithèques	мартышки
Guinea Baboon	Guineapavian	Babouin de Guinée	гвинейский павиан
Hairy-eared Dwarf Lemur	Büschelohriger Katzenmaki	Chirogale aux oreilles velues	
Hairy Hedgehogs	Haarigel	Gymnures	крысиные ежи
Hairy-nosed Wombat	Haarnasenwombat	Wombat à narines poilues	широколобый вомбат
Hairy Saki	Zottelschweifaffe	Saki à perruque	саки-монах
Hairy-tailed Mole	Haarschwanzmaulwurf	Taupe à queue chevelue	волосохвостый крот
Haitian Solenodon	Haiti-Schlitzrüßler	Solenodon	гаитийский щелезуб
Hamadryas Baboon	Mantelpavian	Hamadryas	гамадрил
Hare Wallabies	Hasenkänguruhs	Lièvres wallabies	заячьи кенгуру
Hedgehogs	Igel	Érinacéidés	ежи
Himalayan Entellus Langur	Berghulman	Semnopithèque de l'Himalaya	гималайский гульман
– Water Shrew	Himalajawasserspitzmaus	Chimarrogale de l'Himalaya	гималайская водяная белозубка
Honey Glider	Gleithörnchenbeutler	Phalanger volant	сумчатая летяга
– Phalanger	Honigbeutler	Souris à miel	пяткоход
Hoolock Gibbon	Hulock	Hoolock	гулок
Hottentot Golden Mole	Hottentotten-Goldmull		готтентотский крот
House Shrew	Moschusspitzmaus		бурая мускусная белозубка
Howler Monkeys	Brüllaffen	Singes hurleurs	ревуны
Humboldt's Woolly Monkey	Wollaffe	Lagotriche de Humboldt	шерстистая обезьяна Гумбольдта
Indian Hedgehog	Indischer Igel		индийский еж
Indri	Indri	Indri	короткохвостый индри
Indrisoid Lemurs	Indriartige	Indriinés	индри
Insect-Eaters	Insektenesser	Insectivores	насекомоядные
Italian Hedgehog	Italienischer Igel	Hérisson d'Italie	итальянский еж
Japanese Macaque	Rotgesichtsmakak	Macaque Japonais	японский макак
– Shrew Mole	Japanischer Spitzmull	Taupe des montagnes du Japon	японский землеройковый крот
Jerboa Marsupials	Springbeutelmäuse	Gerboises-Souris marsupiales	сумчатые тушканчики
John's Langur	Nilgirilangur	Semnopithèque des Nilgiris	нилагирийский тонкотел
Kangaroos	Riesenkänguruhs	Grands Kangourous	исполинские кенгуру
Kansu Mole	Kansumaulwurf	Taupe de Kansu	западнокитайский землеройковый крот
Kelaart's Long-clawed Shrew	Kelaarts Langkrallenspitzmaus		когтистая белозубка Келаарта
Kimberley Planigale	Zwergflachkopfbeutelmaus		карликовая плоскоголовая мышевидка
Koala	Koala	Koala	коала
Koala-like Marsupials	Koalaverwandte	Phascolarctinés	сумчатые медведи
Korean Hedgehog	Koreaigel	Hérisson de Corée	корейский еж
Langurs	Languren	Presbytinés	
Large Tree Shrew	Tana	Tana	тупайя тана
Laxmann's Shrew	Maskenspitzmaus	Musaraigne lapone	средняя бурозубка
Leadbeater's Phalanger	Hörnchenkletterbeutler	Opossum de Leadbeater	сумчатая белка
Leaf Monkeys	Schlankaffen	Colobidés	тонкотелые обезьяны
Least Shrew	Nordamerikanische Kleinohrspitzmaus	Petite Musaraigne à queue courte	североамериканская короткоухая бурозубка
Lemur-like Prosimians	Lemurenartige	Lémuriens	лемуровые
Lemurs	Lemuren	Lémurs	лемуры
Lesser Gymnure	Kleiner Rattenigel		малый крысиный еж
– Hedgehog Tenrec	– Igeltanrek		ежевый тенрек Тельфера
– Mouse Lemur	Mausmaki	Chirogale mignon	мышиный маки
– Otter Shrew	Zwerg-Otterspitzmaus	Micropotamogale de Lamotte	карликовая выдровая землеройка
– Shrew	Zwergspitzmaus	Musaraigne pygmée	малая бурозубка
– Slow Loris	Kleiner Plumplori		малый толстый лори
– Tenrecs	Kleintanreks	Microgales	длиннохвостые тенреки
– White-nosed Guenon	Kleine Weißnasenmeerkatze	Hocheur blanc-nez	малая белоносая мартышка
– White-toothed Shrew	Gartenspitzmaus	Musaraigne des jardins	малая белозубка
Lesueur's Rat Kangaroo	Lesueur-Bürstenkänguruh	Rat-Kangourou de Lesueur	кистехвостый кенгуру Лесюера
L'Hoest's Monkey	Vollbartmeerkatze	Cercopithèque de l'Hoest	бородатая мартышка
Lion-tailed Macaque	Wanderu	Macaque Ouanderou	вандеру
Little Northern Dasyure	Zwerg-Fleckenbeutelmarder		североавстралийская сумчатая куница
– Rock Wallaby	Zwergsteinkänguruh	Petit Wallaby de rochers	карликовый каменный кенгуру
– Water Opossum	Dickschwanzbeutelratte		толстохвостый опоссум
Long-beaked Spiny Anteater	Langschnabeligel	Échidné à bec courbe	проехидны
Long-clawed Shrews	Langkrallenspitzmäuse	Pachyures aux griffes longues	
Long-eared Hedgehog	Langohrigel		ушастый еж
Long-fingered Striped Phalanger	Kleiner Streifenbeutler		малый полосатый кускус
Long-haired Spider Monkey	Goldstirn-Klammeraffe	Singe-Araignée à ventre blanc	светлолобая коата
Long-nosed Bandicoot	Langnasenbeutler	Bandicoot à nez long	большой бандикут
– Rat Kangaroo	Langschnauzen-Kaninchenkänguruh	Rat-Kangourou à nez long	крысиный потору
Long-tailed Mole	Langschwanzmaulwurf		длиннохвостый крот
– New Guinea Bandicoot	Langschwänziger Neuguinea-Nasenbeutler		длиннохвостый новогвинейский бандикут
– Tenrec	Langschwanztanrek	Microgale	обыкновенный длиннохвостый тенрек

ENGLISH NAME	GERMAN NAME	FRENCH NAME	RUSSIAN NAME
Lönnberg's Tamarin	Lönnbergtamarin	Tamarin de Lönnberg	тамарин Леннберга
Lorentz's Marsupial Rat	Neuguinea-Spitzhörnchenbeutler		мышевидка Лоренца
Loris-like Prosimians	Loriartige	Lorisidés	лориевые
Lorises	Loris	Lorisidés	лори
Lowland Gorilla	Flachlandgorilla	Gorille de côte	береговая горилла
Lumholtz's Tree Kangaroo	Lumholtz-Baumkänguruh	Kangourou arboricole de Lumholtz	квинслендский древесный кенгуру
Macaques	Makaken	Macaques	макаки
Madagascar Shrew	Madagaskarspitzmaus	Pachyure de Madagascar	мадагаскарская белозубка
Madras Tree-shrew	Elliots Tupaia	Toupaïe d'Elliot	тупайя Эллиота
Mammals	Säugetiere	Mammifères	млекопитающие
Mandrill	Mandrill	Mandrill	собственно мандрил
Maned Tamarins	Löwenäffchen	Singes-Lions	львиные игрунки
Mangabeys	Mangaben	Mangabeys	мангабеи
Mantled Howler Monkey	Mantelbrüllaffe	Hurleur à manteau	
Marmosets	Marmosetten	Ouistitis	
Maroon Leaf Monkey	Maronenlangur	Semnopithèque rubicond	каштановый тонкотел
Marsh Tenrec	Wassertanrek	Limnogale	болотный тенрек
Marsupial Anteater	Ameisenbeutler	Fourmilier marsupial rayé	мурашеед
– Mice	Beutelmäuse	Phascogalinés	мышевидки
– Moles	Beutelmulle	Taupes marsupiales	сумчатые кроты
– Rats	Pinselschwanzbeutler	Rats marsupiaux	
Marsupials	Beuteltiere	Marsupiaux	сумчатые
Masked Shrew	Amerikanische Maskenspitzmaus	Musaraigne cendrée	обыкновенная американская бурозубка
– Titi	Schwarzköpfiger Springaffe	Callicèbe à masque	черноголовый прыгун
Matschie's Tree Kangaroo	Matschie-Baumkänguruh	Kangourou arboricole de Matschie	
Mediterranean Long-tailed Shrew	Mittelmeer-Langschwanzspitzmaus	Musaraigne méditerranéenne	средиземноморская белозубка
– Mole	Blindmaulwurf	Taupe aveugle	слепой крот
– Water Shrew	Sumpfspitzmaus	Musaraigne de Miller	малая кутора
Mentawi Leaf Monkey	Mentawilangur	Semnopithèque de Mentawi	ментавайский тонкотел
Merriam's Desert Shrew	Große Wüstenspitzmaus	Grande Musaraigne du désert	большая пустынная бурозубка
Mexican Mouse Opossum	Schwarzring-Zwergbeutelratte		мексиканская сумчатая крыса
Mindanao Gymnure	Philippinen-Rattenigel		филиппинская малая гимнура
Mole	Maulwurf	Taupe	крот
Mole-like Rice Tenrec	Maulwurfartiger Reistanrek		кротовидный рисовый тенрек
Moles	Maulwürfe	Talpidés	кроты
Mona Monkey	Monameerkatze	Cercopithèque mone	мона
Mongoose Lemur	Mongozmaki	Lémur mongos	монго
Monkeys	Affen	Simiens	обезьяны
Monotremes	Kloakentiere	Monotrèmes	однопроходные
Moon Rat	Großer Haarigel	Gymnure	большой крысиный еж
Moor Macaque	Mohrenmakak	Macaque de Célèbes	черный целебесский макак
Mountain Gorilla	Berggorilla	Gorille de montagne	горная горилла
Mouse Bandicoot	Maus-Nasenbeutler		мышиный бандикут
– Lemurs	Zwergmakis		
– Sminthopsis	Kleine Schmalfußbeutelmaus		обыкновенная сумчатая землеройка
Moustached Monkey	Blaumaulmeerkatze	Moustac	голуболицая мартышка
– Tamarin	Schnurrbarttamarin	Tamarin à Moustaches	усатый тамарин
Murine Opossum	Maus-Zwergbeutelratte		мышиный опоссум
Musky Rat Kangaroo	Moschusrattenkänguruh	Rat musqué Kangourou	мускусный кенгуру
Nail-tail Wallabies	Nagelkänguruhs	Onychogales	когтехвостые кенгуру
Narrow-footed Marsupial Mice	Schmalfußbeutelmäuse	Souris marsupiales	сумчатые землеройки
Native Cats	Fleckenbeutelmarder		сумчатые куницы
Needle-clawed Bushbaby	Kielnagelgalago	Galago mignon et Galago du Congo	
Negro Tamarin	Mohrentamarin	Tamarin nègre	обыкновенный тамарин
New Guinea Bandicoots	Neuguinea-Nasenbeutler	Bandicoots de Nouvelle-Guinée	новогвинейские бандикуты
– – Marsupial Mice	Neuguineabeutelmäuse		новогвинейские мышевидки
– – Mountain Wallaby	Macleay-Buschkänguruh	Wallaby de Macleay	кустовый кенгуру Миклухо-Маклая
– – Pigmy Flying Phalanger	Neuguinea-Zwerggleitbeutler		новогвинейская порхающая сумчатая мышь
New World Monkeys	Neuweltaffen	Singes du Nouveau Monde	обезьяны Нового света
North African Elephant Shrew	Nordafrikanische Elefantenspitzmaus	Macroscélide de l'Afrique du Nord	североафриканский прыгунчик
Northern Black-and-white Colobus	Nördlicher Guereza	Colobe de l'Abyssinie	гвереца
– Nail-tail Wallaby	Flachnagelkänguruh		обыкновенный когтехвостый кенгуру
– New Guinea Wallaby	Hagen-Buschkänguruh		кустовый кенгуру Хагена
– Planigale	Nördliche Flachkopfbeutelmaus		северная плоскоголовая мышевидка
– Water Shrew	– Wasserspitzmaus	Musaraigne palustre	болотная бурозубка

ENGLISH NAME	GERMAN NAME	FRENCH NAME	RUSSIAN NAME
North-western Marsupial Mole	Kleiner Beutelmull	Petite Taupe marsupiale	малый сумчатый крот
Old World Moles	Altweltmaulwürfe	Talpinés	собственно кроты
– – Monkeys	Hundsaffen		мартышковые
– – Simian Primates	Schmalnasen	Catarhiniens	узконосые обезьяны
Olivaceous Woolly Monkey	Grauer Wollaffe	Lagotriche olive	серая шерстистая обезьяна
Opossums	Beutelratten	Opossums d'Amérique	сумчатые крысы
Orabussu Titi	Grauer Springaffe	Callicèbe arabassu	серый прыгун
Orang-Utan	Orang-Utan	Orang-outan	оранг-утан
Otter Shrew	Otterspitzmaus	Potamogale	выдровая землеройка
Owl-faced Guenon	Hamlynmeerkatze	Cercopithèque à tête de Hibou	
Pacific Water Shrew	Pazifische Wasserspitzmaus	Musaraigne des marais	бурозубка Бендайра
Pademelons	Filander	Thylogales	малые кустовые кенгуру
Pale-headed Saki	Blaßkopfsaki	Saki à tête pâle	бледный саки
Parma Wallaby	Parmakänguruh		кенгуру парма
Patas	Schwarznasen-Husarenaffe	Patas	черноносый гусар
Pearson's Long-clawed Shrew	Pearsons Langkrallenspitzmaus		когтистая белозубка Пирсона
Pen-tailed Phalanger	Federschwanzbeutler		перохвостая сумчатая соня
– Tree Shrew	Federschwanz	Ptilocerque	перохвостая тупайя
Peruvian Mountain Woolly Monkey	Gelbschwanzwollaffe		желтохвостая шерстистая обезьяна
– Rat Opossum	Peru-Opossummaus		перуанская первокрыса
Peters' Elephant Shrew	Rotschulterrüsselhündchen	Macroscélide de Peters	хоботковая собачка Петерса
Phalangers	Kletterbeutler	Phalangéridés	лазающие сумчатые
Phayre's Leaf Monkey	Phayres Langur	Semnopithèque de Phayre	тонкотел Фейера
Philander Opossum	Gelbe Wollbeutelratte	Opossum laineux	желтая шерстистая сумчатая крыса
Philippine Tarsier	Philippinenkoboldmaki	Tarsier des Philippines	филиппинский долгопят
Philippine Tree Shrews	Philippinentupaias		филиппинские тупайи
Piebald Shrew	Gescheckte Spitzmaus		пегий путорак
Pied Tamarin	Manteläffchen	Tamarin bicolore	пегая игрунка
Pig-footed Bandicoot	Schweinsfuß		хероп
Pigmy Flying Phalanger	Zwerggleitbeutler	Acrobate pygmée	порхающая сумчатая мышь
– Marmoset	Zwergseidenäffchen	Ouistiti mignon	карликовая игрунка
– Marsupial Mouse	Zwergbeutelmaus		карликовая мышевидка
– possum	Neuguinea-Bilchbeutler		новогвинейская сумчатая соня
Pig-tailed Langur	Pagehstumpfnasenaffe		одноцветный симиас
– Macaque	Schweinsaffe	Macaque à queue de cochon	свинохвостый макак
Placental Mammals	Höhere Säugetiere	Eutheriens	высшие млекопитающие
Plain Rock Wallaby	Queensland-Felskänguruh		квинслендский каменный кенгуру
Platypuses	Schnabeltiere	Ornithorhynchidés	утконосы
Potto	Potto	Potto de Bosman	потто
Pouched Mammals	Beutelsäuger	Métathériens	сумчатые звери
Pretty-face Wallaby	Hübschgesichtkänguruh		кустовый кенгуру Парри
Primates	Herrentiere	Primates	приматы
Proboscis Monkey	Nasenaffe	Nasique	обыкновенный носач
Prosimians	Halbaffen	Prosimiens	полуобезьяны
Pruner's Hedgehog	Pruners Igel	Hérisson de Pruner	еж Прунера
Purple-faced Langur	Weißbartlangur	Semnopithèque blanchâtre	белобородый тонкотел
Pyrenean Desman	Pyrenäendesman	Desman des Pyrénées	пиренейская выхухоль
Queensland Ring Tail	Wander-Ringelschwanzbeutler		обыкновенный кольцехвостый кускус
Quokka	Kurzschwanzkänguruh	Kangourou à queue courte	короткохвостый кустовый кенгуру
Rabbit Bandicoot	Kaninchen-Nasenbeutler		сумчатые зайцы
Rat Kangaroos	Rattenkänguruhs	Rats-Kangourous	кенгуровые крысы
– Opossums	Opossummäuse	Caenolestidés	ценолестовые сумчатые
Rat-tailed Opossum	Nacktschwanzbeutelratte	Opossum à queue de rat	голохвостый опоссум
Red-backed Saki	Rotrückensaki	Saki capucin	красноспинный саки
– Squirrel Monkey	Gelbes Totenköpfchen	Sapajou à dos rouge	желтая саймири
Red-bellied Guenon	Rotbauchmeerkatze	Hocheur à ventre rouge	краснобрюхая мартышка
– Lemur	Rotbauchmaki	Lèmur à ventre rouge	рыжебрюхий маки
– White-lipped Tamarin	Rotbauchtamarin	Tamarin labié	краснобрюхий тамарин
Red-capped Tamarin	Rotkappentamarin	Tamarin à calotte rousse	красногривый тамарин
Red Colobus	Roter Stummelaffe	Colobe bai	красный толстотел
Red-fronted Lemur	Rotstirnmaki	Lémur à front rouge	рыжелобый маки
Red Guenon	Husarenaffe	Patas	обыкновенный гусар
Red-handed Tamarin	Rothandtamarin	Tamarin aux mains rousses	краснорукий тамарин
Red Howler Monkey	Roter Brüllaffe	Hurleur roux	рыжий ревун
– Kangaroo	Rotes Riesenkänguruh	Kangourou roux	рыжий исполинский кенгуру
Red-legged Pademelon	Rotbeinfilander		обожженный кенгуру
Red-mantled Tamarin	Rotmanteltamarin	Tamarin à manteau rouge	
Red-necked Pademelon	Rothalsfilander		падемелон
– Wallaby	Bennettkänguruh		кустовый кенгуру Беннетта
Red Spider Monkey	Panama-Klammeraffe	Singe-Araignée rouge	панамская коата
Red-tailed Phascogale	Kleiner Pinselschwanzbeutler		малая кистехвостая мышевидка
Red Titi	Roter Springaffe	Callicèbe roux	красный прыгун

ENGLISH NAME	GERMAN NAME	FRENCH NAME	RUSSIAN NAME
Red-toothed Shrews	Rotzahnspitzmäuse	Soricinés	землеройки-бурозубки
Red Uakari	Roter Uakari	Ouakari rubicond	красный уакори
Rhesus Macaque	Rhesusaffe	Macaque rhésus	макак-резус
Rice Tenrecs	Reistanreks	Oryzoryctinés	рисовые тенреки
Ring-tailed Lemur	Katta	Lémur catta	катта
– Phalangers	Ringelschwanz-Kletterbeutler	Ringtails	кольцехвостые кускусы
– Rock Wallaby	Ringschwanz-Felskänguruh		желтоногий каменный кенгуру
Rio Napo Tamarin	Rio-Napo-Tamarin	Tamarin de Rio Napo	
Rock Elephant Shrew	Klippen-Elefantenspitzmaus		скалистый прыгунчик
Rock-haunting Ring Tail	Felsen-Ringelschwanzbeutler		кускус Даля
Rock Wallabies	Felskänguruhs	Wallabies de rochers	каменные кенгуру
Roloway Monkey	Roloway	Roloway	
Roman Mole	Römischer Maulwurf	Taupe romaine	римский крот
Ruffed Lemur	Vari	Lémur vrai	вари
Rufous-bellied Pademelon	Rotbauchfilander	Wallaby de Billardier	рыжебрюхий кустовый кенгуру
Rufous-handed Howler Monkey	Rothandbrüllaffe	Hurleur à mains rousses	краснорукий ревун
Rufous Rat Kangaroo	Rotes Rattenkänguruh	Rat-Kangourou rougeâtre	рыжая кенгуровая крыса
Ruhe's Baboon	Webbipavian	Babouin de Ruhe	
Russian Desman	Russischer Desman	Desman de Moscovie	обыкновенная выхухоль
Ruwenzori Otter Shrew	Ruwenzori-Otterspitzmaus	Micropotamogale du Mont Ruwenzori	
Sakis	Schweifaffen	Sakis moines	
Sandy Wallaby	Flinkes Känguruh	Wallaby agile	проворный кустовый кенгуру
Santarém Marmoset	Langohrseidenäffchen	Ouistiti de Santarém	длинноухая игрунка
Savi's Pigmy Shrew	Etruskerspitzmaus	Pachyure étrusque	белозубка-малютка
Scaly-tailed Phalanger	Schuppenschwanzkusu		чешуйчатохвостый кузу
Sclater's Hedgehog	Sclaters Igel	Hérisson de Sclater	еж Склатера
Senegal Bushbaby	Senegalgalago	Galago du Sénégal	сенегальский галаго
Short Bare-tailed Opossums	Spitzmausbeutelratten		землеройковые сумчатые крысы
Short-beaked Spiny Anteater	Kurzschnabeligel	Échidné à bec droit	ехидны
Short-eared Brush-tailed Phalanger	Hundskusu	Phalanger de montagne	собачий кузу
– Elephant Shrew	Kurzohrrüsselspringer		обыкновенный слоновый прыгунчик
– Rock Wallaby	Kurzohr-Felskänguruh		короткоухий каменный кенгуру
Short-nosed Bandicoots	Kurznasenbeutler		курносые бандикуты
– Rat Kangaroos	Bürstenkänguruhs	Rats-Kangourous à nez court	кистехвостые кенгуру
Short-snouted Elephant Shrew	Kurznasen-Elefantenspitzmaus		коротконосый прыгунчик
Short-tailed Indri	Indris	Indris	индри
–Moupin Shrew	Asiatische Kurzschwanzspitzmaus		азиатская короткохвостая бурозубка
– Shrew	Kurzschwanzspitzmaus	Grande Musaraigne à queue courte	североамериканская коротко-хвостая бурозубка
Shrew Hedgehog	Spitzmausigel	Neotétracus	землеройковый еж
– Mole	Spitzmausmaulwurf	Musaraigne-Taupe	ушастый крот
Shrews	Spitzmäuse	Soricidés	землеройки
Siamang	Siamang	Siamang	обыкновенный сиаманг
Sifakas	Sifakas	Propithèques	сифаки
Sikkim Large-clawed Shrew	Sikkim-Großklauen-Spitzmaus	Musaraigne de Sikkim	сиккимская когтистая бурозубка
Silvered Leaf Monkey	Haubenlangur		гривистый тонкотел
Silvery Marmoset	Silberäffchen	Ouistiti melanure	серебристая игрунка
Slender Loris	Schlanklori	Loris grêle	тонкий лори
Slow Loris	Plumplori	– paresseux	толстый лори
Smooth-tailed Tree Shrews	Bergtupaias		горные тупайи
Snub-nosed Monkeys	Stumpfnasenaffen	Rhinopithèques	ринопитеки
Solenodons	Schlitzrüßler	Solénodontidés	щелезубы
Sooty Mangabey	Halsbandmangabe	Mangabey à collier blanc	воротничковый мангабей
South American Mouse Opossums	Zwergbeutelratten	Souris-Opossums	карликовые сумчатые крысы
Southern Black-and-white Colobus	Südlicher Guereza	Colobe à longs poils	королевский толстотел
– Short-nosed Bandicoot	Kleiner Kurznasenbeutler		малый курносый бандикут
South-western Pigmy Phalanger	Dünnschwanz-Schlafbeutler		тонкохвостая сумчатая соня
Speckled Marsupial Mouse	Sprenkelbeutelmaus		крапчатая мышевидка
Spectacled Hare Wallaby	Brillen-Hasenkänguruh		очковый заячий кенгуру
Spider Monkeys	Klammeraffen	Singes-Araignées	коаты
Spiny Hedgehogs	Echte Igel	Hérissons	настоящие ежи
– New Guinea Bandicoots	Stachelnasenbeutler		остроносые бандикуты
Sportive Lemur	Kleiner Wieselmaki	Lépilémur à queue rouge	малый куний маки
Spotted Cuscus	Tüpfelkuskus	Couscous tacheté	пятнистый кускус
Squirrel Monkeys	Totenkopfäffchen	Saimiris	саймири
Star-nosed Mole	Sternmull	Condylure étoilé	звездорыл
Streaked Tenrec	Halbborstenigel	Hémicentètes	полутенреки
Striped Marsupial Rats	Streifenbeutelmäuse		полосатые мышевидки
– Native Cat	Streifenbeutelmarder		полосатая сумчатая куница
– Opossums	Streifenkletterbeutler	Phalangers au pelage rayé	полосатые кускусы

ENGLISH NAME	GERMAN NAME	FRENCH NAME	RUSSIAN NAME
– Phalanger	Großer Streifenbeutler	Phalanger au pelage rayé	большой полосатый кускус
– Ring Tail	Streifen-Ringelschwanzbeutler		желтый кускус
Stuhlmann's Elephant Shrew	Dunkles Rüsselhündchen	Macroscélide de Stuhlmann	темная хоботковая собачка
– Golden Mole		Taupe dorée de Stuhlmann	
Stump-tailed Macaque	Bärenmakak	Macaque brun	медвежий макак
Sunda Island Leaf Monkey	Mützenlangur		чубастый тонкотел
Swamp Guenon	Schwarzgrüne Meerkatze	Cercopithèque noir et vert	чернозеленая мартышка
Szechuan Burrowing Shrew	Stummelschwanzspitzmaus		куцая белозубка
– Water Shrew	Gebirgsbachspitzmaus	Nectogale élégant	тибетская водяная белозубка
Tailless Tenrec	Großer Tanrek	Tanrec	обыкновенный тенрек
Tamarins	Tamarins	Tamarins	тамарины
Tamaulipan Mole	Tamaulipasmaulwurf	Taupe de Tamaulipas	
Tammar	Tammar	Wallaby de l'Ile d'Eugène	кенгуру дерби
Tarsiers	Koboldmakis	Tarsidés	долгопяты
Tasmanian Barred Bandicoot	Tasmanischer Langnasenbeutler		бандикут Гунна
– Devil	Beutelteufel	Sarcophile satanique	сумчатый дьявол
– Echidna	Tasmanien-Kurzschnabeligel	Échidné de Tasmanie	тасманийская ехидна
– Rat Kangaroo	Tasmanienbürstenkänguruh	Rat-Kangourou de Tasmanie	тасманийский кистехвостый кенгуру
– Wolf	Beutelwolf	Loup marsupial	сумчатый волк
Tenrecs	Tanreks	Tanrecs	тенреки
Thick-tailed Bushbaby	Riesengalago	Galago à queue trouffue	толстохвостый галаго
Tiger Cat	Fleckschwanzbeutelmarder	Chat marsupial	исполинская сумчатая куница
Titi Monkeys	Springaffen	Titis	обезьяны-прыгуны
Tonkin Snub-nosed Monkey	Tonkinstumpfnase	Rhinopithèque de Tonkin	авункулярный ринопитек
Toque Monkey	Ceylon-Hutaffe	Macaque couronné	цейлонский макак
Townsend's Mole	Townsends Maulwurf	Taupe de Townsend	крот Тоунсенда
Tree Kangaroos	Baumkänguruhs	Kangourous arboricoles	древесные кенгуру
– Shrews	Spitzhörnchen	Tupaiidés	тупайи
True's Shrew Mole	Trues Spitzmull	Taupe de True	
Typical Lemurs	Mittelgroße Lemuren	Lémurinés	средние лемуры
Uakaris	Kurzschwanzaffen	Ouakaris	уакори
Uganda Armoured Shrew	Ugandapanzerspitzmaus		угандская броненоска
Unalaska Shrew	Unalaskaspitzmaus		алеутская бурозубка
Verreaux's Sifaka	Larvensifaka	Propithèque de Verreaux	сифака Верро
Virginian Opossum	Virginisches Nordopossum	Opossum de Virginie	виргинский опоссум
Wallaroo	Bergkänguruh	Wallaroo	горный кенгуру
Weasel Lemurs	Wieselmakis	Lépilémur	тонкотелые маки
Weddell's Tamarin	Weißlippentamarin	Tamarin de Weddell	
Weeper Capuchin	Brauner Kapuziner	Sapajou brun	бурый капуцин
Western American Moles	Westamerikanische Maulwürfe	Taupes d'Amérique de l'Ouest	западноамериканские кроты
– Dasyure	Schwarzschwanzbeutelmarder		чернохвостая сумчатая куница
– European Hedgehog	Braunbrustigel	Hérisson d'Europe de l'Ouest	бурогрудый еж
– Hare Wallaby	Zottel-Hasenkänguruh		косматый заячий кенгуру
– Needle-clawed Bushbaby	Westlicher Kielnagelgalago	Galago mignon	
– Tarsier	Sundakoboldmaki	Tarsier de Horsfield	сундский долгопят
White-eared Marmoset	Weißohrseidenäffchen	Ouistiti oreillard	белоухая игрунка
White-footed Tamarin	Weißfußäffchen	Pinché aux pieds blancs	белоногая игрунка
White-fronted Capuchin	Weißstirnkapuziner	Sapajou à front blanc	белолобый капуцин
– Leaf Monkey	Weißstirnlangur	Semnopithèque à front blanc	гололобый тонкотел
– Lemur	Weißkopfmaki	Maki à front blanc	белолобый маки
– Marmoset	Weißgesichtseidenäffchen	Ouistiti à tête blanche	белолицая игрунка
White-handed Gibbon	Lar	Gibbon lar	лар
White-headed Saki	Weißkopfsaki	Saki à tête blanche	белоголовый саки
White-mantled Snub Nose	Weißmantelstumpfnase	Rhinopithèque jaune doré	бреличевский ринопитек
White-necked marmoset	Weißnackenseidenäffchen	Ouistiti à col blanc	белошейная игрунка
White-nosed Saki	Weißnasensaki	Saki à nez blanc	белоносый саки
White-shouldered Marmoset	Weißschulterseidenäffchen	Ouistiti à camail	белоплечая игрунка
White-tailed Rabbit Bandicoot	Kleiner Kaninchen-Nasenbeutler		малый сумчатый заяц
White Tamarin	Weißer Tamarin	Tamarin blanc	белый тамарин
White-throated Capuchin	Kapuziner	Sapajou capucin	обыкновенный капуцин
– Guenon	Weißkehlmeerkatze	Cercopithèque à gorge blanche	белогорлая мартышка
White-toothed Shrews	Weißzahnspitzmäuse	Musaraignes à dents blanches	землеройки-белозубки
Wombats	Wombats	Wombats	вомбаты
Woolly Indris	Wollmaki	Avahi lanigère	авахи
– Monkeys	Wollaffen	Singes laineux	шерстистые обезьяны
– Opossums	Wollbeutelratten	Opossums laineux	шерстистые сумчатые крысы
– Spider Monkey	Spinnenaffe	Éroïde	обыкновенная паукообразная обезьяна
Yapok	Schwimmbeutler	Yapock	водяной опоссум
Yellow Baboon	Gelber Babuin	Babouin cynocéphale	бабуин
Yellow-bellied Glider	Großer Gleithörnchenbeutler		большая сумчатая летяга
Yellow-footed Marsupial Mouse	Gelbfußbeutelmaus		желтоногая мышевидка
Yellow-legged Marmoset	Gelbfußäffchen	Ouistiti aux pieds jaunes	желтоногая игрунка

II. German–English–French–Russian

Unterartnamen werden meist aus den Artnamen durch Voranstellen von Eigenschaftswörtern oder geographischen Bezeichnungen gebildet. In diesem Teil des Tierwörterbuchs sind so gebildete deutsche Unterartnamen sowie die wissenschaftlichen Unterartnamen in der Regel nicht aufgeführt.

GERMAN NAME	ENGLISH NAME	FRENCH NAME	RUSSIAN NAME
Aalstrichwallaby	Black-striped Wallaby		полосатый кустовый кенгуру
Acrobates	Pigmy Flying Phalangers	Acrobate pygmée	порхающие сумчатые мыши
– pulchellus	New Guinea Pigmy Flying Phalanger		новогвинейская порхающая сумчатая мышь
– pygmaeus	Pigmy Flying Phalanger	Acrobate pygmeé	австралийская порхающая сумчатая мышь
Aepyprymnus rufescens	Rufous Rat Kangaroo	Rat-Kangourou rougeâtre	рыжая кенгуровая крыса
Aethechinus algirus	Algerian Hedgehog	Hérisson d'Algérie	алжирский еж
– frontalis	Cape Hedgehog		капский еж
– sclateri	Sclater's Hedgehog	Hérisson de Sclater	еж Склатера
Affen	Monkeys and Apes	Simiens	обезьяны
Afrikanische Riesenspitzmaus	African Forest Shrew	Musaraigne géante	гигантская белозубка
– Waldspitzmaus	Forest Shrew		африканская лесная белозубка
Akrobatenmaki	Greater Sportive Lemur	Lépilémur mustélin	большой куний маки
Algerischer Igel	Algerian Hedgehog	Hérisson d'Algérie	алжирский еж
Almiqui	Cuban Solenodon		кубинский щелезуб
Alouatta	Howler Monkeys	Hurleurs	ревуны
– belzebul	Rufous-handed Howler Monkey	Hurleur à mains rousses	красноруклй ревун
– caraya	Black Howler Monkey	– noir	черный ревун
– fusca	Brown Howler Monkey	– brun	бурый ревун
– palliata	Mantled Howler Monkey	– à manteau	
– seniculus	Red Howler Monkey	– roux	рыжий ревун
– villosa	Guatemalan Howler Monkey	– de Guatemala	гватемальский ревун
Alouattinae	Howler Monkeys	Singes hurleurs	ревуны
Alpenspitzmaus	Alpine Shrew	Musaraigne des montagnes	альпийская бурозубка
Altweltaffen	Old World Simian Primates	Catarhiniens	узконосые обезьяны
Altweltmaulwürfe	– – Moles	Talpinés	собственно кроты
Amblysomus hottentotus	Hottentot Golden Mole		готтентотский крот
Ameisenbeutler	Marsupial Anteater	Fourmilier marsupial rayé	мурашеед
Ameisenigel	Echidnas	Échidnés	ехидны
Amerikanisch-Asiatische Maulwürfe	American and Asian Moles	Taupes d'Asie et d'Amérique du Nord	американско-азиатские кроты
Amerikanische Maskenspitzmaus	Masked Shrew	Musaraigne cendrée	обыкновенная американская бурозубка
Amerikanischer Spitzmaus-maulwurf	American Shrew Mole	Taupe de Gibbs	американский землеройковый крот
Amerikanische Zwergspitzmaus	– Pigmy Shrew	Musaraigne pygmée d'Amérique	североамериканская карликовая бурозубка
Amurigel	Amur Hedgehog	Hérisson d'Amur	амурский еж
Anathana ellioti	Madras Tree-shrew	Toupaïe d'Elliot	тупайя Эллиота
Anourosorex squamipes	Szechuan Burrowing Shrew		куцая белозубка
Antechinomys	Jerboa Marsupials	Gerboises-Souris marsupiales	сумчатые тушканчики
– laniger	Eastern Jerboa Marsupial		восточноавстралийский сумчатый тушканчик
– spenceri	Central Jerboa Marsupial		среднеавстралийский сумчатый тушканчик
Antechinus	Broad-footed Marsupial Mice	Rats marsupiaux	
– apicalis	Speckled Marsupial Mouse		крапчатая мышевидка
– flavipes	Yellow-footed Marsupial Mouse		желтоногая мышевидка
– macdonnellensis	Fat-tailed Marsupial Mouse		толстохвостая мышевидка
– maculatus	Pigmy Marsupial Mouse		карликовая мышевидка
Anubispavian	Anubis Baboon	Papion anubis	анубис
Aotus trivirgatus	Douroucouli	Singe de nuit	мирикина
Aotinae	Night and Titi Monkeys	Aotinés	мирикины
Apella	Brown Capuchin	Sapajou apelle	капуцин-фавн
Arctocebus calabarensis	Angwantibo	Angwantibo	медвежий маки
Aschgraue Zwergbeutelratte	Ashy Opossum		пепельная сумчатая крыса
Asiatische Kurzschwanzspitzmaus	Short-tailed Moupin Shrew		азиатская короткохвостая бурозубка
– Wasserspitzmäuse	Asian Water Shrews	Chimarrogales	
Assamrhesus	Assamese Macaque	Macaque d'Assam	горный резус
Atelerix	Central African Hedgehogs	Hérissons d'Afrique centrale	среднеафриканские ежи
– albiventris		Hérisson à ventre blanc	белобрюхий еж
– pruneri	Pruner's Hedgehog	– de Pruner	еж Прунера
Ateles	Spider Monkeys	Singes-Araignées	коаты
Ateles belzebuth	Long-haired Spider Monkey	Singe-Araignée à ventre blanc	светлолобая коата
– fusciceps	Brown-headed Spider Monkey	– à tête brune	буроголовая коата

GERMAN NAME	ENGLISH NAME	FRENCH NAME	RUSSIAN NAME
– *geoffroyi*	Central American Spider Monkey	– aux Mains noirs	коата Жоффруа
– *paniscus*	Black Spider Monkey	– noir	черная коата
Atelinae	Spider and Woolly Monkeys	Atélinés	шерстистые обезьяны и коаты
Äthiopischer Igel	Ethiopian Hedgehog	Hérisson du désert	абиссинский еж
Atopogale cubana	Cuban Solenodon	Almiqui	кубинский щелезуб
Australien-Kurzschnabeligel	Australian Echidna	Échidné de l'Australie	австралийская ехидна
Australischer Zwerggleitbeutler	Pigmy Flying Phalanger	Acrobate pygmée	австралийская порхающая сумчатая мышь
Avahi	Woolly Indri	Avahi lanigère	авахи
Avahi laniger	Woolly Indri	Avahi lanigère	авахи
Aye-Aye	Aye-Aye	Aye-Aye	мадагаскарская руконожка
Babuine	Baboons	Babouins	
Bänderkänguruh	Banded Hare Wallaby	Wallaby rayé	поперечнополосатый кенгуру
Bänder-Langnasenbeutler	Eastern Barred Bandicoot		полосатый бандикут
Bandikuts	Bandicoots	Bandicoots	сумчатые барсуки
Bären-Baumkänguruh	Black Tree Kangaroo		кенгуру-медведь
Bärenkuskus	Bear Phalanger	Couscous des Célèbes	черный кускус
Bärenmakak	Stump-tailed Macaque	Macaque brun	медвежий макак
Bärenmaki	Angwantibo	Angwantibo	медвежий маки
Bärenpavian	Chacma Baboon	Chacma	чакма
Bartäffchen	Moustached Tamarin	Tamarin à Moustaches	усатый тамарин
Bartaffe	Lion-tailed Macaque	Macaque Ouanderou	вандеру
Barton-Langschnabeligel	Barton's Echidna	Échidné de Barton	проехидна Бартона
Bartsakis	Bearded Sakis		
Baumkänguruhs	Tree Kangaroos	Kangourous arboricoles	древесные кенгуру
Bennett-Baumkänguruh	Bennett's Tree Kangaroo	Kangourou arboricole de Bennett	древесный кенгуру Беннетта
Bennettkänguruh	Red-necked Wallaby		кустовый кенгуру Беннетта
Berberaffe	Barbary Ape	Magot	магот
Berggorilla	Mountain Gorilla	Gorille de montagne	горная горилла
Berghulman	Himalayan Entellus Langur	Semnopithèque de l'Himalaya	гималайский гульман
Bergkänguruh	Wallaroo	Wallaroo	горный кенгуру
Bergrhesus	Assamese Macaque	Macaque d'Assam	горный резус
Bergspitzmaus	Mediterranean Water Shrew	Musaraigne de Miller	малая кутора
Bergtupaias	Smooth-tailed Tree Shrews		горные тупайи
Bettongia	Short-nosed Rat Kangaroos	Rats-Kangourous à nez court	кистехвостые кенгуру
– *cuniculus*	Tasmanian Rat Kangaroo	Rat-Kangourou de Tasmanie	тасманийский кистехвостый кенгуру
– *gaimardi*	Gaimard's Rat Kangaroo	Rat-Kangourou de Gaimard	австралийский кистехвостый кенгуру
– *lesueur*	Lesueur's Rat Kangaroo	– de Lesueur	кистехвостый кенгуру Лесюера
Beutelbär	Koala	Koala	коала
Beutelmarder	Dasyures	Dasyurinés	сумчатые куницы
Beutelmaulwürfe	Marsupial Moles	Notoryctidés	сумчатые кроты
Beutelmäuse	– Mice	Phascogalinés	мышевидки
Beutelmulle	– Moles	Taupes marsupiales	сумчатые кроты
Beutelratten	Opossums	Opossums d'Amérique	сумчатые крысы
Beutelsäuger	Pouched Mammals	Métathériens	сумчатые звери
Beutelteufel	Tasmanian Devil	Sarcophile satanique	сумчатый дьявол
Beuteltiere	Marsupials	Marsupiaux	сумчатые
Beutelwolf	Tasmanian Wolf	Loup marsupial	сумчатый волк
Biberspitzmäuse	Asian Water Shrews	Chimarrogales	
Bindenwollbeutelratte	Woolly Opossum	Opossum laineux	полосатая шерстистая сумчатая крыса
Blarina brevicauda	Short-tailed Shrew	Grande Musaraigne à queue courte	североамериканская короткохвостая бурозубка
Blarinella quadraticauda	– Moupin Shrew		азиатская короткохвостая бурозубка
Blaßkopfsaki	Pale-headed Saki	Saki à tête pâle	бледный саки
Blaumaulmeerkatze	Moustached Monkey	Moustac	голуболицая мартышка
Blindmaulwurf	Mediterranean Mole	Taupe aveugle	слепой крот
Borneowasserspitzmaus			индонезийская водяная белозубка
Borstenigel	Tenrecs	Tenrecinés	настоящие тенреки
Brachyteles arachnoides	Woolly Spider Monkey	Éroïde	обыкновенная паукообразная обезьяна
Brandts Igel	Brandt's Hedgehog	Hérisson de Brandt	длинноиглый еж
Braunbrustigel	Western European Hedgehog	Hérisson d'Europe de l'Ouest	бурогрудый еж
Brauner Brüllaffe	Brown Howler Monkey	Hurleur brun	бурый ревун
– Kapuziner	Weeper Capuchin	Sapajou brun	бурый капуцин
– Maki	Brown Lemur	Lémur brun	черноголовый маки
– Wollaffe	– Woolly Monkey	Lagotriche de Castelnau	бурая шерстистая обезьяна
Braune Stumpfnase	– Snub-nosed Langur	Rhinopithèque brun	биэтовский ринопитек
Braunkopfklammeraffe	Brown-headed Spider Monkey	Singe-Araignée à tête brune	буроголовая коата
Braunrückentamarin	– Tamarin	Tamarin à tête brune	буроспинный тамарин
Brazzameerkatze	De Brazza's Monkey	Cercopithèque de Brazza	бразовская мартышка
Breitfußbeutelmäuse	Broad-footed Marsupial Mice	Rats marsupiaux	
Breitfußmaulwurf	– Mole		калифорнийский крот

GERMAN NAME	ENGLISH NAME	FRENCH NAME	RUSSIAN NAME
Breitkopfkänguruh	Broad-faced Rat Kangaroo		широколицый потору
Breitnasenaffen, Breitnasen	New World Monkeys	Singes du Nouveau Monde	обезьяны нового света
Breitschnauzenhalbmaki	Broad-nosed Gentle Lemur	Hapalémur à nez large	широконосый полумаки
Brillen-Hasenkänguruh	Spectacled Hare Wallaby		очковый заячий кенгуру
Brillenlangur	Dusky Leaf Monkey	Semnopithèque obscur	очковый тонкотел
Bruijn-Langschnabeligel	Bruijn's Echidna	Échidné de Bruijn	проехидна Бруйна
Brüllaffen	Howler Monkeys	Hurleurs	ревуны
Bubu-Langschnabeligel	Bubu-Echidna		проехидна острова Бубу
Budeng	Budeng	Budeng	
Burramys parvus	Burramys Pygmy Possum	Souris-Opossum de Burramy	
Bürsten-Felskänguruh	Brush-tailed Rock Wallaby		кистехвостый каменный кенгуру
Bürstenkänguruhs	Short-nosed Rat Kangaroos	Rats-Kangourous à nez court	кистехвостые кенгуру
Bürstenmaulwurf	Hairy-tailed Mole	Taupe à queue chevelue	волосохвостый крот
Buschbabies	Galagos	Galagidés	галаги
Büschelohriger Katzenmaki	Hairy-eared Dwarf Lemur	Chirogale aux oreilles velues	
Buschschwanzbeutelratten	Brush-tailed Possums		кистехвостые сумчатые крысы
Buschschwanztupaias	– Tree Shrews	Tupaiinés	собственно тупайи
Buschwaldgalago	Allen's Bushbaby	Galago d'Allen	
Cacajao	Uakaris	Ouakaris	уакори
– calvus	Bald Uakari	Ouakari chauve	лысый уакори
– melanocephalus	Black-headed Uakari	– à tête noire	черноголовый уакори
– roosevelti	Black Uakari	– de Roosevelt	черный уакори
– rubicundus	Red Uakari	– rubicond	красный уакори
Caenolestes fuliginosus	Ecuador Rat Possum		бурая первокрыса
Caenolestidae	Rat Opossums	Caenolestidés	ценолестовые сумчатые
Callicebus	Titi Monkeys	Titis	обезьяны-прыгуны
– cupreus	Red Titi	Callicèbe roux	красный прыгун
– moloch	Orabussu Titi	– orabassu	серый прыгун
– personatus	Masked Titi	– à masque	черноголовый прыгун
– torquatus	Collared Titi	– à fraise	
Callimico goeldii	Goeldi's Monkey	Tamarin de Goeldi	
Callimiconidae	– Monkeys	Tamarins de Goeldi	
Callithricidae	Marmosets and Tamarins	Callithricidés	когтистые обезьяны
Callithrix	Marmosets	Ouistitis	
– albicollis	White-necked Marmoset	Ouistiti à col blanc	белошейная игрунка
– argentata	Silvery Marmoset	– melanure	серебристая игрунка
– aurita	White-eared Marmoset	– oreillard	белоухая игрунка
– chrysoleucos	Yellow-legged Marmoset	– aux pieds jaunes	желтоногая игрунка
– flaviceps	Buff-headed Marmoset	– à tête jaune	желтоголовая игрунка
– humeralifer	White-shouldered Marmoset	– à camail	белоплечая игрунка
– jacchus	Common Marmoset	Ouistiti	обыкновенная игрунка
– leucocephala	White-fronted Marmoset	– à tête blanche	белолицая игрунка
– penicillata	Black-pencilled Marmoset	– à pinceau noir	кисточковая игрунка
– pygmaea	Pigmy Marmoset	– mignon	карликовая игрунка
– santaremensis	Santarém Marmoset	– de Santarém	длинноухая игрунка
Caloprymnus campestris	Desert Rat Kangaroo	Rat-Kangourou du désert	степная кенгуровая крыса
Caluromys	Woolly Opossums	Opossums laineux	шерстистые сумчатые крысы
– irrupta	– Opossum	Opossum laineux	полосатая шерстистая сумчатая крыса
– laniger	Woolly Opossum	Opossum laineux	рыжая шерстистая сумчатая крыса
– philander	Philander Opossum	Opossum laineux	желтая шерстистая сумчатая крыса
Campbells Meerkatze	Campbell's Guenon	Mone de Campbell	мартышка Кампбелла
Catarrhina	Old World Simian Primates	Catarhiniens	узконосые обезьяны
Cebidae	New World Monkeys	Cébidés	капуцинообразные обезьяны
Cebinae	Capuchin Monkeys	Cébinés	цебусовые обезьяны
Ceboidea	New World Monkeys	Singes du Nouveau Monde	широконосые обезьяны
Cebus	Capuchins	Sapajous	капуцины
– albifrons	White-fronted Capuchin	Sapajou à front blanc	белолобый капуцин
– apella	Brown Capuchin	– apelle	капуцин-фавн
– capucinus	White-throated Capuchin	– capucin	обыкновенный капуцин
– nigrivittatus	Weeper Capuchin	– brun	бурый капуцин
Celebeskoboldmaki	Eastern Tarsier	Tarsier spectre	долгопят-привидение
Celebesmakaken	Moor Macaques	Macaques de Célèbes	целебесские макаки
Ceramnasenbeutler	Ceram Long-nosed Bandicoot		
Cercaërtus	Dormouse Possums	Phalangers Loirs	сумчатые сони
– concinnus	South-western Pigmy Phalanger		тонкохвостая сумчатая соня
– nanus	Dormouse Possum		обыкновенная сумчатая соня
Cercocebus	Mangabeys	Mangabeys	мангабеи
– albigena	Grey-cheeked Mangabey	Mangabey à gorge blanche	гривистый мангабей
– aterrimus	Black Mangabey	– noir	бородатый мангабей
– galeritus	Agile Mangabey	– à ventre doré	чубастый мангабей
– torquatus	Sooty Mangabey	– à collier blanc	воротничковый мангабей
Cercopithecidae	Guenon-like Monkeys	Cercopithecidés	мартышки
Cercopithecoidea	Old World Monkeys		мартышковые

GERMAN NAME	ENGLISH NAME	FRENCH NAME	RUSSIAN NAME
Cercopithecus	Guenons	Cercopithèques	мартышки
– aethiops	Grass Monkey	Singe vert	зеленая мартышка
– cephus	Moustached Monkey	Moustac	голуболицая мартышка
– diana	Diana Monkey	Cercopithèque diane	диана
Cercopithecus erythrogaster	Red-bellied Guenon	Hocheur à ventre rouge	краснобрюхая мартышка
– hamlyni	Owl-faced Guenon	Cercopithèque à tête de Hibou	
– lhoesti	L'Hoest's Monkey	Cercopithèque de l'Hoest	бородатая мартышка
– mitis	Diademed Guenon	– diadème	
– mona	Mona Monkey	– mone	мона
– neglectus	De Brazza's Monkey	– de Brazza	бразовская мартышка
– nictitans	Greater White-nosed Guenon	– hocheur	большая белоносая мартышка
– nigroviridis	Swamp Guenon	– noir et vert	чернозеленая мартышка
– petaurista	Lesser White-nosed Guenon	Hocheur blanc-nez	малая белоносая мартышка
– pogonias	Crowned Guenon	Cercopithèque pogonias	чубатая мартышка
– talapoin	Dwarf Guenon	Talapoin	крошечная мартышка
Ceylon-Hutaffe	Toque Monkey	Macaque couronné	цейлонский макак
Chaeropus ecaudatus	Pig-footed Bandicoot		хероп
Cheirogaleinae	Dwarf Lemurs	Chirogaléinés	крысиные маки
Cheirogaleus	Dwarf Lemurs	Chirogales	
– major	Greater Dwarf Lemur	Chirogale de Milius	
– medius	Fat-tailed Dwarf Lemur		
– trichotis	Hairy-eared Dwarf Lemur	Chirogale aux oreilles velues	
Chile-Opossummaus	Chilean Rat Opossum		чилийская первокрыса
Chimarrogale	Asian Water Shrews	Chimarrogales	
– phaeura			индонезийская водяная белозубка
– platycephala	Himalayan Water Shrew	Chimarrogale de l'Himalaya	гималайская водяная белозубка
Chironectes minimus	Yapok	Yapock	водяной опоссум
Chiropotes	Bearded Sakis		
– albinasa	White-nosed Saki	Saki à nez blanc	белоносый саки
– chiropotes	Red-backed Saki	– capucin	красноспинный саки
– satanas	Black Saki	– noir	чертов саки
Chrysochloridae	Golden Moles	Chrysochloridés	златокроты
Chrysochloris	Golden Moles	Taupes dorées	собственно златокроты
– asiatica	Cape Golden Mole	Taupe dorée du Cap	капский златокрот
– stuhlmanni	Stuhlmann's Golden Mole	– – de Stuhlmann	
Chrysospalax trevelyani	Giant Golden Mole	Grande Taupe dorée	исполинский златокрот
Coahuilamaulwurf	Coahuilan Mole	Taupe de Coahuila	
Colobidae	Leaf Monkeys	Colobidés	тонкотелые обезьяны
Colobus	Colobus Monkeys	Colobes	толстотелы
– abyssinicus	Northern Black-and-white Colobus	Colobe de l'Abyssinie	гвереца
– badius	Red Colobus	– bai	красный толстотел
– polykomos	Southern Black-and-white Colobus	– à longs poils	королевский толстотел
– verus	Green Colobus	– vrai	ванбенеденовский толстотел
Columbischer Bergwollaffe			колумбийская шерстистая обезьяна
Condylura cristata	Star-nosed Mole	Condylure étoilé	звездорыл
Condylurinae	– Moles	Condylures étoilés	звездорылы
Coquerels Zwergmaki	Coquerel's Mouse Lemur	Mirza de Coquerel	
Crocidura caudata	Mediterranean Long-tailed Shrew	Musaraigne méditerranéenne	средиземноморская белозубка
– flavescens	Giant Musk Shrew		большая белозубка
– leucodon	Bicolor White-toothed Shrew	Crocidure leucode	белобрюхая белозубка
– russula	Common European White-toothed Shrew	Musareigne musette	бурая белозубка
– smithi	Desert Musk Shrew	Crocidure du désert	пустынная белозубка
– suaveolens	Lesser White-toothed Shrew	Musaraigne des jardins	малая белозубка
Crocidurinae	White-toothed Shrews	Musaraignes à dents blanches	землеройки-белозубки
Cryptochloris wintoni	De Winton's Golden Mole	Taupe dorée de Winton	златокрот
Cryptotis parva	Least Shrew	Petite Musaraigne à queue courte	североамериканская короткоухая бурозубка
Cynopithecus niger	Celebes Crested Macaque	Cynopithèque nègre	хохлатый павиан
Dactylopsila	Striped Phalangers	Phalangers au pelage rayé	полосатые кускусы
– palpator	Long-fingered Striped Phalanger		малый полосатый кускус
– trivirgata	Striped Phalanger	Phalanger au pelage rayé	большой полосатый кускус
Dasogale fontoynonti	Fontoynont's Hedgehog Tenrec		ежевый тенрек Фонтуанона
Dasycercus	Crest-tailed Marsupial Mice		
– cristicauda	Crest-tailed Marsupial Mouse		гребнехвостая мышевидка
Dasyuridae	Flesh-eating Marsupials	Dasyuridés	хищные сумчатые
Dasyurinae	Dasyures	Dasyurinés	сумчатые куницы
Dasyuroides byrnei	Crest-tailed Marsupial Rat		гребешковая мышевидка
Dasyurus	Dasyures or Native "Cats"		собственно сумчатые куницы
– geoffroyi	Western Dasyure		чернохвостая сумчатая куница
– hallucatus	Little Northern Dasyure		североавстралийская сумчатая куница

GERMAN NAME	ENGLISH NAME	FRENCH NAME	RUSSIAN NAME
– *maculatus*	Tiger Cat	Chat marsupial	исполинская сумчатая куница
– *quoll*	Eastern Dasyure		крапчатая сумчатая куница
Daubentonia madagascariensis	Aye-Aye	Aye-Aye	мадагаскарская руконожка
Daubentoniidae	Aye-Ayes	Daubentoniidés	руконожки
Demidoffgalago	Demidoff's Bushbaby	Galago de Demidoff	галаго Демидова
Dendrogale	Smooth-tailed Tree Shrews		горные тупайи
Dendrolagus	Tree Kangaroos	Kangourous arboricoles	древесные кенгуру
Dendrolagus dorianus	Doria's Tree Kangaroo	Kangourou arboricole de Doria	древесный кенгуру Дория
– *goodfellowi*	Goodfellow's Tree Kangaroo	– – de Goodfellow	
– *lumholtzi*	Lumholtz's Tree Kangaroo	– – de Lumholtz	квинслендский древесный кенгуру
– *matschiei*	Matschie's Tree Kangaroo	– – de Matschie	
– *ursinus*	Black Tree Kangaroo		кенгуру-медведь
Derbykänguruh	Tammar	Wallaby de l'Ile d'Eugène	кенгуру дерби
Desmana moschata	Russian Desman	Desman de Moscovie	обыкновенная выхухоль
Desmane	Desmans	Desmaninés	выхухоли
Desmaninae	Desmans	Desmaninés	выхухоли
Diademmeerkatze	Diademed Guenon	Cercopithèque diadème	
Diademsifaka	– Sifaka	Propithèque diadème	белолобый сифака
Dianameerkatze	Diana Monkey	Cercopithèque diane	диана
Dickschwanzbeutelratte	Little Water Opossum		толстохвостый опоссум
Dickschwänzige Schmalfußbeutelmaus	Fat-tailed Sminthopsis		толстохвостая сумчатая землеройка
Dickschwanz-Schlafbeutler	Dormouse Possum		обыкновенная сумчатая соня
Dickschwanzspitzmäuse		Pachyures	многозубые белозубки
Didelphidae	Opossums	Opossums d'Amérique	сумчатые крысы
Didelphis	Common and Azara's Opossums	Opossums	опоссумы
– *marsupialis*	– Opossum	Opossum commun	обыкновенный опоссум
– *paraguayensis*	Azara's Opossum		казака
Diplomesodon pulchellum	Piebald Shrew		пегий путорак
Distoechurus pennatus	Pen-tailed Phalanger		перохвостая сумчатая соня
Doppelkammbeutelmaus	Crest-tailed Marsupial Rat		гребешковая мышевидка
Dorcopsis hageni	Northern New Guinea Wallaby		кустовый кенгуру Хагена
– *macleayi*	New Guinea Mountain Wallaby	Wallaby de Macleay	кустовый кенгуру Миклухо-Маклая
Doria-Baumkänguruh	Doria's Tree Kangaroo	Kangourou arboricole de Doria	древесный кенгуру Дория
Dorré-Hasenkänguruh		Lièvre wallaby de l'Île de Dorré	
Drill	Drill	Drill	дрил
Dschelada	Gelada Baboon	Gelada	джелада
Dunkelfüßige Waldspitzmaus	Dark-footed Forest Shrew		темнопалая белозубка
Dunkles Rüsselhündchen	Stuhlmann's Elephant Shrew	Macroscélide de Stuhlmann	темная хоботковая собачка
Dünnschwanz-Schlafbeutler	South-western Pigmy Phalanger		тонкохвостая сумчатая соня
Echinops telfairi	Lesser Hedgehog Tenrec		ежевый тенрек Тельфера
Echinosorex gymnurus	Moon Rat	Gymnure	большой крысиный еж
Echinosoricinae	Hairy Hedgehogs	Gymnures	крысиные ежи
Echte Diademmeerkatze		Cercopithèque diadème	
– Igel	Spiny Hedgehogs	Hérissons	настоящие ежи
– Katzenmakis	Dwarf Lemurs	Chirogales	
– Makis	Lemurs	Lémur	маки
Echymipera	Spiny New Guinea Bandicoots		остроносые бандикуты
Eierlegende Säugetiere	Egg-laying Mammals	Protothériens	клоачные
Eigentliche Gibbons	Gibbons	Gibbons	собственно гиббоны
– Känguruhs	Kangaroos and Wallabies	Macropodinés	собственно кенгуру
– Kletterbeutler	Phalangers	Phalangerinés	собственно лазающие сумчатые
Ekuador-Opossummaus	Ecuador Rat Opossum		бурая первокрыса
Elefantenspitzmäuse	Elephant Shrews		
Elephantulus	Elephant Shrews		
– *brachyrhynchus*	Short-snouted Elephant Shrew		коротконосый прыгунчик
– *intufi*	Bushveld Elephant Shrew		
– *rozeti*	North African Elephant Shrew	Macroscélide de l'Afrique du Nord	североафриканский прыгунчик
– *rupestris*	Rock Elephant Shrew		скалистый прыгунчик
Elliots Tupaia	Madras Tree-shrew	Toupaïe d'Elliot	тупайя Эллиота
Erdtanrek		Géogale	земляной тенрек
Eremitalpa granti	Grant's Desert Golden Mole	Taupe dorée de Grant	пустынный крот
Erinaceidae	Hedgehogs	Érinacéidés	ежи
Erinaceinae	Spiny Hedgehogs	Hérissons	настоящие ежи
Erinaceus	Common Hedgehogs	– communs	обыкновенные ежи
– *amurensis*	Amur Hedgehog	Hérisson de l'Amour	амурский еж
– *europaeus*	Western European Hedgehog	– d'Europe de l'Ouest	бурогрудый еж
– *koreanus*	Korean Hedgehog	– de Corée	корейский еж
– *roumanicus*	Eastern European Hedgehog	– d'Europe de l'Est	белогрудый еж
Erythrocebus patas	Red Guenon	Patas	обыкновенный гусар
Etruskerspitzmaus	Savi's Pigmy Shrew	Pachyure étrusque	белозубка-малютка
Eudromicia caudata	Pigmy possum		новогвинейская сумчатая соня
Eulenkopfmeerkatze	Owl-faced Guenon	Cercopithèque à tête de Hibou	

GERMAN NAME	ENGLISH NAME	FRENCH NAME	RUSSIAN NAME
Europäischer Maulwurf	Common Eurasian Mole	Taupe commune	обыкновенный крот
Eutheria	Placental Mammals	Eutheriens	высшие млекопитающие
Everetts Spitzhörnchen	Philippines Tree Shrew		тупайя Эверетта
Faunaffe	Brown Capuchin	Sapajou apelle	капуцин-фавн
Federschwanz	Pen-tailed Tree Shrew	Ptilocerque	перохвостая тупайя
Federschwanzbeutler	– Phalanger		перохвостая сумчатая соня
Federschwanztupaias	– Tree Shrews	Ptilocerques	перохвостые тупайи
Feldspitzmaus	Bicolor White-toothed Shrew	Crocidure leucode	белобрюхая белозубка
Felsen-Ringelschwanzbeutler	Rock-haunting Ring Tail		кускус Даля
Felskänguruhs	Rock Wallabies	Wallabies de rochers	каменные кенгуру
Feroculus feroculus	Kelaart's Long-clawed Shrew		когтистая белозубка Келаарта
Festland-Bürstenkänguruh	Gaimard's Rat Kangaroo	Rat-Kangourou de Gaimard	австралийский кистехвостый кенгуру
Fettschwänzige Breitfußbeutelmaus	Fat-tailed Marsupial Mouse		толстохвостая мышевидка
Fettschwanzmaki	– Dwarf Lemur		
Filander	Pademelon	Thylogale	малые кустовые кенгуру
Fingertier	Aye-Aye	Aye-Aye	мадагаскарская руконожка
Flachkopfbeutelmäuse	Flat-skulled Marsupials		плоскоголовые мышевидки
Flachlandgorilla	Lowland Gorilla	Gorille de côte	береговая горилла
Flachnagelkänguruh	Northern Nail-tail Wallaby		обыкновенный когтехвостый кенгуру
Fleckenbeutelmarder	Native Cats		собственно сумчатые куницы
Fleckschwanzbeutelmarder	Tiger Cat	Chat marsupial	исполинская сумчатая куница
Flinkes Känguruh	Sandy Wallaby	Wallaby agile	проворный кустовый кенгуру
Fontoynonts Igeltanrek	Fontoynont's Hedgehog Tenrec		ежевый тенрек Фонтуанона
Formosamakak	Formosa Macaque	Macaque de Formosa	тайванский резус
Formosarhesus	Formosa Macaque	Macaque de Formosa	тайванский резус
Fuchskusu	Brush-tailed Phalanger	Phalanger-Renard	лисий кузу
Gabelstreifiger Katzenmaki	Fork-marked Dwarf Lemur	Phaner à fourche	вильчатый маки
Galagidae	Galagos	Galagidés	галаги
Galago	Galagos	Galagos	галаго
– *alleni*	Allen's Bushbaby	Galago d'Allen	
– *crassicaudatus*	Thick-tailed Bushbaby	– à queue trouffue	толстохвостый галаго
– *demidovii*	Demidoff's Bushbaby	– de Demidoff	галаго Демидова
– *elegantulus*	Western Needle-clawed Bushbaby	– mignon	
– *inustus*	Eastern Needle-clawed Bushbaby	– du Congo	
– *senegalensis*	Senegal Bushbaby	– du Sénégal	сенегальский галаго
Galagos	Galagos	Galagos	галаго
Galemys pyrenaicus	Pyrenean Desman	Desman des Pyrénées	пиренейская выхухоль
Gartenspitzmaus	Lesser White-toothed Shrew	Musaraigne des jardins	малая белозубка
Gebirgsbachspitzmaus	Szechuan Water Shrew	Nectogale élégant	тибетская водяная белозубка
Geflecktes Rüsselhündchen	Checkered Elephant Shrew		пятнистая хоботковая собачка
Gehaubter Kapuziner	Brown Capuchin	Sapajou apelle	капуцин-фавн
Gelber Babuin	Yellow Baboon	Babouin cynocéphale	бабуин
Gelbes Totenköpfchen	Red-backed Squirrel Monkey	Sapajou à dos rouge	желтая саймири
Gelbe Wollbeutelratte	Philander Opossum	Opossum laineux	желтая шерстистая сумчатая крыса
Gelbfußäffchen	Yellow-legged Marmoset	Ouistiti aux pieds jaunes	желтоногая игрунка
Gelbfußbeutelmaus	Yellow-footed Marsupial Mouse		желтоногая мышевидка
Gelbfußkänguruh	Ring-tailed Rock Wallaby		желтоногий каменный кенгуру
Gelbgrüne Meerkatze		Callitriche	желтозеленая мартышка
Gelbkopfbüscheläffchen	Buff-headed Marmoset	Ouistiti à tête jaune	желтоголовая игрунка
Gelbschenkelgalago		Galago du Kilimandjaro	
Gelbschwanzäffchen	Yellow-legged Marmoset	Ouistiti aux pieds jaunes	желтоногая игрунка
Gelbschwanzwollaffe	Peruvian Mountains Woolly Monkey		желтохвостая шерстистая обезьяна
Geoffroy-Klammeraffe	Central American Spider Monkey	Singe-Araignée aux Mains noirs	коата Жоффруа
Geoffroy-Perückenäffchen	Geoffroy's Tamarin	Pinché de Geoffroy	игрунка Жоффруа
Geogale aurita		Géogale	земляной тенрек
Gescheckte Spitzmaus	Piebald Shrew		пегий путорак
Gewöhnliches Spitzhörnchen	Common Tree-shrew	Toupaïe	обыкновенная тупайя
Gibbons	Gibbons	Gibbons	гиббоны
Gilbert-Kaninchenkänguruh	Gilbert's Rat Kangaroo	Rat-Kangourou de Gilbert	потору Джильберта
Gleithörnchenbeutler	Honey Gliders	Petaurus	сумчатые летяги
Glironia	Bushy-tailed Opossums		кистехвостые сумчатые крысы
Goelditamarin	Goeldi's Monkey	Tamarin de Goeldi	
Goldgelbes Löwenäffchen	Golden Lion Marmoset	Petit singe-lion	розалия
Goldkopflöwenäffchen	Golden-headed Tamarin	Singe-lion à tête dorée	желтоголовая львиная игрунка
Goldkopfsaki	– Saki	Saki à tête dorée	золотистый саки
Goldmanteltamarin	Golden-mantled Tamarin	Tamarin à manteau doré	
Goldmulle	Golden Moles	Chrysochloridés	златокроты
Goldstirn-Klammeraffe	Long-haired Spider Monkey	Singe-Araignée à ventre blanc	светлолобая коата
Goldstumpfnase	Snub-nosed Monkey	Rhinopithèque de Roxellane	рокселланов ринопитек
Golduakari	Red Uakari	Ouakari rubicond	красный уакари

GERMAN NAME	ENGLISH NAME	FRENCH NAME	RUSSIAN NAME
Goodfellow-Baumkänguruh	Goodfellow's Tree Kangaroo	Kangourou arboricole de Goodfellow	
Gorilla	Gorilla	Gorille	горилла
Gorilla gorilla	Gorilla	Gorille	горилла
Grauarmmakak			серорукий целебесский макак
Grauer Halbmaki	Grey Gentle Lemur	Hapalémur	серый полумаки
– Springaffe	Orabussu Titi	Callicèbe arabassu	серый прыгун
Graues Riesenkänguruh	Grey Kangaroo	Kangourou géant	серый исполинский кенгуру
Graue Wüstenspitzmaus	Crawford's Desert Shrew	Musaraigne du désert	серая пустынная бурозубка
Graugrüne Meerkatze	Grivet Monkey	Grivet	серозеленая мартышка
Große Otterspitzmaus	Otter Shrew	Potamogale	выдровая землеройка
Großer Beutelmull	Greater Marsupial Mole	Grande Taupe marsupiale	сумчатый крот
– Gleithörnchenbeutler	Yellow-bellied Glider		большая сумчатая летяга
Großer Haarigel	Moon Rat	Gymnure	большой крысиный еж
– Igeltanrek	Greater Hedgehog Tenrec		обыкновенный ежевый тенрек
– Kaninchen-Nasenbeutler	Rabbit Bandicoot		обыкновенный сумчатый заяц
– Katzenmaki	Greater Dwarf Lemur	Chirogale de Milius	
– Kurznasenbeutler	Brindled Bandicoot		большой курносый бандикут
– Langnasenbeutler	Long-nosed Bandicoot	Bandicoot à nez long	большой бандикут
– Neuguinea-Nasenbeutler	New Guinea Bandicoot	Bandicoot de Nouvelle-Guinée	большой новогвинейский бандикут
– Pinselschwanzbeutler	Black-tailed Phascogale		кистехвостая мышевидка
– Rattenigel	Moon Rat	Gymnure	большой крысиный еж
– Streifenbeutler	Striped Phalanger	Phalanger au pelage rayé	большой полосатый кускус
– Tanrek	Tailless Tenrec	Tanrec	обыкновенный тенрек
– Wieselmaki	Greater Sportive Lemur	Lépilémur mustélin	большой куний маки
Großes Rattenkänguruh	Rufous Rat Kangaroo	Rat-Kangourou rougeâtre	рыжая кенгуровая крыса
Große Weißnasenmeerkatze	Greater White-nosed Guenon	Cercopithèque hocheur	большая белоносая мартышка
– Wüstenspitzmaus	Merriam's Desert Shrew	Grande Musaraigne du désert	большая пустынная бурозубка
Grüne Meerkatze	Grass Monkey	Singe vert	зеленая мартышка
Grüner Pavian	Anubis Baboon	Papion anubis	анубис
– Stummelaffe	Green Colobus	Colobe vrai	ванбенеденовский толстотел
Guatemalabrüllaffe	Guatemalan Howler Monkey	Hurleur de Guatemala	гватемальский ревун
Guerezas	Black-and-white Colobuses	Guérézas	
Guineapavian	Guinea Baboon	Babouin de Guinée	гвинейский павиан
Gymnobelideus leadbeateri	Leadbeater's Phalanger	Opossum de Leadbeater	сумчатая белка
Haarigel	Hairy Hedgehogs	Gymnures	крысиные ежи
Haarnasenwombat	Hairy-nosed Wombat	Wombat à narines poilues	широколобый вомбат
Haarschwanzmaulwurf	Hairy-tailed Mole	Taupe à queue chevelue	волосохвостый крот
Hagen-Buschkänguruh	Northern New Guinea Wallaby		кустовый кенгуру Хагена
Haiti-Schlitzrüßler	Haitian Solenodon	Solenodon	гаитийский щелезуб
Halbaffen	Prosimians	Prosimiens	полуобезьяны
Halbborstenigel	Streaked Tenrec	Hemicentete	полосатый тенрек
Halbmakis	Gentle Lemurs	Hapalémur	полумаки
Halsbandmangabe	Sooty Mangabey	Mangabey à collier blanc	воротничковый мангабей
Hamlynmeerkatze	Owl-faced Guenon	Cercopithèque à tête de Hibou	
Hanuman	Entellus Langur	Houleman	гульман
Hapalemur	Gentle Lemurs	Hapalémur	полумаки
– *griseus*	Grey Gentle Lemur	Hapalémur	серый полумаки
– *simus*	Broad-nosed Gentle Lemur	– à nez large	широконосый полумаки
Hasenkänguruhs	Hare Wallabies	Lièvres wallabies	заячьи кенгуру
Haubenlangur	Silvered Leaf Monkey		гривистый тонкотел
Haubenmangabe	Agile Mangabey	Mangabey à ventre doré	чубастый мангабей
Hausspitzmaus	Common European White-toothed Shrew	Musaraigne musette	бурая белозубка
Hemicentetes	Streaked Tenrec	Hemicentetes	полутенреки
– *nigriceps*	Streaked Tenrec		черноголовый тенрек
– *semispinosus*	Streaked Tenrec	Hemicentete	полосатый тенрек
Hemiechinus	Eared Hedgehog		ушастые ежи
– *auritus*	Long-eared Hedgehog		ушастый еж
Herrentiere	Primates	Primates	приматы
Himalajawasserspitzmaus	Himalayan Water Shrew	Chimarrogale de l'Himalaya	гималайская водяная белозубка
Höhere Säugetiere	Placental Mammals	Eutheriens	высшие млекопитающие
Hominoidea	Apes and Men	Hominoïdés	человекообразные обезьяны
Honigbeutler	Honey Phalanger	Souris à miel	пяткоход
Hörnchenkletterbeutler	Leadbeater's Possum	Opossum de Leadbeater	сумчатая белка
Hottentotten-Goldmull	Hottentot Golden Mole		готтентотский крот
Hübschgesichtkänguruh	Pretty-face Wallaby		кустовый кенгуру Парри
Hulman	Entellus Langur	Houleman	гульман
Hulock	Hoolock Gibbon	Hoolock	гулок
Hundsaffen	Old World Monkeys		мартышковые
Hundskusu	Short-eared Brush-tailed Phalanger	Phalanger de montagne	собачий кузу
Husarenaffe	Red Guenon	Patas	обыкновенный гусар
Hutaffen	Bonnet and Toque Monkeys	Macaques bonnets et Couronnés	

GERMAN NAME	ENGLISH NAME	FRENCH NAME	RUSSIAN NAME
Hylobates	Gibbons	Gibbons	собственно гиббоны
– *agilis*	Dark-handed Gibbon	Gibbon agile	быстрый гиббон
– *concolor*	Black Gibbon	– noir	одноцветный гиббон
– *hoolock*	Hoolock Gibbon	Hoolock	гулок
– *lar*	White-handed Gibbon	Gibbon lar	лар
– *moloch*	Grey Gibbon	– cendré	серебристый гиббон
Hylobatidae	Gibbons	Hylobatidae	гиббоны
Hylomys suillus	Lesser Gymnure		малый крысиный еж
Hypsiprymnodon moschatus	Musky Rat Kangaroo	Rat musqué Kangourou	мускусный кенгуру
Hypsiprymnodontinae	– – Kangaroos	Rats musqués Kangourous	мускусные сумчатки
Igel	Hedgehogs	Hérrissons	ежи
Indischer Hutaffe	Bonnet Monkey	Macaque bonnet chinois	индийский макак
– Igel	Indian Hedgehog		индийский еж
Indri	Indri	Indri	индри
Indri indri	Indri	Indri	индри
Indriartige	Indrisoid Lemurs	Indriinés	индри
Indriidae	Indrisoid Lemurs	Indriinés	индри
Inneraustralische Springbeutelmaus	Central Jerboa Marsupial		среднеавстралийский сумчатый тушканчик
Insectivora	Insect-Eaters	Insectivores	насекомоядные
Insektenesser	Insect-Eaters	Insectivores	насекомоядные
Irmawallaby	Black-gloved Wallaby	Wallaby d'Irma	кенгуру ирма
Japanischer Spitzmull	Japanese Shrew Mole	Taupe des montagnes du Japon	японский землеройковый крот
Japanmakak	– Macaque	Macaque Japonais	японский макак
Javaneraffe	Crab-eating Macaque	– de Buffon	яванский макак
Kahlkopfuakari	Bald Uakari	Ouakari chauve	лысый уакори
Kaiserschnurrbarttamarin	Emperor Tamarin	Tamarin empereur	императорский усатый тамарин
Kalifornischer Maulwurf	Broad-footed Mole		калифорнийский крот
Kammschwanzbeutelmäuse	Crest-tailed Marsupial Mice		
Känguruhs	Wallabies and Kangaroos	Kangourous	кенгуру
Kaninchenkänguruhs	Long-nosed Rat Kangaroos	Rats-Kangourous à nez long	потору
Kaninchen-Nasenbeutler	Rabbit Bandicoots	Bandicoots-Lapins	сумчатые зайцы
Kansumaulwurf	Kansu Mole	Taupe de Kansu	западнокитайский землеройковый крот
Kapgoldmull	Cape Golden Mole	– dorée du Cap	капский златокрот
Kapigel	Cape Hedgehog		капский еж
Kapuziner	White-throated Capuchin	Sapajou capucin	обыкновенный капуцин
Kapuzineraffen	Capuchin Monkeys	Cébinés	цебусовые обезьяны
Kapuzinerartige	New World Monkeys	Cébidés	капуцинообразные обезьяны
Katta	Ring-tailed Lemur	Lémur catta	катта
Katzenmakis	Dwarf Lemurs	Chirogaléinés	крысиные маки
Kelaarts Langkrallenspitzmaus	Kelaart's Long-clawed Shrew		когтистая белозубка Келаарта
Kielnagelgalago	Needle-clawed Bushbaby	Galago mignon et Galago du Congo	
Klammeraffen	Spider Monkeys	Singes-Araignées	коаты
Klammerschwanzaffen	Spider and Woolly Monkeys	Atélinés	шерстистые обезьяны и коаты
Kleideraffe	Douc Langur	Douc	немейский тонкотел
Kleiner Beutelmull	North-western Marsupial Mole	Petite Taupe marsupiale	малый сумчатый крот
– Igeltanrek	Lesser Hedgehog Tenrec		ежевый тенрек Тельфера
– Kaninchen-Nasenbeutler	White-tailed Rabbit Bandicoot		малый сумчатый заяц
– Kurznasenbeutler	Southern Short-nosed Bandicoot		малый курносый бандикут
– Pinselschwanzbeutler	Red-tailed Phascogale		малая кистехвостая мышевидка
– Plumplori	Lesser Slow Loris		малый толстый лори
– Rattenigel	– Gymnure		малый крысиный еж
– Streifenbeutler	Long-fingered Striped Phalanger		малый полосатый кускус
– Wieselmaki	Sportive Lemur	Lépilémur à queue rouge	малый куний маки
Kleine Schmalfußbeutelmaus	Mouse Sminthopsis		обыкновенная сумчатая землеройка
– Weißnasenmeerkatze	Lesser White-nosed Guenon	Hocheur blanc-nez	малая белоносая мартышка
Kleinohrigel	Common Hedgehogs	Hérissons communs	обыкновенные ежи
Kleintanreks	Lesser Tenrecs	Microgales	длиннохвостые тенреки
Kletterbeutler	Phalangers	Phalangéridés	лазающие сумчатые
Klippen-Elefantenspitzmaus	Rock Elephant Shrew		скалистый прыгунчик
Kloakentiere	Monotremes	Monotrèmes	однопроходные
Koala	Koala	Koala	коала
Koalaverwandte	Koala-like Marsupials	Phascolarctinés	сумчатые медведи
Koboldmakis	Tarsiers	Tarsidés	долгопяты
Komba	Thick-tailed Bushbaby	Galago à queue trouffue	толстохвостый галаго
Kongopanzerspitzmaus	Congo Armoured Shrew		конголезская броненоска
Kongowimperspitzmaus			конголезская белозубка
Koreaigel	Korean Hedgehog	Hérisson de Corée	корейский еж
Krallenaffen, Krallenäffchen	Marmosets and Tamarins	Callithricidés	когтистые обезьяны
Kretaigel			критский еж
Kronenmaki	Crowned Lemur	Lémur couronné	хохлатый монго
Kronenmeerkatze	– Guenon	Cercopithèque pogonias	чубатая мартышка
Kuba-Schlitzrüßler	Cuban Solenodon	Almiqui	кубинский щелезуб

GERMAN NAME	ENGLISH NAME	FRENCH NAME	RUSSIAN NAME
Kurzkopfgleitbeutler	Honey Glider	Phalanger volant	короткоголовая сумчатая летяга
Kurznagelkänguruh	Bridled Nail-tail Wallaby		уздечковый когтехвостый кенгуру
Kurznasenbeutler	Short-nosed Bandicoots		курносые бандикуты
Kurznasen-Elefantenspitzmaus	Short-snouted Elephant Shrew		коротконосый прыгунчик
Kurzohr-Felskänguruh	Short-eared Rock Wallaby		короткоухий каменный кенгуру
Kurzohrrüsselspringer	– Elephant Shrew		обыкновенный слоновый прыгунчик
Kurzschnabeligel	Short-beaked Spiny Ant-eater	Échidné à bec droit	ехидны
Kurzschwanzaffen	Uakaris	Ouakaris	уакори
Kurzschwanzkänguruh	Quokka	Kangourou à queue courte	короткохвостый кустовый кенгуру
Kurzschwanzspitzmaus	Short-tailed Shrew	Grande Musaraigne à queue courte	североамериканская коротко-хвостая бурозубка
Kuskuse	Cuscuses	Couscous	кускусы
Küstenmaulwurf	Coast Mole	Taupe de côte	тихоокеанский крот
Kusus	Brush-tailed Phalangers	Opossums d'Australie	кузу
Lagorchestes	Hare Wallabies	Lièvres wallabies	заячьи кенгуру
– *conspicillatus*	Spectacled Hare Wallaby		очковый заячий кенгуру
– *hirsutus*	Western Hare Wallaby		косматый заячий кенгуру
Lagorchestes leporoides	Brown Hare Wallaby		обыкновенный заячий кенгуру
Lagostrophus fasciatus	Banded Hare Wallaby	Wallaby rayé	поперечнополосатый кенгуру
Lagothrix	Woolly Monkeys	Singes laineux	шерстистые обезьяны
– *flavicauda*	Peruvian Mountain Woolly Monkey		желтохвостая шерстистая обезьяна
– *lagotricha*	Humboldt's Woolly Monkey	Lagotriche de Humboldt	шерстистая обезьяна Гумбольдта
Langkrallenspitzmäuse	Long-clawed Shrews	Pachyures aux griffes longues	
Langnasenbeutler	Long-nosed Bandicoots	Bandicoots	бандикуты
Langohr-Hasenkänguruh	Brown Hare Wallaby		обыкновенный заячий кенгуру
Langohrigel	Long-eared Hedgehog		ушастый еж
Langohrseidenäffchen	Santarem Marmoset	Ouistiti de Santarém	длинноухая игрунка
Langschnabeligel	Long-beaked Spiny Ant-eater	Echidné à bec courbe	проехидны
Langschnauzen-Kaninchen-känguruh	Long-nosed Rat Kangaroo	Rat-Kangourou à nez long	крысиный потору
Langschwänziger Neuguinea-Nasenbeutler	Long-tailed New Guinea Bandicoot		длиннохвостый новогвинейский бандикут
Langschwanzmaulwurf	– Mole		длиннохвостый крот
Langschwanztanrek	– Tenrec	Microgale	обыкновенный длиннохвостый тенрек
Languren	Langurs	Presbytinés	
Lar	White-handed Gibbon	Gibbon lar	лар
Larvensifaka	Verreaux's Sifaka	Propithèque de Verreaux	сифака Верро
Lasiorhinus latifrons	Hairy-nosed Wombat	Wombat à narines poilues	широколобый вомбат
Lemur	Lemurs	Lémur	маки
– *catta*	Ring-tailed Lemur	– catta	катта
– *fulvus*	Brown Lemur	– brun	черноголовый маки
– *macaco*	Black Lemur	– macaco	черный маки
– *mongoz*	Mongoose Lemur	– mongos	монго
– *rubriventer*	Red-bellied Lemur	– à ventre rouge	рыжебрюхий маки
– *variegatus*	Ruffed Lemur	– vrai	вари
Lemuren	Lemurs	Lémuridés	лемуры
Lemurenartige	Lemur-like Prosimians	Lémuriens	лемуровые
Lemuren-Ringelschwanzbeutler	Brush-tipped Ring Tail		лемуровый кускус
Lemuridae	Lemurs	Lémuridés	лемуры
Lemuriformes	Lemur-like Prosimians	Lémuriens	лемуровые
Lemurinae	Typical Lemurs	Lémurinés	средние лемуры
Leontideus	Maned Tamarins	Singes-Lions	львиные игрунки
– *chrysomelas*	Golden-headed Tamarin	Singe-lion à tête dorée	желтоголовая львиная игрунка
– *chrysopygus*	Golden-rumped Tamarin	– – queue jaune	краснозадая львиная игрунка
– *rosalia*	Golden Lion Marmoset	Petit singe-lion	розалия
Lepilemur	Weasel Lemurs	Lépilémur	тонкотелые маки
– *mustelinus*	Greater Sportive Lemur	– mustélin	большой куний маки
– *ruficaudatus*	Sportive Lemur	– à queue rouge	малый куний маки
Lestoros inca	Peruvian Rat Opossum		перуанская первокрыса
Lesueur-Bürstenkänguruh	Lesueur's Rat Kangaroo	Rat-Kangourou de Lesueur	кистехвостый кенгуру Лесюера
Limnogale mergulus	Marsh Tenrec	Limnogale	болотный тенрек
Lisztäffchen	Cotton-head Tamarin	Pinché	эдипова игрунка
Lönnbergtamarin	Lönnberg's Tamarin	Tamarin de Lönnberg	тамарин Леннберга
Loriartige	Loris-like Prosimians	Lorisidés	лориевые
Loris	Lorises	Lorisidés	лори
Loris tardigradus	Slender Loris	Loris grêle	тонкий лори
Lorisidae	Lorises	Lorisidés	лори
Lorisiformes	Lori-like Prosimians	Lorisidés	лориевые
Löwenäffchen	Maned Tamarins	Singes-Lions	львиные игрунки
Löwenmakak			бирманский макак
Lumholtz-Baumkänguruh	Lumholtz's Tree Kangaroo	Kangourou arboricole de Lumholtz	квинслендский древесный кенгуру

GERMAN NAME	ENGLISH NAME	FRENCH NAME	RUSSIAN NAME
Luteolina crassicaudata	Little Water Opossum		толстохвостый опоссум
Macaca	Macaques	Macaques	макаки
– arctoides	Stump-tailed Macaque	Macaque brun	медвежий макак
– assamensis	Assamese Macaque	– d'Assam	горный резус
– cyclopis	Formosa Macaque	– de Formosa	тайванский резус
– fuscata	Japanese Macaque	– Japonais	японский макак
– irus	Crab-eating Macaque	– de Buffon	яванский макак
– maura	Moor Macaque	– des Célèbes	черный целебесский макак
– mulatta	Rhesus Macaque	– rhésus	макак-резус
– nemestrina	Pig-tailed Macaque	– à queue de cochon	свинохвостый макак
– radiata	Bonnet Monkey	– bonnet chinois	индийский макак
– silenus	Lion-tailed Macaque	– Ouanderou	вандеру
– sinica	Toque Monkey	– couronné	цейлонский макак
– sylvana	Barbary Ape	Magot	магот
Macleay-Buschkänguruh	New Guinea Mountain Wallaby	Wallaby de Macleay	кустовый кенгуру Миклухо-Маклая
Macropus	Kangaroos	Grands Kangourous	исполинские кенгуру
– giganteus	Grey Kangaroo	Kangourou géant	серый исполинский кенгуру
– robustus	Wallaroo	Wallaroo	горный кенгуру
– rufus	Red Kangaroo	Kangourou roux	рыжий исполинский кенгуру
Macropodidae	Wallabies and Kangaroos	Kangourous	кенгуру
Macropodinae	Kangaroos and Wallabies	Macropodinés	собственно кенгуру
Macroscelides proboscideus	Short-eared Elephant Shrew		обыкновенный слоновый прыгунчик
Macroscelididae	Elephant Shrews	Macroscélidés	прыгунчики
Macrotis	Rabbit Bandicoots	Bandicoots-Lapins	сумчатые зайцы
– lagotis	– Bandicoot		обыкновенный сумчатый заяц
– leucura	White-tailed Rabbit Bandicoot		малый сумчатый заяц
Madagaskarspitzmaus	Madagascar Shrew	Pachyure de Madagascar	мадагаскарская белозубка
Madeirafluß-Totenköpfchen			мадерская саймири
Magot	Barbary Ape	Magot	магот
Makaken	Macaques	Macaques	макаки
Makis	Lemurs	Lémuridés	лемуры
Mammalia	Mammals	Mammifères	млекопитающие
Mandrill	Mandrill	Mandrill	собственно мандрил
Mandrillus leucophaeus	Drill	Drill	дрил
– sphinx	Mandrill	Mandrill	собственно мандрил
Mangaben	Mangabeys	Mangabeys	мангабеи
Manteläffchen	Pied Tamarin	Tamarin bicolore	пегая игрунка
Mantelbrüllaffe	Mantled Howler Monkey	Hurleur à manteau	
Mantelmangabe	Grey-cheeked Mangabey	Mangabey à gorge blanche	гривистый мангабей
Mantelpavian	Hamadryas Baboon	Hamadryas	гамадрил
Marikina	Bare-faced Tamarins	Marikinas	
Marmosa	South American Mouse Opossums	Souris-Opossums	карликовые сумчатые крысы
– cinerea	Ashy Opossum		пепельная сумчатая крыса
– mexicana	Mexican Mouse Opossum		мексиканская сумчатая крыса
– murina	Murine Opossum		мышиный опоссум
Marmosetten	Marmosets	Ouistitis	
Maronenlangur	Maroon Leaf Monkey	Semnopithèque rubicond	каштановый тонкотел
Marsupialia	Marsupials	Marsupiaux	сумчатые
Maskenspitzmaus	Laxmann's Shrew	Musaraigne lapone	средняя бурозубка
Maskentiti	Masked Titi	Callicèbe à masque	черноголовый прыгун
Matschie-Baumkänguruh	Matschie's Tree Kangaroo	Kangourou arboricole de Matschie	
Maulwurfartiger Reistanrek	Mole-like Rice Tenrec		кротовидный рисовый тенрек
Maulwürfe	Moles	Taupes	кроты
Maulwurfspitzmaus			кротовая белозубка
Mausgleitbeutler	Pigmy Flying Phalangers	Acrobate pygmée	порхающие сумчатые мыши
Mausmaki	Lesser Mouse Lemur	Chirogale mignon	мышиный маки
Maus-Nasenbeutler	Mouse Bandicoot		мышиный бандикут
Maus-Zwergbeutelratte	Murine Opossum		мышиный опоссум
Meerkatzen	Guenons	Cercopithèques	мартышки
Meerkatzenartige	Guenon-like Monkeys	Cercopithecidés	мартышки
Menschenaffen	Great Apes	Pongidés	собственно человекообразные обезьяны
Menschenartige	Apes and Men	Hominoïdés	человекообразные обезьяны
Mentawilangur	Mentawi Leaf Monkey	Semnopithèque de Mentawi	ментавайский тонкотел
Metachirops opossum	Four-eyed Opossum	Quatre-oeil	опоссум квика
Metachirus nudicaudatus	Rat-tailed Opossum	Opossum à queue de rat	голохвостый опоссум
Metatheria	Pouched Mammals	Métathériens	сумчатые звери
Microcebus	Mouse Lemurs		
– coquereli	Coquerel's Mouse Lemur	Mirza de Coquerel	
– murinus	Lesser Mouse Lemur	Chirogale mignon	мышиный маки
Microgale	Tenrecs	Microgales	длиннохвостые тенреки
– longicauda	Long-tailed Tenrec	Microgale	обыкновенный длиннохвостый тенрек

GERMAN NAME	ENGLISH NAME	FRENCH NAME	RUSSIAN NAME
Microperoryctes murina	Mouse Bandicoot		мышиный бандикут
Micropotamogale lamottei	Lesser Otter Shrew	Micropotamogale de Lamotte	карликовая выдровая землеройка
– ruwenzorii	Ruwenzori Otter Shrew	Micropotamogale du Mont Ruwenzori	
Microsorex hoyi	American Pigmy Shrew	Musaraigne pygmée d'Amérique	североамериканская карликовая бурозубка
Mirikina	Douroucouli	Singe de nuit	мирикина
Mittelafrikanische Igel	Central African Hedgehogs	Hérissons d'Afrique centrale	среднеафриканские ежи
Mittelgroße Lemuren	Typical Lemurs	Lémurinés	средние лемуры
Mittelmeerigel	Algerian Hedgehog	Hérisson d'Algérie	алжирский еж
Mittelmeer-Langschwanzspitzmaus	Mediterranean Long-tailed Shrew	Musaraigne méditerranéenne	средиземноморская белозубка
Mittlerer Gleithörnchenbeutler	Honey Glider		средняя сумчатая летяга
– Katzenmaki	Fat-tailed Dwarf Lemur		
Moholi	Senegal Bushbaby	Galago du Sénégal	сенегальский галаго
Mohrenmakak	Moor Macaque	Macaque des Célèbes	черный целебесский макак
Mohrenmaki	Black Lemur	Lémur macaco	черный маки
Mohrentamarin	Negro Tamarin	Tamarin nègre	обыкновенный тамарин
Monameerkatze	Mona Monkey	Cercopithèque mone	мона
Mönchsaffe	Hairy Saki	Saki à perruque	саки-монах
Mondnagelkänguruh	Crescent Nail-tail Wallaby		полулунный когтехвостый кенгуру
Mongozmaki	Mongoose Lemur	Lémur mongos	монго
Monodelphis	Short Bare-tailed Opossums		землеройковые сумчатые крысы
Monotremata	Monotremes	Monotrémes	однопроходные
Moorspitzmaus	Northern Water Shrew	Musaraigne palustre	болотная бурозубка
Moschusrattenkänguruh	Musky Rat Kangaroo	Rat musqué Kangourou	мускусный кенгуру
Moschusspitzmaus	House Shrew		бурая мускусная белозубка
Mulgara	Crest-tailed Marsupial Mouse		гребнехвостая мышевидка
Murexia	New Guinea Marsupial Mice		новогвинейские мышевидки
Mützenlangur	Sunda Island Leaf Monkey		чубастый тонкотел
Myoictis melas	Striped Native Cat		полосатая сумчатая куница
Myrmecobiidae	Marsupial Anteaters	Fourmiliers Marsupiaux rayés	мурашееды
Myrmecobius fasciatus	– Anteater	Fourmilier marsupial rayé	мурашеед
Nachtaffe	Douroucouli	Singe de nuit	мирикина
Nacht- und Springaffen	Night and Titi Monkeys	Aotinés	мирикины
Nacktbrustkänguruh	Desert Rat Kangaroo	Rat-Kangourou du désert	степная кенгуровая крыса
Nacktgesichttamarins	Bare-faced Tamarins	Marikinas	
Nacktnasenwombat	Common Wombat	Wombat à narines dénudées	медвежий вомбат
Nacktschwanzbeutelratte	Rat-tailed Opossum	Opossum à queue de rat	голохвостый опоссум
Nagelkänguruhs	Nail-tail Wallabies	Onychogales	когтехвостые кенгуру
Nasalis larvatus	Proboscis Monkey	Nasique	обыкновенный носач
Nasenaffe	Proboscis Monkey	Nasique	обыкновенный носач
Nasenbeutler	Bandicoots	Bandicoots	сумчатые барсуки
Nectogale elegans	Szechuan Water Shrew	Nectogale élégant	тибетская водяная белозубка
Neomys anomalus	Mediterranean Water Shrew	Musaraigne de Miller	малая кутора
– fodiens	European Water Shrew	– aquatique	обыкновенная кутора
Neotetracus sinensis	Shrew Hedgehog	Neotétracus	землеройковый еж
Neuguineabeutelmäuse	New Guinea Marsupial Mice		новогвинейские мышевидки
Neuguinea-Bilchbeutler	Pigmy possum		новогвинейская сумчатая соня
Neuguineafilander	Bruijn's Pademelon	Wallaby de Bruijn	аруанский кустовый кенгуру
Neuguinea-Nasenbeutler	New Guinea Bandicoots	Bandicoots de Nouvelle-Guinée	новогвинейские бандикуты
Neuguinea-Spitzhörnchenbeutler	Lorentz's Marsupial Rat		мышевидка Лоренца
Neuguinea-Zwerggleitbeutler	New Guinea Pigmy Flying Phalanger		новогвинейская порхающая сумчатая мышь
Neurotrichus gibbsi	American Shrew Mole	Taupe de Gibbs	американский землеройковый крот
Neuweltaffen	New World Monkeys	Singes du Nouveau Monde	обезьяны Нового света
Nilgirilangur	John's Langur	Semnopithèque des Nilgiris	нилагирийский тонкотел
Nisnas	Nisnas Monkey	Nisnas	белоносый гусар
Nordafrikanische Elefantenspitzmaus	North African Elephant Shrew	Macroscélide de l'Afrique du Nord	североафриканский прыгунчик
Nordamerikanische Kleinohrspitzmaus	Least Shrew	Petite Musaraigne à queue courte	североамериканская короткоухая бурозубка
Nördliche Flachkopfbeutelmaus	Northern Planigale		северная плоскоголовая мышевидка
Nördlicher Guereza	Northern Black-and-white Colobus	Colobe de l'Abyssinie	гвереца
Nördliche Wasserspitzmaus	– Water Shrew	Musaraigne palustre	болотная бурозубка
Nordopossum	Common Opossum	Opossum commun	обыкновенный опоссум
Notiosorex crawfordi	Crawford's Desert Shrew	Musaraigne du désert	серая пустынная бурозубка
– gigas	Merriam's Desert Shrew	Grand Musaraigne du désert	большая пустынная бурозубка
Notoryctes	Marsupial Moles	Taupes marsupiales	сумчатые кроты
– caurinus	North-western Marsupial Mole	Petite Taupe marsupiale	малый сумчатый крот
– typhlops	Greater Marsupial Mole	Grande Taupe marsupiale	сумчатый крот
Notoryctidae	Marsupial Moles	Notoryctidés	сумчатые кроты
Nycticebus coucang	Slow Loris	Loris paresseux	толстый лори

GERMAN NAME	ENGLISH NAME	FRENCH NAME	RUSSIAN NAME
– *pygmaeus*	Lesser Slow Loris		малый толстый лори
Oedipomidas	Crested Bare-faced Tamarins	Pinchés	пинче
– *geoffroyi*	Geoffroy's Tamarin	Pinché de Geoffroy	игрунка Жоффруа
– *leucopus*	White-footed Tamarin	– aux pieds blancs	белоногая игрунка
– *oedipus*	Cotton-head Tamarin	Pinché	эдипова игрунка
Ohrenbeuteldachse	Rabbit Bandicoots	Bandicoot-Lapins	сумчатые зайцы
Ohrenigel	Eared Hedgehog		ушастые ежи
Ohrenspitzmaus-Maulwürfe	Shrew Moles	Musaraignes-taupes	ушастые кроты
Onychogalea	Nail-tail Wallabies	Onychogale	когтехвостые кенгуру
– *fraenata*	Bridled Nail-tail Wallaby		уздечковый когтехвостый кенгуру
– *lunata*	Crescent Nail-tail Wallaby		полулунный когтехвостый кенгуру
– *unguifer*	Northern Nail-tail Wallaby		обыкновенный когтехвостый кенгуру
Opossummäuse	Rat Opossums	Caenolestidés	ценолестовые сумчатые
Opossums	Common and Azara's Opossums	Opossums	опоссумы
Orang-Utan	Orang-Utan	Orang-outan	оранг-утан
Ornithorhynchidae	Platypuses	Ornithorhynchidés	утконосы
Ornithorhynchus anatinus	Duck-billed Platypus	Ornithorhynque	утконос
Oryzorictes	Rice Tenrecs	Oryzoryctes	рисовые тенреки
– *talpoides*	Mole-like Rice Tenrec		кротовидный рисовый тенрек
Oryzorictinae	Rice Tenrecs	Oryzoryctinés	рисовые тенреки
Ostamerikanischer Maulwurf	Eastern American Mole	Taupe à queue glabre	восточноамериканский крот
Ostgorilla	Mountain Gorilla	Gorille de montagne	горная горилла
Ostigel	Eastern European Hedgehog	Hérisson d'Europe de l'Est	белогрудый еж
Östlicher Kielnagelgalago	– Needle-clawed Bushbaby	Galago du Congo	
Östliche Springbeutelmaus	Eastern Jerboa Marsupial		восточноавстралийский сумчатый тушканчик
Ostmaulwurf	–Mole	Taupe d'Europe de l'Est	короткохвостый крот
Ost-Ringelschwanzbeutler	Queensland Ring Tail		обыкновенный кольцехвостый кускус
Otterspitzmäuse	Otter Shrews	Potamogalidés	выдровые землеройки
Pademelons	Pademelons	Thylogales	малые кустовые кенгуру
Pagehstumpfnasenaffe	Pig-tailed Langur		одноцветный симиас
Panamaperückenäffchen	Geoffroy's Tamarin	Pinché de Geoffroy	игрунка Жоффруа
Panzerspitzmäuse	Armoured Shrews		белозубки-броненоски
Papio	Baboons	Papions	павианы
– *anubis*	Anubis Baboon	Papion anubis	анубис
– *cynocephalus*	Yellow Baboon	Babouin cynocéphale	бабуин
– *hamadryas*	Hamadryas Baboon	Hamadryas	гамадрил
– *papio*	Guinea Baboon	Babouin de Guinée	гвинейский павиан
– *ursinus*	Chacma Baboon	Chacma	чакма
Paracrocidura schoutedeni			конголезская белозубка
Paraechinus aethiopicus	Ethiopian Hedgehog	Hérisson du désert	абиссинский еж
– *hypomelas*	Brandt's Hedgehog	– de Brandt	длинноиглый еж
– *micropus*	Indian Hedgehog		индийский еж
Paraguayanisches Opossum	Azara's Opossum		казака
Parascalops breweri	Hairy-tailed Mole	Taupe à queue chevelue	волосохвостый крот
Parmakänguruh	Parma Wallaby		кенгуру.парма
Patas	Patas	Patas	черноносый гусар
Paviane	Baboons	Papions	павианы
Pazifischer Maulwurf	Coast Mole	Taupe de côte	тихоокеанский крот
Pazifische Wasserspitzmaus	Pacific Water Shrew	Musaraigne des marais	бурозубка Бендайра
Pearsons Langkrallenspitzmaus	Pearson's Long-clawed Shrew		когтистая белозубка Пирсона
Peradorcas concinna	Little Rock Wallaby	Petit Wallaby de rochers	карликовый каменный кенгуру
Perameles	Long-nosed Bandicoots	Bandicoots	бандикуты
– *fasciata*	Eastern Barred Bandicoot		полосатый бандикут
– *gunni*	Tasmanian Barred Bandicoot		бандикут Гунна
– *nasuta*	Long-nosed Bandicoot	Bandicoot à nez long	большой бандикут
Peramelidae	Bandicoots	Bandicoots	сумчатые барсуки
Perodicticus potto	Potto	Potto de Bosman	потто
Peroryctes	New Guinea Bandicoots	Bandicoots de Nouvelle-Guinée	новогвинейские бандикуты
– *longicauda*	Long-tailed New Guinea Bandicoot		длиннохвостый новогвинейский бандикут
– *raffrayanus*	New Guinea Bandicoot	Bandicoot de Nouvelle-Guinée	большой новогвинейский бандикут
Perückenäffchen	Crested Bare-faced Tamarins	Pinchés	пинче
Peru-Opossummaus	Peruvian Rat Opossum		перуанская первокрыса
Petaurus	Honey Gliders	Phalangers volants	сумчатые летяги
– *australis*	Yellow-bellied Glider		большая сумчатая летяга
– *breviceps*	Honey Glider		короткоголовая сумчатая летяга
– *norfolcensis*	Honey Glider		средняя сумчатая летяга
Petrodromus sultan	Forest Elephant Shrew		лесная хоботковая крыса
– *tetradactylus*	Four-toed Elephant Shrew		четырехпалая хоботковая крыса
Petrogale	Rock Wallabies	Wallabies de rochers	каменные кенгуру
– *brachyotis*	Short-eared Rock Wallaby		короткоухий каменный кенгуру

GERMAN NAME	ENGLISH NAME	FRENCH NAME	RUSSIAN NAME
– *inornata*	Plain Rock Wallaby		квинслендский каменный кенгуру
– *penicillata*	Brush-tailed Rock Wallaby		кистехвостый каменный кенгуру
– *xanthopus*	Ring-tailed Rock Wallaby		желтоногий каменный кенгуру
Phalanger	Phalangers	Phalangerinés	собственно лазающие сумчатые
Phalanger	Cuscuses	Couscous	кускусы
– *maculatus*	Spotted Cuscus	Couscous tacheté	пятнистый кускус
– *orientalis*	Grey Cuscus	– gris	серый кускус
– *ursinus*	Bear Phalanger	– des Célèbes	черный кускус
Phalangeridae	Phalangers	Phalangéridés	лазающие сумчатые
Phalangerinae	Phalangers	Phalangerinés	собственно лазающие сумчатые
Phaner furcifer	Fork-marked Dwarf Lemur	Phaner à fourche	вильчатый маки
Phascogale	Marsupial Rats	Rats marsupiaux	
– *calura*	Red-tailed Phascogale		малая кистехвостая мышевидка
– *lorentzi*	Lorentz's Marsupial Rat		мышевидка Лоренца
– *tapoatafa*	Black-tailed Phascogale		кистехвостая мышевидка
Phascogalinae	Marsupial Mice	Phascogalinés	мышевидки
Phascolarctinae	Koala-like Marsupials	Phascolarctinés	сумчатые медведи
Phascolarctos cinereus	Koala	Koala	коала
Phascolosorex	Striped Marsupial Rats		полосатые мышевидки
Phayres Langur	Phayre's Leaf Monkey	Semnopithèque de Phayre	тонкотел Фейера
Philippinenkoboldmaki	Philippine Tarsier	Tarsier des Philippines	филиппинский долгопят
Philippinen-Rattenigel	Mindanao Gymnure	Gymnure des Philippines	филиппинская малая гимнура
Philippinentupaias	Philippines Tree Shrew	Toupaïe des Philippines	филиппинские тупайи
Pinchéäffchen	Crested Bare-faced Tamarins	Pinchés	пинче
Pinselschwanzbeutler	Marsupial Rats	Rats marsupiaux	
Pinselschwanzkänguruh	Brush-tailed Rock Wallaby		кистехвостый каменный кенгуру
Pithecia	Sakis	Sakis moines	
– *monacha*	Hairy Saki	Saki à perruque	саки-монах
– *pithecia*	Pale-headed Saki	Saki à tête pâle	бледный саки
Pitheciinae	Sakis and Uakaris	Pithécinés	саки
Planigale	Flat-skulled Marsupials		плоскоголовые мышевидки
– *ingrami*	Northern Planigale		северная плоскоголовая мышевидка
– *subtilissima*	Kimberley Planigale		карликовая плоскоголовая мышевидка
Platyrrhina	New World Monkeys	Singes du Nouveau Monde	обезьяны Нового света
Plumpbeutler	Wombats	Wombats	вомбаты
Plumplori	Slow Loris	Loris paresseux	толстый лори
Podogymnura truei	Mindanao Gymnure		филиппинская малая гимнура
Pongo pygmaeus	Orang-Utan	Orang-outan	оранг-утан
Pongidae	Great Apes	Pongidés	собственно человекообразные обезьяны
Potamogale velox	Otter Shrew	Potamogale	выдровая землеройка
Potamogalidae	– Shrews	Potamogalidés	выдровые землеройки
Potoroops platyops	Broad-faced Rat Kangaroo		широколицый потору
Potoroinae	Rat Kangaroos	Rats-Kangourous	кенгуровые крысы
Potorous	Long-nosed Rat Kangaroos	– à nez long	потору
– *gilberti*	Gilbert's Rat Kangaroo	Rat-Kangourou de Gilbert	потору Джильберта
– *tridactylus*	Long-nosed Rat Kangaroo	– à nez long	крысиный потору
Potto	Potto	Potto de Bosman	потто
Praesorex goliath	African Forest Shrew	Musaraigne géante	гигантская белозубка
Presbytis	Langurs	Presbytinés	
– *aygula*	Sunda Island Leaf Monkey		чубастый тонкотел
– *cristatus*	Silvered Leaf Monkey		гривистый тонкотел
– *entellus*	Entellus Langur	Houleman	гульман
– *francoisi*	François' Monkey	Semnopithèque de François	тонкинский тонкотел
– *frontatus*	White-fronted Leaf Monkey	– à front blanc	гололобый тонкотел
– *johni*	John's Langur	– des Nilgiris	нилагирийский тонкотел
– *melalophus*	Banded Leaf Monkey	– melalophe	рыжий тонкотел
– *obscurus*	Dusky Leaf Monkey	– obscur	очковый тонкотел
– *phayrei*	Phayre's Leaf Monkey	– de Phayre	тонкотел Фейера
– *pileatus*	Capped Langur		хохлатый тонкотел
– *potenzani*	Mentawi Leaf Monkey	– de Mentawi	ментавайский тонкотел
– *rubicunda*	Maroon Leaf Monkey	– rubicond	каштановый тонкотел
– *senex*	Purple-faced Langur	– blanchâtre	белобородый тонкотел
Primaten	Primates	Primates	приматы
Primates	Primates	Primates	приматы
Propithecus	Sifakas	Propithèques	сифаки
– *diadema*	Diademed Sifaka	Propithèque diadème	белолобый сифака
– *verreauxi*	Verreaux's Sifaka	– de Verreaux	сифака Верро
Prosimiae	Prosimians	Prosimiens	полуобезьяны
Prototheria	Egg-laying Mammals	Protothériens	клоачные
Pruners Igel	Pruner's Hedgehog	Hérisson de Pruner	еж Прунера
Pseudocheirus	Ring-tailed Phalangers	Ringtails	кольцехвостые кускусы
– *archeri*	Striped Ring Tail		желтый кускус
– *dahli*	Rock-haunting Ring Tail		кускус Даля

GERMAN NAME	ENGLISH NAME	FRENCH NAME	RUSSIAN NAME
– *lemuroides*	Brush-tipped Ring Tail		лемуровый кускус
– *peregrinus*	Queensland Ring Tail		обыкновенный кольцехвостый кускус
Ptilocercinae	Pen-tailed Tree Shrews	Ptilocerques	перохвостые тупайи
Ptilocercus lowii	Low's Pen-tailed Tree Shrew	Ptilocerque de Low	
Pygathrix nemaeus	Douc Langur	Douc	немейский тонкотел
Pyrenäendesman	Pyrenean Desman	Desman des Pyrénées	пиренейская выхухоль
Queensland-Felskänguruh	Plain Rock Wallaby		квинслендский каменный кенгуру
Quicka	Four-eyed Opossum	Quatre-oeil	опоссум квика
Quokka	Quokka	Kangourou à queue courte	короткохвостый кустовый кенгуру
Rattenigel	Hairy Hedgehogs	Gymnures	крысиные ежи
Rattenkänguruhs	Rat Kangaroos	Rats-Kangourous	кенгуровые крысы
Raubbeutler	Flesh-eating Marsupials	Dasyuridés	хищные сумчатые
Reistanreks	Rice Tenrecs	Oryzoryctinés	рисовые тенреки
Reiswühler	Rice Tenrecs	Oryzoryctes	рисовые тенреки
Rhesusaffe	Rhesus Macaque	Macaque rhésus	макак-резус
Rhinopithecus	Snub-nosed Monkeys	Rhinopithèques	ринопитеки
– *avunculus*	Tonkin Snub-nosed Monkey	Rhinopithèque de Tonkin	авункулярный ринопитек
– *bieti*	Brown Snub-nosed Langur	– brun	биэтовский ринопитек
– *brelichi*	White-mantled Snub Nose	– jaune doré	бреличевский ринопитек
– *roxellanae*	Snub-nosed Monkey	– de Roxellane	рокселланов ринопитек
Rhynchocyon cirnei	Checkered Elephant Shrew		пятнистая хоботковая собачка
– *petersi*	Peters' Elephant Shrew	Macroscélide de Peters	хоботковая собачка Петерса
– stuhlmanni	Stuhlmann's Elephant Shrew	– de Stuhlmann	темная хоботковая собачка
Rhyncholestes raphanurus	Chilean Rat Opossum		чилийская первокрыса
Rhynchomeles prattorum	Ceram Long-nosed Bandicoot		
Riesenbeutelmarder	Tiger Cat	Chat marsupial	исполинская сумчатая куница
Riesengalago	Thick-tailed Bushbaby	Galago à queue trouffue	толстохвостый галаго
Riesengleitbeutler	Greater Gliding Phalanger	Grand Phalanger volant	исполинская сумчатая летяга
Riesengoldmull	Giant Golden Mole	Grande Taupe dorée	исполинский златокрот
Riesenkänguruhs	Kangaroos	Grands Kangourous	исполинские кенгуру
Riesenwimperspitzmaus	Giant Musk Shrew		большая белозубка
Ringelschwanz-Kletterbeutler	Ring-tailed Phalangers	Ringtails	кольцехвостые кускусы
Ringschwanz-Felskänguruh	– Rock Wallaby		желтоногий каменный кенгуру
Rio-Napo-Tamarin	Rio Napo Tamarin	Tamarin de Rio Napo	
Rohrrüßler	Elephant Shrews	Macroscélidés	прыгунчики
Rollaffen	Capuchins	Sapajou	капуцины
Roloway	Roloway	Roloway	
Römischer Maulwurf	Roman Mole	Taupe romaine	римский крот
Rotbauchfilander	Rufous-bellied Pademelon	Wallaby de Billardier	рыжебрюхий кустовый кенгуру
Rotbauchmaki	Red-bellied Lemur	Lémur à ventre rouge	рыжебрюхий маки
Rotbauchmeerkatze	Red-bellied Guenon	Hocheur à ventre rouge	краснобрюхая мартышка
Rotbauchtamarin	Red-bellied White-lipped Tamarin	Tamarin labié	краснобрюхий тамарин
Rotbeinfilander	Red-legged Pademelon		обожженный кенгуру
Roter Brüllaffe	Red Howler Monkey	Hurleur roux	рыжий ревун
– Langur	Banded Leaf Monkey	Semnopithèque melalophe	рыжий тонкотел
– Springaffe	Red Titi	Callicèbe roux	красный прыгун
– Stummelaffe	– Colobus	Colobe bai	красный толстотел
– Uakari	– Uakari	Ouakari rubicond	красный уакори
Rotes Rattenkänguruh	Rufous Rat Kangaroo	Rat-Kangourou rougeâtre	рыжая кенгуровая крыса
– Riesenkänguruh	Red Kangaroo	Kangourou roux	рыжий исполинский кенгуру
Rote Wollbeutelratte	Woolly Opossum	Opossum laineux	рыжая шерстистая сумчатая крыса
Rotgesichtsmakak	Japanese Macaque	Macaque Japonais	японский макак
Rothalsfilander	Red-necked Pademelon		падемелон
Rothandbrüllaffe	Rufous-handed Howler Monkey	Hurleur à mains rousses	краснорукий ревун
Rothandtamarin	Red-handed Tamarin	Tamarin aux mains rousses	краснорукий тамарин
Rotkappentamarin	Red-capped Tamarin	– à calotte rousse	красногривый тамарин
Rotmanteltamarin	Red-mantled Tamarin	– à manteau rouge	
Rotnackenpademelon	Red-necked Pademelon		падемелон
Rotrückensaki	Red-backed Saki	Saki capucin	красноспинный саки
Rotschulterrüsselhündchen	Peters' Elephant Shrew	Macroscélide de Peters	хоботковая собачка Петерса
Rotschwänziger Wieselmaki	Sportive Lemur	Lépilémur à queue rouge	малый куний маки
Rotsteißlöwenäffchen	Golden-rumped Tamarin	Singe-lion à queue jaune	краснозадая львиная игрунка
Rotstirnmaki	Red-fronted Lemur	Lémur à front rouge	рыжелобый маки
Rotzahnspitzmäuse	Red-toothed Shrews	Soricinés	землеройки-бурозубки
Rückenstreifkänguruh	Black-striped Wallaby		полосатый кустовый кенгуру
Rüsselbeutler	Honey Phalangers	Tarsipidinés	сумчатые медоеды
Rüsselhündchen			хоботковые собачки
Rüsselratten			хоботковые крысы
Rüsselspringer	Elephant Shrews	Macroscélidés	прыгунчики
Russischer Desman	Russian Desman	Desman de Moscovie	обыкновенная выхухоль
Ruwenzori-Otterspitzmaus	Ruwenzori Otter Shrew	Micropotamogale du Mont Ruwenzori	
Saguinus	Tamarins	Tamarins	тамарины

GERMAN NAME	ENGLISH NAME	FRENCH NAME	RUSSIAN NAME
– *bicolor*	Pied Tamarin	Tamarin bicolore	пегая игрунка
– *fuscicollis*	Brown-headed Tamarin	– à tête brune	буроспинный тамарин
– *graellsi*	Rio Napo Tamarin	– de Rio Napo	
– *illigeri*	Red-mantled Tamarin	– à manteau rouge	
– *imperator*	Emperor Tamarin	– empereur	императорский усатый тамарин
– *labiatus*	Red-bellied White-lipped Tamarin	– labié	краснобрюхий тамарин
– *melanoleucus*	White Tamarin	– blanc	белый тамарин
– *midas*	Red-handed Tamarin	– aux mains rousses	краснорукий тамарин
– *mystax*	Moustached Tamarin	– à Moustaches	усатый тамарин
– *nigricollis*	Black-and-red Tamarin	– rouge et noir	черноспинный тамарин
– *pileatus*	Red-capped Tamarin	– à calotte rousse	красногривый тамарин
– *pluto*	Lönnberg's Tamarin	– de Lönnberg	тамарин Леннберга
– *tamarin*	Negro Tamarin	– nègre	обыкновенный тамарин
– *tripartitus*	Golden-mantled Tamarin	– à manteau doré	
– *weddelli*	Weddell's Tamarin	– de Weddell	
Saimiri	Squirrel Monkeys	Saimiris	саймири
– *boliviensis*	Black-headed Squirrel Monkey	Sapajou à tête noire	черноголовая саймири
– *oerstedi*	Red-backed Squirrel Monkey	– à dos rouge	желтая саймири
– *sciureus*	Common Squirrel Monkey	– jaune	саймири-белка
Sakiaffen	Sakis and Uakaris	Pithécinés	саки
Sakis	Sakis	Sakis moines	саки
Sandwallaby	Sandy Wallaby		проворный кустовый кенгуру
Sansibar-Rüsselhündchen			занзибарская хоботковая собачка
Santaremäffchen	Santarém Marmoset	Ouistiti de Santarém	длинноухая игрунка
Sarcophilus harrisi	Tasmanian Devil	Sarcophile satanique	сумчатый дьявол
Satansaffe	Black Saki	Saki noir	чертов саки
Säugetiere	Mammals	Mammifères	млекопитающие
Scalopinae	American and Asian Moles	Taupes d'Asie et d'Amérique du Nord	американско-азиатские кроты
Scalopus aquaticus	Eastern American Mole	Taupe à queue glabre	восточноамериканский крот
– *inflatus*	Tamaulipan Mole	Taupe de Tamaulipas	
– *montanus*	Coahuilan Mole	– – Coahuila	
Scapanulus oweni	Kansu Mole	– – Kansu	западнокитайский землеройковый крот
Scapanus	Western American Moles	Taupes d'Amérique de l'Ouest	западноамериканские кроты
Scapanus latimanus	Broad-footed Mole		калифорнийский крот
– *orarius*	Coast Mole	Taupe de côte	тихоокеанский крот
– *townsendi*	Townsend's Mole	– – Townsend	крот Тоунсенда
Scaptonyx fuscicaudus	Long-tailed Mole		длиннохвостый крот
Scharlachgesicht	Bald Uakari	Ouakari chauve	лысый уакори
Schlafbeutler	Dormouse Possums	Phalangers Loirs	сумчатые сони
Schlankaffen	Leaf Monkeys	Colobidés	тонкотелые обезьяны
Schlanklori	Slender Loris	Loris grêle	тонкий лори
Schlitzrüßler	Solenodons	Solénodontidés	щелезубы
Schmalfußbeutelmäuse	Narrow-footed Marsupial Mice	Souris marsupiales	сумчатые землеройки
Schmalnasen	Old World Simian Primates	Catarhiniens	узконосые обезьяны
Schnabeltier	Duck-billed Platypus	Ornithorhynque	утконос
Schnurrbarttamarin	Moustached Tamarin	Tamarin à Moustaches	усатый тамарин
Schoinobates volans	Greater Gliding Phalanger	Grand Phalanger volant	исполинская сумчатая летяга
Schönwallaby	Pretty-face Wallaby		кустовый кенгуру Парри
Schopfgibbon	Black Gibbon	Gibbon noir	одноцветный гиббон
Schopflangur	Capped Langur		хохлатый тонкотел
Schopfmakak	Celebes Crested Macaque	Cynopithèque nègre	хохлатый павиан
Schopfmangabe	Black Mangabey	Mangabey noir	бородатый мангабей
Schuppenschwanzkusu	Scaly-tailed Phalanger		чешуйчатохвостый кузу
Schwarzer Brüllaffe	Black Howler Monkey	Hurleur noir	черный ревун
– Klammeraffe	– Spider Monkey	Singe-Araignée noir	черная коата
– Uakari	– Uakari	Ouakari de Roosevelt	черный уакори
Schwarzgrüne Meerkatze	Swamp Guenon	Cercopithèque noir et vert	черноземная мартышка
Schwarzköpfiger Springaffe	Masked Titi	Callicèbe à masque	черноголовый прыгун
Schwarzköpfiges Totenköpfchen	Black-headed Squirrel Monkey	Sapajou à tête noire	черноголовая саймири
Schwarzkopfmaki	Brown Lemur	Lémur brun	черноголовый маки
Schwarzkopftanrek	Streaked Tenrec		черноголовый тенрек
Schwarzkopfuakari	Black-headed Uakari	Ouakari à tête noire	черноголовый уакори
Schwarzmakak	Moor Macaque	Macaque des Célèbes	черный целебесский макак
Schwarznasen-Husarenaffe	Patas	Patas	черноносый гусар
Schwarzpinseläffchen	Black-pencilled Marmoset	Ouistiti à pinceau noir	кисточковая игрунка
Schwarzring-Zwergbeutelratte	Mexican Mouse Opossum		мексиканская сумчатая крыса
Schwarzrückentamarin	Black-and-red Tamarin	Tamarin rouge et noir	черноспинный тамарин
Schwarzschulteropossum	Woolly Opossum	Opossum laineux	полосатая шерстистая сумчатая крыса
Schwarzschwanzbeutelmarder	Western Dasyure		чернохвостая сумчатая куница
Schweifaffen	Sakis	Sakis moines	
Schweinsaffe	Pig-tailed Macaque	Macaque à queue de cochon	свинохвостый макак
Schweinsfuß	Pig-footed Bandicoot		хероп
Schwimmbeutelratten	Yapoks	Yapocks	плавуны

GERMAN NAME	ENGLISH NAME	FRENCH NAME	RUSSIAN NAME
Schwimmbeutler	Yapok	Yapock	водяной опоссум
Sclaters Igel	Sclater's Hedgehog	Hérisson de Sclater	еж Склатера
Scutisorex congicus	Congo Armoured Shrew		конголезская броненоска
– somereni	Uganda Armoured Shrew		угандская броненоска
Scutisoricinae	Armoured Shrews		белозубки-броненоски
Senegalgalago	Senegal Bushbaby	Galago du Sénégal	сенегальский галаго
Setifer setosus	Greater Hedgehog Tenrec		обыкновенный ежевый тэнрек
Setonix brachyurus	Quokka	Kangourou à queue courte	короткохвостый кустовый кенгуру
Siamang	Siamang	Siamang	обыкновенный сиаманг
Sifakas	Sifakas	Propithèques	сифаки
Sikkim-Großklauen-Spitzmaus	Sikkim Large-clawed Shrew	Musaraigne de Sikkim	сиккимская когтистая бурозубка
Silberäffchen	Silvery Marmoset	Ouistiti melanure	серебристая игрунка
Silbergibbon	Grey Gibbon	Gibbon cendré	серебристый гиббон
Simiae	Monkeys, Apes and Men	Simiens	обезьяны
Simias concolor	Pig-tailed Langur		одноцветный симиас
Sminthopsis	Narrow-footed Marsupial Mice	Souris marsupiales	сумчатые землеройки
– crassicaudata	Fat-tailed Sminthopsis		толстохвостая сумчатая землеройка
– murina	Mouse Sminthopsis		обыкновенная сумчатая землеройка
Solenodon paradoxus	Haitian Solenodon	Solenodon	гаитийский щелезуб
Solenodontidae	Solenodons	Solénodontidés	щелезубы
Solisorex pearsoni	Pearson's Long-clawed Shrew		когтистая белозубка Пирсона
Sorex alpinus	Alpine Shrew	Musaraigne des montagnes	альпийская бурозубка
– araneus	Common Shrew	– carrelet	обыкновенная бурозубка
– bendirii	Pacific Water Shrew	– des marais	бурозубка Бендайра
– caecutiens	Laxmann's Shrew	– lapone	средняя бурозубка
– cinereus	Masked Shrew	– cendrée	обыкновенная американская бурозубка
– hydrodromus	Unalaska Shrew		алеутская бурозубка
– minutus	Lesser Shrew	– pygmée	малая бурозубка
– palustris	Northern Water Shrew	– palustre	болотная бурозубка
Soricidae	Shrews	Soricidés	землеройки
Soricinae	Red-toothed Shrews	Soricinés	землеройки-бурозубки
Soricoidea	Shrews and Moles	Musaraignes et Taupes	землеройковые
Soriculus nigrescens	Sikkim Large-clawed Shrew	Musaraigne de Sikkim	сиккимская когтистая бурозубка
Sphinxpavian	Guinea Baboon	Babouin de Guinée	гвинейский павиан
Spinnenaffe	Woolly Spider Monkey	**Éroïde**	обыкновенная паукообразная обезьяна
Spitzhörnchen	Tree Shrews	Tupaiidés	тупайевые
Spitzmausartige	Shrews and Moles	Musaraignes et Taupes	землеройковые
Spitzmausbeutelratten	Short Bare-tailed Opossums		землеройковые сумчатые крысы
Spitzmäuse	Shrews	Soricidés	землеройки
Spitzmausigel	Shrew Hedgehog	Neotétracus	землеройковый еж
Spitzmausmaulwurf	– Mole	Musaraigne-Taupe	ушастый крот
Spitzmull	American Shrew Mole	Taupe de Gibbs	американский землеройковый крот
Sprenkelbeutelmaus	Speckled Marsupial Mouse		крапчатая мышевидка
Springaffen	Titi Monkeys	Titis	обезьяны-прыгуны
Springbeutelmäuse	Jerboa Marsupials	Gerboises-Souris marsupiales	сумчатые тушканчики
Springbeutler	Wallabies and Kangaroos	Kangourous	кенгуру
Springtamarin	Goeldi's Monkey	Tamarin de Goeldi	
Stacheligel	Spiny Hedgehogs	Hérissons	настоящие ежи
Stachelnasenbeutler	– New Guinea Bandicoots		остроносые бандикуты
Sternmull	Star-nosed Mole	Condylure étoilé	звездорыл
Sternnasenmaulwürfe	– Moles	Condylures étoilés	звездорылы
Streifenbeutelmarder	Striped Native Cat		полосатая сумчатая куница
Streifenbeutelmäuse	–Marsupial Rats		полосатые мышевидки
Streifenkletterbeutler	–Possums (or- Phalangers)	Phalangers au pelage rayé	полосатые кускусы
Streifenphalanger	–Possums (or- Phalangers)	Phalangers au pelage rayé	полосатые кускусы
Streifen-Ringelschwanzbeutler	– Ring Tail		желтый кускус
Streifentanrek	Streaked Tenrec	Hemicentete	полосатый тенрек
Stummelaffen	Colobus Monkeys	Colobes	толстотелы
Stummelschwanzspitzmaus	Szechuan Burrowing Shrew		куцая белозубка
Stumpfnasenaffen	Snub-nosed Monkeys	Rhinopithèques	ринопитеки
Südliche Kammschwanzbeutelmaus	Crest-tailed Marsupial Mouse		гребнехвостая мышевидка
Südlicher Guereza	Southern Black-and-white Colobus	Colobe à longs poils	королевский толстотел
Südopossum	Azara's Opossum		казака
Sumpfmeerkatze	Swamp Guenon	Cercopithèque noir et vert	чернозеленая мартышка
Sumpfspitzmaus	Mediterranean Water Shrew	Musaraigne de Miller	малая кутора
Sumpfwallaby	Black-tailed Wallaby	Wallaby bicolore	чернохвостый кустовый кенгуру
Suncus cafer	Dark footed Forest Shrew		темнопалая белозубка
– etruscus	Savi's Pigmy Shrew	Pachyure étrusque	белозубка-малютка

GERMAN NAME	ENGLISH NAME	FRENCH NAME	RUSSIAN NAME
– *madagascariensis*	Madagascar Shrew	– de Madagascar	мадагаскарская белозубка
– *murinus*	House Shrew		бурая мускусная белозубка
– *varius*	Forest Shrew		африканская лесная белозубка
Sundakoboldmaki	Western Tarsier	Tarsier de Horsfield	сундский долгопят
Symphalangus klossi	Dwarf Siamang	Siamang de Kloss	карликовый сиаманг
– *syndactylus*	Siamang	Siamang	обыкновенный сиаманг
Tachyglossidae	Echidnas	Échidnés	ехидны
Tachyglossus	Short-beaked Spiny Ant-eater	Échidné à bec droit	ехидны
– ***aculeatus***	Australian Echidna	– de l'Australie	австралийская ехидна
– ***setosus***	Tasmanian Echidna	– de Tasmanie	тасманийская ехидна
Talpa caeca	Mediterranean Mole	Taupe aveugle	слепой крот
– ***europaea***	Common Eurasian Mole	– commune	обыкновенный крот
– ***micrura***	Eastern Mole	– d'Europe de l'Est	короткохвостый крот
– ***romana***	Roman Mole	– romaine	римский крот
Talpidae	Moles	**Talpidés**	кроты
Talpinae	Old World Moles	**Talpinés**	собственно кроты
Tamarins	Tamarins	**Tamarins**	тамарины
Tamaulipasmaulwurf	Tamaulipan Mole	Taupe de Tamaulipas	
Tammar	Tammar	**Wallaby de l'Île d'Eugène**	кенгуру дерби
Tana	Large Tree Shrew	**Tana**	тупайя тана
Tana tana	Large Tree Shrew	**Tana**	тупайя тана
Tanreks	Tenrecs	**Tanrecs**	тенреки
Tarsiidae	Tarsiers	**Tarsidés**	долгопяты
Tarsiiformes	Tarsier-like Prosimians	**Tarsiens**	долгопятовые
Tarsipedinae	Honey Phalangers	**Souris à miel**	сумчатые медоеды
Tarsipes spenserae	– Phalanger	**Souris à miel**	пяткоход
Tarsius	Tarsiers	**Tarsiers**	долгопяты
– ***bancanus***	Western Tarsier	**Tarsier de Horsfield**	сундский долгопят
– ***spectrum***	Eastern Tarsier	**– spectre**	долгопят-привидение
– ***syrichta***	Philippine Tarsier	**– des Philippines**	филиппинский долгопят
Tasmanien-Bürstenkänguruh	Tasmanian Rat Kangaroo	Rat-Kangourou de Tasmanie	тасманийский кистехвостый кенгуру
Tasmanien-Kurzschnabeligel	– Echidna	Échidné de Tasmanie	тасманийская ехидна
Tasmanischer Langnasenbeutler	– Barred Bandicoot		бандикут Гунна
– Teufel	– Devil	Sarcophile satanique	сумчатый дьявол
Tenrec ecaudatus	Tailless Tenrec	Tanrec	обыкновенный тенрек
Tenrecidae	Tenrecs	Tenrecidés	тенреки
Tenrecinae	Tenrecs	Tenrecinés	настоящие тенреки
Theropithecus gelada	Gelada Baboon	Gelada	джелада
Thylacininae	Tasmanian Wolves	**Thylacinés**	сумчатые волки
Thylacinus cynocephalus	Tasmanian Wolf	**Loup marsupial**	сумчатый волк
Thylacis	Short-nosed Bandicoots		курносые бандикуты
– *macrourus*	Brindled Bandicoot		большой курносый бандикут
– *obesolus*	Southern Short-nosed Bandicoot		малый курносый бандикут
Thylogale	Pademelons	Thylogales	малые кустовые кенгуру
– *billardierii*	Rufous-bellied Pademelon	Wallaby de Billardier	рыжебрюхий кустовый кенгуру
– *bruijni*	Bruijn's Pademelon	– – Bruijn	аруанский кустовый кенгуру
– *stigmatica*	Red-legged Pademelon		обожженный кенгуру
– *thetis*	Red-necked Pademelon		падемелон
Tibetanische Wasserspitzmaus	Szechuan Water Shrew	Nectogale élégant	тибетская водяная белозубка
Titis	Titi Monkeys	Titis	обезьяны-прыгуны
Tonkinlangur	François' Monkey	Semnopithèque de François	тонкинский тонкотел
Tonkinstumpfnase	Tonkin Snub-nosed Monkey	Rhinopithèque de Tonkin	авункулярный ринопитек
Totenkopfäffchen, Totenköpfchen	Common Squirrel Monkey	Sapajou jaune	саймири-белка
Townsends Maulwurf	Townsend's Mole	Taupe de Townsend	крот Тоунсенда
Trichosurus	Brush-tailed Phalangers	Opossums d'Australie	кузу
– *caninus*	Short-eared Brush-tailed Phalanger	Phalanger de montagne	собачий кузу
– *vulpecula*	Brush-tailed Phalanger	Phalanger-Renard	лисий кузу
Trockenland-Elefantenspitzmaus	Bushveld Elephant Shrew		
Trues Spitzmull	True's Shrew Mole	Taupe de True	
Tschakma	Chacma Baboon	Chacma	чакма
Tupaia	Tree Shrews	Toupaïes	тупайи
– *glis*	Common Tree-shrew	Toupaïe	обыкновенная тупайя
Tupaias	Tree Shrews	Toupaïes	тупайи
Tupaiidae	Tree Shrews	Tupaiidés	тупайи
Tupaiiformes	Tree Shrews	Tupaiidés	тупайевые
Tupaiinae	Brush-tailed Tree Shrews	Tupaiinés	собственно тупайи
Tüpfelbeutelmarder	Eastern Dasyure		крапчатая сумчатая куница
Tüpfelkuskus	Spotted Cuscus	Couscous tacheté	пятнистый кускус
Uakaris	Uakaris	Ouakaris	уакори
Ugandapanzerspitzmaus	Uganda Armoured Shrew		угандская броненоска
Unalaskaspitzmaus	Unalaska Shrew		алеутская бурозубка
Ungka	Dark-handed Gibbon	Gibbon agile	быстрый гиббон
Urogale everetti	Philippines Tree Shrew	Tupaïe des Philippines	тупайя Эверетта
Uropsilinae	Shrew Moles	Musaraignes-taupes	ушастые кроты

GERMAN NAME	ENGLISH NAME	FRENCH NAME	RUSSIAN NAME
Uropsilus soricipes	– Mole	Musaraigne-Taupe	ушастый крот
Urotrichus pilirostris	True's Shrew Mole	Taupe de True	
– talpoides	Japanese Shrew Mole	– des montagnes du Japon	японский землеройковый крот
Vari	Ruffed Lemur	Lémur vrai	вари
Vieraugenbeutelratte	Four-eyed Opossum	Quatre-oeil	опоссум квика
Vierzehenrüsselratte	Four-toed Elephant Shrew		четырехпалая хоботковая крыса
Virginisches Opossum	Common Opossum	Opossum commun	обыкновенный опоссум
Vollbartmeerkatze	L'Hoest's Monkey	Cercopithèque de l'Hoest	бородатая мартышка
Vombatidae	Wombats	Wombats	вомбаты
Vombatus ursinus	Common Wombat	Wombat à narines dénudées	медвежий вомбат
Waldrüsselratte	Forest Elephant Shrew		лесная хоботковая крыса
Waldspitzmaus	Common Shrew	Musaraigne carrelet	обыкновенная бурозубка
Wallabia	Brush Wallabies	Wallabies	кустовые кенгуру
– agilis	Sandy Wallaby	Wallaby agile	проворный кустовый кенгуру
– bicolor	Black-tailed Wallaby	– bicolore	чернохвостый кустовый кенгуру
– canguru	Pretty-face Wallaby		кустовый кенгуру Парри
– dorsalis	Black-striped Wallaby		полосатый кустовый кенгуру
– eugenii	Tammar	Wallaby de l'Île d'Eugène	кенгуру дерби
– irma	Black-gloved Wallaby	– d'Irma	кенгуру ирма
– parma	Parma Wallaby		кенгуру парма
– rufogrisea	Red-necked Wallaby		кустовый кенгуру Беннетта
Wallabys	Brush Wallabies	Wallabies	кустовые кенгуру
Wallaruh	Wallaroo	Wallaroo	горный кенгуру
Wander-Ringelschwanzbeutler	Queensland Ring Tail		обыкновенный кольцехвостый кускус
Wanderu	Lion-tailed Macaque	Macaque Ouanderou	вандеру
Wasserspitzmaus	European Water Shrew	Musaraigne aquatique	обыкновенная кутора
Wassertanrek	Marsh Tenrec	Limnogale	болотный тенрек
Webbipavian	Ruhe's Baboon	Babouin de Ruhe	
Weißbartlangur	Purple-faced Langur	Semnopithèque blanchâtre	белобородый тонкотел
Weißbauchigel		Hérisson à ventre blanc	белобрюхий еж
Weißbrustigel	Eastern European Hedgehog	– d'Europe de l'Est	белогрудый еж
Weißbüscheläffchen	Common Marmoset	Ouistiti	обыкновенная игрунка
Weißer Tamarin	White Tamarin	Tamarin blanc	белый тамарин
Weißfußäffchen	White-footed Tamarin	Pinché aux pieds blancs	белоногая игрунка
Weißgesichtseidenäffchen	White-fronted Marmoset	Ouistiti à tête blanche	белолицая игрунка
Weißhandgibbon	White-handed Gibbon	Gibbon lar	лар
Weißkehlmeerkatze	White-throated Guenon	Cercopithèque à gorge blanche	белогорлая мартышка
Weißkopfaffe	White-headed Saki	Saki à tête blanche	белоголовый саки
Weißkopfmaki	White-fronted Lemur	Maki à front blanc	белолобый маки
Weißkopfsaki	White-headed Saki	Saki à tête blanche	белоголовый саки
Weißlippentamarin	Weddell's Tamarin	Tamarin de Weddell	
Weißmantelstumpfnase	White-mantled Snub Nose	Rhinopithèque jaune doré	бреличевский ринопитек
Weißnackenseidenäffchen	White-necked Marmoset	Ouistiti à col blanc	белошейная игрунка
Weißnasen-Husarenaffe	Nisnas	Nisnas	белоносый гусар
Weißnasensaki	White-nosed Saki	Saki à nez blanc	белоносый саки
Weißohrseidenäffchen	White-eared Marmoset	Ouistiti oreillard	белоухая игрунка
Weißschulteraffe	White-throated Capuchin	Sapajou capucin	обыкновенный капуцин
Weißschulterseidenäffchen	White-shouldered Marmoset	Ouistiti à camail	белоплечая игрунка
Weißstirnkapuziner	White-fronted Capuchin	Sapajou à front blanc	белолобый капуцин
Weißstirnlangur	White-fronted Leaf Monkey	Semnopithèque à front blanc	гололобый тонкотел
Weißzahnspitzmäuse	White-toothed Shrews	Musaraignes à dents blanches	землеройки-белозубки
Westamerikanische Maulwürfe	Western American Moles	Taupes d'Amérique de l'Ouest	западноамериканские кроты
Westgorilla	Lowland Gorilla	Gorille de côte	береговая горилла
Westigel	Western European Hedgehog	Hérisson d'Europe de l'Ouest	
Westlicher Kielnagelgalago	– Needle-clawed Bushbaby	Galago mignon	
Wieselmakis	Weasel Lemurs	Lépilémur	тонкотелые маки
Wimperspitzmäuse	White-toothed Shrews	Musaraignes à dents blanches	землеройки-белозубки
Winton-Goldmull	De Winton's Golden Mole	Taupe dorée de Winton	златокрот
Witwenaffe	Collared Titi	Callicèbe à fraise	
Wollaffe	Woolly Monkey	Singe laineux	шерстистая обезьяна
Wollbeutelratten	Woolly Opossums	Opossums Laineaux	шерстистые сумчатые крысы
Wollkuskus	Grey Cuscus	Couscous gris	серый кускус
Wollmaki	Woolly Indri	Avahi lanigère	авахи
Wombats	Wombats	Wombats	вомбаты
Wüstengoldmull	Grant's Desert Golden Mole	Taupe dorée de Grant	пустынный крот
Wüstenwimperspitzmaus	Desert Musk Shrew	Crocidure du désert	пустынная белозубка
Wyulda squamicaudata	Scaly-tailed Phalanger		чешуйчатохвостый кузу
Yapok	Yapok	Yapock	водяной опоссум
Zaglossus	Long-beaked Spiny Ant-eater	Echidné à bec courbe	проехидны
– bartoni	Barton's Echidna	– de Barton	проехидна Бартона
– bruijni	Bruijn's Echidna	– de Bruijn	проехидна Бруйна
– bubuensis	Bubu Echidna		проехидна острова Бубу
Zati	Bonnet and Toque Monkeys	Macaques bonnets et Couronnés	
Zottel-Hasenkänguruh	Western Hare Wallaby		косматый заячий кенгуру
Zottelschweifaffe	Hairy Saki	Saki à perruque	саки-монах

GERMAN NAME	ENGLISH NAME	FRENCH NAME	RUSSIAN NAME
Zügelkänguruh	Bridled Nail-tail Wallaby		уздечковый когтехвостый кенгуру
Zweifarbenäffchen	Pied Tamarin	Tamarin bicolore	
Zwergbeutelmaus	Pigmy Marsupial Mouse		карликовая мышевидка
Zwergbeutelratten	South American Mouse Opossums	Souris-Opossums	карликовые сумчатые крысы
Zwergflachkopfbeutelmaus	Kimberley Planigale		карликовая плоскоголовая мышевидка
Zwerg-Fleckenbeutelmarder	Little Northern Dasyure		североавстралийская сумчатая куница
Zwerggalago	Demidoff's Bushbaby	Galago de Demidoff	галаго Демидова
Zwerggleitbeutler	Pigmy Flying Phalangers	Acrobate pygmée	порхающие сумчатые мыши
Zwergmakis	Mouse Lemurs		
Zwergmeerkatze	Dwarf Guenon	Talapoin	крошечная мартышка
Zwerg-Otterspitzmaus	Lesser Otter Shrew	Micropotamogale de Lamotte	карликовая выдровая землеройка
Zwergseidenäffchen	Pigmy Marmoset	Ouistiti mignon	карликовая игрунка
Zwergsiamang	Dwarf Siamang	Siamang de Kloss	карликовый сиаманг
Zwergspitzmaus	Lesser Shrew	Musaraigne pygmée	малая бурозубка
Zwergsteinkänguruh	Little Rock Wallaby	Petit Wallaby de rochers	карликовый каменный кенгуру

III. French—German—English—Russian

Dans la plupart des cas, les noms des sous-espèces sont formés en ajoutant au nom de l'espèce un adjectif ou une désignation géographique. Dans cette partie du dictionnaire zoologique, les noms français des sous-espèces formés de cette manière ne seront en général pas indiqués.

FRENCH NAME	GERMAN NAME	ENGLISH NAME	RUSSIAN NAME
Acrobate pygmée	Australischer Zwerggleitbeutler	Pigmy Flying Phalanger	австралийская порхающая сумчатая мышь
Almiqui	Kuba-Schlitzrüßler	Cuban Solenodon	кубинский щелезуб
Angwantibo	Bärenmaki	Angwantibo	медвежий маки
Aotinés	Nacht- und Springaffen	Night and Titi Monkeys	мирикины
Atélinés	Klammerschwanzaffen	Spider and Woolly Monkeys	шерстистые обезьяны и коаты
Avahi lanigère	Wollmaki	Woolly Indri	авахи
Aye-Aye	Fingertier	Aye-Aye	мадагаскарская руконожка
Babouin cynocéphale	Gelber Babuin	Yellow Baboon	бабуин
– de Guinée	Guineapavian	Guinea Baboon	гвинейский павиан
– de Ruhe	Webbipavian	Ruhe's Baboon	
Bandicoot à nez long	Großer Langnasenbeutler	Long-nosed Bandicoot	большой бандикут
– de Nouvelle-Guinée	Großer Neuguinea-Nasenbeutler	New Guinea Bandicoot	большой новогвинейский бандикут
Bandicoots	Nasenbeutler	Bandicoots	сумчатые барсуки
– de Nouvelle-Guinée	Neuguinea-Nasenbeutler	New Guinea Bandicoots	новогвинейские бандикуты
Bandicoots-Lapins	Kaninchen-Nasenbeutler	Rabbit Bandicoots	сумчатые зайцы
Budeng	Budeng	Budeng	
Caenolestidés	Opossummäuse	Rat Opossums	ценолестовые сумчатые
Callicèbe à fraise	Witwenaffe	Collared Titi	
– à masque	Schwarzköpfiger Springaffe	Masked Titi	черноголовый прыгун
– arabassu	Grauer Springaffe	Orabussu Titi	серый прыгун
– roux	Roter Springaffe	Red Titi	красный прыгун
Callitriche	Gelbgrüne Meerkatze		желтозеленая мартышка
Callithricidés	Krallenaffen	Marmosets and Tamarins	когтистые обезьяны
Catarhiniens	Schmalnasen	Old World Simian Primates	узконосые обезьяны
Cébidés	Kapuzinerartige	New World Monkeys	капуциноообразные обезьяны
Cébinés	Kapuzineraffen	Capuchin Monkeys	цебусовые обезьяны
Cercopithecidés	Meerkatzenartige	Guenon-like Monkeys	мартышки
Cercopithèque à gorge blanche	Weißkehlmeerkatze	White-throated Guenon	белогорлая мартышка
– à tête de Hibou	Hamlynmeerkatze	Owl-faced Guenon	
– de Brazza	Brazzameerkatze	De Brazza's Monkey	бразовская мартышка
– de Gray	Grays Kronenmeerkatze	Gray's Guenon	мартышка Грея
– de l'Hoest	Vollbartmeerkatze	L'Hoest's Monkey	бородатая мартышка
– diadème	Diademmeerkatze	Diademed Guenon	
– diane	Dianameerkatze	Diana Monkey	диана
Cercopithèque hocheur	Große Weißnasenmeerkatze	Greater White-nosed Guenon	большая белоносая мартышка
– mone	Monameerkatze	Mona Monkey	мона
– noir et vert	Schwarzgrüne Meerkatze	Swamp Guenon	черноэеленая мартышка
– pogonias	Kronenmeerkatze	Crowned Guenon	чубатая мартышка
Cercopithèques	Meerkatzen	Guenons	мартышки
Chacma	Bärenpavian	Chacma Baboon	чакма
Chat marsupial	Fleckschwanzbeutelmarder	Tiger Cat	исполинская сумчатая куница
Chimarrogale de l'Himalaya	Himalajawasserspitzmaus	Himalayan Water Shrew	гималайская водяная белозубка

FRENCH NAME	GERMAN NAME	ENGLISH NAME	RUSSIAN NAME
Chimarrogales	Asiatische Wasserspitzmäuse	Asian Water Shrews	
Chirogale aux oreilles velues	Büschelohriger Katzenmaki	Hairy-eared Dwarf Lemur	
– de Milius	Großer Katzenmaki	Greater Dwarf Lemur	
Chirogaléinés	Katzenmakis	Dwarf Lemurs	крысиные маки
Chirogale mignon	Mausmaki	Lesser Mouse Lemur	мышиный маки
Chirogales	Echte Katzenmakis	Dwarf Lemurs	
Chrysochloridés	Goldmulle	Golden Moles	златокроты
Colobe à longs poils	Südlicher Guereza	Southern Black-and-white Colobus	королевский толстотел
– bai	Roter Stummelaffe	Red Colobus	красный толстотел
– de l'Abyssinie	Nördlicher Guereza	Northern Black-and-white Colobus	гвереца
Colobes	Stummelaffen	Colobus Monkeys	толстотелы
Colobe vrai	Grüner Stummelaffe	Green Colobus	ванбенеденовский толстотел
Colobidés	Schlankaffen	Leaf Monkeys	тонкотелые обезьяны
Condylure étoilé	Sternmull	Star-nosed Mole	звездорыл
Couscous	Kuskuse	Cuscuses	кускусы
– de Célèbes	Bärenkuskus	Bear Phalanger	черный кускус
Crocidure du désert	Wüstenwimperspitzmaus	Desert Musk Shrew	пустынная белозубка
– leucode	Feldspitzmaus	Bicolor White-toothed Shrew	белобрюхая белозубка
Cynopithèque nègre	Schopfmakak	Celebes Crested Macaque	хохлатый павиан
Dasyuridés	Raubbeutler	Flesh-eating Marsupials	хищные сумчатые
Dasyurinés	Beutelmarder	Dasyures	собственно сумчатые куницы
Daubentoniidés	Fingertiere	Aye-Ayes	руконожки
Desman de Moscovie	Russischer Desman	Russian Desman	обыкновенная выхухоль
– des Pyrénées	Pyrenäendesman	Pyrenean Desman	пиренейская выхухоль
Desmaninés	Desmane	Desmans	выхухоли
Douc	Kleideraffe	Douc Langur	немейский тонкотел
Drill	Drill	Drill	дрил
Échidné à bec courbe	Langschnabeligel	Long-beaked Spiny Anteater	проехидны
– à bec droit	Kurzschnabeligel	Short-beaked Spiny Anteater	ехидны
– de Barton	Barton-Langschnabeligel	Barton's Echidna	проехидна Бартона
– de Bruijn	Bruijn-Langschnabeligel	Bruijn's Echidna	проехидна Бруйна
– de l'Australie	Australien-Kurzschnabeligel	Australian Echidna	австралийская ехидна
– de Tasmanie	Tasmanien-Kurzschnabeligel	Tasmanian Echidna	тасманийская ехидна
Échidnés	Ameisenigel	Echidnas	ехидны
Érinacéidés	Igel	Hedgehogs	ежи
Éroïde	Spinnenaffe	Woolly Spider Monkey	обыкновенная паукообразная обезьяна
Euthériens	Höhere Säugetiere	Placental Mammals	высшие млекопитающие
Fourmilier marsupial rayé	Ameisenbeutler	Marsupial Anteater	мурашеед
Galagidés	Galagos	Galagos	галаги
Galago à queue trouffue	Riesengalago	Thick-tailed Bushbaby	толстохвостый галаго
– d'Allen	Buschwaldgalago	Allen's Bushbaby	
– de Demidoff	Zwerggalago	Demidoff's Bushbaby	галаго Демидова
– du Congo	Östlicher Kielnagelgalago	Eastern Needle-clawed Bushbaby	
– du Kilimandjaro	Gelbschenkelgalago		
– du Sénégal	Senegalgalago	Senegal Bushbaby	сенегальский галаго
– mignon	Westlicher Kielnagelgalago	Western Needle-clawed Bushbaby	
Gelada	Dschelada	Gelada Baboon	джелада
Géogale	Erdtanrek		земляной тенрек
Gerboises-Souris marsupiales	Springbeutelmäuse	Jerboa Marsupials	сумчатые тушканчики
Gibbon agile	Ungka	Dark-handed Gibbon	быстрый гиббон
– cendré	Silbergibbon	Grey Gibbon	серебристый гиббон
– lar	Lar	White-handed Gibbon	лар
– noir	Schopfgibbon	Black Gibbon	одноцветный гиббон
Gibbons	Eigentliche Gibbons	Gibbons	собственно гиббоны
Gorille	Gorilla	Gorilla	горилла
– de côte	Flachlandgorilla	Lowland Gorilla	береговая горилла
– de montagne	Berggorilla	Mountain Gorilla	горная горилла
Grande Musaraigne à queue courte	Kurzschwanzspitzmaus	Short-tailed Shrew	североамериканская короткохвостая бурозубка
– – du désert	Große Wüstenspitzmaus	Merriam's Desert Shrew	большая пустынная бурозубка
– Taupe dorée	Riesengoldmull	Giant Golden Mole	исполинский златокрот
– – marsupiale	Großer Beutelmull	Greater Marsupial Mole	сумчатый крот
Grand Phalanger volant	Riesengleitbeutler	– Gliding Phalanger	исполинская сумчатая летяга
Grands Kangourous	Riesenkänguruhs	Kangaroos	исполинские кенгуру
Grivet	Graugrüne Meerkatze	Grivet	серозеленая мартышка
Guérézas	Guerezas	Black-and-white Colobuses	
Gymnure	Großer Haarigel	Moon Rat	большой крысиный еж
Gymnures	Haarigel	Hairy Hedgehogs	крысиные ежи
Hamadryas	Mantelpavian	Hamadryas Baboon	гамадрил
Hapalémur à nez large	Breitschnauzenhalbmaki	Broad-nosed Gentle Lemur	широконосый полумаки
Hapalémurs	Halbmakis	Gentle Lemurs	полумаки
Hémicentètes	Halbborstenigel	Streaked Tenrec	полутенреки

FRENCH NAME	GERMAN NAME	ENGLISH NAME	RUSSIAN NAME
Hérisson à ventre blanc	Weißbauchigel		белобрюхий еж
– d'Algérie	Algerischer Igel	Algerian Hedgehog	алжирский еж
– de l'Amour	Amurigel	Amur Hedgehog	амурский еж
– de Brandt	Brandts Igel	Brandt's Hedgehog	длинноиглый еж
– de Corée	Koreaigel	Korean Hedgehog	корейский еж
– d'Espagne	Spanischer Kleinohrigel	Spanish Hedgehog	испанский еж
– d'Europe de l'Est	Weißbrustigel	Eastern European Hedgehog	белогрудый еж
– d'Europe de l'Ouest	Braunbrustigel	Western European Hedgehog	бурогрудый еж
– d'Italie	Italienischer Igel	Italian Hedgehog	итальянский еж
– du désert	Äthiopischer Igel	Ethiopian Hedgehog	абиссинский еж
Hérissons	Echte Igel	Spiny Hedgehogs	настоящие ежи
– communs	Kleinohrigel	Common Hedgehogs	обыкновенные ежи
– d'Afrique centrale	Mittelafrikanische Igel	Central African Hedgehogs	среднеафриканские ежи
Hocheur à ventre rouge	Rotbauchmeerkatze	Red-bellied Guenon	краснобрюхая мартышка
– blanc-nez	Kleine Weißnasenmeerkatze	Lesser White-nosed Guenon	малая белоносая мартышка
Hominoïdés	Menschenartige	Apes and Men	человекообразные обезьяны
Hoolock	Hulock	Hoolock Gibbon	гулок
Houleman	Hulman	Entellus Langur	гульман
Hurleur à mains rousses	Rothandbrüllaffe	Rufous-handed Howler Monkey	краснорукий ревун
– à manteau	Mantelbrüllaffe	Mantled Howler Monkey	
– brun	Brauner Brüllaffe	Brown Howler Monkey	бурый ревун
– de Guatemala	Guatemalabrüllaffe	Guatemalan Howler Monkey	гватемальский ревун
– noir	Schwarzer Brüllaffe	Black Howler Monkey	черный ревун
– roux	Roter Brüllaffe	Red Howler Monkey	рыжий ревун
Hurleurs	Brüllaffen	Howler Monkeys	ревуны
Hylobatidés	Gibbons	Gibbons	гиббоны
Indri	Indri	Indri	короткохвостый индри
Indriinés	Indriartige	Indrisoid Lemurs	индри
Insectivores	Insektenesser	Insect-Eaters	насекомоядные
Kangourou à queue courte	Kurzschwanzkänguruh	Quokka	короткохвостый кустовый кенгуру
– arboricole de Bennett	Bennett-Baumkänguruh	Bennett's Tree Kangaroo	древесный кенгуру Беннетта
– – de Doria	Doria-Baumkänguruh	Doria's Tree Kangaroo	древесный кенгуру Дория
– – de Goodfellow	Goodfellow-Baumkänguruh	Goodfellow's Tree Kangaroo	
– – de Lumholtz	Lumholtz-Baumkänguruh	Lumholtz's Tree Kangaroo	квинслендский древесный кенгуру
– – de Matschie	Matschie-Baumkänguruh	Matschie's Tree Kángaroo	
– géant	Graues Riesenkänguruh	Grey Kangaroo	серый исполинский кенгуру
– roux	Rotes Riesenkänguruh	Red Kangaroo	рыжий исполинский кенгуру
Kangourous	Känguruhs	Kangaroos	кенгуру
– arboricoles	Baumkänguruhs	Tree Kangaroos	древесные кенгуру
Koala	Koala	Koala	коала
Lagotriche de Castelnau	Brauner Wollaffe	Brown Woolly Monkey	бурая шерстистая обезьяна
– de Humboldt	Wollaffe	Humboldt's Woolly Monkey	шерстистая обезьяна Гумбольдта
– grison	Grauer Wollaffe	Grey Woolly Monkey	серая шерстистая обезьяна
– olive	Grauer Wollaffe	Olivaceous Woolly Monkey	серая шерстистая обезьяна
Lémur	Echte Makis	Lemurs	маки
– à front rouge	Rotstirnmaki	Red-fronted Lemur	рыжелобый маки
– à ventre rouge	Rotbauchmaki	Red-bellied Lemur	рыжебрюхий маки
– brun	Schwarzkopfmaki	Brown Lemur	черноголовый маки
– catta	Katta	Ring-tailed Lemur	катта
– couronné	Kronenmaki	Crowned Lemur	хохлатый монго
Lémuridés	Lemuren	Lemurs	лемуры
Lémuriens	Lemurenartige	Lemur-like Prosimians	лемуровые
Lémurinés	Mittelgroße Lemuren	Typical Lemurs	средние лемуры
Lémur macaco	Mohrenmaki	Black Lemur	черный маки
– mongos	Mongozmaki	Mongoose Lemur	монго
– vrai	Vari	Ruffed Lemur	вари
Lépilémur	Wieselmakis	Weasel Lemurs	тонкотелые маки
– à queue rouge	Kleiner Wieselmaki	Sportive Lemur	малый куний маки
– mustélin	Großer Wieselmaki	Greater Sportive Lemur	большой куний маки
Lièvres wallabies	Hasenkänguruhs	Hare Wallabies	заячьи кенгуру
Limnogale	Wassertanrek	Marsh Tenrec	болотный тенрек
Loris grêle	Schlanklori	Slender Loris	тонкий лори
Lorisidés	Loris	Lorises	лори
Loris paresseux	Plumplori	Slow Loris	толстый лори
Loup marsupial	Beutelwolf	Tasmanian Wolf	сумчатый волк
Macaque à queue de cochon	Schweinsaffe	Pig-tailed Macaque	свинохвостый макак
– bonnet chinois	Indischer Hutaffe	Bonnet Monkey	индийский макак
– brun	Bärenmakak	Stump-tailed Macaque	медвежий макак
– couronné	Ceylon-Hutaffe	Toque Monkey	цейлонский макак
– d'Assam	Bergrhesus	Assamese Macaque	горный резус
– de Buffon	Javaneraffe	Crab-eating Macaque	яванский макак
– de Célèbes	Mohrenmakak	Moor Macaque	черный целебесский макак
Macaque de Formosa	Formosamakak	Formosa Macaque	тайванский резус
– des Philippines	Philippinenmakak		филиппинский макак

FRENCH NAME	GERMAN NAME	ENGLISH NAME	RUSSIAN NAME
– Japonais	Rotgesichtsmakak	Japanese Macaque	японский макак
– Ouanderou	Wanderu	Lion-tailed Macaque	вандеру
– rhésus	Rhesusaffe	Rhesus Macaque	макак-резус
Macaques	Makaken	Macaques	макаки
– de Célèbes	Celebesmakaken	Moor Macaques	целебесские макаки
Macropodinés	Eigentliche Känguruhs	Kangaroos and Wallabies	собственно кенгуру
Macroscélide de l'Afrique du Nord	Nordafrikanische Elefanten-spitzmaus	North African Elephant Shrew	североафриканский прыгунчик
– de Peters	Rotschulterrüsselhündchen	Peters' Elephant Shrew	хоботковая собачка Петерса
– de Stuhlmann	Dunkles Rüsselhündchen	Stuhlmann's Elephant Shrew	темная хоботковая собачка
Macroscélidés	Rüsselspringer	Elephant Shrews	прыгунчики
Magot	Magot	Barbary Ape	магот
Maki à front blanc	Weißkopfmaki	White-fronted Lemur	белолобый маки
Mammifères	Säugetiere	Mammals	млекопитающие
Mandrill	Mandrill	Mandrill	собственно мандрил
Mangabey à collier blanc	Halsbandmangabe	Sooty Mangabey	воротничковый мангабей
– à gorge blanche	Mantelmangabe	Grey-cheeked Mangabey	гривистый мангабей
– à ventre doré	Haubenmangabe	Agile Mangabey	чубастый мангабей
– noir	Schopfmangabe	Black Mangabey	бородатый мангабей
Mangabeys	Mangaben	Mangabeys	мангабеи
Marikinas	Nacktgesichttamarins	Bare-faced Tamarins	
Marsupiaux	Beuteltiere	Marsupials	сумчатые
Métathériens	Beutelsäuger	Pouched Mammals	сумчатые звери
Microgales	Kleintanreks	Lesser Tenrecs	длиннохвостые тенреки
Micropotamogale du Mont Ruwenzori	Ruwenzori-Otterspitzmaus	Ruwenzori Otter Shrew	
– de Lamotte	Zwerg-Otterspitzmaus	Lesser Otter Shrew	карликовая выдровая землеройка
Mirza de Coquerel	Coquerels Zwergmaki	Coquerel's Mouse Lemur	
Monotrèmes	Kloakentiere	Monotremes	однопроходные
Moustac	Blaumaulmeerkatze	Moustached Monkey	голуболицая мартышка
Musaraigne aquatique	Wasserspitzmaus	European Water Shrew	обыкновенная кутора
– carrelet	Waldspitzmaus	Common Shrew	обыкновенная бурозубка
– cendrée	Amerikanische Maskenspitzmaus	Masked Shrew	обыкновенная американская бурозубка
– de Miller	Sumpfspitzmaus	Mediterranean Water Shrew	малая кутора
– de Sikkim	Sikkim-Großklauen-Spitzmaus	Sikkim Large-clawed Shrew	сиккимская когтистая бурозубка
– des jardins	Gartenspitzmaus	Lesser White-toothed Shrew	малая белозубка
– des marais	Pazifische Wasserspitzmaus	Pacific Water Shrew	бурозубка Бендайра
– des montagnes	Alpenspitzmaus	Alpine Shrew	альпийская бурозубка
– du désert	Graue Wüstenspitzmaus	Crawford's Desert Shrew	серая пустынная бурозубка
– géante	Afrikanische Riesenspitzmaus	African Forest Shrew	гигантская белозубка
– lapone	Maskenspitzmaus	Laxmann's Shrew	средняя бурозубка
– méditerranéenne	Mittelmeer-Langschwanz-spitzmaus	Mediterranean Long-tailed Shrew	средиземноморская белозубка
– musette	Hausspitzmaus	Common European White-toothed Shrew	бурая белозубка
– palustre	Nördliche Wasserspitzmaus	Northern Water Shrew	болотная бурозубка
– pygmée	Zwergspitzmaus	Lesser Shrew	малая бурозубка
– pygmée d'Amérique	Amerikanische Zwergspitzmaus	American Pigmy Shrew	североамериканская карликовая бурозубка
Musaraignes	Waldspitzmäuse	Shrews	бурозубки
– à dents blanches	Weißzahnspitzmäuse	White-toothed Shrews	землеройки-белозубки
Musaraigne-taupe	Spitzmausmaulwurf	Shrew Mole	ушастый крот
Nasique	Nasenaffe	Proboscis Monkey	обыкновенный носач
Nectogale élégant	Gebirgsbachspitzmaus	Szechuan Water Shrew	тибетская водяная белозубка
Neotétracus	Spitzmausigel	Shrew Hedgehog	землеройковый еж
Nisnas	Weißnasen-Husarenaffe	Nisnas	белоносый гусар
Notoryctidés	Beutelmulle	Marsupial Moles	сумчатые кроты
Onychogales	Nagelkänguruhs	Nail-tail Wallabies	когтехвостые кенгуру
Opossum à queue de rat	Nacktschwanzbeutelratte	Rat-tailed Opossum	голохвостый опоссум
– commun	Nordopossum	Common Opossum	обыкновенный опоссум
– de Leadbeater	Hörnchenkletterbeutler	Leadbeater's Possum	сумчатая белка
– laineux	Wollbeutelratte	Woolly Opossum	шерстистая сумчатая крыса
Opossums	Opossums	Common and Azara's Opossums	опоссумы
– d'Amérique	Beutelratten	Opossums	сумчатые крысы
– d'Australie	Kusus	Brush-tailed Phalangers	кузу
Orang-outan	Orang-Utan	Orang-Utan	оранг-утан
Ornithorhynque	Schnabeltier	Duck-billed Platypus	утконос
Oryzoryctes	Reiswühler	Rice Tenrecs	рисовые тенреки
Oryzoryctinés	Reistanreks	Rice Tenrecs	рисовые тенреки
Ouakari à tête noire	Schwarzkopfuakari	Black-headed Uakari	черноголовый уакори
– chauve	Scharlachgesicht	Bald Uakari	лысый уакори
– de Roosevelt	Schwarzer Uakari	Black Uakari	черный уакори
– rubicond	Roter Uakari	Red Uakari	красный уакори
Ouakaris	Uakaris	Uakaris	уакори
Ouistiti	Weißbüscheläffchen	Common Marmoset	обыкновенная игрунка

FRENCH NAME	GERMAN NAME	ENGLISH NAME	RUSSIAN NAME
– à camail	Weißschulterseidenäffchen	White-shouldered Marmoset	белоплечая игрунка
Ouistiti à col blanc	Weißnackenseidenäffchen	White-necked Marmoset	белошейная игрунка
– à pinceau noir	Schwarzpinseläffchen	Black-pencilled Marmoset	кисточковая игрунка
– à tête blanche	Weißgesichtseidenäffchen	White-fronted Marmoset	белолицая игрунка
– à tête jaune	Gelbkopfbüscheläffchen	Buff-headed Marmoset	желтоголовая игрунка
– aux pieds jaunes	Gelbfußäffchen	Yellow-legged Marmoset	желтоногая игрунка
– de Santarém	Langohrseidenäffchen	Santarém Marmoset	длинноухая игрунка
– melanure	Silberäffchen	Silvery Marmoset	серебристая игрунка
– mignon	Zwergseidenäffchen	Pigmy Marmoset	карликовая игрунка
– oreillard	Weißohrseidenäffchen	White-eared Marmoset	белоухая игрунка
Ouistitis	Marmosetten	Marmosets	
Pachyure étrusque	Etruskerspitzmaus	Savi's Pigmy Shrew	белозубка-малютка
– de Madagascar	Madagaskarspitzmaus	Madagascar Shrew	мадагаскарская белозубка
Pachyures	Dickschwanzspitzmäuse		многозубые белозубки
– aux griffes longues	Langkrallenspitzmäuse	Long-clawed Shrews	
Papion anubis	Grüner Pavian	Anubis Baboon	анубис
Papions	Paviane	Baboons	павианы
Patas	Husarenaffe	Red Guenon	обыкновенный гусар
Petaurus	Gleithörnchenbeutler	Honey Gliders	сумчатые летяги
Petite Musaraigne à queue courte	Nordamerikanische Kleinohrspitzmaus	Least Shrew	североамериканская короткоухая бурозубка
– Taupe marsupiale	Kleiner Beutelmull	North-western Marsupial Mole	малый сумчатый крот
Petit singe-lion	Goldgelbes Löwenäffchen	Golden Lion Marmoset	розалия
– Wallaby de rochers	Zwergsteinkänguruh	Little Rock Wallaby	карликовый каменный кенгуру
Phalanger à pelage rayé	Streifenkletterbeutler	Striped Phalanger	полосатые кускусы
– de montagne	Hundskusu	Short-eared Brush-tailed Phalanger	собачий кузу
Phalangéridés	Kletterbeutler	Phalangers	лазающие сумчатые
Phalangérinés	Eigentliche Kletterbeutler	Phalangers	собственно лазающие сумчатые
Phalanger-Renard	Fuchskusu	Brush-tailed Phalanger	лисий кузу
Phalangers Loirs	Schlafbeutler	Dormouse Possums	сумчатые сони
Phalanger volant	Gleitbeutler	Honey Glider	короткоголовая сумчатая летяга
Phaner à fourche	Gabelstreifiger Katzenmaki	Fork-marked Dwarf Lemur	вильчатый маки
Phascogalinés	Beutelmäuse	Marsupial Mice	мышевидки
Phascolarctinés	Koalaverwandte	Koala-like Marsupials	сумчатые медведи
Pinché	Perückenäffchen	Cotton-head Tamarin	эдипова игрунка
– aux pieds blancs	Weißfußäffchen	White-footed Tamarin	белоногая игрунка
– de Geoffroy	Panamaperückenäffchen	Geoffroy's Tamarin	игрунка Жоффруа
Pithécinés	Sakiaffen	Sakis and Uakaris	саки
Pongidés	Menschenaffen	Great Apes	собственно человекообразные обезьяны
Potamogale	Große Otterspitzmaus	Otter Shrew	выдровая землеройка
Potamogalidés	Otterspitzmäuse	Otter Shrews	выдровые землеройки
Potto de Bosman	Potto	Potto	потто
Presbytinés	Languren	Langurs	
Primates	Herrentiere	Primates	приматы
Propithèque de Verreaux	Larvensifaka	Verreaux's Sifaka	сифака Верро
– diadème	Diademsifaka	Diademed Sifaka	белолобый сифака
Propithèques	Sifakas	Sifakas	сифаки
Prosimiens	Halbaffen	Prosimians	полуобезьяны
Protothériens	Eierlegende Säugetiere	Egg-laying Mammals	клоачные
Ptilocerques	Federschwanztupaias	Pen-tailed Tree Shrews	перохвостые тупайи
Quatre-oeil	Vieraugenbeutelratte	Four-eyed Opossum	опоссум квика
Rat-Kangourou de Gaimard	Festland-Bürstenkänguruh	Gaimard's Rat Kangaroo	австралийский кистехвостый кенгуру
– de Gilbert	Gilbert-Kaninchenkänguruh	Gilbert's Rat Kangaroo	потору Джильберта
– de Lesueur	Lesueur-Bürstenkänguruh	Lesueur's Rat Kangaroo	кистехвостый кенгуру Лесюера
– de Tasmanie	Tasmanienbürstenkänguruh	Tasmanian Rat Kangaroo	тасманийский кистехвостый кенгуру
– du désert	Nacktbrustkänguruh	Desert Rat Kangaroo	степная кенгуровая крыса
– rougeâtre	Rotes Rattenkänguruh	Rufous Rat Kangaroo	рыжая кенгуровая крыса
Rat musqué Kangourou	Moschusrattenkänguruh	Musky Rat Kangaroo	мускусный кенгуру
Rats-Kangourous	Rattenkänguruhs	Rat Kangaroos	кенгуровые крысы
– à nez court	Bürstenkänguruhs	Short-nosed Rat Kangaroos	кистехвостые кенгуру
– à nez long	Kaninchenkänguruhs	Long-nosed Rat Kangaroos	потору
Rats marsupiaux	Breitfußbeutelmäuse	Broad-footed Marsupial Mice	
Rhinopithèque brun	Braune Stumpfnase	Brown Snub-nosed Langur	биэтовский ринопитек
– de Roxellane	Goldstumpfnase	Snub-nosed Monkey	рокселланов ринопитек
– de Tonkin	Tonkinstumpfnase	Tonkin Snub-nosed Monkey	авункулярный ринопитек
– jaune doré	Weißmantelstumpfnase	White-mantled Snub Nose	бреличевский ринопитек
Rhinopithèques	Stumpfnasenaffen	Snub-nosed Monkeys	ринопитеки
Roloway	Roloway	Roloway	
Saimiris	Totenkopfäffchen	Squirrel Monkeys	саймири
Saki à nez blanc	Weißnasensaki	White-nosed Saki	белоносый саки
– à perruque	Zottelschweifaffe	Hairy Saki	саки-монах
– à tête blanche	Weißkopfsaki	White-headed Saki	белоголовый саки

FRENCH NAME	GERMAN NAME	ENGLISH NAME	RUSSIAN NAME
– à tête dorée	Goldkopfsaki	Golden-headed Saki	золотистый саки
– capucin	Rotrückensaki	Red-backed Saki	красноспинный саки
– noir	Satansaffe	Black Saki	чертов саки
Sakis moines	Schweifaffen	Sakis	
Sapajou	Kapuziner	Capuchins	капуцины
– à dos rouge	Gelbes Totenköpfchen	Red-backed Squirrel Monkey	желтая саймири
– à front blanc	Weißstirnkapuziner	White-fronted Capuchin	белолобый капуцин
– apelle	Apella	Brown Capuchin	капуцин-фавн
– à tête noire	Schwarzköpfiges Totenköpfchen	Black-headed Squirrel Monkey	черноголовая саймири
– brun	Brauner Kapuziner	Weeper Capuchin	бурый капуцин
– capucin	Kapuziner	White throated Capuchin	обыкновенный капуцин
– jaune	Totenkopfäffchen	Common Squirrel Monkey	саймири-белка
Sarcophile satanique	Beutelteufel	Tasmanian Devil	сумчатый дьявол
Semnopithèque à front blanc	Weißstirnlangur	White-fronted Leaf Monkey	гололобый тонкотел
– blanchâtre	Weißbartlangur	Purple-faced Langur	белобородый тонкотел
– de François	Tonkinlangur	François's Monkey	тонкинский тонкотел
– de l'Himalaya	Berghulman	Himalayan Entellus Langur	гималайский гульман
– de Mentawi	Mentawilangur	Mentawi Leaf Monkey	ментавайский тонкотел
– de Phayre	Phayres Langur	Phayre's Leaf Monkey	тонкотел Фейера
– des Nilgiris	Nilgirilangur	John's Langur	нилагирийский тонкотел
– melalophe	Roter Langur	Banded Leaf Monkey	рыжий тонкотел
– obscur	Brillenlangur	Dusky Leaf Monkey	очковый тонкотел
– rubicond	Maronenlangur	Maroon Leaf Monkey	каштановый тонкотел
Siamang	Siamang	Siamang	обыкновенный сиаманг
– de Kloss	Zwergsiamang	Dwarf Siamang	карликовый сиаманг
Simiens	Affen	Monkeys, Apes, and Men	обезьяны
Singe-Araignée à tête brune	Braunkopfklammeraffe	Brown-headed Spider Monkey	буроголовая коата
– à ventre blanc	Goldstirn-Klammeraffe	Long-haired Spider Monkey	светлолобая коата
– aux mains noires	Geoffroy-Klammeraffe	Central American Spider Monkey	коата Жоффруа
– noir	Schwarzer Klammeraffe	Black Spider Monkey	черная коата
– rouge	Panama-Klammeraffe	Red Spider Monkey	панамская коата
Singe de nuit	Nachtaffe	Douroucouli	мирикина
Singe-lion à queue jaune	Rotsteißlöwenäffchen	Golden-rumped Tamarin	краснозадая львиная игрунка
– à tête dorée	Goldkopflöwenäffchen	Golden-headed Tamarin	желтоголовая львиная игрунка
Singes-Araignées	Klammeraffen	Spider Monkeys	коаты
Singes du Nouveau Monde	Neuweltaffen	New World Monkeys	обезьяны Нового света
– hurleurs	Brüllaffen	Howler Monkeys	ревуны
– laineux	Wollaffen	Woolly Monkeys	шерстистые обезьяны
Singes-Lions	Löwenäffchen	Maned Tamarins	львиные игрунки
Singe vert	Grüne Meerkatze	Grass Monkey	зеленая мартышка
Solénodon	Haiti-Schlitzrüßler	Haitian Solenodon	гаитийский щелезуб
Solénodontidés	Schlitzrüßler	Solenodons	щелезубы
Soricidés	Spitzmäuse	Shrews	землеройки
Soricinés	Rotzahnspitzmäuse	Red-toothed Shrews	землеройки-бурозубки
Souris à miel	Honigbeutler	Honey Phalanger	пяткоход
– marsupiales	Schmalfußbeutelmäuse	Narrow-footed Marsupial Mice	сумчатые землеройки
Souris-Opossum de Burramy		Burramys Pigmy Possum	
Souris-Opossums	Zwergbeutelratten	South American Mouse Opossums	карликовые сумчатые крысы
Talapoin	Zwergmeerkatze	Dwarf Guenon	крошечная мартышка
Talpidés	Maulwürfe	Moles	кроты
Talpinés	Altweltmaulwürfe	Old World Moles	собственно кроты
Tamarin à calotte rousse	Rotkappentamarin	Red-capped Tamarin	красногривый тамарин
– à manteau doré	Goldmanteltamarin	Golden-mantled Tamarin	
– à manteau rouge	Rotmanteltamarin	Red-mantled Tamarin	
– à moustaches	Schnurrbarttamarin	Moustached Tamarin	усатый тамарин
– à tête brune	Braunrückentamarin	Brown-headed Tamarin	буроспинный тамарин
– aux mains rousses	Rothandtamarin	Red-handed Tamarin	краснорукий тамарин
– bicolore	Manteläffchen	Pied Tamarin	пегая игрунка
– blanc	Weißer Tamarin	White Tamarin	белый тамарин
– de Goeldi	Springtamarin	Goeldi's Monkey	
– de Lönnberg	Lönnbergtamarin	Lönnberg's Tamarin	тамарин Леннберга
– de Martins	Martins-Manteläffchen		
– de Rio Napo	Rio-Napo-Tamarin	Rio Napo Tamarin	
– de Weddell	Weißlippentamarin	Weddell's Tamarin	
– empereur	Kaiserschnurrbarttamarin	Emperor Tamarin	императорский усатый тамарин
– labié	Rotbauchtamarin	Red-bellied White-lipped Tamarin	краснобрюхий тамарин
– nègre	Mohrentamarin	Negro Tamarin	обыкновенный тамарин
– ocre	Ockermanteläffchen		
– rouge et noir	Schwarzrückentamarin	Black-and-red Tamarin	черноспинный тамарин
Tamarins	Tamarins	Tamarins	тамарины
Tana	Tana	Large Tree Shrew	тупайя тана
Tanrec	Großer Tanrek	Tailless Tenrec	обыкновенный тенрек
Tanrecs	Tanreks	Tenrecs	тенреки
Tarsidés	Koboldmakis	Tarsiers	долгопяты
Tarsiens	Koboldmakis	Tarsier-like Prosimians	долгопятовые

FRENCH NAME	GERMAN NAME	ENGLISH NAME	RUSSIAN NAME
Tarsier de Horsfield	Sundakoboldmaki	Western Tarsier	сундский долгопят
– des Philippines	Philippinenkoboldmaki	Philippine Tarsier	филиппинский долгопят
Tarsiers	Koboldmakis	Tarsiers	долгопяты
Tarsier spectre	Celebeskoboldmaki	Eastern Tarsier	долгопят-привидение
Tarsipedinés	Rüsselbeutler	Honey Phalangers	сумчатые медоеды
Taupe	Maulwurf	Mole	крот
– à queue chevelue	Haarschwanzmaulwurf	Hairy-tailed Mole	волосохвостый крот
Taupe à queue glabre	Ostamerikanischer Maulwurf	Eastern American Mole	восточноамериканский крот
– aveugle	Blindmaulwurf	Mediterranean Mole	слепой крот
– commune	Europäischer Maulwurf	Common Eurasian Mole	обыкновенный крот
– de Coahuila	Coahuilamaulwurf	Coahuilan Mole	
– de côte	Pazifischer Maulwurf	Coast Mole	тихоокеанский крот
– de Gibbs	Amerikanischer Spitzmausmaulwurf	American Shrew Mole	американский землеройковый крот
– de Kansu	Kansumaulwurf	Kansu Mole	западнокитайский землеройковый крот
– des montagnes du Japon	Japanischer Spitzmull	Japanese Shrew Mole	японский землеройковый крот
– de Tamaulipas	Tamaulipasmaulwurf	Tamaulipan Mole	
– de Townsend	Townsends Maulwurf	Townsend's Mole	крот Тоунсенда
– de True	Trues Spitzmull	True's Shrew Mole	
– d'Europe de l'Est	Ostmaulwurf	Eastern Mole	короткохвостый крот
– dorée de Grant	Wüstengoldmull	Grant's Desert Golden Mole	пустынный крот
– dorée de Stuhlmann		Stuhlmann's Golden Mole	
– – de Winton	Winton-Goldmull	De Winton's Golden Mole	златокрот
– – du Cap	Kapgoldmull	Cape Golden Mole	капский златокрот
– romaine	Römischer Maulwurf	Roman Mole	римский крот
Taupes d'Amérique de l'Ouest	Westamerikanische Maulwürfe	Western American Moles	западноамериканские кроты
– d'Asie et d'Amérique du Nord	Amerikanisch-Asiatische Maulwürfe	American and Asian Moles	американско-азиатские кроты
– dorées		Golden Moles	собственно златокроты
– marsupiales	Beutelmulle	Marsupial Moles	сумчатые кроты
Tenrecidés	Tanreks	Tenrecs	тенреки
Tenrecinés	Borstenigel	Tenrecs	настоящие тенреки
Thylacinés	Beutelwölfe	Tasmanian Wolves	сумчатые волки
Thylogales	Pademelons	Pademelons	малые кустовые кенгуру
Titis	Springaffen	Titi Monkeys	обезьяны-прыгуны
Toupaïe	Gewöhnliches Spitzhörnchen	Common Tree-shrew	обыкновенная тупайя
– d'Elliot	Elliots Tupaia	Madras Tree-shrew	тупайя Эллиота
Toupaïes	Tupaias	Tree Shrews	тупайи
Tupaiidés	Spitzhörnchen	Tree Shrews	тупайи
Tupaiinés	Buschschwanztupaias	Brush-tailed Tree Shrews	собственно тупайи
Wallabies	Wallabys	Brush Wallabies	кустовые кенгуру
– de rochers	Felskänguruhs	Rock Wallabies	каменные кенгуру
Wallaby agile	Flinkes Känguruh	Sandy Wallaby	проворный кустовый кенгуру
– bicolore	Sumpfwallaby	Black-tailed Wallaby	чернохвостый кустовый кенгуру
– de Billardier	Rotbauchfilander	Rufous-bellied Pademelon	рыжебрюхий кустовый кенгуру
– de Bruijn	Neuguineafilander	Bruijn's Pademelon	аруанский кустовый кенгуру
– de Grey	Östliches Irmawallaby	Grey's Wallaby	восточный кенгуру ирма
– de l'Île d'Eugène	Tammar	Tammar	кенгуру дерби
– de Macleay	Macleay-Buschkänguruh	New Guinea Mountain Wallaby	кустовый кенгуру Миклухо-Маклая
– d'Irma	Irmawallaby	Black-gloved Wallaby	кенгуру ирма
– rayé	Bänderkänguruh	Banded Hare Wallaby	поперечнополосатый кенгуру
Wallaroo	Bergkänguruh	Wallaroo	горный кенгуру
Wombat à narines dénudées	Nacktnasenwombat	Common Wombat	медвежий вомбат
– à narines poilues	Haarnasenwombat	Hairy-nosed Wombat	широколобый вомбат
Wombats	Wombats	Wombats	вомбаты
Yapock	Schwimmbeutler	Yapok	водяной опоссум

IV. Russian—German—English—French

Названия подвидов отличаются от видовых чаще всего лишь дополнительным прилагательным, главным образом географического характера. Такие русские названия подвидов как правило не включены в данную часть зоологического словаря.

RUSSIAN NAME	GERMAN NAME	ENGLISH NAME	FRENCH NAME
абиссинский еж	Äthiopischer Igel	Ethiopian Hedgehog	Hérisson du désert
авахи	Wollmaki	Woolly Indri	Avahi lanigère
австралийская ехидна	Australien-Kurzschnabeligel	Australian Echidna	Échidné de l'Australie
австралийская порхающая сумчатая мышь	Australischer Zwerggleitbeutler	Pigmy Flying Phalanger	Acrobate pygmée

RUSSIAN NAME	GERMAN NAME	ENGLISH NAME	FRENCH NAME
австралийский кистехвостый кенгуру	Festland-Bürstenkänguruh	Gaimard's Rat Kangaroo	Rat-Kangourou de Gaimard
авункулярный ринопитек	Tonkinstumpfnase	Tonkin Snub-nosed Monkey	Rhinopithèque de Tonkin
азиатская короткохвостая бурозубка	Asiatische Kurzschwanz-spitzmaus	Short-tailed Moupin Shrew	
алеутская бурозубка	Unalaskaspitzmaus	Unalaska Shrew	
алжирский еж	Algerischer Igel	Algerian Hedgehog	Hérisson d'Algérie
альпийская бурозубка	Alpenspitzmaus	Alpine Shrew	Musaraigne des montagnes
американский землеройковый крот	Amerikanischer Spitzmaus-maulwurf	American Shrew Mole	Taupe de Gibbs
американско-азиатские кроты	Amerikanisch-Asiatische Maulwürfe	American and Asian Moles	Taupes d'Asie et d'Amérique du Nord
амурский еж	Amurigel	Amur Hedgehog	Hérisson de l'Amour
анубис	Grüner Pavian	Anubis Baboon	Papion anubis
аруанский кустовый кенгуру	Neuguineafilander	Bruijn's Pademelon	Wallaby de Bruijn
африканская лесная белозубка	Afrikanische Waldspitzmaus	Forest Shrew	
бабуин	Gelber Babuin	Yellow Baboon	Babouin cynocéphale
бандикут Гунна	Tasmanischer Langnasenbeutler	Tasmanian Barred Bandicoot	
бандикуты	Langnasenbeutler	Long-nosed Bandicoots	Bandicoots
белобородый тонкотел	Weißbartlangur	Purple-faced Langur	Semnopithèque blanchâtre
белобрюхая белозубка	Feldspitzmaus	Bicolor White-toothed Shrew	Crocidure leucode
белобрюхий еж	Weißbauchigel		Hérisson à ventre blanc
белоголовый саки	Weißkopfsaki	White-headed Saki	Saki à tête blanche
белогорлая мартышка	Weißkehlmeerkatze	White-throated Guenon	Cercopithèque à gorge blanche
белогрудый еж	Weißbrustigel	Eastern European Hedgehog	Hérisson d'Europe de l'Est
белозубка-малютка	Etruskerspitzmaus	Savi's Pigmy Shrew	Pachyure étrusque
белозубки-броненоски	Panzerspitzmäuse	Armoured Shrews	
белолицая игрунка	Weißgesichtseidenäffchen	White-fronted Marmoset	Ouistiti à tête blanche
белолобый капуцин	Weißstirnkapuziner	– Capuchin	Sapajou à front blanc
белолобый маки	Weißkopfmaki	– Lemur	Maki à front blanc
белолобый сифака	Diademsifaka	Diademed Sifaka	Propithèque diadème
белоногая игрунка	Weißfußäffchen	White-footed Tamarin	Pinché aux pieds blancs
белоносый саки	Weißnasensaki	White-nosed Saki	Saki à nez blanc
белоплечая игрунка	Weißschulterseidenäffchen	White-shouldered Marmoset	Ouistiti à camail
белоухая игрунка	Weißohrseidenäffchen	White-eared Marmoset	– oreillard
белошейная игрунка	Weißnackenseidenäffchen	White-necked marmoset	– à col blanc
белый тамарин	Weißer Tamarin	White Tamarin	Tamarin blanc
береговая горилла	Flachlandgorilla	Lowland Gorilla	Gorille de côte
бирманский макак	Löwenmakak		
биэтовский ринопитек	Braune Stumpfnase	Brown Snub-nosed Langur	Rhinopithèque brun
бледный саки	Blaßkopfsaki	Pale-headed Saki	Saki à tête pâle
болотная бурозубка	Nördliche Wasserspitzmaus	Northern Water Shrew	Musaraigne palustre
болотный тенрек	Wassertanrek	Marsh Tenrec	Limnogale
большая белозубка	Riesenwimperspitzmaus	Giant Musk Shrew	
большая белоносая мартышка	Große Weißnasenmeerkatze	Greater White-nosed Guenon	Cercopithèque hocheur
большая пустынная бурозубка	– Wüstenspitzmaus	Merriam's Desert Shrew	Grande Musaraigne du désert
большая сумчатая летяга	Großer Gleithörnchenbeutler	Yellow-bellied Glider	
большой бандикут	– Langnasenbeutler	Long-nosed Bandicoot	Bandicoot à nez long
большой крысиный еж	– Haarigel	Moon Rat	Gymnure
большой куний макиа	– Wieselmaki	Greater Sportive Lemur	Lépilémur mustélin
большой курносый бкндикут	– Kurznasenbeutler	Brindled Bandicoot	
большой новогвинейсйи бандикут	– Neuguinea-Nasenbeutler	New Guinea Bandicoot	Bandicoot de Nouvelle-Guinée
большой полосатый кускус	– Streifenbeutler	Striped Phalanger	Phalanger au pelage rayé
бородатая мартышка	Vollbartmeerkatze	L'Hoest's Monkey	Cercopithèque de l'Hoest
бородатый мангабей	Schopfmangabe	Black Mangabey	Mangabey noir
бразовская мартышка	Brazzameerkatze	De Brazza's Monkey	Cercopithèque de Brazza
бреличевский ринопитек	Weißmantelstumpfnase	White-mantled Snub Nose	Rhinopithèque jaune doré
бурая белозубка	Hausspitzmaus	Common European White-toothed Shrew	Musaraigne musette
бурая мускусная белозубка	Moschusspitzmaus	House Shrew	
бурая первокрыса	Ekuador-Opossummaus	Ecuador Rat Opossum	
бурая шерстистая обезьяна	Brauner Wollaffe	Brown Woolly Monkey	Lagotriche de Castelnau
буроголовая коата	Braunkopfklammeraffe	Brown-headed Spider Monkey	Singe-Araignée à tête brune
бурогрудый еж	Braunbrustigel	Western European Hedgehog	Hérisson d'Europe de l'Ouest
бурозубка Бендайра	Pazifische Wasserspitzmaus	Pacific Water Shrew	Musaraigne des marais
бурозубки	Waldspitzmäuse	Shrews	Musaraignes
буроспинный тамарин	Braunrückentamarin	Brown-headed Tamarin	Tamarin à tête brune
бурый капуцин	Brauner Kapuziner	Weeper Capuchin	Sapajou brun
бурый ревун	– Brüllaffe	Brown Howler Monkey	Hurleur brun
быстрый гиббон	Ungka	Dark-handed Gibbon	Gibbon agile
ванбенеденовский толстотел	Grüner Stummelaffe	Green Colobus	Colobe vrai
вандеру	Wanderu	Lion-tailed Macaque	Macaque Ouanderou
вари	Vari	Ruffed Lemur	Lémur vrai
вильчатый маки	Gabelstreifiger Katzenmaki	Fork-marked Dwarf Lemur	Phaner à fourche
водяной опоссум	Schwimmbeutler	Yapok	Yapock
волосохвостый крот	Haarschwanzmaulwurf	Hairy-tailed Mole	Taupe à queue chevelue

RUSSIAN NAME	GERMAN NAME	ENGLISH NAME	FRENCH NAME
вомбаты	Wombats	Wombats	Wombats
воротничковый мангабей	Halsbandmangabe	Sooty Mangabey	Mangabey à collier blanc
восточноавстралийский сумчатый тушканчик	Östliche Springbeutelmaus	Eastern Jerboa Marsupial	
восточноамериканский крот	Ostamerikanischer Maulwurf	– American Mole	Taupe à queue glabre
восточный кенгуру ирма	Östliches Irmawallaby	Grey's Wallaby	Wallaby de Grey
выдровая землеройка	Große Otterspitzmaus	Otter Shrew	Potamogale
выдровые землеройки	Otterspitzmäuse	Otter Shrews	Potamogalidés
высшие млекопитающие	Höhere Säugetiere	Placental Mammals	Eutheriens
выхухоли	Desmane	Desmans	Desmaninés
галаго	Galago	Galago	Galago
галаго Демидова	Zwerggalago	Demidoff's Bushbaby	Galago de Demidoff
гаитийский щелезуб	Haiti-Schlitzrüßler	Haitian Solenodon	Solénodon
гамадрил	Mantelpavian	Hamadryas Baboon	Hamadryas
гватемальский ревун	Guatemalabrüllaffe	Guatemalan Howler Monkey	Hurleur de Guatemala
гвереца	Nördlicher Guereza	Northern Black-and-white Colobus	Colobe de l'Abyssinie
гвинейский павиан	Guineapavian	Guinea Baboon	Babouin de Guinée
гиббоны	Gibbons	Gibbons	Hylobatidae
гигантская белозубка	Afrikanische Riesenspitzmaus	African Forest Shrew	Musaraigne géante
гималайская водяная белозубка	Himalajawasserspitzmaus	Himalayan Water Shrew	Chimarrogale de l'Himalaya
гималайский гульман	Berghulman	– Entellus Langur	Semnopithèque de l'Himalaya
гололобый тонкотел	Weißstirnlangur	White-fronted Leaf Monkey	– à front blanc
голохвостый опоссум	Nacktschwanzbeutelratte	Rat-tailed Opossum	Opossum à queue de rat
голуболицая мартышка	Blaumaulmeerkatze	Moustached Monkey	Moustac
горилла	Gorilla	Gorilla	Gorille
горные тупайи	Bergtupaias	Smooth-tailed Tree Shrews	
горный кенгуру	Bergkänguruh	Wallaroo	Wallaroo
горный резус	Bergrhesus	Assamese Macaque	Macaque d'Assam
готтентотский крот	Hottentotten-Goldmull	Hottentot Golden Mole	
гребешковая мышевидка	Doppelkammbeutelmaus	Crest-tailed Marsupial Rat	
гребнехвостая мышевидка	Südliche Kammschwanzbeutelmaus	– – Mouse	
гривистый мангабей	Mantelmangabe	Grey-cheeked Mangabey	Mangabey à gorge blanche
гривистый тонкотел	Haubenlangur	Silvered Leaf Monkey	
гулок	Hulock	Hoolock Gibbon	Hoolock
гульман	Hulman	Entellus Langur	Houleman
джелада	Dschelada	Gelada Baboon	Gelada
диана	Dianameerkatze	Diana Monkey	Cercopithèque diane
длинноиглый еж	Brandts Igel	Brandt's Hedgehog	Hérisson de Brandt
длинноухая игрунка	Langohrseidenäffchen	Santarém Marmoset	Ouistiti de Santarém
длиннохвостые тенреки	Kleintanreks	Lesser Tenrecs	Microgales
длиннохвостый крот	Langschwanzmaulwurf	Long-tailed Mole	
длиннохвостый новогвинейский бандикут	Langschwänziger Neuguinea-Nasenbeutler	– New Guinea Bandicoot	
долгопятовые	Koboldmakis	Tarsier-like Prosimians	Tarsiens
долгопят-привидение	Celebeskoboldmaki	Eastern Tarsier	Tarsier spectre
долгопяты	Koboldmakis	Tarsiers	Tarsidés
древесные кенгуру	Baumkänguruhs	Tree Kangaroos	Kangourous arboricoles
древесный кенгуру Беннетта	Bennett-Baumkänguruh	Bennett's Tree Kangaroo	Kangourou arboricole de Bennett
древесный кенгуру Дория	Doria-Baumkänguruh	Doria's Tree Kangaroo	– – de Doria
дрил	Drill	Drill	Drill
еж Прунера	Pruners Igel	Pruner's Hedgehog	Hérisson de Pruner
еж Склатера	Sclaters Igel	Sclater's Hedgehog	– de Sclater
ежевый тенрек Тельфера	Kleiner Igeltanrek	Lesser Hedgehog Tenrec	
ежевый тенрек Фонтуанона	Fontoynonts Igeltanrek	Fontoynont's Hedgehog Tenrec	
ежи	Igel	Hedgehogs	Érinacéidés
ехидны	Ameisenigel	Echidnas	Échidnés
желтая саймири	Gelbes Totenköpfchen	Red-backed Squirrel Monkey	Sapajou à dos rouge
желтая шерстистая сумчатая крыса	Gelbe Wollbeutelratte	Philander Opossum	Opossum laineux
желтоголовая игрунка	Gelbkopfbüscheläffchen	Buff-headed Marmoset	Ouistiti à tête jaune
желтоголовая львиная игрунка	Goldkopflöwenäffchen	Golden-headed Tamarin	Singe-lion à tête dorée
желтозеленая мартышка	Gelbgrüne Meerkatze		Callitriche
желтоногая игрунка	Gelbfußäffchen	Yellow-legged Marmoset	Ouistiti aux pieds jaunes
желтоногая мышевидка	Gelbfußbeutelmaus	Yellow-footed Marsupial Mouse	
желтоногий каменный кенгуру	Ringschwanz-Felskänguruh	Ring-tailed Rock Wallaby	
желтохвостая шерстистая обезьяна	Gelbschwanzwollaffe	Peruvian Mountain Woolly Monkey	
желтый кускус	Streifen-Ringelschwanzbeutler	Striped Ring Tail	
закавказский еж	Transkaukasischer Igel	Transcaucasian Hedgehog	
занзибарская хоботковая собачка	Sansibar-Rüsselhündchen		
западноамериканские кроты	Westamerikanische Maulwürfe	Western American Moles	Taupes d'Amérique de l'Ouest
западноевропейский еж	Westeuropäischer Igel	– European Hedgehog	
западнокитайский землеройковый крот	Kansumaulwurf	Kansu Mole	Taupe de Kansu
заячьи кенгуру	Hasenkänguruhs	Hare Wallabies	Lièvres wallabies

RUSSIAN NAME	GERMAN NAME	ENGLISH NAME	FRENCH NAME
звездорыл	Sternmull	Star-nosed Mole	Condylure étoilé
зеленая мартышка	Grüne Meerkatze	Grass Monkey	Singe vert
землеройки	Spitzmäuse	Shrews	Soricidés
землеройки-белозубки	Weißzahnspitzmäuse	White-toothed Shrews	Musaraignes à dents blanches
землеройки-бурозубки	Rotzahnspitzmäuse	Red-toothed Shrews	Soricinés
землеройковые	Spitzmausartige	Shrews and Moles	Musaraignes et Taupes
землеройковые сумчатые крысы	Spitzmausbeutelratten	Short Bare-tailed Opossums	
землеройковый еж	Spitzmausigel	Shrew Hedgehog	Neotétracus
земляной тенрек	Erdtanrek		Géogale
златокрот	Winton-Goldmull	De Winton's Golden Mole	Taupe dorée de Winton
златокроты	Goldmulle	Golden Moles	Chrysochloridés
золотистый саки	Goldkopfsaki	Golden-headed Saki	Saki à tête dorée
игрунка Жоффруа	Panamaperückenäffchen	Geoffroy's Tamarin	Pinché de Geoffroy
императорский усатый тамарин	Kaiserschnurrbarttamarin	Emperor Tamarin	Tamarin empereur
индийский еж	Indischer Igel	Indian Hedgehog	
индийский макак	– Hutaffe	Bonnet Monkey	Macaque bonnet chinois
индонезийская водяная белозубка	Borneowasserspitzmaus		
индри	Indris	Indrises	Indris
исполинская сумчатая куница	Fleckschwanzbeutelmarder	Tiger Cat	Chat marsupial
исполинская сумчатая летяга	Riesengleitbeutler	Greater Gliding Phalanger	Grand Phalanger volant
исполинские кенгуру	Riesenkänguruhs	Kangaroos	Grands Kangourous
исполинский златокрот	Riesengoldmull	Giant Golden Mole	Grande Taupe dorée
казака	Südopossum	Azara's Opossum	
калифорнийский крот	Kalifornischer Maulwurf	Broad-footed Mole	
каменные кенгуру	Felskänguruhs	Rock Wallabies	Wallabies de rochers
капский еж	Kapigel	Cape Hedgehog	
капский златокрот	Kapgoldmull	– Golden Mole	Taupe dorée du Cap
капуцинообразные обезьяны	Kapuzinerartige	New World Monkeys	Cébidés
капуцин-фавн	Apella	Brown Capuchin	Sapajou apelle
капуцины	Kapuziner	Capuchins	Sapajou
карликовая выдровая землеройка	Zwerg-Otterspitzmaus	Lesser Otter Shrew	Micropotamogale de Lamotte
карликовая игрунка	Zwergseidenäffchen	Pigmy Marmoset	Ouistiti mignon
карликовая мышевидка	Zwergbeutelmaus	– Marsupial Mouse	
карликовая плоскоголовая мышевидка	Zwerg-Flachkopf-Beutelmaus	Kimberley Planigale	
карликовые сумчатые крысы	Zwergbeutelratten	South American Mouse Opossums	Souris-Opossums
карликовый каменный кенгуру	Zwergsteinkänguruh	Little Rock Wallaby	Petit Wallaby de rochers
карликовый сиаманг	Zwergsiamang	Dwarf Siamang	Siamang de Kloss
катта	Katta	Ring-tailed Lemur	Lémur catta
каштановый тонкотел	Maronenlangur	Maroon Leaf Monkey	Semnopithèque rubicond
квинслендский древесный кенгуру	Lumholtz-Baumkänguruh	Lumholtz's Tree Kangaroo	Kangourou arboricole de Lumholtz
квинслендский каменный кенгуру	Queensland-Felskänguruh	Plain Rock Wallaby	
кенгуровые крысы	Rattenkänguruhs	Rat Kangaroos	Rats-Kangourous
кенгуру	Känguruhs	Wallabies and Kangaroos	Kangourous
кенгуру дерби	Tammar	Tammar	Wallaby de l'Île d'Eugène
кенгуру ирма	Irmawallaby	Black-gloved Wallaby	– d'Irma
кенгуру-медведь	Bären-Baumkänguruh	Black Tree Kangaroo	
кенгуру парма	Parmakänguruh	Parma Wallaby	
кистехвостая мышевидка	Großer Pinselschwanzbeutler	Black-tailed Phascogale	
кистехвостые кенгуру	Bürstenkänguruhs	Short-nosed Rat Kangaroos	Rats-Kangourous à nez court
кистехвостые сумчатые крысы	Buschschwanzbeutelratten	Brush-tailed Opossums	
кистехвостый каменный кенгуру	Bürsten-Felskänguruh	– Rock Wallaby	
кистехвостый кенгуру Лесюера	Lesueur-Bürstenkänguruh	Lesueur's Rat Kangaroo	Rat-Kangourou de Lesueur
кисточковая игрунка	Schwarzpinseläffchen	Black-pencilled Marmoset	Ouistiti à pinceau noir
клоачные	Eierlegende Säugetiere	Egg-laying Mammals	Protothériens
коала	Koala	Koala	Koala
коата Жоффруа	Geoffroy-Klammeraffe	Central American Spider Monkey	Singe-Araignée aux Mains noirs
коаты	Klammeraffen	Spider Monkeys	Singes-Araignées
когтехвостые кенгуру	Nagelkänguruhs	Nail-tail Wallabies	Onychogales
когтистая белозубка Келаарта	Kelaarts Langkrallenspitzmaus	Kelaart's Long-clawed Shrew	
когтистая белозубка Пирсона	Pearsons Langkrallenspitzmaus	Pearson's Long-clawed Shrew	
когтистые обезьяны	Krallenaffen	Marmosets and Tamarins	Callithricidés
колумбийская шерстистая обезьяна	Columbischer Bergwollaffe		
кольцехвостые кускусы	Ringelschwanz-Kletterbeutler	Ring-tailed Phalangers	Ringtails
конголезская белозубка	Kongowimperspitzmaus		
конголезская броненоска	Kongopanzerspitzmaus	Congo Armoured Shrew	
корейский еж	Koreaigel	Korean Hedgehog	Hérisson de Corée
королевский толстотел	Südlicher Guereza	Southern Black-and-white Colobus	Colobe à longs poils
короткоголовая сумчатая летяга	Kurzkopfgleitbeutler	Honey Glider	Phalanger volant
коротконосый прыгунчик	Kurznasen-Elefantenspitzmaus	Short-snouted Elephant Shrew	
короткоухий каменный кенгуру	Kurzohr-Felskänguruh	Short-eared Rock Wallaby	
короткохвостый индри	Indri	Indris	Indri

RUSSIAN NAME	GERMAN NAME	ENGLISH NAME	FRENCH NAME
короткохвостый крот	Ostmaulwurf	Eastern Mole	Taupe d'Europe de l'Est
короткохвостый кустовый кенгуру	Kurzschwanzkänguruh	Quokka	Kangourou à queue courte
косматый заячий кенгуру	Zottel-Hasenkänguruh	Western Hare Wallaby	
крапчатая мышевидка	Sprenkelbeutelmaus	Speckled Marsupial Mouse	
крапчатая сумчатая куница	Tüpfelbeutelmarder	Eastern Dasyure	
краснобрюхая мартышка	Rotbauchmeerkatze	Red-bellied Guenon	Hocheur à ventre rouge
краснобрюхий тамарин	Rotbauchtamarin	– White-lipped Tamarin	Tamarin labié
красногривый тамарин	Rotkappentamarin	Red-capped Tamarin	– à calotte rousse
краснозадая львиная игрунка	Rotsteißlöwenäffchen	Golden-rumped Tamarin	Singe-lion à queue jaune
красторукий ревун	Rothandbrüllaffe	Rufous-handed Howler Monkey	Hurleur à mains rousses
красторукий тамарин	Rothandtamarin	Red-handed Tamarin	Tamarin aux mains rousses
красноспинный саки	Rotrückensaki	Red-backed Saki	Saki capucin
красный прыгун	Roter Springaffe	Red Titi	Callicèbe roux
красный толстотел	– Stummelaffe	– Colobus	Colobe bai
красный уакари	– Uakari	– Uakari	Ouakari rubicond
крот Тоунсенда	Townsends Maulwurf	Townsend's Mole	Taupe de Townsend
кротовая белозубка	Maulwurfspitzmaus		
кротовидный рисовый тенрек	Maulwurfartiger Reistanrek	Mole-like Rice Tenrec	
кроты	Maulwürfe	Moles	Talpidés
крошечная мартышка	Zwergmeerkatze	Dwarf Guenon	Talapoin
крысиные ежи	Haarigel	Hairy Hedgehogs	Gymnures
крысиные маки	Katzenmakis	Dwarf Lemurs	Chirogaléinés
крысиный потору	Langschnauzen-Kaninchen-känguruh	Long-nosed Rat Kangaroo	Rat-Kangourou à nez long
кубинский щелезуб	Kuba-Schlitzrüßler	Cuban Solenodon	Almiqui
кузу	Kusus	Brush-tailed Phalangers	Opossums d'Australie
курносые бандикуты	Kurznasenbeutler	Short-nosed Bandicoots	
кускус Даля	Felsen-Ringelschwanzbeutler	Rock-haunting Ring Tail	
кускусы	Kuskuse	Cuscuses	Couscous
кустовые кенгуру	Wallabys	Brush Wallabies	Wallabies
кустовый кенгуру Беннетта	Bennettkänguruh	Red-necked Wallaby	
кустовый кенгуру Миклухо-Маклая	Macleay-Buschkänguruh	New Guinea Mountain Wallaby	Wallaby de Macleay
кустовый кенгуру Парри	Hübschgesichtkänguruh	Pretty-face Wallaby	
кустовый кенгуру Хагена	Hagen-Buschkänguruh	Northern New Guinea Wallaby	
куцая белозубка	Stummelschwanzspitzmaus	Szechuan Burrowing Shrew	
лазающие сумчатые	Kletterbeutler	Phalangers	Phalangéridés
лар	Lar	White-handed Gibbon	Gibbon lar
лемуровые	Lemurenartige	Lemur-like Prosimians	Lémuriens
лемуровый кускус	Lemuren-Ringelschwanzbeutler	Brush-tipped Ring Tail	
лемуры	Lemuren	Lemurs	Lémuridés
лесная хоботковая крыса	Waldrüsselratte	Forest Elephant Shrew	
лисий кузу	Fuchskusu	Brush-tailed Phalanger	Phalanger-Renard
лори	Loris	Lorises	Lorisidés
лориевые	Loriartige	Loris-like Prosimians	Lorisidés
лысый уакари	Scharlachgesicht	Bald Uakari	Ouakari chauve
львиные игрунки	Löwenäffchen	Maned Tamarins	Singes-Lions
магот	Magot	Barbary Ape	Magot
мадагаскарская белозубка	Madagaskarspitzmaus	Madagascar Shrew	Pachyure de Madagascar
мадагаскарская руконожка	Fingertier	Aye-Aye	Aye-Aye
мадерская саймири	Madeirafluß-Totenköpfchen		
макаки	Makaken	Macaques	Macaques
макак-резус	Rhesusaffe	Rhesus Macaque	Macaque rhésus
маки	Echte Makis	Lemurs	Lémur
малая белозубка	Gartenspitzmaus	Lesser White-toothed Shrew	Musaraigne des jardins
малая белоносая мартышка	Kleine Weißnasenmeerkatze	– White-nosed Guenon	Hocheur blanc-nez
малая бурозубка	Zwergspitzmaus	– Shrew	Musaraigne pygmée
малая кистехвостая мышевидка	Kleiner Pinselschwanzbeutler	Red-tailed Phascogale	
малая кутора	Sumpfspitzmaus	Mediterranean Water Shrew	Musaraigne de Miller
малые кустовые кенгуру	Filander	Pademelons	Thylogales
малый крысиный еж	Kleiner Rattenigel	Lesser Gymnure	
малый куний маки	– Wieselmaki	Sportive Lemur	Lépilémur à queue rouge
малый курносый бандикут	– Kurznasenbeutler	Southern Short-nosed Bandicoot	
малый полосатый кускус	– Streifenbeutler	Long-fingered Striped Phalanger	
малый сумчатый заяц	– Kaninchen-Nasenbeutler	White-tailed Rabbit Bandicoot	
малый сумчатый крот	– Beutelmull	North-western Marsupial Mole	Petite Taupe marsupiale
малый толстый лори	– Plumplori	Lesser Slow Loris	
мангабеи	Mangaben	Mangabeys	Mangabeys
мартышки	Meerkatzen	Guenons	Cercopithèques
мартышковые	Hundsaffen	Old World Monkeys	
медвежий вомбат	Nacktnasenwombat	Common Wombat	Wombat à narines dénudées
медвежий макак	Bärenmakak	Stump-tailed Macaque	Macaque brun
медвежий маки	Bärenmaki	Angwantibo	Angwantibo
мексиканская сумчатая крыса	Schwarzring-Zwergbeutelratte	Mexican Mouse Opossum	
ментавайский тонкотел	Mentawilangur	Mentawi Leaf Monkey	Semnopithèque de Mentawi

RUSSIAN NAME	GERMAN NAME	ENGLISH NAME	FRENCH NAME
мирикина	Nachtaffe	Douroucouli	Singe de nuit
млекопитающие	Säugetiere	Mammals	Mammifères
многозубые белозубки	Dickschwanzspitzmäuse		Pachyures
мона	Monameerkatze	Mona Monkey	Cercopithèque mone
монго	Mongozmaki	Mongoose Lemur	Lémur mongos
муравеед	Ameisenbeutler	Marsupial Anteater	Fourmilier marsupial rayé
мускусный кенгуру	Moschusrattenkänguruh	Musky Rat Kangaroo	Rat musqué Kangourou
мышевидка Лоренца	Neuguinea-Spitzhörnchen-beutler	Lorentz's Marsupial Rat	
мышевидки	Beutelmäuse	Marsupial Mice	Phascogalinés
мышиный бандикут	Maus-Nasenbeutler	Mouse Bandicoot	
мышиный маки	Mausmaki	Lesser Mouse Lemur	Chirogale mignon
мышиный опоссум	Maus-Zwergbeutelratte	Murine Opossum	
насекомоядные	Insektenesser	Insect-Eaters	Insectivores
настоящие ежи	Echte Igel	Spiny Hedgehogs	Hérissons
настоящие тенреки	Borstenigel	Tenrecs	Tenrecinés
немейский тонкотел	Kleideraffe	Douc Langur	Douc
нилагирийский тонкотел	Nilgirilangur	John's Langur	Semnopithèque des Nilgiris
новогвинейская порхающая сумчатая мышь	Neuguinea-Zwerggleitbeutler	New Guinea Pigmy Flying Phalanger	
новогвинейская сумчатая соня	Neuguinea-Bilchbeutler	Pigmy Possum	
новогвинейские бандикуты	Neuguinea-Nasenbeutler	New Guinea Bandicoots	Bandicoots de Nouvelle-Guinée
новогвинейские мышевидки	Neuguineabeutelmäuse	– – Marsupial Mice	
обезьяны	Affen	Monkeys, Apes, and Men	Simiens
обезьяны Нового света	Neuweltaffen	New World Monkeys	Singes du Nouveau Monde
обезьяны-прыгуны	Springaffen	Titi Monkeys	Titis
обожженный кенгуру	Rotbeinfilander	Red-legged Pademelon	
обыкновенная американская бурозубка	Amerikanische Maskenspitzmaus	Masked Shrew	Musaraigne cendrée
обыкновенная бурозубка	Waldspitzmaus	Common Shrew	– carrelet
обыкновенная выхухоль	Russischer Desman	Russian Desman	Desman de Moscovie
обыкновенная игрунка	Weißbüscheläffchen	Common Marmoset	Ouistiti
обыкновенная кутора	Wasserspitzmaus	European Water Shrew	Musaraigne aquatique
обыкновенная паукообразная обезьяна	Spinnenaffe	Woolly Spider Monkey	Éroïde
обыкновенная сумчатая земле-ройка	Kleine Schmalfußbeutelmaus	Mouse Sminthopsis	
обыкновенная сумчатая соня	Dickschwanz-Schlafbeutler	Dormouse Possum	
обыкновенная тупайя	Gewöhnliches Spitzhörnchen	Common Tree-shrew	Toupaïe
обыкновенные ежи	Kleinohrigel	– Hedgehogs	Hérissons communs
обыкновенный гусар	Husarenaffe	Red Guenon	Patas
обыкновенный длиннохвостый тенрек	Langschwanztanrek	Long-tailed Tenrec	Microgale
обыкновенный ежевый тенрек	Großer Igeltanrek	Greater Hedgehog Tenrec	
обыкновенный заячий кенгуру	Langohr-Hasenkänguruh	Brown Hare Wallaby	
обыкновенный капуцин	Kapuziner	White-throated Capuchin	Sapajou capucin
обыкновенный когтехвостый кенгуру	Flachnagelkänguruh	Northern Nail-tail Wallaby	
обыкновенный кольцехвостый кускус	Wander-Ringelschwanzbeutler	Queensland Ring Tail	
обыкновенный крот	Europäischer Maulwurf	Common Eurasian Mole	Taupe commune
обыкновенный носач	Nasenaffe	Proboscis Monkey	Nasique
обыкновенный опоссум	Nordopossum	Common Opossum	Opossum commun
обыкновенный сиаманг	Siamang	Siamang	Siamang
обыкновенный слоновый прыгунчик	Kurzohrrüsselspringer	Short-eared Elephant Shrew	
обыкновенный сумчатый заяц	Großer Kaninchen-Nasenbeutler	Rabbit Bandicoot	
обыкновенный тамарин	Mohrentamarin	Negro Tamarin	Tamarin nègre
обыкновенный тенрек	Großer Tanrek	Tailless Tenrec	Tanrec
однопроходные	Kloakentiere	Monotremes	Monotrèmes
одноцветный гиббон	Schopfgibbon	Black Gibbon	Gibbon noir
одноцветный симиас	Pagehstumpfnasenaffe	Pig-tailed Langur	
опоссум квика	Vieraugenbeutelratte	Four-eyed Opossum	Quatre-oeil
опоссумы	Opossums	Common and Azara's Opossums	Opossums
оранг-утан	Orang-Utan	Orang-Utan	Orang-outan
остроносые бандикуты	Stachelnasenbeutler	Spiny New Guinea Bandicoots	
очковый заячий кенгуру	Brillen-Hasenkänguruh	Spectacled Hare Wallaby	
очковый тонкотел	Brillenlangur	Dusky Leaf Monkey	Semnopithèque obscur
павианы	Paviane	Baboons	Papions
падемелон	Rothalsfilander	Red-necked Pademelon	
пегая игрунка	Manteläffchen	Pied Tamarin	Tamarin bicolore
пегий путорак	Gescheckte Spitzmaus	Piebald Shrew	
пепельная сумчатая крыса	Aschgraue Zwergbeutelratte	Ashy Opossum	
перохвостая сумчатая соня	Federschwanzbeutler	Pen-tailed Phalanger	
перохвостая тупайя	Lows Federschwanz	– Tree Shrew	
перуанская первокрыса	Peru-Opossummaus	Peruvian Rat Opossum	

RUSSIAN NAME	GERMAN NAME	ENGLISH NAME	FRENCH NAME
пинче	Perückenäffchen	Crested Bare-faced Tamarins	Pinchés
пиренейская выхухоль	Pyrenäendesman	Pyrenean Desman	Desman des Pyrénées
плавуны	Schwimmbeutelratten	Yapoks	Yapocks
плоскоголовые мышевидки	Flachkopfbeutelmäuse	Flat-skulled Marsupials	
полосатая сумчатая куница	Streifenbeutelmarder	Striped Native Cat	
полосатая шерстистая сумчатая крыса	Bindenwollbeutelratte	Woolly Opossum	Opossum laineux
полосатые кускусы	Streifenkletterbeutler	Striped Phalanger	Phalanger à pelage rayé
полосатые мышевидки	Streifenbeutelmäuse	– Marsupial Rats	
полосатый бандикут	Bänder-Langnasenbeutler	Eastern Barred Bandicoot	
полосатый кустовый кенгуру	Rückenstreifkänguruh	Black-striped Wallaby	
полосатый тенрек	Halbborstenigel	Streaked Tenrec	Hémicentète
полулунный когтехвостый кенгуру	Mondnagelkänguruh	Crescent Nail-tail Wallaby	
полумаки	Halbmakis	Gentle Lemurs	Hapalémur
полуобезьяны	Halbaffen	Prosimians	Prosimiens
полутенреки	Halbborstenigel	Streaked Tenrecs	Hémicentètes
поперечнополосатый кенгуру	Bänderkänguruh	Banded Hare Wallaby	Wallaby rayé
порхающие сумчатые мыши	Zwerggleitbeutler	Pigmy Flying Phalangers	Acrobate pygmée
потору	Kaninchenkänguruhs	Long-nosed Rat Kangaroos	Rats-Kangourous à nez long
потору Джильберта	Gilbert-Kaninchenkänguruh	Gilbert's Rat Kangaroo	Rat-Kangourou de Gilbert
потто	Potto	Potto	Potto de Bosman
приматы	Herrentiere	Primates	Primates
проворный кустовый кенгуру	Flinkes Känguruh	Sandy Wallaby	Wallaby agile
проехидна Бартона	Barton-Langschnabeligel	Barton's Echidna	Échidné de Barton
проехидна Бруйна	Bruijn-Langschnabeligel	Bruijn's Echidna	– de Bruijn
проехидна острова Бубу	Bubu-Langschnabeligel	Bubu-Echidna	
проехидны	Langschnabeligel	Long-beaked Spiny Anteater	Échidné à bec courbe
прыгунчики	Rüsselspringer	Elephant Shrews	Macroscélidés
пустынная белозубка	Wüstenwimperspitzmaus	Desert Musk Shrew	Crocidure du désert
пустынный крот	Wüstengoldmull	Grant's Desert Golden Mole	Taupe dorée de Grant
пяткоход	Honigbeutler	Honey Phalanger	Souris à miel
пятнистая хоботковая собачка	Geflecktes Rüsselhündchen	Checkered Elephant Shrew	
пятнистый кускус	Tüpfelkuskus	Spotted Cuscus	Couscous tacheté
ревуны	Brüllaffen	Howler Monkeys	Singes hurleurs
римский крот	Römischer Maulwurf	Roman Mole	Taupe romaine
ринопитеки	Stumpfnasenaffen	Snub-nosed Monkeys	Rhinopithèques
рисовые тенреки	Reistanreks	Rice Tenrecs	Oryzoryctinés
розалия	Goldgelbes Löwenäffchen	Golden Lion Marmoset	Petit singe-lion
рокселланов ринопитек	Goldstumpfnase	Snub-nosed Monkey	Rhinopithèque de Roxellane
руконожки	Fingertiere	Aye-Ayes	Daubentoniidés
рыжая кенгуровая крыса	Rotes Rattenkänguruh	Rufous Rat Kangaroo	Rat-Kangourou rougeâtre
рыжая шерстистая сумчатая крыса	Rote Wollbeutelratte	Woolly Opossum	Opossum laineux
рыжебрюхий кустовый кенгуру	Rotbauchfilander	Rufous-bellied Pademelon	Wallaby de Billardier
рыжебрюхий маки	Rotbauchmaki	Red-bellied Lemur	Lémur à ventre rouge
рыжелобый маки	Rotstirnmaki	Red-fronted Lemur	– à front rouge
рыжий исполинский кенгуру	Rotes Riesenkänguruh	Red Kangaroo	Kangourou roux
рыжий ревун	Roter Brüllaffe	– Howler Monkey	Hurleur roux
рыжий тонкотел	– Langur	Banded Leaf Monkey	Semnopithèque melalophe
саки	Sakiaffen	Sakis and Uakaris	Pithécinés
саки-монах	Zottelschweifaffe	Hairy Saki	Saki ä perruque
саймири	Totenkopfäffchen	Squirrel Monkeys	Saimiris
саймири-белка	Totenkopfäffchen	Common Squirrel Monkey	Sapajou jaune
светлолобая коата	Goldstirn-Klammeraffe	Long-haired Spider Monkey	Singe-Araignée à ventre blanc
свинохвостый макак	Schweinsaffe	Pig-tailed Macaque	Macaque à queue de cochon
северная плоскоголовая мышевидка	Nördliche Flachkopfbeutelmaus	Northern Planigale	
североавстралийская сумчатая куница	Zwerg-Fleckenbeutelmarder	Little Northern Dasyure	
североамериканская карликовая бурозубка	Amerikanische Zwergspitzmaus	American Pigmy Shrew	Musaraigne pygmée d'Amérique
североамериканская короткоухая бурозубка	Nordamerikanische Kleinohrspitzmaus	Least Shrew	Petite Musaraigne à queue courte
североамериканская короткохвостая бурозубка	Kurzschwanzspitzmaus	Short-tailed Shrew	Grande Musaraigne à queue courte
североафриканский прыгунчик	Nordafrikanische Elefantenspitzmaus	North African Elephant Shrew	Macroscélide de l'Afrique du Nord
сенегальский галаго	Senegalgalago	Senegal Bushbaby	Galago du Sénégal
серая пустынная бурозубка	Graue Wüstenspitzmaus	Crawford's Desert Shrew	Musaraigne du désert
серебристая игрунка	Silberäffchen	Silvery Marmoset	Ouistiti melanure
серебристый гиббон	Silbergibbon	Grey Gibbon	Gibbon cendré
серорукий целебесский макак	Grauarmmakak		
серый исполинский кенгуру	Graues Riesenkänguruh	Grey Kangaroo	Kangourou géant
серый кускус	Wollkuskus	– Cuscus	
серый полумаки	Grauer Halbmaki	– Gentle Lemur	Hapalémur

RUSSIAN NAME	GERMAN NAME	ENGLISH NAME	FRENCH NAME
серый прыгун	– Springaffe	Orabussu Titi	Callicèbe arabassu
сиаманг	Siamang	Siamang	Siamang
сиккимская когтистая бурозубка	Sikkim-Großklauen-Spitzmaus	Sikkim Large-clawed Shrew	Musaraigne de Sikkim
сифака Верро	Larvensifaka	Verreaux's Sifaka	Propithèque de Verreaux
сифаки	Sifakas	Sifakas	Propithèques
слепой крот	Blindmaulwurf	Mediterranean Mole	Taupe aveugle
скалистый прыгунчик	Klippen-Elefantenspitzmaus	Rock Elephant Shrew	
собачий кузу	Hundskusu	Short-eared Brush-tailed Phalanger	Phalanger de montagne
собственно гиббоны	Eigentliche Gibbons	Gibbons	Gibbons
собственно златокроты	Goldmulle	Golden Moles	Taupes dorées
собственно кенгуру	Eigentliche Känguruhs	Kangaroos and Wallabies	Macropodinés
собственно кроты	Altweltmaulwürfe	Old World Moles	Talpinés
собственно лазающие сумчатые	Eigentliche Kletterbeutler	Phalangers	Phalangerinés
собственно мандрил	Mandrill	Mandrill	Mandrill
собственно сумчатые куницы	Beutelmarder	Dasyures	Dasyurinés
собственно тупайи	Buschschwanztupaias	Brush-tailed Tree Shrews	Tupaiinés
собственно человекообразные обезьяны	Menschenaffen	Great Apes	Pongidés
средиземноморская белозубка	Mittelmeer-Langschwanz-spitzmaus	Mediterranean Long-tailed Shrew	Musaraigne méditerranéenne
среднеавстралийский сумчатый тушканчик	Inneraustralische Springbeutel-maus	Central Jerboa Marsupial	
среднеафриканские ежи	Mittelafrikanische Igel	– African Hedgehogs	Hérissons d'Afrique centrale
средние лемуры	Mittelgroße Lemuren	Typical Lemurs	Lémurinés
средняя бурозубка	Maskenspitzmaus	Laxmann's Shrew	Musaraigne lapone
средняя сумчатая летяга	Mittlerer Gleithörnchenbeutler	Honey Glider	
степная кенгуровая крыса	Nacktbrustkänguruh	Desert Rat Kangaroo	Rat-Kangourou du désert
сумчатая белка	Hörnchenkletterbeutler	Leadbeater's Possum	Opossum de Leadbeater
сумчатые	Beuteltiere	Marsupials	Marsupiaux
сумчатые барсуки	Nasenbeutler	Bandicoots	Bandicoots
сумчатые волки	Beutelwölfe	Tasmanian Wolves	Loups marsupiaux
сумчатые зайцы	Kaninchen-Nasenbeutler	Rabbit Bandicoots	Bandicoots-Lapins
сумчатые звери	Beutelsäuger	Pouched Mammals	Métathériens
сумчатые землеройки	Schmalfußbeutelmäuse	Narrow-footed Marsupial Mice	Souris marsupiales
сумчатые кроты	Beutelmulle	Marsupial Moles	Taupes marsupiales
сумчатые крысы	Beutelratten	Opossums	Opossums d'Amérique
сумчатые куницы	Fleckenbeutelmarder	Native Cats	
сумчатые летяги	Gleithörnchenbeutler	Honey Gliders	Petaurus
сумчатые медведи	Koalaverwandte	Koala-like Marsupials	Phascolarctinés
сумчатые медоеды	Rüsselbeutler	Honey Phalangers	Souris à miel
сумчатые сони	Schlafbeutler	Dormouse Possums	Phalangers Loirs
сумчатые тушканчики	Springbeutelmäuse	Jerboa Marsupials	Gerboises-Souris marsupiales
сумчатый волк	Beutelwolf	Tasmanian Wolf	Loup marsupial
сумчатый дьявол	Beutelteufel	– Devil	Sarcophile satanique
сумчатый крот	Großer Beutelmull	Greater Marsupial Mole	Grande Taupe marsupiale
сундский долгопят	Sundakoboldmaki	Western Tarsier	Tarsier de Horsfield
тайванский резус	Formosamakak	Formosa Macaque	Macaque de Formosa
тамарин Леннберга	Lönnbergtamarin	Lönnberg's Tamarin	Tamarin de Lönnberg
тамарины	Tamarins	Tamarins	Tamarins
тасманийская ехидна	Tasmanien-Kurzschnabeligel	Tasmanian Echidna	Échidné de Tasmanie
тасманийский кистехвостый кенгуру	Tasmanienbürstenkänguruh	– Rat Kangaroo	Rat-Kangourou de Tasmanie
темная хоботковая собачка	Dunkles Rüsselhündchen	Stuhlmann's Elephant Shrew	Macroscélide de Stuhlmann
темнопалая белозубка	Dunkelfüßige Waldspitzmaus	Dark-footed Forest Shrew	
тенреки	Tanreks	Tenrecs	Tanrecs
тибетская водяная белозубка	Gebirgsbachspitzmaus	Szechuan Water Shrew	Nectogale élégant
тихоокеанский крот	Pazifischer Maulwurf	Coast Mole	Taupe de côte
толстотелы	Stummelaffen	Colobus Monkeys	Colobes
толстохвостая мышевидка	Fettschwänzige Breitfuß-beutelmaus	Fat-tailed Marsupial Mouse	
толстохвостая сумчатая землеройка	Dickschwänzige Schmalfuß-beutelmaus	– Sminthopsis	
толстохвостый галаго	Riesengalago	Thick-tailed Bushbaby	Galago à queue trouffue
толстохвостый опоссум	Dickschwanzbeutelratte	Little Water Opossum	
толстый лори	Plumplori	Slow Loris	Loris paresseux
тонкий лори	Schlanklori	Slender Loris	– grêle
тонкинский тонкотел	Tonkinlangur	François' Monkey	Semnopithèque de François
тонкотел Фейера	Phayres Langur	Phayre's Leaf Monkey	– de Phayre
тонкотелые маки	Wieselmakis	Weasel Lemurs	Lépilémur
тонкотелые обезьяны	Schlankaffen	Leaf Monkeys	Colobidés
тонкохвостая сумчатая соня	Dünnschwanz-Schlafbeutler	South-western Pigmy Phalanger	
тупайевые	Spitzhörnchen	Tree Shrews	Tupaiidés
тупайи	Spitzhörnchen	Tree Shrews	Tupaiidés
тупайя тана	Tana	Large Tree Shrew	Tana
тупайя Эверетта	Everetts Spitzhörnchen	Philippine Tree Shrew	

RUSSIAN NAME	GERMAN NAME	ENGLISH NAME	FRENCH NAME
тупайя Эллиота	Elliots Tupaia	Madras Tree-shrew	Toupaïe d'Elliot
уакори	Kurzschwanzaffen	Uakaris	Ouakaris
угандская броненоска	Ugandapanzerspitzmaus	Uganda Armoured Shrew	
уздечковый когтехвостый кенгуру	Kurznagelkänguruh	Bridled Nail-tail Wallaby	
узконосые обезьяны	Schmalnasen	Old World Simian Primates	Catarhiniens
усатый тамарин	Schnurrbarttamarin	Moustached Tamarin	Tamarin à Moustaches
утконос	Schnabeltier	Duck-billed Platypus	Ornithorhynque
ушастые ежи	Ohrenigel	Eared Hedgehog	
ушастые кроты	Ohrenspitzmaus-Maulwürfe	Shrew Moles	Musaraignes-taupes
ушастый еж	Langohrigel	Long-eared Hedgehog	
ушастый крот	Spitzmausmaulwurf	Shrew Mole	Musaraigne-taupe
филиппинская малая гимнура	Philippinen-Rattenigel	Mindanao Gymnure	
филиппинские тупайи	Philippinentupaias	Philippine Tree Shrew	
филиппинский долгопят	Philippinenkoboldmaki	Philippine Tarsier	Tarsier des Philippines
филиппинский макак	Philippinenmakak		Macaque des Philippines
хероп	Schweinsfuß	Pig-footed Bandicoot	
хищные сумчатые	Raubbeutler	Flesh-eating Marsupials	Dasyuridés
хоботковая собачка Петерса	Rotschulterrüsselhündchen	Peters' Elephant Shrew	Macroscélide de Peters
хоботковые крысы	Rüsselratten		
хоботковые собачки	Rüsselhündchen		
хохлатый монго	Kronenmaki	Crowned Lemur	Lémur couronné
хохлатый павиан	Schopfmakak	Celebes crested Macaque	Cynopithèque nègre
хохлатый тонкотел	Schopflangur	Capped Langur	
цебусовые обезьяны	Kapuzineraffen	Capuchin Monkeys	Cébinés
цейлонский макак	Ceylon-Hutaffe	Toque Monkey	Macaque couronné
целебесские макаки	Celebesmakaken	Moor Macaques	Macaques de Célèbes
ценолестовые сумчатые	Opossummäuse	Rat Opossums	Caenolestidés
чакма	Bärenpavian	Chacma Baboon	Chacma
человекообразные обезьяны	Menschenaffen	Apes	Pongidés
черная коата	Schwarzer Klammeraffe	Black Spider Monkey	Singe-Araignée noir
черноголовая саймири	Schwarzköpfiges Totenköpfchen	Black-headed Squirrel Monkey	Sapajou à tête noire
черноголовый маки	Schwarzkopfmaki	Brown Lemur	Lémur brun
черноголовый прыгун	Schwarzköpfiger Springaffe	Masked Titi	Callicèbe à masque
черноголовый тенрек	Schwarzkopftanrek	Streaked Tenrec	
черноголовый уакори	Schwarzkopfuakari	Black-headed Uakari	Ouakari à tête noire
черноземеная мартышка	Schwarzgrüne Meerkatze	Swamp Guenon	Cercopithèque noir et vert
черноспинный тамарин	Schwarzrückentamarin	Black-and-red Tamarin	Tamarin rouge et noir
чернохвостая сумчатая куница	Schwarzschwanzbeutelmarder	Western Dasyure	
чернохвостый кустовый кенгуру	Sumpfwallaby	Black-tailed Wallaby	Wallaby bicolore
черный кускус	Bärenkuskus	Bear Phalanger	Couscous de Célèbes
черный маки	Mohrenmaki	Black Lemur	Lémur macaco
черный ревун	Schwarzer Brüllaffe	– Howler Monkey	Hurleur noir
черный уакори	– Uakari	– Uakari	Ouakari de Roosevelt
черный целебесский макак	Mohrenmakak	Moor Macaque	Macaque de Célèbes
чертов саки	Satansaffe	Black Saki	Saki noir
четырехпалая хоботковая крыса	Vierzehenrüsselratte	Four-toed Elephant Shrew	
чшеуйчатохвостый кузу	Schuppenschwanzkusu	Scaly-tailed Phalanger	
чилийская первокрыса	Chile-Opossummaus	Chilean Rat Opossum	
чубастый мангабей	Haubenmangabe	Agile Mangabey	Mangabey à ventre doré
чубастый тонкотел	Mützenlangur	Sunda Island Leaf Monkey	
чубатая мартышка	Kronenmeerkatze	Crowned Guenon	Cercopithèque pogonias
шерстистая обезьяна Гумбольдта	Wollaffe	Humboldt's Woolly Monkey	Lagotriche de Humboldt
шерстистые обезьяны	Wollaffen	Woolly Monkeys	Singes laineux
шерстистые обезьяны и коаты	Klammerschwanzaffen	Spider and Woolly Monkeys	Atélinés
шерстистые сумчатые крысы	Wollbeutelratten	Woolly Opossums	Opossums laineux
широколицый потору	Breitkopfkänguruh	Broad-faced Rat Kangaroo	
широколобый вомбат	Haarnasenwombat	Hairy-nosed Wombat	Wombat à narines poilues
широконосые обезьяны	Breitnasenaffen	New World Monkeys	Singes du Nouveau Monde
широконосый полумаки	Breitschnauzenhalbmaki	Broad-nosed Gentle Lemur	Hapalémur à nez large
щелезубы	Schlitzrüßler	Solenodons	Solénodontidés
дипова игрунка	Lisztäffchen	Cotton-head Tamarin	Pinché
яванский макак	Javaneraffe	Crab-eating Macaque	Macaque de Buffon
японский землеройковый крот	Japanischer Spitzmull	Japanese Shrew Mole	Taupe des montagnes du Japon
японский макак	Rotgesichtsmakak	– Macaque	Macaque Japonais

Conversion Tables of Metric to U.S. and British Systems

U.S. Customary to Metric		*Metric to U.S. Customary*	
Length			
To convert	*Multiply by*	*To convert*	*Multiply by*
in. to mm.	25.4	mm. to in.	0.039
in. to cm.	2.54	cm. to in.	0.394
ft. to m.	0.305	m. to ft.	3.281
yd. to m.	0.914	m. to yd.	1.094
mi. to km.	1.609	km. to mi.	0.621
Area			
sq. in. to sq. cm.	6.452	sq. cm. to sq. in.	0.155
sq. ft. to sq. mi.	0.093	sq. m. to sq. ft.	10.764
sq. yd. to sq. m.	0.836	sq. m. to sq. yd.	1.196
sq. mi. to ha.	258.999	ha. to sq. mi.	0.004
Volume			
cu. in. to cc.	16.387	cc. to cu. in.	0.061
cu. ft. to cu. m.	0.028	cu. m. to cu. ft.	35.315
cu. yd. to cu. m.	0.765	cu. m. to cu. yd.	1.308
Capacity (liquid)			
fl. oz. to liter	0.03	liter to fl. oz.	33.815
qt. to liter	0.946	liter to qt.	1.057
gal. to liter	3.785	liter to gal.	0.264
Mass (weight)			
oz. avdp. to g.	28.35	g. to oz. avdp.	0.035
lb. avdp. to kg.	0.454	kg. to lb. avdp.	2.205
ton to t.	0.907	t. to ton	1.102
l. t. to t.	1.016	t. to l. t.	0.984

Abbreviations

U.S. Customary

avdp.—avoirdupois
ft.—foot, feet
gal.—gallon(s)
in.—inch(es)
lb.—pound(s)
l. t.—long ton(s)
mi.—mile(s)
oz.—ounce(s)
qt.—quart(s)
sq.—square
yd.—yard(s)

Metric

cc.—cubic centimeter(s)
cm.—centimeter(s)
cu.—cubic
g.—gram(s)
ha.—hectare(s)
kg.—kilogram(s)
m.—meter(s)
mm.—millimeter(s)
t.—metric ton(s)

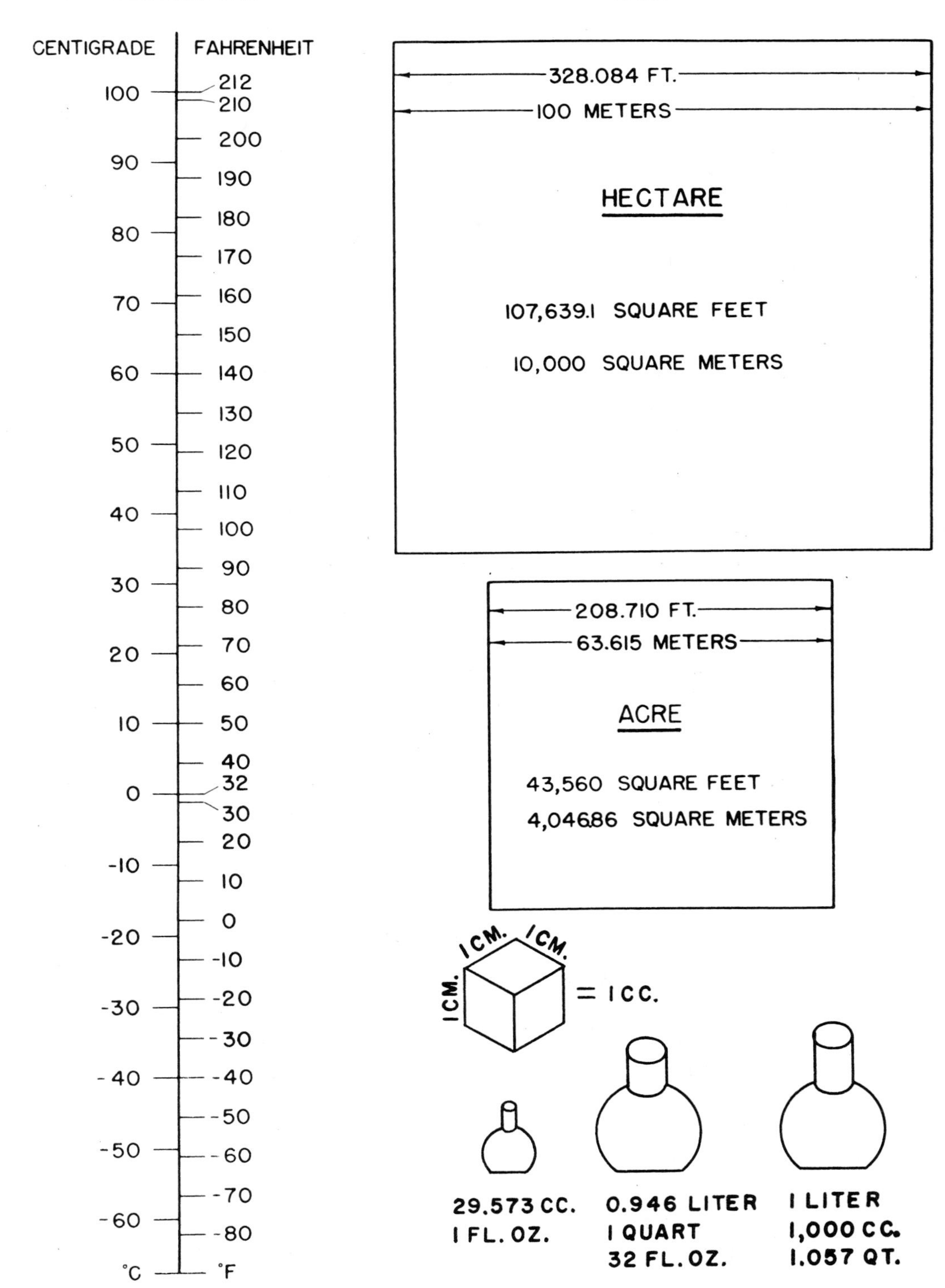
TEMPERATURE
CENTIGRADE
FAHRENHEIT
100
212
210
200
90
190
180
80
170
70
160
150
60
140
130
50
120
110
40
100
90
30
80
70
20
60
10
50
40
32
0
30
20
-10
10
0
-20
-10
-20
-30
-30
-40
-40
-50
-50
-60
-70
-60
-80
°C
°F
AREA
328.084 FT.
100 METERS
HECTARE
107,639.1 SQUARE FEET
10,000 SQUARE METERS
208.710 FT.
63.615 METERS
ACRE
43,560 SQUARE FEET
4,046.86 SQUARE METERS
1 CM.
1 CM.
1 CM.
= 1 CC.
29.573 CC.
1 FL. OZ.
0.946 LITER
1 QUART
32 FL. OZ.
1 LITER
1,000 CC.
1.057 QT.

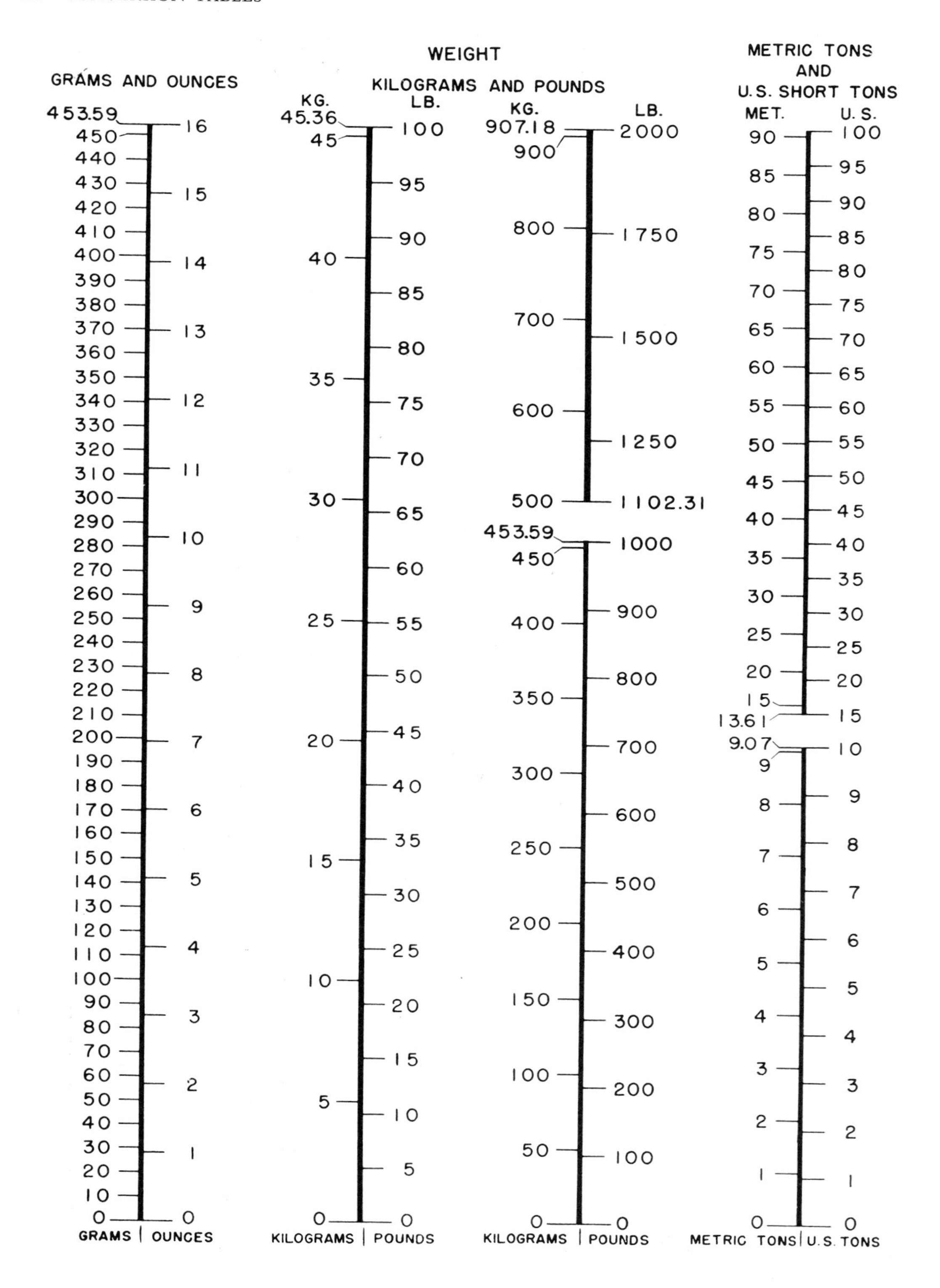
WEIGHT
GRAMS AND OUNCES
453.59
16
GRAMS | OUNCES
KILOGRAMS AND POUNDS
KG.
LB.
45.36
100
KILOGRAMS | POUNDS
KG.
LB.
907.18
2000
500
1102.31
453.59
1000
KILOGRAMS | POUNDS
METRIC TONS
AND
U.S. SHORT TONS
MET.
U.S.
90
100
13.61
15
9.07
10
METRIC TONS | U.S. TONS

LENGTH: MILLIMETERS AND INCHES

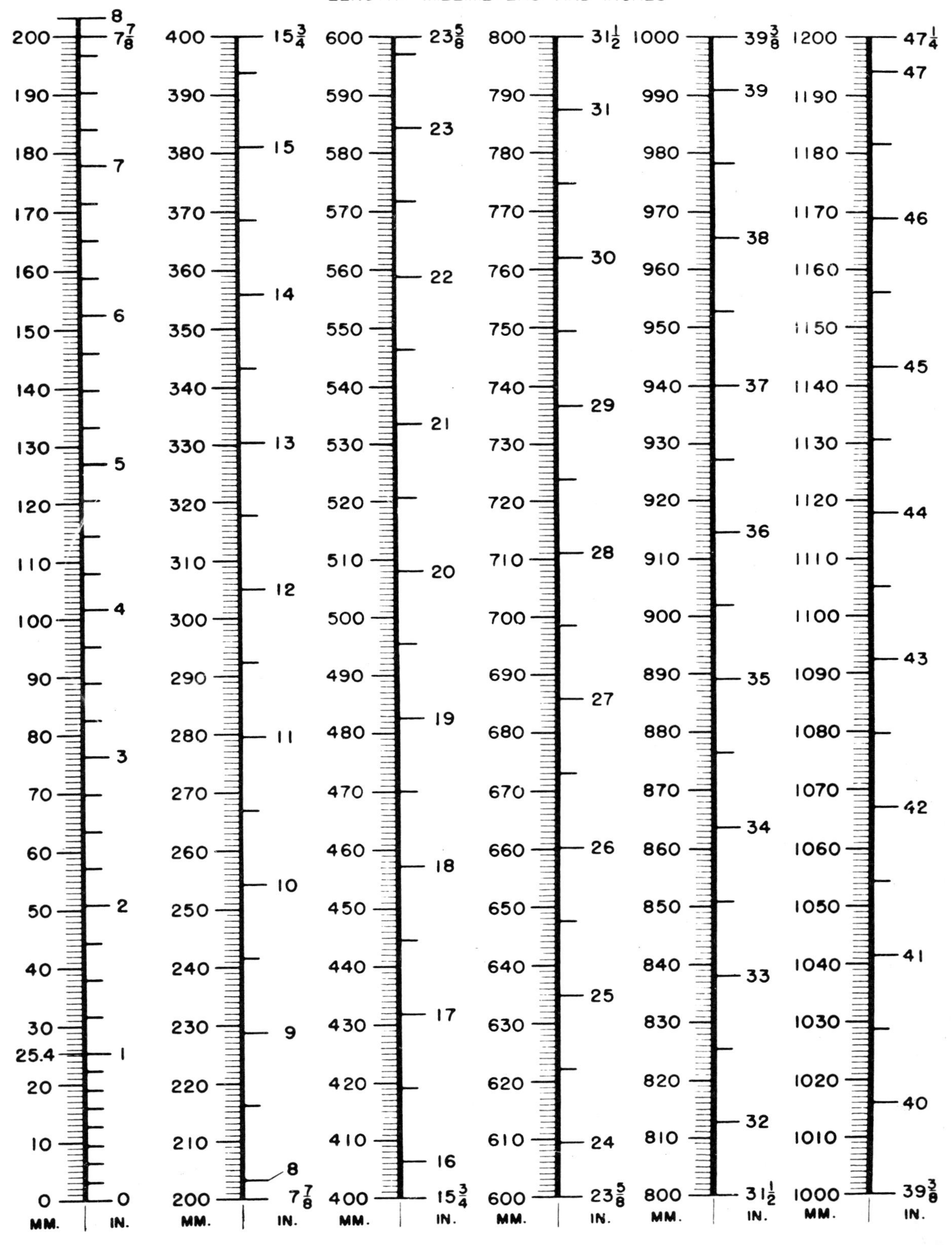

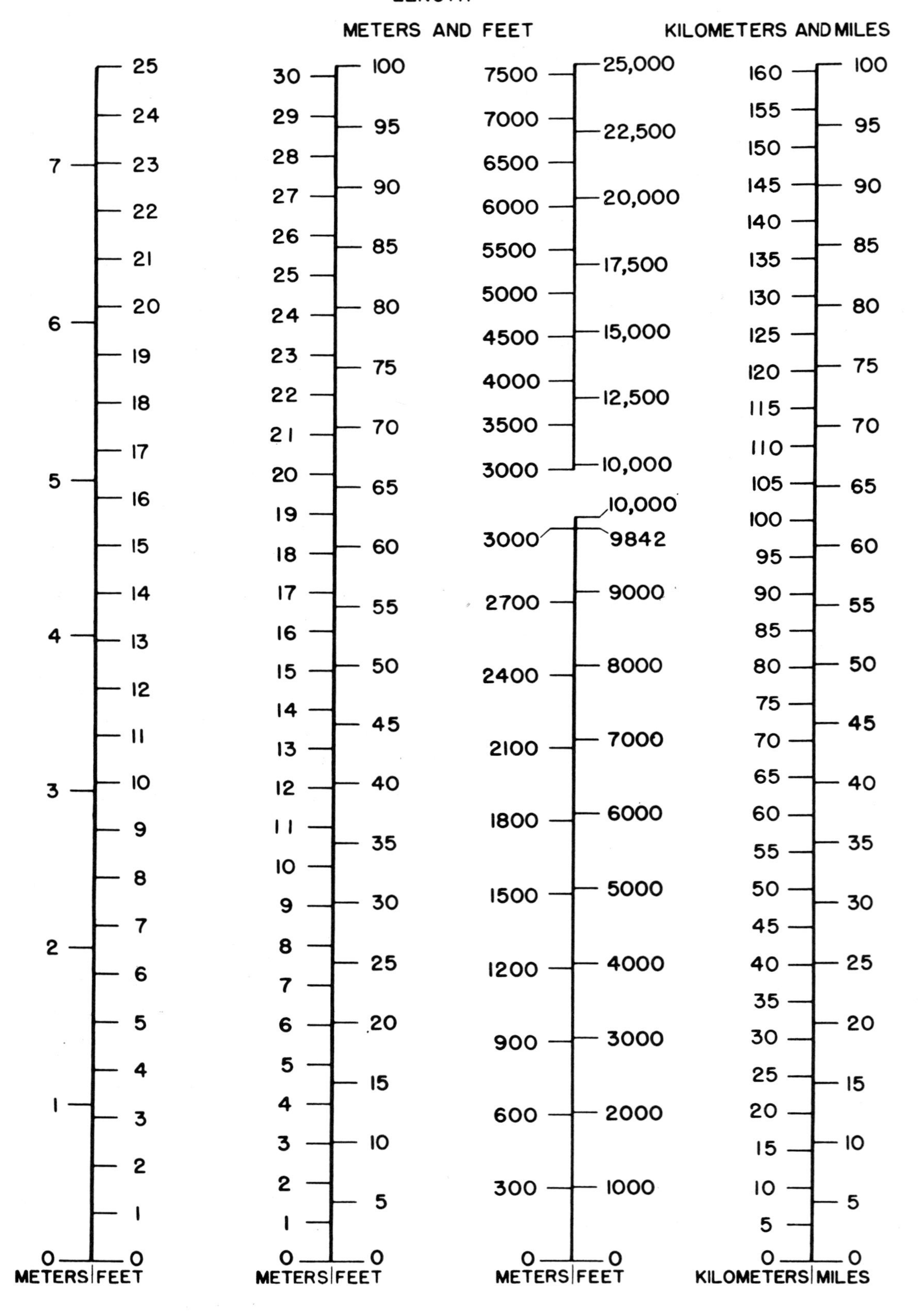
LENGTH
METERS AND FEET
KILOMETERS AND MILES
0 1 2 3 4 5 6 7
METERS
0 1 2 3 4 5 6 7 8 9 10 11 12 13 14 15 16 17 18 19 20 21 22 23 24 25
FEET
0 1 2 3 4 5 6 7 8 9 10 11 12 13 14 15 16 17 18 19 20 21 22 23 24 25 26 27 28 29 30
METERS
0 5 10 15 20 25 30 35 40 45 50 55 60 65 70 75 80 85 90 95 100
FEET
3000 3500 4000 4500 5000 5500 6000 6500 7000 7500
10,000 12,500 15,000 17,500 20,000 22,500 25,000
0 300 600 900 1200 1500 1800 2100 2400 2700 3000
METERS
0 1000 2000 3000 4000 5000 6000 7000 8000 9000 9842 10,000
FEET
0 5 10 15 20 25 30 35 40 45 50 55 60 65 70 75 80 85 90 95 100 105 110 115 120 125 130 135 140 145 150 155 160
KILOMETERS
0 5 10 15 20 25 30 35 40 45 50 55 60 65 70 75 80 85 90 95 100
MILES

Supplementary Readings

These references of books and articles published in scientific journals deal with animals and topics that are covered in this volume. Some of these were the original sources on which the content of this book is based. These titles are intended as an aid to readers who are interested in additional information and more detailed coverage of the subjects contained in this book.*

MAMMALS

Allen, Glover. 1942. *Extinct and Vanishing Mammals of the Western Hemisphere.* American Committee for International Wildlife Protection.

Arey, L. B. 1965. *Developmental Anatomy,* 7th ed. W. B. Saunders Company, Philadelphia.

Blair, W. F. and others. 1957. *Vertebrates of the United States.* McGraw-Hill Book Co., Inc., New York.

Bourlière. F. 1954. *The Natural History of Mammals.* Alfred A. Knopf, New York.

Burt, W. H. and R. P. Grossenheider. 1952. *A Field Guide to the Mammals.* Houghton Mifflin Company, Boston.

Cahalane, Victor H. 1947. *Mammals of North America.* The Macmillan Company, New York.

Carter, T. D., J. E. Hill, and G. H. H. Tate. 1945. *Mammals of the Pacific World.* The Macmillan Company, New York.

Cockrum, E. L. 1962. *Introduction to Mammalogy.* Ronald Press Co., New York.

— . 1962. *Laboratory and Field Manual for Introduction to Mammalogy.* Ronald Press Co., New York.

Colbert, E. H. 1955. *Evolution of the Vertebrates.* Wiley, New York.

Davis, D. E. and F. B. Golley. 1963. *Principles in Mammalogy.* Van Nostrand Reinhold Company, New York.

Drimmer, F., ed. 1954. *The Animal Kingdom,* Vols. I & II. Greystone Press.

Flower, W. H. 1885. *An Introduction to the Osteology of the Mammalia,* 3rd ed. The Macmillan Company, New York. (Republished by Dover Press, New York, 1962.)

Flower, W. H. and R. Lydekker. 1891. *An Introduction to the Study of Mammals, Living and Extinct.* Adam and Charles Black, London.

Gregory, W. K. 1951. *Evolution Emerging.* 2 vols. The Macmillan Company, New York.

Hafez, E. S. E. 1969. *The Behavior of Domestic Animals,* 2nd ed. Williams and Wilkins, Baltimore.

Hall, E. Raymond and K. R. Kelson. 1959. *The Mammals of North America.* 2 vols. Ronald Press Co., New York.

Halstead, L. B. 1968. *The Pattern of Vertebrate Evolution.* W. H. Freeman and Co., Inc., San Francisco.

Hamilton, William John. 1939. *American Mammals.* McGraw-Hill Book Co., Inc., New York.

— . 1943. *The Mammals of Eastern United States.* Comstock Publishing Company, Ithaca, New York.

Harper, Francis. 1945. *Extinct and Vanishing Mammals of the Old World.* American Committee for International Wildlife Protection.

Herrick, C. J. 1948. *Brain of Rats and Men.* University of Chicago Press, Chicago.

Hvass, H. 1961. *Mammals of the World.* Methuen.

Kon, Ş. K. and A. T. Cowie, 1961. *Milk: The Mammary Gland and its Secretions.* Academic Press, New York.

Matthews, L. Harrison. 1952. *British Mammals.* Collins.

Murie, Olaus J. 1954. *A Field Guide to Animal Tracks.* Houghton Mifflin Co., Boston.

National Geographic Book Service. 1960. *Wild Animals of North America.* The National Geographic Society.

Palmer, E. L. 1957. *Fieldbook of Mammals.* E. P. Dutton and Co., New York.

Palmer, Ralph S. 1954. *The Mammal Guide.* Doubleday, New York.

Peyer, B. 1968. *Comparative Odontology.* University of Chicago Press, Chicago.

Romer, A. S. 1970. *The Vertebrate Body,* 4th ed. W. B. Saunders Company, Philadelphia.

Sanderson, Ivan T. 1955. *Living Mammals of the World.* Doubleday, New York.

Scheele, W. E. 1955. *The First Mammals.* World Publishing, New York.

Scott, William B. 1962. *A History of Land Mammals in the Western Hemisphere.* Hafner.

Simpson, G. G. 1966. *The Meaning of Evolution.* Yale University Press, New Haven.

Smythe, R. H. 1961. *Animal Vision.* Thomas.

Spinage, C. A. 1963. *Animals of East Africa.* Houghton Mifflin Co., Boston.

Tate, George H. H. 1947. *Mammals of Eastern Asia.* The Macmillan Company, New York.

Thompson, D. W. 1942. *On Growth and Form,* 2nd ed. Cambridge University Press, Cambridge, England.

Tinbergen, N. 1965. *Animal Behavior.* Time-Life Books, New York.

Troughton, Ellis. 1947. *Furred Animals of Australia.* Scribner, New York.

Walker, E. P. 1968. *Mammals of the World,* 2nd ed. 3 vols. Johns Hopkins Press, Baltimore.

Young, J. Z. 1957. *The Life of Mammals.* Oxford Univeristy Press, London.

Allen, G. M. 1939. A Checklist of African Mammals. *Bulletin of the Museum of Comparative Zoology of Harvard* 83:1–763.

Barnett, C. H., R. J. Harrison, and J. D. W. Tomlinson. 1958. Variations in the Venous Systems of Mammals. *Biology Review* 33:442–487.

Collies, N. E. 1944. Aggressive Behavior among Vertebrate Animals. *Physiological Zoology* 17:83–123.

Crompton, A. W. 1963. The Evolution of the Mammalian Jaw. *Evolution* 17(4):431–439.

Elias, H. and S. Bortner. 1957. On the Phylogeny of Hair. *American Museum Novitiates* 1820:1–15.

Hardy, J. D. 1961. Physiology of Temperate Regulation. *Physiological Review* 41:521–606.

Mitchell, P. C. 1906. On the Intestinal Tract of Mammals. *Transactions of the Zoological Society of London* 17:437–536.

Simpson, G. G. 1945. The Principles of Classification and a Classification of Mammals. *Bulletin, American Museum of Natural History* 85:1–350.

Snider, R. S. 1958. The Cerebellum. *Scientific American* 199(2):84–90.

MARSUPIALS, MONOTREMES AND INSECTIVORES

Burrel, H. 1927. *The Platypus.* Sydney.

Crowcroft, Peter. 1957. *The Life of the Shrew.* Reinhardt, London.

Fleay, David. 1944. *We Breed the Platypus.* Robertson and Mullens.

Godfrey, Gillian and P. Crowcroft. 1960. *The Life of the Mole.* Museum Press, London.

Hartman, C. G. 1952. *Possums.* University of Texas Press, Austin.

Kohn, Bernice. 1964. *Marvelous Mammals: Monotremes and Marsupials.* Prentice-Hall, Englewood Cliffs, New Jersey.

*Supplementary Readings prepared by John B. Brown.

Clemens, W. A. 1968. Origin and Early Evolution of Marsupials. *Evolution* 22(1):1-18.

Evans, F. Gayner. 1942. The Osteology and Relationships of the Elephant Shrews. *Bulletin, American Museum of Natural History* 80:83-125.

PRIMATES: GENERAL

Altmann, S. A. 1962. The Social Behavior of Anthropoid Primates: An Analysis of Some Recent Concepts, pp. 277-285. In E. L. Buss, ed., *The Roots of Behavior.* Harper and Row, New York.
— ., ed. 1967. *Social Communication Among Primates.* University of Chicago Press, Chicago.

Benchley, B. 1942. *My Friends, the Apes.* Little, Brown and Co., Boston.
Boulenger, E. G. *Apes and Monkeys.* Robert M. McBride, London.
Bourne, Geoffrey Howard. 1971. *The Ape People.* Putnam, New York.
Buettner-Janusch, John, ed. 1963, 1964. *Evolutionary and Genetic Biology of the Primates.* 2 vols. Academic Press, New York.
— . 1966. *Origins of Man: Physical Anthropology.* Wiley, New York.

Carpenter, C. R. 1964. *Naturalistic Behavior of Non-human Primates.* Pennsylvania State University Press, University Park.
Chance, M. R. A. 1961. The Nature and Special Features of the Instinctive Social Bond of Primates, pp. 17-33. In S. L. Washburn, ed., *The Social Life of Early Man.* Viking Fund Publication No. 31, Aldine Publishing Co., Chicago.

De Vore, I., ed. 1965. *Primate Behavior: Field Studies of Monkeys and Apes.* Holt, Rinehart and Winston, New York.
— and S. Eimerl. 1965. *The Primates.* Time, Inc., New York.

Elliot, D. G. 1913. *A Review of the Primates.* 3 vols. Monographs of the American Museum of Natural History.

Forbes, Henry Ogg. 1896-97. *A Handbook to the Primates.* 2 vols. E. Lloyd, London.

Gavan, J. A., ed. 1955. *The Nonhuman Primates and Human Evolution.* Wayne State University Press, Detroit.

Hall, K. R. L. 1963. Some Problems in the Analysis and Comparison of Monkey and Ape Behavior, pp. 273-300. In S. L. Washburn, ed., *Classification and Human Evolution.* Viking Fund Publication No. 37, Wenner-Gren Foundation, New York.
— . 1964. Aggression in Monkey and Ape Societies, pp. 51-64. In J. D. Carthy and F. J. Ebling, eds., *The Natural History of Aggression.* Academic Press, New York.
Harlow, H. F. 1959. Basic Social Capacity of Primates, pp. 40-52. In J. N. Spuhler, ed., *The Evolution of Man's Capacity for Culture.* Wayne State University Press, Detroit.
Hartmann, R. 1885. *Anthropoid Apes.* Kegan, Paul Trench and Co., London.
Hayes, C. 1951. *The Ape in Our House.* Harper, New York.
Hill, W. C. O. 1953, 1955, 1957, 1960, 1962, 1966. *Primates: Comparative Anatomy and Taxonomy,* Vols. I-VI. Interscience Publishers, New York.
Hofner, H., A. H. Schultz and D. Stark, eds. 1956. *Primatologia,* Vol I, S. Karger, Basel.
Hooton, Earnest Albert. 1937. *Apes, Men and Morons.* G. P. Putnam's Sons, New York.
— . 1942. *Man's Poor Relations.* Doubleday, New York.
— . 1946. *Up from the Ape.* The Macmillan Company, New York.
Hubrecht, Ambrosius Arnold Willem. 1897. *The Descent of the Primates.* Lectures delivered on the occasion of the sesquicentennial celebration of Princeton University. C. Scribner's Sons, New York.

James, W. W. 1960. *The Jaws and Teeth of Primates.* Pitman Medical Publishing Co., London.
Jay, P. C. 1968. *Primates: Studies in Adaptation and Variability.* Holt, Rinehart and Winston, New York.

Kawamura, S. and J. Itani, eds. *Monkeys and Apes.* Chuohoron-Sha, Tokyo.
Klüver, H. 1933. *Behavior Mechanisms in Monkeys.* University of Chicago Press, Chicago.
Köhler, Wolfgang. 1959. *The Mentality of Apes.* Random House, New York.

Lancaster, J. B. and R. B. Lee. 1965. The Annual Reproductive Cycle in Monkeys and Apes, pp. 486-513. In De Vore, ed., *Primate Behavior.* Holt, Rinehart and Winston, New York.
Lanyon, W. E. L. and W. N. Tavolga, eds. 1960. *Animal Sounds and Communication.* American Institute of Biological Sciences, Washington D.C.
Le Gros Clark, Sir Wilfrid Edward. 1934. *Early Forerunners of Man: A Morphological Study of the Evolutionary Origin of the Primates.* W. Wood and Co., Baltimore.
— . 1960. *The Antecedents of Man: An Introduction to the Evolution of the Primates.* Quadrangle Books, Chicago.
— . 1966. *History of the Primates,* 5th ed. University of Chicago Press, Chicago.
Lehrman, R. L. 1961. *The Long Road to Man.* Basic Books, Inc., New York.

Morris, D. 1968. *Primate Ethology.* Aldine Publishing Co., Chicago.
Morris, Ramona and Desmond. 1966. *Men and Apes.* Hutchinson and Co., London.

Napier, John Russell. 1970. *The Roots of Mankind.* Smithsonian Institution Press, Washington.
— and P.H. 1968. *Handbook of Living Primates,* Academic Press, Inc., New York.
— and — . 1970. *Conference, Systematics of the Old World Monkeys, Wartenstein Castle, 1969.* Academic Press, London.
Nissen, H. W. 1951. Social Behavior in Primates. In E. P. Stone, ed., *Comparative Psychology,* 3rd ed. Prentice-Hall, Englewood Cliffs, New Jersey.
Noback, Charles R. and William Montagna. 1970. *The Primate Brain.* Appleton-Century-Crofts, New York.
Noback, C. R. and N. Moskowitz. 1963. The Primate Nervous System: Functional and Structural Aspects in Phylogeny, pp. 131-177. In Buettner-Janusch, ed., *Evolutionary and Genetic Biology of Primates.* Academic Press, New York.

Reynolds, V. 1967. *The Apes.* E. P. Dutton, New York.
Rohles, F. H., ed. 1969. *Circadian Rhythms in Non-human Primates.* S. Karger, Basel, New York.
Rosenblum, Leonard A. 1970. *Primate Behavior: Developments in Field and Laboratory Research,* Vol. I. Academic Press, New York.
Ruch, T. C. 1941. *Bibliographica Primatologica: A Classified Bibliography of Primates other than Man.* Thomas, Springfield, Illinois.

Sanderson, I. T. 1957. *The Monkey Kingdom.* Hanover House, Garden City, New York.
Schrier, A. M., ed. 1965, 1966. *Behavior of Non-human Primates,* Vols. I & II. Academic Press, Inc., New York.
Southwick, C. H., ed. 1963. *Primate Social Behavior.* Van Nostrand Reinhold Company, New York.

Washburn, Sherwood Larned. 1963. *Classification and Human Evolution.* Aldine Publishing Co., Chicago.
— . 1966. Conflict in Primate Society, pp. 3-15. In A. de Reuck and J. Knight, eds., *Conflict in Society.* Little, Brown and Co., Boston.
Williams, Leonard. 1965. *Samba and the Monkey Mind.* Bodley Head, London.
— . 1967. *Man and Monkey.* Lippincott, Philadelphia.

Yerkes, R. M. 1916. *The Mental Life of Monkeys and Apes: A Study of Ideational Behavior.* H. Holt and Company, Cambridge, Boston, New York.
— . 1925. *Almost Human.* The Century Company, New York and London.

Zoological Society of London. 1962. The Primates. *Symposia,* No. 10. Academic Press, Inc., New York.
Zuckerman, Solly. 1932. *The Social Life of Monkeys and Apes.* Harcourt, Brace and World, New York.
— . 1933. *Functional Affinities of Man, Monkeys and Apes: A Study of the Bearings of Physiology and Behavior on the Taxonomy and Phylogeny of Lemurs, Monkeys, Apes and Man.* K. Paul Trench, Turbner, London.

Andrew, R. J. 1963. Evolution of Facial Expressions. *Science* 142(3595).

— . 1963. Evolution of Vocalization in Monkeys and Apes. *Symposia of the Zoological Society of London* 10:89-101.
— . 1963. The Origin and Evolution of the Calls and Facial Expressions of the Primates. *Behavior* 20:1-109.
Ashton, E. H. and C. E. Osnard. 1964. Locomotor Patterns in Primates. *Proceedings of the Zoological Society of London* 142:1-28.
Avis, Virginia. 1962. Brachiation: The Crucial Issue for Man's Ancestry. *Southwestern Journal of Anthropology* 18:119-148.

Bishop, A. 1962. Hand Control in Lower Primates. *Annals of the New York Academy of Sciences* 102:316-337.
Bolwig, N. 1964. Facial Expressions in Primates with remarks on a Parallel Development in Certain Carnivores. *Behavior* 22:167-193.
Buettner-Janusch, J. and R. L. Hill. 1965. Molecules and Monkeys. *Science* 147(3660):836-842.
— and others. 1962. The Relatives of Man. *Annals of the New York Academy of Sciences* (Vol. 102, Article 2) December 1962.

Carpenter, C. R. 1942. Societies of Monkeys and Apes. *Biology Symposium* 8:177-204.
Chance, M. R. A. 1963. The Social Bond of the Primates. *Primates* 14(4):1-21.
— and A. P. Mead. 1956. Social Behavior and Primate Evolution. *Symposia of the Society of Experimental Biology* 7:395-439.
Crook, J. H. 1966. The Evolution of Primate Societies. *Nature* (London) 210:1200-1203.

Erichson, G. E. 1963. Brachiation in New World Monkeys and in Anthropoid Apes. *Symposia of the Zoological Society of London* 10:135-164.

Hall, K. R. L. 1965. Social Organization of the Old World Monkeys and Apes. *Symposia of the Zoological Society of London* 14:265-289.
Hill, J. P. 1932. The Developmental History of the Primates. *Philosophical Transactions of the Royal Society of Britain* 221:45-178.

Imanischi, K. 1960. Social Organization of Subhuman Primates in their Natural Habitat. *Current Anthropology* 1:393-407.
Kortlandt, A. and M. Kooij. 1963. Protohominid Behavior in Primates. *Symposia of the Zoological Society of London* 10:61-88.

Napier, J. R. 1960. Studies of the Hands of Living Primates. *Proceedings of the Zoological Society of London* 134:647-657.
— . 1961. Prehensibility and Opposability in the Hands of Primates. *Symposia of the Zoological Society of London* 5:115-132.
— . 1962. The Evolution of the Hand. *Scientific American* 207:56-62.
— . 1963. Brachiation and Brachiators. *Symposia of the Zoological Society of London* 10:183-194.
— and N. A. Barnicot, eds. 1963. The Primates. *Symposia of the Zoological Society of London* (no. 10).

Oxnard, C. E. 1963. Locomotor Adaptations in the Primate Forelimb. *Symposia of the Zoological Society of London* 10:165-182.

Sahlins, M. D. 1959. The Social Life of Monkeys, Apes and Primitive Man. *Human Biology* 31:54-73.
Schultz, A. H. 1936. Characters Common to Higher Primates and Characteristics Specific for Man. *Quarterly Review of Biology* 11:259-283, 425-455.
— . 1937. Proportions of Long Bones in Man and Apes. *Human Biology* 9:281-328.
Simons, E. L. 1964. The Early Relatives of Man. *Scientific American* 211(1):50-62.
Southwick, C. H. 1962. Patterns of Intergroup Social Behavior in Primates, with Special Reference to Rhesus and Howling Monkeys. *Annals of the New York Academy of Sciences* 102:436-454.
Szalsy, F. S. 1968. The Beginnings of Primates. *Evolution* 22(1):19-36.

Van Hoof, J. A. R. A. M. 1962. Facial Expressions in Higher Primates. *Symposia of the Zoological Society of London* 8:97-125.
Van Valen, L. and R. E. Sloan. 1965. The Earliest Primates. *Science* 150:743-45.

Washburn, S. L., P. Jay and J. B. Lancaster. 1965. Field Studies of Old World Monkeys and Apes. *Science* 150:1541-1547.
Wiener, A. S. and J. Moor-Jankowski. Blood Groups in Anthropoid Apes and Baboons. *Science* 142:(3588)67-69.

Woollard, H. H. 1927. The Retina of Primates. *Proceedings of the Zoological Society of London* 1927:1-17.

Zuckerman, S. 1930. The Menstrual Cycle of the Primates. *Proceedings of the Zoological Society of London* 1930:691-754.

PRIMATES: STUDIES OF SPECIFIC GROUPS

Altman, Stuart A. and Jeanne Altmann. 1971. *Baboon Ecology.* University of Chicago Press, Chicago.

Ballantyne, R. M. *The Gorilla Hunters.* Collins Clear-type Press, London.
Bingham, Harold Clyde. 1932. *Gorillas in a Native Habitat.* Carnegie Institution of Washington, Washington.

Carpenter, Clarence Ray. 1934. *A Field Study of the Behavior and Social Relations of the Howling Monkeys.* Johns Hopkins Press, Baltimore.

Davis, Delbert Dwight. 1951. *The Baculum of the Gorilla.* Chicago Natural History Museum, Chicago.
De Vore, I. 1962. *The Social Behavior and Organization of Baboon Troops.* Unpublished Ph.D. Thesis, University of Chicago.
— and K. R. L. Hall. 1965. Baboon Ecology, pp. 20-52. In I. De Vore, ed., *Primate Behavior: Field Studies of Monkeys and Apes.* Holt, Rinehart and Winston, Inc., New York.
— and S. L. Washburn. 1963. Baboon Ecology and Human Evolution, pp. 335-367. In F. Clark Howell and F. Bourlière, eds., *African Ecology and Human Evolution.* Viking Fund Publications on Anthropology, No. 36. Aldine Publishing Co., Chicago.

Gregory, W. K., ed. 1950. *The Anatomy of the Gorilla.* Columbia University Press, New York.

Harrisson, B. 1963. *Orang-Utan.* Doubleday, New York.

Jolly, A. 1967. *Lemur Behavior.* University of Chicago Press, Chicago.

Kummer, Hans. 1968. *Social Organization of Hamadryas Baboons, a Field Study.* Karger, Basel, New York.

Malinow, M. R., ed. 1968. *Biology of the Howler Monkey (Alouatta caraya).* S. Karger, Basel.

Petter, J. 1965. The Lemurs of Madagascar, pp. 292-319. In I. De Vore, ed., *Primate Behavior: Field Studies of Monkeys and Apes.* Holt, Rinehart and Winston, Inc., New York.

Richards, P. W. 1965. *The Ecology and Social Behavior of the Ceylon Gray Langur.* Unpublished Ph.D. thesis, University of California.
Rosenblum, L. A. and R. W. Cooper, eds. 1968. *The Squirrel Monkey.* Academic Press, New York.

Schaller, G. B. 1963. *The Mountain Gorilla: Ecology and Behavior.* University of Chicago Press, Chicago.
— . 1964. *The Year of the Gorilla.* University of Chicago Press, Chicago.
Simonds, P. E. 1965. The Bonnet Macaque in South India, pp. 175-196. In I. De Vore, ed., *Primate Behavior: Field Studies of Monkeys and Apes.* Holt, Rinehart and Winston, Inc., New York.
Southwick, C. H., M. A. Beg and M. R. Siddiqi. 1965. Rhesus Monkeys in North India, pp. 111-159. In I. De Vore, ed., *Primate Behavior: Field Studies of Monkeys and Apes.* Holt, Rinehart and Winston, Inc., New York.

Vagtborg, H., ed., 1965. *The Baboon in Medical Research.* University of Texas Press, Austin.

Yerkes, Robert Mearns. 1929. *The Mind of the Gorilla.* Clark University, Worcester, Mass.
— . 1934. *The Great Apes: A Study of Anthropoid Life.* H. Milford, Oxford University Press, London.
— and A. W. Yerkes. 1929. *The Great Apes.* Yale University Press, New Haven.

Akeley, Carl E. 1922. Hunting Gorillas in Central Africa. *World's Work* 44:169-183, 307-318, 393-399, 525-533.
Altmann, S. A. 1959. Field Observations on a Howling Monkey Society. *Journal of Mammalogy* 40:317-330.

—. 1962. A Field Study of the Sociobiology of Rhesus Monkeys, *Macaca mulatta*. *Annals of the New York Academy of Sciences* 102(2):338–435.

Avon, F. 1963. Drills and Mandrills. *Animal Life* 13.

Bernstein, I. S. 1964. A Field Study of the Activities of Howler Monkeys. *Animal Behavior* 12:92–97.

—. 1965. Activity Patterns in a *Cebus* Monkey Group. *Folia Primatologica* 3:211–224.

Bode, N. C. 1952. Sakis, Elves of the Amazon. *Zoonooz* 25 No. 4:5–6.

Bolwig, N. 1959. A Study of the Behavior of the Chacma Baboon. *Behavior* 14:136–163.

Booth, A. H. 1954. A Note on the Colobus Monkeys of the Gold and Ivory Coast. *Annual Magazine of Natural History* 7(12):857–860.

—. 1955. Speciation in the Mona Monkeys. *Journal of Mammology* 36:434–449.

—. 1957. Observations on the Natural History of the Olive Colobus Monkey, *Procolobus verus* (Van Beneden). *Proceedings of the Zoological Society of London* 129:421–430.

Bourlière, F., A. Petter-Rousseaux and J. J. Petter. 1962. Regular Breeding in Captivity of the Lesser Mouse Lemur. *International Zoological Yearbook*, Vol. III (C. Jarvis and D. Morris, eds.). Zoological Society of London.

Buettner-Janusch, J. 1964. The Breeding of Galagos in Captivity and Some Notes on Behavior. *Folia Primatologica* 2:93–110.

Carpenter, C. R. 1935. Behavior of Red Spider Monkeys in Panama. *Journal of Mammalogy* 16:171–180.

—. 1940. A Field Study in Siam of the Behavior and Social Relations of the Gibbon *(Hylobates lar)*. *Comparative Psychology Monographs*, Vol. XVI, No. 5.

Chance, M. R. A. 1956. Social Structure of a Colony of *Macaca mulatta*. *British Journal of Animal Behavior* 4:1–13.

Chapman, F. M. 1937. My Monkey Neighbors on Barro Colorado. *Natural History of New York* 40:471–479.

Crook, J. H. 1966. Gelada Baboon Herd Structure and Movement: A Comparative Report. *Symposia of the Zoological Society of London* 18:237–258.

Dart, Raymond A. 1963. The Carnivorous Propensity of Baboons. *Symposia of the Zoological Society of London* 10:49–56.

Davenport, R. K. 1967. The Orang-Utan in Sabah. *Folia Primitologica* 5:247–263.

Ditmars, R. L. 1933. Development of the Silky Marmoset. *Bulletin of the New York Zoological Society* 36(6):175–176.

Emlen, J. T. 1962. The Display of the Gorilla. *Proceedings of the American Philosophical Society* 106:516–619.

— and G. B. Schaller. 1960. Distribution and Status of the Mountain Gorilla *(Gorilla gorilla beringei)*. *Zoologica* 45(1):41–53.

Fitzgerald, A. 1935. Rearing of Marmosets in Captivity. *Journal of Mammalogy* 16:181–188.

Fooden, J. 1967. Identification of the Stump-tailed Monkey, *Macaca speciosa*, I. Geoffroy, 1826. *Folia Primitologica* 5:153–164.

Fossey, D. 1970. Making Friends with Mountain Gorillas. *National Geographic Magazine* 137(1):48–67.

Furuya, Y. 1961–62. On the Ecological Survey of the Wild Crab-eating Monkeys in Malaya. *Primates* 3(1):75–76.

—. 1961–62. The Social Life of the Silvered Leaf Monkeys *(Trachypithecus cristatus)*. *Primates* 3(2):41–60.

Gee, E. P. 1961. The Distribution and Feeding Habits of the Golden Langur, *Presbytis geei gee* (Khajuria, 1956). *Journal of the Bombay Natural History Society* 58:1–12.

Groves, C. P. 1967. Ecology and Taxonomy of the Gorilla. *Nature*, London 890–893.

Haddow, A. J. 1952. Field and Laboratory Studies on an African Monkey *Cercopithecus ascanius schmidti* Matschi. *Proceedings of the Zoological Society of London* 122:297–394.

—. 1956. The Blue Monkey Group in Uganda. *Uganda Wild Life and Sport* 1:22–26.

Hall, K. R. L. 1963. Variations in the Ecology of the Chacma Baboon, *Papio ursinus*. *Symposia of the Zoological Society of London* 10:1–28.

—. 1965. Behavior and Ecology of the Wild Patas Monkeys, *Erythrocebus patas*, in Uganda. *Journal of the Zoological Society of London* 148:15–87.

— and J. S. Gartlan. 1965. Ecology and Behavior of the Vervet Monkey, *Ceropithecus aethiops* Lolui Island, Lake Victoria. *Proceedings of the Zoological Society of London* 145:37–56.

Harlow, H. F. 1959. Love in Infant Monkeys. *Scientific American* 200(6):68–74.

— and M. K. Harlow. 1963. Social Deprivation in Monkeys. *Scientific American* 207(5):136–146.

Harrison, J. L. 1955. Apes and Monkeys of Malaya (including the Slow Loris). *Malayan Museum Pamphlet*, 9.

Harrisson, Barbara. 1960. A Study of Orang-Utan Behavior in Semi-wild State. *Sarawak Museum Journal* 9:422–447.

Hartman, C. G. 1938. Some Observations on the Bonnet Macaque. *Journal of Mammalogy* 19:468–474.

Hershkovitz, P. 1963. A Systematic and Zoogeographic Account of the Monkeys of the Genus *Callicebus* (Cebidae) of the Amazons and Orinoco River Basins. *Mammalia* 27:1–79.

Hill, W. C. Osman. 1934. A Monograph on the Purple-faced Leaf Monkeys *(Pithecus vetulus)*. *Ceylon Journal of Science* (B) 19:23–88.

Hinde, R. G. and T. E. Rowell. 1962. Communication by Postures and Facial Expression in the Rhesus Monkey *(Macaca mulatta)*. *Proceedings of the Zoological Society of London* 138:1–21.

Hoogstraal, H. 1947. The Inside Story of the Tarsier. *Chicago Natural History Museum Bulletin* 18, Nos. 11 and 12.

Imanishi, K. 1957. Social Behavior in Japanese Monkeys, *Macaca Fuscata*. *Psychologia* 1:47–54.

Itani, J., K. Tokuda, Y. Furuya, K. Kano, and Y. Shin. 1963. The Social Construction of Natural Troops of Japanese Monkeys in Takasakiyama. *Primates* 4(3):1–42.

Itani, J. 1963. Vocal Communication of the Wild Japanese Monkey. *Primates* 4(2):11–66.

Jolly, A. 1966. Lemur Social Behavior and Primate Intelligence. *Science* 153(3735):501–506.

Jolly, C. J. 1966. Introduction to the Pithecoidea with Notes on their use as Laboratory Animals. *Symposia of the Zoological Society of London* 17:427–457.

Kaufmann, J. H. 1962. Ecology and Social Behavior of the Coati, *Nasua narica*, on Barro Colorado Island, Panama. *University of California Publications in Zoology* 60(3):95–222.

—. 1965. Studies on the Behavior of Captive Tree-shrews. *Folia Primitologica* 3:50–74.

Kawai, M. and H. Mizuhara. 1959. An Ecological Study of the Wild Mountain Gorilla. *Primates* 2:1–42.

Keith, A. 1896. An Introduction to the Study of Anthropoid Apes IV, the Gibbon. *Natural Science* 9:372–379.

Kellog, R. and E. A. Goldman. 1944. Review of the Spider Monkeys. *Proceedings of the United States National Museum* 96:1–45.

Kern, J. A. 1965. The Proboscis Monkey. *Animals* 6(9):522–526.

Kummer, H. and F. Kurt. 1963. Social Units of a Free-living Population of Hamadryas Baboons. *Folia Primatologica* 1:4–19.

Lyon, M. W. 1907. Notes on the Slow Lemurs. *Proceedings of the United States National Museum* 31:527–539.

—. 1913. Treeshrews: An Account of the Mammalian Family Tupaiidae. *Proceedings of the United States National Museum* 45:1–188.

Mason, W. A. 1966. Social Organization of the South American Monkey, *Callicebus mollock:* A Preliminary Report. *Tulane Studies in Zoology* 13:23–28.

Miyadi, D. 1964. Social Life of Japanese Monkeys. *Science* 143:783–786.

Nolte, Angela. 1955. Field Observations on the Daily Routine and Social Behavior of Common Indian Monkeys with Special Reference to the Bonnet Monkey. *(Macaca radiata* Geoffroy). *Journal of the Bombay Natural History Society* 53:177–184.

Osborn, R. 1963. Observations on the Behavior of the Mountain Gorilla. *Symposia of the Zoological Society of London* 10:29–37.

Owen, R. 1866. On the Aye-Aye. *Transactions of the Zoological Society of London* 5:33–101.

Petter, J. J. 1962. Ecological and Behavioral Studies of Madagascan

Lemurs in the Field. *Annals of the New York Academy of Sciences* 102:267-281.

Pitman, C. R. S. 1937. The Gorillas of the Kayonsa Region, Western Kigezi, Southwest Uganda. *Simthsonian Institution Annual Report,* Washington, D.C.

Pocock, R. I. 1917. The Genera of Hapalidae (Marmosets). *Annual Magazine of Natural History* (8)20:247-258.

— . 1917. Lemurs of the *Hapalemur* Group. *Annual Magazine of Natural History* (8)19:343-352.

— . 1925. The External Characters of the Catarrhine Monkeys and Apes. *Proceedings of the American Philosophical Society* 1925:1479-1579.

— . 1927. The Gibbons of the Genus *Hylobates. Proceedings of the Zoological Society of London* 719-741.

Pournelle, G. H. 1959. Allen's Monkey. *Zoonooz* 32, No. 10.

— . 1965. The Gorilla—Its Status in Captivity. *Zoonooz* 38; No. 9:9-10.

Rowell, T. E. 1966. Forest Living Baboons in Uganda. *Journal of Zoology in London* 149:344-364.

Schaller, G. B. 1961. The Orangutan in Sarawak. *Zoologica* 46(2):73-82.

Schwarz, E. 1928. The Species of the Genus *Cercocebus. Annual Magazine of Natural History* (10)1:644-670.

— . 1929. On the Local Races and Distribution of the Black and White Colobus Monkeys. *Proceedings of the Zoological Society of London:* 585-598.

— . 1931. On the African Long-tailed Lemurs or Galagos. *Annual Magazine of Natural History* (10)7:41-66.

— . 1931. On the African Short-tailed Lemurs or Pottos. *Annual Magazine of Natural History* (10)8:249-256.

Sorenson, M. W. and C. H. Conaway. 1966. Observations on the Social Behavior of Treeshrews in Captivity. *Folia Primatologica* 4:124-145.

Stott, K. and G. J. Selsor. 1961. Observations of the Maroon Leaf Monkey in North Borneo. *Mammalia* 25:184-189.

Struhsaker, T. 1967. Behavior of Vervet Monkeys *(Cercopithecus aethiops). University of California Publications in Zoology* 82:1-74.

— . 1967. Social Structure among Vervet Monkeys *(Cercopithecus aethiops). Behavior* 29:6-121.

Tappen, N. C. 1960. Problems of Distribution and Adaptation of the African Monkeys. *Current Anthropology* 1(2):91-120.

Tokuda, K. 1961-62. A Study on the Sexual Behavior in the Japanese Monkey Troop. *Primates* 3(2):1-40.

Vandenbergh, J. G. 1963. Feeding, Activity and Social Behavior of the Treeshrew, *Tupaia glis,* in a Large Outdoor Enclosure. *Folia Primatologica* 1:199-207.

Van Valen, L. 1965. Treeshrews, Primates and Fossils. *Evolution* 19(2):137-151.

Washburn, S. L. 1944. The Genera of Malaysian Langurs. *Journal of Mammalogy* 25:289-294.

— and I. De Vore. 1961. The Social Life of Baboons. *Scientific American* (Vol. 204, No. 6) June 1961.

Wharton, C. H. 1950. The Tarsier in Captivity. *Journal of Mammalogy* 31:260-269.

Williams, L. 1967. Breeding Humboldt's Woolly Monkey *L. lagotricha* at Murraytown Wolly Monkey Sanctuary. *International Zoo Year Book,* Vol. VII (ed., C. Jarvis), Zoological Society of London.

Yoshiba, K. 1964. The Orang-Utan in North Borneo. *Primates* 5(1-2):11-26.

Picture Credits

Artists: P. Barruel (p. 177, 178, 179, 180, 283/284). B. Bertram (p. 101, 131, 132, 133, 134, 135, 136, 137, 138). H. Diller, from Haltenorth and Diller, *Afrikas Wild,* with the permission of the University Press, Wolf and Son, Munich (p. 248, 249, 250, 251, 252, 253). W. Eigener (p. 89/90, 327, 328, 337, 338, 339, 340, 375, 376, 386, 387, 388, 389, 390, 391, 392, 393, 394, 411/412, 453, 454, 455, 456, 476, 512). K. Grossman (p. 25, 26, 247, 254, 318). E. Hudecek-Neubauer (p. 52, 167, 237, 238). F. Reimann (p. 223/224). W. Weber (p. 41, 189, 190, Frontispiece). R. Zieger (p. 62, 71, 72, 99, 100). Scientific responsibility for the artwork rests with: Prof. Dr. H. Dathe (Reimann), Prof. Dr. H. Dathe and Prof. Dr. G. Stein (Zieger), Dr. Th. Haltenorth (H. Diller), Dr. D. Heinemann (Barruel, Eigener, Weber), Dr. F. Hückinghaus (p. 25, 26), B. Marlow (Bertram), Prof. E. Thenius (Hudecek-Neubauer).

Color Photographs: Agfa/PIP (p. 385). Boz/PIP (p. 146, top). Danesch (p. 61). Dominis/Life (p. 31/32). FPG/PIP (p. 350). Good/PIP (p. 264). A. Grassman (p. 509). Grzimek (p. 51, 102 bottom, 144, 263, 511, 521, 524). Morse/Life (p. 474, 475). Müller/PIP (p. 349). A. van den Nieuwenhuizen (p. 265, 266, 317). Van Nostrand/PIP (p. 446). Paysan (p. 145). Reuhs/Grzimek (p. 168). Alan Root/Grzimek (p. 42 upper/lower, 102 upper, 103 upper/lower, 104, 104/105, 146 lower, 445). Siegel (p. 439). Steinemann, Basel Zoo (p. 143). Zellman, Berlin Zoo (p. 440, 510).

The cover picture was taken by Mrs. A. Grassmann and shows the male orang-utan "Jocki" of the Berlin Zoo. Color plate, p. 522/523 was taken from the *Serie Säugetiere* by Dr. Erna Mohr, *Sammlung Nafurkundlicher Tafeln,* published by Kroner Verlag. The color plates on p. 247 and p. 439 were made available to us by C. H. Boehringer & son, Ingelheim, and Carl Zeiss, Oberkochem, Württemberg, Germany.

Line drawings: Erich Diller and Jörg Kühn. Fig. 1-3, p. 34, according to O. Kuhn; Fig. 2-1—2-5, pp. 46, 47 and Fig. 11-8, p. 203, according to Brehm; Fig. 5-4, p. 75, Fig. 5-5, p. 75 according to Marlow; Fig. 6-4, p. 108 according to Walker; Fig. 9-1, p. 150, Fig. 9-2 and 9-3, p. 152, Fig. 9-4—9-7, pp. 148-149, Fig. 9-20, p. 173 according to Hediger; Fig. 11-12, p. 213 according to photograph (Herter); Fig. 11-27, p. 240 according to Eisentraut; Fig. 15-3, p. 331, Fig. 15-7, p. 335, Fig. 16-4, p. 367, Fig. 17-16, p. 437 according to Hill; Figs. 15-11, 15-12, 15-13, p. 346, Fig. 15-14, p. 347 according to photographs (Nolte); Fig. 18-4, p. 461 according to Napier; Fig. 11-29, p. 241, Fig. 14-1, p. 313 according to photographs; Fig. 21-2, p. 505 according to photograph (Berlin Zoo; all other line drawings are based on materials supplied by the authors.

Index

*Heavy type indicates the main entry, an asterisk * indicates an illustration, and m indicates a distribution map.*

Abbreviations and Symbols

C, °C . . . . Celsius, degrees centigrade

C.S.I.R.O. . . . Commonwealth Scientific and Industrial Res. Org. (Australia)

f . . . . following (page)

ff . . . . following (pages)

L . . . . total length (from tip of nose [bill] to end of tail)

I.R.S.A.C. . . . Institute for Scientific Res. in Central Africa, Congo

I.U.C.N. . . . . Intern. Union for Conserv. of Nature and Natural Resources

BH . . . . body height

HRL . . . . head-rump length (from nose to base of tail or end of body)

N, N- . . . . . . North, Northern, North-

NE, NE- . . . . Northeast, Northeastern, Northeast-

E, E- . . . . East, Eastern, East-

S, S- . . . . South, Southern, South-

TL . . . . tail length

SE, SE- . . . . . Southeast, Southeastern, Southeast-

SW, SW- . . . Southwest, Southwestern, Southwest-

W, W- . . . . . West, Western, West-

♂ . . . . male

♂♂ . . . . males

♀ . . . . female

♀♀ . . . . females

♂♀ . . . . pair

\+ . . . . extinct

$\frac{2 \cdot 1 \cdot 2 \cdot 3}{2 \cdot 1 \cdot 2 \cdot 3}$ . . . tooth formula, explanation in Volume X

▷ . . . . following (opposite page) color plate

▷▷ . . . . Color plate or double color plate on the page following the next

▷▷▷ . . . . Third color plate or double color plate (etc.)

⌽ . . . . Endangered species and subspecies